Concepts of Modern Physics

Arthur Beiser
and
Kok Wai Cheah

Mc Graw Hill

CONCEPTS OF MODERN PHYSICS

By Arthur Beiser, Kok Wai Cheah

International Edition 2020

1 2 3 4 5 6 7 8 9 0 MHE-KOREA 20 20

When ordering this titie, use ISBN 979-11-321-0226-7

This edition is for sale in Korea only.

Printed in Korea

Contents

CHAPTER 2

Particle Properties of Waves 69

CHAPTER 3

Wave Properties of Particles 109

CHAPTER 8
Molecules 285

CHAPTER 9
Statistical Mechanics 315

CHAPTER 11

CHAPTER 12

Preface

Modern physics began in 1900 with Max Planck's discovery of the role of energy quantization in blackbody radiation, a revolutionary idea soon followed by Albert Einstein's equally revolutionary theory of relativity and quantum theory of light. Students today must wonder why the label "modern" remains attached to this branch of physics. Yet it is not really all that venerable: my father was born in 1900, for instance, and when I was learning modern physics most of its founders, including Einstein, were still alive; I even had the privilege of meeting a number of them, including Heisenberg, Pauli, and Dirac. Few aspects of contemporary science—indeed, of contemporary life—are unaffected by the insights into matter and energy provided by modern physics, which continues as an active discipline as it enters its second century.

This book is intended to be used with a one-semester course in modern physics for students who have already had basic physics and calculus courses. Relativity and quantum ideas are considered first to provide a framework for understanding the physics of atoms and nuclei. The theory of the atom is then developed with emphasis on quantum-mechanical notions. Next comes a discussion of the properties of aggregates of atoms, which includes a look at statistical mechanics. Finally atomic nuclei and elementary particles are examined.

The balance in this book leans more toward ideas than toward experimental methods and practical applications, because I believe that the beginning student is better served by a conceptual framework than by a mass of details. For a similar reason the sequence of topics follows a logical rather than strictly historical order. The merits of this approach have led to the extensive worldwide use of the five previous editions of *Concepts of Modern Physics,* including translations into a number of other languages, since the first edition appeared nearly forty years ago.

Wherever possible, important subjects are introduced on an elementary level, which enables even relatively unprepared students to understand what is going on from the start and also encourages the development of physical intuition in readers in whom the mathematics (rather modest) inspires no terror. More material is included than can easily be covered in one semester. Both factors give scope to an instructor to fashion the type of course desired, whether a general survey, a deeper inquiry into selected subjects, or a combination of both.

Like the text, the exercises are on all levels, from the quite easy (for practice and reassurance) to those for which real thought is needed (for the joy of discovery). The exercises are grouped to correspond to sections of the text with answers to the odd-numbered exercises given at the back of the book. In addition, a Student Solutions Manual has been prepared by Craig Watkins that contains solutions to the odd-numbered exercises.

Because the ideas of modern physics represented totally new directions in thought when first proposed, rather than extensions of previous knowledge, the story of their development is exceptionally interesting. Although there is no room here for a full account, bits and pieces are included where appropriate, and thirty-nine brief biographies of important contributors are sprinkled through the text to help provide a human persepctive. Many books on the history of modern physics are available for those who wish to go further into this subject; those by Abraham Pais and by Emilio Segré, themselves distinguished physicists, are especially recommended.

For this edition of *Concepts of Modern Physics* the treatments of special relativity, quantum mechanics, and elementary particles received major revisions. In addition, numerous smaller changes and updates were made throughout the book, and several new topics were added, for instance Einstein's derivation of the Planck radiation law. There is more material on aspects of astrophysics that nicely illustrate important elements of modern physics, which for this reason are discussed where relevant in the text rather than being concentrated in a single chapter.

Many students, although able to follow the arguments in the book, nevertheless may have trouble putting their knowledge to use. To help them, each chapter has a selection of worked examples. Together with those in the *Solutions Manual,* over 350 solutions are thus available to problems that span all levels of difficulty. Understanding these solutions should bring the unsolved even-numbered exercises within reach.

In revising *Concepts of Modern Physics* for the sixth edition I have had the benefit of constructive criticism from the following reviewers, whose generous assistance was of great value: Steven Adams, *Widener University;* Amitava Bhattacharjee, *The University of Iowa;* William E. Dieterle, *California University of Pennsylvania*; Nevin D. Gibson, *Denison University;* Asif Khand Ker, *Millsaps College;* Teresa Larkin-Hein, *American University;* Jorge A. López, *University of Texas at El Paso*; Carl A. Rotter, *West Virginia University;* and Daniel Susan, *Texas A&M University–Kingsville*. I am also grateful to the following reviewers of previous editions for their critical reviews and comments: Donald R. Beck, *Michigan Technological University*; Ronald J. Bieniek, *University of Missouri–Rolla*; Lynn R. Cominsky, *Sonoma State University*; Brent Cornstubble, *United States Military Academy*; Richard Gass, *University of Cincinnati*; Nicole Herbot, *Arizona State University*; Vladimir Privman, *Clarkson University*; Arnold Strassenberg, *State University of New York–Stony Brook*; the students at *Clarkson* and *Arizona State Universities* who evaluated an earlier edition from their point of view; and Paul Sokol of *Pennsylvania State University* who supplied a number of excellent exercises. I am especially indebted to Craig Watkins of *Massachusetts Institute of Technology* who went over the manuscript with a meticulous and skeptical eye and who checked the answers to all the exercises. Finally, I want to thank my friends at McGraw-Hill for their skilled and enthusiastic help throughout the project.

Arthur Beiser

The primary aim of the revision of *Concepts of Modern Physics* is to update some of the text to be in-line with the Modern Physics course. The revision includes the addition of more fundamental topics and updated examples to provide students with a good foundation of the subject. Rearrangement of topics has been made – Chapter 1, for example, has been revamped – Lorentz Transformation and Spacetime were moved from Appendix into the main text. Doppler Effect, Momentum and Energy are now aligned into a sequence to make them more coherent.

An introductory chapter has been added to highlight some of the developing processes arisen from theoretical and experimental results, as well as those that had involved in these processes. The introductory chapter provides an historical perspective on the beginnings of modern physics; starting from classical physics of the 1890's, with the realization on waves and particles, conservation laws and forces, and the atomic theory of matter.

I am delighted to have this opportunity to get involved in revising this book. Modern physics is a bridge for the young undergraduates to get across into 20th century physics. In preparing this revision, I'd like to thank Anthony Tzu Liang Chan for his expert advice on Computational Physics, and Suet Ying Ching, Kin Long Chan and Shumei Chen for helping in gathering all the information and preparing the drafts of the new applications.

Prof. Kok Wai Cheah
Head and Chair Professor in Department of Physics,
Director of Institute of Advanced Materials,
Hong Kong Baptist University;
Fellow of Institute of Physics, UK;
Visiting Professor at Fudan University, Shanghai, People's Republic of China;
National Sun Yat-sen University, Kaohsiung, Taiwan, Republic of China;
Jinan University, Guangzhou, People's Republic of China

INTRODUCTION

1 19TH CENTURY PHYSICS

By the 19th century, modern science and technology began to shape themselves. Many of the famous physicists that we have come to know as pioneers of modern science and technology appeared in this century, for example, Michael Faraday, Ampere, Hertz, and James Watts. By the late 19th century, giants like J. J. Thomson, Marie Curie, and James Maxwell were there showing the way into modern physics. They further inspired pioneers of modern physics such as Einstein, Bohr, Rutherford, and Planck, among others. Modern Physics is in fact a narration of their story, told in physical manner that allow us to understand how this physical world works—to the best of our current understanding. With better experimental techniques, greater computation power, we have been pushing the frontier of physics. So the physics we describe now is not the final version; it is just the latest version.

1.1 THE INDUSTRIAL REVOLUTION

The industrial revolution in Europe is referred to the period when mankind began widely using machines to do large-scale manufacturing. Energy conversion and efficiently harvesting this energy to drive machines was singularly lacking until then. James Watt was the first to recognize steam power and how to harness this power to drive manufacturing machines. The technology in energy conversion and the engineering of machines needed a much better understanding of science. In fact it was through diligent study of the process of nature, a deeper understanding of its physics that allowed the industrial revolution to begin.

One of the most important developments in physics was the formal quantification of energy; James Joule (1818–1889) demonstrated the mechanical equivalent of heat through a series of mechanical and electrical experiments. The result was published in 1845 and because of this fundamental contribution, the SI unit of energy is Joule. It was from his work that Hermann Helmholtz (1821–1894) proposed the conservation of energy. This has had a profound implication and it is being used in all energy conversion processes; from mechanical processes such as the collision of simple billiard balls to the fusion and creation of elementary particles. In fact the conservation law was used to prove that light has particle properties.

The energy derived from different sources can be represented by:

Thermal: $E = k_B\ T$ where k_B is the Boltzmann constant and T is the absolute temperature

Kinetic: $E = \frac{1}{2}mv^2$ where m is the object mass and v is the velocity

Mass/Energy: $E = m_0c^2$ where m_0 is the rest mass and c is the velocity of light

1.2 UNIFICATION OF ELECTRICITY AND MAGNETISM

Electricity and magnetism until the 19th century was considered as two separate physical forces. It was in 1873 when James Clerk Maxwell who, in fact, extended the ideas and results proposed by Faraday, Ampere and others, formally showed that they are in fact a single force. Interestingly, Maxwell built his model based on the existence of a medium that supposedly conducted electromagnetic waves: aether which was subsequently shown to not exist. He treated the process quite mechanically yet he was able to derive four equations that completely described the electromagnetic wave characteristics, and from them he could also derive the speed of light. He then showed that light was not a separate entity but part of the electromagnetic spectrum. These four equations which are known now as Maxell Equations show that:

i. A current in a wire would generate a circular magnetic field.
ii. A loop wire cutting through a magnetic field would induce a current.
iii. Both electric charges and magnetic poles repel the same charge/pole and attract opposite charges/poles.
iv. The force between two charges/poles obeys the inverse square law with respect to the distance between them.

The four Maxwell Equations expressed in differential form are:

Displacement current: $div\ D = \ \pi\rho$; $\boldsymbol{\rho}$ is the charge density

Magnetic induction: $\mathrm{div}\ B = 0$

Magnetic intensity: $curl\ H = 4\pi\ j + \frac{1}{c}\frac{\partial D}{\partial t}$; j is the current density

Electric intensity: $curl\ E = -\frac{1}{c}\frac{\partial B}{\partial t}$; c is the speed of light

These four equations first appeared in his paper "On physical lines of force" published in *Philosophical Magazine*, 1861–1862. It is available at http://ow.ly/scEDk.

It is worth doing a brief summary on the discoveries made by others: Michael Faraday (1791–1867) made several critical discoveries; among them, the most important is electric induction which later extends to magnetic induction. This is the well-known Faraday's Law and is described mathematically in Maxwell Equations. The other two more notable discoveries are the Faraday Effect in which the polarization of light can be rotated by an externally applied magnetic field. The Faraday Cage which shows that external charge has no effect on any object inside a conductor i.e., electromagnetic shielding.

Andre-Marie Ampere (1775–1836) produced a mathematical model that describes the work done by Hans Christian Ørsted. Ampere's Law states that the mutual action of two lengths of current-carrying wire is proportional to their lengths and to the intensities of their currents. This is the basis of the inverse square law.

The eventual realization that electricity and magnetism is basically one force—electromagnetism—was one step towards understanding the fundamental forces of nature.

1.3 NEWTONIAN PHYSICS

Sir Isaac Newton (1642–1727) was probably the most influential physicist until the early 20th century. Although, strictly speaking, he was an 18th century person, his influence on classical physics reaches to today. He laid the foundation for classical mechanics which is often called Newtonian Mechanics for obvious reasons. He summarized and expanded the formulation of kinetic motions, on Earth as well as those of celestial objects, in his laws of motion and gravitation law. These laws are being used daily to determine phenomena from the flight of a ball to the trajectories of space probes traversing the space between Earth and other planets. In fact we very much rely on these four laws in detecting exo-planetary systems so we can arguable say that Newton, after more than 300 years, has helped to show that our Solar System is not unique in this universe! Besides the kinetic laws, he also has major contributions in optics and inventing calculus and the power series. We will briefly describe some of his important works.

1.3.1 LAWS OF MOTION

In 1687 Newton published his main work, *Principia Mathematica*, within which he stated the three laws of motion:

(i) The First Law: It states that the kinetic status of an object does not change unless an external force acts on it. This means that if it is at rest then it stays at rest. If it is in uniform motion then it stays in uniform motion until an external force acts on it. The law implies the presence of reference frames (called inertial frames).

(ii) The Second Law: It states that an applied force, $\boldsymbol{F}$, that is acting on an object, is equal to the rate of change of its momentum, $\boldsymbol{p}$, with time. This can be expressed as

$$F = \frac{dp}{dt} = \frac{d(mv)}{dt}$$

In classical physics constant mass is assumed. Thus, m can be taken out of the derivative operator. Therefore, the derivative is operating on velocity only viz it is acceleration, the equation then can be written in the familiar form

$$F = ma$$

The second law describes the dynamics of force from simple harmonic motions, which are being used among many mechanic models for classical models of atomic oscillation in crystals and electron oscillation in condensed states, e.g. plasmon. It is also being used in the Drude model which describes the condensed state of metal, although it is being used to describe other matter in a modified form.

(iii) The Third Law: It states that for every action there is an equal and opposite reaction. This means that when a force is exerted onto an object there is a force that is equal but opposite in direction exerting back.

One example is the recoil of a rifle; when a bullet is fired there is a force propelling the bullet forward. At the same time there is an equal but opposite force exerted back onto the rifle and this is felt by the shooter.

This law is also behind rocket propulsion in which the reaction force from the fuel burning propels the rocket upwards.

Delta rocket launch
Source: ©McGraw-Hill Education/ SuperStock

1.3.2 GRAVITATIONAL LAW

From the discoveries of Copernicus, Galileo, Kepler and others, by the 17th century it was quite clear that the Earth was revolving the Sun, the Moon is resolving the Earth and the other planets, Mars, Venus, and Jupiter were all also revolving around the Sun. From Kepler's three laws, the orbiter characteristics were quite clearly described.

However, these laws can best be termed as observation laws describing the orbital motion of the planets. It did not explain why these celestial bodies attract each other and what force(s) keep them together. It was Isaac Newton's law of universal gravitation that described the gravitational attraction between any two bodies. However, the fundamental cause that gives rise to gravitation force remains unknown. Nevertheless, Newton's theory achieved full support with the success in predicting the presence of Neptune from the orbital motion of Uranus. Although subsequent observation of Mercury's orbit irregularity put the validity of the theory in some doubt; it was generally agreed that the theory is sound and can be used to demonstrate the dominance of gravity in celestial bodies; for example the tidal effect due to Earth-Moon and Jupiter-Europa systems fully shown the power of gravitational force in the macro-world.

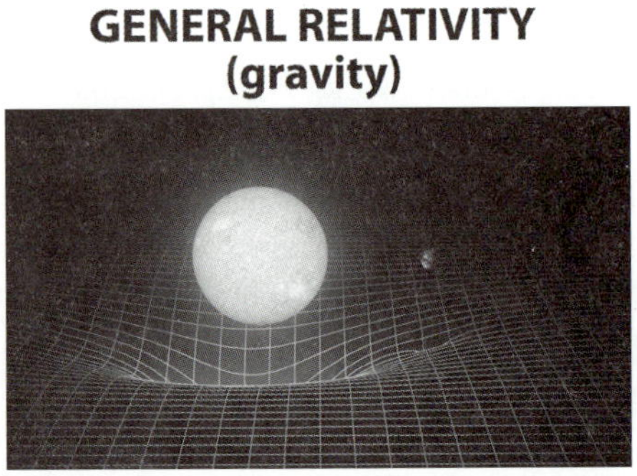

Massive Sun warps spacetime and the Earth follows the straight possible track through curved spacetime.
Source: © McGraw-Hill Education/T. Pyle/Caltech/MIT/LIGO Lab

The discrepancy of Mercury's orbit had to wait until Albert Einstein's General Relativity Theory came before it could be fully explained. In general relativity, gravitation is caused by spacetime curvature; mass curves spacetime and it is the curved spacetime that causes objects "falling" into each other.

2. NEW UNDERSTANDING AND DISCOVERIES

By the end of the 19th century, physics experiments were getting more sophisticated resulting in greater accuracy, broader temperature ranges, better computations. This led to more in depth exploration of the laws of nature which in turn bring in Modern Physics.

2.1 THE WORLD OF CLASSICAL PHYSICS

Classical physics is sometimes called Newtonian physics. This reflects the importance of the three laws of motion and universal gravitation law proposed by Newton. They basically summarize the principles behind the dynamics of the macro-world that we live in. In fact there are two other aspects of the physical world that were being looked at: Heat and waves. The dynamics of heat or thermodynamics was intensely being investigated as there was clearly a relationship with energy transfer but this relationship was not so obvious at all. Among the phenomena being investigated was blackbody radiation and it was this area that led to the eventual discovery of energy quantization.

The investigation of waves led to the search for the medium in which electromagnetic waves propagate. Maxwell's equations have described the properties of electromagnetic waves, and at the turn of the 20th century there was much research in the properties of electromagnetic waves like speed of the wave, propagating medium and others. The discovery that there was no medium needed for electromagnetic waves to propagate, and that there was in fact a coordinate transformation non-invariant, eventually led to the development of the Theory of Special Relativity.

There were also other experiments which led to new discoveries and understanding.

2.2 NEW EXPERIMENTS

A brief narration is given here highlighting some of the new physical phenomena observed: In the process of investigating the electrical resistance of mercury, H. K. Onnes, at very low temperature (4.2K) observed the sudden drop in resistance. This observation led to the discovery of superconductivity and superfluid transition. Marie and Joule Curie were studying the radioactive properties of uranium and were extracting it from tar. Their work led to discovering radioacitivity and the radio-active decay of elements. Other properties of gaseous elements were studied using gas discharge and the work gave rise to the discovery of X-rays, and the cathode ray. The observation of discrete spectral lines from the absorption spectra of the discharged gases paved the way to the realization of discrete electron energy levels in the atomic structure.

All these discoveries and new theories basically form the basis of modern physics. Sections below provide some introductory scenes of these discoveries:

2.3 BLACKBODY RADIATION

By the end of the 19th century, there were several equations and laws that described thermal radiation; for example, Stefan-Boltzmann Law gives the total radiative intensity emitted by a blackbody while Wien's Displacement Law provides the radiative spectrum of a hot object. But none of them could provide a full description of the full emission characteristics across the temperature spectrum. It was Max Planck who re-visited the fundamental principles of thermal radiation and proposed what is now known as Planck's Law on blackbody radiation in which the radiation energy is quantized. The smallest quantized energy is called a photon and when proposed was immediately used by Einstein, de Broglie, Heisenberg and others in their new insight into the world of atoms and elementary particles.

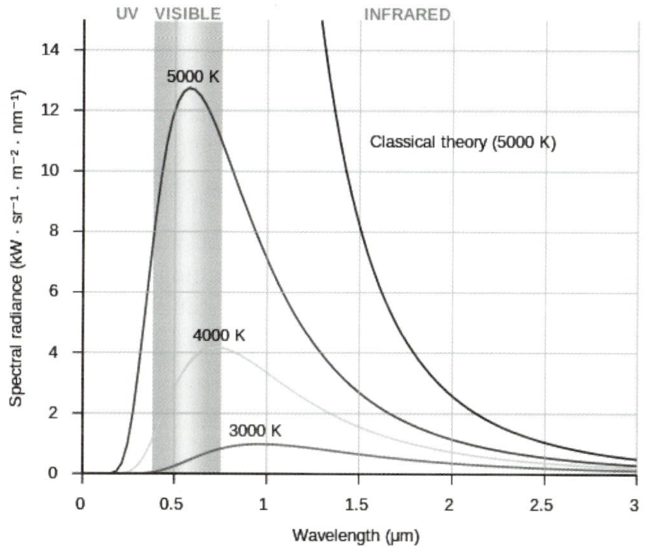

Blackbody Radiation Curve; the peak shifts as function of the body temperature
Source: © McGraw-Hill Education

2.4 STABILITY OF ATOMS

In 1808, John Dalton proposed his idea on the composition of matter in his atomic theory. In his theory the smallest divisible unit is called the atom and they are invariant and stable. In fact this follows the teaching of Socrates who declared that matters of heaven are invariants and stable. In 1896, Henri Becquerel discovered the natural radiation from Uranium salts and this radiation had similar penetrating power as X-rays which was only discovered a year earlier. Marie Curie decided to investigate further with her husband Pierre Curie and their works showed that atoms were not necessary stable. The radiation from these atoms would lead to the transmutation change of the atoms. A few years later J. J. Thompson discovered electrons while investigating cathode rays and Rutherford and others unraveled the structure of atom. Thus began a new field in Physics—Particle Physics.

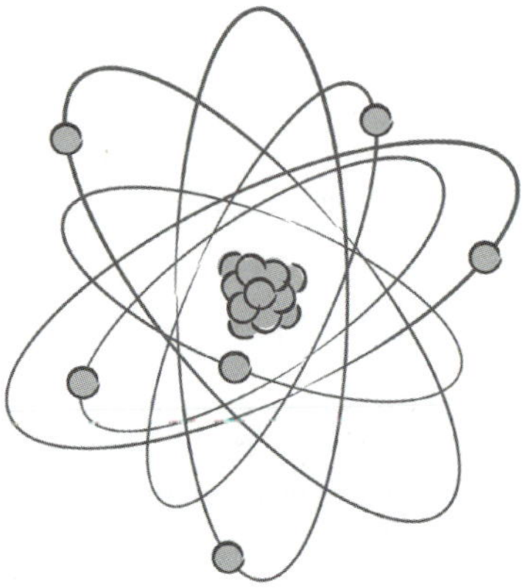

Atomic structure
Source: © McGraw-Hill Education

2.5 INVARIANCE IN COORDINATE TRANSFORMATION

All physical laws are expected to behave exactly the same irrespective of which reference frame they are. Therefore it is natural that we expect Maxwell equations would behave in the same manner. However, it turns out that there was apparent non-invariance in Galilean transformations. While two like charges repel each other in a rest frame, it was found that when using Galilean transformations to transform them to a moving frame they attracted each other because moving charges generated a mutually attractive force. This is obviously not correct. It was Einstein who showed that coordinate transformation under close to the speed of light required a correction term which incorporated in the Lorentz Transformation. He included the Lorentz Transformation into his Theory of Special Relativity describing the space-time relationship of matter. This will be discussed in section 3.1 and in greater detail in Chapter 1.

3 THE ATOMIC WORLD

In the atomic world, or more precisely sub-atomic world, classical physics was found to be inadequate to describe various behaviors. Some obvious examples were that the Galilean transformation fails to maintain the invariance of many physical laws, that the apparent instability of atoms could not be explained properly, and spacetime phenomenon is different from that of the macro-world.

3.1 RELATIVITY

By using the Lorentz Transformation, Albert Einstein was able to show that invariance can be maintained for all physical laws in all frames of reference. In the Theory of Special Relativity, he proposed the idea of the speed of light being independent of all frames of reference and that the velocity of any object is relative to it. This satisfies all coordinate transformation under the Lorentz Transformation which can be approximated to Galilean Transformation when the object velocity is significantly smaller than that of light. The concept of time and space is also different; time can slow down, distance shrinks and mass gets heavier as the speed of an object approaches the speed of light. This theory is being used every day in calculating where you are when you use GPS (Global Position System) and gravity variation which is measured using atomic clocks in the satellites.

In the General Theory of Relativity, Einstein brought gravity into the theory in which he proposed how spacetime is warped by masses such as planets and stars. The warped spacetime "bends" light as it moves across the area of a mass. The General Theory of Relativity also predicts the dynamics of the universe, that is, how the universe evolves, and recently cosmologists have been using it to investigate dark matter and dark energy which have yet to be identified.

A GPS satellite orbiting the earth.
Source: © McGraw-Hill Education/Shutterstock/Andrey Armyagov

3.2 PROPERTIES OF ATOMS

By the end of the 19th century and from the beginning of the 20th century it was realized that atoms are not completely stable; some elements emit radioactivity and decay or transmute eventually into another element. In fact, each element can have several isotopes i.e. their atomic weights are different because they carry different numers of neutrons, but still having the same number of protons. Hence, the same number of electrons. Since their atomic and chemical properties are ruled by the number of electrons, so they are the same element but with different weight.

The study of their radioactivity and their nucleus have led to the discovery of the atomic structure and properties of the atomic nucleus.

Critical experiments by Rutherford augmented by Bohr's atomic model have shown that the atomic and sub-atomic is quantum in nature and the quantum property blurs into the continuum property of the macro-world with the increasingly large number of atoms being considered. The quantum property of the atomic world has since then branched into other areas of science besides Physics; there are quantum chemistry which has also given rise to several Nobel laureates, theoretical biology and biophysics which gives new insight into bio-molecules' interaction which is so important in understanding the nature of new viruses such as SARS and avian flu.

3.3 STRUCTURE OF ATOMS—NUCLEAR PHYSICS

Rutherford has shown that the atomic structure is nothing that we envisaged at the turn of the 20th century; the nucleus of an atom is compact and dense, electrons are orbiting around the nucleus at a relatively distance from the nucleus and are in discrete orbits.

With heavier elements, their nuclei are increasing complex and also seemingly less stable. This is the beginning of Nuclear Physics that leads to the discovery of two major fundamental forces: strong and weak nuclear forces. These two forces are responsible for fusion, fission and radioactivity. The understanding of these two forces led to the development of the atomic bomb and the hydrogen bomb. Nuclear power stations

Cooling towers of a nuclear power station, Inner Mongolia, China
Source: © McGraw-Hill Education/Rodrigo A Torres/Glow Images

Tokamak Equipment for experiments with high temperature plasma.
Source: © NFRI

have, however, been developed for a peaceful application of nuclear fission. Harvesting nuclear fusion turns out to be very challenging especially in confining the fusion plasma. In the effort to understand the nature of fundamental forces, Salam, Glashow, and Weinberg put forward a theory that united electromagnetism with the nuclear weak force and it gained them a Nobel Prize in Physics in 1979. With this theory, three of the four fundamental forces, strong, weak nuclear and electromagnetic forces are united. Gravity remains the only force showing no obvious 'link' with the rest of the fundamental forces.

3.4 PARTICLE PHYSICS AND COSMOLOGY

Another new branch of Physics also emerges from the study of the atomic nucleus—Particle Physics. This is the study of sub-atomic particles in their own right, their creation, annihilation, and interaction. At the same time, Einstein's General Theory of Relativity and discovery of cosmic background radiation has led to strong belief that the universe was created from a Big Bang. Sub-atomic particles were the first to form from this Big Bang. So the search for "ancient" particles and the emission they produced has become part of Cosmology research today.

Huge accelerators and cyclotrons were built to study them. By increasing the kinetic energy of the high speed particles, the total energy at the moment of collision increases, and therefore more and more exotic elementary particles are created. These high energy exotic particles would have appeared soon after the Big Bang. So in order to verify the Big Bang theory, higher and higher energy accelerators are built to create collisional energy close to that estimated Big Bang moment. In March 2013 one of fundamental particles, the Higgs boson, predicted by the standard model was discovered. This confirmation allows us to move forward in our theory and understanding in this physical world.

To study the state of the universe and its evolution, a plethora of telescopes on the ground as well as in space have been deployed to study these sub-atomic particles travelling across the universe with immense energy—cosmic rays. These telescopes detect radio waves, X-rays, probing different physical processes that are still happening in this universe.

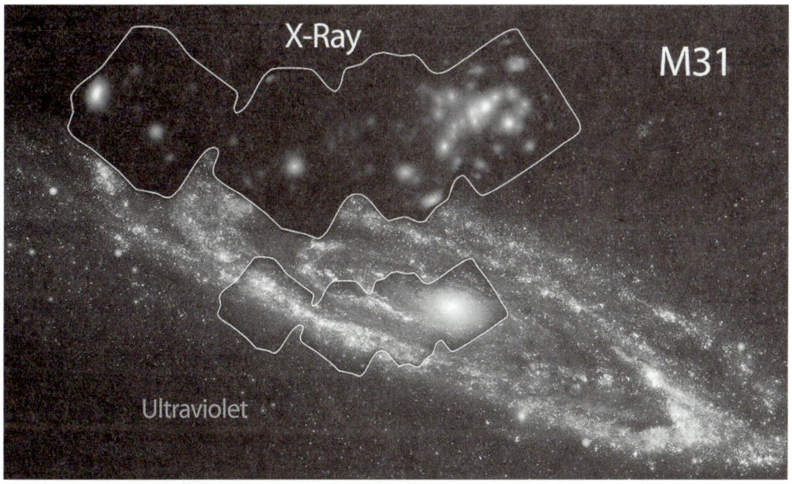

This composite image combines the X-ray image taken by the NuSTAR telescope with the ultraviolet image of Andromeda galaxy(M31) from the GALEX space telescope.
Source: © NASA/JPL-Caltech/GSFC

4 THE QUANTUM CONNECTION

The word quantum and its world started when Planck showed that energy comes in a discrete form and the smallest possible discrete energy unit is called quantum. This was a revelation for many physicists at that moment; some of the most well-known examples were:

(i) Einstein used quantized energy to explain the photo-electric effect; discrete photon energy is responsible for the observed photo-emission of electrons from solids, and the fact that the threshold photon energy is the inherent property of the elements indicates that each element has its unique electronic structure. Among the elemental properties obtained from this understanding, work function is probably most well-used. It is used in semiconductor devices, electron microscopy/spectroscopy. Einstein was awarded with the Nobel Prize for the explanation of the photo-electric effect in 1921.

(ii) Niels Bohr used the quantum energy concept to explain the scattering work done by Ernest Rutherford who showed that the atomic structure is composed of tight, condensed positive charged the nucleus is surrounded by electrons orbiting around it at a relatively large distance. Among the then points of concern is the apparent stability of electrons orbiting round a positive charged nucleus without decaying uncontrollably into the center. Bohr proposed that there were stable/stationery orbits around the nucleus with energies $E = nh\nu$, where h is the Planck constant, ν is the photon frequency and n is called the quantum number representing the orbits; the lowest orbit has $n = 1$. His atomic model was able to explain the discrete emission lines from the gas discharge. He received his Nobel Prize in 1922.

(iii) Quantization of energy leads to the realization of the particle properties of light. This was demonstrated by Compton (Nobel Prize 1927). However it was de Broglie (Nobel Prize 1929) who showed that all elementary particles have both wave/particle properties. This fact is both revealing and intriguing; up to now there is no satisfactory explanation why elementary particles should possess this dual property.

4.1 FROM ATOM TO QUANTUM

When Rutherford showed what the atomic structure was composed of, Bohr based on the then newly proposed knowledge of the quantization of energy to propose how electrons orbit around the nucleus in fixed orbits with discrete energy levels. This began quantum mechanics describing the dynamics of elementary particles. The evolution of quantum mechanics to its present-day form started with Heisenberg, Dirac, and Schrodinger. The format which we are familiar with is different in its presentation from the old quantum mechanics that was first proposed by Bohr—the Copenhagen interpretation. The essence of the principle does not change but the interpretation and mathematical format is different.

Some of those moments are worth noting down again; the mathematics of early quantum mechanics was in matrix format and it was not easy to relate to or understand. It was Schrodinger who came up with a new formulation with a format very similar to those representing harmonic motions. This received a big welcome from other physicists and was immediately adopted and is now known as the Schrodinger Equation. For his contribution he was awarded the Nobel Prize in 1933.

4.2 THE CONSEQUENCE OF QUANTUM MECHANICS

The description of elementary particles in quantum mechanics does not intuitively relate to our common senses; the theory describes the particles as possessing wave/particle properties which cannot be separated. The theory also describes that position and momentum are intrinsically related; Heisenberg's Uncertainty Principle and determining for example the positions of electrons orbiting around a nucleus is a probability determination.

These interpretations are not accepted completely by all scientists. Einstein for example was very skeptical of the non-deterministic nature of the theory. Hence, his famous remark: God does not play dice.

Subsequent study in the quantum phenomena led to the full development of quantum mechanics. Many fundamental properties of matter hitherto not known or understood can now be explained such as the natural line width observed in emission or absorption explained by Heisenberg's Uncertainty Principle. The realization that there is a class of matter situated between conductor and insulator now known as semiconductors allows us to develop electronics, optoelectronics and photonics. Quantisation of light also gives rise to new physics to describe its property leading to the emergence of Quantum Optics (Roy Glauber, Nobel Prize 2005).

Since interaction at the atomic scale involves quantum mechanics, therefore modeling at this level inevitably requires vigorous quantum treatment; Paul Dirac in 1929 pointed that while the underlying mathematics for a large part of Physics and whole of Chemistry is known, its applications are too complex to obtain a solution. Nevertheless the rise in computational power, modeling of molecular structures using quantum mechanics becomes possible. The Hartree-Fock method was first introduced in the 1930s to perform some physics modelling and it was adopted for molecular systems. Dirac quite accurately pointed that modeling in the physical world remained a challenge and it was in Chemistry that quantum mechanics provided the most initial benefit. There have been two Chemistry Nobel Prizes given to quantum chemistry; in 1998 and 2013. These show how important quantum chemistry is to the understanding of chemistry at the atomic level.

4.3 COMPUTATIONAL PHYSICS

4.3.1 The Emergence of Computational Physics

The invention of solid state electronics has led to our ability to build large scale computation electronics like the computer. With increasing computational power, physicists have begun to use it to solve many complex dynamic problems numerically. Very often analytical solutions cannot be obtained due to the complexity of a dynamic system e.g. classical fluid, electrons in organic compound or plasma, and the motion of stars in a galaxy. Analytically we are able to solve up to two-body problems but beyond two bodies we are unable to solve them analytically and have to rely on a computer to do so. In this section we will make a brief introduction to computational physics so that Physics undergraduates have some appreciation of what are involved in doing computational modeling.

Mathematical formulation of a dynamic system based on physical principles that govern it can be very complex because the motions of many particles, e.g. ions in a fluid or gas molecules in air, are very often non-linear in nature sometimes even with positive feedback that can lead to chaotic scenes. There are many such complex systems in nature e.g. atmospheric turbulence which gives us the weather, the magnetohydrodynamics behavior of the plasma in nuclear fusion and high intensity optical fields in microcavities.

Supercomputers that can simulate such complex problems can be constructed by connecting computers using a high speed network. Such computer clusters can distribute their data and the computational workload among different computers, also known as parallel computing. Recent modelling work on galactic/stellar evolution and microbe evolution demonstrate how powerful this computational power can be. The top 10 computers listed in November 2013 by Top 500 Supercomputer Sites are shown in Table 1. The speed of a computer system is typically measured in floating point operations per second (FLOPS). To give you an idea, the speed of recent handheld

Table 1 Top 10 Supercomputers in the World (as of November 2013)

Rank	Site	System Name	Cores	Speed (TFLOPS)
1	National Super Computer Center in Guangzhou China	**Tianhe-2 (Milky Way-2)** - TH-IVB-FEP Cluster, Intel Xeon E5-2692 12C 2.200GHz, Intel Xeon Phi 31S1P NUDT	3,120,000	33,862.7
2	Oak Ridge National Laboratory United States	**Titan** - Cray XK7, Opteron 6274 16C 2.200GHz, NVIDIA K20x Cray Inc.	560,640	17,590.0
3	Lawrence Livermore National Laboratory United States	**Sequoia** - BlueGene/Q, Power BQC 16C 1.60 GHz, IBM	1,572,864	17,173.2
4	RIKEN Advanced Institute for Computational Science Japan	**K computer**, SPARC64 VIIIfx 2.0GHz, Fujitsu	705,024	10,510.0
5	Argonne National Laboratory United States	**Mira** - BlueGene/Q, Power BQC 16C 1.60GHz, IBM	786,432	8,586.6
6	Swiss National Supercomputing Centre Switzerland	**Piz Daint** - Cray XC30, Xeon E5-2670 8C 2.600GHz, NVIDIA K20x Cray Inc.	115,984	6,271.0
7	Texas Advanced Computing Center, University of Texas at Austin United States	**Stampede** - PowerEdge C8220, Xeon E5-2680 8C 2.700GHz, Intel Xeon Phi SE10P Dell	462,462	5,168.1
8	Forschungszentrum Juelich Germany	**JUQUEEN** - BlueGene/Q, Power BQC 16C 1.600GHz, IBM	458,752	5,008.9
9	Lawrence Livermore National Laboratory United States	**Vulcan** - BlueGene/Q, Power BQC 16C 1.600GHz, IBM	393,216	4,293.3
10	Leibniz Rechenzentrum Germany	**SuperMUC** - iDataPlex DX360M4, Xeon E5-2680 8C 2.70GHz, IBM	147,456	2,897.0

Source: Top 500 Supercomputer Sites

devices with multiple computing cores has a few GFLOPS (giga or 109 FLOPS), while the speed of a recent PC is a few 10s to a 100 GFLOPS. The speed of the top supercomputers is already beyond PFLOPS (peta or 1015 FLOPS). Lately, it is recognized that the capability of the GPU (graphics processing unit) to handle large data blocks simultaneously is not only beneficial for graphics display but also for high performance parallel computing. The latest supercomputers start to incorporate GPUs (see NVIDIA in the table below) to enhance their speed.

However using computers to numerically solve all these complex systems is not a simple direct translation of the physical laws or principles into a computer program. Most mathematical equations are either Ordinary Differential Equations (ODE) or Partial Differential Equations (PDE). Since ODE and PDE are in continuous form and a computer can only work in discrete steps therefore the transformation of the continuous form into discrete form is necessary. The resolution of these steps has to be carefully defined so as not to render too large an approximation that leads to meaningless results. While if the step is too fine, the computation time will take too long.

Because computational physics encompasses a broad class of problems, it is generally divided amongst the different mathematical problems it numerically solves, or the methods it applies.

4.3.2 COMPUTATIONAL METHODS

There are several computational methods being developed to solve the mathematical formulation in the Physical theories and many variations have arisen from these methods. Below is a short list of the more well-known methods and the specific equations they solve:

i. Ordinary differential equations

Runge–Kutta methods are most commonly used. These methods were developed by German mathematicians C. Runge and M. W. Kutta around 1900 and they are used for the approximation of solutions of ordinary differential equations.

A simple example is a first-order differential equation that could be describing a heat flow problem:

$$\frac{dy}{dx} = f(x, y)$$

The discretization of the differential equation means that each increment in x and y directions are expressed as:

$$y = y_0 + \Delta y \quad \text{and} \quad x = x_0 + \Delta x$$

where Δy and Δx are finite increments along the y and x-axis respectively.

Then the differential equation can be translated into a difference equation in the form:

$$y_1 = y\,(x_0) + \Delta y \approx y\,(x_0) + f\,(x_0, y_0) + \Delta x$$

The difference equation is equivalent to performing a Taylor expansion of $y\,(x)$ around the point $x = x_0$ keeping the terms up the first order derivative. Higher order variations of the method are second (RK2) and fourth-order formulae (RK4), for example.

ii. Partial differential equations

There are several computational methods being used to solve partial differential equations. The finite difference method is a high dimensional extension of the discretization demonstrated above. On a uniform mesh of grids, finite difference equations are defined to approximate the partial differential equation. The finite element method is probably most widely met because it is used extensively in engineering problems involving cars, oil pipelines, fluid dynamics, and weather prediction. In each of these problems, the model consists of meshes that can have different sizes; each mesh represents the state of the physical system in time at that particular space. A finer mesh is used for rapid changes in state and a coarser mesh is used for more stable states. This in fact reduces the computational time without compromising the accuracy. The spectral method expresses the solution of the partial differential equation as a Fourier transform. This allows analysis of the solutions in terms of wavelength and frequency, and efficient algorithms of Fast Fourier Transform (FFT) can be used.

iii. Integration

High dimensional integration is frequently encountered in statistical physics, which studies many-particle systems with large degrees of freedom. Since it is practically impossible to perform exact integration or examine all the possible states of such systems, Monte Carlo integration are typically employed. The Monte Carlo method explores the states of the system in a probabilistic manner that resembles gambling in a casino, hence, the name Monte Carlo was coined. This method is also useful to model stochastic processes where a system is subject to some random forces, for example Brownian motion and diffusion.

Applications of Monte Carlo simulations have spread beyond Physics. In the commercial world, risk probability in business is being modelled by Monte Carlo methods. They are also being extensively used in complex optimization problems in daily life, such as the well-known traveling salesman problem. In mathematics, Monte Carlo is also essential in solving multi-dimensional definitive integrals that have complicated boundary conditions.

The above three classes of problems only provide a brief idea on the approaches taken to solve many of the physical formulations numerically. It is not meant to be a comprehensive review and if you are interested in computational physics, you should certainly look up for references in this field.

In computational modeling there are several commonly used computer languages: FORTRAN, C, Java and Python. FORTRAN is a commonly used language for scientific research with well-developed software packages and mathematical libraries that very often are freely available worldwide. However, there is a growing popularity towards C and C++ for numerical computations. Java and Python are more recently developed. An advantage of developing a computer program in Java is that it is portable as it can run on multiple computer platforms. Python uses very readable expressions that are almost English-like, thus is easy to learn and understand.

If writing dedicated programs is not too appealing to you, there are commercial software packages that can solve the mathematical formulation for you. Some examples are MatLab and Mathematica which can be used for both undergraduate teaching and research. There also are packages such as COMSOL designed for various physics and engineering applications.

4.3.3 Examples

The advancement in computational power is such that Nobel Prizes were given to several scientists in Chemistry who have all used theories developed from physics and/or chemistry and translated them into computational models that can give us detailed insight on the dynamics of molecules, chemical reactions, and processes.

In 1998 Walter Kohn and John A. Pople were awarded the Nobel Prize in Chemistry for the development of the Density Functional Theory (by Kohn who is a physicist!) and computational methods in quantum chemistry (by Pople). Conventional derivation of molecular properties calculates the motions of all the electrons in molecules; this is too complicated even for a computer to do. Kohn showed that a reasonable approximation can be obtained if the average number of electrons per unit volume is known (hence the name density functional); this leads to a considerably simpler computational method. Therefore, more complex molecules such as enzyme reactions can be modelled. Pople devised computational programs that use quantum mechanics to describe, at the molecular level, chemical reactions. Thus, the combination of these two developments allows us to perform very detail theoretical (quantum) chemistry at the molecular level that hitherto could not be done.

In 2013, Martin Karplus, Michael Levitt, and Arieh Warshel shared the Nobel Prize in Chemistry for the development of multi-scale models for complex chemical systems. Using large scale computational power, they first used quantum mechanics to model how atoms of a drug couple to the active site of a protein, then used classical physics to simulate the rest of the protein. Therefore, they were able to study theoretically how chemical reactions took place almost atom by atom.

Here we present one example in basic computational simulation that describes the dynamic characteristics of some complex systems:

Ising model

The Ising model is a model that describes the magnetic moment in a solid. Usually, the magnetic moment originates from the electron spin within an atom. In its simplest form, the atoms are arranged in a square lattice and the direction of the individual spin can either point up or down (Fig. 1). There are interactions between the magnetic moment of different atoms and the Ising model aims to study how the spin of each atom flips due to such interactions and under the influence of temperature. Fig. 1 illustrates how a solid can undergo a transition from a paramagnetic state (the spins

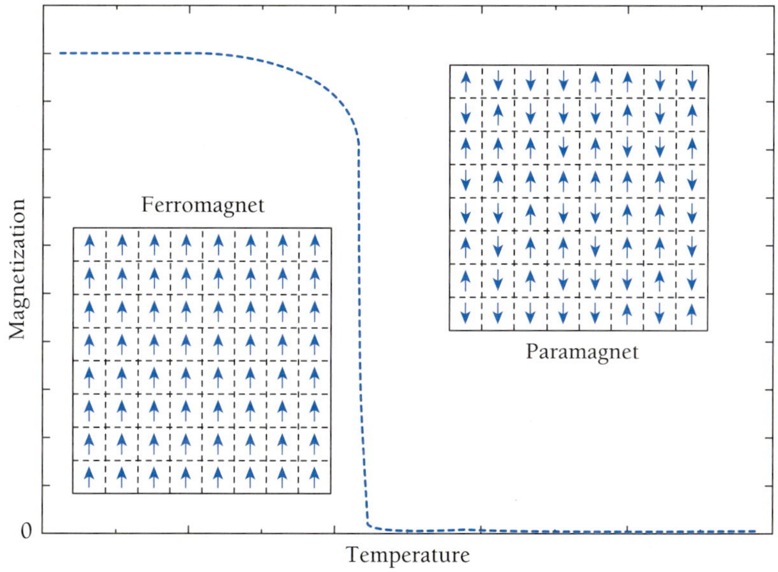

Figure 1

are randomly oriented) to a ferromagnetic state (all the spins are aligned in one direction, non-zero magnetization) as temperature decreases. This book does not deal in detail with regard to the physics and the computational method used. We will only point out that the model seeks to work out the process of the spin flips as the system reaches thermal equilibrium. There are competition forces as the system settles down towards thermal equilibrium. The Ising model can be utilized to study the properties of strong permanent magnets found in many applications in modern technology such as computer hard disks and wind turbine generators. In fact this is a thermodynamic problem that becomes increasingly complex as a more realistic 3-D model of a solid is considered.

CHAPTER 1

Relativity

According to the theory of relativity, nothing can travel faster than light. Although today's spacecraft can exceed 10 km/s, they are far from this ultimate speed limit.

In 1905 a young physicist of twenty-six named Albert Einstein showed how measurements of time and space are affected by motion between an observer and what is being observed. To say that Einstein's theory of relativity revolutionized science is no exaggeration. Relativity connects space and time, matter and energy, electricity and magnetism—links that are crucial to our understanding of the physical universe. From relativity have come a host of remarkable predictions, all of which have been confirmed by experiment. For all their profundity, many of the conclusions of relativity can be reached with only the simplest of mathematics.

1.1 SPECIAL RELATIVITY

All motion is relative; the speed of light in free space is the same for all observers

When such quantities as length, time interval, and mass are considered in elementary physics, no special point is made about how they are measured. Since a standard unit exists for each quantity, who makes a certain determination would not seem to matter—everybody ought to get the same result. For instance, there is no question of principle involved in finding the length of an airplane when we are on board. All we have to do is put one end of a tape measure at the airplane's nose and look at the number on the tape at the airplane's tail.

But what if the airplane is in flight and we are on the ground? It is not hard to determine the length of a distant object with a tape measure to establish a baseline, a surveyor's transit to measure angles, and a knowledge of trigonometry. When we measure the moving airplane from the ground, though, we find it to be shorter than it is to somebody in the airplane itself. To understand how this unexpected difference arises we must analyze the process of measurement when motion is involved.

Frames of Reference

The first step is to clarify what we mean by motion. When we say that something is moving, what we mean is that its position relative to something else is changing. A passenger moves relative to an airplane; the airplane moves relative to the earth; the earth moves relative to the sun; the sun moves relative to the galaxy of stars (the Milky Way) of which it is a member; and so on. In each case a **frame of reference** is part of the description of the motion. To say that something is moving always implies a specific frame of reference.

An **inertial frame of reference** is one in which Newton's first law of motion holds. In such a frame, an object at rest remains at rest and an object in motion continues to move at constant velocity (constant speed and direction) if no force acts on it. Any frame of reference that moves at constant velocity relative to an inertial frame is itself an inertial frame.

All inertial frames are equally valid. Suppose we see something changing its position with respect to us at constant velocity. Is it moving or are we moving? Suppose we are in a closed laboratory in which Newton's first law holds. Is the laboratory moving or is it at rest? These questions are meaningless because all constant-velocity motion is relative. There is no universal frame of reference that can be used everywhere, no such thing as "absolute motion."

The **theory of relativity** deals with the consequences of the lack of a universal frame of reference. **Special relativity,** which is what Einstein published in 1905, treats problems that involve inertial frames of reference. **General relativity**, published by Einstein a decade later, describes the relationship between gravity and the geometrical structure of space and time. The special theory has had an enormous impact on much of physics, and we shall concentrate on it here.

Postulates of Special Relativity

Two postulates underlie special relativity. The first, the **principle of relativity**, states:

> The laws of physics are the same in all inertial frames of reference.

This postulate follows from the absence of a universal frame of reference. If the laws of physics were different for different observers in relative motion, the observers could find from these differences which of them were "stationary" in space and which were "moving." But such a distinction does not exist, and the principle of relativity expresses this fact.

The second postulate is based on the results of many experiments:

> The speed of light in free space has the same value in all inertial frames of reference.

This speed is 2.998×10^8 m/s to four significant figures.

To appreciate how remarkable these postulates are, let us look at a hypothetical experiment basically no different from actual ones that have been carried out in a number of ways. Suppose I turn on a searchlight just as you fly past in a spacecraft at a speed of 2×10^8 m/s (Fig. 1.1). We both measure the speed of the light waves from the searchlight using identical instruments. From the ground I find their speed to be 3×10^8 m/s as usual. "Common sense" tells me that you ought to find a speed of $(3 - 2) \times 10^8$ m/s, or only 1×10^8 m/s, for the same light waves. But you also find their speed to be 3×10^8 m/s, even though to me you seem to be moving parallel to the waves at 2×10^8 m/s.

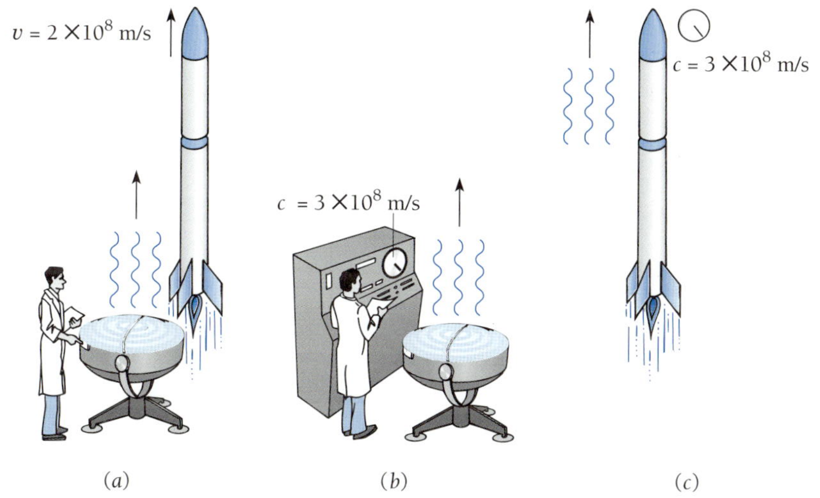

Figure 1.1 The speed of light is the same to all observers.

About Scientists

Albert A. Michelson (1852–1931) was born in Germany but came to the United States at the age of two with his parents, who settled in Nevada. He attended the U.S. Naval Academy at Annapolis where, after two years of sea duty, he became a science instructor. To improve his knowledge of optics, in which he wanted to specialize, Michelson went to Europe and studied in Berlin and Paris. Then he left the Navy to work first at the Case School of Applied Science in Ohio, then at Clark University in Massachusetts, and finally at the University of Chicago, where he headed the physics department from 1892 to 1929. Michelson's speciality was high-precision measurement, and for many decades his successive figures for the speed of light were the best available. He redefined the meter in terms of wavelengths of a particular spectral line and devised an interferometer that could determine the diameter of a star (stars appear as points of light in even the most powerful telescopes).

Michelson's most significant achievement, carried out in 1887 in collaboration with Edward Morley, was an experiment to measure the motion of the earth through the "ether," a hypothetical medium pervading the universe in which light waves were supposed to occur. The notion of the ether was a hangover from the days before light waves were recognized as electromagnetic, but nobody at the time seemed willing to discard the idea that light propagates relative to some sort of universal frame of reference.

To look for the earth's motion through the ether, Michelson and Morley used a pair of light beams formed by a half-silvered mirror, as in Fig. 1.2. One light beam is directed to a mirror along a path perpendicular to the ether current, and the other goes to a mirror along a path parallel to the ether current. Both beams end up at the same viewing screen. The clear glass plate ensures that both beams pass through the same thicknesses of air and glass. If the transit times of the two beams are the same, they will arrive at the screen in phase and will interfere constructively. An ether current due to the earth's motion parallel to one of the beams, however, would cause the beams to have different transit times and the result would be destructive interference at the screen. This is the essence of the experiment.

Although the experiment was sensitive enough to detect the expected ether drift, to everyone's surprise none was found. The negative result had two consequences. First, it showed that the ether does not exist and so there is no such thing as "absolute motion" relative to the ether: all motion is relative to a specified frame of reference, not to a universal one. Second, the result showed that the speed of light is the same for all observers, which is not true of waves that need a material medium in which to occur (such as sound and water waves).

The Michelson-Morley experiment set the stage for Einstein's 1905 special theory of relativity, a theory that Michelson himself was reluctant to accept. Indeed, not long before the flowering of relativity and quantum theory revolutionized physics, Michelson announced that "physical discoveries in the future are a matter of the sixth decimal place." This was a common opinion of the time. Michelson received a Nobel Prize in 1907, the first American to do so.

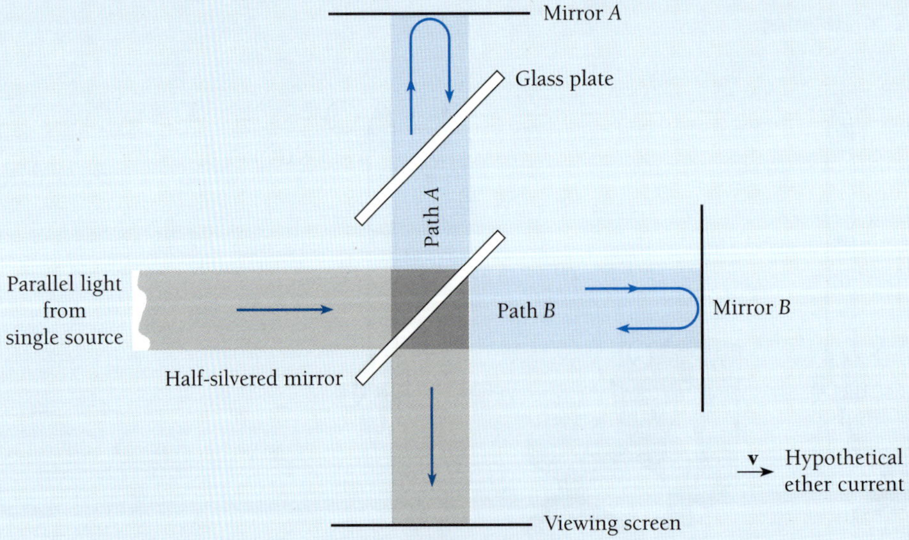

Figure 1.2 The Michelson-Morley experiment.

There is only one way to account for these results without violating the principle of relativity. It must be true that measurements of space and time are not absolute but depend on the relative motion between an observer and what is being observed. If I were to measure from the ground the rate at which your clock ticks and the length of your meter stick, I would find that the clock ticks more slowly than it did at rest on the ground and that the meter stick is shorter in the direction of motion of the spacecraft. To you, your clock and meter stick are the same as they were on the ground before you took off. To me they are different because of the relative motion, different in such a way that the speed of light you measure is the same 3×10^8 m/s I measure. Time intervals and lengths are relative quantities, but the speed of light in free space is the same to all observers.

Before Einstein's work, a conflict had existed between the principles of mechanics, which were then based on Newton's laws of motion, and those of electricity and magnetism, which had been developed into a unified theory by Maxwell. Newtonian mechanics had worked well for over two centuries. Maxwell's theory not only covered all that was then known about electric and magnetic phenomena but had also predicted that electromagnetic waves exist and identified light as an example of them. However, the equations of Newtonian mechanics and those of electromagnetism differ in the way they relate measurements made in one inertial frame with those made in a different inertial frame.

Einstein showed that Maxwell's theory is consistent with special relativity whereas Newtonian mechanics is not, and his modification of mechanics brought these branches of physics into accord. As we will find, relativistic and Newtonian mechanics agree for relative speeds much lower than the speed of light, which is why Newtonian mechanics seemed correct for so long. At higher speeds Newtonian mechanics fails and must be replaced by the relativistic version.

1.2 THE LORENTZ TRANSFORMATION

Prospective perception from different reference frames

Suppose we are in an inertial frame of reference S and find the coordinates of some event that occurs at the time t are x, y, z. An observer located in a different inertial frame S9 which is moving with respect to S at the constant velocity $\mathbf{v}$ will find that the same event occurs at the time t' and has the coordinates x', y', z'. (In order to simplify our work, we shall assume that $\mathbf{v}$ is in the $+x$ direction, as in Fig. 1.3.) How are the measurements x, y, z, t related to x', y', z', t'?

Galilean Transformation

Before special relativity, transforming measurements from one inertial system to another seemed obvious. If clocks in both systems are started when the origins of S and S' coincide, measurements in the x direction made is S will be greater than those made in S' by the amount vt, which is the distance S' has moved in the x direction. That is,

$$x' = x - vt \tag{1.1}$$

There is no relative motion in the y and z directions, and so

$$y' = y \tag{1.2}$$

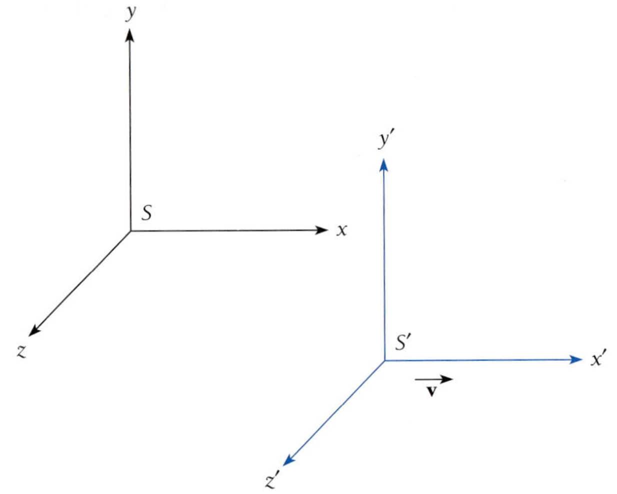

Figure 1.3 Frame S' moves in the $+x$ direction with the speed v relative to frame S. The Lorentz transformation must be used to convert measurements made in one of these frames to their equivalents in the other.

$$z' = z \tag{1.3}$$

In the absence of any indication to the contrary in our everyday experience, we further assume that

$$t' = t \tag{1.4}$$

The set of Eqs. (1.1) to (1.4) is known as the **Galilean transformation.**

To convert velocity components measured in the S frame to their equivalents in the S' frame according to the Galilean transformation, we simply differentiate x', y', and z' with respect to time:

$$v'_x = \frac{dx'}{dt'} = v_x - v \tag{1.5}$$

$$v'_y = \frac{dy'}{dt'} = v_y \tag{1.6}$$

$$v'_z = \frac{dz'}{dt'} = v_z \tag{1.7}$$

Although the Galilean transformation and the corresponding velocity transformation seem straightforward enough, they violate both of the postulates of special relativity. The first postulate calls for the same equations of physics in both the S and S' inertial frames, but the equations of electricity and magnetism become very different when the Galilean transformation is used to convert quantities measured in one frame into their equivalents in the other. The second postulate calls for the same value of the speed of light c whether determined in S or S'. If we measure the speed of light in the x direction in the S system to be c, however, in the S' system it will be

$$c' = c - v$$

according to Eq. (1.5). Clearly a different transformation is required if the postulates of special relativity are to be satisfied. We would expect both time dilation and length contraction to follow naturally from this new transformation.

Lorentz Transformation

A reasonable guess about the nature of the correct relationship between x and x' is

$$x' = k(x - \boldsymbol{v}t) \tag{1.8}$$

Here k is a factor that does not depend upon either x or t but may be a function of $\boldsymbol{v}$. The choice of Eq. (1.8) follows from several considerations:

1 It is linear in x and x', so that a single event in frame S corresponds to a single event in frame S', as it must.
2 It is simple, and a simple solution to a problem should always be explored first.
3 It has the possibility of reducing to Eq. (1.1), which we know to be correct in ordinary mechanics.

Because the equations of physics must have the same form in both S and S', we need only change the sign of $\boldsymbol{v}$ (in order to take into account the difference in the direction of relative motion) to write the corresponding equation for x in terms of x' and t':

$$x = k(x' + \boldsymbol{v}t') \tag{1.9}$$

The factor k must be the same in both frames of reference since there is no difference between S and S' other than in the sign of $\boldsymbol{v}$.

As in the case of the Galilean transformation, there is nothing to indicate that there might be differences between the corresponding coordinates y, y' and z, z' which are perpendicular to the direction of $\boldsymbol{v}$. Hence we again take

$$y' = y \tag{1.10}$$

$$z' = z \tag{1.11}$$

The time coordinates t and t', however, are *not* equal. We can see this by substituting the value of x' given by Eq. (1.8) into Eq. (1.9). This gives

$$x = k^2(x - \boldsymbol{v}t) + k\boldsymbol{v}t'$$

from which we find that

$$t' = kt + \left(\frac{1 - k^2}{k\boldsymbol{v}}\right)x \tag{1.12}$$

Equations (1.8) and (1.10) to (1.12) constitute a coordinate transformation that satisfies the first postulate of special relativity.

The second postulate of relativity gives us a way to evaluate k. At the instant $t = 0$, the origins of the two frames of reference S and S' are in the same place, according to our initial conditions, and $t' = 0$ then also. Suppose that a flare is set off at the common origin of S and S' at $t = t' = 0$, and the observers in each system measure the speed with which the flare's light spreads out. Both observers must find the same speed c (Fig. 1.4), which means that in the S frame

$$x = ct \tag{1.13}$$

and in the S' frame

$$x' = ct' \tag{1.14}$$

Substituting for x' and t' in Eq. (1.14) with the help of Eqs. (1.8) and (1.12) gives

$$k(x - \boldsymbol{v}t) = ckt + \left(\frac{1-k^2}{k\boldsymbol{v}}\right)cx$$

and solving for x,

$$x = \frac{ckt + \boldsymbol{v}kt}{k - \left(\dfrac{1-k^2}{k\boldsymbol{v}}\right)c} = ct\left[\frac{k + \dfrac{\boldsymbol{v}}{c}k}{k - \left(\dfrac{1-k^2}{k\boldsymbol{v}}\right)c}\right] = ct\left[\frac{1 + \dfrac{\boldsymbol{v}}{c}}{1 - \left(\dfrac{1}{k^2} - 1\right)\dfrac{c}{\boldsymbol{v}}}\right]$$

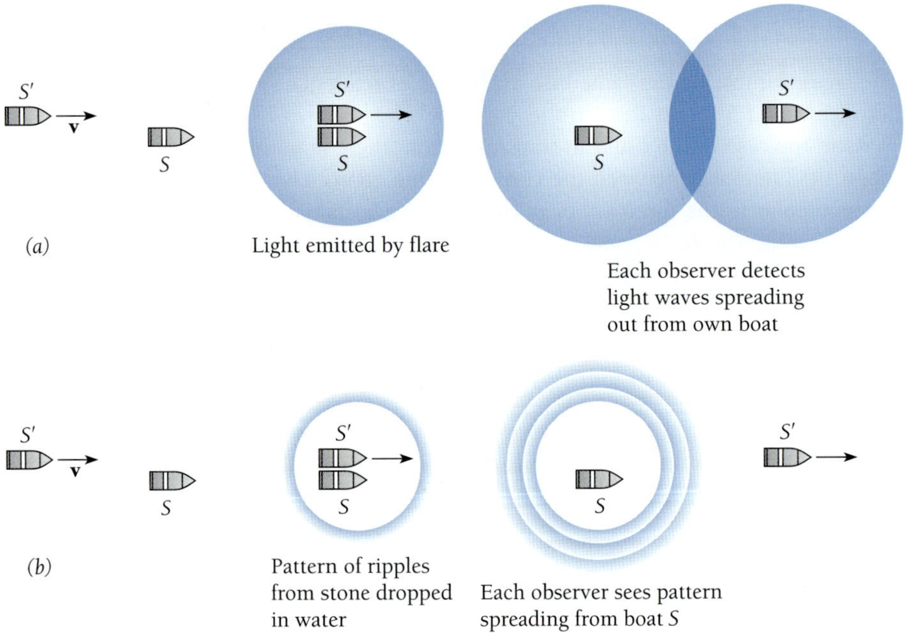

Figure 1.4 (*a*) Inertial frame S' is a boat moving at speed $\boldsymbol{v}$ in the $+x$ direction relative to another boat, which is the inertial frame S. When $t = t' = 0$, S' is next to S, and $x = x_0 = 0$. At this moment a flare is fired from one of the boats. An observer on boat S detects light waves spreading out at speed c from his boat. An observer on boat S' also detects light waves spreading out at speed c from her boat, even though S' is moving to the right relative to S. (*b*) If instead a stone were dropped in the water at $t = t' = 0$, the observers would find a pattern of ripples spreading out around S at different speeds relative to their boats. The difference between (*a*) and (*b*) is that water, in which the ripples move, is itself a frame of reference whereas space, in which light moves, is not.

This expression for x will be the same as that given by Eq. (1.13), namely, $x = ct$, provided that the quantity in the brackets equals 1. Therefore

$$\frac{1+\dfrac{v}{c}}{1-\left(\dfrac{1}{k^2}-1\right)\dfrac{c}{v}} = 1$$

and

$$k = \frac{1}{\sqrt{1 - v^2/c^2}} \tag{1.15}$$

Finally we put this value of k in Eqs. (1.11) and (1.15). Now we have the complete transformation of measurements of an event made in S to the corresponding measurements made in S':

Lorentz transformation

$$x' = \frac{x - vt}{\sqrt{1 - v^2/c^2}} \tag{1.16}$$

$$y' = y \tag{1.17}$$

$$z' = z \tag{1.18}$$

$$t' = \frac{t - \dfrac{vx}{c^2}}{\sqrt{1 - v^2/c^2}} \tag{1.19}$$

These equations comprise the **Lorentz transformation.** They were first obtained by the Dutch physicist H.A. Lorentz, who showed that the basic formulas of electromagnetism are the same in all inertial frames only when Eqs. (1.41) to (1.44) are used. It was not until several years later that Einstein discovered their full significance. It is obvious that the Lorentz transformation reduces to the Galilean transformation when the relative velocity v is small compared with the velocity of light c.

Example 1.1

Derive the relativistic length contraction using the Lorentz transformation.

Solution

Let us consider a rod lying along the x' axis in the moving frame S'. An observer in this frame determines the coordinates of its ends to be x'_1 and x'_2, and so the proper length of the rod is

$$L_0 = x'_2 - x'_1$$

In order to find $L = x_2 - x_1$, the length of the rod as measured in the stationary frame S at the time t, we make use of Eq. (1.16) to give

$$x'_1 = \frac{x_1 - vt}{\sqrt{1 - v^2/c^2}} \qquad x'_2 = \frac{x_2 - vt}{\sqrt{1 - v^2/c^2}}$$

Hence $$L = x_2 - x_1 = (x'_2 - x'_1)\sqrt{1 - v^2/c^2} = L_0\sqrt{1 - v^2/c^2}$$

This is the same as Eq. (1.27)

About Scientists

Hendrik A. Lorentz (1853–1928) was born in Arnhem, Holland, and studied at the University of Leyden. At nineteen he returned to Arnhem and taught at the high school there while preparing a doctoral thesis that extended Maxwell's theory of electromagnetism to cover the details of the refraction and reflection of light. In 1878 he became professor of theoretical physics at Leyden, the first such post in Holland, where he remained for thirty-four years until he moved to Haarlem. Lorentz went on to reformulate and simplify Maxwell's theory and to introduce the idea that electromagnetic fields are created by electric charges on the atomic level. He proposed that the emission of light by atoms and various optical phenomena could be traced to the motions and interactions of atomic electrons. The discovery in 1896 by Pieter Zeeman, a student of his, that the spectral lines of atoms that radiate in a magnetic field are split into components of slightly different frequency confirmed Lorentz's work and led to a Nobel Prize for both of them in 1902.

The set of equations that enables electromagnetic quantities in one frame of reference to be transformed into their values in another frame of reference moving relative to the first were found by Lorentz in 1895, although their full significance was not realized until Einstein's theory of special relativity ten years afterward. Lorentz (and, independently, the Irish physicist G. F. Fitzgerald) suggested that the negative result of the Michelson-Morley experiment could be understood if lengths in the direction of motion relative to an observer were contracted. Subsequent experiments showed that although such contractions do occur, they are not the real reason for the Michelson-Morley result, which is that there is no "ether" to serve as a universal frame of reference.

Inverse Lorentz Transformation

In Example 1.9 the coordinates of the ends of the moving rod were measured in the stationary frame S at the same time t, and it was easy to use Eq. (1.16) to find L in terms of L_0 and $\boldsymbol{v}$. If we want to examine time dilation, though, Eq. (1.19) is not convenient, because t_1 and t_2, the start and finish of the chosen time interval, must be measured when the moving clock is at the respective *different* positions x_1 and x_2. In situations of this kind it is easier to use the **inverse Lorentz transformation**, which converts measurements made in the moving frame S' to their equivalents in S.

To obtain the inverse transformation, primed and unprimed quantities in Eqs. (1.16) to (1.19) are exchanged, and $\boldsymbol{v}$ is replaced by $-\boldsymbol{v}$:

Inverse Lorentz transformation

$$x = \frac{x' + vt'}{\sqrt{1 - v^2/c^2}} \tag{1.20}$$

$$y = y' \tag{1.21}$$

$$z = z' \tag{1.22}$$

$$t = \frac{t' + \dfrac{vx'}{c^2}}{\sqrt{1 - v^2/c^2}} \tag{1.23}$$

Example 1.2

Derive the formula for time dilation using the inverse Lorentz transformation.

Solution

Let us consider a clock at the point x' in the moving frame S'. When an observer in S' finds that the time is t'_1, an observer in S will find it to be t_1, where, from Eq. (1.23),

$$t_1 = \frac{t'_1 + \dfrac{vx'}{c^2}}{\sqrt{1 - v^2/c^2}}$$

After a time interval of t_0 (to him), the observer in the moving system finds that the time is now t'_2 according to his clock. That is,

$$t_0 = t'_2 - t'_1$$

The observer in S, however, measures the end of the same time interval to be

$$t_2 = \frac{t'_2 + \dfrac{vx'}{c^2}}{\sqrt{1 - v^2/c^2}}$$

so to her the duration of the interval t is

$$t = t_2 - t_1 = \frac{t'_2 - t'_1}{\sqrt{1 - v^2/c^2}} = \frac{t_0}{\sqrt{1 - v^2/c^2}}$$

This is what we found earlier with the help of a light-pulse clock.

Velocity Addition

Special relativity postulates that the speed of light c in free space has the same value for all observers, regardless of their relative motion."Common sense" (which means here the Galilean transformation) tells us that if we throw a ball forward at 10 m/s from a car moving at 30 m/s, the ball's speed relative to the road will be 40 m/s, the sum of the two speeds. What if we switch on the car's headlights when its speed is $\mathbf{v}$? The same reasoning suggests that their light, which is emitted from the reference frame S' (the car) in the direction of its motion relative to another frame S (the road), ought to have a speed of $c + \mathbf{v}$ as measured in S. But this violates the above postulate, which has had ample experimental verification. Common sense is no more reliable as a guide in science than it is elsewhere, and we must turn to the Lorentz transformation equations for the correct scheme of velocity addition.

Suppose something is moving relative to both S and S'. An observer in S measures its three velocity components to be

$$V_x = \frac{dx}{dt} \qquad V_y = \frac{dy}{dt} \qquad V_z = \frac{dz}{dt}$$

while to an observer in S' they are

$$V'_x = \frac{dx'}{dt'} \qquad V'_y = \frac{dy'}{dt'} \qquad V'_z = \frac{dz'}{dt'}$$

By differentiating the inverse Lorentz transformation equations for x, y, z, and t, we obtain

$$dx = \frac{dx' + v\,dt'}{\sqrt{1 - v^2/c^2}} \qquad dy = dy' \qquad dz = dz' \qquad dt = \frac{dt' + \dfrac{v\,dx'}{c^2}}{\sqrt{1 - v^2/c^2}}$$

and so
$$V_x = \frac{dx}{dt} = \frac{dx' + v\,dt'}{dt' + \dfrac{v\,dx'}{c^2}} = \frac{\dfrac{dx'}{dt'} + v}{1 + \dfrac{v}{c^2}\dfrac{dx'}{dt'}}$$

Relativistic velocity transformation

$$V_x = \frac{V'_x + v}{1 + \dfrac{vV'_x}{c^2}} \tag{1.24}$$

Similarly,
$$V_y = \frac{V'_y\sqrt{1 - v^2/c^2}}{1 + \dfrac{vV'_x}{c^2}} \tag{1.25}$$

$$V_z = \frac{V'_z\sqrt{1 - v^2/c^2}}{1 + \dfrac{vV'_x}{c^2}} \tag{1.26}$$

If $V'_x = c$, that is, if light is emitted in the moving frame S' in its direction of motion relative to S, an observer in frame S will measure the speed

$$V_x = \frac{V'_x + v}{1 + \dfrac{vV'_x}{c^2}} = \frac{c + v}{1 + \dfrac{vc}{c^2}} = \frac{c(c + v)}{c + v} = c$$

Thus observers in the car and on the road both find the same value for the speed of light, as they must.

Example 1.3

Spacecraft Alpha is moving at $0.90c$ with respect to the earth. If spacecraft Beta is to pass Alpha at a relative speed of $0.50c$ in the same direction, what speed must Beta have with respect to the earth?

Solution

According to the Galilean transformation, Beta would need a speed relative to the earth of $0.90c + 0.50c = 1.40c$, which we know is impossible. According to Eq. (1.24), however, with $V'_x = 0.50c$ and $v = 0.90c$, the required speed is only

$$V_x = \frac{V'_x + v}{1 + \dfrac{vV'_x}{c^2}} = \frac{0.50c + 0.90c}{1 + \dfrac{(0.90c)(0.50c)}{c^2}} = 0.97c$$

which is less than c. It is necessary to go less than 10 percent faster than a spacecraft traveling at $0.90c$ in order to pass it at a relative speed of $0.50c$.

Simultaneity

The relative character of time as well as space has many implications. Notably, events that seem to take place simultaneously to one observer may not be simultaneous to another observer in relative motion, and vice versa.

Let us examine two events—the setting off of a pair of flares, say—that occur at the same time t_0 to somebody on the earth but at the different locations x_1 and x_2. What does the pilot of a spacecraft in flight see? To her, the flare at x_1 and t_0 appears at the time

$$t_1' = \frac{t_0 - vx_1/c^2}{\sqrt{1 - v^2/c^2}}$$

according to Eq. (1.19), while the flare at x_2 and t_0 appears at the time

$$t_2' = \frac{t_0 - vx_2/c^2}{\sqrt{1 - v^2/c^2}}$$

Hence two events that occur simultaneously to one observer are separated by a time interval of

$$t_2' - t_1' = \frac{v(x_1 - x_2)/c^2}{\sqrt{1 - v^2/c^2}}$$

to an observer moving at the speed $\boldsymbol{v}$ relative to the other observer. Who is right? The question is, of course, meaningless: both observers are "right" since each simply measures what he or she sees.

Because simultaneity is a relative concept and not an absolute one, physical theories that require simultaneity in events at different locations cannot be valid. For instance, saying that total energy is conserved in an isolated system does not rule out a process in which an amount of energy ΔE vanishes at one place while an equal amount of energy ΔE comes into being somewhere else with no actual transport of energy from one place to the other. Because simultaneity is relative, some observers of the process will find energy not being conserved. To rescue conservation of energy in the light of special relativity, then, we have to say that, when energy disappears somewhere and appears elsewhere, it has actually flowed from the first location to the second. Thus energy is conserved *locally* everywhere, not merely when an isolated system is considered—a much stronger statement of this principle.

1.3 ELECTRICITY AND MAGNETISM

Relativity is the bridge

One of the puzzles that set Einstein on the trail of special relativity was the connection between electricity and magnetism, and the ability of his theory to clarify the nature of this connection is one of its triumphs.

Because the moving charges (usually electrons) whose interactions give rise to many of the magnetic forces familiar to us have speeds far smaller than c, it is not obvious that the operation of an electric motor, say, is based on a relativistic effect. The idea becomes less implausible, however, when we reflect on the strength of electric forces. The electric attraction between the electron and proton in a hydrogen atom, for instance, is 10^{39} times greater than the gravitational attraction between them. Thus even a small change in the character of these forces due to relative motion, which is what magnetic forces represent, may have large consequences. Furthermore, although the effective speed of an individual electron in a current-carrying wire (<1 mm/s) is less than that of a tired caterpillar, there may be 10^{20} or more moving electrons per centimeter in such a wire, so the total effect may be considerable.

Although the full story of how relativity links electricity and magnetism is mathematically complex, some aspects of it are easy to appreciate. An example is the origin of the magnetic force between two parallel currents. An important point is that, like the speed of light,

Electric charge is relativistically invariant.

A charge whose magnitude is found to be Q in one frame of reference is also Q in all other frames.

Let us look at the two idealized conductors shown in Fig. 1.5*a*. They contain equal numbers of positive and negative charges at rest that are equally spaced. Because the conductors are electrically neutral, there is no force between them.

Figure 1.5*b* shows the same conductors when they carry currents i_I and i_{II} in the same direction. The positive charges move to the right and the negative charges move to the left, both at the same speed $\boldsymbol{v}$ as seen from the laboratory frame of reference. (Actual currents in metals consist of flows of negative electrons only, of course, but the electrically equivalent model here is easier to analyze and the results are the same.) Because the charges are moving, their spacing is smaller than before by the factor $\sqrt{1 - v^2/c^2}$. Since $\boldsymbol{v}$ is the same for both sets of charges, their spacings shrink by the same amounts, and both conductors remain neutral to an observer in the laboratory. However, the conductors now attract each other. Why?

Let us look at conductor II from the frame of reference of one of the negative charges in conductor I. Because the negative charges in II appear at rest in this frame, their spacing is not contracted, as in Fig. 1.5*c*. On the other hand, the positive charges in II now have the velocity $2\boldsymbol{v}$, and their spacing is accordingly contracted to a greater extent than they are in the laboratory frame. Conductor II therefore appears to have a net positive charge, and an attractive force acts on the negative charge in I.

Next we look at conductor II from the frame of reference of one of the positive charges in conductor I. The positive charges in II are now at rest, and the negative charges there move to the left at the speed $2\boldsymbol{v}$. Hence the negative charges are closer together than the positive ones, as in Fig. 1.5*d*, and the entire conductor appears negatively charged. An attractive force therefore acts on the positive charges in I.

Identical arguments show that the negative and positive charges in II are attracted to I. Thus all the charges in each conductor experience forces directed toward the other conductor. To each charge, the force on it is an "ordinary" electric force that arises because the charges of opposite sign in the other conductor are closer together than the charges of the same sign, so the other conductor appears to have a net charge. From the laboratory frame the situation is less straightforward. Both conductors are electrically neutral in this frame, and it is natural to explain their mutual attraction by attributing it to a special "magnetic" interaction between the currents.

A similar analysis explains the repulsive force between parallel conductors that carry currents in opposite directions. Although it is convenient to think of magnetic forces as being different from electric ones, they both result from a single electromagnetic interaction that occurs between charged particles.

Clearly a current-carrying conductor that is electrically neutral in one frame of reference might not be neutral in another frame. How can this observation be reconciled with charge invariance? The answer is that we must consider the entire circuit of which

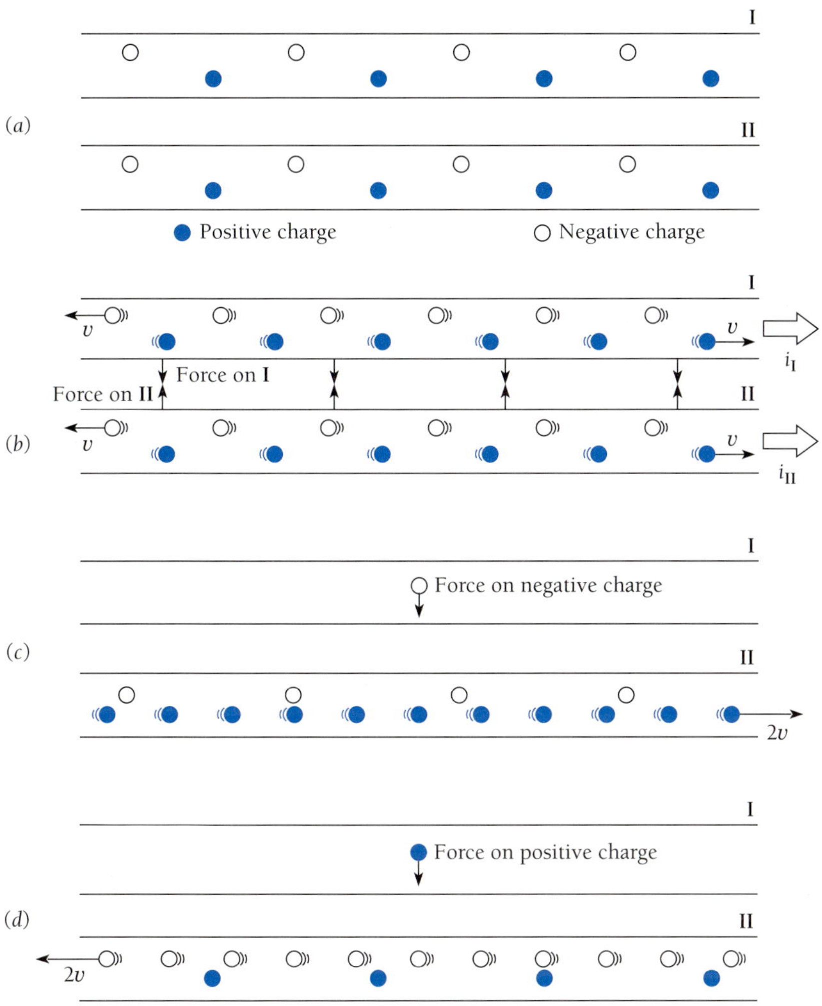

Figure 1.5 How the magnetic attraction between parallel currents arises. (*a*) Idealized parallel conductors that contain equal numbers of positive and negative charges. (*b*) When the conductors carry currents, the spacing of their moving charges undergoes a relativistic contraction as seen from the laboratory. The conductors attract each other when i_I and i_{II} are in the same direction. (*c*) As seen by a negative charge in I, the negative charges in II are at rest whereas the positive charges are in motion. The contracted spacing of the latter leads to a net positive charge in II that attracts the negative charge in I. (*d*) As seen by a positive charges in I, the positive charges in II are at rest whereas the negative charges are in motion. The contracted spacing of the latter leads to a net negative charge on II that attrats the positive charge in I. The contracted spacings in *b*, *c*, and *d* are greatly exaggerated.

the conductor is a part. Because the circuit must be closed for a current to occur in it, for every current element in one direction that a moving observer finds to have, say, a positive charge, there must be another current element in the opposite direction which the same observer finds to have a negative charge. Hence magnetic forces always act between different parts of the same circuit, even though the circuit as a whole appears electrically neutral to all observers.

The preceding discussion considered only a particular magnetic effect. All other magnetic phenomena can also be interpreted on the basis of Coulomb's law, charge invariance, and special relativity, although the analysis is usually more complicated.

1.4 LENGTH CONTRACTION

Faster means shorter

Measurements of lengths as well as of time intervals are affected by relative motion. The length L of an object in motion with respect to an observer always appears to the observer to be shorter than its length L_0 when it is at rest with respect to him. This contraction occurs only in the direction of the relative motion. The length L_0 of an object in its rest frame is called its **proper length.** (We note that in Fig. 1.9 the clock is moving perpendicular to $\mathbf{v}$, hence $L = L_0$ there.)

The length contraction can be derived in a number of ways. Perhaps the simplest is based on time dilation and the principle of relativity. Let us consider what happens to unstable particles called muons that are created at high altitudes by fast cosmic-ray particles (largely protons) from space when they collide with atomic nuclei in the earth's atmosphere. A muon has a mass 207 times that of the electron and has a charge of either $+e$ or $-e$; it decays into an electron or a positron after an average lifetime of 2.2 μs (2.2×10^{-6} s).

Cosmic-ray muons have speeds of about 2.994×10^8 m/s ($0.998c$) and reach sea level in profusion—one of them passes through each square centimeter of the earth's surface on the average slightly more often than once a minute. But in $t_0 = 2.2\ \mu$s, their average lifetime, muons can travel a distance of only

$$vt_0 = (2.994 \times 10^8 \text{ m/s})(2.2 \times 10^{-6} \text{ s}) = 6.6 \times 10^2 \text{ m} = 0.66 \text{ km}$$

before decaying, whereas they are actually created at altitudes of 6 km or more.

To resolve the paradox, we note that the muon lifetime of $t_0 = 2.2\ \mu$s is what an observer at rest with respect to a muon would find. Because the muons are hurtling toward us at the considerable speed of $0.998c$, their lifetimes are extended in our frame of reference by time dilation to

$$t = \frac{t_0}{\sqrt{1 - v^2/c^2}} = \frac{2.2 \times 10^{-6} \text{ s}}{\sqrt{1 - (0.998c)^2/c^2}} = 34.8 \times 10^{-6} \text{ s} = 34.8\ \mu\text{s}$$

The moving muons have lifetimes almost 16 times longer than those at rest. In a time interval of 34.8 μs, a muon whose speed is $0.998c$ can cover the distance

$$vt = (2.994 \times 10^8 \text{ m/s})(34.8 \times 10^{-6} \text{ s}) = 1.04 \times 10^4 \text{ m} = 10.4 \text{ km}$$

Although its lifetime is only $t_0 = 2.2\ \mu$s in its own frame of reference, a muon can reach the ground from altitudes of as much as 10.4 km because in the frame in which these altitudes are measured, the muon lifetime is $t = 34.8\ \mu$s.

What if somebody were to accompany a muon in its descent at $v = 0.998c$, so that to him or her the muon is at rest? The observer and the muon are now in the same frame of reference, and in this frame the muon's lifetime is only 2.2 μs. To the observer, the muon can travel only 0.66 km before decaying. The only way to account for the arrival of the muon at ground level is if the distance it travels, from the point of view of an observer in the moving frame, is shortened by virtue of its motion (Fig. 1.6). The principle of relativity tells us the extent of the shortening—it must be by the same factor of $\sqrt{1 - v^2/c^2}$ that the muon lifetime is extended from the point of view of a stationary observer.

We therefore conclude that an altitude we on the ground find to be h_0 must appear in the muon's frame of reference as the lower altitude

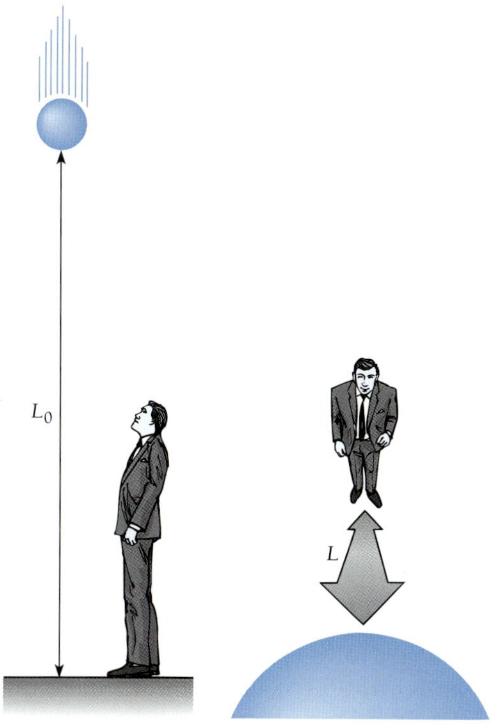

As found by observer on the ground, the muon altitude is L_0.

As found by an observer moving with the muon, the ground is L below it, which is a shorter distance than L_0.

Figure 1.6 Muon decay as seen by different observers. The muon size is greatly exaggerated here; in fact, the muon seems likely to be a point particle with no extension in space.

$$h = h_0\sqrt{1 - v^2/c^2}$$

In our frame of reference the muon can travel $h_0 = 10.4$ km because of time dilation. In the muon's frame of reference, where there is no time dilation, this distance is abbreviated to

$$h = (10.4 \text{ km})\sqrt{1 - (0.998c)^2/c^2} = 0.66 \text{ km}$$

As we know, a muon traveling at $0.998c$ goes this far in 2.2 μs.

The relativistic shortening of distances is an example of the general contraction of lengths in the direction of motion:

Length contraction

$$L = L_0\sqrt{1 - v^2/c^2} \tag{1.27}$$

Figure 1.7 is a graph of L/L_0 versus v/c. Clearly the length contraction is most significant at speeds near that of light. A speed of 1000 km/s seems fast to us, but it only results in a shortening in the direction of motion to 99.9994 percent of the proper length of an object moving at this speed. On the other hand, something traveling at nine-tenths the speed of light is shortened to 44 percent of its proper length, a significant change.

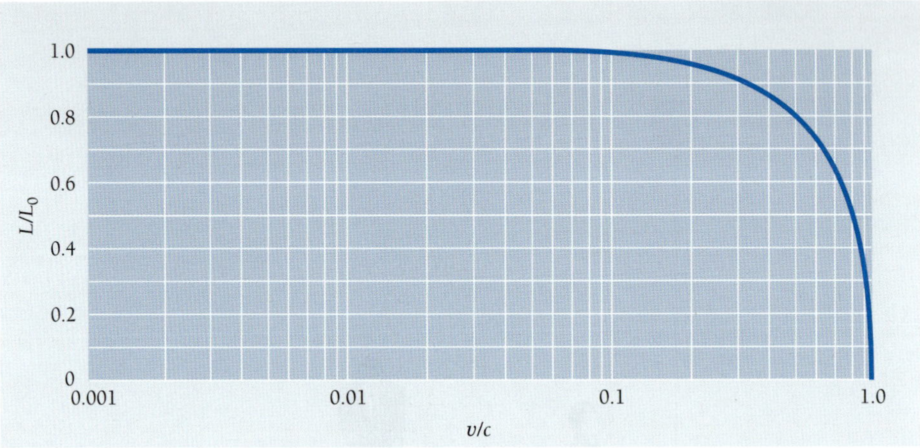

Figure 1.7 Relativistic length contraction. Only lengths in the direction of motion are affected. The horizontal scale is logarithmic.

Like time dilation, the length contraction is a reciprocal effect. To a person in a spacecraft, objects on the earth appear shorter than they did when he or she was on the ground by the same factor of $\sqrt{1 - v^2/c^2}$ that the spacecraft appears shorter to somebody at rest. The proper length L_0 found in the rest frame is the maximum length any observer will measure. As mentioned earlier, only lengths in the direction of motion undergo contraction. Thus to an outside observer a spacecraft is shorter in flight than on the ground, but it is not narrower.

1.5 TIME DILATION

A moving clock ticks more slowly than a clock at rest

Measurements of time intervals are affected by relative motion between an observer and what is observed. As a result, a clock that moves with respect to an observer ticks more slowly than it does without such motion, and all processes (including those of life) occur more slowly to an observer when they take place in a different inertial frame.

If someone in a moving spacecraft finds that the time interval between two events in the spacecraft is t_0, we on the ground would find that the same interval has the longer duration t. The quantity t_0, which is determined by events that occur *at the same place* in an observer's frame of reference, is called the **proper time** of the interval between the events. When witnessed from the ground, the events that mark the beginning and end of the time interval occur at different places, and in consequence the duration of the interval appears longer than the proper time. This effect is called **time dilation** (to dilate is to become larger).

To see how time dilation comes about, let us consider two clocks, both of the particularly simple kind shown in Fig. 1.8. In each clock a pulse of light is reflected back and forth between two mirrors L_0 apart. Whenever the light strikes the lower mirror, an electric signal is produced that marks the recording tape. Each mark corresponds to the tick of an ordinary clock.

One clock is at rest in a laboratory on the ground and the other is in a spacecraft that moves at the speed $\mathbf{v}$ relative to the ground. An observer in the laboratory watches both clocks: does she find that they tick at the same rate?

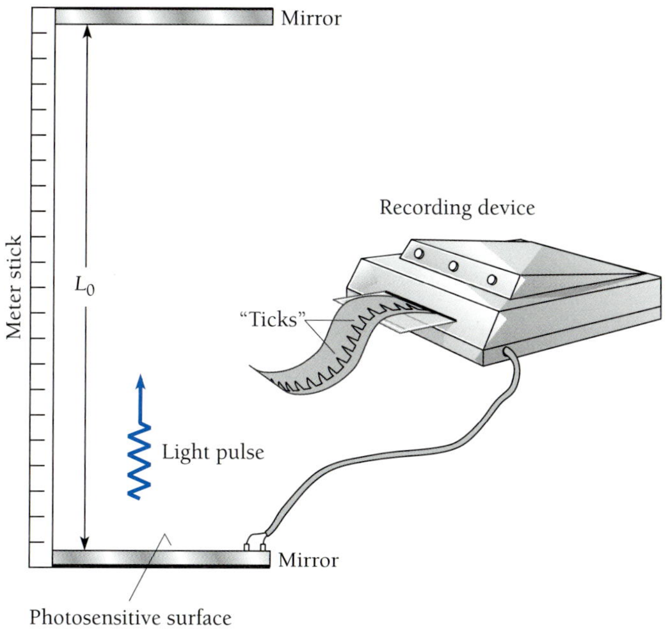

Figure 1.8 A simple clock. Each "tick" corresponds to a round trip of the light pulse from the lower mirror to the upper one and back.

Figure 1.9 shows the laboratory clock in operation. The time interval between ticks is the proper time t_0 and the time needed for the light pulse to travel between the mirrors at the speed of light c is $t_0/2$. Hence $t_0/2 = L_0/c$ and

$$t_0 = \frac{2L_0}{c} \tag{1.28}$$

Figure 1.10 shows the moving clock with its mirrors perpendicular to the direction of motion relative to the ground. The time interval between ticks is t. Because the clock is moving, the light pulse, as seen from the ground, follows a zigzag path. On its way from the lower mirror to the upper one in the time $t/2$, the pulse travels a horizontal distance of $\boldsymbol{v}(t/2)$ and a total distance of $c(t/2)$. Since L_0 is the vertical distance between the mirrors,

$$\left(\frac{ct}{2}\right)^2 = L_0^2 + \left(\frac{\boldsymbol{v}t}{2}\right)^2$$

$$\frac{t^2}{4}(c^2 - \boldsymbol{v}^2) = L_0^2$$

$$t^2 = \frac{4L_0^2}{c^2 - \boldsymbol{v}^2} = \frac{(2L_0)^2}{c^2(1 - \boldsymbol{v}^2/c^2)}$$

$$t = \frac{2L_0/c}{\sqrt{1 - \boldsymbol{v}^2/c^2}} \tag{1.29}$$

But $2L_0/c$ is the time interval t_0 between ticks on the clock on the ground, as in Eq. (1.28), and so

Time dilation

$$t = \frac{t_0}{\sqrt{1 - \boldsymbol{v}^2/c^2}} \tag{1.30}$$

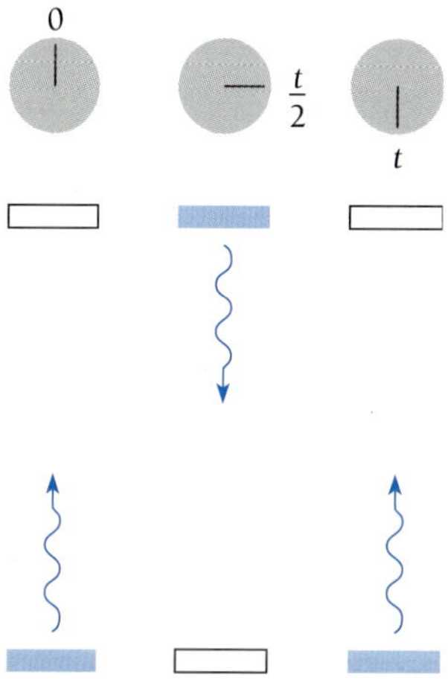

Figure 1.9 A light-pulse clock at rest on the ground as seen by an observer on the ground. The dial represents a conventional clock on the ground.

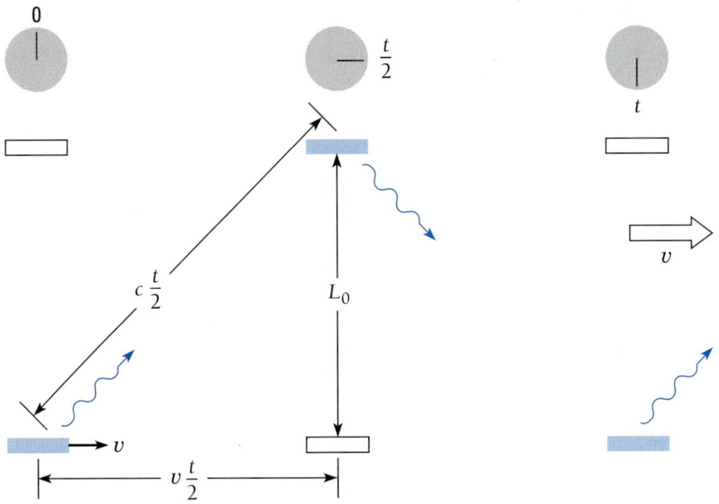

Figure 1.10 A light-pulse clock in a spacecraft as seen by an observer on the ground. The mirrors are parallel to the direction of motion of the spacecraft. The dial represents a conventional clock on the ground.

Here is a reminder of what the symbols in Eq. (1.31) represent:

t_0 = time interval on clock at rest relative to an observer = proper time
t = time interval on clock in motion relative to an observer
v = speed of relative motion
c = speed of light

Because the quantity $\sqrt{1 - v^2/c^2}$ is always smaller than 1 for a moving object, t is always greater than t_0. The moving clock in the spacecraft appears to tick at a slower rate than the stationary one on the ground, as seen by an observer on the ground.

Exactly the same analysis holds for measurements of the clock on the ground by the pilot of the spacecraft. To him, the light pulse of the ground clock follows a zigzag path that requires a total time t per round trip. His own clock, at rest in the spacecraft, ticks at intervals of t_0. He too finds that

$$t = \frac{t_0}{\sqrt{1 - v^2/c^2}}$$

so the effect is reciprocal: *every* observer finds that clocks in motion relative to him tick more slowly than clocks at rest relative to him.

Our discussion has been based on a somewhat unusual clock. Do the same conclusions apply to ordinary clocks that use machinery—spring-controlled escapements, tuning forks, vibrating quartz crystals, or whatever—to produce ticks at constant time intervals? The answer must be yes, since if a mirror clock and a conventional clock in the spacecraft agree with each other on the ground but not when in flight, the disagreement between then could be used to find the speed of the spacecraft independently of any outside frame of reference—which contradicts the principle that all motion is relative.

The Ultimate Speed Limit

The earth and the other planets of the solar system seem to be natural products of the evolution of the sun. Since the sun is a rather ordinary star in other ways, it is not surprising that other stars have been found to have planetary systems around them as well. Life developed here on earth, and there is no known reason why it should not also have done so on some of these planets. Can we expect ever to be able to visit them and meet our fellow citizens of the universe? The trouble is that nearly all stars are very far away—thousands or millions of light-years away. (A light-year, the distance light travels in a year, is 9.46×10^{15} m.) But if we can build a spacecraft whose speed is thousands or millions of times greater than the speed of light c, such distances would not be an obstacle.

Alas, a simple argument based on Einstein's postulates shows that nothing can move faster than c. Suppose you are in a spacecraft traveling at a constant speed $\mathbf{v}$ relative to the earth that is greater than c. As I watch from the earth, the lamps in the spacecraft suddenly go out. You switch on a flashlight to find the fuse box at the front of the spacecraft and change the blown fuse (Fig. 1.11*a*). The lamps go on again.

From the ground, though, I would see something quite different. To me, since your speed $\mathbf{v}$ is greater than c, the light from your flashlight illuminates the *back* of the spacecraft (Fig. 1.11*b*). I can only conclude that the laws of physics are different in your inertial frame from what they are in my inertial frame—which contradicts the principle of relativity. The only way to avoid this contradiction is to assume that nothing can move faster than the speed of light. This assumption has been tested experimentally many times and has always been found to be correct

The speed of light c in relativity is always its value in free space of 3.00×10^8 m/s. In all material media, such as air, water, or glass, light travels more slowly than this, and atomic particles are able to move faster in such media than does light. When an electrically charged particle moves through a transparent substance at a speed exceeding that of light in the substance, a cone of light waves is emitted that corresponds to the bow wave produced by a ship moving through the water faster than water waves do. These light waves are known as **Cerenkov radiation** and form the basis of a method of determining the speeds of such particles. The minimum speed a particle must have to emit Cerenkov radiation is c/n in a medium whose index of refraction is n. Cerenkov radiation is visible as a bluish glow when an intense beam of particles is involved.

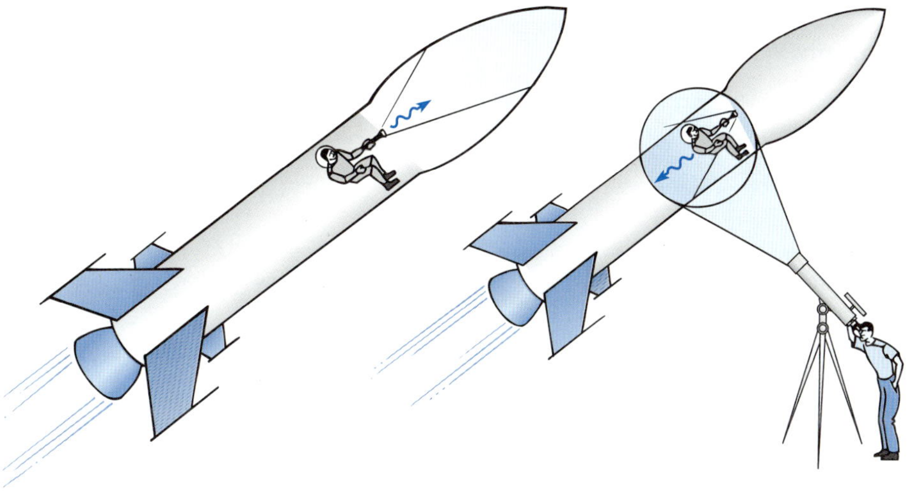

(*a*) (*b*)

Figure 1.11 A person switches on a flashlight in a spacecraft assumed to be moving relative to the earth faster than light. (*a*) In the spacecraft frame, the light goes to the front of the spacecraft. (*b*) In the earth frame, the light goes to the back of the spacecraft. Because observers in the spacecraft and on the earth would see different events, the principle of relativity would be violated. The conclusion is that the spacecraft cannot be moving faster than light relative to the earth (or relative to anything else).

About Scientists

Albert Einstein (1879–1955), bitterly unhappy with the rigid discipline of the schools of his native Germany, went at sixteen to Switzerland to complete his education, and later got a job examining patent applications at the Swiss Patent Office. Then, in 1905, ideas that had been germinating in his mind for years when he should have been paying attention to other matters (one of his math teachers called Einstein a "lazy dog") blossomed into three short papers that were to change decisively the course not only of physics but of modern civilization as well.

The first paper, on the photoelectric effect, proposed that light has a dual character with both particle and wave properties. The subject of the second paper was Brownian motion, the irregular zigzag movement of tiny bits of suspended matter, such as pollen grains in water. Einstein showed that Brownian motion results from the bombardment of the particles by randomly moving molecules in the fluid in which they are suspended. This provided the long-awaited definite link with experiment that convinced the remaining doubters of the molecular theory of matter. The third paper introduced the special theory of relativity.

Although much of the world of physics was originally either indifferent or skeptical, even the most unexpected of Einstein's conclusions were soon confirmed and the development of what is now called modern physics began in earnest. After university posts in Switzerland and Czechoslovakia, in 1913 he took up an appointment at the Kaiser Wilhelm Institute in Berlin that left him able to do research free of financial worries and routine duties. Einstein's interest was now mainly in gravitation, and he started where Newton had left off more than two centuries earlier.

Einstein's general theory of relativity, published in 1916, related gravity to the structure of space and time. In this theory the force of gravity can be thought of as arising from a warping of spacetime around a body of matter so that a nearby mass tends to move toward it, much as a marble rolls toward the bottom of a saucer-shaped hole. From general relativity came a number of remarkable predictions, such as that light should be subject to gravity, all of which were verified experimentally. The later discovery that the universe is expanding fit neatly into the theory. In 1917 Einstein introduced the idea of stimulated emission of radiation, an idea that bore fruit forty years later in the invention of the laser.

The development of quantum mechanics in the 1920s disturbed Einstein, who never accepted its probabilistic rather than deterministic view of events on an atomic scale. "God does not play dice with the world," he said, but for once his physical intuition seemed to be leading him in the wrong direction.

Einstein, by now a world celebrity, left Germany in 1933 after Hitler came to power and spent the rest of his life at the Institute for Advanced Study in Princeton, New Jersey, thereby escaping the fate of millions of other European Jews at the hands of the Germans. His last years were spent in an unsuccessful search for a theory that would bring gravitation and electromagnetism together into a single picture, a problem worthy of his gifts but one that remains unsolved to this day.

Example 1.4

A spacecraft is moving relative to the earth. An observer on the earth finds that, between 1 P.M. and 2 P.M. according to her clock, 3601 s elapse on the spacecraft's clock. What is the spacecraft's speed relative to the earth?

Solution

Here $t_0 = 3600$ s is the proper time interval on the earth and $t = 3601$ s is the time interval in the moving frame as measured from the earth. We proceed as follows:

$$t = \frac{t_0}{\sqrt{1 - v^2/c^2}}$$

$$1 - \frac{v^2}{c^2} = \left(\frac{t_0}{t}\right)^2$$

$$v = c\sqrt{1 - \left(\frac{t_0}{t}\right)^2} = (2.998 \times 10^8 \text{ m/s})\sqrt{1 - \left(\frac{3600 \text{ s}}{3601 \text{ s}}\right)^2}$$

$$= 7.1 \times 10^6 \text{ m/s}$$

Today's spacecraft are much slower than this. For instance, the highest speed of the Apollo 11 spacecraft that went to the moon was only 10,840 m/s, and its clocks differed from those on the earth by less than one part in 10^9. Most of the experiments that have confirmed time dilation made use of unstable nuclei and elementary particles which readily attain speeds not far from that of light.

Apollo 11 lifts off its pad to begin the first human visit to the moon. At its highest speed of 10.8 km/s relative to the earth, its clocks differed from those on the earth by less than one part in a billion.

Although time is a relative quantity, not all the notions of time formed by everyday experience are incorrect. Time does not run backward to *any* observer, for instance. A sequence of events that occur at some particular point at $t_1, t_2, t_3, \ldots$ will appear in the same order to all observers everywhere, though not necessarily with the same time intervals $t_2 - t_1, t_3 - t_2, \ldots$ between each pair of events. Similarly, no distant observer, regardless of his or her state of motion, can see an event before it happens—more precisely, before a nearby observer sees it—since the speed of light is finite and signals require the minimum period of time L/c to travel a distance L. There is no way to peer into the future, although past events may appear different to different observers.

1.6 TWIN PARADOX

A longer life, but it will not seem longer

We are now in a position to understand the famous relativistic effect known as the twin paradox. This paradox involves two identical clocks, one of which remains on the earth while the other is taken on a voyage into space at the speed v and eventually is brought back. It is customary to replace the clocks with the pair of twins Dick and Jane, a substitution that is perfectly acceptable because the processes of life—heartbeats, respiration, and so on—constitute biological clocks of reasonable regularity.

Dick is 20 y old when he takes off on a space voyage at a speed of 0.80c to a star 20 light-years away. To Jane, who stays behind, the pace of Dick's life is slower than hers by a factor of

$$\sqrt{1 - v^2/c^2} = \sqrt{1 - (0.80c)^2/c^2} = 0.60 = 60\%$$

To Jane, Dick's heart beats only 3 times for every 5 beats of her heart; Dick takes only 3 breaths for every 5 of hers; Dick thinks only 3 thoughts for every 5 of hers. Finally Dick returns after 50 years have gone by according to Jane's calendar, but to Dick the trip has taken only 30 y. Dick is therefore 50 y old whereas Jane, the twin who stayed home, is 70 y old (Fig. 1.12).

Where is the paradox? If we consider the situation from the point of view of Dick in the spacecraft, Jane on the earth is in motion relative to him at a speed of 0.80c. Should not Jane then be 50 y old when the spacecraft returns, while Dick is then 70—the precise opposite of what was concluded above?

But the two situations are not equivalent. Dick changed from one inertial frame to a different one when he started out, when he reversed direction to head home, and when he landed on the earth. Jane, however, remained in the same inertial frame during Dick's whole voyage. The time dilation formula applies to Jane's observations of Dick, but not to Dick's observations of her.

To look at Dick's voyage from his perspective, we must take into account that the distance L he covers is shortened to

$$L = L_0\sqrt{1 - v^2/c^2} = (20 \text{ light-years})\sqrt{1 - (0.80c)^2/c^2} = 12 \text{ light-years}$$

To Dick, time goes by at the usual rate, but his voyage to the star has taken $L/v = 15$ y and his return voyage another 15 y, for a total of 30 y. Of course, Dick's life span has not been extended *to him*, because regardless of Jane's 50-y wait, he has spent only 30 y on the roundtrip.

Figure 1.12 An astronaut who returns from a space voyage will be younger than his or her twin who remains on earth. Speeds close to the speed of light (here $v = 0.8c$) are needed for this effect to be conspicuous.

The nonsymmetric aging of the twins has been verified by experiments in which accurate clocks were taken on an airplane trip around the world and then compared with identical clocks that had been left behind. An observer who departs from an inertial system and then returns after moving relative to that system will always find his or her clocks slow compared with clocks that stayed in the system.

Example 1.5

Dick and Jane each send out a radio signal once a year while Dick is away. How many signals does Dick receive? How many does Jane receive?

Solution

On the outward trip, Dick and Jane are being separated at a rate of $0.80c$. With the help of the reasoning used to analyze the doppler effect in Sec. 1.3, we find that each twin receives signals

$$T_1 = t_0\sqrt{\frac{1 + v/c}{1 - v/c}} = (1\ \text{y})\sqrt{\frac{1 + 0.80}{1 - 0.80}} = 3\ \text{y}$$

apart. On the return trip, Dick and Jane are getting closer together at the same rate, and each receives signals more frequently, namely

$$T_2 = t_0\sqrt{\frac{1 - v/c}{1 + v/c}} = (1\ \text{y})\sqrt{\frac{1 - 0.80}{1 + 0.80}} = \frac{1}{3}\ \text{y}$$

apart.

To Dick, the trip to the star takes 15 y, and he receives $15/3 = 5$ signals from Jane. During the 15 y of the return trip, Dick receives $15/(1/3) = 45$ signals from Jane, for a total of 50 signals. Dick therefore concludes that Jane has aged by 50 y in his absence. Both Dick and Jane agree that Jane is 70 y old at the end of the voyage.

To Jane, Dick needs $L_0/v = 25$ y for the outward trip. Because the star is 20 light-years away. Jane on the earth continues to receive Dick's signals at the original rate of one every 3 y for 20 y after Dick has arrived at the star. Hence Jane receives signals every 3 y for 25 y + 20 y = 45 y to give a total of $45/3 = 15$ signals. (These are the 15 signals Dick sent out on the outward trip.) Then, for the remaining 5 y of what is to Jane a 50-y voyage, signals arrive from Dick at the shorter intervals of $1/3$ y for an additional $5/(1/3) = 15$ signals. Jane thus receives 30 signals in all and concludes that Dick has aged by 30 y during the time he was away—which agrees with Dick's own figure. Dick is indeed 20 y younger than his twin Jane on his return.

1.7 RELATIVISTIC MOMENTUM

Redefining an important quantity

In classical mechanics linear momentum $\mathbf{p} = m\mathbf{v}$ is a useful quantity because it is conserved in a system of particles not acted upon by outside forces. When an event such as a collision or an explosion occurs inside an isolated system, the vector sum of the momenta of its particles before the event is equal to their vector sum afterward. We now have to ask whether $\mathbf{p} = m\mathbf{v}$ is valid as the definition of momentum in inertial frames in relative motion, and if not, what a relativistically correct definition is.

To start with, we require that $\mathbf{p}$ be conserved in a collision for all observers in relative motion at constant velocity. Also, we know that $\mathbf{p} = m\mathbf{v}$ holds in classical mechanics, that is, for $v \ll c$. Whatever the relativistically correct $\mathbf{p}$ is, then, it must reduce to $m\mathbf{v}$ for such velocities.

Let us consider an elastic collision (that is, a collision in which kinetic energy is conserved) between two particles A and B, as witnessed by observers in the reference frames S and S' which are in uniform relative motion. The properties of A and B are identical when determined in reference frames in which they are at rest. The frames S and S' are oriented as in Fig. 1.13, with S' moving in the $+x$ direction with respect to S at the velocity $\mathbf{v}$.

Before the collision, particle A had been at rest in frame S and particle B in frame S'. Then, at the same instant, A was thrown in the $+y$ direction at the speed V_A while B was thrown in the $-y'$ direction at the speed V'_B, where

$$V_A = V'_B \tag{1.31}$$

Hence the behavior of A as seen from S is exactly the same as the behavior of B as seen from S'.

When the two particles collide, A rebounds in the $-y$ direction at the speed V_A, while B rebounds in the $+y'$ direction at the speed V'_B. If the particles are thrown from positions Y apart, an observer in S finds that the collision occurs at $y = \frac{1}{2}Y$ and one in S' finds that it occurs at $y' = y = \frac{1}{2}Y$. The round-trip time T_0 for A as measured in frame S is therefore

$$T_0 = \frac{Y}{V_A} \tag{1.32}$$

and it is the same for B in S':

$$T_0 = \frac{Y}{V'_B}$$

In S the speed V_B is found from

$$V_B = \frac{Y}{T} \tag{1.33}$$

where T is the time required for B to make its round trip *as measured in S*. In S', however, B's trip requires the time T_0, where

$$T = \frac{T_0}{\sqrt{1 - v^2/c^2}} \tag{1.34}$$

according to our previous results. Although observers in both frames see the same event, they disagree about the length of time the particle thrown from the other frame requires to make the collision and return.

Replacing T in Eq. (1.33) with its equivalent in terms of T_0, we have

$$V_B = \frac{Y\sqrt{1 - v^2/c^2}}{T_0}$$

From Eq. (1.32),

$$V_A = \frac{Y}{T_0}$$

If we use the classical definition of momentum, $\mathbf{p} = m\mathbf{v}$, then in frame S

$$p_A = m_A V_A = m_A\left(\frac{Y}{T_0}\right)$$

$$p_B = m_B V_B = m_B\sqrt{1 - v^2/c^2}\left(\frac{Y}{T_0}\right)$$

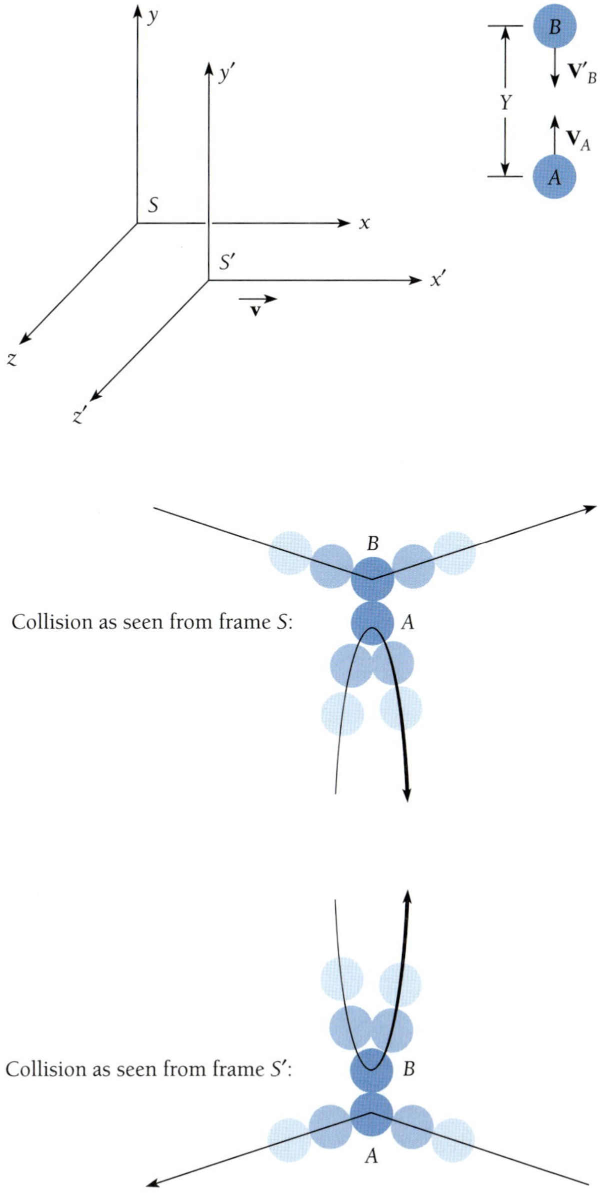

Figure 1.13 An elastic collision as observed in two different frames of reference. The balls are initially Y apart, which is the same distance in both frames since S' moves only in the x direction.

This means that, in this frame, momentum will not be conserved if $m_A = m_B$, where m_A and m_B are the masses as measured in S. However, if

$$m_B = \frac{m_A}{\sqrt{1 - v^2/c^2}} \tag{1.35}$$

then momentum *will* be conserved.

In the collision of Fig. 1.13 both A and B are moving in both frames. Suppose now that V_A and V_B' are very small compared with $\boldsymbol{v}$, the relative velocity of the two frames.

In this case an observer in S will see B approach A with the velocity v, make a glancing collision (since $V'_B \ll v$), and then continue on. In the limit of $V_A = 0$, if m is the mass in S of A when A is at rest, then $m_A = m$. In the limit of $V'_B = 0$, if $m(v)$ is the mass in S of B, which is moving at the velocity v, then $m_B = m(v)$. Hence Eq. (1.35) becomes

$$m(v) = \frac{m}{\sqrt{1 - v^2/c^2}} \tag{1.36}$$

We can see that if linear momentum is defined as

Relativistic momentum

$$\mathbf{p} = \frac{m\mathbf{v}}{\sqrt{1 - v^2/c^2}} \tag{1.37}$$

then conservation of momentum is valid in special relativity. When $v \ll c$, Eq. (1.37) becomes just $\mathbf{p} = m\mathbf{v}$, the classical momentum, as required. Equation (1.37) is often written as

Relativistic momentum

$$\mathbf{p} = \gamma m\mathbf{v} \tag{1.38}$$

where

$$\gamma = \frac{1}{\sqrt{1 - v^2/c^2}} \tag{1.39}$$

In this definition, m is the **proper mass** (or **rest mass**) of an object, its mass when measured at rest relative to an observer. (The symbol γ is the Greek letter gamma.) Figure 1.14 shows how p varies with v/c for both γmv and mv. When v/c is small, mv and γmv are very nearly the same. (For $v = 0.01c$, the difference is only 0.005 percent; for $v = 0.1c$, it is 0.5 percent, still small). As v approaches c, however, the curve for γmv rises more and more steeply (for $v = 0.9c$, the difference is 229 percent). If $v = c$, $p = \gamma mv = \infty$, which is impossible. We conclude that no material object can travel as fast as light.

But what if a spacecraft moving at $v_1 = 0.5c$ relative to the earth fires a projectile at $v_2 = 0.5c$ in the same direction? We on earth might expect to observe the projectile's speed as $v_1 + v_2 = c$. Actually, as discussed in Appendix I to this chapter, velocity addition in relativity is not so simple a process, and we would find the projectile's speed to be only $0.8c$ in such a case.

"Relativistic Mass"

We could alternatively regard the increase in an object's momentum over the classical value as being due to an increase in the object's mass. Then we would call $m_0 = m$ the rest mass of the object and $m = m(v)$ from Eq. (1.43) its relativistic mass, its mass when moving relative to an observer, so that $\mathbf{p} = m\mathbf{v}$. This is the view often taken in the past, at one time even by Einstein. However, as Einstein later wrote, the idea of relativistic mass is "not good" because "no clear definition can be given. It is better to introduce no other mass concept than the 'rest mass' m." In this book the term mass and the symbol m will always refer to proper (or rest) mass, which will be considered relativistically invariant.

Relativistic Second Law

In relativity Newton's second law of motion is given by

Relativistic second law

$$\mathbf{F} = \frac{d\mathbf{p}}{dt} = \frac{d}{dt}(\gamma m\mathbf{v}) \tag{1.40}$$

This is more complicated than the classical formula $\mathbf{F} = m\mathbf{a}$ because γ is a function of v. When $v \ll c$, γ is very nearly equal to 1, and $\mathbf{F}$ is very nearly equal to $m\mathbf{v}$, as it should be.

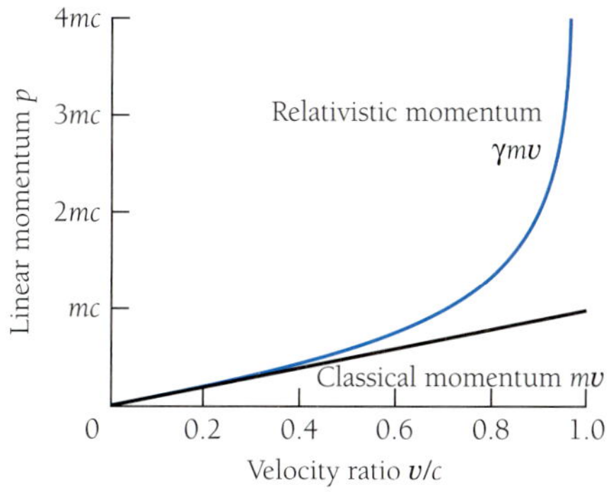

Figure 1.14 The momentum of an object moving at the velocity v relative to an observer. The mass m of the object is its value when it is at rest relative to the observer. The object's velocity can never reach c because its momentum would then be infinite, which is impossible. The relativistic momentum γmv is always correct; the classical momentum mv is valid for velocities much smaller than c.

Example 1.6

Find the acceleration of a particle of mass m and velocity $\mathbf{v}$ when it is acted upon by the constant force $\mathbf{F}$, where $\mathbf{F}$ is parallel to $\mathbf{v}$.

Solution

From Eq. (1.40), since $a = dv/dt$,

$$F = \frac{d}{dt}(\gamma mv) = m\frac{d}{dt}\left(\frac{v}{\sqrt{1 - v^2/c^2}}\right)$$

$$= m\left[\frac{1}{\sqrt{1 - v^2/c^2}} + \frac{v^2/c^2}{(1 - v^2/c^2)^{3/2}}\right]\frac{dv}{dt}$$

$$= \frac{ma}{(1 - v^2/c^2)^{3/2}}$$

We note that F is equal to $\gamma^3 ma$, *not* to γma. Merely replacing m by γm in classical formulas does not always give a relativistically correct result.

The acceleration of the particle is therefore

$$a = \frac{F}{m}(1 - v^2/c^2)^{3/2}$$

Even though the force is constant, the acceleration of the particle decreases as its velocity increases. As $v \rightarrow c$, $a \rightarrow 0$, so the particle can never reach the speed of light, a conclusion we expect.

1.8 MASS AND ENERGY

Where $E_0 = mc^2$ comes from

The most famous relationship Einstein obtained from the postulates of special relativity—how powerful they turn out to be!—concerns mass and energy. Let us see how this relationship can be derived from what we already know.

As we recall from elementary physics, the work W done on an object by a constant force of magnitude F that acts through the distance s, where $\mathbf{F}$ is in the same direction as $\mathbf{s}$, is given by $W = Fs$. If no other forces act on the object and the object starts from rest, all the work done on it becomes kinetic energy KE, so KE $= Fs$. In the general case where F need not be constant, the formula for kinetic energy is the integral

$$\text{KE} = \int_0^s F\,ds$$

In nonrelativistic physics, the kinetic energy of an object of mass m and speed v is KE $= \frac{1}{2}mv^2$. To find the correct relativistic formula for KE we start from the relativistic form of the second law of motion, Eq. (1.40), which gives

$$\text{KE} = \int_0^s \frac{d(\gamma m v)}{dt}\,ds = \int_0^{mv} v\,d(\gamma m v) = \int_0^v v\,d\left(\frac{mv}{\sqrt{1 - v^2/c^2}}\right)$$

Integrating by parts ($\int x\,dy = xy - \int y\,dx$),

$$\begin{aligned}\text{KE} &= \frac{mv^2}{\sqrt{1 - v^2/c^2}} - m\int_0^v \frac{v\,dv}{\sqrt{1 - v^2/c^2}} \\ &= \frac{mv^2}{\sqrt{1 - v^2/c^2}} + \left[mc^2\sqrt{1 - v^2/c^2}\right]_0^v \\ &= \frac{mc^2}{\sqrt{1 - v^2/c^2}} - mc^2\end{aligned}$$

Kinetic energy

$$\text{KE} = \gamma mc^2 - mc^2 = (\gamma - 1)mc^2 \tag{1.41}$$

This result states that the kinetic energy of an object is equal to the difference between γmc^2 and mc^2. Equation (1.41) may be written

Total energy

$$E = \gamma mc^2 = mc^2 + \text{KE} \tag{1.42}$$

If we interpret γmc^2 as the **total energy** E of the object, we see that when it is at rest and KE $= 0$, it nevertheless possesses the energy mc^2. Accordingly mc^2 is called the **rest energy** E_0 of something whose mass is m. We therefore have

$$E = E_0 + \text{KE}$$

where

Rest energy

$$E_0 = mc^2 \tag{1.43}$$

If the object is moving, its total energy is

Total energy

$$E = \gamma mc^2 = \frac{mc^2}{\sqrt{1 - v^2/c^2}} \tag{1.44}$$

Example 1.7

A stationary body explodes into two fragments each of mass 1.0 kg that move apart at speeds of 0.6c relative to the original body. Find the mass of the original body.

Solution

The rest energy of the original body must equal the sum of the total energies of the fragments. Hence

$$E_0 = mc^2 = \gamma m_1c^2 + \gamma m_2c^2 = \frac{m_1c^2}{\sqrt{1 - v_1^2/c^2}} + \frac{m_2c^2}{\sqrt{1 - v_2^2/c^2}}$$

and

$$m = \frac{E_0}{c^2} = \frac{(2)(1.0 \text{ kg})}{\sqrt{1 - (0.60)^2}} = 2.5 \text{ kg}$$

Since mass and energy are not independent entities, their separate conservation principles are properly a single one—the principle of conservation of mass energy. Mass *can* be created or destroyed, but when this happens, an equivalent amount of energy simultaneously vanishes or comes into being, and vice versa. Mass and energy are different aspects of the same thing.

It is worth emphasizing the difference between a *conserved* quantity, such as total energy, and an *invariant* quantity, such as proper mass. Conservation of E means that, in a given reference frame, the total energy of some isolated system remains the same regardless of what events occur in the system. However, the total energy may be different as measured from another frame. On the other hand, the invariance of m means that m has the same value in all inertial frames.

The conversion factor between the unit of mass (the kilogram, kg) and the unit of energy (the joule, J) is c^2, so 1 kg of matter—the mass of this book is about that—has an energy content of $mc^2 = (1 \text{ kg})(3 \times 10^8 \text{ m/s})^2 = 9 \times 10^{16}$ J. This is enough to send a payload of a million tons to the moon. How is it possible for so much energy to be bottled up in even a modest amount of matter without anybody having been aware of it until Einstein's work?

In fact, processes in which rest energy is liberated are very familiar. It is simply that we do not usually think of them in such terms. In every chemical reaction that evolves energy, a certain amount of matter disappears, but the lost mass is so small a fraction of the total mass of the reacting substances that it is imperceptible. Hence the "law" of conservation of mass in chemistry. For instance, only about 6×10^{-11} kg of matter vanishes when 1 kg of dynamite explodes, which is impossible to measure directly, but the more than 5 million joules of energy that is released is hard to avoid noticing.

Example 1.8

Solar energy reaches the earth at the rate of about 1.4 kW per square meter of surface perpendicular to the direction of the sun (Fig. 1.15). By how much does the mass of the sun decrease per second owing to this energy loss? The mean radius of the earth's orbit is 1.5×10^{11} m.

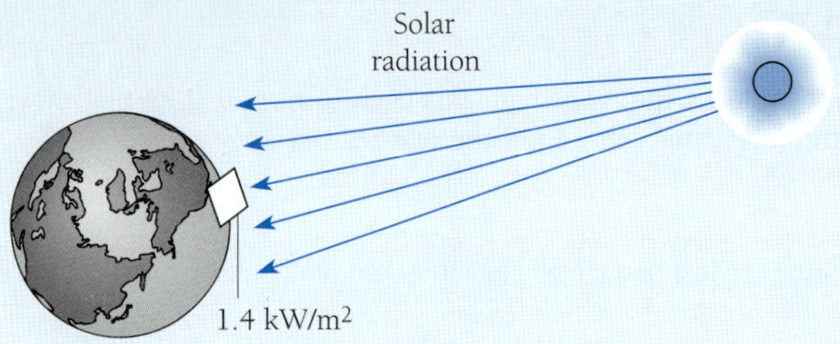

Figure 1.15

Solution

The surface area of a sphere of radius r is $A = 4\pi r^2$. The total power radiated by the sun, which is equal to the power received by a sphere whose radius is that of the earth's orbit, is therefore

$$P = \frac{P}{A}A = \frac{P}{A}(4\pi r^2) = (1.4 \times 10^3\ \text{W/m}^2)(4\pi)(1.5 \times 10^{11}\ \text{m})^2 = 4.0 \times 10^{26}\ \text{W}$$

Thus the sun loses $E_0 = 4.0 \times 10^{26}$ J of rest energy per second, which means that the sun's rest mass decreases by

$$m = \frac{E_0}{c^2} = \frac{4.0 \times 10^{26}\ \text{J}}{(3.0 \times 10^8\ \text{m/s})^2} = 4.4 \times 10^9\ \text{kg}$$

per second. Since the sun's mass is 2.0×10^{30} kg, it is in no immediate danger of running out of matter. The chief energy-producing process in the sun and most other stars is the conversion of hydrogen to helium in its interior. The formation of each helium nucleus is accompanied by the release of 4.0×10^{-11} J of energy, so 10^{37} helium nuclei are produced in the sun per second.

Kinetic Energy at Low Speeds

When the relative speed $\boldsymbol{v}$ is small compared with c, the formula for kinetic energy must reduce to the familiar $\frac{1}{2}m\boldsymbol{v}^2$, which has been verified by experiment at such speeds. Let us see if this is true. The relativistic formula for kinetic energy is

Kinetic energy

$$\text{KE} = \gamma mc^2 - mc^2 = \frac{mc^2}{\sqrt{1 - v^2/c^2}} - mc^2 \tag{1.45}$$

Since $v^2/c^2 \ll 1$, we can use the binomial approximation $(1 + x)^n \approx 1 + nx$, valid for $|x| \ll 1$, to obtain

$$\frac{1}{\sqrt{1 - v^2/c^2}} \approx 1 + \frac{1}{2}\frac{v^2}{c^2} \qquad v \ll c$$

Thus we have the result

$$\text{KE} \approx \left(1 + \frac{1}{2}\frac{v^2}{c^2}\right)mc^2 - mc^2 \approx \frac{1}{2}mv^2 \qquad v \ll c$$

At low speeds the relativistic expression for the kinetic energy of a moving object does indeed reduce to the classical one. So far as is known, the correct formulation of mechanics has its basis in relativity, with classical mechanics representing an approximation that is valid only when $v \ll c$. Figure 1.16 shows how the kinetic energy of a moving object varies with its speed according to both classical and relativistic mechanics.

The degree of accuracy required is what determines whether it is more appropriate to use the classical or to use the relativistic formulas for kinetic energy. For instance, when $v = 10^7$ m/s (0.033c), the formula $\frac{1}{2}mv^2$ understates the true kinetic energy by only 0.08 percent; when $v = 3 \times 10^7$ m/s (0.1c), it understates the true kinetic energy by 0.8 percent; but when $v = 1.5 \times 10^8$ m/s (0.5c), the understatement is a significant 19 percent; and when $v = 0.999c$, the understatement is a whopping 4300 percent. Since 10^7 m/s is about 6310 mi/s, the nonrelativistic formula $\frac{1}{2}mv^2$ is entirely satisfactory for finding the kinetic energies of ordinary objects, and it fails only at the extremely high speeds reached by elementary particles under certain circumstances.

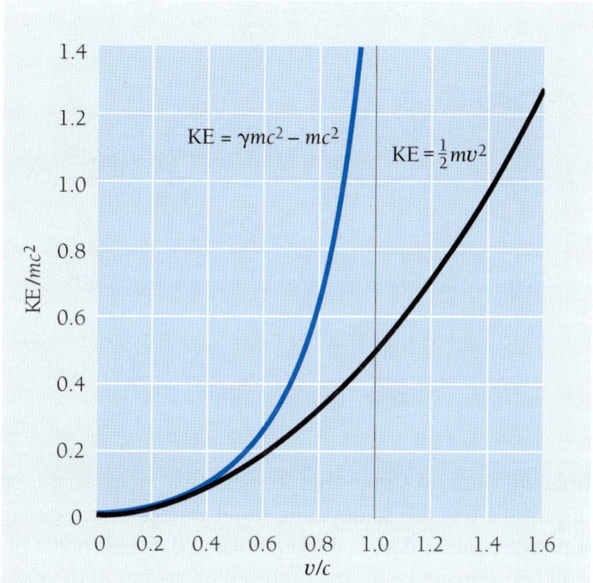

Figure 1.16 A comparison between the classical and relativistic formulas for the ratio between kinetic energy KE of a moving body and its rest energy mc^2. At low speeds the two formulas give the same result, but they diverge at speeds approaching that of light. According to relativistic mechanics, a body would need an infinite kinetic energy to travel with the speed of light, whereas in classical mechanics it would need only a kinetic energy of half its rest energy to have this speed.

1.9 ENERGY AND MOMENTUM

How they fit together in relativity

Total energy and momentum are conserved in an isolated system, and the rest energy of a particle is invariant. Hence these quantities are in some sense more fundamental than velocity or kinetic energy, which are neither. Let us look into how the total energy, rest energy, and momentum of a particle are related.

We begin with Eq. (1.46) for total energy,

Total energy
$$E = \frac{mc^2}{\sqrt{1 - v^2/c^2}} \tag{1.46}$$

and square it to give

$$E^2 = \frac{m^2c^4}{1 - v^2/c^2}$$

From Eq. (1.47) for momentum,

Momentum

$$p = \frac{mv}{\sqrt{1 - v^2/c^2}} \tag{1.47}$$

we find that

$$p^2c^2 = \frac{m^2v^2c^2}{1 - v^2/c^2}$$

Now we subtract p^2c^2 from E^2:

$$E^2 - p^2c^2 = \frac{m^2c^4 - m^2v^2c^2}{1 - v^2/c^2} = \frac{m^2c^4(1 - v^2/c^2)}{1 - v^2/c^2}$$
$$= (mc^2)^2$$

Hence

Energy and momentum

$$E^2 = (mc^2)^2 + p^2c^2 \tag{1.48}$$

which is the formula we want. We note that, because mc^2 is invariant, so is $E^2 - p^2c^2$: this quantity for a particle has the same value in all frames of reference.

For a system of particles rather than a single particle, Eq. (1.48) holds provided that the rest energy mc^2—and hence mass m—is that of the entire system. If the particles in the system are moving with respect to one another, the sum of their individual rest energies may not equal the rest energy of the system. We saw this in Example 1.7 when a stationary body of mass 2.5 kg exploded into two smaller bodies, each of mass 1.0 kg, that then moved apart. If we were inside the system, we would interpret the difference of 0.5 kg of mass as representing its conversion into kinetic energy of the smaller bodies. But seen as a whole, the system is at rest both before and after the explosion, so the *system* did not gain kinetic energy. Therefore the rest energy of the system includes the kinetic energies of its internal motions and it corresponds to a mass of 2.5 kg both before and after the explosion.

In a given situation, the rest energy of an isolated system may be greater than, the same as, or less than the sum of the rest energies of its members. An important case in which the system rest energy is less than the rest energies of its members is that of a system of particles held together by attractive forces, such as the neutrons and protons in an atomic nucleus. The rest energy of a nucleus (except that of ordinary hydrogen, which is a single proton) is less than the total of the rest energies of its constituent particles. The difference is called the *binding energy* of the nucleus. To break a nucleus up completely calls for an amount of energy at least equal to its binding energy. This topic will be explored in detail in Sec. 11.4. For the moment it is interesting to note how large nuclear binding energies are—nearly 10^{12} kJ per kg of nuclear matter is typical. By comparison, the binding energy of water molecules in liquid water is only 2260 kJ/kg; this is the energy needed to turn 1 kg of water at 100°C to steam at the same temperature.

Massless Particles

Can a massless particle exist? To be more precise, can a particle exist which has no rest mass but which nevertheless exhibits such particlelike properties as energy and momentum? In classical mechanics, a particle must have rest mass in order to have energy and momentum, but in relativistic mechanics this requirement does not hold.

From Eqs. (1.47) and (1.46), when $m = 0$ and $v \ll c$, it is clear that $E = p = 0$. A massless particle with a speed less than that of light can have neither energy nor momentum. However, when $m = 0$ and $v = c$, $E = 0/0$ and $p = 0/0$, which are indeterminate: E and p can have any values. Thus Eqs. (1.47) and (1.46) are consistent with the existence of massless particles that possess energy and momentum *provided that they travel with the speed of light.*

Equation (1.48) gives us the relationship between E and p for a particle with $m = 0$:

Massless particle $$E = pc \tag{1.49}$$

The conclusion is not that massless particles necessarily occur, only that the laws of physics do not exclude the possibility as long as $v = c$ and $E = pc$ for them. In fact, a massless particle—the photon—indeed exists and its behavior is as expected, as we shall find in Chap. 2.

Electronvolts

In atomic physics the usual unit of energy is the **electronvolt** (eV), where 1 eV is the energy gained by an electron accelerated through a potential difference of 1 volt. Since $W = QV$,

$$1 \text{ eV} = (1.602 \times 10^{-19} \text{ C})(1.000 \text{ V}) = 1.602 \times 10^{-19} \text{ J}$$

Two quantities normally expressed in electronvolts are the ionization energy of an atom (the work needed to remove one of its electrons) and the binding energy of a molecule (the energy needed to break it apart into separate atoms). Thus the ionization energy of nitrogen is 14.5 eV and the binding energy of the hydrogen molecule H_2 is 4.5 eV. Higher energies in the atomic realm are expressed in **kiloelectronvolts** (keV), where 1 keV $= 10^3$ eV.

In nuclear and elementary-particle physics even the keV is too small a unit in most cases, and the **megaelectronvolt** (MeV) and **gigaelectronvolt** (GeV) are more appropriate, where

$$1 \text{ MeV} = 10^6 \text{ eV} \qquad 1 \text{ GeV} = 10^9 \text{ eV}$$

An example of a quantity expressed in MeV is the energy liberated when the nucleus of a certain type of uranium atom splits into two parts. Each such fission event releases about 200 MeV; this is the process that powers nuclear reactors and weapons.

The rest energies of elementary particles are often expressed in MeV and GeV and the corresponding rest masses in MeV/c^2 and GeV/c^2. The advantage of the latter units is that the rest energy equivalent to a rest mass of, say, 0.938 GeV/c^2 (the rest mass of the proton) is just $E_0 = mc^2 = 0.938$ GeV. If the proton's kinetic energy is 5.000 GeV, finding its total energy is simple:

$$E = E_0 + \text{KE} = (0.938 + 5.000)\ \text{GeV} = 5.938\ \text{GeV}$$

In a similar way the MeV/c and GeV/c are sometimes convenient units of linear momentum. Suppose we want to know the momentum of a proton whose speed is 0.800c. From Eq. (1.47) we have

$$p = \frac{mv}{\sqrt{1 - v^2/c^2}} = \frac{(0.938\ \text{GeV}/c^2)(0.800c)}{\sqrt{1 - (0.800c)^2/c^2}}$$

$$= \frac{0.750\ \text{GeV}/c}{0.600} = 1.25\ \text{GeV}/c$$

Example 1.9

An electron ($m = 0.511$ MeV/c^2) and a photon ($m = 0$) both have momenta of 2.000 MeV/c. Find the total energy of each.

Solution

(*a*) From Eq. (1.48) the electron's total energy is

$$E = \sqrt{m^2c^4 + p^2c^2} = \sqrt{(0.511\ \text{MeV}/c^2)^2c^4 + (2.000\ \text{MeV}/c)^2c^2}$$
$$= \sqrt{(0.511\ \text{MeV})^2 + (2.000\ \text{MeV})^2} = 2.064\ \text{MeV}$$

(*b*) From Eq. (1.49) the photon's total energy is

$$E = pc = (2.000\ \text{MeV}/c)c = 2.000\ \text{MeV}$$

1.10 SPACETIME

How space and time are related

As we have seen, the concepts of space and time are inextricably mixed in nature. A length that one observer can measure with only a meter stick may have to be measured with both a meter stick and a clock by another observer.

A convenient and elegant way to express the results of special relativity is to regard events as occurring in a four-dimensional **spacetime** in which the usual three coordinates x, y, z refer to space and a fourth coordinate ict refers to time, where $i = \sqrt{-1}$. Although we cannot visualize spacetime, it is no harder to deal with mathematically than three-dimensional space.

The reason that ict is chosen as the time coordinate instead of just t is that the quantity

$$s^2 = x^2 + y^2 + z^2 - (ct)^2 \qquad (1.50)$$

is **invariant** under a Lorentz transformation. That is, if an event occurs at x, y, z, t in an inertial frame S and at x', y', z', t' in another inertial frame S', then

$$s^2 = x^2 + y^2 + z^2 - (ct)^2 = x'^2 + y'^2 + z'^2 - (ct')^2$$

Because s^2 is invariant, we can think of a Lorentz transformation merely as a rotation in spacetime of the coordinate axes x, y, z, ict (Fig. 1.17).

The four coordinates x, y, z, ict define a vector in spacetime, and this *four-vector* remains fixed in spacetime regardless of any rotation of the coordinate system—that is, regardless of any shift in point of view from one inertial frame S to another S'.

Another four-vector whose magnitude remains constant under Lorentz transformations has the components $p_x, p_y, p_z, iE/c$. Here p_x, p_y, p_z are the usual components of the linear momentum of a body whose total energy is E. Hence the value of

$$p_x^2 + p_y^2 + p_z^2 - \frac{E^2}{c}$$

is the same in all inertial frames even though p_x, p_y, p_z and E separately may be different. This invariance was noted earlier in connection with Eq. (1.48); we note that $p^2 = p_x^2 + p_y^2 + p_z^2$.

A more mathematically elaborate formulation brings together the electric and magnetic fields **E** and **B** into an invariant quantity called a tensor. This approach to incorporating special relativity into physics has led both to a deeper understanding of natural laws and to the discovery of new phenomena and relationships.

Spacetime Intervals

The statements made at the end of Sec. 1.5 are easy to confirm using the idea of spacetime. Figure 1.18 shows two events plotted on the axes x and ct. Event 1 occurs at x = 0 , t = 0 and event 2 occurs at x = Δx, $t = \Delta t$ The *spacetime interval* Δs between them is defined by

Spacetime interval between events

$$(\Delta s)^2 = (c\Delta t)^2 - (\Delta x)^2 \tag{1.51}$$

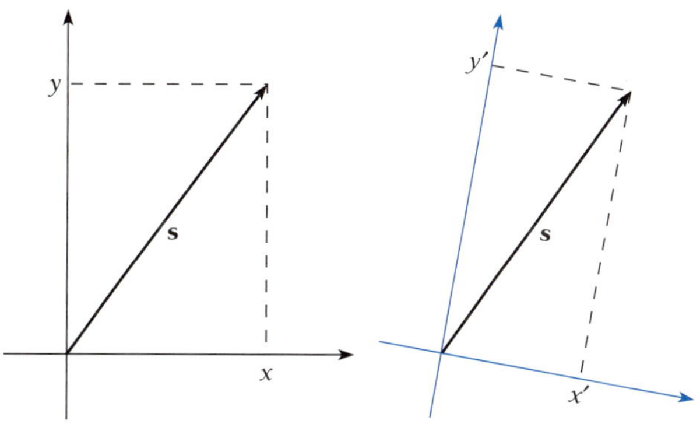

Figure 1.17 Rotating a two-dimensional coordinate system does not change the quantity $s^2 = x^2 + y^2 = x'^2 + y'^2$, where s is the length of the vector **s**. This result can be generalized to the four-dimensional spacetime coordinate system x, y, z, ict.

The virtue of this definition is that $(\Delta s)^2$, like the s^2 of Eq. 1.50, is invariant under Lorentz transformations. If Δx and Δt are the differences in space and time between two events measured in the S frame and $\Delta x'$ and $\Delta t'$ are the same quantities measured in the S' frame,

$$(\Delta s)^2 = (c\Delta t)^2 - (\Delta x)^2 = (c\Delta t')^2 - (\Delta x')^2$$

Therefore whatever conclusions we arrive at in the S frame in which event 1 is at the origin hold equally well in any other frame in relative motion at constant velocity.

Now let us look into the possible relationships between events 1 and 2. Event 2 can be related causally in some way to event 1 provided that a signal traveling slower than the speed of light can connect these events, that is, provided that

$$c\Delta t > |\Delta x|$$

or

Timelike interval

$$(\Delta s)^2 > 0 \tag{1.52}$$

An interval in which $(\Delta s)^2 > 0$ is said to be *timelike*. Every timelike interval that connects event 1 with another event lies within the *light cones* bounded by $x = \pm ct$ in Fig. 1.18. All events that could have affected event 1 lie in the past light cone; all events that event 1 is able to affect lie in the future light cone. (Events connected by timelike intervals need not *necessarily* be related, of course, but it is *possible* for them to be related.)

Conversely, the criterion for there being no causal relationship between events 1 and 2 is that

$$c\Delta t < |\Delta x|$$

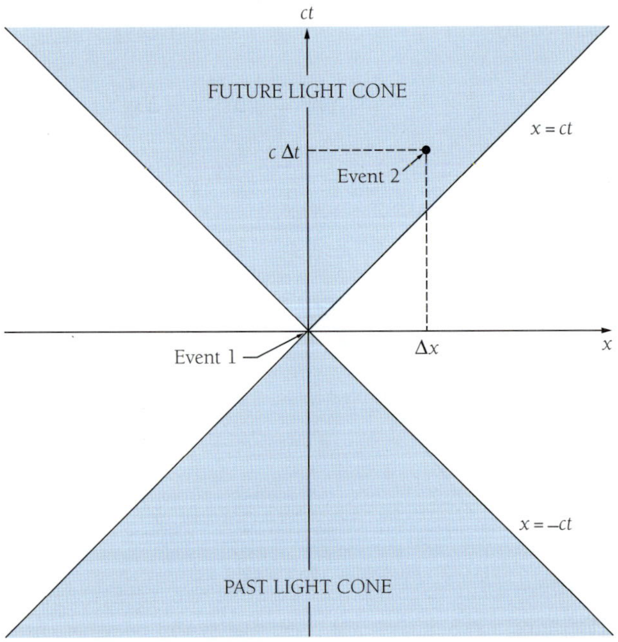

Figure 1.18 The past and future light cones in spacetime of event 1.

or

Spacelike interval $$(\Delta s)^2 < 0 \tag{1.53}$$

An interval in which $(\Delta s)^2 < 0$ is said to be *spacelike*. Every event that is connected with event 1 by a spacelike interval lies outside the light cones of event 1 and neither has interacted with event 1 in the past nor is capable of interacting with it in the future; the two events must be entirely unrelated.

When events 1 and 2 can be connected with a light signal only,

$$c\Delta t = |\Delta x|$$

or

Lightlike interval $$\Delta s = 0 \tag{1.54}$$

An interval in which $\Delta s = 0$ is said to be *lightlike*. Events that can be connected with event 1 by lightlike intervals lie on the boundaries of the light cones.

These conclusions hold in terms of the light cones of event 2 because $(\Delta s)^2$ is invariant; for example, if event 2 is inside the past light cone of event 1, event 1 is inside the future light cone of event 2. In general, events that lie in the future of an event as seen in one frame of reference S lie in its future in every other frame S', and events that lie in the past of an event in S lie in its past in every other frame S'. Thus "future" and "past" have invariant meanings. However, "simultaneity" is an ambiguous concept, because all events that lie outside the past and future light cones of event 1 (that is, all events connected by spacelike intervals with event 1) can appear to occur simultaneously with event 1 in some particular frame of reference.

The path of a particle in spacetime is called its *world line* (Fig. 1.19). The world line of a particle must lie within its light cones.

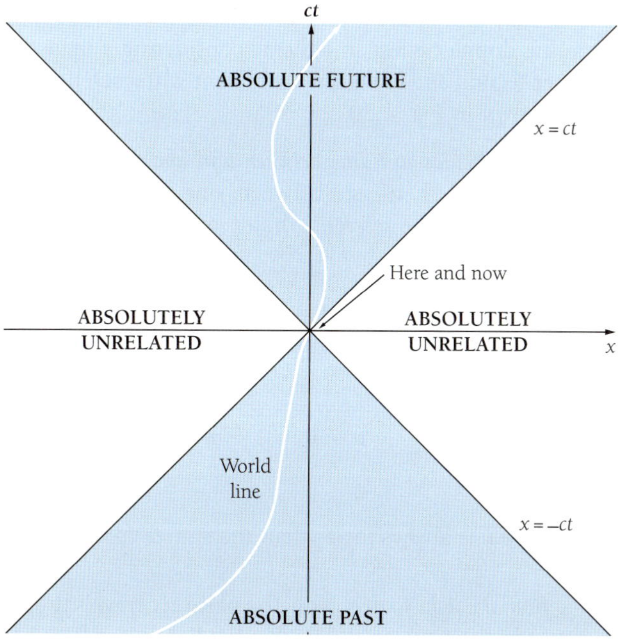

Figure 1.19 The world line of a particle in spacetime.

1.11 DOPPLER EFFECT

Why the universe is believed to be expanding

We are all familiar with the increase in pitch of a sound when its source approaches us (or we approach the source) and the decrease in pitch when the source recedes from us (or we recede from the source). These changes in frequency constitute the **doppler effect,** whose origin is straightforward. For instance, successive waves emitted by a source moving toward an observer are closer together than normal because of the advance of the source; because the separation of the waves is the wavelength of the sound, the corresponding frequency is higher. The relationship between the source frequency ν_0 and the observed frequency ν is

Doppler effect in sound

$$\nu = \nu_0\left(\frac{1 + v/c}{1 - V/c}\right) \tag{1.55}$$

where c = speed of sound
v = speed of observer (+ for motion toward the source, − for motion away from it)
V = speed of the source (+ for motion toward the observer, − for motion away from him)

If the observer is stationary, $v = 0$, and if the source is stationary, $V = 0$.

The doppler effect in sound varies depending on whether the source, or the observer, or both are moving. This appears to violate the principle of relativity: all that should count is the relative motion of source and observer. But sound waves occur only in a material medium such as air or water, and this medium is itself a frame of reference with respect to which motions of source and observer are measurable. Hence there is no contradiction. In the case of light, however, no medium is involved and only relative motion of source and observer is meaningful. The doppler effect in light must therefore differ from that in sound.

We can analyze the doppler effect in light by considering a light source as a clock that ticks ν_0 times per second and emits a wave of light with each tick. We will examine the three situations shown in Fig. 1.20.

1 *Observer moving perpendicular to a line between him and the light source.* The proper time between ticks is $t_0 = 1/\nu_0$, so between one tick and the next the time $t = t_0/\sqrt{1 - v^2/c^2}$ elapses in the reference frame of the observer. The frequency he finds is accordingly

$$\nu(\text{transverse}) = \frac{1}{t} = \frac{\sqrt{1 - v^2/c^2}}{t_0}$$

Transverse doppler effect in light

$$\nu = \nu_0\sqrt{1 - v^2/c^2} \tag{1.56}$$

The observed frequency ν is always lower than the source frequency ν_0.

2 *Observer receding from the light source.* Now the observer travels the distance vt away from the source between ticks, which means that the light wave from a given tick takes vt/c longer to reach him than the previous one. Hence the total time between the arrival of successive waves is

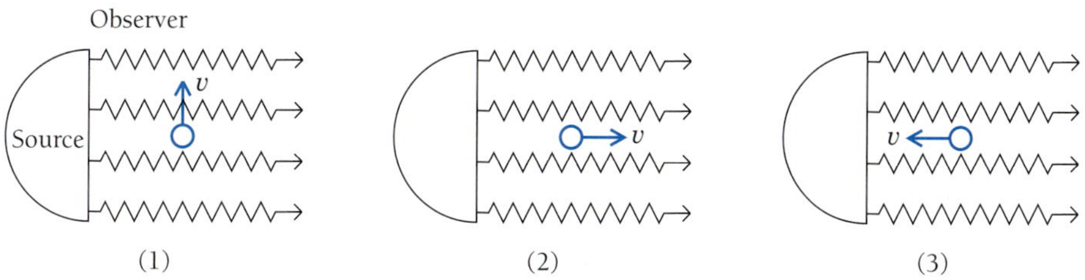

Figure 1.20 The frequency of the light seen by an observer depends on the direction and speed of the observer's motion relative to its source.

$$T = t + \frac{vt}{c} = t_0 \frac{1 + v/c}{\sqrt{1 - v^2/c^2}} = t_0 \frac{\sqrt{1 + v/c}\sqrt{1 + v/c}}{\sqrt{1 + v/c}\sqrt{1 - v/c}} = t_0 \sqrt{\frac{1 + v/c}{1 - v/c}}$$

and the observed frequency is

$$\nu(\text{receding}) = \frac{1}{T} = \frac{1}{t_0}\sqrt{\frac{1 - v/c}{1 + v/c}} = \nu_0 \sqrt{\frac{1 - v/c}{1 + v/c}} \tag{1.57}$$

The observed frequency ν is lower than the source frequency ν_0. Unlike the case of sound waves, which propagate relative to a material medium it makes no difference whether the observer is moving away from the source or the source is moving away from the observer.

3 *Observer approaching the light source*. The observer here travels the distance vt toward the source between ticks, so each light wave takes vt/c less time to arrive than the previous one. In this case $T = t - vt/c$ and the result is

$$\nu(\text{approaching}) = \nu_0 \sqrt{\frac{1 + v/c}{1 - v/c}} \tag{1.58}$$

The observed frequency is higher than the source frequency. Again, the same formula holds for motion of the source toward the observer.

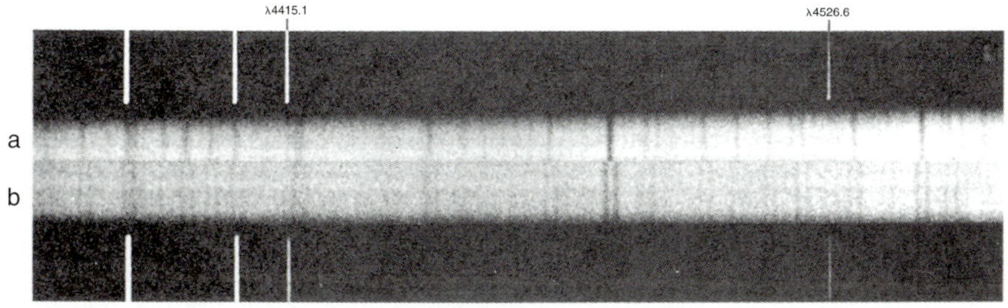

Spectra of the double star Mizar, which consists of two stars that circle their center of mass, taken 2 days apart. In *a* the stars are in line with no motion toward or away from the earth, so their spectral lines are superimposed. In *b* one star is moving toward the earth and the other is moving away from the earth, so the spectral lines of the former are doppler-shifted toward the blue end of the spectrum and those of the latter are shifted toward the red end.

Equations (1.57) and (1.58) can be combined in the single formula

Longitudinal doppler effect in light

$$\nu = \nu_0 \sqrt{\frac{1 + v/c}{1 - v/c}} \tag{1.59}$$

by adopting the convention that v is + for source and observer approaching each other and − for source and observer receding from each other.

Example 1.10

A driver is caught going through a red light. The driver claims to the judge that the color she actually saw was green ($\nu = 5.60 \times 10^{14}$ Hz) and not red ($\nu_0 = 4.80 \times 10^{14}$ Hz) because of the doppler effect. The judge accepts this explanation and instead fines her for speeding at the rate of \$1 for each km/h she exceeded the speed limit of 80 km/h. What was the fine?

Solution

Solving Eq. (1.59) for v gives

$$v = c\left(\frac{\nu^2 - \nu_0^2}{\nu^2 + \nu_0^2}\right) = (3.00 \times 10^8 \text{ m/s})\left[\frac{(5.60)^2 - (4.80)^2}{(5.60)^2 + (4.80)^2}\right]$$

$$= 4.59 \times 10^7 \text{ m/s} = 1.65 \times 10^8 \text{ km/h}$$

since 1 m/s = 3.6 km/h. The fine is therefore $\$(1.65 \times 10^8 - 80) = \$164{,}999{,}920$.

Visible light consists of electromagnetic waves in a frequency band to which the eye is sensitive. Other electromagnetic waves, such as those used in radar and in radio communications, also exhibit the doppler effect in accord with Eq. (1.59). Doppler shifts in radar waves are used by police to measure vehicle speeds, and doppler shifts in the radio waves emitted by a set of earth satellites formed the basis of the highly accurate Transit system of marine navigation.

About Scientists

Edwin Hubble (1889–1953) was born in Missouri and, although always interested in astronomy, pursued a variety of other subjects as well at the University of Chicago. He then went as a Rhodes Scholar to Oxford University in England where he concentrated on law, Spanish, and heavyweight boxing. After two years of teaching at an Indiana high school, Hubble realized what his true vocation was and returned to the University of Chicago to study astronomy.

At Mt. Wilson Observatory in California, Hubble made the first accurate measurements of the distances of spiral galaxies which showed that they are far away in space from our own Milky Way galaxy. It had been known for some time that such galaxies have red shifts in their spectra that indicate motion away from the Milky Way, and Hubble joined his distance figures with the observed red shifts to conclude that the recession speeds were proportional to distance. This implies that the universe is expanding, a remarkable discovery that has led to the modern picture of the universe. Hubble was the first to use the 200-inch telescope, for many years the world's largest, at Mt. Palomar in California, in 1949. In his later work Hubble tried to determine the structure of the universe by finding how the concentration of remote galaxies varies with distance, a very difficult task that only today is being accomplished.

The Expanding Universe

The doppler effect in light is an important tool in astronomy. Stars emit light of certain characteristic frequencies called spectral lines, and motion of a star toward or away from the earth shows up as a doppler shift in these frequencies. The spectral lines of distant galaxies of stars are all shifted toward the low-frequency (red) end of the spectrum and hence are called "red shifts." Such shifts indicate that the galaxies are receding from us and from one another. The speeds of recession are observed to be proportional to distance, which suggests that the entire universe is expanding (Fig. 1.21). This proportionality is called **Hubble's law.**

The expansion apparently began about 13 billion years ago when a very small, intensely hot mass of primeval matter exploded, an event usually called the **Big Bang.** As described in Chap. 13, the matter soon turned into the electrons, protons, and neutrons of which the present universe is composed. Individual aggregates that formed during the expansion became the galaxies of today. Present data suggest that the current expansion will continue forever.

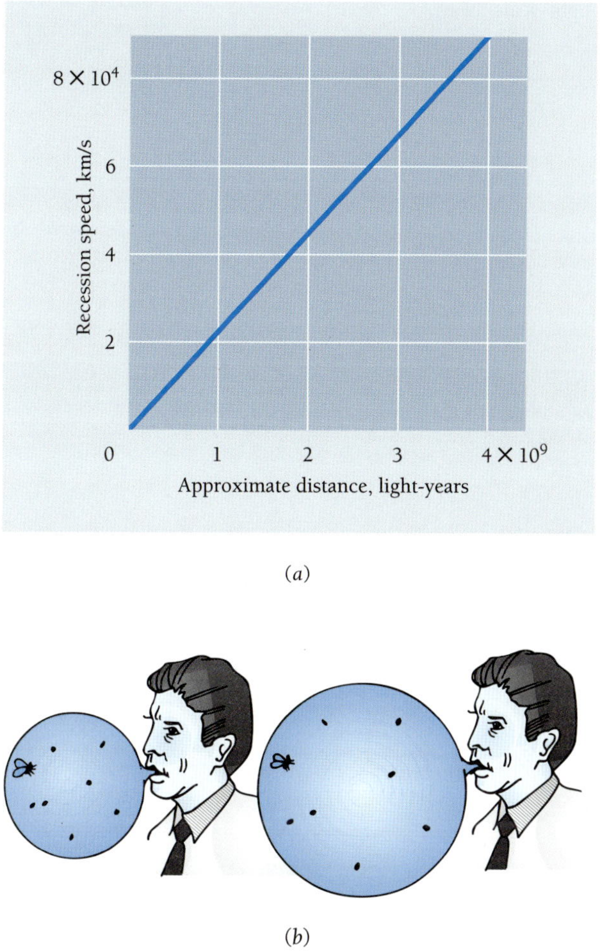

Figure 1.21 (*a*) Graph of recession speed versus distance for distant galaxies. The speed of recession averages about 21 km/s per million light-years. (*b*) Two-dimensional analogy of the expanding universe. As the balloon is inflated, the spots on it become farther apart. A bug on the balloon would find that the farther away a spot is from its location, the faster the spot seems to be moving away; this is true no matter where the bug is. In the case of the universe, the more distant a galaxy is from us, the faster it is moving away, which means that the universe is expanding uniformly.

Example 1.11

A distant galaxy in the constellation Hydra is receding from the earth at 6.12×10^7 m/s. By how much is a green spectral line of wavelength 500 nm (1 nm = 10^{-9} m) emitted by this galaxy shifted toward the red end of the spectrum?

Solution

Since $\lambda = c/\nu$ and $\lambda_0 = c/\nu_0$, from Eq. (1.57) we have

$$\lambda = \lambda_0 \sqrt{\frac{1 + v/c}{1 - v/c}}$$

Here $v = 0.204c$ and $\lambda_0 = 500$ nm, so

$$\lambda = 500 \text{ nm} \sqrt{\frac{1 + 0.204}{1 - 0.204}} = 615 \text{ nm}$$

which is in the orange part of the spectrum. The shift is $\lambda - \lambda_0 = 115$ nm. This galaxy is believed to be 2.9 billion light-years away.

Appendix I to Chapter 1

General Relativity

Special relativity is concerned only with inertial frames of reference, that is, frames that are not accelerated. Einstein's 1916 **general theory of relativity** goes further by including the effects of accelerations on what we observe. Its essential conclusion is that the force of gravity arises from a warping of spacetime around a body of matter (Fig. 1.22). As a result, an object moving through such a region of space in general follows a curved path rather than a straight one, and may even be trapped there.

The **principle of equivalence** is central to general relativity:

> An observer in a closed laboratory cannot distinguish between the effects produced by a gravitational field and those produced by an acceleration of the laboratory.

This principle follows from the experimental observation (to better than 1 part in 10^{12}) that the inertial mass of an object, which governs the object's acceleration when a force acts on it, is always equal to its gravitational mass, which governs the gravitational force another object exerts on it. (The two masses are actually proportional; the constant of proportionality is set equal to 1 by an appropriate choice of the constant of gravitation G.)

Gravity and Light

It follows from the principle of equivalence that light should be subject to gravity. If a light beam is directed across an accelerated laboratory, as in Fig. 1.23, its path relative to the laboratory will be curved. This means that, if the light beam is subject to the

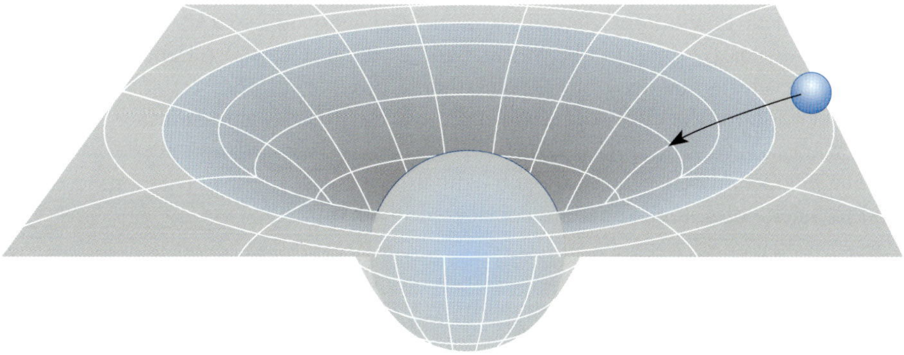

Figure 1.22 General relativity pictures gravity as a warping of spacetime due to the presence of a body of matter. An object nearby experiences an attractive force as a result of this distortion, much as a marble rolls toward the bottom of a depression in a rubber sheet. To paraphrase J. A. Wheeler, spacetime tells mass how to move, and mass tells spacetime how to curve.

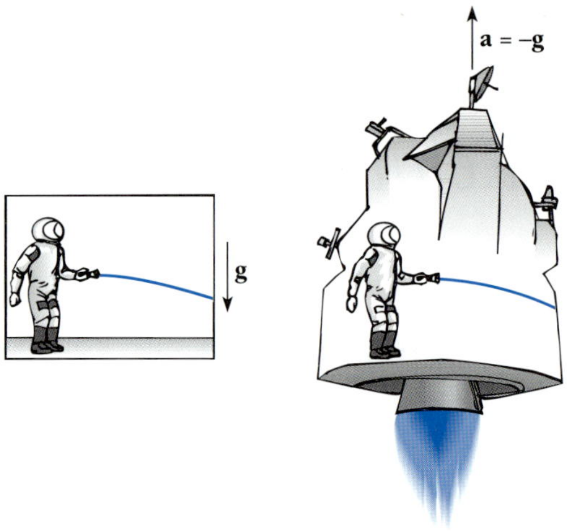

Figure 1.23 According to the principle of equivalence, events that take place in an accelerated laboratory cannot be distinguished from those which take place in a gravitational field. Hence the deflection of a light beam relative to an observer in an accelerated laboratory means that light must be similarly deflected in a gravitational field.

gravitational field to which the laboratory's acceleration is equivalent, the beam would follow the same curved path.

According to general relativity, light rays that graze the sun should have their paths bent toward it by 0.005°—the diameter of a dime seen from a mile away. This prediction was first confirmed in 1919 by photographs of stars that appeared in the sky near the sun during an eclipse, when they could be seen because the sun's disk was covered by the moon. The photographs were then compared with other photographs of the same part of the sky taken when the sun was in a distant part of the sky (Fig. 1.24). Einstein became a world celebrity as a result.

Because light is deflected in a gravitational field, a dense concentration of mass—such as a galaxy of stars—can act as a lens to produce multiple images of a distant light source located behind it (Fig. 1.25). A **quasar,** the nucleus of a young galaxy, is brighter than 100 billion stars but is no larger than the solar system. The first observation of gravitational lensing was the discovery in 1979 of what seemed to be a pair of nearby quasars but was actually a single one whose light was deviated by an intervening massive object. Since then a number of other gravitational lenses have been found; the effect occurs in radio waves from distant sources as well as in light waves.

The interaction between gravity and light also gives rise to the gravitational red shift and to black holes, topics that are considered in Chap. 2.

Other Findings of General Relativity

A further success of general relativity was the clearing up of a long-standing puzzle in astronomy. The perihelion of a planetary orbit is the point in the orbit nearest the sun. Mercury's orbit has the peculiarity that its perihelion shifts (precesses) about 1.6° per century (Fig. 1.26). All but 43″ (1″ = 1 arc second = $\frac{1}{3600}$ of a

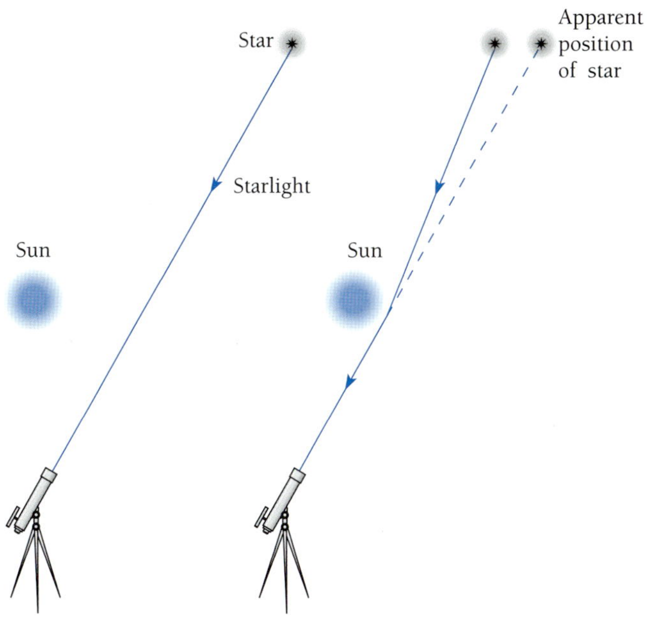

Figure 1.24 Starlight passing near the sun is deflected by its strong gravitational field. The deflection can be measured during a solar eclipse when the sun's disk is obscured by the moon.

degree) of this shift is due to the attractions of other planets, and for a while the discrepancy was used as evidence for an undiscovered planet called Vulcan whose orbit was supposed to lie inside that of Mercury. When gravity is weak, general relativity gives very nearly the same results as Newton's formula $F = Gm_1m_2/r^2$. But Mercury is close to the sun and so moves in a strong gravitational field, and Einstein was able to show from general relativity that a precession of 43″ per century was to be expected for its orbit.

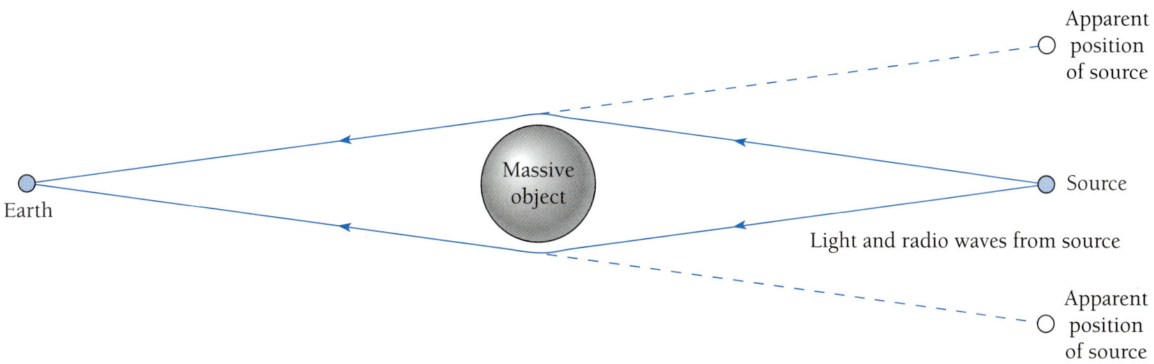

Figure 1.25 A gravitational lens. Light and radio waves from a source such as a quasar are deviated by a massive object such as a galaxy so that they seem to come from two or more identical sources. A number of such gravitational lenses have been identified.

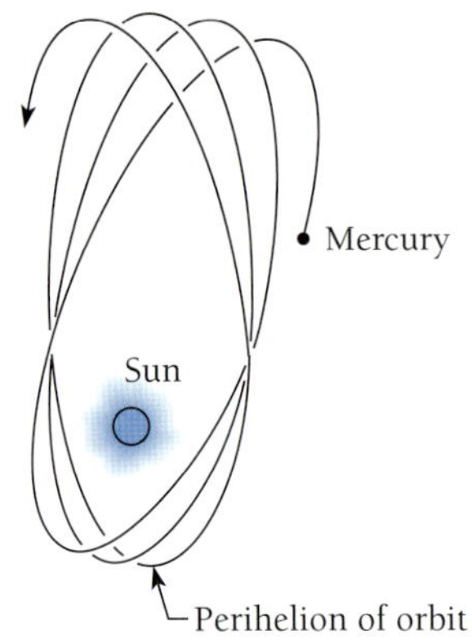

Figure 1.26 The precession of the perihelion of Mercury's orbit.

The existence of **gravitational waves** that travel with the speed of light was the prediction of general relativity that had to wait the longest to be verified. To visualize gravitational waves, we can think in terms of the model of Fig. 1.17 in which two-dimensional space is represented by a rubber sheet distorted by masses embedded in it. If one of the masses vibrates, waves will be sent out in the sheet that set other masses in vibration. A vibrating electric charge similarly sends out electromagnetic waves that excite vibrations in other charges.

A big difference between the two kinds of waves is that gravitational waves are extremely weak, so that despite much effort none have as yet been directly detected. However, in 1974 strong evidence for gravitational waves was found in the behavior of a system of two nearby stars, one a pulsar, that revolve around each other. A **pulsar** is a very small, dense star, composed mainly of neutrons, that spins rapidly and sends out flashes of light and radio waves at a regular rate, much as the rotating beam of a lighthouse does (see Sec. 9.11). The pulsar in this particular binary system emits pulses every 59 milliseconds (ms), and it and its companion (probably another neutron star) have an orbital period of about 8 h. According to general relativity, such a system should give off gravitational waves and lose energy as a result, which would reduce the orbital period as the stars spiral in toward each other. A change in orbital period means a change in the arrival times of the pulsar's flashes, and in the case of the observed binary system the orbital period was found to be decreasing at 75 ms per year. This is so close to the figure that general relativity predicts for the system that there seems to be no doubt that gravitational radiation is responsible. The 1993 Nobel Prize in physics was awarded to Joseph Taylor and Russell Hulse for this work.

Much more powerful sources of gravitational waves ought to be such events as two black holes colliding and supernova explosions in which the remnant star cores collapse into neutron stars (again, see Sec. 9.11). A gravitational wave that passes through a body of matter will cause distortions to ripple through it due to fluctuations in the gravitational field. Because gravitational forces are feeble—the electric attraction between a proton and an electron is over 10^{39} times greater than the gravitational attraction between them—such distortions at the earth induced by gravitational waves from a supernova in our galaxy (which occurs an average of once every 30 years or so) would amount to only about 1 part in 10^{18}, even less for a more distant supernova. This corresponds to a change in, say, the height of a person by well under the diameter of an atomic nucleus, yet it seems to be detectable—just—with current technology.

In one method, a large metal bar cooled to a low temperature to minimize the random thermal motions of its atoms is monitored by sensors for vibrations due to gravitational waves. In another method, an interferometer similar to the one shown in Fig. 1.2 with a laser as the light source is used to look for changes in the lengths of the arms to which the mirrors are attached. Instruments of both kinds are operating, thus far with no success.

A really ambitious scheme has been proposed that would use six spacecraft in orbit around the sun placed in pairs at the corners of a triangle whose sides are 5 million kilometers (km) long. Lasers, mirrors, and sensors in the spacecraft would detect changes in their spacings resulting from the passing of a gravitational wave. It may only be a matter of time before gravitational waves will be providing information about a variety of cosmic disturbances on the largest scale.

Global Positioning System (GPS)

The Global Positioning System (GPS) is a satellite-based navigation system which can consists of 24 orbiting satellites (see image below) each of which travel one circle around the Earth every 12 hours. These satellites are all in orbits at an altitude of 11000 miles. Signals from these satellites are collected by GPS receiver systems, and these signals contain important position information and time information for the GPS receiver to calculate the distance between these satellites. Only use three of these satellites, the information is enough for triangulating the longitude and latitude information of the GPS receiver on the ground. If we use four or more GPS satellites, altitude position is known as well. Hence, the GPS system has the function of real-time, three-dimensional positioning all over the world, and now it has been widely used in mapping, police, science area and so on.

Space satellite orbiting the earth.
Source: © McGraw-Hill Education/shutterstock/Andrey Armyagov

The satellites can be adjusted by land-based radar systems. Therefore, the exact location and orbit of the satellites should be known in advance. To get accurate time, atomic clock is used for synchronizing the timer in satellites and the GPS receivers. At exactly the same time determined by atomic clock, receivers and satellites will generate a code and then collect the codes from each other. The difference between the time satellites generated code and the time receiver received the code can now be determined. This time interval is the travel time of the code, which is used for calculating the distance between the receivers and satellites.

However, some errors exist in GPS because of the relativistic effect. One of them is relativistic effects, which includes the time dilations, gravitational frequency shift and so on. According to the theory of relativity, the clocks on the satellites are affected by their speed. Typically these atomic clocks will be slower than those on the stationary ground. Using Special relativity theory, if the atomic clocks have the same velocity as the orbital speeds of GPS satellites, which is about 4 km/s, a delay of about 7μs/day will be occurred. This error distance due to time dilation effect has been verified in GPS.

The second error effect is gravitational frequency shift. Due to general relativity, the atomic clock should be slower when it is closer to a massive object. Since most of the GPS receivers are much closer to earth ground than GPS satellites, the clocks in satellites will go faster. In general, this shift causes the clocks in satellites to be faster by 45.9μs/day. With the time dilations effect, the total time error is about 38μs/day, and without relativity theory correction, an error distance of 10km/day is accumulated. Therefore, to avoid the frequency shift, the frequency standard on board each satellite is given a rate offset before launching, making it run slightly slower than the desired frequency on Earth.

EXERCISES

But be ye doers of the word, and not hearers only, deceiving your own selves. —James I:22

1.1 Special Relativity

1. If the speed of light were smaller than it is, would relativistic phenomena be more or less conspicuous than they are now?
2. It is possible for the electron beam in a television picture tube to move across the screen at a speed faster than the speed of light. Why does this not contradict special relativity?

1.2 The Lorentz Transformation

3. An observer detects two explosions, one that occurs near her at a certain time and another that occurs 2.00 ms later 100 km away. Another observer finds that the two explosions occur at the same place. What time interval separates the explosions to the second observer?
4. An observer detects two explosions that occur at the same time, one near her and the other 100 km away. Another observer finds that the two explosions occur 160 km apart. What time interval separates the explosions to the second observer?
5. A spacecraft moving in the $+x$ direction receives a light signal from a source in the xy plane. In the reference frame of the fixed stars, the speed of the spacecraft is v and the signal arrives at an angle θ to the axis of the spacecraft. (*a*) With the help of the Lorentz transformation find the angle θ' at which the signal arrives in the reference frame of the spacecraft. (*b*) What would you conclude from this result about the view of the stars from a porthole on the side of the spacecraft?
6. A body moving at $0.500c$ with respect to an observer disintegrates into two fragments that move in opposite directions relative to their center of mass along the same line of motion as the original body. One fragment has a velocity of $0.600c$ in the backward direction relative to the center of mass and the other has a velocity of $0.500c$ in the forward direction. What velocities will the observer find?
7. A man on the moon sees two spacecraft, A and B, coming toward him from opposite directions at the respective speeds of $0.800c$ and $0.900c$. (*a*) What does a man on A measure for the speed with which he is approaching the moon? For the speed with which he is approaching B? (*b*) What does a man on B measure for the speed with which he is approaching the moon? For the speed with which he is approaching A?
8. An electron whose speed relative to an observer in a laboratory is $0.800c$ is also being studied by an observer moving in the same direction as the electron at a speed of $0.500c$ relative to the laboratory. What is the kinetic energy (in MeV) of the electron to each observer?

1.4 Length Contraction

9. An astronaut whose height on the earth is exactly 6 ft is lying parallel to the axis of a spacecraft moving at $0.90c$ relative to the earth. What is his height as measured by an observer in the same spacecraft? By an observer on the earth?
10. An astronaut is standing in a spacecraft parallel to its direction of motion. An observer on the earth finds that the spacecraft speed is $0.60c$ and the astronaut is 1.3 m tall. What is the astronaut's height as measured in the spacecraft?
11. How much time does a meter stick moving at $0.100c$ relative to an observer take to pass the observer? The meter stick is parallel to its direction of motion.
12. A meter stick moving with respect to an observer appears only 500 mm long to her. What is its relative speed? How long does it take to pass her? The meter stick is parallel to its direction of motion.
13. A spacecraft antenna is at an angle of 10° relative to the axis of the spacecraft. If the spacecraft moves away from the earth at a speed of $0.70c$, what is the angle of the antenna as seen from the earth?

1.5 Time Dilation

14. An observer on a spacecraft moving at $0.700c$ relative to the earth finds that a car takes 40.0 min to make a trip. How long does the trip take to the driver of the car?
15. An athlete has learned enough physics to know that if he measures from the earth a time interval on a moving spacecraft, what he finds will be greater than what somebody on the spacecraft would measure. He therefore proposes to set a world record for the 100-m dash by having his time taken by an observer on a moving spacecraft. Is this a good idea?
16. An airplane is flying at 300 m/s (672 mi/h). How much time must elapse before a clock in the airplane and one on the ground differ by 1.00 s?
17. Two observers, A on earth and B in a spacecraft whose speed is 2.00×10^8 m/s, both set their watches to the same time when the ship is abreast of the earth. (*a*) How much time must elapse by A's reckoning before the watches differ by 1.00 s? (*b*) To A, B's watch seems to run slow. To B, does A's watch seem to run fast, run slow, or keep the same time as his own watch?
18. The Apollo 11 spacecraft that landed on the moon in 1969 traveled there at a speed relative to the earth of 1.08×10^4 m/s. To an observer on the earth, how much longer than his own day was a day on the spacecraft?
19. How fast must a spacecraft travel relative to the earth for each day on the spacecraft to correspond to 2 d on the earth?
20. A certain particle has a lifetime of 1.00×10^{-7} s when measured at rest. How far does it go before decaying if its speed is $0.99c$ when it is created?

1.6 Twin Paradox

21. A woman leaves the earth in a spacecraft that makes a round trip to the nearest star, 4 light-years distant, at a speed of 0.9*c*. How much younger is she upon her return than her twin sister who remained behind?

22. Twin *A* makes a round trip at 0.6*c* to a star 12 light-years away, while twin *B* stays on the earth. Each twin sends the other a signal once a year by his own reckoning. (*a*) How many signals does *A* send during the trip? How many does *B* send? (*b*) How many signals does *A* receive? How many does *B* receive?

1.7 Relativistic Momentum

23. All definitions are arbitrary, but some are more useful than others. What is the objection to defining linear momentum as $\mathbf{p} = m\mathbf{v}$ instead of the more complicated $\mathbf{p} = \gamma m\mathbf{v}$?

24. (*a*) An electron's speed is doubled from 0.2*c* to 0.4*c*. By what ratio does its momentum increase? (*b*) What happens to the momentum ratio when the electron's speed is doubled again from 0.4*c* to 0.8*c*?

25. Verify that

$$\left(\frac{1}{\sqrt{1 - v^2/c^2}}\right)^2 = 1 + \frac{p^2}{m^2c^2}$$

1.8 Mass and Energy

26. A certain quantity of ice at 0°C melts into water at 0°C and in so doing gains 1.00 kg of mass. What was its initial mass?

27. Dynamite liberates about 5.4×10^6 J/kg when it explodes. What fraction of its total energy content is this?

28. How many joules of energy per kilogram of rest mass are needed to bring a spacecraft from rest to a speed of 0.90*c*?

29. At what speed does the kinetic energy of a particle equal its rest energy?

30. Verify that, for $E \gg E_0$,

$$\frac{v}{c} \approx 1 - \frac{1}{2}\left(\frac{E_0}{E}\right)^2$$

31. An electron has a kinetic energy of 0.100 MeV. Find its speed according to classical and relativistic mechanics.

32. (*a*) The speed of a proton is increased from 0.20*c* to 0.40*c*. By what factor does its kinetic energy increase? (*b*) The proton speed is again doubled, this time to 0.80*c*. By what factor does its kinetic energy increase now?

33. A particle has a kinetic energy 20 times its rest energy. Find the speed of the particle in terms of *c*.

34. (*a*) Derive a formula for the minimum kinetic energy needed by a particle of rest mass *m* to emit Cerenkov radiation in a medium of index of refraction *n*. [*Hint*: Start from Eqs. (1.47) and (1.49).] (*b*) Use this formula to find KE_{min} for an electron in a medium of $n = 1.5$.

35. How much work (in MeV) must be done to increase the speed of an electron from 1.2×10^8 m/s to 2.4×10^8 m/s?

36. Prove that $\frac{1}{2}\gamma m v^2$, does *not* equal the kinetic energy of a particle moving at relativistic speeds.

37. A moving electron collides with a stationary electron and an electron-positron pair comes into being as a result (a positron is a positively charged electron). When all four particles have the same velocity after the collision, the kinetic energy required for this process is a minimum. Use a relativistic calculation to show that $KE_{min} = 6mc^2$, where *m* is the rest mass of the electron.

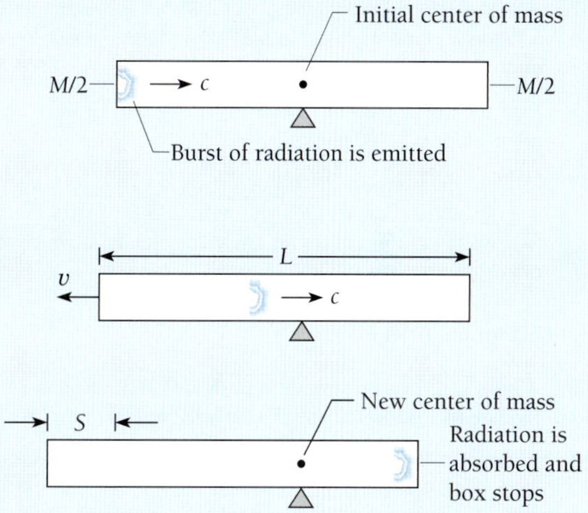

Figure 1.27 The box has moved the distance *S* to the left when it stops.

38. An alternative derivation of the mass-energy formula $E_0 = mc^2$, also given by Einstein, is based on the principle that the location of the center of mass (CM) of an isolated system cannot be changed by any process that occurs inside the system. Figure 1.27 shows a rigid box of length *L* that rests on a frictionless surface; the mass *M* of the box is equally divided between its two ends. A burst of electromagnetic radiation of energy E_0 is emitted by one end of the box. According to classical physics, the radiation has the momentum $p = E_0/c$, and when it is emitted, the box recoils with the speed $v \approx E_0/Mc$ so that the total momentum of the system remains zero. After a time $t \approx L/c$ the radiation reaches the other end of the box and is absorbed there, which brings the box to a stop after having moved the distance *S*. If the CM of the box is to remain in its original place, the radiation must have transferred mass from one end to the other. Show that this amount of mass is $m = E_0/c^2$.

1.9 Energy and Momentum

39. In its own frame of reference, a proton takes 5 min to cross the Milky Way galaxy, which is about 10^5 light-years in diameter. (*a*) What is the approximate energy of the proton in electronvolts? (*b*) About how long would the proton take to cross the galaxy as measured by an observer in the galaxy's reference frame?

40. Find the SI equivalents of the mass unit MeV/c^2 and the momentum unit MeV/*c*.

41. Find the momentum (in MeV/c) of an electron whose speed is $0.600c$.

42. What is the energy of a photon whose momentum is the same as that of a proton whose kinetic energy is 10.0 MeV?

43. Find the momentum of an electron whose kinetic energy equals its rest energy of 511 keV.

44. Find the total energy and kinetic energy (in GeV) and the momentum (in GeV/c) of a proton whose speed is $0.900c$. The mass of the proton is 0.938 GeV/c^2.

45. Find the speed and momentum (in GeV/c) of a proton whose total energy is 3.500 GeV.

46. Verify that $v/c = pc/E$.

47. A particle has a kinetic energy of 62 MeV and a momentum of 335 MeV/c. Find its mass (in MeV/c^2) and speed (as a fraction of c).

48. Find the total energy of a neutron (m = 0.940 GeV/c^2) whose momentum is 1.200 GeV/c.

49. (*a*) Find the mass (in GeV/c^2) of a particle whose total energy is 4.00 GeV and whose momentum is 1.45 GeV/c. (*b*) Find the total energy of this particle in a reference frame in which its momentum is 2.00 GeV/c.

1.11 Doppler Effect

50. A spacecraft receding from the earth at $0.97c$ transmits data at the rate of 1.00×10^4 pulses/s. At what rate are they received?

51. A galaxy in the constellation Ursa Major is receding from the earth at 15,000 km/s. If one of the characteristic wavelengths of the light the galaxy emits is 550 nm, what is the corresponding wavelength measured by astronomers on the earth?

52. The frequencies of the spectral lines in light from a distant galaxy are found to be two-thirds as great as those of the same lines in light from nearby stars. Find the recession speed of the distant galaxy.

53. A spacecraft receding from the earth emits radio waves at a constant frequency of 10^9 Hz. If the receiver on earth can measure frequencies to the nearest hertz, at what spacecraft speed can the difference between the relativistic and classical doppler effects be detected? For the classical effect, assume the earth is stationary.

54. A car moving at 150 km/h (93 mi/h) is approaching a stationary police car whose radar speed detector operates at a frequency of 15 GHz. What frequency change is found by the speed detector?

55. If the angle between the direction of motion of a light source of frequency ν_0 and the direction from it to an observer is θ, the frequency ν the observer finds is given by

$$\nu = \nu_0 \frac{\sqrt{1 - v^2/c^2}}{1 - (v/c) \cos \theta}$$

where v is the relative speed of the source. Show that this formula includes Eqs. (1.56) and (1.58) as special cases.

56. (*a*) Show that when $v \ll c$, the formulas for the doppler effect both in light and in sound for an observer approaching a source, and vice versa, all reduce to $\nu \approx \nu_0(1 + v/c)$, so that $\Delta\nu/\nu \approx v/c$. [*Hint*: For $x \ll 1$, $1/(1 + x) \approx 1 - x$.] (*b*) What do the formulas for an observer receding from a source, and vice versa, reduce to when $v \ll c$?

CHAPTER 2

Particle Properties of Waves

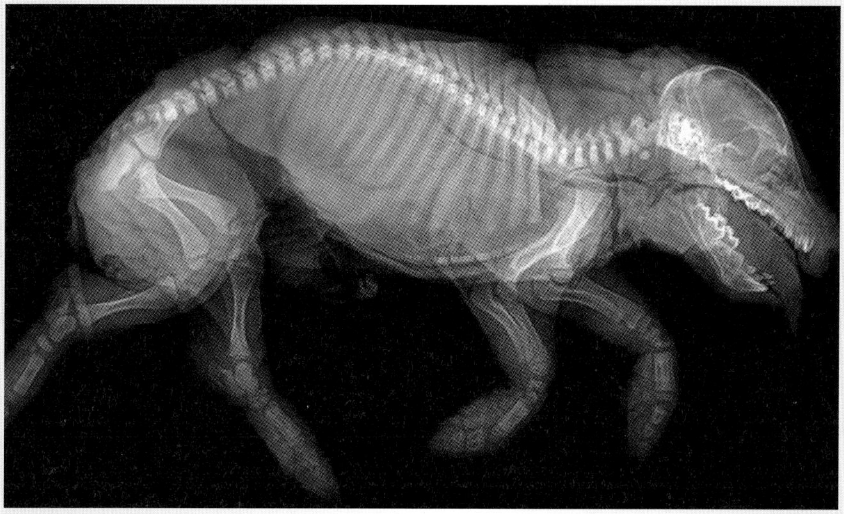

The penetrating ability of x-rays enabled them to reveal the skeleton of a fetal pig. They consist of high-energy photons.

In our everyday experience there is nothing mysterious or ambiguous about the concepts of **particle** and **wave.** A stone dropped into a lake and the ripples that spread out from its point of impact apparently have in common only the ability to carry energy and momentum from one place to another. Classical physics, which mirrors the "physical reality" of our sense impressions, treats particles and waves as separate components of that reality. The mechanics of particles and the optics of waves are traditionally independent disciplines, each with its own chain of experiments and principles based on their results.

The physical reality we perceive has its roots in the microscopic world of atoms and molecules, electrons and nuclei, but in this world there are neither particles nor waves in our sense of these terms. We regard electrons as particles because they possess charge and mass and behave according to the laws of particle mechanics in such familiar devices as television picture tubes. We shall see, however, that it is just as correct to interpret a moving electron as a wave manifestation as it is to interpret it as a particle manifestation. We regard electromagnetic waves as waves because under suitable circumstances they exhibit diffraction, interference, and polarization. Similarly, we shall see that under other circumstances electromagnetic waves behave as though they consist of streams of particles. Together with special relativity, the wave-particle duality is central to an understanding of modern physics, and in this book there are few arguments that do not draw upon either or both of these fundamental ideas.

2.1 ELECTROMAGNETIC WAVES

Coupled electric and magnetic oscillations that move with the speed of light and exhibit typical wave behavior

In 1864 the British physicist James Clerk Maxwell made the remarkable suggestion that accelerated electric charges generate linked electric and magnetic disturbances that can travel indefinitely through space. If the charges oscillate periodically, the disturbances are waves whose electric and magnetic components are perpendicular to each other and to the direction of propagation, as in Fig. 2.1.

From the earlier work of Faraday, Maxwell knew that a changing magnetic field can induce a current in a wire loop. Thus a changing magnetic field is equivalent in its effects to an electric field. Maxwell proposed the converse: a changing electric field has a magnetic field associated with it. The electric fields produced by electromagnetic induction are easy to demonstrate because metals offer little resistance to the flow of charge. Even a weak field can lead to a measurable current in a metal. Weak magnetic fields are much harder to detect, however, and Maxwell's hypothesis was based on a symmetry argument rather than on experimental findings.

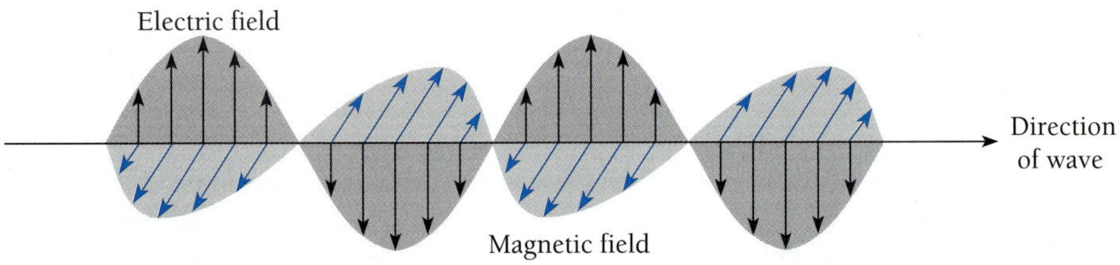

Figure 2.1 The electric and magnetic fields in an electromagnetic wave vary together. The fields are perpendicular to each other and to the direction of propagation of the wave.

About Scientists

James Clerk Maxwell (1831–1879) was born in Scotland shortly before Michael Faraday discovered electromagnetic induction. At nineteen he entered Cambridge University to study physics and mathematics. While still a student, he investigated the physics of color vision and later used his ideas to make the first color photograph. Maxwell became known to the scientific world at twenty-four when he showed that the rings of Saturn could not be solid or liquid but must consist of separate small bodies. At about this time Maxwell became interested in electricity and magnetism and grew convinced that the wealth of phenomena Faraday and others had discovered were not isolated effects but had an underlying unity of some kind. Maxwell's initial step in establishing that unity came in 1856 with the paper "On Faraday's Lines of Force," in which he developed a mathematical description of electric and magnetic fields.

Maxwell left Cambridge in 1856 to teach at a college in Scotland and later at King's College in London. In this period he expanded his ideas on electricity and magnetism to create a single comprehensive theory of electromagnetism. The fundamental equations he arrived at remain the foundations of the subject today. From these equations Maxwell predicted that electromagnetic waves should exist that travel with the speed of light, described the properties the waves should have, and surmised that light consisted of electromagnetic waves. Sadly, he did not live to see his work confirmed in the experiments of the German physicist Heinrich Hertz.

Maxwell's contributions to kinetic theory and statistical mechanics were on the same profound level as his contributions to electromagnetic theory. His calculations showed that the viscosity of a gas ought to be independent of its pressure, a surprising result that Maxwell, with the help of his wife, confirmed in the laboratory. They also found that the viscosity was proportional to the absolute temperature of the gas. Maxwell's explanation for this proportionality gave him a way to estimate the size and mass of molecules, which until then could only be guessed at. Maxwell shares with Boltzmann credit for the equation that gives the distribution of molecular energies in a gas.

In 1865 Maxwell returned to his family's home in Scotland. There he continued his research and also composed a treatise on electromagnetism that was to be the standard text on the subject for many decades. It was still in print a century later. In 1871 Maxwell went back to Cambridge to establish and direct the Cavendish Laboratory, named in honor of the pioneering physicist Henry Cavendish. Maxwell died of cancer at the age of forty-eight in 1879, the year in which Albert Einstein was born. Maxwell had been the greatest theoretical physicist of the nineteenth century; Einstein was to be the greatest theoretical physicist of the twentieth century. (By a similar coincidence, Newton was born in the year of Galileo's death.)

If Maxwell was right, electromagnetic (em) waves must occur in which constantly varying electric and magnetic fields are coupled together by both electromagnetic induction and the converse mechanism he proposed. Maxwell was able to show that the speed c of electromagnetic waves in free space is given by

$$c = \frac{1}{\sqrt{\epsilon_0 \mu_0}} = 2.998 \times 10^8 \text{ m/s}$$

where ϵ_0 is the electric permittivity of free space and μ_0 is its magnetic permeability. This is the same as the speed of light waves. The correspondence was too great to be accidental, and Maxwell concluded that light consists of electromagnetic waves.

During Maxwell's lifetime the notion of em waves remained without direct experimental support. Finally, in 1888, the German physicist Heinrich Hertz showed that em waves indeed exist and behave exactly as Maxwell had predicted. Hertz generated the waves by applying an alternating current to an air gap between two metal balls. The width of the gap was such that a spark occurred each time the current reached a peak. A wire loop with a small gap was the detector; em waves set up oscillations in the loop that produced sparks in the gap. Hertz determined the wavelength and speed of the waves he generated, showed that they have both electric and magnetic components, and found that they could be reflected, refracted, and diffracted.

Light is not the only example of an em wave. Although all such waves have the same fundamental nature, many features of their interaction with matter depend upon

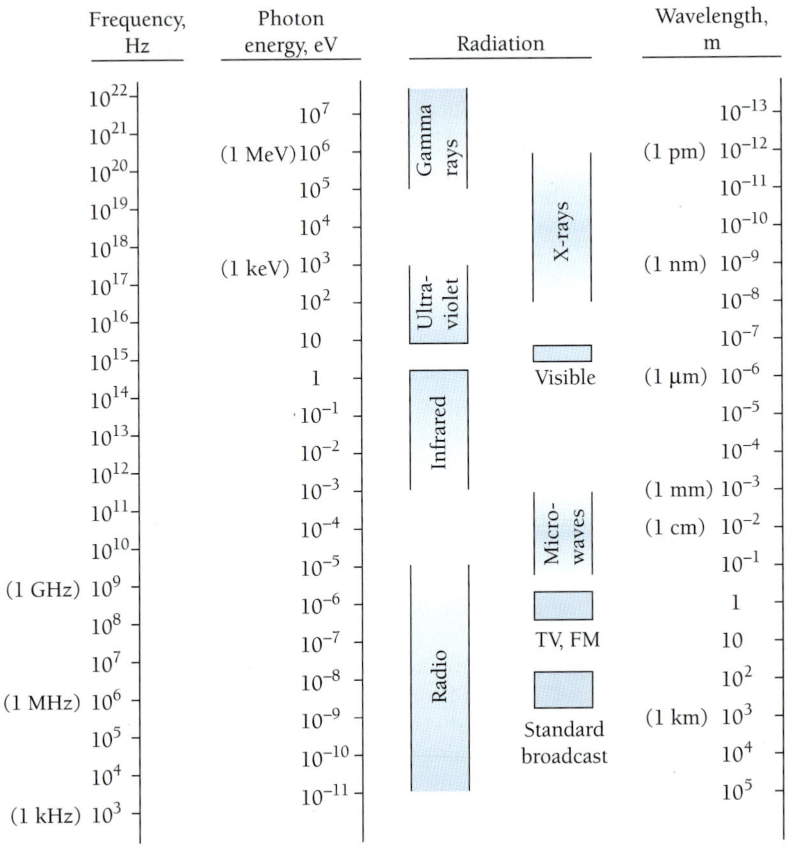

Figure 2.2 The spectrum of electromagnetic radiation.

their frequencies. Light waves, which are em waves the eye responds to, span only a brief frequency interval, from about 4.3×10^{14} Hz for red light to about 7.5×10^{14} Hz for violet light. Figure 2.2 shows the em wave spectrum from the low frequencies used in radio communication to the high frequencies found in x-rays and gamma rays.

A characteristic property of all waves is that they obey the **principle of superposition:**

> When two or more waves of the same nature travel past a point at the same time, the instantaneous amplitude there is the sum of the instantaneous amplitudes of the individual waves.

Instantaneous amplitude refers to the value at a certain place and time of the quantity whose variations constitute the wave. ("Amplitude" without qualification refers to the maximum value of the wave variable.) Thus the instantaneous amplitude of a wave in a stretched string is the displacement of the string from its normal position; that of a water wave is the height of the water surface relative to its normal level; that of a sound wave is the change in pressure relative to the normal pressure. Since the electric and magnetic fields in a light wave are related by $E = cB$, its instantaneous amplitude can be taken as either E or B. Usually E is used, since it is the electric fields of light waves whose interactions with matter give rise to nearly all common optical effects.

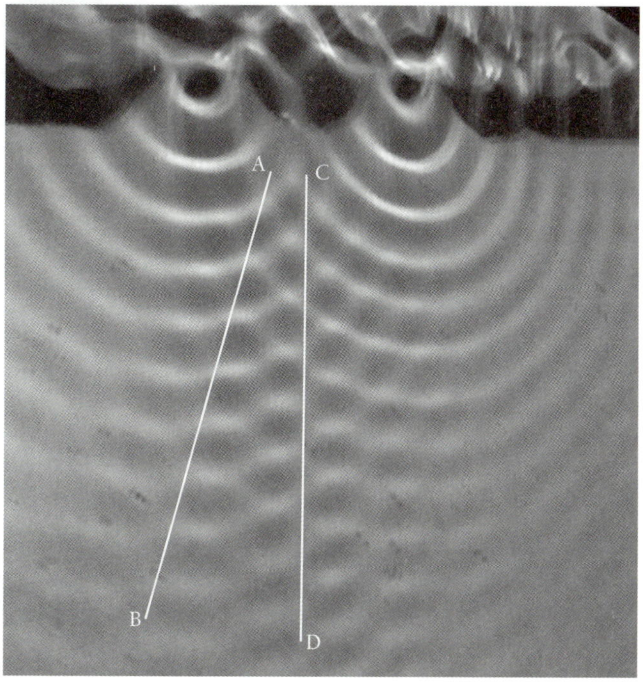

The interference of water waves. Constructive interference occurs along the line *AB* and destructive interference occurs along the line *CD*.

When two or more trains of light waves meet in a region, they **interfere** to produce a new wave there whose instantaneous amplitude is the sum of those of the original waves. Constructive interference refers to the reinforcement of waves with the same phase to produce a greater amplitude, and destructive interference refers to the partial or complete cancellation of waves whose phases differ (Fig. 2.3). If the original waves have different frequencies, the result will be a mixture of constructive and destructive interference, as in Fig. 3.4.

The interference of light waves was first demonstrated in 1801 by Thomas Young, who used a pair of slits illuminated by monochromatic light from a single source (Fig. 2.4). From each slit secondary waves spread out as though originating at the slit; this is an example of **diffraction,** which, like interference, is a characteristic wave phenomenon. Owing to interference, the screen is not evenly lit but shows a pattern of alternate bright and dark lines. At those places on the screen where the path lengths from the two slits differ by an odd number of half wavelengths ($\lambda/2$, $3\lambda/2$, $5\lambda/2$, . . .), destructive interference occurs and a dark line is the result. At those places where the path lengths are

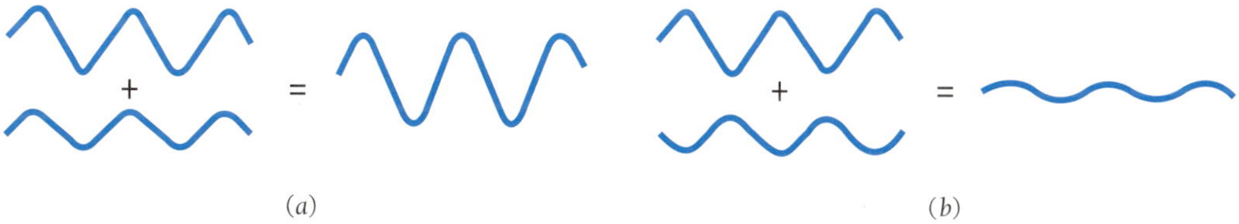

Figure 2.3 (*a*) In constructive interference, superposed waves in phase reinforce each other. (*b*) In destructive interference, waves out of phase partially or completely cancel each other.

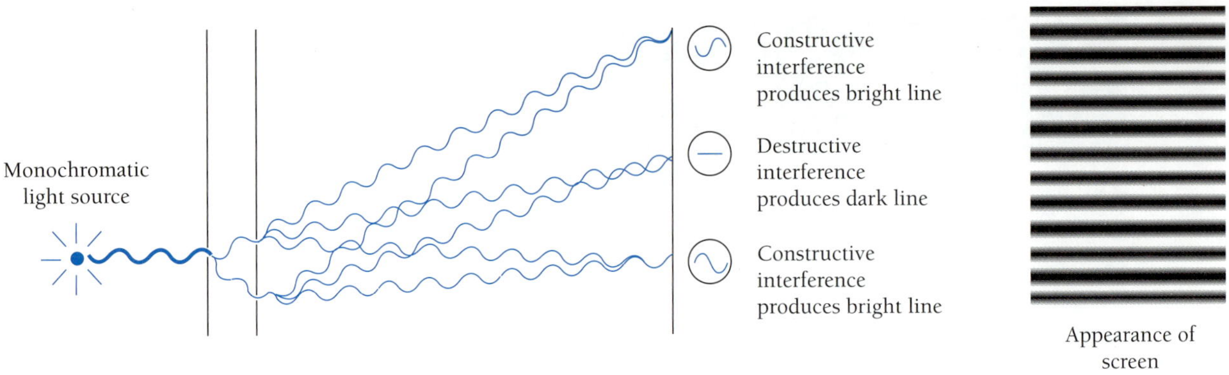

Figure 2.4 Origin of the interference pattern in Young's experiment. Constructive interference occurs where the difference in path lengths from the slits to the screen is θ, λ, 2λ, ... Destructive interference occurs where the path difference is $\lambda/2$, $3\lambda/2$, $5\lambda/2$, ...

equal or differ by a whole number of wavelengths (λ, 2λ, 3λ, . . .), constructive interference occurs and a bright line is the result. At intermediate places the interference is only partial, so the light intensity on the screen varies gradually between the bright and dark lines.

Interference and diffraction are found only in waves—the particles we are familiar with do not behave in those ways. If light consisted of a stream of classical particles, the entire screen would be dark. Thus Young's experiment is proof that light consists of waves. Maxwell's theory further tells us what kind of waves they are: electromagnetic. Until the end of the nineteenth century the nature of light seemed settled forever.

2.2 BLACKBODY RADIATION

Only the quantum theory of light can explain its origin

Following Hertz's experiments, the question of the fundamental nature of light seemed clear: light consisted of em waves that obeyed Maxwell's theory. This certainty lasted only a dozen years. The first sign that something was seriously amiss came from attempts to understand the origin of the radiation emitted by bodies of matter.

We are all familiar with the glow of a hot piece of metal, which gives off visible light whose color varies with the temperature of the metal, going from red to yellow to white as it becomes hotter and hotter. In fact, other frequencies to which our eyes do not respond are present as well. An object need not be so hot that it is luminous for it to be radiating em energy; *all* objects radiate such energy continuously whatever their temperatures, though which frequencies predominate depends on the temperature. At room temperature most of the radiation is in the infrared part of the spectrum and hence is invisible.

The ability of a body to radiate is closely related to its ability to absorb radiation. This is to be expected, since a body at a constant temperature is in thermal equilibrium with its surroundings and must absorb energy from them at the same rate as it emits energy. It is convenient to consider as an ideal body one that absorbs *all* radiation incident upon it, regardless of frequency. Such a body is called a **blackbody.**

The point of introducing the idealized blackbody in a discussion of thermal radiation is that we can now disregard the precise nature of whatever is radiating, since all blackbodies behave identically. In the laboratory a blackbody can be approximated by a hollow object with a very small hole leading to its interior (Fig. 2.5). Any radiation striking the hole enters the cavity, where it is trapped by reflection back and forth until it is absorbed. The cavity walls are constantly emitting and absorbing radiation, and it is in the properties of this radiation **(blackbody radiation)** that we are interested.

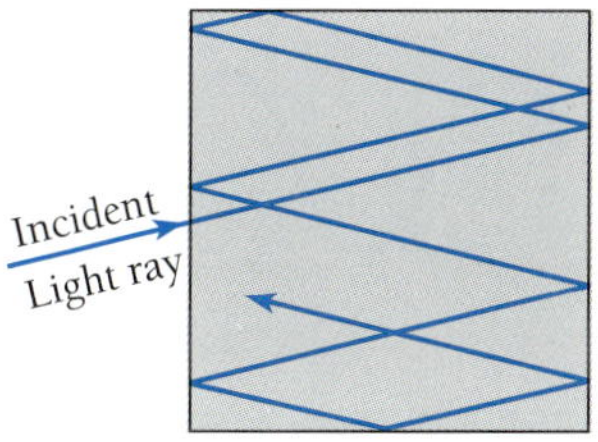

Figure 2.5 A hole in the wall of a hollow object is an excellent approximation of a blackbody.

Experimentally we can sample blackbody radiation simply by inspecting what emerges from the hole in the cavity. The results agree with everyday experience. A blackbody radiates more when it is hot than when it is cold, and the spectrum of a hot blackbody has its peak at a higher frequency than the peak in the spectrum of a cooler one. We recall the behavior of an iron bar as it is heated to progressively higher temperatures: at first it glows dull red, then bright orange-red, and eventually it becomes "white hot." The spectrum of blackbody radiation is shown in Fig. 2.6 for two temperatures.

The Ultraviolet Catastrophe

Why does the blackbody spectrum have the shape shown in Fig. 2.6? This problem was examined at the end of the nineteenth century by Lord Rayleigh and James Jeans. The details of their calculation are given in Chap. 9. They started by considering the radiation inside a cavity of absolute temperature T whose walls are perfect reflectors to be a series of standing em waves (Fig. 2.7). This is a three-dimensional generalization of standing waves in a stretched string. The condition

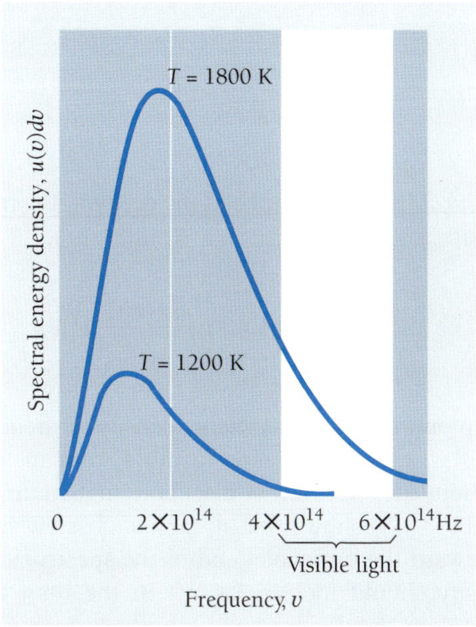

Figure 2.6 Blackbody spectra. The spectral distribution of energy in the radiation depends only on the temperature of the body. The higher the temperature, the greater the amount of radiation and the higher the frequency at which the maximum emission occurs. The dependence of the latter frequency on temperature follows a formula called Wien's displacement law, which is discussed in Sec. 9.6.

The color and brightness of an object heated until it glows, such as the filament of this light bulb, depends upon its temperature, which here is about 3000 K. An object that glows white is hotter than it is when it glows red, and it gives off more light as well.

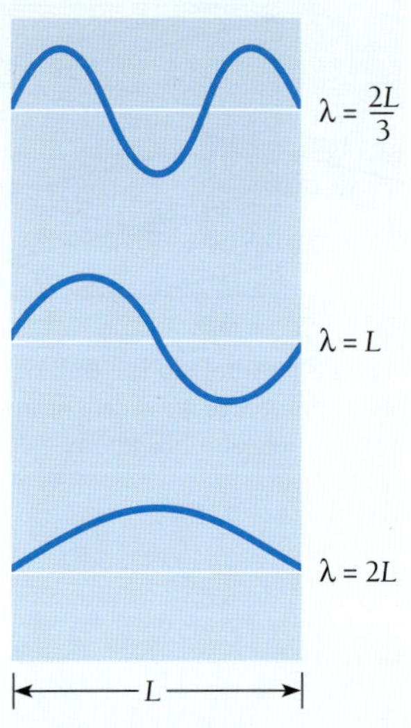

Figure 2.7 Em radiation in a cavity whose walls are perfect reflectors consists of standing waves that have nodes at the walls, which restricts their possible wavelengths. Shown are three possible wavelengths when the distance between opposite walls is L.

for standing waves in such a cavity is that the path length from wall to wall, whatever the direction, must be a whole number of half-wavelengths, so that a node occurs at each reflecting surface. The number of independent standing waves $G(\nu)d\nu$ in the frequency interval between ν and $d\nu$ per unit volume in the cavity turned out to be

Density of standing waves in cavity

$$G(\nu)d\nu = \frac{8\pi\nu^2 d\nu}{c^3} \tag{2.1}$$

This formula is independent of the shape of the cavity. As we would expect, the higher the frequency ν, the shorter the wavelength and the greater the number of possible standing waves.

The next step is to find the average energy per standing wave. According to the **theorem of equipartition of energy,** a mainstay of classical physics, the average energy per degree of freedom of an entity (such as a molecule of an ideal gas) that is a member of a system of such entities in thermal equilibrium at the temperature T is $\frac{1}{2}kT$. Here k is **Boltzmann's constant:**

Boltzmann's constant

$$k = 1.381 \times 10^{-23}\ \text{J/K}$$

A degree of freedom is a mode of energy possession. Thus a monatomic ideal gas molecule has three degrees of freedom, corresponding to kinetic energy of motion in three independent directions, for an average total energy of $\frac{3}{2}kT$.

A one-dimensional harmonic oscillator has two degrees of freedom, one that corresponds to its kinetic energy and one that corresponds to its potential energy. Because each standing wave in a cavity originates in an oscillating electric charge in the cavity wall, two degrees of freedom are associated with the wave and it should have an average energy of $2(\frac{1}{2})kT$:

Classical average energy per standing wave

$$\overline{\epsilon} = kT \tag{2.2}$$

The total energy $u(\nu)\, d\nu$ per unit volume in the cavity in the frequency interval from ν to $\nu + d\nu$ is therefore

Rayleigh-Jeans formula

$$u(\nu)\, d\nu = \overline{\epsilon}G(\nu)\, d\nu = \frac{8\pi kT}{c^3}\nu^2\, d\nu \tag{2.3}$$

This radiation rate is proportional to this energy density for frequencies between ν and $\nu + d\nu$. Equation (2.3), the **Rayleigh-Jeans formula,** contains everything that classical physics can say about the spectrum of blackbody radiation.

Even a glance at Eq. (2.3) shows that it cannot possibly be correct. As the frequency ν increases toward the ultraviolet end of the spectrum, this formula predicts that the energy density should increase as ν^2. In the limit of infinitely high frequencies, $u(\nu)\, d\nu$ therefore should also go to infinity. In reality, of course, the energy density (and radiation rate) falls to 0 as $\nu \to \infty$ (Fig. 2.8). This discrepancy became known as the **ultraviolet catastrophe** of classical physics. Where did Rayleigh and Jeans go wrong?

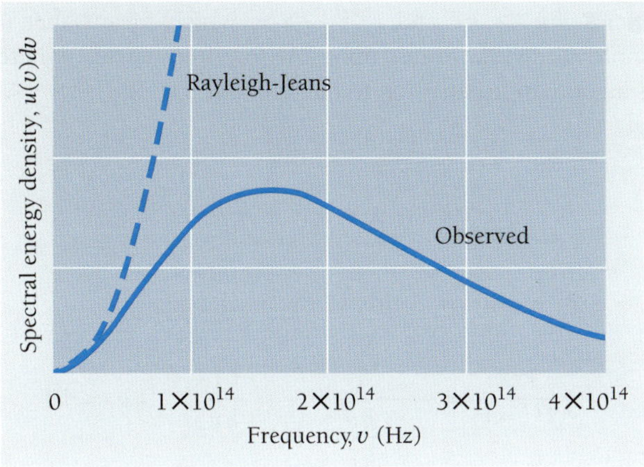

Figure 2.8 Comparison of the Rayleigh-Jeans formula for the spectrum of the radiation from a blackbody at 1500 K with the observed spectrum. The discrepancy is known as the ultraviolet catastrophe because it increases with increasing frequency. This failure of classical physics led Planck to the discovery that radiation is emitted in quanta whose energy is $h\nu$.

Planck Radiation Formula

In 1900 the German physicist Max Planck used "lucky guesswork" (as he later called it) to come up with a formula for the spectral energy density of blackbody radiation:

Planck radiation formula

$$u(\nu)\,d\nu = \frac{8\pi h}{c^3}\frac{\nu^3\,d\nu}{e^{h\nu/kT}-1} \tag{2.4}$$

Here h is a constant whose value is

Planck's constant

$$h = 6.626 \times 10^{-34}\ \mathrm{J \cdot s}$$

About Scientists

Max Planck (1858–1947) was born in Kiel and educated in Munich and Berlin. At the University of Berlin he studied under Kirchhoff and Helmholtz, as Hertz had done earlier. Planck realized that blackbody radiation was important because it was a fundamental effect independent of atomic structure, which was still a mystery in the late nineteenth century, and worked at understanding it for six years before finding the formula the radiation obeyed. He "strived from the day of its discovery to give it a real physical interpretation." The result was the discovery that radiation is emitted in energy steps of $h\nu$. Although this discovery, for which he received the Nobel Prize in 1918, is now considered to mark the start of modern physics, Planck himself remained skeptical for a long time of the physical reality of quanta. As he later wrote, "My vain attempts to somehow reconcile the elementary quantum with classical theory continued for many years and cost me great effort. . . . Now I know for certain that the quantum of action has a much more fundamental significance than I originally suspected."

Like many physicists, Planck was a competent musician (he sometimes played with Einstein) and in addition enjoyed mountain climbing. Although Planck remained in Germany during the Hitler era, he protested the Nazi treatment of Jewish scientists and lost his presidency of the Kaiser Wilhelm Institute as a result. In 1945 one of his sons was implicated in a plot to kill Hitler and was executed. After World War II the Institute was renamed after Planck and he was again its head until his death.

At high frequencies, $h\nu \gg kT$ and $e^{h\nu/kT} \rightarrow \infty$, which means that $u(\nu)\, d\nu \rightarrow 0$ as observed. No more ultraviolet catastrophe. At low frequencies, where the Rayleigh-Jeans formula is a good approximation to the data (see Fig. 2.8), $h\nu \ll kT$ and $h\nu/kT \ll 1$. In general,

$$e^x = 1 + x + \frac{x^2}{2!} + \frac{x^3}{3!} + \cdots$$

If x is small, $e^x \approx 1 + x$, and so for $h\nu/kT \ll 1$ we have

$$\frac{1}{e^{h\nu/kT} - 1} \approx \frac{1}{1 + \dfrac{h\nu}{kT} - 1} \approx \frac{kT}{h\nu} \qquad h\nu \ll kT$$

Thus at low frequencies Planck's formula becomes

$$u(\nu)\, d\nu \approx \frac{8\pi h}{c^3}\nu^3 \left(\frac{kT}{h\nu}\right) d\nu \approx \frac{8\pi kT}{c^3}\nu^2\, d\nu$$

which is the Rayleigh-Jeans formula. Planck's formula is clearly at least on the right track; in fact, it has turned out to be completely correct.

Next Planck had the problem of justifying Eq. (2.4) in terms of physical principles. A new principle seemed needed to explain his formula, but what was it? After several weeks of "the most strenuous work of my life," Planck found the answer: The oscillators in the cavity walls could not have a continuous distribution of possible energies ϵ but must have only the specific energies

Oscillator energies $$\epsilon_n = nh\nu \qquad n = 0, 1, 2, \ldots \tag{2.5}$$

An oscillator emits radiation of frequency ν when it drops from one energy state to the next lower one, and it jumps to the next higher state when it absorbs radiation of frequency ν. Each discrete bundle of energy $h\nu$ is called a **quantum** (plural **quanta**) from the Latin for "how much."

With oscillator energies limited to $nh\nu$, the average energy per oscillator in the cavity walls—and so per standing wave—turned out to be not $\bar{\epsilon} = kT$ as for a continuous distribution of oscillator energies, but instead

Actual average energy per standing wave $$\bar{\epsilon} = \frac{h\nu}{e^{h\nu/kT} - 1} \tag{2.6}$$

This average energy leads to Eq. (2.4). Blackbody radiation is further discussed in Chap. 9.

Example 2.1

Assume that a certain 660-Hz tuning fork can be considered as a harmonic oscillator whose vibrational energy is 0.04 J. Compare the energy quanta of this tuning fork with those of an atomic oscillator that emits and absorbs orange light whose frequency is 5.00×10^{14} Hz.

Solution

(*a*) For the tuning fork,

$$h\nu_1 = (6.63 \times 10^{-34}\ \text{J} \cdot \text{s})\ (660\ \text{s}^{-1}) = 4.38 \times 10^{-31}\ \text{J}$$

The total energy of the vibrating tines of the fork is therefore about 10^{29} times the quantum energy $h\nu$. The quantization of energy in the tuning fork is obviously far too small to be observed, and we are justified in regarding the fork as obeying classical physics.

(*b*) For the atomic oscillator,

$$h\nu_2 = (6.63 \times 10^{-34}\ \text{J} \cdot \text{s})\ (5.00 \times 10^{14}\ \text{s}^{-1}) = 3.32 \times 10^{-19}\ \text{J}$$

In electronvolts, the usual energy unit in atomic physics,

$$h\nu_2 = \frac{3.32 \times 10^{-19}\ \text{J}}{1.60 \times 10^{-19}\ \text{J/eV}} = 2.08\ \text{eV}$$

This is a significant amount of energy on an atomic scale, and it is not surprising that classical physics fails to account for phenomena on this scale.

The concept that the oscillators in the cavity walls can interchange energy with standing waves in the cavity only in quanta of $h\nu$ is, from the point of view of classical physics, impossible to understand. Planck regarded his quantum hypothesis as an "act of desperation" and, along with other physicists of his time, was unsure of how seriously to regard it as an element of physical reality. For many years he held that, although the energy transfers between electric oscillators and em waves apparently are quantized, em waves themselves behave in an entirely classical way with a continuous range of possible energies.

2.3 PHOTOELECTRIC EFFECT

The energies of electrons liberated by light depend on the frequency of the light

During his experiments on em waves, Hertz noticed that sparks occurred more readily in the air gap of his transmitter when ultraviolet light was directed at one of the metal balls. He did not follow up this observation, but others did. They soon discovered that the cause was electrons emitted when the frequency of the light was sufficiently high. This phenomenon is known as the **photoelectric effect** and the emitted electrons are called **photoelectrons.** It is one of the ironies of history that the same work to demonstrate that light consists of em waves also gave the first hint that this was not the whole story.

Figure 2.9 shows how the photoelectric effect was studied. An evacuated tube contains two electrodes connected to a source of variable voltage, with the metal plate whose surface is irradiated as the anode. Some of the photoelectrons that emerge from this surface have enough energy to reach the cathode despite its negative polarity, and they constitute the measured current. The slower photoelectrons are repelled before they get to the cathode. When the voltage is increased to a certain value V_0, of the order of several volts, no more photoelectrons arrive, as indicated by the current dropping to zero. This extinction voltage corresponds to the maximum photoelectron kinetic energy.

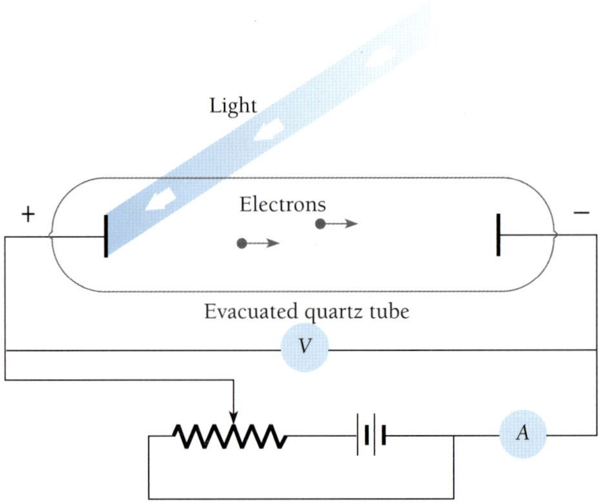

Figure 2.9 Experimental observation of the photoelectric effect.

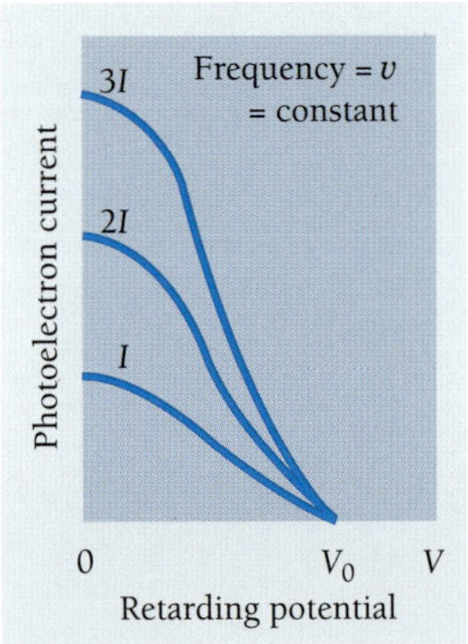

Figure 2.10 Photoelectron current is proportional to light intensity I for all retarding voltages. The stopping potential V_0, which corresponds to the maximum photoelectron energy, is the same for all intensities of light of the same frequency ν.

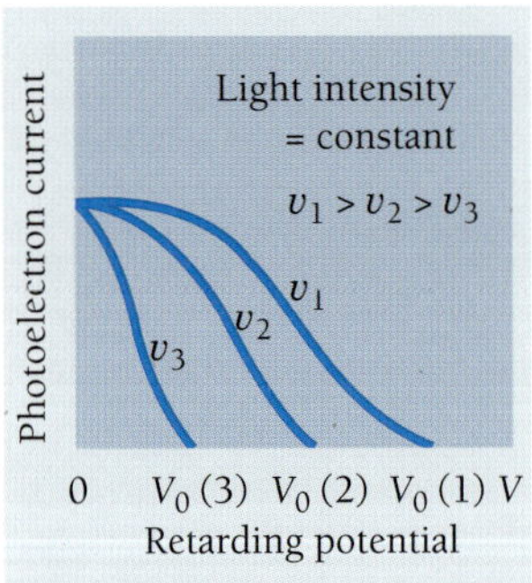

Figure 2.11 The stopping potential V_0, and hence the maximum photoelectron energy, depends on the frequency of the light. When the retarding potential is $V = 0$, the photoelectron current is the same for light of a given intensity regardless of its frequency.

The existence of the photoelectric effect is not surprising. After all, light waves carry energy, and some of the energy absorbed by the metal may somehow concentrate on individual electrons and reappear as their kinetic energy. The situation should be like water waves dislodging pebbles from a beach. But three experimental findings show that no such simple explanation is possible.

1 Within the limits of experimental accuracy (about 10^{-9} s), there is no time interval between the arrival of light at a metal surface and the emission of photoelectrons. However, because the energy in an em wave is supposed to be spread across the wavefronts, a period of time should elapse before an individual electron accumulates enough energy (several eV) to leave the metal. A detectable photoelectron current results when 10^{-6} W/m^2 of em energy is absorbed by a sodium surface. A layer of sodium 1 atom thick and 1 m^2 in area contains about 10^{19} atoms, so if the incident light is absorbed in the uppermost atomic layer, each atom receives energy at an average rate of 10^{-25} W. At this rate over a month would be needed for an atom to accumulate energy of the magnitude that photoelectrons from a sodium surface are observed to have.

2 A bright light yields more photoelectrons than a dim one of the same frequency, but the electron energies remain the same (Fig. 2.10). The em theory of light, on the contrary, predicts that the more intense the light, the greater the energies of the electrons.

3 The higher the frequency of the light, the more energy the photoelectrons have (Fig. 2.11). Blue light results in faster electrons than red light. At frequencies below a certain critical frequency ν_0, which is characteristic of each particular metal, no electrons are emitted. Above ν_0 the photoelectrons range in energy from 0 to a maximum value that increases linearly with increasing frequency (Fig. 2.12). This observation, also, cannot be explained by the em theory of light.

Quantum Theory of Light

When Planck's derivation of his formula appeared, Einstein was one of the first—perhaps the first—to understand just how radical the postulate of energy quantization of oscillators was: "It was as if the ground was pulled from under one." A few years later, in 1905, Einstein realized that the photoelectric effect could be understood if the

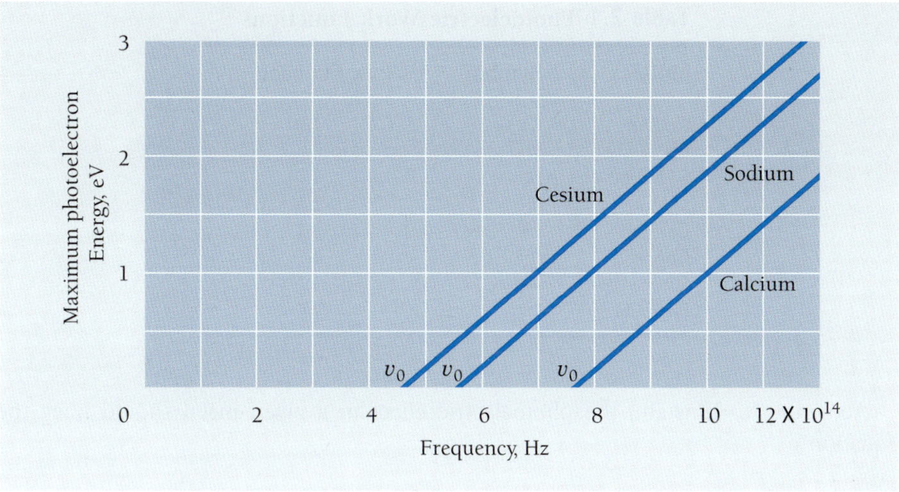

Figure 2.12 Maximum photoelectron kinetic energy KE_{max} versus frequency of incident light for three metal surfaces.

energy in light is not spread out over wavefronts but is concentrated in small packets, or **photons.** (The term photon was coined by the chemist Gilbert Lewis in 1926.) Each photon of light of frequency ν has the energy $h\nu$, the same as Planck's quantum energy. Planck had thought that, although energy from an electric oscillator apparently had to be given to em waves in separate quanta of $h\nu$ each, the waves themselves behaved exactly as in conventional wave theory. Einstein's break with classical physics was more drastic: Energy was not only given to em waves in separate quanta but was also carried by the waves in separate quanta.

The three experimental observations listed above follow directly from Einstein's hypothesis. (1) Because em wave energy is concentrated in photons and not spread out, there should be no delay in the emission of photoelectrons. (2) All photons of frequency ν have the same energy, so changing the intensity of a monochromatic light beam will change the number of photoelectrons but not their energies. (3) The higher the frequency ν, the greater the photon energy $h\nu$ and so the more energy the photoelectrons have.

What is the meaning of the critical frequency ν_0 below which no photoelectrons are emitted? There must be a minimum energy ϕ for an electron to escape from a particular metal surface or else electrons would pour out all the time. This energy is called the **work function** of the metal, and is related to ν_0 by the formula

Work function
$$\phi = h\nu_0 \tag{2.7}$$

The greater the work function of a metal, the more energy is needed for an electron to leave its surface, and the higher the critical frequency for photoelectric emission to occur.

Some examples of photoelectric work functions are given in Table 2.1. To pull an electron from a metal surface generally takes about half as much energy as that needed to pull an electron from a free atom of that metal (see Fig. 7.10); for instance, the ionization energy of cesium is 3.9 eV compared with its work function of 1.9 eV. Since the visible spectrum extends from about 4.3 to about 7.5 $\times$ 10^{14} Hz, which corresponds to quantum energies of 1.7 to 3.3 eV, it is clear from Table 2.1 that the photoelectric effect is a phenomenon of the visible and ultraviolet regions.

Table 2.1 Photoelectric Work Functions

Metal	Symbol	Work Function, eV
Cesium	Cs	1.9
Potassium	K	2.2
Sodium	Na	2.3
Lithium	Li	2.5
Calcium	Ca	3.2
Copper	Cu	4.7
Silver	Ag	4.7
Platinum	Pt	6.4

According to Einstein, the photoelectric effect in a given metal should obey the equation

Photoelectric effect
$$h\nu = \mathrm{KE}_{\max} + \phi \tag{2.8}$$

where $h\nu$ is the photon energy, $\mathrm{KE}_{\max}$ is the maximum photoelectron energy (which is proportional to the stopping potential), and ϕ is the minimum energy needed for an electron to leave the metal. Because $\phi = h\nu_0$, Eq. (2.8) can be rewritten (Fig. 2.13)

$$h\nu = \mathrm{KE}_{\max} + h\nu_0$$

$$\mathrm{KE}_{\max} = h\nu - h\nu_0 = h(\nu - \nu_0) \tag{2.9}$$

This formula accounts for the relationships between $\mathrm{KE}_{\max}$ and ν plotted in Fig. 2.12 from experimental data. If Einstein was right, the slopes of the lines should all be equal to Planck's constant h, and this is indeed the case.

All light-sensitive detectors, including the eye and the one used in this video camera, are based on the absorption of energy from photons of light by electrons in the atoms the light falls on.

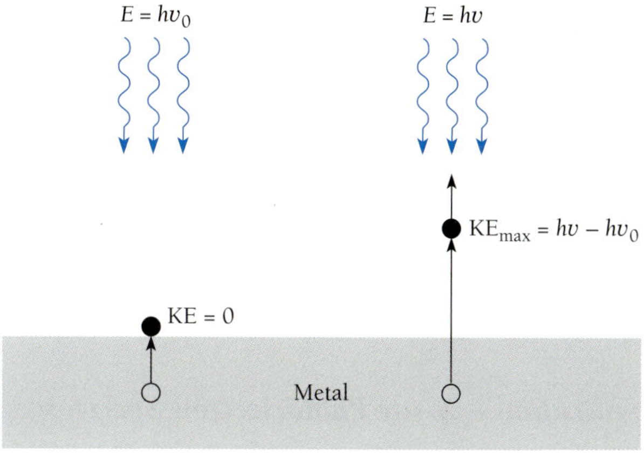

Figure 2.13 If the energy $h\nu_0$ (the work function of the surface) is needed to remove an electron from a metal surface, the maximum electron kinetic energy will be $h\nu - h\nu_0$ when light of frequency ν is directed at the surface.

In terms of electronvolts, the formula $E = h\nu$ for photon energy becomes

Photon energy

$$E = \left(\frac{6.626 \times 10^{-34}\ \text{J}\cdot\text{s}}{1.602 \times 10^{-19}\ \text{J/eV}}\right)\nu = (4.136 \times 10^{-15})\nu\ \text{eV} \tag{2.10}$$

If we are given instead the wavelength λ of the light, then since $\nu = c/\lambda$ we have

Photon energy

$$E = \frac{(4.136 \times 10^{-15}\ \text{eV}\cdot\text{s})(2.998 \times 10^{8}\ \text{m/s})}{\lambda} = \frac{1.240 \times 10^{-6}\ \text{eV}\cdot\text{m}}{\lambda} \tag{2.11}$$

Example 2.2

Ultraviolet light of wavelength 350 nm and intensity 1.00 W/m^2 is directed at a potassium surface. (*a*) Find the maximum KE of the photoelectrons. (*b*) If 0.50 percent of the incident photons produce photoelectrons, how many are emitted per second if the potassium surface has an area of 1.00 cm^2?

Solution

(*a*) From Eq. (2.11) the energy of the photons is, since 1 nm = 1 nanometer = 10^{-9} m,

$$E_p = \frac{1.24 \times 10^{-6}\ \text{eV}\cdot\text{m}}{(350\ \text{nm})(10^{-9}\ \text{m/nm})} = 3.5\ \text{eV}$$

Table 2.1 gives the work function of potassium as 2.2 eV, so

$$\text{KE}_{\text{max}} = h\nu - \phi = 3.5\ \text{eV} - 2.2\ \text{eV} = 1.3\ \text{eV}$$

(*b*) The photon energy in joules is 5.68×10^{-19} J. Hence the number of photons that reach the surface per second is

$$n_p = \frac{E/t}{E_p} = \frac{(P/A)(A)}{E_p} = \frac{(1.00\ \text{W/m}^2)\,(1.00 \times 10^{-4}\ \text{m}^2)}{5.68 \times 10^{-19}\ \text{J/photon}} = 1.76 \times 10^{14}\ \text{photons/s}$$

The rate at which photoelectrons are emitted is therefore

$$n_e = (0.0050)n_p = 8.8 \times 10^{11}\ \text{photoelectrons/s}$$

Application – X-ray Photoelectron Spectroscopy

X-ray photoelectron spectroscopy (XPS) is a surface sensitive quantitative spectro scopic technique that can detect and quantify the elements that exist in materials (except for H and He). XPS was first developed in the 1960s by Kai Siegbahn and his research group at University of Uppsala, Sweden.[1] As it has the ability to analyze conductors, insulators and chemical bonding, XPS can be an ideal technique for a wide variety of applications such as analyzing the surface chemistry of materials, identifying stains and discolorations, analyzing the composition of materials and determining contaminant sources.

XPS is an ultra-high vacuum surface analysis technique with a sampling volume that extends from the surface to a depth of 5-10nm. The basic principle is that a sample will be irradiated by the monochromatic X-ray, and the sample will be excited to have the emission of photoelectrons. A typical XPS result is an XPS spectrum that is an plot of the number of photoelectrons detected against the binding energy of the electrons detected. As each element produces a different set of XPS peaks at specific binding energy values, it can identify each element that exists in or on the surface of materials.

Thermionic Emission

Einstein's interpretation of the photoelectric effect is supported by studies of thermionic emission. Long ago it was discovered that the presence of a very hot object increases the electric conductivity of the surrounding air. Eventually the reason for this effect was found to be the emission of electrons from such an object. Thermionic emission makes possible the operation of such devices as television picture tubes, in which metal filaments or specially coated cathodes at high temperature supply dense streams of electrons.

The emitted electrons evidently obtain their energy from the thermal agitation of the particles of the metal, and we would expect the electrons to need a certain minimum energy to escape. This minimum energy can be determined for many surfaces, and it is always close to the photoelectric work function for the same surfaces. In photoelectric emission, photons of light provide the energy required by an electron to escape, while in thermionic emission heat does so.

1. K. Siegbahn, Et. Al., Nova Acta Regiae Soc.Sci., Ser. IV, Vol. 20 (1967)

2.4 WHAT IS LIGHT?

Both wave and particle

The concept that light travels as a series of little packets is directly opposed to the wave theory of light (Fig. 2.14). Both views have strong experimental support, as we have seen. According to the wave theory, light waves leave a source with their energy spread out continuously through the wave pattern. According to the quantum theory, light consists of individual photons, each small enough to be absorbed by a single electron. Yet, despite the particle picture of light it presents, the quantum theory needs the frequency of the light to describe the photon energy.

Which theory are we to believe? A great many scientific ideas have had to be revised or discarded when they were found to disagree with new data. Here, for the first time, two different theories are needed to explain a single phenomenon. This situation is not the same as it is, say, in the case of relativistic versus newtonian mechanics, where one turns out to be an approximation of the other. The connection between the wave and quantum theories of light is something else entirely.

To appreciate this connection, let us consider the formation of a double-slit interference pattern on a screen. In the wave model, the light intensity at a place on the screen depends on $\overline{E^2}$, the average over a complete cycle of the square of the instantaneous magnitude E of the em wave's electric field. In the particle model, this intensity depends instead on $Nh\nu$, where N is the number of photons per second per unit area that reach the same place on the screen. Both descriptions must give the same value for the intensity, so N is proportional to $\overline{E^2}$. If N is large enough, somebody looking at the screen would see the usual double-slit interference pattern and would have no reason to doubt the wave model. If N is small—perhaps so small that only one photon at a time reaches the screen—the observer would find a series of apparently random flashes and would assume that he or she is watching quantum behavior.

If the observer keeps track of the flashes for long enough, though, the pattern they form will be the same as when N is large. Thus the observer is entitled to conclude that the *probability* of finding a photon at a certain place and time depends on the value of $\overline{E^2}$ there. If we regard each photon as somehow having a wave associated with it, the intensity of this wave at a given place on the screen determines the likelihood that a photon will arrive there. When it passes through the slits, light is behaving as a wave does. When it strikes the screen, light is behaving as a particle does. Apparently light travels as a wave but absorbs and gives off energy as a series of particles.

We can think of light as having a dual character. **The wave theory and the quantum theory complement each other.** Either theory by itself is only part of the story and can explain only certain effects. A reader who finds it hard to understand how light can be both a wave and a stream of particles is in good company: shortly before his death, Einstein remarked that "All these fifty years of conscious brooding have brought me no nearer to the answer to the question, 'What are light quanta?'" The "true nature" of light includes both wave and particle characters, even though there is nothing in everyday life to help us visualize that.

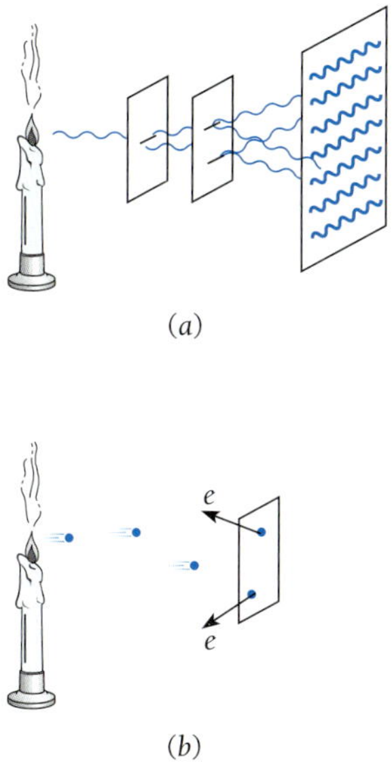

Figure 2.14 (*a*) The wave theory of light explains diffraction and interference, which the quantum theory cannot account for. (*b*) The quantum theory explains the photoelectric effect, which the wave theory cannot account for.

2.5 X-RAYS

They consist of high-energy photons

The photoelectric effect provides convincing evidence that photons of light can transfer energy to electrons. Is the inverse process also possible? That is, can part or all of the kinetic energy of a moving electron be converted into a photon? As it happens, the

About Scientists

Wilhelm Konrad Roentgen (1845–1923) was born in Lennep, Germany, and studied in Holland and Switzerland. After periods at several German universities, Roentgen became professor of physics at Würzburg where, on November 8, 1895, he noticed that a sheet of paper coated with barium platinocyanide glowed when he switched on a nearby cathode-ray tube that was entirely covered with black cardboard. In a cathode-ray tube, electrons are accelerated in a vacuum by an electric field, and it was the impact of these electrons on the glass end of the tube that produced the penetrating "x" (since their nature was then unknown) rays that caused the salt to glow. Roentgen said of his discovery that, when people heard of it, they would say, "Roentgen has probably gone crazy." In fact, x-rays were an immediate sensation, and only two months later were being used in medicine. They also stimulated research in new directions; Becquerel's discovery of radioactivity followed within a year. Roentgen received the first Nobel Prize in physics in 1902. He refused to benefit financially from his work and died in poverty in the German inflation that followed the end of World War I.

inverse photoelectric effect not only does occur but had been discovered (though not understood) before the work of Planck and Einstein.

In 1895 Wilhelm Roentgen found that a highly penetrating radiation of unknown nature is produced when fast electrons impinge on matter. These **x-rays** were soon found to travel in straight lines, to be unaffected by electric and magnetic fields, to pass readily through opaque materials, to cause phosphorescent substances to glow, and to expose photographic plates. The faster the original electrons, the more penetrating the resulting x-rays, and the greater the number of electrons, the greater the intensity of the x-ray beam.

Not long after this discovery it became clear that x-rays are em waves. Electromagnetic theory predicts that an accelerated electric charge will radiate em waves, and a rapidly moving electron suddenly brought to rest is certainly accelerated. Radiation produced under these circumstances is given the German name **bremsstrahlung** ("braking radiation"). Energy loss due to bremsstrahlung is more important for electrons than for heavier particles because electrons are more violently accelerated when passing near nuclei in their paths. The greater the energy of an electron and the greater the atomic number of the nuclei it encounters, the more energetic the bremsstrahlung.

In 1912 a method was devised for measuring the wavelengths of x-rays. A diffraction experiment had been recognized as ideal, but as we recall from physical optics, the spacing between adjacent lines on a diffraction grating must be of the same order of magnitude as the wavelength of the light for satisfactory results, and gratings cannot be ruled with the minute spacing required by x-rays. Max von Laue realized that the wavelengths suggested for x-rays were comparable to the spacing between adjacent atoms in crystals. He therefore proposed that crystals be used to diffract x-rays, with their regular lattices acting as a kind of three-dimensional grating. In experiments carried out the following year, wavelengths from 0.013 to 0.048 nm were found, 10^{-4} of those in visible light and hence having quanta 10^4 times as energetic.

Electromagnetic radiation with wavelengths from about 0.01 to about 10 nm falls into the category of x-rays. The boundaries of this category are not sharp: the shorter-wavelength end overlaps gamma rays and the longer-wavelength end overlaps ultraviolet light (see Fig. 2.2).

Figure 2.15 is a diagram of an x-ray tube. A cathode, heated by a filament through which an electric current is passed, supplies electrons by thermionic emission. The high potential difference V maintained between the cathode and a metallic target accelerates the electrons toward the latter. The face of the target is at an angle relative to the electron beam, and the x-rays that leave the target pass through the side of the tube. The tube is evacuated to permit the electrons to get to the target unimpeded.

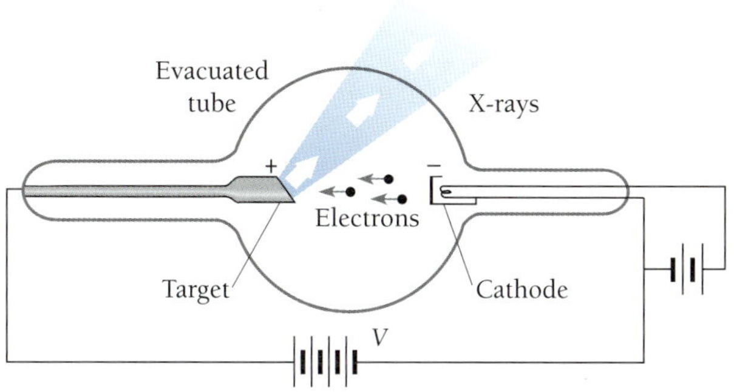

Figure 2.15 An x-ray tube. The higher the accelerating voltage V, the faster the electrons and the shorter the wavelengths of the x-rays.

In modern x-ray tubes like these, circulating oil carries heat away from the target and releases it to the outside air through a heat exchanger. The use of x-rays as a diagnostic tool in medicine is based upon the different extents to which different tissues absorb them. Because of its calcium content, bone is much more opaque to x-rays than muscle, which in turn is more opaque than fat. To enhance contrast, "meals" that contain barium are given to patients to better display their digestive systems, and other compounds may be injected into the bloodstream to enable the condition of blood vessels to be studied.

As mentioned earlier, classical electromagnetic theory predicts bremsstrahlung when electrons are accelerated, which accounts in general for the x-rays produced by an x-ray tube. However, the agreement between theory and experiment is not satisfactory in certain important respects. Figures 2.16 and 2.17 show the x-ray spectra that result when tungsten and molybdenum targets are bombarded by electrons at several different accelerating potentials. The curves exhibit two features electromagnetic theory cannot explain:

1 In the case of molybdenum, intensity peaks occur that indicate the enhanced production of x-rays at certain wavelengths. These peaks occur at specific wavelengths for each target material and originate in rearrangements of the electron structures of the target atoms after having been disturbed by the bombarding electrons. This phenomenon will be discussed in Sec. 7.9; the important thing to note at this point is the presence of x-rays of specific wavelengths, a decidedly nonclassical effect, in addition to a continuous x-ray spectrum.

2 The x-rays produced at a given accelerating potential V vary in wavelength, but none has a wavelength shorter than a certain value λ_{min}. Increasing V decreases λ_{min}. At a particular V, λ_{min} is the *same* for both the tungsten and molybdenum targets. Duane and Hunt found experimentally that λ_{min} is inversely proportional to V; their precise relationship is

X-ray production

$$\lambda_{min} = \frac{1.24 \times 10^{-6}}{V} \text{ V} \cdot \text{m} \tag{2.12}$$

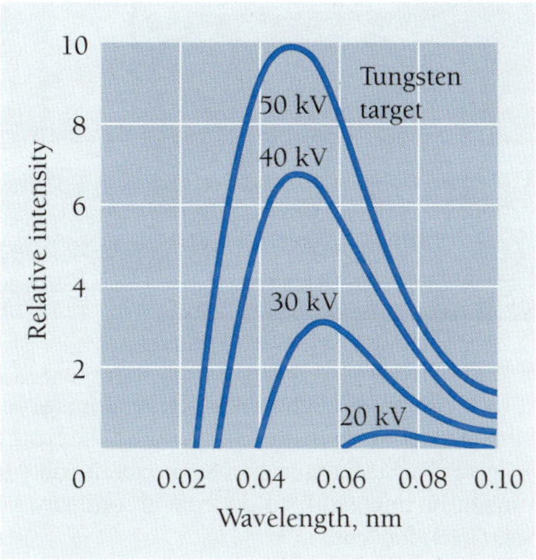

Figure 2.16 X-ray spectra of tungsten at various accelerating potentials.

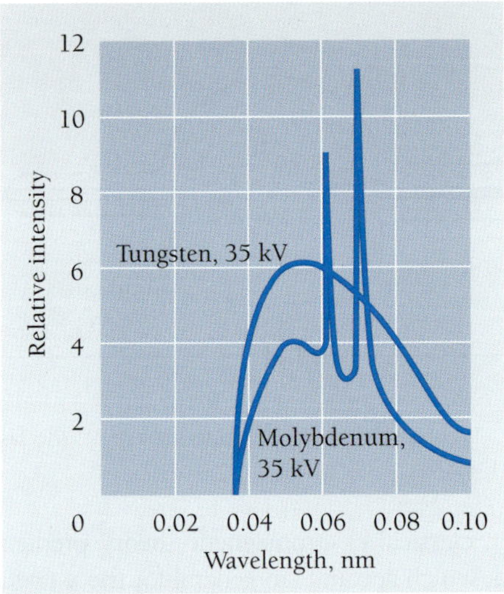

Figure 2.17 X-ray spectra of tungsten and molybdenum at 35 kV accelerating potential.

The second observation fits with the quantum theory of radiation. Most of the electrons that strike the target undergo numerous glancing collisions, with their energy going simply into heat. (This is why the targets in x-ray tubes are made from high-melting-point metals such as tungsten, and a means of cooling the target is usually employed.) A few electrons, though, lose most or all of their energy in single collisions with target atoms. This is the energy that becomes x-rays.

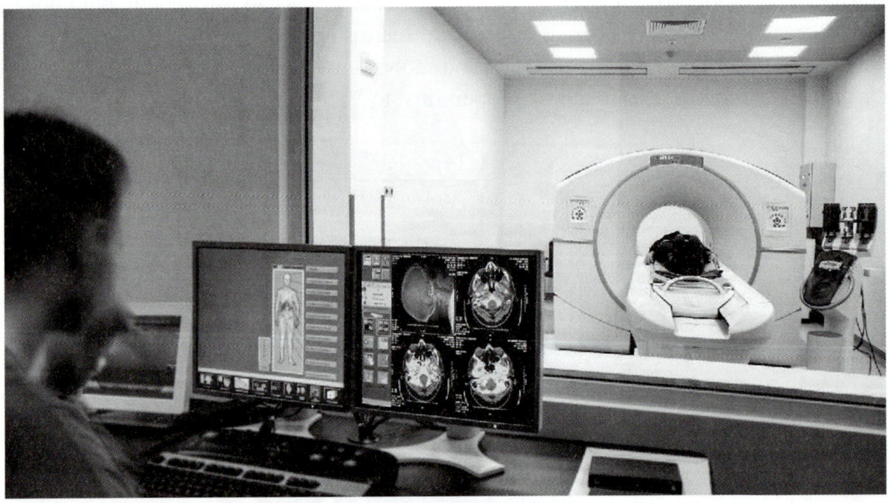

In a CT (computerized tomography) scanner, a series of x-ray exposures of a patient taken from different directions are combined by a computer to give cross-sectional images of the parts of the body being examined. In effect, the tissue is sliced up by the computer on the basis of the x-ray exposures, and any desired slice can be displayed. This technique enables an abnormality to be detected and its exact location established, which might be impossible to do from an ordinary x-ray picture. (The word tomography comes from tomos, Greek for "cut.")

X-ray production, then, except for the peaks mentioned in observation 1 above, represents an inverse photoelectric effect. Instead of photon energy being transformed into electron KE, electron KE is being transformed into photon energy. A short wavelength means a high frequency, and a high frequency means a high photon energy $h\nu$.

Since work functions are only a few electronvolts whereas the accelerating potentials in x-ray tubes are typically tens or hundreds of thousands of volts, we can ignore the work function and interpret the short wavelength limit of Eq. (2.12) as corresponding to the case where the entire kinetic energy KE $= Ve$ of a bombarding electron is given up to a single photon of energy $h\nu_{max}$. Hence

$$Ve = h\nu_{max} = \frac{hc}{\lambda_{min}}$$

$$\lambda_{min} = \frac{hc}{Ve} = \frac{1.240 \times 10^{-6}}{V} \text{ V} \cdot \text{m}$$

which is the Duane-Hunt formula of Eq. (2.12)—and, indeed, the same as Eq. (2.11) except for different units. It is therefore appropriate to regard x-ray production as the inverse of the photoelectric effect.

Example 2.3

Find the shortest wavelength present in the radiation from an x-ray machine whose accelerating potential is 50,000 V.

Solution

From Eq. (2.12) we have

$$\lambda_{min} = \frac{1.24 \times 10^{-6} \text{ V} \cdot \text{m}}{5.00 \times 10^{4} \text{ V}} = 2.48 \times 10^{-11} \text{ m} = 0.0248 \text{ nm}$$

This wavelength corresponds to the frequency

$$\nu_{max} = \frac{c}{\lambda_{min}} = \frac{3.00 \times 10^{8} \text{ m/s}}{2.48 \times 10^{-11} \text{ m}} = 1.21 \times 10^{19} \text{ Hz}$$

2.6 X-RAY DIFFRACTION

How x-ray wavelengths can be determined

A crystal consists of a regular array of atoms, each of which can scatter em waves. The mechanism of scattering is straightforward. An atom in a constant electric field becomes polarized since its negatively charged electrons and positively charged nucleus experience forces in opposite directions. These forces are small compared with the forces holding the atom together, and so the result is a distorted charge distribution equivalent to an electric dipole. In the presence of the alternating electric field of an em wave of frequency ν, the polarization changes back and forth with the same frequency ν. An oscillating electric dipole is thus created at the expense of some of the energy of the incoming wave. The oscillating dipole in turn radiates em waves of frequency ν, and these secondary waves go out in all directions except along the dipole axis. (In an assembly of atoms exposed to unpolarized radiation, the latter restriction does not apply since the contributions of the individual atoms are random.)

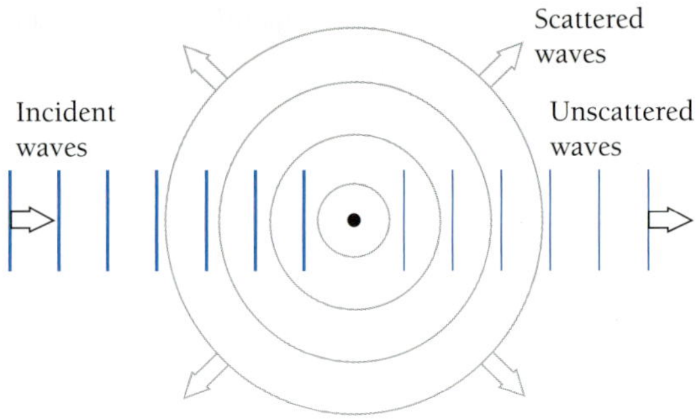

Figure 2.18 The scattering of electromagnetic radiation by a group of atoms. Incident plane waves are reemitted as spherical waves.

In wave terminology, the secondary waves have spherical wave fronts in place of the plane wave fronts of the incoming waves (Fig. 2.18). The scattering process, then, involves atoms that absorb incident plane waves and reemit spherical waves of the same frequency.

A monochromatic beam of x-rays that falls upon a crystal will be scattered in all directions inside it. However, owing to the regular arrangement of the atoms, in certain directions the scattered waves will constructively interfere with one another while in others they will destructively interfere. The atoms in a crystal may be thought of as defining families of parallel planes, as in Fig. 2.19, with each family having a characteristic separation between its component planes. This analysis was suggested in 1913 by W. L Bragg, in honor of whom the above planes are called **Bragg planes.**

The conditions that must be fulfilled for radiation scattered by crystal atoms to undergo constructive interference may be obtained from a diagram like that in Fig. 2.20. A beam containing x-rays of wavelength λ is incident upon a crystal at an angle θ with a family of Bragg planes whose spacing is d. The beam goes past atom A in the first plane and atom B in the next, and each of them scatters part of the beam in random directions. Constructive interference takes place only between those scattered rays that are parallel and whose paths differ by exactly λ, 2λ, 3λ, and so on. That is, the path difference must be $n\lambda$, where n is an integer. The only rays scattered by A and B for which this is true are those labeled I and II in Fig. 2.20.

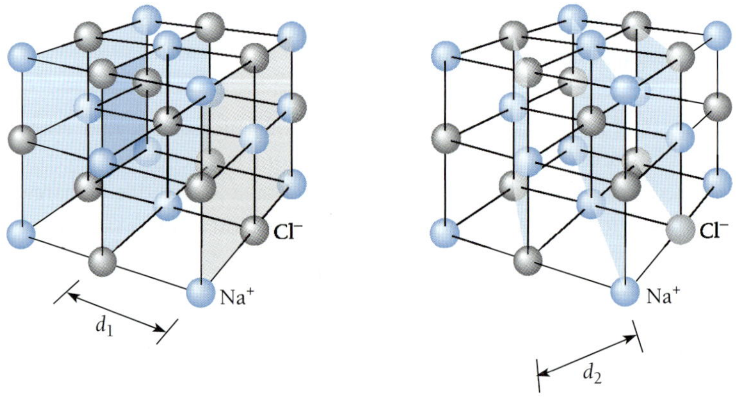

Figure 2.19 Two sets of Bragg planes in a NaCl crystal.

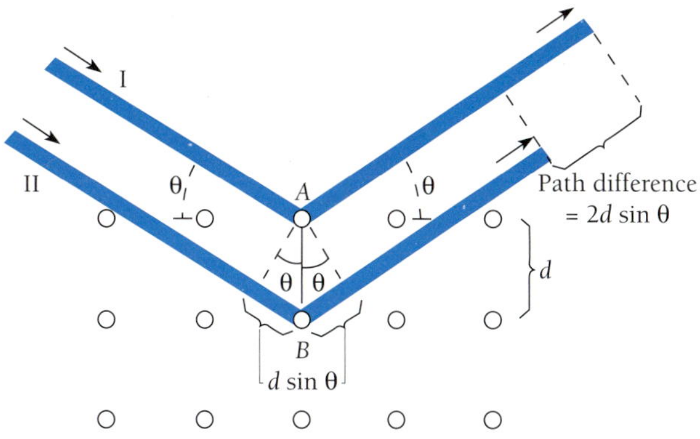

Figure 2.20 X-ray scattering from a cubic crystal.

The interference pattern produced by the scattering of x-rays from ions in a crystal of NaCl. The bright spots correspond to the directions where x-rays scattered from various layers in the crystal interfere constructively. The cubic pattern of the NaCl lattice is suggested by the fourfold symmetry of the pattern. The large central spot is due to the unscattered x-ray beam.

The first condition on I and II is that their common scattering angle be equal to the angle of incidence θ of the original beam. (This condition, which is independent of wavelength, is the same as that for ordinary specular reflection in optics: angle of incidence = angle of reflection.) The second condition is that

$$2d \sin \theta = n\lambda \qquad n = 1, 2, 3, \ldots \tag{2.13}$$

since ray II must travel the distance $2d \sin \theta$ farther than ray I. The integer n is the **order** of the scattered beam.

The schematic design of an x-ray spectrometer based upon Bragg's analysis is shown in Fig. 2.21. A narrow beam of x-rays falls upon a crystal at an angle θ, and a detector is placed so that it records those rays whose scattering angle is also θ. X-rays reaching the detector therefore obey the first Bragg condition. As θ is varied, the detector will

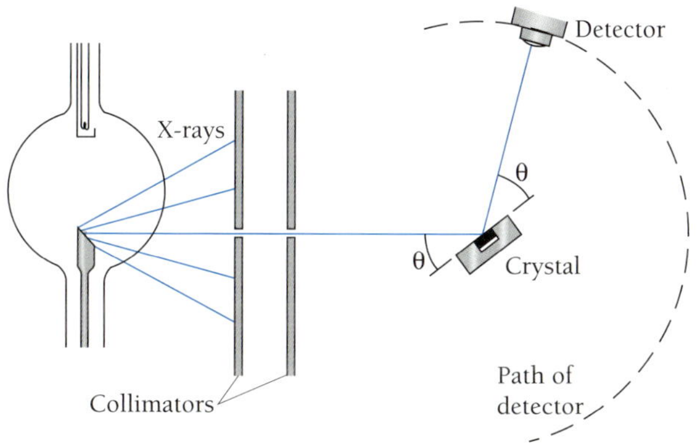

Figure 2.21 X-ray spectrometer.

record intensity peaks corresponding to the orders predicted by Eq. (2.13). If the spacing d between adjacent Bragg planes in the crystal is known, the x-ray wavelength λ may be calculated.

2.7 COMPTON EFFECT

Further confirmation of the photon model

According to the quantum theory of light, photons behave like particles except for their lack of rest mass. How far can this analogy be carried? For instance, can we consider a collision between a photon and an electron as if both were billiard balls?

Figure 2.22 shows such a collision: an x-ray photon strikes an electron (assumed to be initially at rest in the laboratory coordinate system) and is scattered away from its original direction of motion while the electron receives an impulse and begins to move. We can think of the photon as losing an amount of energy in the collision that is the same as the kinetic energy KE gained by the electron, although actually separate photons are involved. If the initial photon has the frequency ν associated with it, the scattered photon has the lower frequency ν', where

$$\text{Loss in photon energy} = \text{gain in electron energy}$$

$$h\nu - h\nu' = \text{KE} \tag{2.14}$$

From Chap. 1 we recall that the momentum of a massless particle is related to its energy by the formula

$$E = pc \tag{1.25}$$

Since the energy of a photon is $h\nu$, its momentum is

Photon momentum

$$p = \frac{E}{c} = \frac{h\nu}{c} \tag{2.15}$$

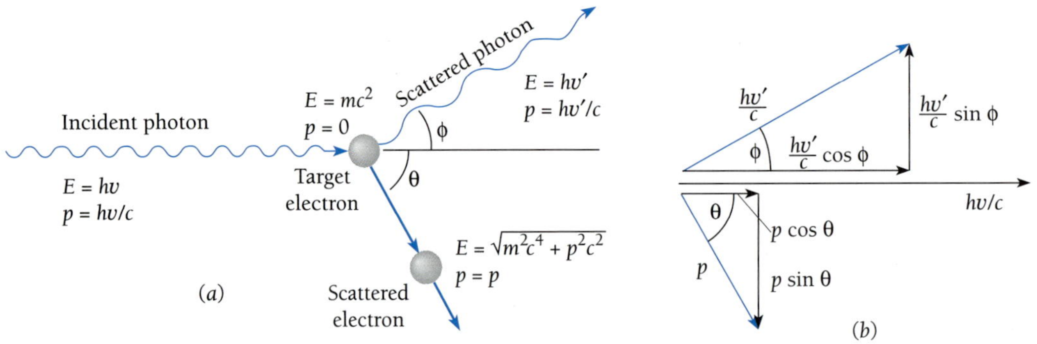

Figure 2.22 (*a*) The scattering of a photon by an electron is called the Compton effect. Energy and momentum are conserved in such an event, and as a result the scattered photon has less energy (longer wavelength) than the incident photon. (*b*) Vector diagram of the momenta and their components of the incident and scattered photons and the scattered electron.

Momentum, unlike energy, is a vector quantity that incorporates direction as well as magnitude, and in the collision momentum must be conserved in each of two mutually perpendicular directions. (When more than two bodies participate in a collision, momentum must be conserved in each of three mutually perpendicular directions.) The directions we choose here are that of the original photon and one perpendicular to it in the plane containing the electron and the scattered photon (Fig. 2.22).

The initial photon momentum is $h\nu/c$, the scattered photon momentum is $h\nu'/c$, and the initial and final electron momenta are respectively 0 and p. In the original photon direction

$$\text{Initial momentum} = \text{final momentum}$$

$$\frac{h\nu}{c} + 0 = \frac{h\nu'}{c} \cos\phi + p\cos\theta \tag{2.16}$$

and perpendicular to this direction

$$\text{Initial momentum} = \text{final momentum}$$

$$0 = \frac{h\nu'}{c} \sin\phi - p\sin\theta \tag{2.17}$$

The angle ϕ is that between the directions of the initial and scattered photons, and θ is that between the directions of the initial photon and the recoil electron. From Eqs. (2.14), (2.16), and (2.17) we can find a formula that relates the wavelength difference between initial and scattered photons with the angle ϕ between their directions, both of which are readily measurable quantities (unlike the energy and momentum of the recoil electron).

The first step is to multiply Eqs. (2.16) and (2.17) by c and rewrite them as

$$pc\cos\theta = h\nu - h\nu'\cos\phi$$

$$pc\sin\theta = h\nu'\sin\phi$$

By squaring each of these equations and adding the new ones together, the angle θ is eliminated, leaving

$$p^2c^2 = (h\nu)^2 - 2(h\nu)(h\nu')\cos\phi + (h\nu')^2 \tag{2.18}$$

Next we equate the two expressions for the total energy of a particle

$$E = \text{KE} + mc^2 \tag{1.20}$$

$$E = \sqrt{m^2c^4 + p^2c^2} \tag{1.24}$$

from Chap. 1 to give

$$(\text{KE} + mc^2)^2 = m^2c^4 + p^2c^2$$

$$p^2c^2 = \text{KE}^2 + 2mc^2\,\text{KE}$$

Since

$$\text{KE} = h\nu - h\nu'$$

we have

$$p^2c^2 = (h\nu)^2 - 2(h\nu)(h\nu') + (h\nu')^2 + 2mc^2(h\nu - h\nu') \tag{2.19}$$

Substituting this value of p^2c^2 in Eq. (2.18), we finally obtain

$$2mc^2(h\nu - h\nu') = 2(h\nu)(h\nu')(1 - \cos\phi) \tag{2.20}$$

This relationship is simpler when expressed in terms of wavelength λ. Dividing Eq. (2.20) by $2h^2c^2$,

$$\frac{mc}{h}\left(\frac{\nu}{c} - \frac{\nu'}{c}\right) = \frac{\nu}{c}\frac{\nu'}{c}(1 - \cos\phi)$$

and so, since $\nu/c = 1/\lambda$ and $\nu'/c = 1/\lambda'$,

$$\frac{mc}{h}\left(\frac{1}{\lambda} - \frac{1}{\lambda'}\right) = \frac{1 - \cos\phi}{\lambda\lambda'}$$

Compton effect

$$\lambda' - \lambda = \frac{h}{mc}(1 - \cos\phi) \tag{2.21}$$

Equation (2.21) was derived by Arthur H. Compton in the early 1920s, and the phenomenon it describes, which he was the first to observe, is known as the **Compton effect.** It constitutes very strong evidence in support of the quantum theory of radiation.

Equation (2.21) gives the change in wavelength expected for a photon that is scattered through the angle ϕ by a particle of rest mass m. This change is independent of the wavelength λ of the incident photon. The quantity

Compton wavelength

$$\lambda_C = \frac{h}{mc} \tag{2.22}$$

is called the **Compton wavelength** of the scattering particle. For an electron $\lambda_C = 2.426 \times 10^{-12}$ m, which is 2.426 pm (1 pm = 1 picometer = 10^{-12} m). In terms of λ_C, Eq. (2.21) becomes

Compton effect

$$\lambda' - \lambda = \lambda_C(1 - \cos\phi) \tag{2.23}$$

The Compton wavelength gives the scale of the wavelength change of the incident photon. From Eq. (2.23) we note that the greatest wavelength change possible corresponds to $\phi = 180°$, when the wavelength change will be twice the Compton wavelength λ_C. Because $\lambda_C = 2.426$ pm for an electron, and even less for other particles owing to their larger rest masses, the maximum wavelength change in the Compton effect is 4.852 pm. Changes of this magnitude or less are readily observable only in x-rays: the shift in wavelength for visible light is less than 0.01 percent of the initial wavelength, whereas for x-rays of $\lambda = 0.1$ nm it is several percent. The Compton effect is the chief means by which x-rays lose energy when they pass through matter.

About Scientists

Arthur Holly Compton (1892–1962), a native of Ohio, was educated at College of Wooster and Princeton. While at Washington University in St. Louis he found that x-rays increase in wavelength when scattered, which he explained in 1923 on the basis of the quantum theory of light. This work convinced remaining doubters of the reality of photons.

After receiving the Nobel Prize in 1927, Compton, now at the University of Chicago, studied cosmic rays and helped establish that they are fast charged particles (today known to be atomic nuclei, largely protons) that circulate in space and are not high-energy gamma rays as many had thought. He did this by showing that cosmic-ray intensity varies with latitude, which makes sense only if they are ions whose paths are influenced by the earth's magnetic field. During World War II Compton was one of the leaders in the development of the atomic bomb.

Example 2.4

X-rays of wavelength 10.0 pm are scattered from a target. (*a*) Find the wavelength of the x-rays scattered through 45°. (*b*) Find the maximum wavelength present in the scattered x-rays. (*c*) Find the maximum kinetic energy of the recoil electrons.

Solution

(*a*) From Eq. (2.23), $\lambda' - \lambda = \lambda_C(1 - \cos\phi)$, and so

$$\begin{aligned}\lambda' &= \lambda + \lambda_C(1 - \cos 45^\circ)\\ &= 10.0\text{ pm} + 0.293\lambda_C\\ &= 10.7\text{ pm}\end{aligned}$$

(*b*) $\lambda' - \lambda$ is a maximum when $(1 - \cos\phi) = 2$, in which case

$$\lambda' = \lambda + 2\lambda_C = 10.0\text{ pm} + 4.9\text{ pm} = 14.9\text{ pm}$$

(*c*) The maximum recoil kinetic energy is equal to the difference between the energies of the incident and scattered photons, so

$$\text{KE}_{\text{max}} = h(\nu - \nu') = hc\left(\frac{1}{\lambda} - \frac{1}{\lambda'}\right)$$

where λ' is given in (*b*). Hence

$$\begin{aligned}\text{KE}_{\text{max}} &= \frac{(6.626 \times 10^{-34}\text{ J}\cdot\text{s})(3.00 \times 10^{8}\text{ m/s})}{10^{-12}\text{ m/pm}}\left(\frac{1}{10.0\text{ pm}} - \frac{1}{14.9\text{ pm}}\right)\\ &= 6.54 \times 10^{-15}\text{ J}\end{aligned}$$

which is equal to 40.8 keV.

The experimental demonstration of the Compton effect is straightforward. As in Fig. 2.23, a beam of x-rays of a single, known wavelength is directed at a target, and the wavelengths of the scattered x-rays are determined at various angles ϕ. The results, shown in Fig. 2.24, exhibit the wavelength shift predicted by Eq. (2.21), but at each angle the scattered x-rays also include many that have the initial wavelength. This is not hard to understand. In deriving Eq. (2.21) it was assumed that the scattering particle is able to move freely, which is reasonable since many of the electrons in matter

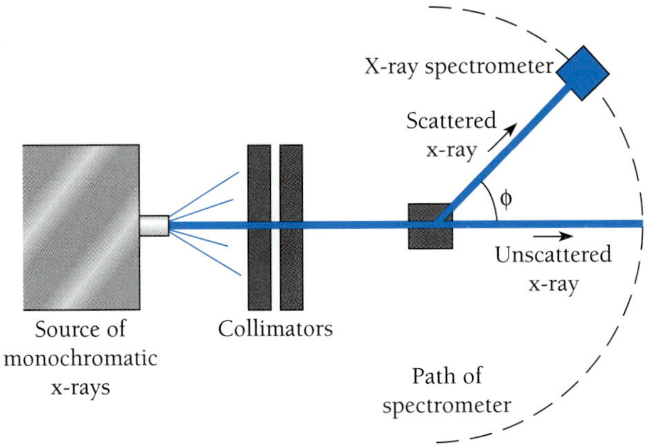

Figure 2.23 Experimental demonstration of the Compton effect.

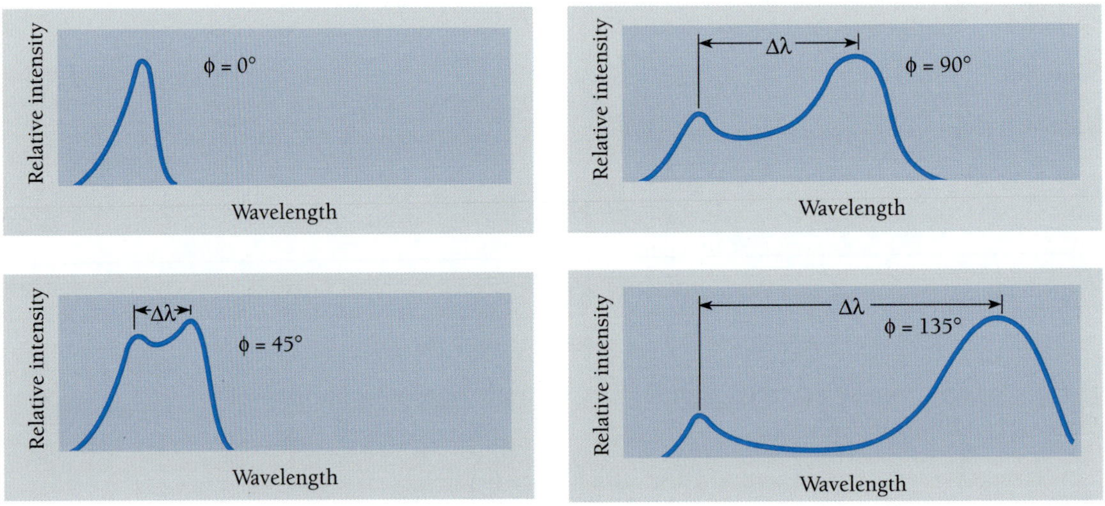

Figure 2.24 Experimental confirmation of Compton scattering. The greater the scattering angle, the greater the wavelength change, in accord with Eq. (2.21).

are only loosely bound to their parent atoms. Other electrons, however, are very tightly bound and when struck by a photon, the entire atom recoils instead of the single electron. In this event the value of m to use in Eq. (2.21) is that of the entire atom, which is tens of thousands of times greater than that of an electron, and the resulting Compton shift is accordingly so small as to be undetectable.

2.8 PAIR PRODUCTION

Energy into matter

As we have seen, in a collision a photon can give an electron all of its energy (the photoelectric effect) or only part (the Compton effect). It is also possible for a photon to materialize into an electron and a positron, which is a positively charged electron. In this process, called **pair production**, electromagnetic energy is converted into matter.

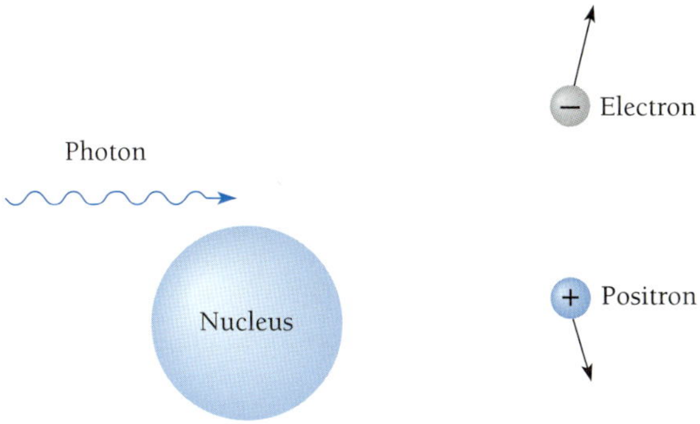

Figure 2.25 In the process of pair production, a photon of sufficient energy materializes into an electron and a positron.

No conservation principles are violated when an electron-positron pair is created near an atomic nucleus (Fig. 2.25). The sum of the charges of the electron ($q = -e$) and of the positron ($q = +e$) is zero, as is the charge of the photon; the total energy, including rest energy, of the electron and positron equals the photon energy; and linear momentum is conserved with the help of the nucleus, which carries away enough photon momentum for the process to occur. Because of its relatively enormous mass, the nucleus absorbs only a negligible fraction of the photon energy. (Energy and linear momentum could not both be conserved if pair production were to occur in empty space, so it does not occur there.)

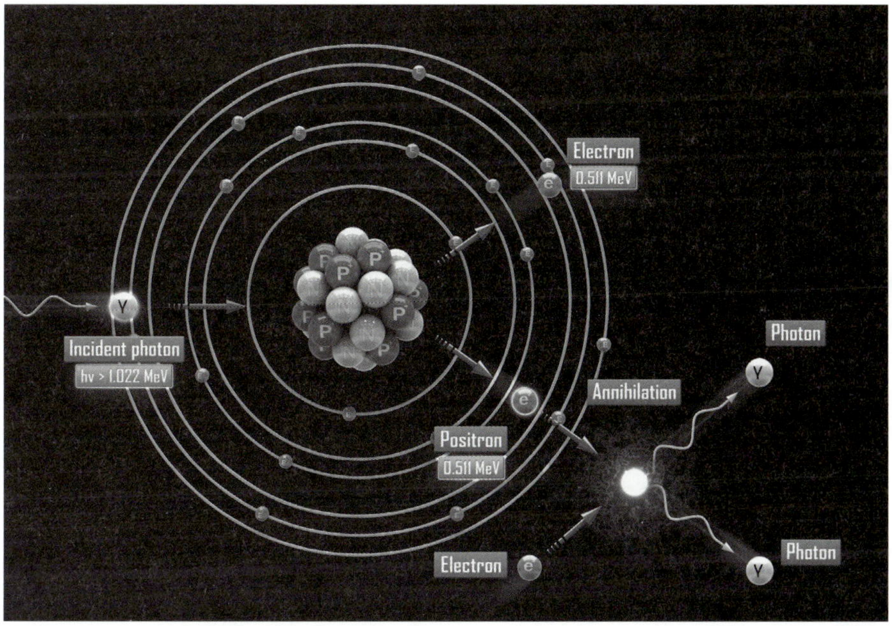

Particles and their corresponding energies in the pair production. A positron created by the pair production or from radioligands annihilates after its collision with an electron nearby and releases gamma rays. By detecting gamma rays from this pair annihilation, a 3D image within a patient's body is constructed from positron emission tomography(PET).

The rest energy mc^2 of an electron or positron is 0.51 MeV, hence pair production requires a photon energy of at least 1.02 MeV. Any additional photon energy becomes kinetic energy of the electron and positron. The corresponding maximum photon wavelength is 1.2 pm. Electromagnetic waves with such wavelengths are called **gamma rays,** symbol γ, and are found in nature as one of the emissions from radioactive nuclei and in cosmic rays.

The inverse of pair production occurs when a positron is near an electron and the two come together under the influence of their opposite electric charges. Both particles vanish simultaneously, with the lost mass becoming energy in the form of two gamma-ray photons:

$$e^+ + e^- \rightarrow \gamma + \gamma$$

The total mass of the positron and electron is equivalent to 1.02 MeV, and each photon has an energy $h\nu$ of 0.51 MeV plus half the kinetic energy of the particles relative to their center of mass. The directions of the photons are such as to conserve both energy and linear momentum, and no nucleus or other particle is needed for this **pair annihilation** to take place.

Example 2.5

Show that pair production cannot occur in empty space.

Solution

From conservation of energy,

$$h\nu = 2\gamma mc^2$$

where $h\nu$ is the photon energy and γmc^2 is the total energy of each member of the electron-position pair. Figure 2.26 is a vector diagram of the linear momenta of the photon, electron, and positron. The angles θ are equal in order that momentum be conserved in the transverse direction. In the direction of motion of the photon, for momentum to be conserved it must be true that

$$\frac{h\nu}{c} = 2p \cos \theta$$

$$h\nu = 2pc \cos \theta$$

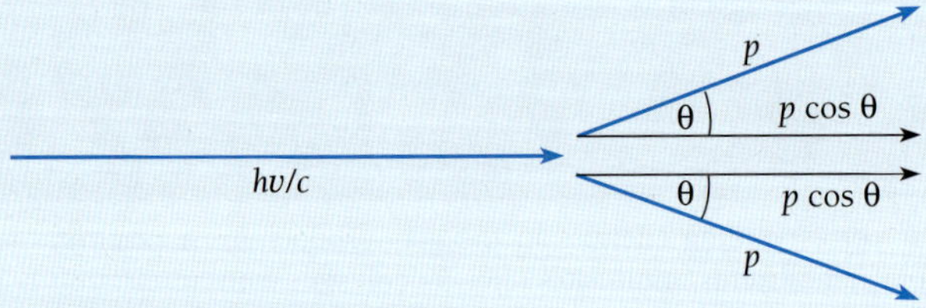

Figure 2.26 Vector diagram of the momenta involved if a photon were to materialize into an electron-positron pair in empty space. Because such an event cannot conserve both energy and momentum, it does not occur. Pair production always involves an atomic nucleus that carries away part of the initial photon momentum.

Since $p = \gamma m v$ for the electron and positron,

$$h\nu = 2\gamma mc^2\left(\frac{v}{c}\right)\cos\,\theta$$

Because $v/c < 1$ and $\cos\,\theta \leq 1$,

$$h\nu < 2\gamma mc^2$$

But conservation of energy requires that $h\nu = 2\gamma mc^2$. Hence it is impossible for pair production to conserve both energy and momentum unless some other object is involved in the process to carry away part of the initial photon momentum.

Example 2.6

An electron and a positron are moving side by side in the $+x$ direction at $0.500c$ when they annihilate each other. Two photons are produced that move along the x axis. (*a*) Do both photons move in the $+x$ direction? (*b*) What is the energy of each photon?

Solution

(*a*) In the center-of-mass (CM) system (which is the system moving with the original particles), the photons move off in opposite directions to conserve momentum. They must also do so in the lab system because the speed of the CM system is less than the speed c of the photons.
(*b*) Let p_1 be the momentum of the photon moving in the $+x$ direction and p_2 be the momentum of the photon moving in the $-x$ direction. Then conservation of momentum (in the lab system) gives

$$p_1 - p_2 = 2\gamma mv = \frac{2(mc^2)(v/c^2)}{\sqrt{1 - v/c^2}}$$

$$= \frac{2(0.511\text{ MeV}/c^2)(c^2)(0.500c)/c^2}{\sqrt{1 - (0.500)^2}} = 0.590\text{ MeV}/c$$

Conservation of energy gives

$$p_1c + p_2c = 2\gamma mc^2 = \frac{2mc^2}{\sqrt{1 - v^2/c^2}} = \frac{2(0.511\text{ MeV})}{\sqrt{1 - (0.500)^2}} = 1.180\text{ MeV}$$

and so

$$p_1 + p_2 = 1.180\text{ MeV}/c$$

Now we add the two results and solve for p_1 and p_2:

$$(p_1 - p_2) + (p_1 + p_2) = 2p_1 = (0.590 + 1.180)\text{ MeV}/c$$

$$p_1 = 0.885\text{ MeV}/c$$

$$p_2 = (p_1 + p_2) - p_1 = 0.295\text{ MeV}/c$$

The photon energies are accordingly

$$E_1 = p_1c = 0.885\text{ MeV} \qquad E_2 = p_2c = 0.295\text{ MeV}$$

Photon Absorption

The three chief ways in which photons of light, x-rays, and gamma rays interact with matter are summarized in Fig. 2.27. In all cases photon energy is transferred to electrons which in turn lose energy to atoms in the absorbing material.

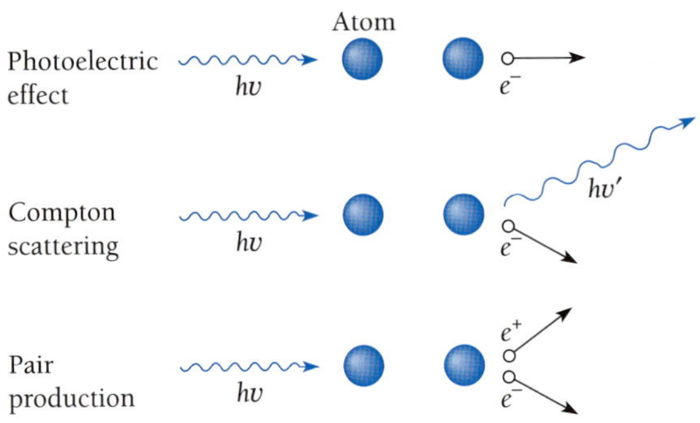

Figure 2.27 X- and gamma rays interact with matter chiefly through the photoelectric effect, Compton scattering, and pair production. Pair production requires a photon energy of at least 1.02 MeV.

At low photon energies the photoelectric effect is the chief mechanism of energy loss. The importance of the photoelectric effect decreases with increasing energy, to be succeeded by Compton scattering. The greater the atomic number of the absorber, the higher the energy at which the photoelectric effect remains significant. In the lighter elements, Compton scattering becomes dominant at photon energies of a few tens of keV, whereas in the heavier ones this does not happen until photon energies of nearly 1 MeV are reached (Fig. 2.28).

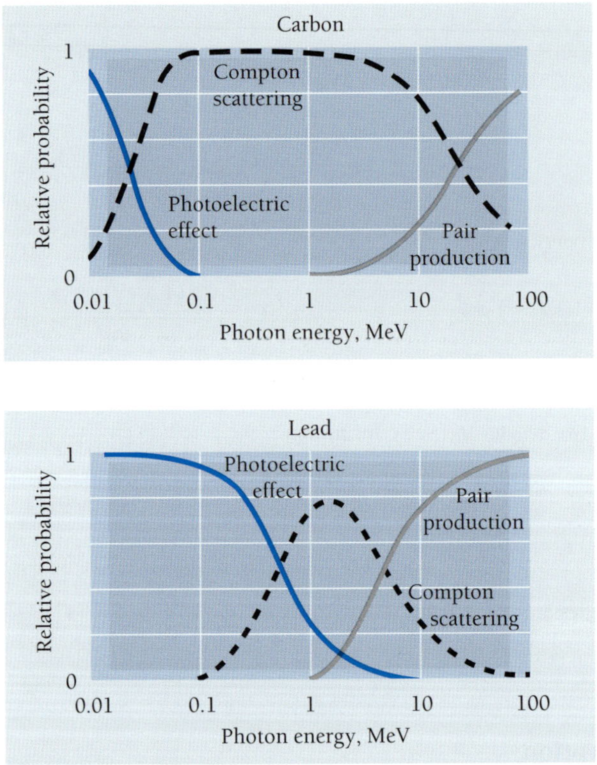

Figure 2.28 The relative probabilities of the photoelectric effect, Compton scattering, and pair production as functions of energy in carbon (a light element) and lead (a heavy element).

Pair production becomes increasingly likely the more the photon energy exceeds the threshold of 1.02 MeV. The greater the atomic number of the absorber, the lower the energy at which pair production takes over as the principal mechanism of energy loss by gamma rays. In the heaviest elements, the crossover energy is about 4 MeV, but it is over 10 MeV for the lighter ones. Thus gamma rays in the energy range typical of radioactive decay interact with matter largely through Compton scattering.

The intensity I of an x- or gamma-ray beam is equal to the rate at which it transports energy per unit cross-sectional area of the beam. The fractional energy $-dI/I$ lost by the beam in passing through a thickness dx of a certain absorber is found to be proportional to dx:

$$-\frac{dI}{I} = \mu\, dx \tag{2.24}$$

The proportionality constant μ is called the **linear attenuation coefficient** and its value depends on the energy of the photons and on the nature of the absorbing material. Integrating Eq. (2.24) gives

Radiation intensity

$$I = I_0 e^{-\mu x} \tag{2.25}$$

The intensity of the radiation decreases exponentially with absorber thickness x. Figure 2.29 is a graph of the linear attenuation coefficient for photons in lead as a function of photon energy. The contribution to μ of the photoelectric effect, Compton scattering, and pair production are shown.

We can use Eq. (2.25) to relate the thickness x of the absorber needed to reduce the intensity of an x- or gamma-ray beam by a given amount to the attenuation coefficient μ. If the ratio of the final and initial intensities is I/I_0,

$$\frac{I}{I_0} = e^{-\mu x} \qquad \frac{I_0}{I} = e^{\mu x} \qquad \ln\frac{I_0}{I} = \mu x$$

Absorber thickness

$$x = \frac{\ln (I_0/I)}{\mu} \tag{2.26}$$

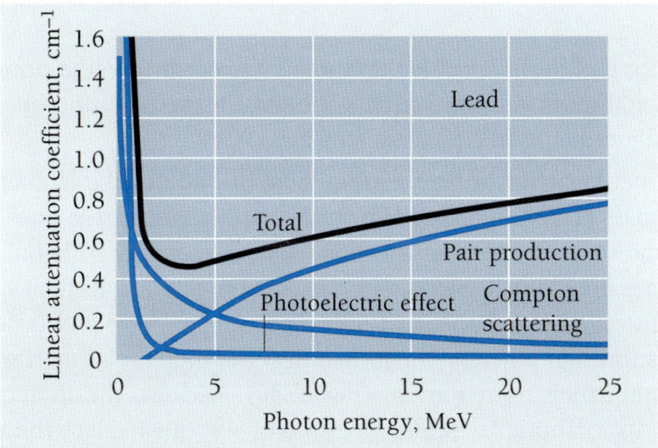

Figure 2.29 Linear attentuation coefficients for photons in lead.

Example 2.7

The linear attenuation coefficient for 2.0-MeV gamma rays in water is $4.9\ \text{m}^{-1}$. (*a*) Find the relative intensity of a beam of 2.0-MeV gamma rays after it has passed through 10 cm of water. (*b*) How far must such a beam travel in water before its intensity is reduced to 1 percent of its original value?

Solution

(*a*) Here $\mu x = (4.9\ \text{m}^{-1})(0.10\ \text{m}) = 0.49$ and so, from Eq. (2.25)

$$\frac{I}{I_0} = e^{-\mu x} = e^{-0.49} = 0.61$$

The intensity of the beam is reduced to 61 percent of its original value after passing through 10 cm of water.
(*b*) Since $I_0/I = 100$, Eq. (2.26) yields

$$x = \frac{\ln(I_0/I)}{\mu} = \frac{\ln 100}{4.9\ \text{m}^{-1}} = 0.94\ \text{m}$$

2.9 PHOTONS AND GRAVITY

Although they lack rest mass, photons behave as though they have gravitational mass

In Sec. 1.10 we learned that light is affected by gravity by virtue of the curvature of spacetime around a mass. Another way to approach the gravitational behavior of light follows from the observation that, although a photon has no rest mass, it nevertheless interacts with electrons as though it had the inertial mass

Photon "mass"
$$m = \frac{p}{v} = \frac{h\nu}{c^2} \tag{2.27}$$

(We recall that, for a photon, $p = h\nu/c$ and $v = c$.) According to the principle of equivalence, gravitational mass is always equal to inertial mass, so a photon of frequency ν ought to act gravitationally like a particle of mass $h\nu/c^2$.

The gravitational behavior of light can be demonstrated in the laboratory. When we drop a stone of mass m from a height H near the earth's surface, the gravitational pull of the earth accelerates it as it falls and the stone gains the energy mgH on the way to the ground. The stone's final kinetic energy $\frac{1}{2}mv^2$ is equal to mgH, so its final speed is $\sqrt{2gH}$.

All photons travel with the speed of light and so cannot go any faster. However, a photon that falls through a height H can manifest the increase of mgH in its energy by an increase in frequency from ν to ν' (Fig. 2.30). Because the frequency change is extremely small in a laboratory-scale experiment, we can neglect the corresponding change in the photon's "mass" $h\nu/c^2$.

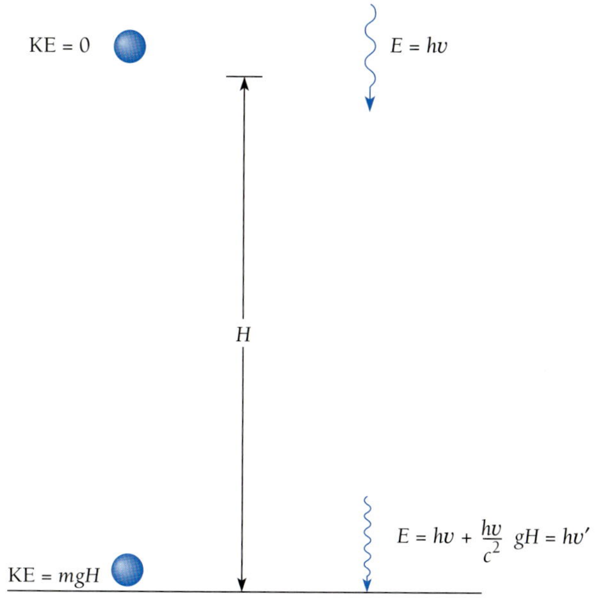

Figure 2.30 A photon that falls in a gravitational field gains energy, just as a stone does. This gain in energy is manifested as an increase in frequency from ν to ν'.

Hence,

$$\text{final photon energy} = \text{initial photon energy} + \text{increase in energy}$$

$$h\nu' = h\nu + mgH$$

and so

$$h\nu' = h\nu + \left(\frac{h\nu}{c^2}\right) gH$$

Photon energy after falling through height H

$$h\nu' = h\nu\left(1 + \frac{gH}{c^2}\right) \tag{2.28}$$

Example 2.8

The increase in energy of a fallen photon was first observed in 1960 by Pound and Rebka at Harvard. In their work H was 22.5 m. Find the change in frequency of a photon of red light whose original frequency is 7.3×10^{14} Hz when it falls through 22.5 m.

Solution

From Eq. (2.28) the change in frequency is

$$\nu' - \nu = \left(\frac{gH}{c^2}\right)\nu$$

$$= \frac{(9.8 \text{ m/s}^2)(22.5 \text{ m})(7.3 \times 10^{14} \text{ Hz})}{(3.0 \times 10^8 \text{ m/s})^2} = 1.8 \text{ Hz}$$

Pound and Rebka actually used gamma rays of much higher frequency, as described in Exercise 53.

Gravitational Red Shift

An interesting astronomical effect is suggested by the gravitational behavior of light. If the frequency associated with a photon moving toward the earth increases, then the frequency of a photon moving away from it should decrease.

The earth's gravitational field is not particularly strong, but the fields of many stars are. Suppose a photon of initial frequency ν is emitted by a star of mass M and radius R, as in Fig. 2.31. The potential energy of a mass m on the star's surface is

$$\text{PE} = -\frac{GMm}{R}$$

where the minus sign is required because the force between M and m is attractive. The potential energy of a photon of "mass" $h\nu/c^2$ on the star's surface is therefore

$$\text{PE} = -\frac{GMh\nu}{c^2R}$$

and its total energy E, the sum of PE and its quantum energy $h\nu$, is

$$E = h\nu - \frac{GMh\nu}{c^2R} = h\nu\left(1 - \frac{GM}{c^2R}\right)$$

At a larger distance from the star, for instance at the earth, the photon is beyond the star's gravitational field but its total energy remains the same. The photon's energy is now entirely electromagnetic, and

$$E = h\nu'$$

where ν' is the frequency of the arriving photon. (The potential energy of the photon in the earth's gravitational field is negligible compared with that in the star's field.) Hence

$$h\nu' = h\nu\left(1 - \frac{GM}{c^2R}\right)$$

$$\frac{\nu'}{\nu} = 1 - \frac{GM}{c^2R}$$

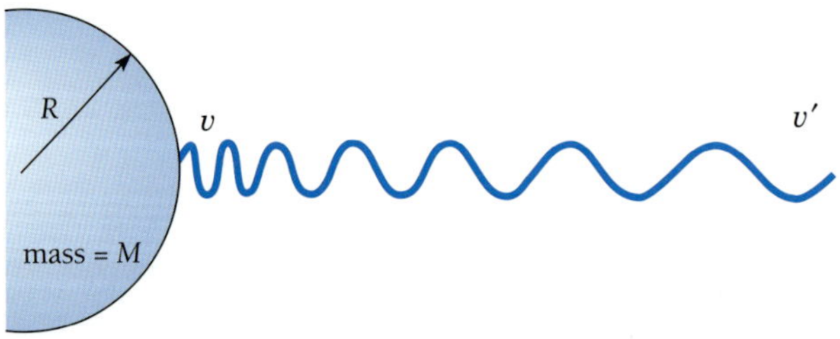

Figure 2.31 The frequency of a photon emitted from the surface of a star decreases as it moves away from the star.

and the relative frequency change is

Gravitational red shift

$$\frac{\Delta\nu}{\nu} = \frac{\nu - \nu'}{\nu} = 1 - \frac{\nu'}{\nu} = \frac{GM}{c^2R} \tag{2.29}$$

The photon has a *lower* frequency at the earth, corresponding to its loss in energy as it leaves the field of the star.

A photon in the visible region of the spectrum is thus shifted toward the red end, and this phenomenon is accordingly known as the **gravitational red shift.** It is different from the doppler red shift observed in the spectra of distant galaxies due to their apparent recession from the earth, a recession that seems to be due to a general expansion of the universe.

As we shall learn in Chap. 4, when suitably excited, the atoms of every element emit photons of certain specific frequencies only. The validity of Eq. (2.29) can therefore be checked by comparing the frequencies found in stellar spectra with those in spectra obtained in the laboratory. For most stars, including the sun, the ratio M/R is too small for a gravitational red shift to be apparent. However, for a class of stars known as **white dwarfs,** it is just on the limit of measurement—and has been observed. A white dwarf is an old star whose interior consists of atoms whose electron structures have collapsed and so it is very small: a typical white dwarf is about the size of the earth but has the mass of the sun.

Black Holes

An interesting question is, what happens if a star is so dense that $GM/c^2R \geq 1$? If this is the case, then from Eq. (2.29) we see that no photon can ever leave the star, since to do so requires more energy than its initial energy $h\nu$. The red shift would, in effect, have then stretched the photon wavelength to infinity. A star of this kind cannot radiate and so would be invisible—a **black hole** in space.

In a situation in which gravitational energy is comparable with total energy, as for a photon in a black hole, general relativity must be applied in detail. The correct criterion for a star to be a black hole turns out to be $GM/c^2R \geq \frac{1}{2}$. The **Schwarzschild radius** R_S of a body of mass M is defined as

Quasars and Galaxies

In even the most powerful telescope, a **quasar** appears as a sharp point of light, just as a star does. Unlike stars, quasars are powerful sources of radio waves; hence their name, a contraction of *quast-stellar radio sources*. Hundreds of quasars have been discovered, and there seem to be many more. Though a typical quasar is smaller than the solar system, its energy output may be thousands of times the output of our entire Milky Way galaxy.

Most astronomers believe that at the heart of every quasar is a black hole whose mass is at least that of 100 million suns. As nearby stars are pulled toward the black hole, their matter is squeezed and heated to produce the observed radiation. While being swallowed, a star may liberate 10 times as much energy as it would have given off had it lived out a normal life. A diet of a few stars a year seems enough to keep a quasar going at the observed rates. It is possible that quasars are the cores of newly formed galaxies. Did all galaxies once undergo a quasar phase? Nobody can say as yet, but there is evidence that all galaxies, including the Milky Way, contain massive black holes at their centers.

Schwarzschild radius

$$R_S = \frac{2GM}{c^2} \tag{2.30}$$

The body is a black hole if all its mass is inside a sphere with this radius. The boundary of a black hole is called its **event horizon.** The escape speed from a black hole is equal to the speed of the light c at the Schwarzschild radius, hence nothing at all can ever leave a black hole. For a star with the sun's mass, R_S is 3 km, a quarter of a million times smaller than the sun's present radius. Anything passing near a black hole will be sucked into it, never to return to the outside world.

Since it is invisible, how can a black hole be detected? A black hole that is a member of a double-star system (double stars are quite common) will reveal its presence by its gravitational pull on the other star; the two stars circle each other. In addition, the intense gravitational field of the black hole will attract matter from the other star, which will be compressed and heated to such high temperatures that x-rays will be emitted profusely. One of a number of invisible objects that astronomers believe on this basis to be black holes is known as Cygnus X-1. Its mass is perhaps 8 times that of the sun, and its radius may be only about 10 km. The region around a black hole that emits x-rays should extend outward for several hundred kilometers.

Only very heavy stars end up as black holes. Lighter stars evolve into white dwarfs and neutron stars, which as their name suggests consist largely of neutrons (see Sec. 9.11). But as time goes on, the strong gravitational fields of both white dwarfs and neutron stars attract more and more cosmic dust and gas. When they have gathered up enough mass, they too will become black holes. If the universe lasts long enough, then everything in it may be in the form of black holes.

Black holes are also believed to be at the cores of galaxies. Again, the clues come from the motions of nearby bodies and from the amount and type of radiation emitted. Stars close to a galactic center are observed to move so rapidly that only the gravitational pull of an immense mass could keep them in their orbits instead of flying off. How immense? As much as a billion times the sun's mass. And, as in the case of black holes that were once stars, radiation pours out of galactic centers so copiously that only black holes could be responsible.

EXERCISES

"Why," said the Dodo, "the best way to explain it is to do it." —Lewis Carroll, *Alice's Adventures in Wonderland*

2.2 Blackbody Radiation

1. If Planck's constant were smaller than it is, would quantum phenomena be more or less conspicuous than they are now?
2. Express the Planck radiation formula in terms of wavelength.

2.3 Photoelectric Effect

3. Is it correct to say that the maximum photoelectron energy KE_{max} is proportional to the frequency ν of the incident light? If not, what would a correct statement of the relationship between KE_{max} and ν be?
4. Compare the properties of particles with those of waves. Why do you think the wave aspect of light was discovered earlier than its particle aspect?
5. Find the energy of a 700-nm photon.
6. Find the wavelength and frequency of a 100-MeV photon.
7. A 1.00-kW radio transmitter operates at a frequency of 880 kHz. How many photons per second does it emit?
8. Under favorable circumstances the human eye can detect 1.0×10^{-18} J of electromagnetic energy. How many 600-nm photons does this represent?

9. Light from the sun arrives at the earth, an average of 1.5 $\times 10^{11}$ m away, at the rate of 1.4×10^3 W/m^2 of area perpendicular to the direction of the light. Assume that sunlight is monochromatic with a frequency of 5.0×10^{14} Hz. (*a*) How many photons fall per second on each square meter of the earth's surface directly facing the sun? (*b*) What is the power output of the sun, and how many photons per second does it emit? (*c*) How many photons per cubic meter are there near the earth?

10. A detached retina is being "welded" back in place using 20-ms pulses from a 0.50-W laser operating at a wavelength of 632 nm. How many photons are in each pulse?

11. The maximum wavelength for photoelectric emission in tungsten is 230 nm. What wavelength of light must be used in order for electrons with a maximum energy of 1.5 eV to be ejected?

12. The minimum frequency for photoelectric emission in copper is 1.1×10^{15} Hz. Find the maximum energy of the photoelectrons (in electronvolts) when light of frequency 1.5×10^{15} Hz is directed on a copper surface.

13. What is the maximum wavelength of light that will cause photoelectrons to be emitted from sodium? What will the maximum kinetic energy of the photoelectrons be if 200-nm light falls on a sodium surface?

14. A silver ball is suspended by a string in a vacuum chamber and ultraviolet light of wavelength 200 nm is directed at it. What electrical potential will the ball acquire as a result?

15. 1.5 mW of 400-nm light is directed at a photoelectric cell. If 0.10 percent of the incident photons produce photoelectrons, find the current in the cell.

16. Light of wavelength 400 nm is shone on a metal surface in an apparatus like that of Fig. 2.9. The work function of the metal is 2.50 eV. (*a*) Find the extinction voltage, that is, the retarding voltage at which the photoelectron current disappears. (*b*) Find the speed of the fastest photoelectrons.

17. A metal surface illuminated by 8.5×10^{14} Hz light emits electrons whose maximum energy is 0.52 eV. The same surface illuminated by 12.0×10^{14} Hz light emits electrons whose maximum energy is 1.97 eV. From these data find Planck's constant and the work function of the surface.

18. The work function of a tungsten surface is 5.4 eV. When the surface is illuminated by light of wavelength 175 nm, the maximum photoelectron energy is 1.7 eV. Find Planck's constant from these data.

19. Show that it is impossible for a photon to give up all its energy and momentum to a free electron. This is the reason why the photoelectric effect can take place only when photons strike bound electrons.

2.5 X-Rays

20. What voltage must be applied to an x-ray tube for it to emit x-rays with a minimum wavelength of 30 pm?

21. Electrons are accelerated in television tubes through potential differences of about 10 kV. Find the highest frequency of the electromagnetic waves emitted when these electrons strike the screen of the tube. What kind of waves are these?

2.6 X-Ray Diffraction

22. The smallest angle of Bragg scattering in potassium chloride (KCl) is 28.4° for 0.30-nm x-rays. Find the distance between atomic planes in potassium chloride.

23. The distance between adjacent atomic planes in calcite ($CaCO_3$) is 0.300 nm. Find the smallest angle of Bragg scattering for 0.030-nm x-rays.

24. Find the atomic spacing in a crystal of rock salt (NaCl), whose structure is shown in Fig. 2.19. The density of rock salt is 2.16 $\times 10^3$ kg/m^3 and the average masses of the Na and Cl atoms are respectively 3.82×10^{-26} kg and 5.89×10^{-26} kg.

2.7 Compton Effect

25. What is the frequency of an x-ray photon whose momentum is 1.1×10^{-23} kg · m/s?

26. How much energy must a photon have if it is to have the momentum of a 10-MeV proton?

27. In Sec. 2.7 the x-rays scattered by a crystal were assumed to undergo no change in wavelength. Show that this assumption is reasonable by calculating the Compton wavelength of a Na atom and comparing it with the typical x-ray wavelength of 0.1 nm.

28. A monochromatic x-ray beam whose wavelength is 55.8 pm is scattered through 46°. Find the wavelength of the scattered beam.

29. A beam of x-rays is scattered by a target. At 45° from the beam direction the scattered x-rays have a wavelength of 2.2 pm. What is the wavelength of the x-rays in the direct beam?

30. An x-ray photon whose initial frequency was 1.5×10^{19} Hz emerges from a collision with an electron with a frequency of 1.2×10^{19} Hz. How much kinetic energy was imparted to the electron?

31. An x-ray photon of initial frequency 3.0×10^{19} Hz collides with an electron and is scattered through 90°. Find its new frequency.

32. Find the energy of an x-ray photon which can impart a maximum energy of 50 keV to an electron.

33. At what scattering angle will incident 100-keV x-rays leave a target with an energy of 90 keV?

34. (*a*) Find the change in wavelength of 80-pm x-rays that are scattered 120° by a target. (*b*) Find the angle between the directions of the recoil electron and the incident photon. (*c*) Find the energy of the recoil electron.

35. A photon of frequency ν is scattered by an electron initially at rest. Verify that the maximum kinetic energy of the recoil electron is $KE_{max} = (2h^2\nu^2/mc^2)/(1 + 2h\nu/mc^2)$.

36. In a Compton-effect experiment in which the incident x-rays have a wavelength of 10.0 pm, the scattered x-rays at a certain angle have a wavelength of 10.5 pm. Find the momentum (magnitude and direction) of the corresponding recoil electrons.

37. A photon whose energy equals the rest energy of the electron undergoes a Compton collision with an electron. If the electron moves off at an angle of 40° with the original photon direction, what is the energy of the scattered photon?

38. A photon of energy E is scattered by a particle of rest energy E_0. Find the maximum kinetic energy of the recoiling particle in terms of E and E_0.

2.8 Pair Production

39. A positron collides head on with an electron and both are annihilated. Each particle had a kinetic energy of 1.00 MeV. Find the wavelength of the resulting photons.

40. A positron with a kinetic energy of 2.000 MeV collides with an electron at rest and the two particles are annihilated. Two photons are produced; one moves in the same direction as the incoming positron and the other moves in the opposite direction. Find the energies of the photons.

41. Show that, regardless of its initial energy, a photon cannot undergo Compton scattering through an angle of more than 60° and still be able to produce an electron-positron pair. (*Hint:* Start by expressing the Compton wavelength of the electron in terms of the maximum photon wavelength needed for pair production.)

42. (*a*) Verify that the minimum energy a photon must have to create an electron-positron pair in the presence of a stationary nucleus of mass M is $2mc^2(1 + m/M)$, where m is the electron rest mass. (*b*) Find the minimum energy needed for pair production in the presence of a proton.

43. (*a*) Show that the thickness $x_{1/2}$ of an absorber required to reduce the intensity of a beam of radiation by a factor of 2 is given by $x_{1/2} = 0.693/\mu$. (*b*) Find the absorber thickness needed to produce an intensity reduction of a factor of 10.

44. (*a*) Show that the intensity of the radiation absorbed in a thickness x of an absorber is given by $I_0\mu x$ when $\mu x \ll 1$. (*b*) If $\mu x = 0.100$, what is the percentage error in using this formula instead of Eq. (2.25)?

45. The linear absorption coefficient for 1-MeV gamma rays in lead is 78 m^{-1}. Find the thickness of lead required to reduce by half the intensity of a beam of such gamma rays.

46. The linear absorption coefficient for 50-keV x-rays in sea-level air is 5.0×10^{-3} m^{-1}. By how much is the intensity of a beam of such x-rays reduced when it passes through 0.50 m of air? Through 5.0 m of air?

47. The linear absorption coefficients for 2.0-MeV gamma rays are 4.9 m^{-1} in water and 52 m^{-1} in lead. What thickness of water would give the same shielding for such gamma rays as 10 mm of lead?

48. The linear absorption coefficient of copper for 80-keV x-rays is 4.7×10^4 m^{-1}. Find the relative intensity of a beam of 80-keV x-rays after it has passed through a 0.10-mm copper foil.

49. What thickness of copper is needed to reduce the intensity of the beam in Exercise 48 by half?

50. The linear absorption coefficients for 0.05-nm x-rays in lead and in iron are, respectively, 5.8×10^4 m^{-1} and 1.1×10^4 m^{-1}. How thick should an iron shield be in order to provide the same protection from these x-rays as 10 mm of lead?

2.9 Photons and Gravity

51. The sun's mass is 2.0×10^{30} kg and its radius is 7.0×10^8 m. Find the approximate gravitational red shift in light of wavelength 500 nm emitted by the sun.

52. Find the approximate gravitational red shift in 500-nm light emitted by a white dwarf star whose mass is that of the sun but whose radius is that of the earth, 6.4×10^6 m.

53. As discussed in Chap. 12, certain atomic nuclei emit photons in undergoing transitions from "excited" energy states to their "ground" or normal states. These photons constitute gamma rays. When a nucleus emits a photon, it recoils in the opposite direction. (*a*) The $^{57}_{27}$Co nucleus decays by K capture to $^{57}_{26}$Fe, which then emits a photon in losing 14.4 keV to reach its ground state. The mass of a $^{57}_{26}$Fe atom is 9.5×10^{-26} kg. By how much is the photon energy reduced from the full 14.4 keV available as a result of having to share energy and momentum with the recoiling atom? (*b*) In certain crystals the atoms are so tightly bound that the entire crystal recoils when a gamma-ray photon is emitted, instead of the individual atom. This phenomenon is known as the **Mössbauer effect.** By how much is the photon energy reduced in this situation if the excited $^{57}_{26}$Fe nucleus is part of a 1.0-g crystal? (*c*) The essentially recoil-free emission of gamma rays in situations like that of *b* means that it is possible to construct a source of virtually monoenergetic and hence monochromatic photons. Such a source was used in the experiment described in Sec. 2.9. What is the original frequency and the change in frequency of a 14.4-keV gamma-ray photon after it has fallen 20 m near the earth's surface?

54. Find the Schwarzschild radius of the earth, whose mass is 5.98×10^{24} kg.

55. The gravitational potential energy U relative to infinity of a body of mass m at a distance R from the center of a body of mass M is $U = -GmM/R$. (*a*) If R is the radius of the body of mass M, find the escape speed v_e of the body, which is the minimum speed needed to leave it permanently. (*b*) Obtain a formula for the Schwarzschild radius of the body by setting $v_e = c$, the speed of light, and solving for R. (Of course, a relativistic calculation is correct here, but it is interesting to see what a classical calculation produces.)

Wave Properties of Particles

In a scanning electron microscope, an electron beam that scans a specimen causes secondary electrons to be ejected in numbers that vary with the angle of the surface. A suitable data display suggests the three-dimensional form of the specimen. The high resolution of this image of an ant is a consequence of the wave nature of moving electrons.

Looking back, it may seem odd that two decades passed between the 1905 discovery of the particle properties of waves and the 1924 speculation that particles might show wave behavior. It is one thing, however, to suggest a revolutionary concept to explain otherwise mysterious data and quite another to suggest an equally revolutionary concept without a strong experimental mandate. The latter is just what Louis de Broglie did in 1924 when he proposed that moving objects have wave as well as particle characteristics. So different was the scientific climate at the time from that around the turn of the century that de Broglie's ideas soon received respectful attention, whereas the earlier quantum theory of light of Planck and Einstein had been largely ignored despite its striking empirical support. The existence of de Broglie waves was experimentally demonstrated by 1927, and the duality principle they represent provided the starting point for Schrödinger's successful development of quantum mechanics in the previous year.

3.1 DE BROGLIE WAVES

A moving body behaves in certain ways as though it has a wave nature

A photon of light of frequency ν has the momentum

$$p = \frac{h\nu}{c} = \frac{h}{\lambda}$$

since $\lambda\nu = c$. The wavelength of a photon is therefore specified by its momentum according to the relation

Photon wavelength

$$\lambda = \frac{h}{p} \tag{3.1}$$

De Broglie suggested that Eq. (3.1) is a completely general one that applies to material particles as well as to photons. The momentum of a particle of mass m and velocity v is $p = \gamma mv$, and its **de Broglie wavelength** is accordingly

De Broglie wavelength

$$\lambda = \frac{h}{\gamma mv} \tag{3.2}$$

About Scientists

Louis de Broglie (1892–1987), although coming from a French family long identified with diplomacy and the military and initially a student of history, eventually followed his older brother Maurice in a career in physics. His doctoral thesis in 1924 contained the proposal that moving bodies have wave properties that complement their particle properties: these "seemingly incompatible conceptions can each represent an aspect of the truth. . . . They may serve in turn to represent the facts without ever entering into direct conflict." Part of de Broglie's inspiration came from Bohr's theory of the hydrogen atom, in which the electron is supposed to follow only certain orbits around the nucleus. "This fact suggested to me the idea that electrons . . . could not be considered simply as particles but that periodicity must be assigned to them also." Two years later Erwin Schrödinger used the concept of de Broglie waves to develop a general theory that he and others applied to explain a wide variety of atomic phenomena. The existence of de Broglie waves was confirmed in diffraction experiments with electron beams in 1927, and in 1929 de Broglie received the Nobel Prize.

The greater the particle's momentum, the shorter its wavelength. In Eq. (3.2) γ is the relativistic factor

$$\gamma = \frac{1}{\sqrt{1 - v^2/c^2}}$$

As in the case of em waves, the wave and particle aspects of moving bodies can never be observed at the same time. We therefore cannot ask which is the "correct" description. All that can be said is that in certain situations a moving body resembles a wave and in others it resembles a particle. Which set of properties is most conspicuous depends on how its de Broglie wavelength compares with its dimensions and the dimensions of whatever it interacts with.

Example 3.1

Find the de Broglie wavelengths of (*a*) a 46-g golf ball with a velocity of 30 m/s, and (*b*) an electron with a velocity of 10^7 m/s.

Solution

(*a*) Since $v \ll c$, we can let $\gamma = 1$. Hence

$$\lambda = \frac{h}{mv} = \frac{6.63 \times 10^{-34}\ \text{J} \cdot \text{s}}{(0.046\ \text{kg})(30\ \text{m/s})} = 4.8 \times 10^{-34}\ \text{m}$$

The wavelength of the golf ball is so small compared with its dimensions that we would not expect to find any wave aspects in its behavior.

(*b*) Again $v \ll c$, so with $m = 9.1 \times 10^{-31}$ kg, we have

$$\lambda = \frac{h}{mv} = \frac{6.63 \times 10^{-34}\ \text{J} \cdot \text{s}}{(9.1 \times 10^{-31}\ \text{kg})(10^7\ \text{m/s})} = 7.3 \times 10^{-11}\ \text{m}$$

The dimensions of atoms are comparable with this figure—the radius of the hydrogen atom, for instance, is 5.3×10^{-11} m. It is therefore not surprising that the wave character of moving electrons is the key to understanding atomic structure and behavior.

Example 3.2

Find the kinetic energy of a proton whose de Broglie wavelength is 1.000 fm = 1.000×10^{-15} m, which is roughly the proton diameter.

Solution

A relativistic calculation is needed unless pc for the proton is much smaller than the proton rest energy of $E_0 = 0.938$ GeV. To find out, we use Eq. (3.2) to determine pc:

$$pc = (\gamma m v)c = \frac{hc}{\lambda} = \frac{(4.136 \times 10^{-15}\ \text{eV} \cdot \text{s})(2.998 \times 10^8\ \text{m/s})}{1.000 \times 10^{-15}\ \text{m}} = 1.240 \times 10^9\ \text{eV}$$

$$= 1.24\ \text{GeV}$$

Since $pc > E_0$ a relativistic calculation is required. From Eq. (1.24) the total energy of the proton is

$$E = \sqrt{E_0^2 + p^2c^2} = \sqrt{(0.938\ \text{GeV})^2 + (1.240\ \text{GeV})^2} = 1.555\ \text{GeV}$$

The corresponding kinetic energy is

$$KE = E - E_0 = (1.555 - 0.938) \text{ GeV} = 0.617 \text{ GeV} = 617 \text{ MeV}$$

De Broglie had no direct experimental evidence to support his conjecture. However, he was able to show that it accounted in a natural way for the energy quantization—the restriction to certain specific energy values—that Bohr had had to postulate in his 1913 model of the hydrogen atom. (This model is discussed in Chap. 4.) Within a few years Eq. (3.2) was verified by experiments involving the diffraction of electrons by crystals. Before we consider one of these experiments, let us look into the question of what kind of wave phenomenon is involved in the matter waves of de Broglie.

3.2 WAVES OF WHAT?

Waves of probability

In water waves, the quantity that varies periodically is the height of the water surface. In sound waves, it is pressure. In light waves, electric and magnetic fields vary. What is it that varies in the case of matter waves?

The quantity whose variations make up matter waves is called the **wave function,** symbol Ψ (the Greek letter psi). The value of the wave function associated with a moving body at the particular point x, y, z in space at the time t is related to the likelihood of finding the body there at the time.

About Scientists

Max Born (1882–1970) grew up in Breslau, then a German city but today part of Poland, and received a doctorate in applied mathematics at Göttingen in 1907. Soon afterward he decided to concentrate on physics, and was back in Göttingen in 1909 as a lecturer. There he worked on various aspects of the theory of crystal lattices, his "central interest" to which he often returned in later years. In 1915, at Planck's recommendation, Born became professor of physics in Berlin where, among his other activities, he played piano to Einstein's violin. After army service in World War I and a period at Frankfurt University, Born was again in Göttingen, now as professor of physics. There a remarkable center of theoretical physics developed under his leadership: Heisenberg and Pauli were among his assistants and Fermi, Dirac, Wigner, and Goeppert were among those who worked with him, just to name future Nobel Prize winners. In those days, Born wrote, "There was complete freedom of teaching and learning in German universities, with no class examinations, and no control of students. The University just offered lectures and the student had to decide for himself which he wished to attend."

Born was a pioneer in going from "the bright realm of classical physics into the still dark and unexplored underworld of the new quantum mechanics;" he was the first to use the latter term. From Born came the basic concept that the wave function Ψ of a particle is related to the probability of finding it. He began with an idea of Einstein, who "sought to make the duality of particles (light quanta or photons) and waves comprehensible by interpreting the square of the optical wave amplitude as probability density for the occurrence of photons. This idea could at once be extended to the Ψ-function: $|\Psi|^2$ must represent the probability density for electrons (or other particles). To assert this was easy; but how was it to be proved? For this purpose atomic scattering processes suggested themselves." Born's development of the quantum theory of atomic scattering (collisions of atoms with various particles) not only verified his "new way of thinking about the phenomena of nature" but also founded an important branch of theoretical physics.

Born left Germany in 1933 at the start of the Nazi period, like so many other scientists. He became a British subject and was associated with Cambridge and then Edinburg universities until he retired in 1953. Finding the Scottish climate harsh and wishing to contribute to the democratization of postwar Germany, Born spent the rest of his life in Bad Pyrmont, a town near Göttingen. His textbooks on modern physics and on optics were standard works on these subjects for many years.

The wave function Ψ itself, however, has no direct physical significance. There is a simple reason why Ψ cannot by interpreted in terms of an experiment. The probability that something be in a certain place at a given time must lie between 0 (the object is definitely not there) and 1 (the object is definitely there). An intermediate probability, say 0.2, means that there is a 20% chance of finding the object. But the amplitude of a wave can be negative as well as positive, and a negative probability, say -0.2, is meaningless. Hence Ψ by itself cannot be an observable quantity.

This objection does not apply to $|\Psi|^2$, the square of the absolute value of the wave function, which is known as **probability density:**

> The probability of experimentally finding the body described by the wave function Ψ at the point x, y, z, at the time t is proportional to the value of $|\Psi|^2$ there at t.

A large value of $|\Psi|^2$ means the strong possibility of the body's presence, while a small value of $|\Psi|^2$ means the slight possibility of its presence. As long as $|\Psi|^2$ is not actually 0 somewhere, however, there is a definite chance, however small, of detecting it there. This interpretation was first made by Max Born in 1926.

There is a big difference between the probability of an event and the event itself. Although we can speak of the wave function Ψ that describes a particle as being spread out in space, this does not mean that the particle itself is thus spread out. When an experiment is performed to detect electrons, for instance, a whole electron is either found at a certain time and place or it is not; there is no such thing as a 20 percent of an electron. However, it is entirely possible for there to be a 20 percent chance that the electron be found at that time and place, and it is this likelihood that is specified by $|\Psi|^2$.

W. L. Bragg, the pioneer in x-ray diffraction, gave this loose but vivid interpretation: "The dividing line between the wave and particle nature of matter and radiation is the moment 'now.' As this moment steadily advances through time it coagulates a wavy future into a particle past. . . . Everything in the future is a wave, everything in the past is a particle." If "the moment 'now' " is understood to be the time a measurement is performed, this is a reasonable way to think about the situation. (The philosopher Søren Kierkegaard may have been anticipating this aspect of modern physics when he wrote, "Life can only be understood backwards, but it must be lived forwards.")

Alternatively, if an experiment involves a great many identical objects all described by the same wave function Ψ, the *actual density* (number per unit volume) of objects at x, y, z at the time t is proportional to the corresponding value of $|\Psi|^2$. It is instructive to compare the connection between Ψ and the density of particles it describes with the connection discussed in Sec. 2.4 between the electric field E of an electromagnetic wave and the density N of photons associated with the wave.

While the wavelength of the de Broglie waves associated with a moving body is given by the simple formula $\lambda = h/\gamma m v$, to find their amplitude Ψ as a function of position and time is often difficult. How to calculate Ψ is discussed in Chap. 5 and the ideas developed there are applied to the structure of the atom in Chap. 6. Until then we can assume that we know as much about Ψ as each situation requires.

3.3 DESCRIBING A WAVE

A general formula for waves

How fast do de Broglie waves travel? Since we associate a de Broglie wave with a moving body, we expect that this wave has the same velocity as that of the body. Let us see if this is true.

If we call the de Broglie wave velocity v_p, we can apply the usual formula

$$v_p = \nu\lambda$$

to find v_p. The wavelength λ is simply the de Broglie wavelength $\lambda = h/\gamma mv$. To find the frequency, we equate the quantum expression $E = h\nu$ with the relativistic formula for total energy $E = \gamma mc^2$ to obtain

$$h\nu = \gamma mc^2$$

$$\nu = \frac{\gamma mc^2}{h}$$

The de Broglie wave velocity is therefore

De Broglie phase velocity

$$v_p = \nu\lambda = \left(\frac{\gamma mc^2}{h}\right)\left(\frac{h}{\gamma mv}\right) = \frac{c^2}{v} \tag{3.3}$$

Because the particle velocity v must be less than the velocity of light c, the de Broglie waves always travel faster than light! In order to understand this unexpected result, we must look into the distinction between **phase velocity** and **group velocity.** (Phase velocity is what we have been calling wave velocity.)

Let us begin by reviewing how waves are described mathematically. For simplicity we consider a string stretched along the x axis whose vibrations are in the y direction, as in Fig. 3.1, and are simple harmonic in character. If we choose $t = 0$ when the displacement y of the string at $x = 0$ is a maximum, its displacement at any future time t at the same place is given by the formula

$$y = A \cos 2\pi\nu t \tag{3.4}$$

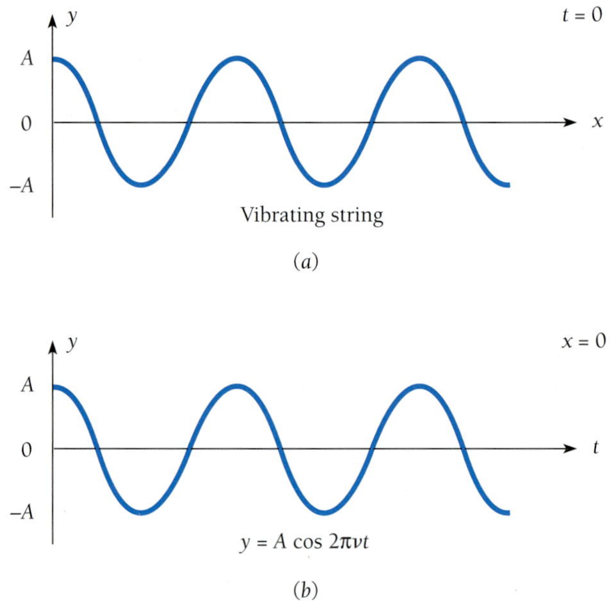

Figure 3.1 (*a*) The appearance of a wave in a stretched string at a certain time. (*b*) How the displacement of a point on the string varies with time.

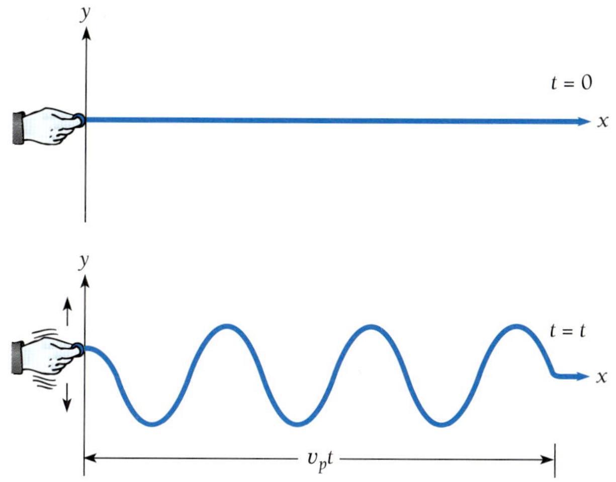

Figure 3.2 Wave propagation.

where A is the amplitude of the vibrations (that is, their maximum displacement on either side of the x axis) and ν their frequency.

Equation (3.4) tells us what the displacement of a single point on the string is as a function of time t. A complete description of wave motion in a stretched string, however, should tell us what y is at *any* point on the string at *any* time. What we want is a formula giving y as a function of both x and t.

To obtain such a formula, let us imagine that we shake the string at $x = 0$ when $t = 0$, so that a wave starts to travel down the string in the $+x$ direction (Fig. 3.2). This wave has some speed v_p that depends on the properties of the string. The wave travels the distance $x = v_pt$ in the time t, so the time interval between the formation of the wave at $x = 0$ and its arrival at the point x is x/v_p. Hence the displacement y of the string at x at any time t is exactly the same as the value of y at $x = 0$ *at the earlier time* $t - x/v_p$. By simply replacing t in Eq. (3.4) with $t - x/v_p$, then, we have the desired formula giving y in terms of both x and t:

Wave formula
$$y = A \cos 2\pi\nu\left(t - \frac{x}{v_p}\right) \tag{3.5}$$

As a check, we note that Eq. (3.5) reduces to Eq. (3.4) at $x = 0$.

Equation (3.5) may be rewritten

$$y = A \cos 2\pi\left(\nu t - \frac{\nu x}{v_p}\right)$$

Since the wave speed v_p is given by $v_p = \nu\lambda$ we have

Wave formula
$$y = A \cos 2\pi\left(\nu t - \frac{x}{\lambda}\right) \tag{3.6}$$

Equation (3.6) is often more convenient to use than Eq. (3.5).

Perhaps the most widely used description of a wave, however, is still another form of Eq. (3.5). The quantities **angular frequency** ω and **wave number** k are defined by the formulas

Angular frequency $$\omega = 2\pi\nu \tag{3.7}$$

Wave number $$k = \frac{2\pi}{\lambda} = \frac{\omega}{v_p} \tag{3.8}$$

The unit of ω is the radian per second and that of k is the radian per meter. Angular frequency gets its name from uniform circular motion, where a particle that moves around a circle ν times per second sweeps out $2\pi\nu$ rad/s. The wave number is equal to the number of radians corresponding to a wave train 1 m long, since there are 2π rad in one complete wave.

In terms of ω and k, Eq. (3.5) becomes

Wave formula $$y = A \cos (\omega t - kx) \tag{3.9}$$

In three dimensions k becomes a vector $\mathbf{k}$ normal to the wave fronts and x is replaced by the radius vector $\mathbf{r}$. The scalar product $\mathbf{k} \cdot \mathbf{r}$ is then used instead of kx in Eq. (3.9).

3.4 PHASE AND GROUP VELOCITIES

A group of waves need not have the same velocity as the waves themselves

The amplitude of the de Broglie waves that correspond to a moving body reflects the probability that it will be found at a particular place at a particular time. It is clear that de Broglie waves cannot be represented simply by a formula resembling Eq. (3.9), which describes an indefinite series of waves all with the same amplitude A. Instead, we expect the wave representation of a moving body to correspond to a **wave packet,** or **wave group,** like that shown in Fig. 3.3, whose waves have amplitudes upon which the likelihood of detecting the body depends.

A familiar example of how wave groups come into being is the case of **beats.** When two sound waves of the same amplitude but of slightly different frequencies are produced simultaneously, the sound we hear has a frequency equal to the average of the two original frequencies and its amplitude rises and falls periodically. The amplitude fluctuations occur as many times per second as the difference between the two original frequencies. If the original sounds have frequencies of, say, 440 and 442 Hz, we will hear a fluctuating sound of frequency 441 Hz with two loudness peaks, called beats, per second. The production of beats is illustrated in Fig. 3.4.

A way to mathematically describe a wave group, then, is in terms of a superposition of individual waves of different wavelengths whose interference with one another results in the variation in amplitude that defines the group shape. If the velocities of the waves are the same, the velocity with which the wave group travels is the common phase velocity. However, if the phase velocity varies with wavelength, the different individual waves do not proceed together. This situation is called **dispersion.** As a result the wave group has a velocity different from the phase velocities of the waves that make it up. This is the case with de Broglie waves.

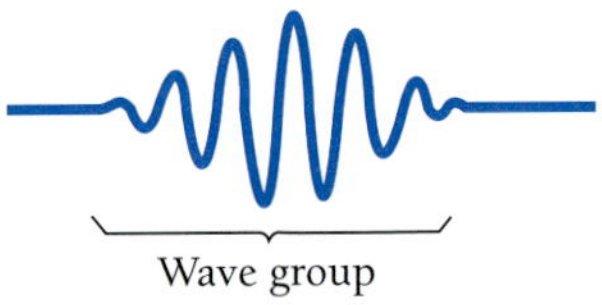

Figure 3.3 A wave group.

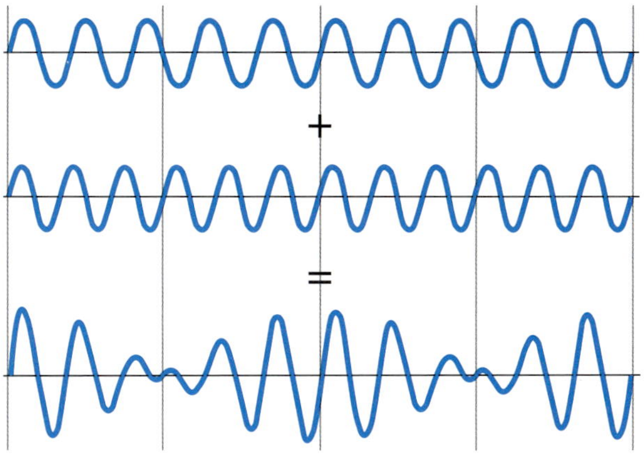

Figure 3.4 Beats are produced by the superposition of two waves with different frequencies.

It is not hard to find the velocity v_g with which a wave group travels. Let us suppose that the wave group arises from the combination of two waves that have the same amplitude A but differ by an amount $\Delta\omega$ in angular frequency and an amount Δk in wave number. We may represent the original waves by the formulas

$$y_1 = A \cos (\omega t - kx)$$
$$y_2 = A \cos [(\omega + \Delta\omega)t - (k + \Delta k)x]$$

The resultant displacement y at any time t and any position x is the sum of y_1 and y_2. With the help of the identity

$$\cos \alpha + \cos \beta = 2 \cos \tfrac{1}{2}(\alpha + \beta) \cos \tfrac{1}{2}(\alpha - \beta)$$

and the relation

$$\cos(-\theta) = \cos \theta$$

we find that

$$\begin{aligned} y &= y_1 + y_2 \\ &= 2A \cos \tfrac{1}{2}[(2\omega + \Delta\omega)t - (2k + \Delta k)x] \cos \tfrac{1}{2}(\Delta\omega\, t - \Delta k\, x) \end{aligned}$$

Since $\Delta\omega$ and Δk are small compared with ω and k respectively,

$$2\omega + \Delta\omega \approx 2\omega$$
$$2k + \Delta k \approx 2k$$

and so

Beats $$y = 2A \cos (\omega t - kx) \cos \left(\frac{\Delta\omega}{2} t - \frac{\Delta k}{2} x \right) \tag{3.10}$$

Equation (3.10) represents a wave of angular frequency ω and wave number k that has superimposed upon it a modulation of angular frequency $\frac{1}{2}\Delta\omega$ and of wave number $\frac{1}{2}\Delta k$.

The effect of the modulation is to produce successive wave groups, as in Fig. 3.4. The phase velocity v_p is

Phase velocity

$$v_p = \frac{\omega}{k} \tag{3.11}$$

and the velocity v_g of the wave groups is

Group velocity

$$v_g = \frac{\Delta\omega}{\Delta k} \tag{3.12}$$

When ω and k have continuous spreads instead of the two values in the preceding discussion, the group velocity is instead given by

Group velocity

$$v_g = \frac{d\omega}{dk} \tag{3.13}$$

Depending on how phase velocity varies with wave number in a particular situation, the group velocity may be less or greater than the phase velocities of its member waves. If the phase velocity is the same for all wavelengths, as is true for light waves in empty space, the group and phase velocities are the same.

The angular frequency and wave number of the de Broglie waves associated with a body of mass m moving with the velocity v are

$$\omega = 2\pi\nu = \frac{2\pi\gamma mc^2}{h}$$

Angular frequency of de Broglie waves

$$= \frac{2\pi mc^2}{h\sqrt{1 - v^2/c^2}} \tag{3.14}$$

$$k = \frac{2\pi}{\lambda} = \frac{2\pi\gamma mv}{h}$$

Wave number of de Broglie waves

$$= \frac{2\pi mv}{h\sqrt{1 - v^2/c^2}} \tag{3.15}$$

Both ω and k are functions of the body's velocity v.

The group velocity v_g of the de Broglie waves associated with the body is

$$v_g = \frac{d\omega}{dk} = \frac{d\omega/dv}{dk/dv}$$

Now

$$\frac{d\omega}{dv} = \frac{2\pi mv}{h(1 - v^2/c^2)^{3/2}}$$

$$\frac{dk}{dv} = \frac{2\pi m}{h(1 - v^2/c^2)^{3/2}}$$

Electron Microscopes

The wave nature of moving electrons is the basis of the electron microscope, the first of which was built in 1932. The resolving power of any optical instrument, which is limited by diffraction, is proportional to the wavelength of whatever is used to illuminate the specimen. In the case of a good microscope that uses visible light, the maximum useful magnification is about 500×; higher magnifications give larger images but do not reveal any more detail. Fast electrons, however, have wavelengths very much shorter than those of visible light and are easily controlled by electric and magnetic fields because of their charge. X-rays also have short wavelengths, but it is not (yet?) possible to focus them adequately.

In an electron microscope, current-carrying coils produce magnetic fields that act as lenses to focus an electron beam on a specimen and then produce an enlarged image on a fluorescent screen or photographic plate (Fig. 3.5). To prevent the beam from being scattered and thereby blurring the image, a thin specimen is used and the entire system is evacuated.

The technology of magnetic "lenses" does not permit the full theoretical resolution of electron waves to be realized in practice. For instance, 100-keV electrons have wavelengths of 0.0037 nm, but the actual resolution they can provide in an electron microscope may be only about 0.1 nm. However, this is still a great improvement on the ~200-nm resolution of an optical microscope, and magnifications of over 1,000,000× have been achieved with electron microscopes.

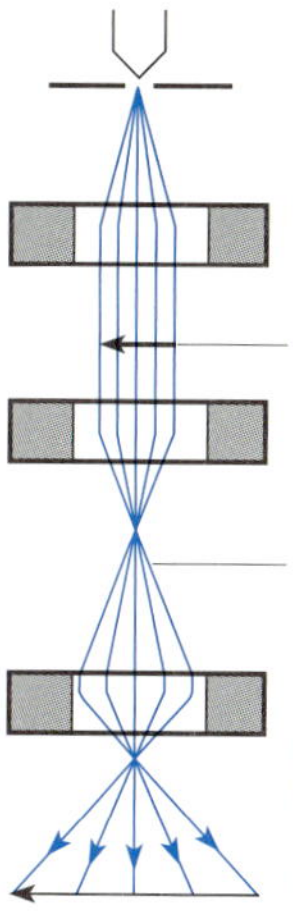

Figure 3.5 Because the wavelengths of the fast electrons in an electron microscope are shorter than those of the light waves in an optical microscope, the electron microscope can produce sharp images at higher magnifications. The electron beam in an electron microscope is focused by magnetic fields.

An electron microscope.

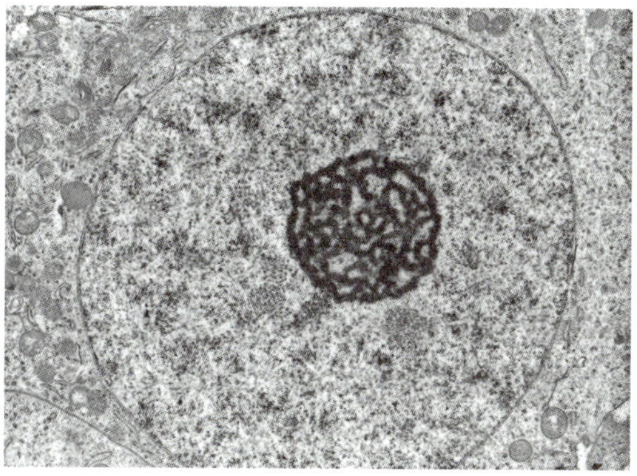

Transmission electron microscope micrograph showing the ultrastructure of a nucleus with a very prominent nucleolus and a remarkable nuclear envelope.

and so the group velocity turns out to be

De Broglie group velocity

$$v_g = v \tag{3.16}$$

The de Broglie wave group associated with a moving body travels with the same velocity as the body.

The phase velocity v_p of de Broglie waves is, as we found earlier,

De Broglie phase velocity

$$v_p = \frac{\omega}{k} = \frac{c^2}{v} \tag{3.3}$$

This exceeds both the velocity of the body v and the velocity of light c, since $v < c$. However, v_p has no physical significance because the motion of the wave group, not the motion of the individual waves that make up the group, corresponds to the motion of the body, and $v_g < c$ as it should be. The fact that $v_p > c$ for de Broglie waves therefore does not violate special relativity.

Example 3.3

An electron has a de Broglie wavelength of 2.00 pm = 2.00×10^{-12} m. Find its kinetic energy and the phase and group velocities of its de Broglie waves.

Solution

(*a*) The first step is to calculate pc for the electron, which is

$$pc = \frac{hc}{\lambda} = \frac{(4.136 \times 10^{-15}\ \text{eV}\cdot\text{s})(3.00 \times 10^8\ \text{m/s})}{2.00 \times 10^{-12}\ \text{m}} = 6.20 \times 10^5\ \text{eV}$$

$$= 620\ \text{keV}$$

The rest energy of the electron is $E_0 = 511$ keV, so

$$\text{KE} = E - E_0 = \sqrt{E_0^2 + (pc)^2} - E_0 = \sqrt{(511\ \text{keV})^2 + (620\ \text{keV})^2} - 511\ \text{keV}$$

$$= 803\ \text{keV} - 511\ \text{keV} = 292\ \text{keV}$$

(*b*) The electron velocity can be found from

$$E = \frac{E_0}{\sqrt{1 - v^2/c^2}}$$

to be

$$v = c\sqrt{1 - \frac{E_0^2}{E^2}} = c\sqrt{1 - \left(\frac{511\ \text{keV}}{803\ \text{keV}}\right)^2}$$

Hence the phase and group velocities are respectively

$$v_p = \frac{c^2}{v} = \frac{c^2}{0.771c} = 1.30c$$

$$v_g = v = 0.771c$$

3.5 PARTICLE DIFFRACTION

An experiment that confirms the existence of de Broglie waves

A wave effect with no analog in the behavior of Newtonian particles is diffraction. In 1927 Clinton Davisson and Lester Germer in the United States and G. P. Thomson in England independently confirmed de Broglie's hypothesis by demonstrating that electron beams are diffracted when they are scattered by the regular atomic arrays of crystals. (All three received Nobel Prizes for their work. J. J. Thomson, G. P.'s father, had earlier won a Nobel Prize for verifying the particle nature of the electron: the wave-particle duality seems to have been the family business.) We shall look at the experiment of Davisson and Germer because its interpretation is more direct.

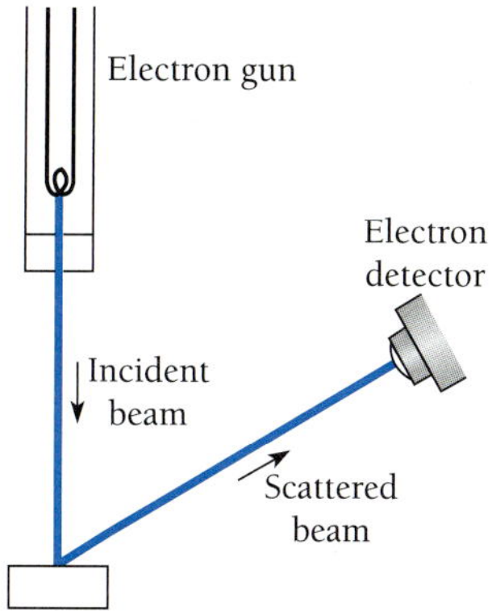

Figure 3.6 The Davisson-Germer experiment.

Davisson and Germer were studying the scattering of electrons from a solid using an apparatus like that sketched in Fig. 3.6. The energy of the electrons in the primary beam, the angle at which they reach the target, and the position of the detector could all be varied. Classical physics predicts that the scattered electrons will emerge in all directions with only a moderate dependence of their intensity on scattering angle and even less on the energy of the primary electrons. Using a block of nickel as the target, Davisson and Germer verified these predictions.

In the midst of their work an accident occurred that allowed air to enter their apparatus and oxidize the metal surface. To reduce the oxide to pure nickel, the target was baked in a hot oven. After this treatment, the target was returned to the apparatus and the measurements resumed.

Now the results were very different. Instead of a continuous variation of scattered electron intensity with angle, distinct maxima and minima were observed whose positions depended upon the electron energy! Typical polar graphs of electron intensity after the accident are shown in Fig. 3.7. The method of plotting is such that the intensity at any angle is proportional to the distance of the curve at that angle from the point of scattering. If the intensity were the same at all scattering angles, the curves would be circles centered on the point of scattering.

Two questions come to mind immediately: What is the reason for this new effect? Why did it not appear until after the nickel target was baked?

De Broglie's hypothesis suggested that electron waves were being diffracted by the target, much as x-rays are diffracted by planes of atoms in a crystal. This idea received

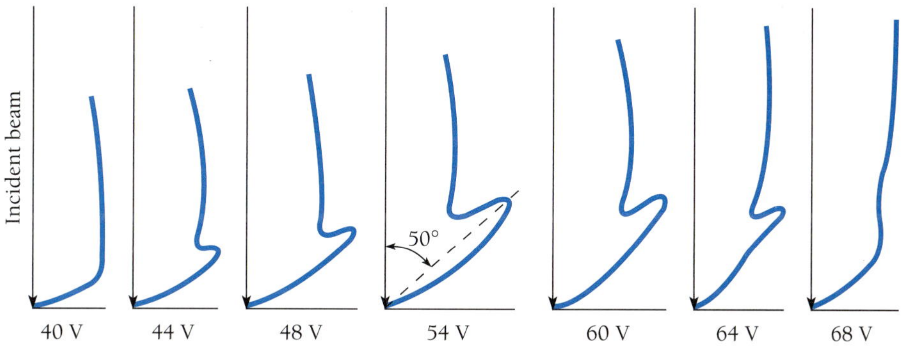

Figure 3.7 Results of the Davisson-Germer experiment, showing how the number of scattered electrons varied with the angle between the incoming beam and the crystal surface. The Bragg planes of atoms in the crystal were not parallel to the crystal surface, so the angles of incidence and scattering relative to one family of these planes were both 65° (see Fig. 3.8).

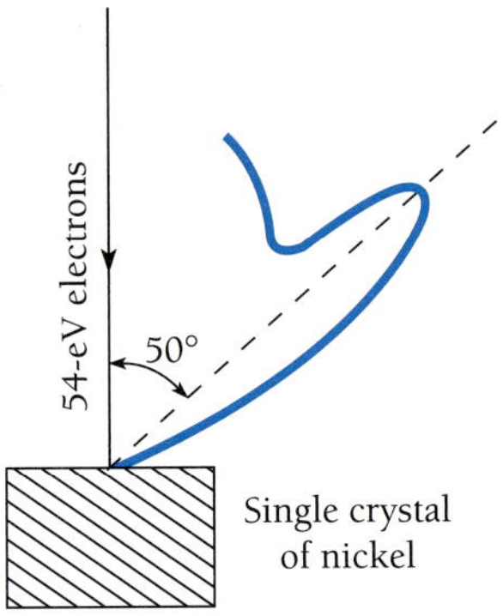

Figure 3.8 The diffraction of the de Broglie waves by the target is responsible for the results of Davisson and Germer.

support when it was realized that heating a block of nickel at high temperature causes the many small individual crystals of which it is normally composed to form into a single large crystal, all of whose atoms are arranged in a regular lattice.

Let us see whether we can verify that de Broglie waves are responsible for the findings of Davisson and Germer. In a particular case, a beam of 54-eV electrons was directed perpendicularly at the nickel target and a sharp maximum in the electron distribution occurred at an angle of 50° with the original beam. The angles of incidence and scattering relative to the family of Bragg planes shown in Fig. 3.8 are both 65°. The spacing of the planes in this family, which can be measured by x-ray diffraction, is 0.091 nm. The Bragg equation for maxima in the diffraction pattern is

$$n\lambda = 2d \sin \theta \tag{2.13}$$

Here $d = 0.091$ nm and $\theta = 65°$. For $n = 1$ the de Broglie wavelength λ of the diffracted electrons is

$$\lambda = 2d \sin \theta = (2)(0.091 \text{ nm})(\sin 65°) = 0.165 \text{ nm}$$

Now we use de Broglie's formula $\lambda = h/\gamma m v$ to find the expected wavelength of the electrons. The electron kinetic energy of 54 eV is small compared with its rest energy mc^2 of 0.51 MeV, so we can let $\gamma = 1$. Since

$$\text{KE} = \tfrac{1}{2}mv^2$$

the electron momentum mv is

$$\begin{aligned} mv &= \sqrt{2m\text{KE}} \\ &= \sqrt{(2)(9.1 \times 10^{-31} \text{ kg})(54 \text{ eV})(1.6 \times 10^{-19} \text{ J/eV})} \\ &= 4.0 \times 10^{-24} \text{ kg} \cdot \text{m/s} \end{aligned}$$

The electron wavelength is therefore

$$\lambda = \frac{h}{mv} = \frac{6.63 \times 10^{-34} \text{ J} \cdot \text{s}}{4.0 \times 10^{-24} \text{ kg} \cdot \text{m/s}} = 1.66 \times 10^{-10} \text{ m} = 0.166 \text{ nm}$$

which agrees well with the observed wavelength of 0.165 nm. The Davisson-Germer experiment thus directly verifies de Broglie's hypothesis of the wave nature of moving bodies.

Analyzing the Davisson-Germer experiment is actually less straightforward than indicated above because the energy of an electron increases when it enters a crystal by an amount equal to the work function of the surface. Hence the electron speeds in the experiment were greater inside the crystal and the de Broglie wavelengths there shorter than the values outside. Another complication arises from interference between waves diffracted by different families of Bragg planes, which restricts the occurrence of maxima to certain combinations of electron energy and angle of incidence rather than merely to any combination that obeys the Bragg equation.

Electrons are not the only bodies whose wave behavior can be demonstrated. The diffraction of neutrons and of whole atoms when scattered by suitable crystals has been observed, and in fact neutron diffraction, like x-ray and electron diffraction, has been used for investigating crystal structures.

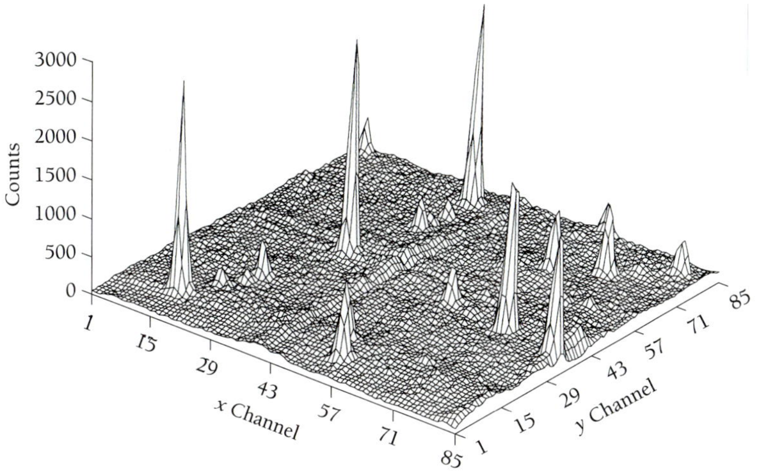

Neutron diffraction by a quartz crystal. The peaks represent directions in which constructive interference occurred. *(Courtesy Frank J. Rotella and Arthur J. Schultz, Argonne National Laboratory)*

Neutron Scattering

The wave nature of particles are explored in some of useful analytical tools; this allows the examination of fine structures of materials either at their surface using, for instance, a Scanning Electron Microscope, or at sub-surface with a Transmission Electron Microscope, or through the process known as Neutron Scattering. The latter has an advantage over electron microscopy because neutron has zero charge and therefore its scattering is not affected by the charges of electrons and protons. Since neutrons are electrically neutral, they can penetrate matter deeper than electrically charged particles of comparable kinetic energy; thus they can be used to probe bulk properties of materials. Neutron scattering almost always has elastic and inelastic component. Depending on the research nature of the experiment, most focuses on either the elastic or the inelastic scattering.

While neutrons are electrically neutral, they do interact with atomic nuclei and magnetic fields from unpaired electrons. These interactions cause pronounced interference and energy transfer effects in scattering experiments. Furthermore, neutron scattering and absorption cross sections vary widely from isotope to isotope. The scattering can be incoherent or coherent which is isotope dependent.

To obtain a 'monochromatic' neutron beam, a precise energy and de Broglie wavelength, of the neutron beam is important as this would give a precise velocity. Monochromaticity is achieved either with a crystal monochromator or a time of flight spectrometer. In the time-of-flight technique, neutrons are sent through a sequence of two rotating slits, so that only neutrons of a particular velocity are selected.

Neutron scattering also has an advantage over X-ray diffraction; in X-ray diffraction the scattering cross sections systematically increase with atomic number. This means that elements with smaller atomic numbers such as hydrogen, lithium and oxygen are not easily detected. In neutron scattering, hydrogen has the highest neutron scattering cross section and important elements like carbon and oxygen are very visible. This makes neutron scattering a very useful tool for material analysis and has found applications in crystallography, biophysics, physical chemistry, and materials research. Fig. 3.9 depicts schematically a typical neutron scattering facility.

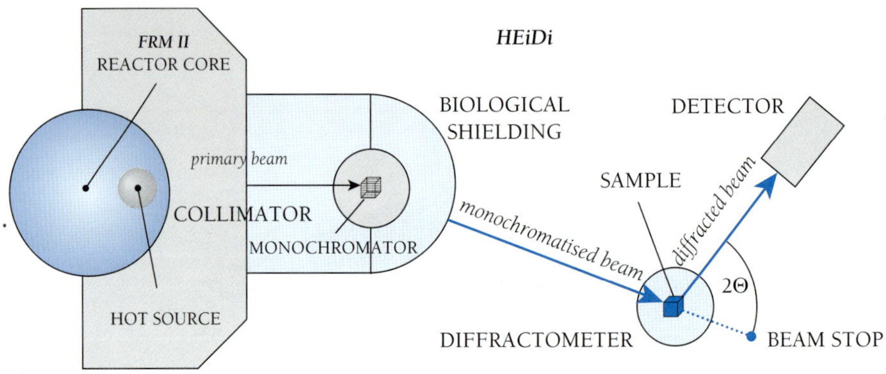

Figure 3.9 Schematic diagram of a neutron scattering used to determine the atomic and/or magnetic structure of a material.

3.6 PARTICLE IN A BOX

Why the energy of a trapped particle is quantized

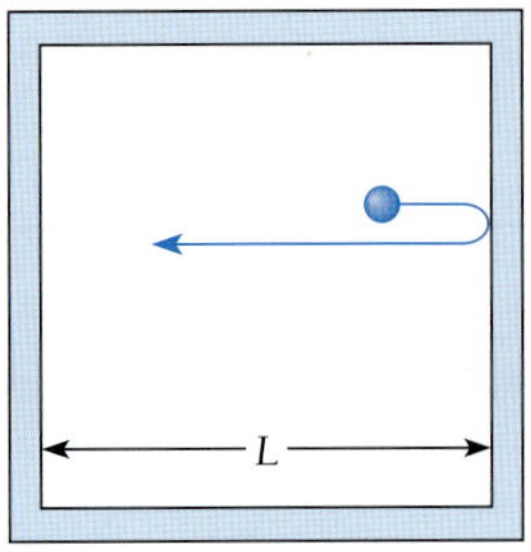

Figure 3.10 A particle confined to a box of width L. The particle is assumed to move back and forth along a straight line between the walls of the box.

The wave nature of a moving particle leads to some remarkable consequences when the particle is restricted to a certain region of space instead of being able to move freely.

The simplest case is that of a particle that bounces back and forth between the walls of a box, as in Fig. 3.10. We shall assume that the walls of the box are infinitely hard, so the particle does not lose energy each time it strikes a wall, and that its velocity is sufficiently small so that we can ignore relativistic considerations. Simple as it is, this model situation requires fairly elaborate mathematics in order to be properly analyzed, as we shall learn in Chap. 5. However, even a relatively crude treatment can reveal the essential results.

From a wave point of view, a particle trapped in a box is like a standing wave in a string stretched between the box's walls. In both cases the wave variable (transverse displacement for the string, wave function Ψ for the moving particle) must be 0 at the walls, since the waves stop there. The possible de Broglie wavelengths of the particle in the box therefore are determined by the width L of the box, as in Fig. 3.11. The longest wavelength is specified by $\lambda = 2L$, the next by $\lambda = L$, then $\lambda = 2L/3$, and so forth. The general formula for the permitted wavelengths is

De Broglie wavelengths of trapped particle

$$\lambda_n = \frac{2L}{n} \qquad n = 1, 2, 3, \ldots \tag{3.17}$$

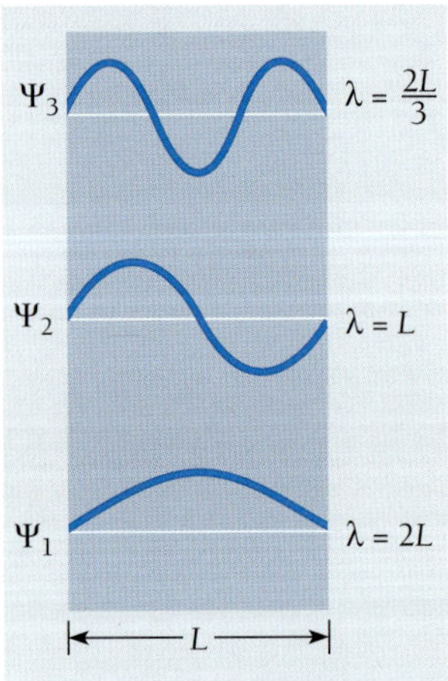

Figure 3.11 Wave functions of a particle trapped in a box L wide.

Because $mv = h/\lambda$, the restrictions on de Broglie wavelength λ imposed by the width of the box are equivalent to limits on the momentum of the particle and, in turn, to limits on its kinetic energy. The kinetic energy of a particle of momentum mv is

$$\text{KE} = \tfrac{1}{2}mv^2 = \frac{(mv)^2}{2m} = \frac{h^2}{2m\lambda^2}$$

The permitted wavelengths are $\lambda_n = 2L/n$, and so, because the particle has no potential energy in this model, the only energies it can have are

Particle in a box
$$E_n = \frac{n^2h^2}{8mL^2} \qquad n = 1, 2, 3, \ldots \tag{3.18}$$

Each permitted energy is called an **energy level**, and the integer n that specifies an energy level E_n is called its **quantum number.**

We can draw three general conclusions from Eq. (3.18). These conclusions apply to *any* particle confined to a certain region of space (even if the region does not have a well-defined boundary), for instance an atomic electron held captive by the attraction of the positively charged nucleus.

1 A trapped particle cannot have an arbitrary energy, as a free particle can. The fact of its confinement leads to restrictions on its wave function that allow the particle to have only certain specific energies and no others. Exactly what these energies are depends on the mass of the particle and on the details of how it is trapped.

2 A trapped particle cannot have zero energy. Since the de Broglie wavelength of the particle is $\lambda = h/mv$, a speed of $v = 0$ means an infinite wavelength. But there is no way to reconcile an infinite wavelength with a trapped particle, so such a particle must have at least some kinetic energy. The exclusion of $E = 0$ for a trapped particle, like the limitation of E to a set of discrete values, is a result with no counterpart in classical physics, where all non-negative energies, including zero, are allowed.

3 Because Planck's constant is so small—only 6.63×10^{-34} J · s—quantization of energy is conspicuous only when m and L are also small. This is why we are not aware of energy quantization in our own experience. Two examples will make this clear.

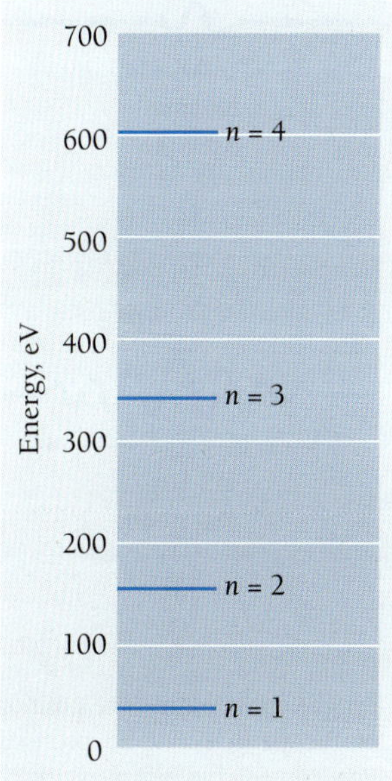

Figure 3.12 Energy levels of an electron confined to a box 0.1 nm wide.

Example 3.4

An electron is in a box 0.10 nm across, which is the order of magnitude of atomic dimensions. Find its permitted energies.

Solution

Here $m = 9.1 \times 10^{-31}$ kg and $L = 0.10$ nm $= 1.0 \times 10^{-10}$ m, so that the permitted electron energies are

$$E_n = \frac{(n^2)(6.63 \times 10^{-34}\ \text{J} \cdot \text{s})^2}{(8)(9.1 \times 10^{-31}\ \text{kg})(1.0 \times 10^{-10}\ \text{m})^2} = 6.0 \times 10^{-18}n^2\ \text{J}$$
$$= 38n^2\ \text{eV}$$

The minimum energy the electron can have is 38 eV, corresponding to $n = 1$. The sequence of energy levels continues with $E_2 = 152$ eV, $E_3 = 342$ eV, $E_4 = 608$ eV, and so on (Fig. 3.12). If such a box existed, the quantization of a trapped electron's energy would be a prominent feature of the system. (And indeed energy quantization is prominent in the case of an atomic electron.)

Example 3.5

A 10-g marble is in a box 10 cm across. Find its permitted energies.

Solution

With $m = 10$ g $= 1.0 \times 10^{-2}$ kg and $L = 10$ cm $= 1.0 \times 10^{-1}$ m,

$$E_n = \frac{(n^2)(6.63 \times 10^{-34}\ \text{J}\cdot\text{s})^2}{(8)(1.0 \times 10^{-2}\ \text{kg})(1.0 \times 10^{-1}\ \text{m})^2}$$
$$= 5.5 \times 10^{-64} n^2\ \text{J}$$

The minimum energy the marble can have is 5.5×10^{-64} J, corresponding to $n = 1$. A marble with this kinetic energy has a speed of only 3.3×10^{-31} m/s and therefore cannot be experimentally distinguished from a stationary marble. A reasonable speed a marble might have is, say, $\frac{1}{3}$ m/s—which corresponds to the energy level of quantum number $n = 10^{30}$! The permissible energy levels are so very close together, then, that there is no way to determine whether the marble can take on only those energies predicted by Eq. (3.18) or any energy whatever. Hence in the domain of everyday experience, quantum effects are imperceptible, which accounts for the success of Newtonian mechanics in this domain.

3.7 UNCERTAINTY PRINCIPLE 1

We cannot know the future because we cannot know the present

To regard a moving particle as a wave group implies that there are fundamental limits to the accuracy with which we can measure such "particle" properties as position and momentum.

To make clear what is involved, let us look at the wave group of Fig. 3.3. The particle that corresponds to this wave group may be located anywhere within the group at a given time. Of course, the probability density $|\Psi|^2$ is a maximum in the middle of the group, so it is most likely to be found there. Nevertheless, we may still find the particle anywhere that $|\Psi|^2$ is not actually 0.

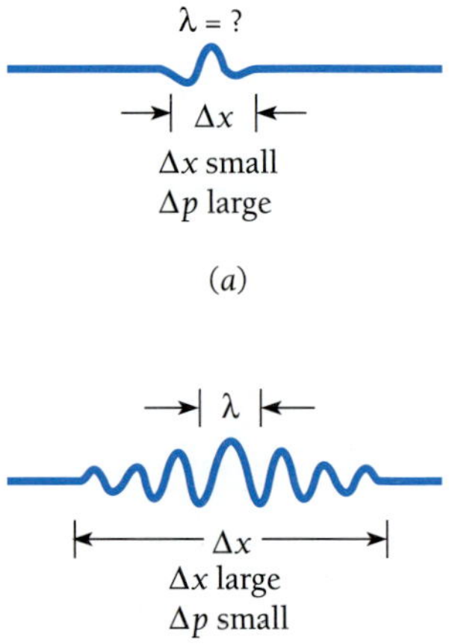

Figure 3.13 (*a*) A narrow de Broglie wave group. The position of the particle can be precisely determined, but the wavelength (and hence the particle's momentum) cannot be established because there are not enough waves to measure accurately. (*b*) A wide wave group. Now the wavelength can be precisely determined but not the position of the particle.

The narrower its wave group, the more precisely a particle's position can be specified (Fig. 3.13*a*). However, the wavelength of the waves in a narrow packet is not well defined; there are not enough waves to measure λ accurately. This means that since $\lambda = h/\gamma m v$, the particle's momentum $\gamma m v$ is not a precise quantity. If we make a series of momentum measurements, we will find a broad range of values.

On the other hand, a wide wave group, such as that in Fig. 3.13*b*, has a clearly defined wavelength. The momentum that corresponds to this wavelength is therefore a precise quantity, and a series of measurements will give a narrow range of values. But where is the particle located? The width of the group is now too great for us to be able to say exactly where the particle is at a given time.

Thus we have the **uncertainty principle:**

It is impossible to know both the exact position and exact momentum of an object at the same time.

This principle, which was discovered by Werner Heisenberg in 1927, is one of the most significant of physical laws.

A formal analysis supports the above conclusion and enables us to put it on a quantitative basis. The simplest example of the formation of wave groups is that given in Sec. 3.4, where two wave trains slightly different in angular frequency ω and wave number k were superposed to yield the series of groups shown in Fig. 3.4. A moving body corresponds to a single wave group, not a series of them, but a single wave group can also be thought of in terms of the superposition of trains of harmonic waves. However, an infinite number of wave trains with different frequencies, wave numbers, and amplitudes is required for an isolated group of arbitrary shape, as in Fig. 3.14.

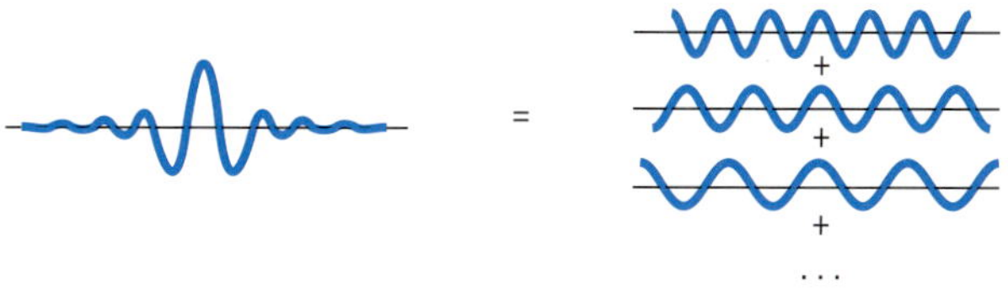

Figure 3.14 An isolated wave group is the result of superposing an infinite number of waves with different wavelengths. The narrower the wave group, the greater the range of wavelengths involved. A narrow de Broglie wave group thus means a well-defined position (Δx smaller) but a poorly defined wavelength and a large uncertainty Δp in the momentum of the particle the group represents. A wide wave group means a more precise momentum but a less precise position.

At a certain time t, the wave group $\Psi(x)$ can be represented by the **Fourier integral**

$$\Psi(x) = \int_0^\infty g(k) \cos kx \, dk \tag{3.19}$$

where the function $g(k)$ describes how the amplitudes of the waves that contribute to $\Psi(x)$ vary with wave number k. This function is called the **Fourier transform** of $\Psi(x)$, and it specifies the wave group just as completely as $\Psi(x)$ does. Figure 3.15 contains graphs of the Fourier transforms of a pulse and of a wave group. For comparison, the Fourier transform of an infinite train of harmonic waves is also included. There is only a single wave number in this case, of course.

Strictly speaking, the wave numbers needed to represent a wave group extend from $k = 0$ to $k = \infty$, but for a group whose length Δx is finite, the waves whose amplitudes $g(k)$ are appreciable have wave numbers that lie within a finite interval Δk. As Fig. 3.15 indicates, the narrower the group, the broader the range of wave numbers needed to describe it, and vice versa.

The relationship between the distance Δx and the wave-number spread Δk depends upon the shape of the wave group and upon how Δx and Δk are defined. The minimum value of the product $\Delta x \, \Delta k$ occurs when the envelope of the group has the familiar bell shape of a Gaussian function. In this case the Fourier transform happens to be a Gaussian function also. If Δx and Δk are taken as the standard deviations of the respective functions $\Psi(x)$ and $g(k)$, then this minimum value is $\Delta x \, \Delta k = \frac{1}{2}$. Because wave groups in general do not have Gaussian forms, it is more realistic to express the relationship between Δx and Δk as

$$\Delta x \, \Delta k \geq \tfrac{1}{2} \tag{3.20}$$

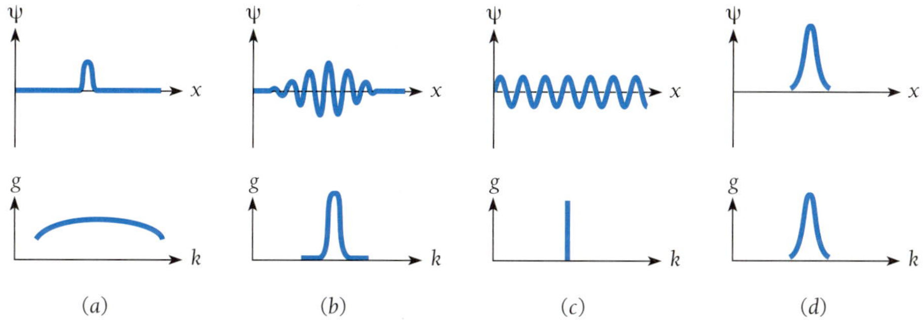

Figure 3.15 The wave functions and Fourier transforms for (*a*) a pulse, (*b*) a wave group, (*c*) a wave train, and (*d*) a Gaussian distribution. A brief disturbance needs a broader range of frequencies to describe it than a disturbance of greater duration. The Fourier transform of a Gaussian function is also a Gaussian function.

Gaussian Function

When a set of measurements is made of some quantity x in which the experimental errors are random, the result is often a **Gaussian distribution** whose form is the bell-shaped curve shown in Fig. 3.15. The **standard deviation** σ of the measurements is a measure of the spread of x values about the mean of x_0, where σ equals the square root of the average of the squared deviations from x_0. If N measurements were made,

Standard deviation
$$\sigma = \sqrt{\frac{1}{N}\sum_{i=1}^{N}(x_1 - x_0)^2}$$

The width of a Gaussian curve at half its maximum value is 2.35σ.

The ***Gaussian function*** $f(x)$ that describes the above curve is given by

Gaussian function
$$f(x) = \frac{1}{\sigma\sqrt{2\pi}} e^{-(x - x_0)^2/2\sigma^2}$$

where $f(x)$ is the probability that the value x be found in a particular measurement. Gaussian functions occur elsewhere in physics and mathematics as well. (Gabriel Lippmann had this to say about the Gaussian function: "Experimentalists think that it is a mathematical theorem while mathematicians believe it to be an experimental fact.")

The probability that a measurement lie inside a certain range of x values, say between x_1 and x_2, is given by the area of the $f(x)$ curve between these limits. This area is the integral

$$P_{x_1x_2} = \int_{x_1}^{x_2} f(x)\, dx$$

An interesting questions is what fraction of a series of measurements has values within a standard deviation of the mean value x_0. In this case $x_1 = x_0 - \sigma$ and $x_2 = x_0 + \sigma$, and

$$P_{x_0\pm\sigma} = \int_{x_0-\sigma}^{x_0+\sigma} f(x)\, dx = 0.683$$

Hence 68.3 percent of the measurements fall in this interval, which is shaded in Fig. 3.16. A similar calculation shows that 95.4 percent of the measurements fall within two standard deviations of the mean value.

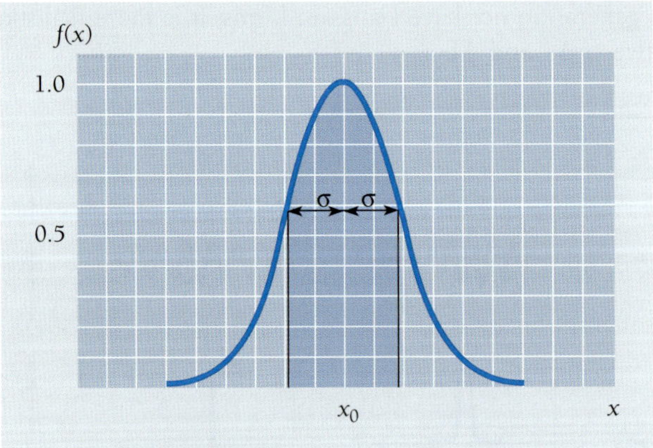

Figure 3.16 A Gaussian distribution. The probability of finding a value of x is given by the Gaussian function $f(x)$. The mean value of x is x_0, and the total width of the curve at half its maximum value is 2.35σ, where σ is the standard deviation of the distribution. The total probability of finding a value of x within a standard deviation of x_0 is equal to the shaded area and is 68.3 percent.

The de Broglie wavelength of a particle of momentum p is $\lambda = h/p$ and the corresponding wave number is

$$k = \frac{2\pi}{\lambda} = \frac{2\pi p}{h}$$

In terms of wave number the particle's momentum is therefore

$$p = \frac{hk}{2\pi}$$

Hence an uncertainty Δk in the wave number of the de Broglie waves associated with the particle results in an uncertainty Δp in the particle's momentum according to the formula

$$\Delta p = \frac{h\,\Delta k}{2\pi}$$

Since $\Delta x\,\Delta k \geq \frac{1}{2}$, $\Delta k \geq 1/(2\Delta x)$ and

Uncertainty principle

$$\Delta x\,\Delta p \geq \frac{h}{4\pi} \tag{3.21}$$

This equation states that the product of the uncertainty Δx in the position of an object at some instant and the uncertainty Δp in its momentum component in the x direction at the same instant is equal to or greater than $h/4\pi$.

If we arrange matters so that Δx is small, corresponding to a narrow wave group, then Δp will be large. If we reduce Δp in some way, a broad wave group is inevitable and Δx will be large.

About Scientists

Werner Heisenberg (1901–1976) was born in Duisberg, Germany, and studied theoretical physics at Munich, where he also became an enthusiastic skier and mountaineer. At Göttingen in 1924 as an assistant to Max Born, Heisenberg became uneasy about mechanical models of the atom: "Any picture of the atom that our imagination is able to invent is for that very reason defective," he later remarked. Instead he conceived an abstract approach using matrix algebra. In 1925, together with Born and Pascual Jordan, Heisenberg developed this approach into a consistent theory of quantum mechanics, but it was so difficult to understand and apply that it had very little impact on physics at the time. Schrödinger's wave formulation of quantum mechanics the following year was much more successful; Schrödinger and others soon showed that the wave and matrix versions of quantum mechanics were mathematically equivalent.

In 1927, working at Bohr's institute in Copenhagen, Heisenberg developed a suggestion by Wolfgang Pauli into the uncertainty principle. Heisenberg initially felt that this principle was a consequence of the disturbances inevitably produced by any measuring process. Bohr, on the other hand, thought that the basic cause of the uncertainties was the wave-particle duality, so that they were built into the natural world rather than solely the result of measurement. After much argument Heisenberg came around to Bohr's view. (Einstein, always skeptical about quantum mechanics, said after a lecture by Heisenberg on the uncertainty principle: "Marvelous, what ideas the young people have these days. But I don't believe a word of it.") Heisenberg received the Nobel Prize in 1932.

Heisenberg was one of the very few distinguished scientists to remain in Germany during the Nazi period. In World War II he led research there on atomic weapons, but little progress had been made by the war's end. Exactly why remains unclear, although there is no evidence that Heisenberg, as he later claimed, had moral qualms about creating such weapons and more or less deliberately dragged his feet. Heisenberg recognized early that "an explosive of unimaginable consequences" could be developed, and he and his group should have been able to have gotten farther than they did. In fact, alarmed by the news that Heisenberg was working on an atomic bomb, the U.S. government sent the former Boston Red Sox catcher Moe Berg to shoot Heisenberg during a lecture in neutral Switzerland in 1944. Berg, sitting in the second row, found himself uncertain from Heisenberg's remarks about how advanced the German program was, and kept his gun in his pocket.

These uncertainties are due not to inadequate apparatus but to the imprecise character in nature of the quantities involved. Any instrumental or statistical uncertainties that arise during a measurement only increase the product $\Delta x\, \Delta p$. Since we cannot know exactly both where a particle is right now and what its momentum is, we cannot say anything definite about where it will be in the future or how fast it will be moving then. *We cannot know the future for sure because we cannot know the present for sure.* But our ignorance is not total: we can still say that the particle is more likely to be in one place than another and that its momentum is more likely to have a certain value than another.

H-Bar

The quantity $h/2\pi$ appears often in modern physics because it turns out to be the basic unit of angular momentum. It is therefore customary to abbreviate $h/2\pi$ by the symbol $\hbar$ ("h-bar"):

$$\hbar = \frac{h}{2\pi} = 1.054 \times 10^{-34}\ \text{J} \cdot \text{s}$$

In the remainder of this book $\hbar$ is used in place of $h/2\pi$. In terms of $\hbar$, the uncertainty principle becomes

Uncertainty principle

$$\Delta x\, \Delta p \geq \frac{\hbar}{2} \tag{3.22}$$

Example 3.6

A measurement establishes the position of a proton with an accuracy of $\pm 1.00 \times 10^{-11}$ m. Find the uncertainty in the proton's position 1.00 s later. Assume $v \ll c$.

Solution

Let us call the uncertainty in the proton's position Δx_0 at the time $t = 0$. The uncertainty in its momentum at this time is therefore, from Eq. (3.22),

$$\Delta p \geq \frac{\hbar}{2\Delta x_0}$$

Since $v \ll c$, the momentum uncertainty is $\Delta p = \Delta(mv) = m\, \Delta v$ and the uncertainty in the proton's velocity is

$$\Delta v = \frac{\Delta p}{m} \geq \frac{\hbar}{2m\, \Delta x_0}$$

The distance x the proton covers in the time t cannot be known more accurately than

$$\Delta x = t\, \Delta v \geq \frac{\hbar t}{2m\, \Delta x_0}$$

Hence Δx is inversely proportional to Δx_0: the *more* we know about the proton's position at $t = 0$, the *less* we know about its later position at $t > 0$. The value of Δx at $t = 1.00$ s is

$$\Delta x \geq \frac{(1.054 \times 10^{-34}\ \text{J} \cdot \text{s})(1.00\ \text{s})}{(2)(1.672 \times 10^{-27}\ \text{kg})(1.00 \times 10^{-11}\ \text{m})}$$

$$\geq 3.15 \times 10^{3}\ \text{m}$$

This is 3.15 km—nearly 2 mi! What has happened is that the original wave group has spread out to a much wider one (Fig. 3.17). This occurred because the phase velocities of the component waves vary with wave number and a large range of wave numbers must have been present to produce the narrow original wave group. See Fig. 3.15.

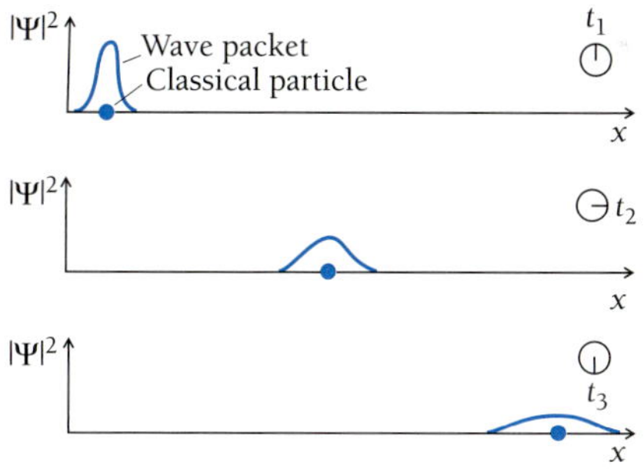

Figure 3.17 The wave packet that corresponds to a moving packet is a composite of many individual waves, as in Fig. 3.13. The phase velocities of the individual waves vary with their wave lengths. As a result, as the particle moves, the wave packet spreads out in space. The narrower the original wavepacket—that is, the more precisely we know its position at that time—the more it spreads out because it is made up of a greater span of waves with different phase velocities.

3.8 UNCERTAINTY PRINCIPLE II

A particle approach gives the same result

The uncertainty principle can be arrived at from the point of view of the particle properties of waves as well as from the point of view of the wave properties of particles.

We might want to measure the position and momentum of an object at a certain moment. To do so, we must touch it with something that will carry the required information back to us. That is, we must poke it with a stick, shine light on it, or perform some similar act. The measurement process itself thus requires that the object be interfered with in some way. If we consider such interferences in detail, we are led to the same uncertainty principle as before even without taking into account the wave nature of moving bodies.

Suppose we look at an electron using light of wavelength λ, as in Fig. 3.18. Each photon of this light has the momentum h/λ. When one of these photons bounces off the electron (which must happen if we are to "see" the electron), the electron's original momentum will be changed. The exact amount of the change Δp cannot be predicted, but it will be of the same order of magnitude as the photon momentum h/λ. Hence

$$\Delta p \approx \frac{h}{\lambda} \tag{3.23}$$

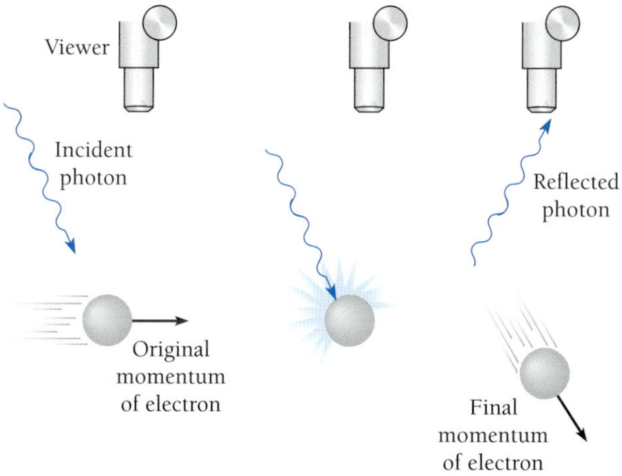

Figure 3.18 An electron cannot be observed without changing its momentum.

The longer the wavelength of the observing photon, the smaller the uncertainty in the electron's momentum.

Because light is a wave phenomenon as well as a particle phenomenon, we cannot expect to determine the electron's location with perfect accuracy regardless of the instrument used. A reasonable estimate of the minimum uncertainty in the measurement might be one photon wavelength, so that

$$\Delta x \geq \lambda \tag{3.24}$$

The shorter the wavelength, the smaller the uncertainty in location. However, if we use light of short wavelength to increase the accuracy of the position measurement, there will be a corresponding decrease in the accuracy of the momentum measurement because the higher photon momentum will disturb the electron's motion to a greater extent. Light of long wavelength will give a more accurate momentum but a less accurate position.

Combining Eqs. (3.23) and (3.24) gives

$$\Delta x \, \Delta p \geq h \tag{3.25}$$

This result is consistent with Eq. (3.22), $\Delta x \, \Delta p \geq \hbar/2$.

Arguments like the preceding one, although superficially attractive, must be approached with caution. The argument above implies that the electron can possess a definite position and momentum at any instant and that it is the measurement process that introduces the indeterminacy in $\Delta x \, \Delta p$. On the contrary, *this indeterminacy is inherent in the nature of a moving body*. The justification for the many "derivations" of this kind is first, they show it is impossible to imagine a way around the uncertainty principle; and second, they present a view of the principle that can be appreciated in a more familiar context than that of wave groups.

3.9 APPLYING THE UNCERTAINTY PRINCIPLE

A useful tool, not just a negative statement

Planck's constant h is so small that the limitations imposed by the uncertainty principle are significant only in the realm of the atom. On such a scale, however, this principle is of great help in understanding many phenomena. It is worth keeping in mind that the lower limit of $\hbar/2$ for $\Delta x \, \Delta p$ is rarely attained. More usually $\Delta x \, \Delta p \geq \hbar$, or even (as we just saw) $\Delta x \, \Delta p \geq h$.

Example 3.7

A typical atomic nucleus is about 5.0×10^{-15} m in radius. Use the uncertainty principle to place a lower limit on the energy an electron must have if it is to be part of a nucleus.

Solution

Letting $\Delta x = 5.0 \times 10^{-15}$ m we have

$$\Delta p \geq \frac{\hbar}{2\Delta x} \geq \frac{1.054 \times 10^{-34} \text{ J} \cdot \text{s}}{(2)(5.0 \times 10^{-15} \text{ m})} \geq 1.1 \times 10^{-20} \text{ kg} \cdot \text{m/s}$$

If this is the uncertainty in a nuclear electron's momentum, the momentum p itself must be at least comparable in magnitude. An electron with such a momentum has a kinetic energy KE many times greater than its rest energy mc^2. From Eq. (1.24) we see that we can let KE $= pc$ here to a sufficient degree of accuracy. Therefore

$$\text{KE} = pc \geq (1.1 \times 10^{-20}\ \text{kg} \cdot \text{m/s})(3.0 \times 10^{8}\ \text{m/s}) \geq 3.3 \times 10^{-12}\ \text{J}$$

Since 1 eV $= 1.6 \times 10^{-19}$ J, the kinetic energy of an electron must exceed 20 MeV if it is to be inside a nucleus. Experiments show that the electrons emitted by certain unstable nuclei never have more than a small fraction of this energy, from which we conclude that nuclei cannot contain electrons. The electron an unstable nucleus may emit comes into being at the moment the nucleus decays (see Secs. 11.3 and 12.5).

Example 3.8

A hydrogen atom is 5.3×10^{-11} m in radius. Use the uncertainty principle to estimate the minimum energy an electron can have in this atom.

Solution

Here we find that with $\Delta x = 5.3 \times 10^{-11}$ m.

$$\Delta p \geq \frac{\hbar}{2\Delta x} \geq 9.9 \times 10^{-25}\ \text{kg} \cdot \text{m/s}$$

An electron whose momentum is of this order of magnitude behaves like a classical particle, and its kinetic energy is

$$\text{KE} = \frac{p^2}{2m} \geq \frac{(9.9 \times 10^{-25}\ \text{kg} \cdot \text{m/s})^2}{(2)(9.1 \times 10^{-31}\ \text{kg})} \geq 5.4 \times 10^{-19}\ \text{J}$$

which is 3.4 eV. The kinetic energy of an electron in the lowest energy level of a hydrogen atom is actually 13.6 eV.

Energy and Time

Another form of the uncertainty principle concerns energy and time. We might wish to measure the energy E emitted during the time interval Δt in an atomic process. If the energy is in the form of em waves, the limited time available restricts the accuracy with which we can determine the frequency ν of the waves. Let us assume that the minimum uncertainty in the number of waves we count in a wave group is one wave. Since the frequency of the waves under study is equal to the number of them we count divided by the time interval, the uncertainty $\Delta\nu$ in our frequency measurement is

$$\Delta\nu \geq \frac{1}{\Delta t}$$

The corresponding energy uncertainty is

$$\Delta E = h\,\Delta\nu$$

and so

$$\Delta E \geq \frac{h}{\Delta t} \qquad \text{or} \qquad \Delta E\, \Delta t \geq h$$

A more precise calculation based on the nature of wave groups changes this result to

Uncertainties in energy and time

$$\Delta E\, \Delta t \geq \frac{\hbar}{2} \tag{3.26}$$

Equation (3.26) states that the product of the uncertainty ΔE in an energy measurement and the uncertainty Δt in the time at which the measurement is made is equal to or greater than $\hbar/2$. This result can be derived in other ways as well and is a general one not limited to em waves.

Example 3.9

An "excited" atom gives up its excess energy by emitting a photon of characteristic frequency, as described in Chap. 4. The average period that elapses between the excitation of an atom and the time it radiates is 1.0×10^{-8} s. Find the inherent uncertainty in the frequency of the photon.

Solution

The photon energy is uncertain by the amount

$$\Delta E \geq \frac{\hbar}{2\Delta t} \geq \frac{1.054 \times 10^{-34}\ \text{J} \cdot \text{s}}{2(1.0 \times 10^{-8}\ \text{s})} \geq 5.3 \times 10^{-27}\ \text{J}$$

The corresponding uncertainty in the frequency of light is

$$\Delta \nu = \frac{\Delta E}{h} \geq 8 \times 10^{6}\ \text{Hz}$$

This is the irreducible limit to the accuracy with which we can determine the frequency of the radiation emitted by an atom. As a result, the radiation from a group of excited atoms does not appear with the precise frequency ν. For a photon whose frequency is, say, 5.0×10^{14} Hz, $\Delta\nu/\nu = 1.6 \times 10^{-8}$. In practice, other phenomena such as the doppler effect contribute more than this to the broadening of spectral lines.

EXERCISES

It is only the first step that takes the effort. —Marquise du Deffand

3.1 De Broglie Waves

1. A photon and a particle have the same wavelength. Can anything be said about how their linear momenta compare? About how the photon's energy compares with the particle's total energy? About how the photon's energy compares with the particle's kinetic energy?

2. Find the de Broglie wavelength of (*a*) an electron whose speed is 1.0×10^8 m/s, and (*b*) an electron whose speed is 2.0×10^8 m/s.

3. Find the de Broglie wavelength of a 1.0-mg grain of sand blown by the wind at a speed of 20 m/s.

4. Find the de Broglie wavelength of the 40-keV electrons used in a certain electron microscope.

5. By what percentage will a nonrelativistic calculation of the de Broglie wavelength of a 100-keV electron be in error?

6. Find the de Broglie wavelength of a 1.00-MeV proton. Is a relativistic calculation needed?

7. The atomic spacing in rock salt, NaCl, is 0.282 nm. Find the kinetic energy (in eV) of a neutron with a de Broglie wavelength of 0.282 nm. Is a relativistic calculation needed? Such neutrons can be used to study crystal structure.

8. Find the kinetic energy of an electron whose de Broglie wavelength is the same as that of a 100-keV x-ray.

9. Green light has a wavelength of about 550 nm. Through what potential difference must an electron be accelerated to have this wavelength?

10. Show that the de Broglie wavelength of a particle of mass m and kinetic energy KE is given by

$$\lambda = \frac{hc}{\sqrt{\mathrm{KE}(\mathrm{KE} + 2mc^2)}}$$

11. Show that if the total energy of a moving particle greatly exceeds its rest energy, its de Broglie wavelength is nearly the same as the wavelength of a photon with the same total energy.

12. (*a*) Derive a relativistically correct formula that gives the de Broglie wavelength of a charged particle in terms of the potential difference V through which it has been accelerated. (*b*) What is the nonrelativistic approximation of this formula, valid for $eV \ll mc^2$?

3.4 Phase and Group Velocities

13. An electron and a proton have the same velocity. Compare the wavelengths and the phase and group velocities of their de Broglie waves.

14. An electron and a proton have the same kinetic energy. Compare the wavelengths and the phase and group velocities of their de Broglie waves.

15. Verify the statement in the text that, if the phase velocity is the same for all wavelengths of a certain wave phenomenon (that is, there is no dispersion), the group and phase velocities are the same.

16. The phase velocity of ripples on a liquid surface is $\sqrt{2\pi S/\lambda\rho}$, where S is the surface tension and ρ the density of the liquid. Find the group velocity of the ripples.

17. The phase velocity of ocean waves is $\sqrt{g\lambda/2\pi}$, where g is the acceleration of gravity. Find the group velocity of ocean waves.

18. Find the phase and group velocities of the de Broglie waves of an electron whose speed is $0.900c$.

19. Find the phase and group velocities of the de Broglie waves of an electron whose kinetic energy is 500 keV.

20. Show that the group velocity of a wave is given by $v_g = d\nu/d(1/\lambda)$.

21. (*a*) Show that the phase velocity of the de Broglie waves of a particle of mass m and de Broglie wavelength λ is given by

$$v_p = c\sqrt{1 + \left(\frac{mc\lambda}{h}\right)^2}$$

(*b*) Compare the phase and group velocities of an electron whose de Broglie wavelength is exactly 1×10^{-13} m.

22. In his original paper, de Broglie suggested that $E = h\nu$ and $p = h/\lambda$, which hold for electromagnetic waves, are also valid for moving particles. Use these relationships to show that the group velocity v_g of a de Broglie wave group is given by dE/dp, and with the help of Eq. (1.48), verify that $v_g = v$ for a particle of velocity v.

3.5 Particle Diffraction

23. What effect on the scattering angle in the Davisson-Germer experiment does increasing the electron energy have?

24. A beam of neutrons that emerges from a nuclear reactor contains neutrons with a variety of energies. To obtain neutrons with an energy of 0.050 eV, the beam is passed through a crystal whose atomic planes are 0.20 nm apart. At what angles relative to the original beam will the desired neutrons be diffracted?

25. In Sec. 3.5 it was mentioned that the energy of an electron entering a crystal increases, which reduces its de Broglie wavelength. Consider a beam of 54-eV electrons directed at a nickel target. The potential energy of an electron that enters the target changes by 26 eV. (*a*) Compare the electron speeds outside and inside the target. (*b*) Compare the respective de Broglie wavelengths.

26. A beam of 50-keV electrons is directed at a crystal and diffracted electrons are found at an angle of 50° relative to the original beam. What is the spacing of the atomic planes of the crystal? A relativistic calculation is needed for λ.

3.6 Particle in a Box

27. Obtain an expression for the energy levels (in MeV) of a neutron confined to a one-dimensional box 1.00×10^{-14} m wide. What is the neutron's minimum energy? (The diameter of an atomic nucleus is of this order of magnitude.)

28. The lowest energy possible for a certain particle trapped in a certain box is 1.00 eV. (*a*) What are the next two higher energies the particle can have? (*b*) If the particle is an electron, how wide is the box?

29. A proton in a one-dimensional box has an energy of 400 keV in its first excited state. How wide is the box?

3.9 Applying the Uncertainty Principle

30. Discuss the prohibition of $E = 0$ for a particle trapped in a box L wide in terms of the uncertainty principle. How does the minimum momentum of such a particle compare with the momentum uncertainty required by the uncertainty principle if we take $\Delta x = L$?

31. The atoms in a solid possess a certain minimum **zero-point energy** even at 0 K, while no such restriction holds for the molecules in an ideal gas. Use the uncertainty principle to explain these statements.

32. Compare the uncertainties in the velocities of an electron and a proton confined in a 1.00-nm box.

33. The position and momentum of a 1.00-keV electron are simultaneously determined. If its position is located to within 0.100 nm, what is the percentage of uncertainty in its momentum?

34. (*a*) How much time is needed to measure the kinetic energy of an electron whose speed is 10.0 m/s with an uncertainty of no more than 0.100 percent? How far will the electron have traveled in this period of time? (*b*) Make the same calculations for a 1.00-g insect whose speed is the same. What do these sets of figures indicate?

35. How accurately can the position of a proton with $v \ll c$ be determined without giving it more than 1.00 keV of kinetic energy?

36. (*a*) Find the magnitude of the momentum of a particle in a box in its *n*th state. (*b*) The minimum change in the particle's momentum that a measurement can cause corresponds to a change of ± 1 in the quantum number n. If $\Delta x = L$, show that $\Delta p \, \Delta x \geq \hbar/2$.

37. A marine radar operating at a frequency of 9400 MHz emits groups of electromagnetic waves 0.0800 μs in duration. The time needed for the reflections of these groups to return indicates the distance to a target. (*a*) Find the length of each group and the number of waves it contains. (*b*) What is the approximate minimum bandwidth (that is, spread of frequencies) the radar receiver must be able to process?

38. An unstable elementary particle called the eta meson has a rest mass of 549 MeV/c^2 and a mean lifetime of 7.00×10^{-19} s. What is the uncertainty in its rest mass?

39. The frequency of oscillation of a harmonic oscillator of mass m and spring constant C is $\nu = \sqrt{C/m}/2\pi$. The energy of the oscillator is $E = p^2/2m + Cx^2/2$, where p is its momentum when its displacement from the equilibrium position is x. In classical physics the minimum energy of the oscillator is $E_{min} = 0$. Use the uncertainty principle to find an expression for E in terms of x only and show that the minimum energy is actually $E_{min} = h\nu/2$ by setting $dE/dx = 0$ and solving for E_{min}.

40. (*a*) Verify that the uncertainty principle can be expressed in the form $\Delta L \, \Delta\theta \geq \hbar/2$, where ΔL is the uncertainty in the angular momentum of a particle and $\Delta\theta$ is the uncertainty in its angular position. (*Hint*: Consider a particle of mass m moving in a circle of radius r at the speed v, for which $L = mvr$.) (*b*) At what uncertainty in L will the angular position of a particle become completely indeterminate?

CHAPTER 4

Atomic Structure

Ti:sapphire lasers, extensively used in numerous science experiments, are near-infrared tunable lasers. The sapphire (Al2O3) crystal doped with titanium is attached under the cooling fins, and the green laser beam from the cylinder on the left is guided by several mirrors toward the crystal. The titanium is pumped with radiation from this green laser, which is a highly efficient method.

Even in the past people began to suspect that matter, despite appearing continuous, had a definite structure on a microscopic level beyond the direct reach of our senses. This suspicion did not take on a more concrete form until a little over a century and a half ago. Since then the existence of atoms and molecules, the ultimate particles of matter in its common forms, has been amply demonstrated, and their own ultimate particles, electrons, protons, and neutrons, have been identified and studied as well. In this chapter and in others to come our chief concern will be the structure of the atom, since it is this structure that is responsible for nearly all the properties of matter that have shaped the world around us.

Every atom consists of a small nucleus of protons and neutrons with a number of electrons some distance away. It is tempting to think that the electrons circle the nucleus as planets do the sun, but classical electromagnetic theory denies the possibility of stable electron orbits. In an effort to resolve this paradox, Niels Bohr applied quantum ideas to atomic structure in 1913 to obtain a model which, despite its inadequacies and later replacement by a quantum-mechanical description of greater accuracy and usefulness, still remains a convenient mental picture of the atom. Bohr's theory of the hydrogen atom is worth examining both for this reason and because it provides a valuable transition to the more abstract quantum theory of the atom.

4.1 THE NUCLEAR ATOM

An atom is largely empty space

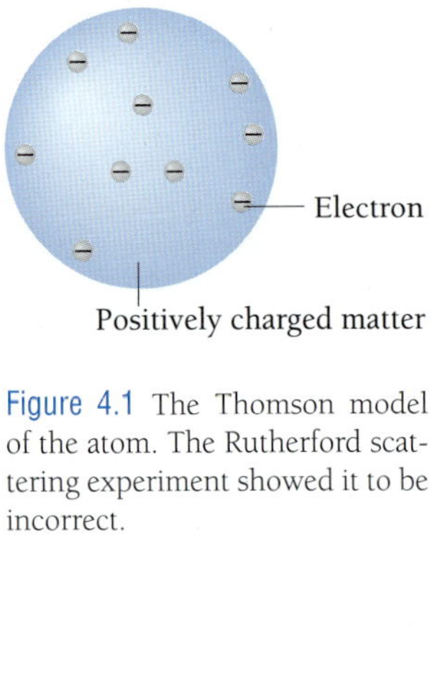

Figure 4.1 The Thomson model of the atom. The Rutherford scattering experiment showed it to be incorrect.

Most scientists of the late nineteenth century accepted the idea that the chemical elements consist of atoms, but they knew almost nothing about the atoms themselves. One clue was the discovery that all atoms contain electrons. Since electrons carry negative charges whereas atoms are neutral, positively charged matter of some kind must be present in atoms. But what kind? And arranged in what way?

One suggestion, made by the British physicist J. J. Thomson in 1898, was that atoms are just positively charged lumps of matter with electrons embedded in them, like raisins in a fruitcake (Fig. 4.1). Because Thomson had played an important role in discovering the electron, his idea was taken seriously. But the real atom turned out to be quite different.

The most direct way to find out what is inside a fruitcake is to poke a finger into it, which is essentially what Hans Geiger and Ernest Marsden did in 1911. At the suggestion of Ernest Rutherford, they used as probes the fast **alpha particles** emitted by certain radioactive elements. Alpha particles are helium atoms that have lost two electrons each, leaving them with a charge of $+2e$.

Geiger and Marsden placed a sample of an alpha-emitting substance behind a lead screen with a small hole in it, as in Fig. 4.2, so that a narrow beam of alpha particles was produced. This beam was directed at a thin gold foil. A zinc sulfide screen, which gives off a visible flash of light when struck by an alpha particle, was set on the other side of the foil with a microscope to see the flashes.

It was expected that the alpha particles would go right through the foil with hardly any deflection. This follows from the Thomson model, in which the electric charge inside an atom is assumed to be uniformly spread through its volume. With only weak electric forces exerted on them, alpha particles that pass through a thin foil ought to be deflected only slightly, 1° or less.

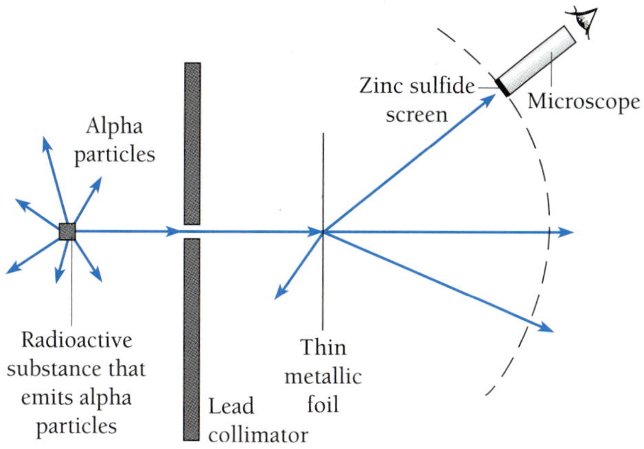

Figure 4.2 The Rutherford scattering experiment.

What Geiger and Marsden actually found was that although most of the alpha particles indeed were not deviated by much, a few were scattered through very large angles. Some were even scattered in the backward direction. As Rutherford remarked, "It was as incredible as if you fired a 15-inch shell at a piece of tissue paper and it came back and hit you."

Alpha particles are relatively heavy (almost 8000 electron masses) and those used in this experiment had high speeds (typically 2×10^7 m/s), so it was clear that powerful forces were needed to cause such marked deflections. The only way to

About Scientists

Ernest Rutherford (1871–1937), a native of New Zealand, was on his family's farm digging potatoes when he learned that he had won a scholarship for graduate study at Cambridge University in England. "This is the last potato I will every dig," he said, throwing down his spade. Thirteen years later he received the Nobel Prize in chemistry.

At Cambridge, Rutherford was a research student under J. J. Thomson, who would soon announce the discovery of the electron. Rutherford's own work was on the newly found phenomenon of radioactivity, and he quickly distinguished between alpha and beta particles, two of the emissions of radioactive materials. In 1898 he went to McGill University in Canada, where he found that alpha particles are the nuclei of helium atoms and that the radioactive decay of an element gives rise to another element. Working with the chemist Frederick Soddy and others, Rutherford traced the successive transformations of radioactive elements, such as uranium and radium, until they end up as stable lead.

In 1907 Rutherford returned to England as professor of physics at Manchester, where in 1911 he showed that the nuclear model of the atom was the only one that could explain the observed scattering of alpha particles by thin metal foils. Rutherford's last important discovery, reported in 1919, was the disintegration of nitrogen nuclei when bombarded with alpha particles, the first example of the artificial transmutation of elements into other elements. After other similar experiments, Rutherford suggested that all nuclei contain hydrogen nuclei, which he called protons. He also proposed that a neutral particle was present in nuclei as well.

In 1919 Rutherford became director of the Cavendish Laboratory at Cambridge, where under his stimulus great strides in understanding the nucleus continued to be made. James Chadwick discovered the neutron there in 1932. The Cavendish Laboratory was the site of the first accelerator for producing high-energy particles. With the help of this accelerator, fusion reactions in which light nuclei unite to form heavier nuclei were observed for the first time.

Rutherford was not infallible: only a few years before the discovery of fission and the building of the first nuclear reactor, he dismissed the idea of practical uses for nuclear energy as "moonshine." He died in 1937 of complications of a hernia and was buried near Newton in Westminster Abbey.

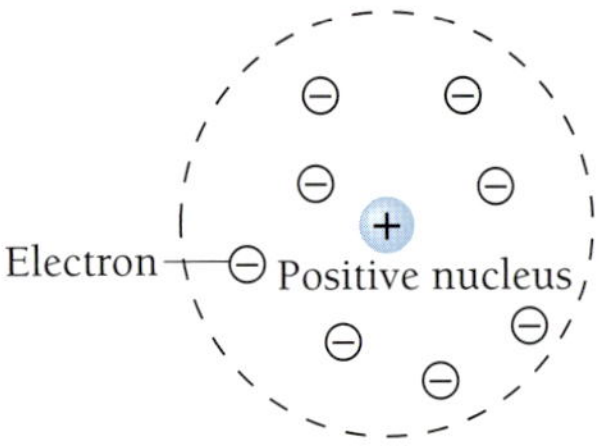

Figure 4.3 The Rutherford model of the atom.

explain the results, Rutherford found, was to picture an atom as being composed of a tiny nucleus in which its positive charge and nearly all its mass are concentrated, with the electrons some distance away (Fig. 4.3). With an atom being largely empty space, it is easy to see why most alpha particles go right through a thin foil. However, when an alpha particle happens to come near a nucleus, the intense electric field there scatters it through a large angle. The atomic electrons, being so light, do not appreciably affect the alpha particles.

The experiments of Geiger and Marsden and later work of a similar kind also supplied information about the nuclei of the atoms that composed the various target foils. The deflection of an alpha particle when it passes near a nucleus depends on the magnitude of the nuclear charge. Comparing the relative scattering of alpha particles by different foils thus provides a way to find the nuclear charges of the atoms involved.

All the atoms of any one element turned out to have the same unique nuclear charge, and this charge increased regularly from element to element in the periodic table. The nuclear charges always turned out to be multiples of $+e$; the number Z of unit positive charges in the nuclei of an element is today called the **atomic number** of the element. We know now that protons, each with a charge $+e$, provide the charge on a nucleus, so the atomic number of an element is the same as the number of protons in the nuclei of its atoms.

Ordinary matter, then, is mostly empty space. The solid wood of a table, the steel that supports a bridge, the hard rock underfoot, all are simply collections of tiny charged particles comparatively farther away from one another than the sun is from the planets. If all the actual matter, electrons and nuclei, in our bodies could somehow be packed closely together, we would shrivel to specks just visible with a microscope.

Rutherford Scattering Formula

The formula that Rutherford obtained for alpha particle scattering by a thin foil on the basis of the nuclear model of the atom is

Rutherford scattering formula

$$N(\theta) = \frac{N_i n t Z^2 e^4}{(8\pi\epsilon_0)^2 r^2 \, \mathrm{KE}^2 \sin^4(\theta/2)} \tag{4.1}$$

This formula is derived in the Appendix to this chapter. The symbols in Eq. (4.1) have the following meanings:

$N(\theta)$ = number of alpha particles per unit area that reach the screen at a scattering angle of θ
N_i = total number of alpha particles that reach the screen
n = number of atoms per unit volume in the foil
Z = atomic number of the foil atoms
r = distance of the screen from the foil
KE = kinetic energy of the alpha particles
t = foil thickness

The predictions of Eq. (4.1) agreed with the measurements of Geiger and Marsden, which supported the hypothesis of the nuclear atom. This is why Rutherford is credited

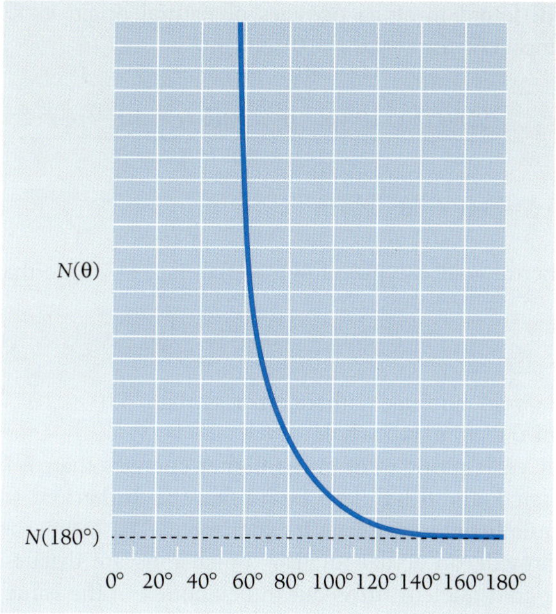

Figure 4.4 Rutherford scattering. $N(\theta)$ is the number of alpha particles per unit area that reach the screen at a scattering angle of θ; $N(180°)$ is this number for backward scattering. The experimental findings follow this curve, which is based on the nuclear model of the atom.

with the "discovery" of the nucleus. Because $N(\theta)$ is inversely proportional to $\sin^4(\theta/2)$ the variation of $N(\theta)$ with θ is very pronounced (Fig. 4.4): only 0.14 percent of the incident alpha particles are scattered by more than 1°.

Nuclear Dimensions

In his derivation of Eq. (4.1) Rutherford assumed that the size of a target nucleus is small compared with the minimum distance R to which incident alpha particles approach the nucleus before being deflected away. Rutherford scattering therefore gives us a way to find an upper limit to nuclear dimensions.

Let us see what the distance of closest approach R was for the most energetic alpha particles employed in the early experiments. An alpha particle will have its smallest R when it approaches a nucleus head on, which will be followed by a 180° scattering. At the instant of closest approach the initial kinetic energy KE of the particle is entirely converted to electric potential energy, and so at that instant

$$\text{KE}_{\text{initial}} = \text{PE} = \frac{1}{4\pi\epsilon_0}\frac{2Ze^2}{R}$$

since the charge of the alpha particle is $2e$ and that of the nucleus is Ze. Hence

Distance of closest approach

$$R = \frac{2Ze^2}{4\pi\epsilon_0 \text{KE}_{\text{initial}}} \tag{4.2}$$

The maximum KE found in alpha particles of natural origin is 7.7 MeV, which is 1.2×10^{-12} J. Since $1/4\pi\epsilon_0 = 9.0 \times 10^9 \text{ N} \cdot \text{m}^2/\text{C}^2$,

$$R = \frac{(2)(9.0 \times 10^9 \text{ N} \cdot \text{m}^2/\text{C}^2)(1.6 \times 10^{-19} \text{ C})^2 Z}{1.2 \times 10^{-12} \text{ J}}$$

$$= 3.8 \times 10^{-16} Z \text{ m}$$

The atomic number of gold, a typical foil material, is $Z = 79$, so that

$$R \text{ (Au)} = 3.0 \times 10^{-14} \text{ m}$$

The radius of the gold nucleus is therefore less than 3.0×10^{-14} m, well under 10^{-4} the radius of the atom as a whole.

In more recent years particles of much higher energies than 7.7 MeV have been artificially accelerated, and it has been found that the Rutherford scattering formula does indeed eventually fail to agree with experiment. These experiments and the information they provide on actual nuclear dimensions are discussed in Chap. 11. The radius of the gold nucleus turns out to be about $\frac{1}{5}$ of the value of R (Au) found above.

Neutron Stars

The density of nuclear matter is about 2.4×10^{17} kg/m^3, which is equivalent to 4 billion tons per cubic inch. As discussed in Sec. 9.11, neutron stars are stars whose atoms have been so compressed that most of their protons and electrons have fused into neutrons, which are the most stable form of matter under enormous pressures. The densities of neutron stars are comparable to those of nuclei: a neutron star packs the mass of one or two suns into a sphere only about 10 km in radius. If the earth were this dense, it would fit into a large apartment house.

4.2 ELECTRON ORBITS

The planetary model of the atom and why it fails

The Rutherford model of the atom, so convincingly confirmed by experiment, pictures a tiny, massive, positively charged nucleus surrounded at a relatively great distance by enough electrons to render the atom electrically neutral as a whole. The electrons cannot be stationary in this model, because there is nothing that can keep them in place against the electric force pulling them to the nucleus. If the electrons are in motion, however, dynamically stable orbits like those of the planets around the sun are possible (Fig. 4.5).

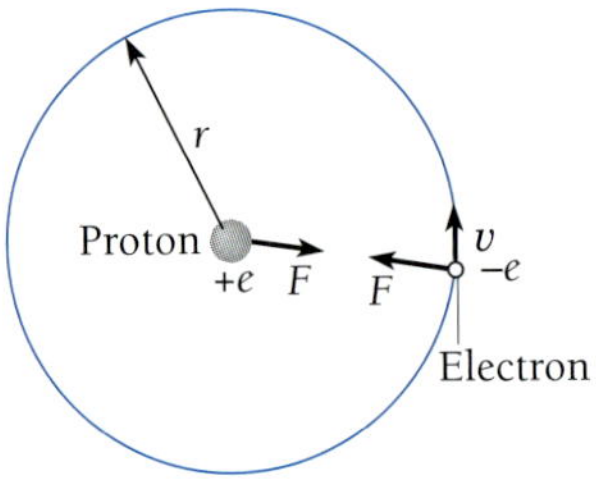

Figure 4.5 Force balance in the hydrogen atom.

Let us look at the classical dynamics of the hydrogen atom, whose single electron makes it the simplest of all atoms. We assume a circular electron orbit for convenience, though it might as reasonably be assumed to be elliptical in shape. The centripetal force

$$F_c = \frac{mv^2}{r}$$

holding the electron in an orbit r from the nucleus is provided by the electric force

$$F_e = \frac{1}{4\pi\epsilon_0}\frac{e^2}{r^2}$$

between them. The condition for a dynamically stable orbit is

$$F_c = F_e$$

$$\frac{mv^2}{r} = \frac{1}{4\pi\epsilon_0}\frac{e^2}{r^2} \tag{4.3}$$

The electron velocity v is therefore related to its orbit radius r by the formula

Electron velocity

$$v = \frac{e}{\sqrt{4\pi\epsilon_0 mr}} \tag{4.4}$$

The total energy E of the electron in a hydrogen atom is the sum of its kinetic and potential energies, which are

$$\text{KE} = \frac{1}{2}mv^2 \qquad \text{PE} = -\frac{e^2}{4\pi\epsilon_0 r}$$

(The minus sign follows from the choice of PE $= 0$ at $r = \infty$, that is, when the electron and proton are infinitely far apart.) Hence

$$E = \text{KE} + \text{PE} = \frac{mv^2}{2} - \frac{e^2}{4\pi\epsilon_0 r}$$

Substituting for v from Eq. (4.4) gives

$$E = \frac{e^2}{8\pi\epsilon_0 r} - \frac{e^2}{4\pi\epsilon_0 r}$$

Total energy of hydrogen atom

$$E = -\frac{e^2}{8\pi\epsilon_0 r} \tag{4.5}$$

The total energy of the electron is negative. This holds for every atomic electron and reflects the fact that it is bound to the nucleus. If E were greater than zero, an electron would not follow a closed orbit around the nucleus.

Actually, of course, the energy E is not a property of the electron alone but is a property of the system of electron + nucleus. The effect of the sharing of E between the electron and the nucleus is considered in Sec. 4.7.

Example 4.1

Experiments indicate that 13.6 eV is required to separate a hydrogen atom into a proton and an electron; that is, its total energy is $E = -13.6$ eV. Find the orbital radius and velocity of the electron in a hydrogen atom.

Solution

Since 13.6 eV = 2.2×10^{-18} J, from Eq. (4.5)

$$r = -\frac{e^2}{8\pi\epsilon_0 E} = -\frac{(1.6 \times 10^{-19}\ \text{C})^2}{(8\pi)(8.85 \times 10^{-12}\ \text{F/m})(-2.2 \times 10^{-18}\ \text{J})}$$
$$= 5.3 \times 10^{-11}\ \text{m}$$

An atomic radius of this magnitude agrees with estimates made in other ways. The electron's velocity can be found from Eq. (4.4):

$$v = \frac{e}{\sqrt{4\pi\epsilon_0 mr}} = \frac{1.6 \times 10^{-19}\ \text{C}}{\sqrt{(4\pi)(8.85 \times 10^{-12}\ \text{F/m})(9.1 \times 10^{-31}\ \text{kg})(5.3 \times 10^{-11}\ \text{m})}}$$
$$= 2.2 \times 10^6\ \text{m/s}$$

Since $v \ll c$, we can ignore special relativity when considering the hydrogen atom.

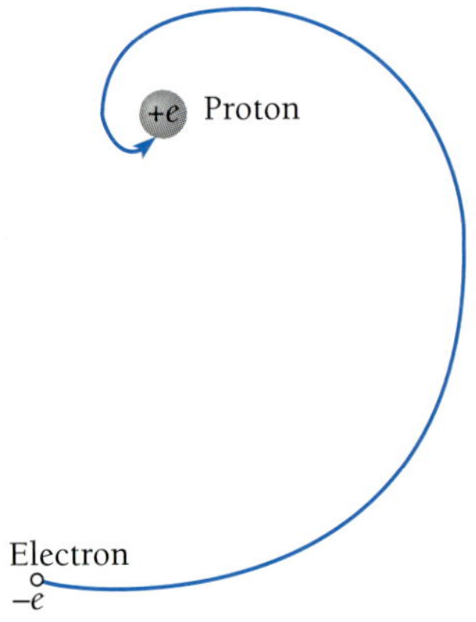

Figure 4.6 An atomic electron should, classically, spiral rapidly into the nucleus as it radiates energy due to its acceleration.

The Failure of Classical Physics

The analysis above is a straightforward application of Newton's laws of motion and Coulomb's law of electric force—both pillars of classical physics—and is in accord with the experimental observation that atoms are stable. However, it is *not* in accord with electromagnetic theory—another pillar of classical physics—which predicts that accelerated electric charges radiate energy in the form of em waves. An electron pursuing a curved path is accelerated and therefore should continuously lose energy, spiraling into the nucleus in a fraction of a second (Fig. 4.6).

But atoms do not collapse. This contradiction further illustrates what we saw in the previous two chapters: The laws of physics that are valid in the macroworld do not always hold true in the microworld of the atom.

Is Rutherford's Analysis Valid?

An interesting question comes up at this point. When he derived his scattering formula, Rutherford used the same laws of physics that prove such dismal failures when applied to atomic stability. Might it not be that this formula is not correct and that in reality the atom does not resemble Rutherford's model of a small central nucleus surrounded by distant electrons? This is not a trivial point. It is a curious coincidence that the quantum-mechanical analysis of alpha particle scattering by thin foils yields precisely the same formula that Rutherford found.

To verify that a classical calculation ought to be at least approximately correct, we note that the de Broglie wavelength of an alpha particle whose speed is 2.0×10^7 m/s is

$$\lambda = \frac{h}{mv} = \frac{6.63 \times 10^{-34}\ \text{J}\cdot\text{s}}{(6.6 \times 10^{-27}\ \text{kg})(2.0 \times 10^7\ \text{m/s})}$$
$$= 5.0 \times 10^{-15}\ \text{m}$$

As we saw in Sec. 4.1, the closest an alpha particle with this wavelength ever gets to a gold nucleus is 3.0×10^{-14} m, which is six de Broglie wavelengths. It is therefore just reasonable to regard the alpha particle as a classical particle in the interaction. We are correct in thinking of the atom in terms of Rutherford's model, though the dynamics of the atomic electrons—which is another matter—requires a nonclassical approach.

Classical physics fails to provide a meaningful analysis of atomic structure because it approaches nature in terms of "pure" particles and "pure" waves. In reality particles and waves have many properties in common, though the smallness of Planck's constant makes the wave-particle duality imperceptible in the macroworld. The usefulness of classical physics decreases as the scale of the phenomena under study decreases, and we must allow for the particle behavior of waves and the wave behavior of particles to understand the atom. In the rest of this chapter we shall see how the Bohr atomic model, which combines classical and modern notions, accomplishes part of the latter task. Not until we consider the atom from the point of view of quantum mechanics, which makes no compromise with the intuitive notions we pick up in our daily lives, will we find a really successful theory of the atom.

4.3 ATOMIC SPECTRA

Each element has a characteristic line spectrum

Atomic stability is not the only thing that a successful theory of the atom must account for. The existence of spectral lines is another important aspect of the atom that finds no explanation in classical physics.

We saw in Chap. 2 that condensed matter (solids and liquids) at all temperatures emits em radiation in which all wavelengths are present, though with different intensities. The observed features of this radiation were explained by Planck without reference to exactly how it was produced by the radiating material or to the nature of the material. From this it follows that we are witnessing the collective behavior of a great many interacting atoms rather than the characteristic behavior of the atoms of a particular element.

At the other extreme, the atoms or molecules in a rarefied gas are so far apart on the average that they only interact during occasional collisions. Under these circumstances we would expect any emitted radiation to be characteristic of the particular atoms or molecules present, which turns out to be the case.

When an atomic gas or vapor at somewhat less than atmospheric pressure is suitably "excited," usually by passing an electric current through it, the emitted radiation has a spectrum which contains certain specific wavelengths only. An idealized arrangement for observing such atomic spectra is shown in Fig. 4.7; actual spectrometers use diffraction

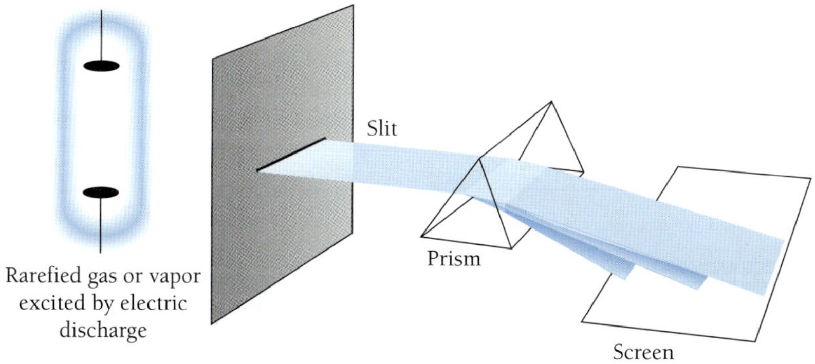

Figure 4.7 An idealized spectrometer.

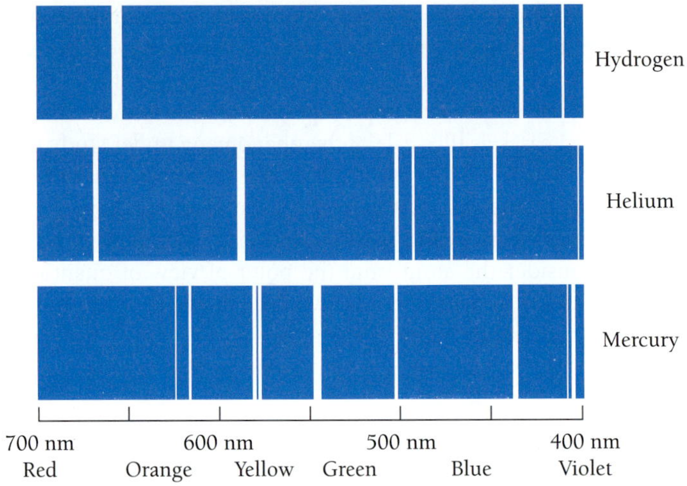

Figure 4.8 Some of the principal lines in the emission spectra of hydrogen, helium, and mercury.

Gas atoms excited by electric currents in these tubes radiate light of wavelengths characteristic of the gas used.

gratings. Figure 4.8 shows the **emission line spectra** of several elements. Every element displays a unique line spectrum when a sample of it in the vapor phase is excited. Spectroscopy is therefore a useful tool for analyzing the composition of an unknown substance.

When white light is passed through a gas, the gas is found to absorb light of certain wavelengths present in its emission spectrum. The resulting **absorption line spectrum** consists of a bright background crossed by dark lines that correspond to the missing wavelengths (Fig. 4.9); emission spectra consist of bright lines on a dark background. The spectrum of sunlight has dark lines in it because the luminous part

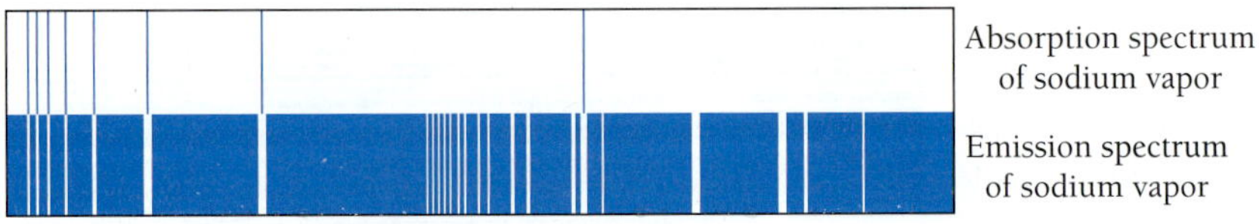

Figure 4.9 The dark lines in the absorption spectrum of an element correspond to bright lines in its emission spectrum.

of the sun, which radiates very nearly like a blackbody heated to 5800 K, is surrounded by an envelope of cooler gas that absorbs light of certain wavelengths only. Most other stars have spectra of this kind.

The number, intensity, and exact wavelengths of the lines in the spectrum of an element depend upon temperature, pressure, the presence of electric and magnetic fields, and the motion of the source. It is possible to tell by examining its spectrum not only what elements are present in a light source but much about their physical state. An astronomer, for example, can establish from the spectrum of a star which elements its atmosphere contains, whether they are ionized, and whether the star is moving toward or away from the earth.

Spectral Series

A century ago the wavelengths in the spectrum of an element were found to fall into sets called **spectral series.** The first such series was discovered by J. J. Balmer in 1885 in the course of a study of the visible part of the hydrogen spectrum. Figure 4.10 shows the **Balmer series.** The line with the longest wavelength, 656.3 nm, is designated H_α, the next, whose wavelength is 486.3 nm, is designated H_β, and so on. As the wave-length decreases, the lines are found closer together and weaker in intensity until the **series limit** at 364.6 nm is reached, beyond which there are no further separate lines but only a faint continuous spectrum. Balmer's formula for the wavelengths of this series is

Balmer
$$\frac{1}{\lambda} = R\left(\frac{1}{2^2} - \frac{1}{n^2}\right) \qquad n = 3, 4, 5, \ldots \tag{4.6}$$

The quantity R, known as the **Rydberg constant,** has the value

Rydberg constant
$$R = 1.097 \times 10^7 \text{ m}^{-1} = 0.01097 \text{ nm}^{-1}$$

The H_α line corresponds to $n = 3$, the H_β line to $n = 4$, and so on. The series limit corresponds to $n = \infty$, so that it occurs at a wavelength of $4/R$, in agreement with experiment.

The Balmer series contains wavelengths in the visible portion of the hydrogen spectrum. The spectral lines of hydrogen in the ultraviolet and infrared regions fall into several other series. In the ultraviolet, the **Lyman series** contains the wavelengths given by the formula

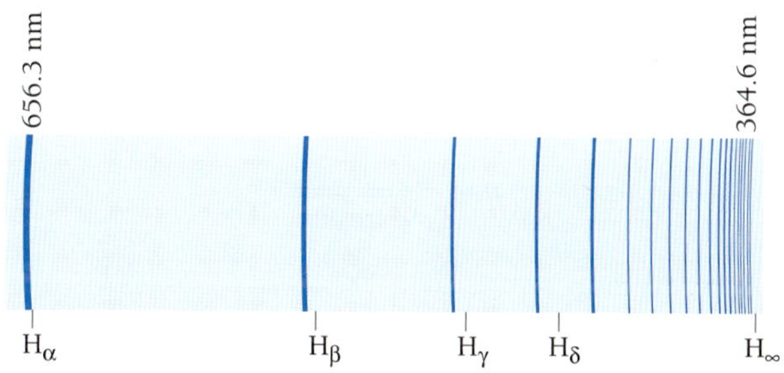

Figure 4.10 The Balmer series of hydrogen. The H_α line is red, the H_β line is blue, the H_γ and H_δ lines are violet, and the other lines are in the near ultraviolet.

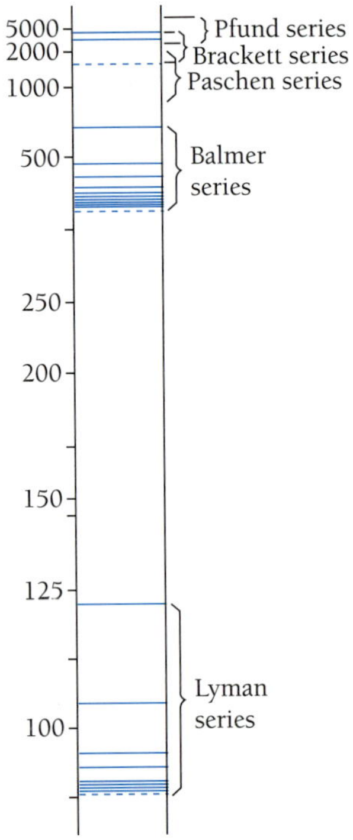

Figure 4.11 The spectral series of hydrogen. The wavelengths in each series are related by simple formulas.

Lyman $$\frac{1}{\lambda} = R\left(\frac{1}{1^2} - \frac{1}{n^2}\right) \qquad n = 2, 3, 4, \ldots \tag{4.7}$$

In the infrared, three spectral series have been found whose lines have the wavelengths specified by the formulas

Paschen $$\frac{1}{\lambda} = R\left(\frac{1}{3^2} - \frac{1}{n^2}\right) \qquad n = 4, 5, 6, \ldots \tag{4.8}$$

Brackett $$\frac{1}{\lambda} = R\left(\frac{1}{4^2} - \frac{1}{n^2}\right) \qquad n = 5, 6, 7, \ldots \tag{4.9}$$

Pfund $$\frac{1}{\lambda} = R\left(\frac{1}{5^2} - \frac{1}{n^2}\right) \qquad n = 6, 7, 8, \ldots \tag{4.10}$$

These spectral series of hydrogen are plotted in terms of wavelength in Fig. 4.11; the Brackett series evidently overlaps the Paschen and Pfund series. The value of R is the same in Eqs. (4.6) to (4.10).

These observed regularities in the hydrogen spectrum, together with similar regularities in the spectra of more complex elements, pose a definitive test for any theory of atomic structure.

4.4 THE BOHR ATOM

Electron waves in the atom

The first theory of the atom to meet with any success was put forward in 1913 by Niels Bohr. The concept of matter waves leads in a natural way to this theory, as de Broglie found, and this is the route that will be followed here. Bohr himself used a different approach, since de Broglie's work came a decade later, which makes his achievement all the more remarkable. The results are exactly the same, however.

We start by examining the wave behavior of an electron in orbit around a hydrogen nucleus. (In this chapter, since the electron velocities are much smaller than c, we will assume that $\gamma = 1$ and for simplicity omit γ from the various equations.) The de Broglie wavelength of this electron is

$$\lambda = \frac{h}{mv}$$

where the electron velocity v is that given by Eq. (4.4):

$$v = \frac{e}{\sqrt{4\pi\epsilon_0 mr}}$$

Hence

Orbital electron wavelength $$\lambda = \frac{h}{e}\sqrt{\frac{4\pi\epsilon_0 r}{m}} \tag{4.11}$$

By substituting 5.3×10^{-11} m for the radius r of the electron orbit (see Example 4.1), we find the electron wavelength to be

$$\lambda = \frac{6.63 \times 10^{-34}\ \text{J}\cdot\text{s}}{1.6 \times 10^{-19}\text{C}} \sqrt{\frac{(4\pi)(8.85 \times 10^{-12}\ \text{C}^2/\text{N}\cdot\text{m}^2)(5.3 \times 10^{-11}\text{m})}{9.1 \times 10^{-31}\ \text{kg}}}$$
$$= 33 \times 10^{-11}\ \text{m}$$

This wavelength is exactly the same as the circumference of the electron orbit,

$$2\pi r = 33 \times 10^{-11}\ \text{m}$$

The orbit of the electron in a hydrogen atom corresponds to one complete electron wave joined on itself (Fig. 4.12)!

The fact that the electron orbit in a hydrogen atom is one electron wavelength in circumference provides the clue we need to construct a theory of the atom. If we consider the vibrations of a wire loop (Fig. 4.13), we find that their wavelengths always fit an integral number of times into the loop's circumference so that each wave joins smoothly with the next. If the wire were perfectly elastic, these vibrations would continue indefinitely. Why are these the only vibrations possible in a wire loop? If a fractional number of wavelengths is placed around the loop, as in Fig. 4.14, destructive

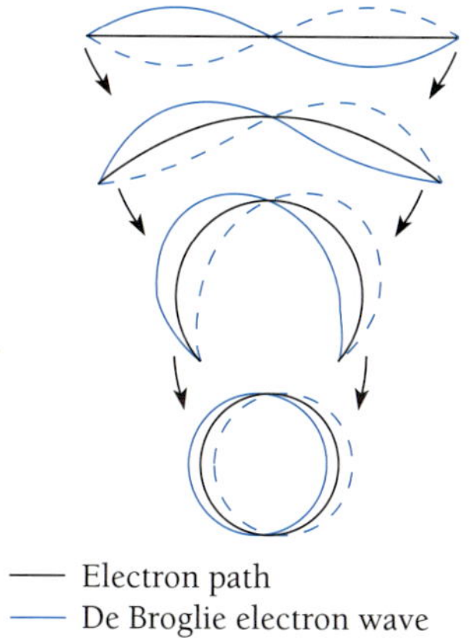

Figure 4.12 The orbit of the electron in a hydrogen atom corresponds to a complete electron de Broglie wave joined on itself.

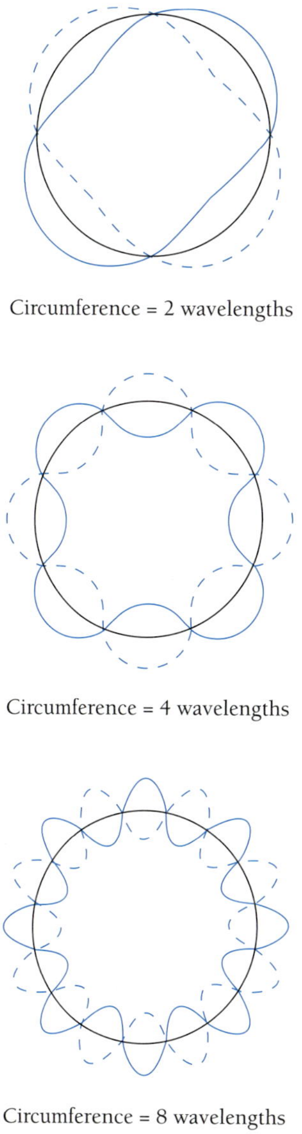

Figure 4.13 Some modes of vibration of a wire loop. In each case a whole number of wavelengths fit into the circumference of the loop.

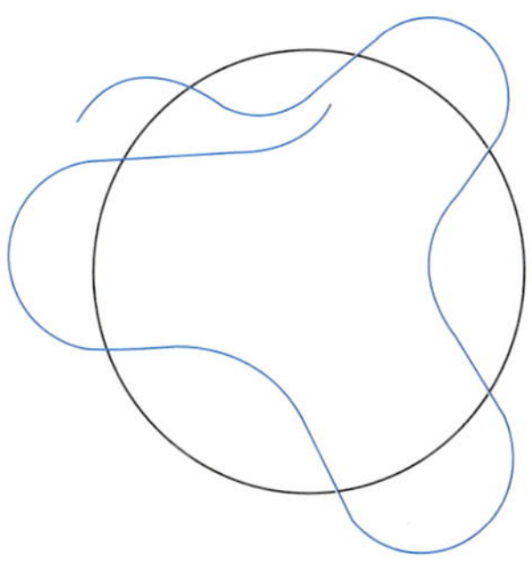

Figure 4.14 A fractional number of wavelengths cannot persist because destructive interference will occur.

About Scientists

Niels Bohr (1884–1962) was born and spent most of his life in Copenhagen, Denmark. After receiving his doctorate at the university there in 1911, Bohr went to England to broaden his scientific horizons. At Rutherford's laboratory in Manchester, Bohr was introduced to the just-discovered nuclear model of the atom, which was in conflict with the existing principles of physics. Bohr realized that it was "hopeless" to try to make sense of the atom in the framework of classical physics alone, and he felt that the quantum theory of light must somehow be the key to understanding atomic structure.

Back in Copenhagen in 1913, a friend suggested to Bohr that Balmer's formula for one set of the spectral lines of hydrogen might be relevant to his quest. "As soon as I saw Balmer's formula the whole thing was immediately clear to me," Bohr said later. To construct his theory, Bohr began with two revolutionary ideas. The first was that an atomic electron can circle its nucleus only in certain orbits, and the other was that an atom emits or absorbs a photon of light when an electron jumps from one permitted orbit to another.

What is the condition for a permitted orbit? To find out, Bohr used as a guide what became known as the correspondence principle: When quantum numbers are very large, quantum effects should not be conspicuous, and the quantum theory must then give the same results as classical physics. Applying this principle showed that the electron in a permitted orbit must have an angular momentum that is a multiple of $\hbar = h/2\pi$. A decade later Louis de Broglie explained this quantization of angular momentum in terms of the wave nature of a moving electron.

Bohr was able to account for all the spectral series of hydrogen, not just the Balmer series, but the publication of the theory aroused great controversy. Einstein, an enthusiastic supporter of the theory (which "appeared to me like a miracle—and appears to me as a miracle even today," he wrote many years later), nevertheless commented on its bold mix of classical and quantum concepts, "One ought to be ashamed of the successes [of the theory] because they have been earned according to the Jesuit maxim, 'Let not thy left hand know what the other doeth.'" Other noted physicists were more deeply disturbed: Otto Stern and Max von Laue said they would quit physics if Bohr were right. (They later changed their minds.) Bohr and others tried to extend his model to many-electron atoms with occasional success—for instance, the correct prediction of the properties of the then-unknown element hafnium—but real progress had to wait for Wolfgang Pauli's exclusion principle of 1925.

In 1916 Bohr returned to Rutherford's laboratory, where he stayed until 1919. Then the Institute of Theoretical Physics was created for him in Copenhagen, and he directed it until his death. The institute was a magnet for quantum theoreticians from all over the world, who were stimulated by the exchange of ideas at regular meetings there. Bohr received the Nobel Prize in 1922. His last important work came in 1939, when he used an analogy between a large nucleus and a liquid drop to explain why nuclear fission, which had just been discovered, occurs in certain nuclei but not in others. During World War II Bohr contributed to the development of the atomic bomb at Los Alamos, New Mexico. After the war, Bohr returned to Copenhagen, where he died in 1962.

interference will occur as the waves travel around the loop, and the vibrations will die out rapidly.

By considering the behavior of electron waves in the hydrogen atom as analogous to the vibrations of a wire loop, then, we can say that

> An electron can circle a nucleus only if its orbit contains an integral number of de Broglie wavelengths.

This statement combines both the particle and wave characters of the electron since the electron wavelength depends upon the orbital velocity needed to balance the pull of the nucleus. To be sure, the analogy between an atomic electron and the standing waves of Fig. 4.13 is hardly the last word on the subject, but it represents an illuminating step along the path to the more profound and comprehensive, but also more abstract, quantum-mechanical theory of the atom.

It is easy to express the condition that an electron orbit contain an integral number of de Broglie wavelengths. The circumference of a circular orbit of radius r is $2\pi r$, and so the condition for orbit stability is

Condition for orbit stability

$$n\lambda = 2\pi r_n \qquad n = 1, 2, 3, \ldots \tag{4.12}$$

where r_n designates the radius of the orbit that contain n wavelengths. The integer n is called the **quantum number** of the orbit. Substituting for λ, the electron wavelength given by Eq. (4.11), yields

$$\frac{nh}{e}\sqrt{\frac{4\pi\epsilon_0 r_n}{m}} = 2\pi r_n$$

and so the possible electron orbits are those whose radii are given by

Orbital radii in Bohr atom

$$r_n = \frac{n^2 h^2 \epsilon_0}{\pi m e^2} \qquad n = 1, 2, 3, \ldots \tag{4.13}$$

The radius of the innermost orbit is customarily called the **Bohr radius** of the hydrogen atom and is denoted by the symbol a_0:

Bohr radius

$$a_0 = r_1 = 5.292 \times 10^{-11}\ \text{m}$$

The other radii are given in terms of a_0 by the formula

$$r_n = n^2 a_0 \tag{4.14}$$

4.5 ENERGY LEVELS AND SPECTRA

A photon is emitted when an electron jumps from one energy level to a lower level

The various permitted orbits involve different electron energies. The electron energy E_n is given in terms of the orbit radius r_n by Eq. (4.5) as

$$E_n = -\frac{e^2}{8\pi\epsilon_0 r_n}$$

Substituting for r_n from Eq (4.13), we see that

Energy levels

$$E_n = -\frac{me^4}{8\epsilon_0^2 h^2}\left(\frac{1}{n^2}\right) = \frac{E_1}{n^2} \qquad n = 1, 2, 3, \ldots \tag{4.15}$$

$$E_1 = -2.18 \times 10^{-18}\ \text{J} = -13.6\ \text{eV}$$

The energies specified by Eq. (4.15) are called the **energy levels** of the hydrogen atom and are plotted in Fig. 4.15. These levels are all negative, which signifies that the electron does not have enough energy to escape from the nucleus. An atomic electron can have only these energies and no others. An analogy might be a person on a ladder, who can stand only on its steps and not in between.

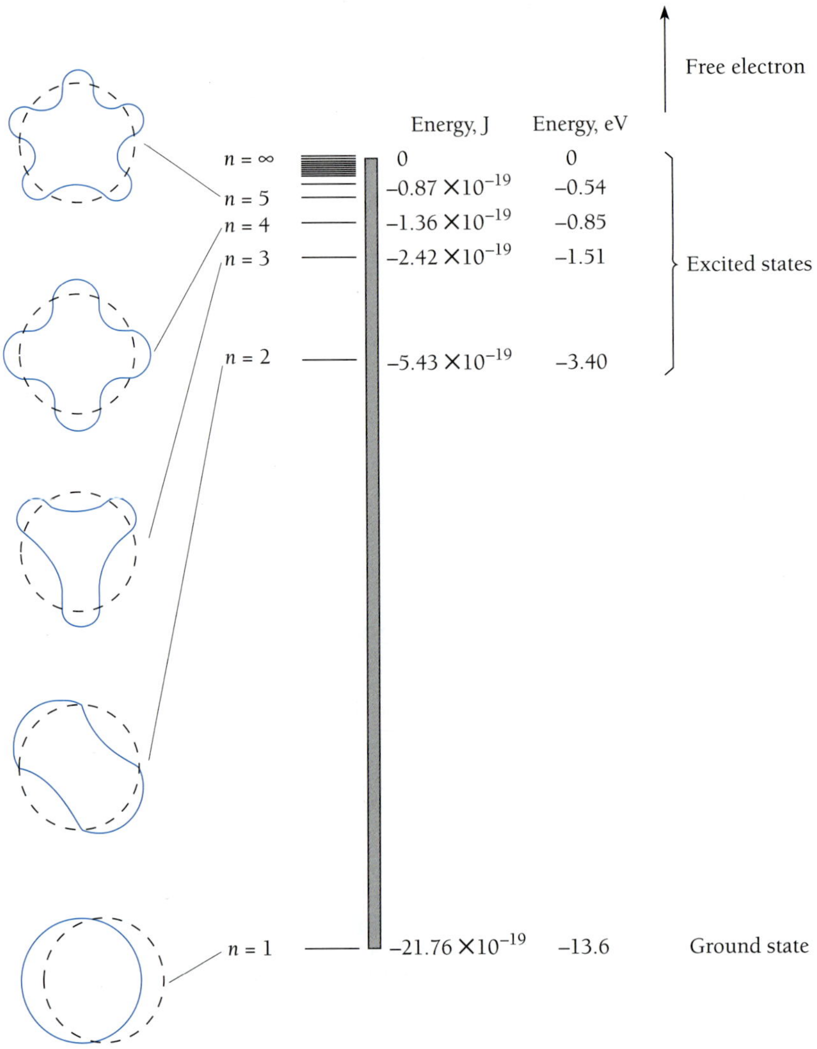

Figure 4.15 Energy levels of the hydrogen atom.

The lowest energy level E_1 is called the **ground state** of the atom, and the higher levels E_2, E_3, E_4, . . . are called **excited states.** As the quantum number n increases, the corresponding energy E_n approaches closer to 0. In the limit of $n = \infty$, $E_\infty = 0$ and the electron is no longer bound to the nucleus to form an atom. A positive energy for a nucleus-electron combination means that the electron is free and has no quantum conditions to fulfill; such a combination does not constitute an atom, of course.

The work needed to remove an electron from an atom in its ground state is called its **ionization energy.** The ionization energy is accordingly equal to $-E_1$, the energy that must be provided to raise an electron from its ground state to an energy of $E = 0$, when it is free. In the case of hydrogen, the ionization energy is 13.6 eV since the ground-state energy of the hydrogen atom is -13.6 eV. Figure 7.10 shows the ionization energies of the elements.

Example 4.2

An electron collides with a hydrogen atom in its ground state and excites it to a state of $n = 3$. How much energy was given to the hydrogen atom in this inelastic (KE not conserved) collision?

Solution

From Eq. (4.15) the energy change of a hydrogen atom that goes from an initial state of quantum number n_i to a final state of quantum number n_f is

$$\Delta E = E_f - E_i = \frac{E_1}{n_f^2} - \frac{E_1}{n_i^2} = E_1\left(\frac{1}{n_f^2} - \frac{1}{n_i^2}\right)$$

Here $n_i = 1$, $n_f = 3$, and $E_1 = -13.6$ eV, so

$$\Delta E = -13.6\left(\frac{1}{3^2} - \frac{1}{1^2}\right) \text{ eV} = 12.1 \text{ eV}$$

Example 4.3

Hydrogen atoms in states of high quantum number have been created in the laboratory and observed in space. They are called **Rydberg atoms.** (*a*) Find the quantum number of the Bohr orbit in a hydrogen atom whose radius is 0.0100 mm. (*b*) What is the energy of a hydrogen atom in this state?

Solution

(*a*) From Eq. (4.14) with $r_n = 1.00 \times 10^{-5}$ m,

$$n = \sqrt{\frac{r_n}{a_0}} = \sqrt{\frac{1.00 \times 10^{-5} \text{ m}}{5.29 \times 10^{-11} \text{ m}}} = 435$$

(*b*) From Eq. (4.15),

$$E_n = \frac{E_1}{n^2} = \frac{-13.6 \text{ eV}}{(435)^2} = -7.19 \times 10^{-5} \text{ eV}$$

Rydberg atoms are obviously extremely fragile and are easily ionized, which is why they are found in nature only in the near-vacuum of space. The spectra of Rydberg atoms range down to radio frequencies and their existence was established from radio telescope data.

Origin of Line Spectra

We must now confront the equations developed above with experimentation. An especially striking observation is that atoms exhibit line spectra in both emission and absorption. Do such spectra follow from our model?

The presence of discrete energy levels in the hydrogen atom suggests the connection. Let us suppose that when an electron in an excited state drops to a lower state, the lost energy is emitted as a single photon of light. According to our model, electrons cannot exist in an atom except in certain specific energy levels. The jump of an electron from one level to another, with the difference in energy between the levels being given off all at once in a photon rather than in some more gradual manner, fits in well with this model.

Quantization in the Atomic World

Sequences of energy levels are characteristic of all atoms, not just those of hydrogen. As in the case of a particle in a box, the confinement of an electron to a region of space leads to restrictions on its possible wave functions that in turn limit the possible energies to well-defined values only. The existence of atomic energy levels is a further example of the quantization, or graininess, of physical quantities on a microscopic scale.

In the world of our daily lives, matter, electric charge, energy, and so forth appear to be continuous. In the world of the atom, in contrast, matter is composed of elementary particles that have definite rest masses, charge always comes in multiples of $+e$ or $-e$, electromagnetic waves of frequency ν appear as streams of photons each with the energy $h\nu$, and stable systems of particles, such as atoms, can possess only certain energies. As we shall find, other quantities in nature are also quantized, and this quantization enters every aspect of how electrons, protons, and neutrons interact to endow the matter around us (and of which we consist) with its familiar properties.

If the quantum number of the initial (higher-energy) state is n_i and the quantum number of the final (lower-energy) state is n_f, we are asserting that

$$\text{Initial energy} - \text{final energy} = \text{photon energy}$$

$$E_i - E_f = h\nu \tag{4.16}$$

where ν is the frequency of the emitted photon. From Eq. (4.15) we have

$$E_i - E_f = E_1\left(\frac{1}{n_i^2} - \frac{1}{n_f^2}\right) = -E_1\left(\frac{1}{n_f^2} - \frac{1}{n_i^2}\right)$$

We recall that E_1 is a negative quantity (-13.6 eV, in fact), so $-E_1$ is a positive quantity. The frequency of the photon released in this transition is therefore

$$\nu = \frac{E_i - E_f}{h} = -\frac{E_1}{h}\left(\frac{1}{n_f^2} - \frac{1}{n_i^2}\right) \tag{4.17}$$

Since $\lambda = c/\nu$, $1/\lambda = \nu/c$ and

Hydrogen spectrum

$$\frac{1}{\lambda} = -\frac{E_1}{ch}\left(\frac{1}{n_f^2} - \frac{1}{n_i^2}\right) \tag{4.18}$$

Equation (4.18) states that the radiation emitted by excited hydrogen atoms should contain certain wavelengths only. These wavelengths, furthermore, fall into definite sequences that depend upon the quantum number n_f of the final energy level of the electron (Fig. 4.16). Since $n_i > n_f$ in each case, in order that there be an excess of energy to be given off as a photon, the calculated formulas for the first five series are

Lyman $\quad n_f = 1: \quad \dfrac{1}{\lambda} = -\dfrac{E_1}{ch}\left(\dfrac{1}{1^2} - \dfrac{1}{n^2}\right) \quad n = 2, 3, 4, \ldots$

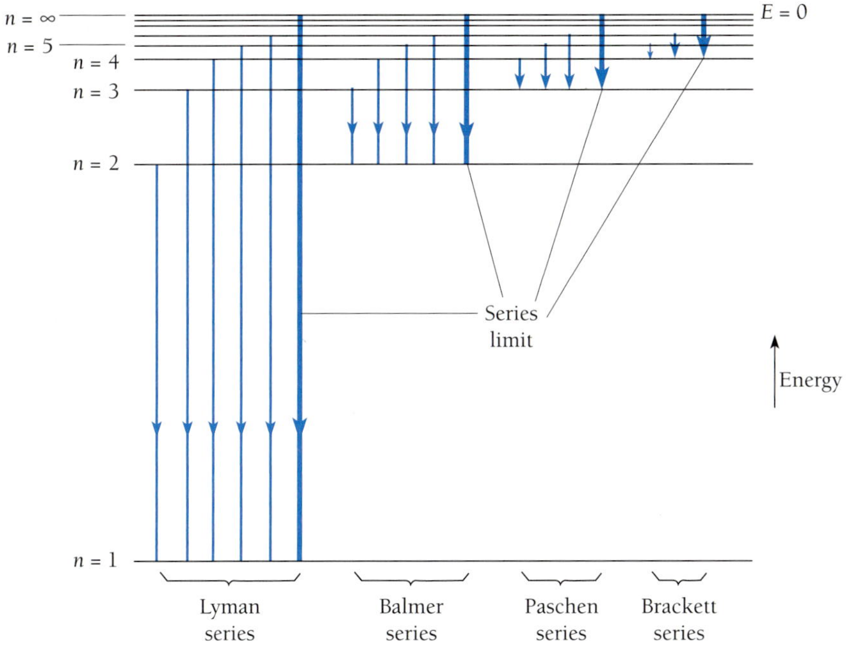

Figure 4.16 Spectral lines originate in transitions between energy levels. Shown are the spectral series of hydrogen. When $n = \infty$, the electron is free.

Balmer $\quad n_f = 2: \quad \frac{1}{\lambda} = -\frac{E_1}{ch}\left(\frac{1}{2^2} - \frac{1}{n^2}\right) \quad n = 3, 4, 5, \ldots$

Paschen $\quad n_f = 3: \quad \frac{1}{\lambda} = -\frac{E_1}{ch}\left(\frac{1}{3^2} - \frac{1}{n^2}\right) \quad n = 4, 5, 6, \ldots$

Brackett $\quad n_f = 4: \quad \frac{1}{\lambda} = -\frac{E_1}{ch}\left(\frac{1}{4^2} - \frac{1}{n^2}\right) \quad n = 5, 6, 7, \ldots$

Pfund $\quad n_f = 5: \quad \frac{1}{\lambda} = -\frac{E_1}{ch}\left(\frac{1}{5^2} - \frac{1}{n^2}\right) \quad n = 6, 7, 8, \ldots$

These sequences are identical in form with the empirical spectral series discussed earlier. The Lyman series corresponds to $n_f = 1$; the Balmer series corresponds to $n_f = 2$; the Paschen series corresponds to $n_f = 3$; the Brackett series corresponds to $n_f = 4$; and the Pfund series corresponds to $n_f = 5$.

Our final step is to compare the value of the constant term in the above equations with that of the Rydberg constant in Eqs. (4.6) to (4.10). The value of the constant term is

$$
\begin{aligned}
-\frac{E_1}{ch} &= \frac{me^4}{8\epsilon_0^2 ch^3} \\
&= \frac{(9.109 \times 10^{-31}\ \text{kg})(1.602 \times 10^{-19}\ \text{C})^4}{(8)(8.854 \times 10^{-12}\ \text{C}^2/\text{N} \cdot \text{m}^2)(2.998 \times 10^8\ \text{m/s})(6.626 \times 10^{-34}\ \text{J} \cdot \text{s})^3} \\
&= 1.097 \times 10^7\ \text{m}^{-1}
\end{aligned}
$$

which is indeed the same as R. Bohr's model of the hydrogen atom is therefore in accord with the spectral data.

Example 4.4

Find the longest wavelength present in the Balmer series of hydrogen, corresponding to the H_α line.

Solution

In the Balmer series the quantum number of the final state is $n_f = 2$. The longest wavelength in this series corresponds to the smallest energy difference between energy levels. Hence the initial state must be $n_i = 3$ and

$$\frac{1}{\lambda} = R\left(\frac{1}{n_f^2} - \frac{1}{n_i^2}\right) = R\left(\frac{1}{2^2} - \frac{1}{3^2}\right) = 0.139R$$

$$\lambda = \frac{1}{0.139R} = \frac{1}{0.139(1.097 \times 10^7\text{m}^{-1})} = 6.56 \times 10^{-7}\text{m} = 656 \text{ nm}$$

This wavelength is near the red end of the visible spectrum.

4.6 CORRESPONDENCE PRINCIPLE

The greater the quantum number, the closer quantum physics approaches classical physics

Quantum physics, so different from classical physics in the microworld beyond reach of our senses, must nevertheless give the same results as classical physics in the macroworld where experiments show that the latter is valid. We have already seen that this basic requirement is true for the wave theory of moving bodies. We shall now find that it is also true for Bohr's model of the hydrogen atom.

According to electromagnetic theory, an electron moving in a circular orbit radiates em waves whose frequencies are equal to its frequency of revolution and to harmonics (that is, integral multiples) of that frequency. In a hydrogen atom the electron's speed is

$$v = \frac{e}{\sqrt{4\pi\epsilon_0 mr}}$$

according to Eq. (4.4), where r is the radius of its orbit. Hence the frequency of revolution f of the electron is

$$f = \frac{\text{electron speed}}{\text{orbit circumference}} = \frac{v}{2\pi r} = \frac{e}{2\pi\sqrt{4\pi\epsilon_0 mr^3}}$$

The radius r_n of a stable orbit is given in terms of its quantum number n by Eq. (4.13) as

$$r_n = \frac{n^2h^2\epsilon_0}{\pi me^2}$$

and so the frequency of revolution is

Frequency of revolution

$$f = \frac{me^4}{8\epsilon_0^2 h^3}\left(\frac{2}{n^3}\right) = \frac{-E_1}{h}\left(\frac{2}{n^3}\right) \tag{4.19}$$

Example 4.5

(*a*) Find the frequencies of revolution of electrons in $n = 1$ and $n = 2$ Bohr orbits. (*b*) What is the frequency of the photon emitted when an electron in an $n = 2$ orbit drops to an $n = 1$ orbit? (*c*) An electron typically spends about 10^{-8} s in an excited state before it drops to a lower state by emitting a photon. How many revolutions does an electron in an $n = 2$ Bohr orbit make in 1.00×10^{-8} s?

Solution

(*a*) From Eq. (4.19),

$$f_1 = \frac{-E_1}{h}\left(\frac{2}{1^3}\right) = \left(\frac{2.18 \times 10^{-18}\text{ J}}{6.63 \times 10^{-34}\text{ J}\cdot\text{s}}\right)(2) = 6.58 \times 10^{15}\text{ rev/s}$$

$$f_2 = \frac{-E_1}{h}\left(\frac{2}{2^3}\right) = \frac{f_1}{8} = 0.823 \times 10^{15}\text{ rev/s}$$

(*b*) From Eq. (4.17),

$$\nu = \frac{-E_1}{h}\left(\frac{1}{n_f^2} - \frac{1}{n_i^2}\right) = \left(\frac{2.18 \times 10^{-18}\text{ J}}{6.63 \times 10^{-34}\text{ J}\cdot\text{s}}\right)\left(\frac{1}{1^2} - \frac{1}{2^2}\right) = 2.47 \times 10^{15}\text{ Hz}$$

This frequency is intermediate between f_1 and f_2.

(*c*) The number of revolutions the electron makes is

$$N = f_2\,\Delta t = (8.23 \times 10^{14}\text{ rev/s})(1.00\times10^{-8}\text{ s}) = 8.23 \times 10^6\text{ rev}$$

The earth takes 8.23 million y to make this many revolutions around the sun.

Under what circumstances should the Bohr atom behave classically? If the electron orbit is so large that we might be able to measure it directly, quantum effects ought not to dominate. An orbit 0.01 mm across, for instance, meets this specification. As we found in Example 4.3, its quantum number is $n = 435$.

What does the Bohr theory predict such an atom will radiate? According to Eq. (4.17), a hydrogen atom dropping from the n_ith energy level to the n_fth energy level emits a photon whose frequency is

$$\nu = \frac{-E_1}{h}\left(\frac{1}{n_f^2} - \frac{1}{n_i^2}\right)$$

Let us write n for the initial quantum number n_i and $n - p$ (where $p = 1, 2, 3, \ldots$) for the final quantum number n_f. With this substitution,

$$\nu = \frac{-E_1}{h}\left[\frac{1}{(n-p)^2} - \frac{1}{n^2}\right] = \frac{-E_1}{h}\left[\frac{2np - p^2}{n^2(n-p)^2}\right]$$

When n_i and n_f are both very large, n is much greater than p, and

$$2np - p^2 \approx 2np$$

$$(n-p)^2 \approx n^2$$

so that

Frequency of photon

$$\nu = \frac{-E_1}{h}\left(\frac{2p}{n^3}\right) \tag{4.20}$$

When $p = 1$, the frequency ν of the radiation is exactly the same as the frequency of rotation f of the orbital electron given in Eq. (4.19). Multiples of this frequency are radiated when $p = 2, 3, 4, \ldots$. Hence both quantum and classical pictures of the hydrogen atom make the same predictions in the limit of very large quantum numbers. When $n = 2$, Eq. (4.19) predicts a radiation frequency that differs from that given by Eq. (4.20) by almost 300 percent. When $n = 10{,}000$, the discrepancy is only about 0.01 percent.

The requirement that quantum physics give the same results as classical physics in the limit of large quantum numbers was called by Bohr the **correspondence principle.** It has played an important role in the development of the quantum theory of matter.

Bohr himself used the correspondence principle in reverse, so to speak, to look for the condition for orbit stability. Starting from Eq. (4.19) he was able to show that stable orbits must have electron orbital angular momenta of

Condition for orbital stability

$$mvr = \frac{nh}{2\pi} \qquad n = 1, 2, 3, \ldots \tag{4.21}$$

Since the de Broglie electron wavelength is $\lambda = h/mv$, Eq. (4.21) is the same as Eq. (4.12), $n\lambda = 2\pi r$, which states that an electron orbit must contain an integral number of wavelengths.

4.7 NUCLEAR MOTION

The nuclear mass affects the wavelengths of spectral lines

Thus far we have been assuming that the hydrogen nucleus (a proton) remains stationary while the orbital electron revolves around it. What must actually happen, of course, is that both nucleus and electron revolve around their common center of mass, which is very close to the nucleus because the nuclear mass is much greater than that of the electron (Fig. 4.17). A system of this kind is equivalent to a single particle of mass m' that revolves around the position of the heavier particle. (This equivalence is

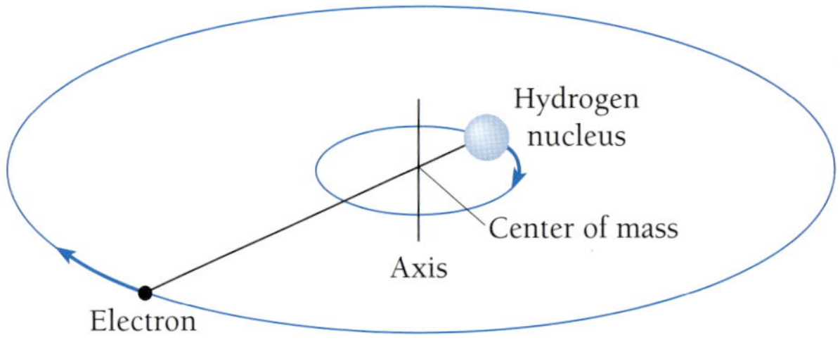

Figure 4.17 Both the electron and nucleus of a hydrogen atom revolve around a common center of mass (not to scale !).

demonstrated in Sec. 8.6.) If m is the electron mass and M the nuclear mass, then m' is given by

Reduced mass

$$m' = \frac{mM}{m + M} \tag{4.22}$$

The quantity m' is called the **reduced mass** of the electron because its value is less than m.

To take into account the motion of the nucleus in the hydrogen atom, then, all we need do is replace the electron with a particle of mass m'. The energy levels of the atom then become

Energy levels corrected for nuclear motion

$$E'_n = -\frac{m'e^4}{8\epsilon_0^2 h^2}\left(\frac{1}{n^2}\right) = \left(\frac{m'}{m}\right)\left(\frac{E_1}{n^2}\right) \tag{4.23}$$

Owing to the motion of the nucleus, all the energy levels of hydrogen are changed by the fraction

$$\frac{m'}{m} = \frac{M}{M + m} = 0.99945$$

This represents an increase of 0.055 percent because the energies E_n, being smaller in absolute value, are therefore less negative.

The use of Eq. (4.23) in place of (4.15) removes a small but definite discrepancy between the predicted wavelengths of the spectral lines of hydrogen and the measured ones. The value of the Rydberg constant R to eight significant figures without correcting for nuclear motion is $1.0973731 \times 10^7\ \text{m}^{-1}$; the correction lowers it to $1.0967758 \times 10^7\ \text{m}^{-1}$.

The notion of reduced mass played an important part in the discovery of **deuterium**, a variety of hydrogen whose atomic mass is almost exactly double that of ordinary hydrogen because its nucleus contains a neutron as well as a proton. About one hydrogen atom in 6000 is a deuterium atom. Because of the greater nuclear mass, the spectral lines of deuterium are all shifted slightly to wavelengths shorter than the corresponding ones of ordinary hydrogen. Thus the H_α line of deuterium, which arises from a transition from the $n = 3$ to the $n = 2$ energy level, occurs at a wavelength of 656.1 nm, whereas the H_α line of hydrogen occurs at 656.3 nm. This difference in wavelength was responsible for the identification of deuterium in 1932 by the American chemist Harold Urey.

Example 4.6

A **positronium** "atom" is a system that consists of a positron and an electron that orbit each other. Compare the wavelengths of the spectral lines of positronium with those of ordinary hydrogen.

Solution

Here the two particles have the same mass m, so the reduced mass is

$$m' = \frac{mM}{m + M} = \frac{m^2}{2m} = \frac{m}{2}$$

where m is the electron mass. From Eq. (4.23) the energy levels of a positronium "atom" are

$$E'_n = \left(\frac{m'}{m}\right)\frac{E_1}{n^2} = \frac{E_1}{2n^2}$$

This means that the Rydberg constant—the constant term in Eq. (4.18)—for positronium is half as large as it is for ordinary hydrogen. As a result the wavelengths in the positronium spectral lines are all twice those of the corresponding lines in the hydrogen spectrum.

Example 4.7

A **muon** is an unstable elementary particle whose mass is $207m_e$ and whose charge is either $+e$ or $-e$. A negative muon (μ^-) can be captured by a nucleus to form a muonic atom. (*a*) A proton captures a μ^-. Find the radius of the first Bohr orbit of this atom. (*b*) Find the ionization energy of the atom.

Solution

(*a*) Here $m = 207m_e$ and $M = 1836m_e$, so the reduced mass is

$$m' = \frac{mM}{m+M} = \frac{(207m_e)(1836m_e)}{207m_e + 1836m_e} = 186m_e$$

According to Eq. (4.13) the orbit radius corresponding to $n = 1$ is

$$r_1 = \frac{h^2\epsilon_0}{\pi m_e e^2}$$

where $r_1 = a_0 = 5.29 \times 10^{-11}$ m. Hence the radius r' that corresponds to the reduced mass m' is

$$r'_1 = \left(\frac{m}{m'}\right) r_1 = \left(\frac{m_e}{186m_e}\right) a_0 = 2.85 \times 10^{-13}\ \text{m}$$

The muon is 186 times closer to the proton than an electron would be, so a muonic hydrogen atom is much smaller than an ordinary hydrogen atom.

(*b*) From Eq. (4.23) we have, with $n = 1$ and $E_1 = -13.6$ eV,

$$E'_1 = \left(\frac{m'}{m}\right) E_1 = 186E_1 = -2.53 \times 10^3\ \text{eV} = -2.53\ \text{keV}$$

The ionization energy is therefore 2.53 keV, 186 times that for an ordinary hydrogen atom.

4.8 ATOMIC EXCITATION

How atoms absorb and emit energy

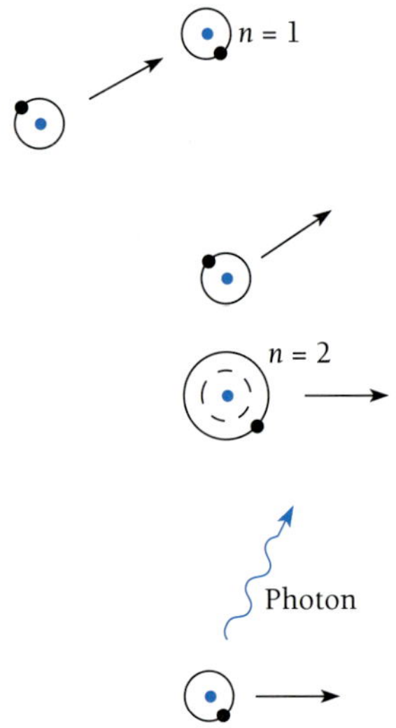

Figure 4.18 Excitation by collision. Some of the available energy is absorbed by one of the atoms, which goes into an excited energy state. The atom then emits a photon in returning to its ground (normal) state.

There are two main ways in which an atom can be excited to an energy above its ground state and thereby become able to radiate. One of these ways is by a collision with another particle in which part of their joint kinetic energy is absorbed by the atom. Such an excited atom will return to its ground state in an average of 10^{-8} s by emitting one or more photons (Fig. 4.18).

To produce a luminous discharge in a rarefied gas, an electric field is established that accelerates electrons and atomic ions until their kinetic energies are sufficient to

Auroras are caused by streams of fast protons and electrons from the sun that excite atoms in the upper atmosphere. The green hues of an auroral display come from oxygen, and the reds originate in both oxygen and nitrogen.

excite atoms they collide with. Because energy transfer is at maximum when the colliding particles have the same mass (see Fig. 12.22), the electrons in such a discharge are more effective than the ions in providing energy to atomic electrons. Neon signs and mercury-vapor lamps are familiar examples of how a strong electric field applied between electrodes in a gas-filled tube leads to the emission of the characteristic spectral radiation of that gas, which happens to be reddish light in the case of neon and bluish light in the case of mercury vapor.

Another excitation mechanism is involved when an atom absorbs a photon of light whose energy is just the right amount to raise the atom to a higher energy level. For example, a photon of wavelength 121.7 nm is emitted when a hydrogen atom in the $n = 2$ state drops to the $n = 1$ state. Absorbing a photon of wavelength 121.7 nm by a hydrogen atom initially in the $n = 1$ state will therefore bring it up to the $n = 2$ state (Fig. 4.19). This process explains the origin of absorption spectra.

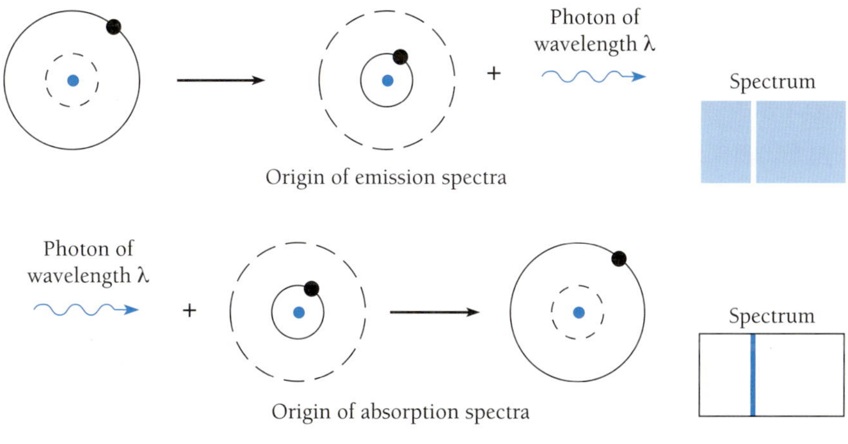

Figure 4.19 How emission and absorption spectral lines originate.

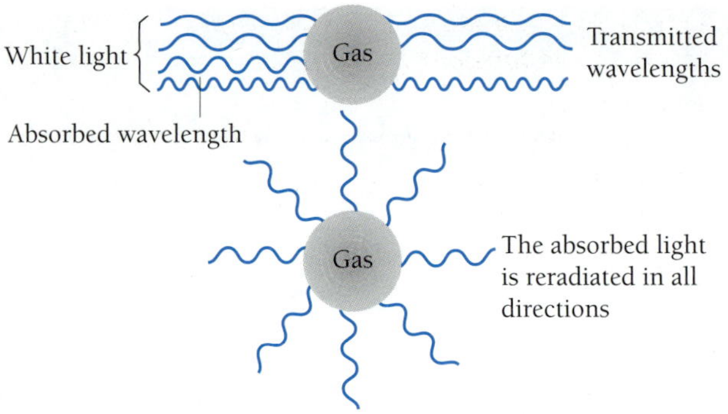

Figure 4.20 The dark lines in an absorption spectrum are never totally dark.

When white light, which contains all wavelengths, is passed through hydrogen gas, photons of those wavelengths that correspond to transitions between energy levels are absorbed. The resulting excited hydrogen atoms reradiate their excitation energy almost at once, but these photons come off in random directions with only a few in the same direction as the original beam of white light (Fig. 4.20). The dark lines in an absorption spectrum are therefore never completely black but only appear so by contrast with the bright background. We expect the lines in the absorption spectrum of any element to coincide with those in its emission spectrum that represent transitions to the ground state, which agrees with observation (see Fig. 4.9).

Franck-Hertz Experiment

Atomic spectra are not the only way to investigate energy levels inside atoms. A series of experiments based on excitation by collision was performed by James Franck and Gustav Hertz (a nephew of Heinrich Hertz) starting in 1914. These experiments demonstrated that atomic energy levels indeed exist and, furthermore, that the ones found in this way are the same as those suggested by line spectra.

Franck and Hertz bombarded the vapors of various elements with electrons of known energy, using an apparatus like that shown in Fig. 4.21. A small potential difference V_0 between the grid and collecting plate prevents electrons having energies less than a certain minimum from contributing to the current I through the ammeter. As the accelerating potential V is increased, more and more electrons arrive at the plate and I rises (Fig. 4.22).

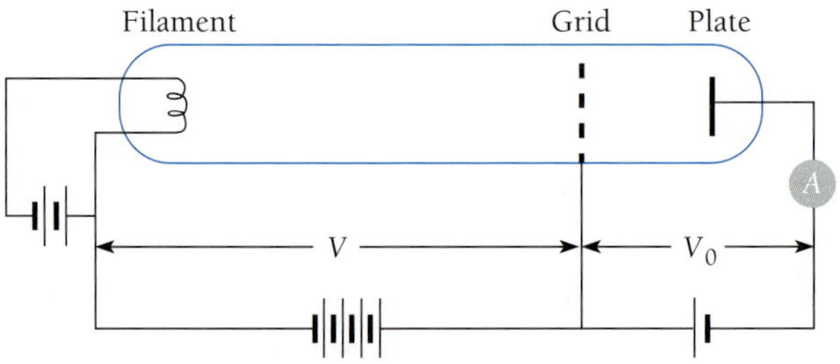

Figure 4.21 Apparatus for the Franck-Hertz experiment.

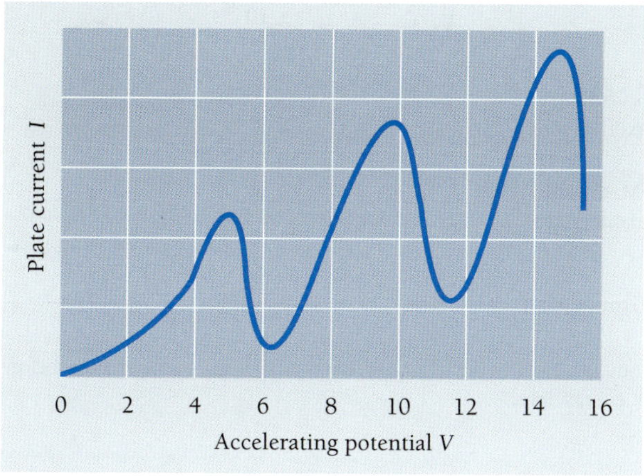

Figure 4.22 Results of the Franck-Hertz experiment, showing critical potentials in mercury vapor.

If KE is conserved when an electron collides with one of the atoms in the vapor, the electron merely bounces off in a new direction. Because an atom is much heavier than an electron, the electron loses almost no KE in the process. After a certain critical energy is reached, however, the plate current drops abruptly. This suggests that an electron colliding with one of the atoms gives up some or all of its KE to excite the atom to an energy level above its ground state. Such a collision is called inelastic, in contrast to an elastic collision in which KE is conserved. The critical electron energy equals the energy needed to raise the atom to its lowest excited state.

Then, as the accelerating potential V is raised further, the plate current again increases, since the electrons now have enough energy left to reach the plate after undergoing an inelastic collision on the way. Eventually another sharp drop in plate current occurs, which arises from the excitation of the same energy level in other atoms by the electrons. As Fig. 4.22 shows, a series of critical potentials for a given atomic vapor is obtained. Thus the higher potentials result from two or more inelastic collisions and are multiples of the lowest one.

To check that the critical potentials were due to atomic energy levels, Franck and Hertz observed the emission spectra of vapors during electron bombardment. In the case of mercury vapor, for example, they found that a minimum electron energy of 4.9 eV was required to excite the 253.6-nm spectral line of mercury—and a photon of 253.6-nm light has an energy of just 4.9 eV. The Franck-Hertz experiments were performed shortly after Bohr announced his theory of the hydrogen atom, and they independently confirmed his basic ideas.

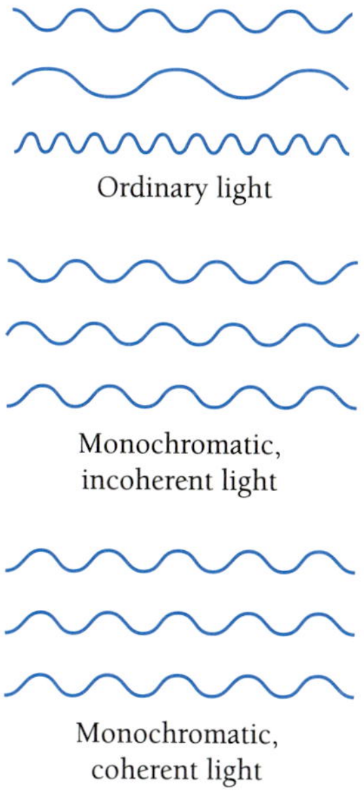

Figure 4.23 A laser produces a beam of light whose waves all have the same frequency (monochromatic) and are in phase with one another (coherent). The beam is also well collimated and so spreads out very little, even over long distances.

4.9 THE LASER

How to produce light waves all in step

The **laser** is a device that produces a light beam with some remarkable properties:

1 The light is very nearly monochromatic.

2 The light is coherent, with the waves all exactly in phase with one another (Fig.4.23).

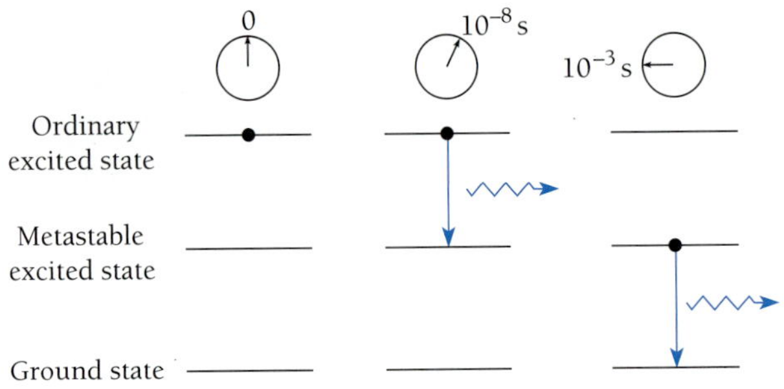

Figure 4.24 An atom can exist in a metastable energy level for a longer time before radiating than it can in an ordinary energy level.

3 A laser beam hardly diverges at all. Such a beam sent from the earth to a mirror left on the moon by the Apollo 11 expedition remained narrow enough to be detected on its return to the earth, a total distance of over three-quarters of a million kilometers. A light beam produced by any other means would have spread out too much for this to be done.
4 The beam is extremly intense, more intense by far than the light from any other source. To achieve an energy density equal to that in some laser beams, a hot object would have to be at a temperature of 10^{30} K.

The last two of these properties follow from the second of them.

The term *laser* stands for *l*ight *a*mplification by *s*timulated *e*mission of *r*adiation. The key to the laser is the presence in many atoms of one or more excited energy levels whose lifetimes may be 10^{-3} s or more instead of the usual 10^{-8} s. Such relatively long-lived states are called **metastable** (temporarily stable); see Fig. 4.24.

Three kinds of transition involving electromagnetic radiation are possible between two energy levels, E_0 and E_1, in an atom (Fig. 4.25). If the atom is initially in the lower state E_0, it can be raised to E_1 by absorbing a photon of energy $E_1 - E_0 = h\nu$. This process is called **stimulated absorption.** If the atom is initially in the upper state E_1, it can drop to E_0 by emitting a photon of energy $h\nu$. This is **spontaneous emission.**

Einstein, in 1917, was the first to point out a third possibility, **stimulated emission,** in which an incident photon of energy $h\nu$ causes a transition from E_1 to E_0. In stimulated emission, the radiated light waves are exactly in phase with the incident ones, so the result is an enhanced beam of coherent light. Einstein showed that stimulated emission has the same probability as stimulated absorption (see Sec. 9.7). That is, a photon of energy $h\nu$ incident on an atom in the upper

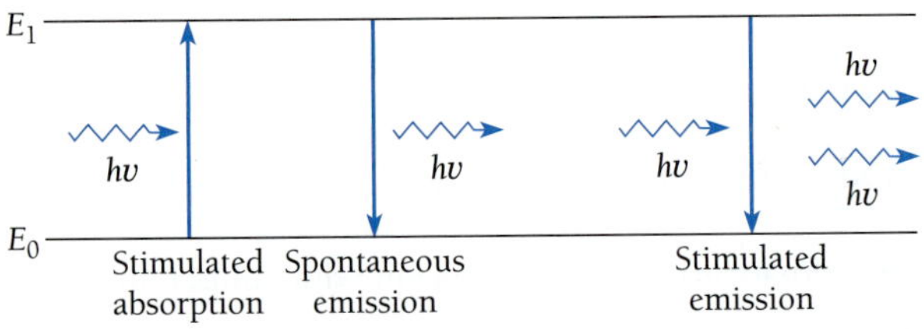

Figure 4.25 Transitions between two energy levels in an atom can occur by stimulated absorption, spontaneous emission, and stimulated emission.

About Scientists

Charles H. Townes (1915–) was born in Greenville, South Carolina, and attended Furman University there. After graduate study at Duke University and the California Institute of Technology, he spent 1939 to 1947 at the Bell Telephone Laboratories designing radar-controlled bombing systems. Townes then joined the physics department of Columbia University. In 1951, while sitting on a park bench, the idea for the **maser** (*m*icrowave *a*mplification by *s*timulated *e*mission of *r*adiation) occurred to him as a way to produce high-intensity microwaves, and in 1953 the first maser began operating. In this device ammonia (NH_3) molecules were raised to an excited vibrational state and then fed into a resonant cavity where, as in a laser, stimulated emission produced a cascade of photons of identical wavelength, here 1.25 cm in the microwave part of the spectrum. "Atomic clocks" of great accuracy are based on this concept, and solid-state maser amplifiers are used in such applications as radioastronomy.

In 1958 Townes and Arthur Schawlow attracted much attention with a paper showing that a similar scheme ought to be possible at optical wavelengths. Slightly earlier Gordon Gould, then a graduate student at Columbia, had come to the same conclusion, but did not publish his calculations at once since that would prevent securing a patent. Gould tried to develop the laser—his term—in private industry, but the Defense Department classified the project (and his original notebooks) as secret and denied him clearance to work on it. Finally, twenty years later, Gould succeeded in establishing his priority and received two patents on the laser, and still later, a third. The first working laser was built by Theodore Maiman at Hughes Research Laboratories in 1960. In 1964, Townes, along with two Russian laser pioneers, Aleksander Prokhorov and Nikolai Basov, was awarded a Nobel Prize. In 1981 Schawlow shared a Nobel Prize for precision spectroscopy using lasers.

Soon after its invention, the laser was spoken of as a "solution looking for a problem" because few applications were then known for it. Today, of course, lasers are widely employed for a variety of purposes.

state E_1 has the same likelihood of causing the emission of another photon of energy $h\nu$ as its likelihood of being absorbed if it is incident on an atom in the lower state E_0.

Stimulated emission involves no novel concepts. An analogy is a harmonic oscillator, for instance, a pendulum, which has a sinusoidal force applied to it whose period is the same as its natural period of vibration. If the applied force is exactly in phase with the pendulum swings, the amplitude of the swings increases. This corresponds to stimulated absorption. However, if the applied force is 180° out of phase with the pendulum swings, the amplitude of the swings *decreases*. This corresponds to stimulated emission.

A **three-level laser,** the simplest kind, uses an assembly of atoms (or molecules) that have a metastable state $h\nu$ in energy above the ground state and a still higher excited state that decays to the metastable state (Fig. 4.26). What we want is more atoms in the metastable state than in the ground state. If we can arrange this and then shine light of frequency ν on the assembly, there will be more stimulated emissions from atoms in the metastable state than stimulated absorptions by atoms in the ground state. The result will be an amplification of the original light. This is the concept that underlies the operation of the laser.

The term **population inversion** describes an assembly of atoms in which the majority are on energy levels above the ground state; normally the ground state is occupied to the greatest extent.

A number of ways exist to produce a population inversion. One of them, called **optical pumping,** is illustrated in Fig. 4.27. Here an external light source is used some of whose photons have the right frequency to raise ground-state atoms to the excited state that decays spontaneously to the desired metastable state.

Why are three levels needed? Suppose there are only two levels, a metastable state $h\nu$ above the ground state. The more photons of frequency ν we pump into the assembly

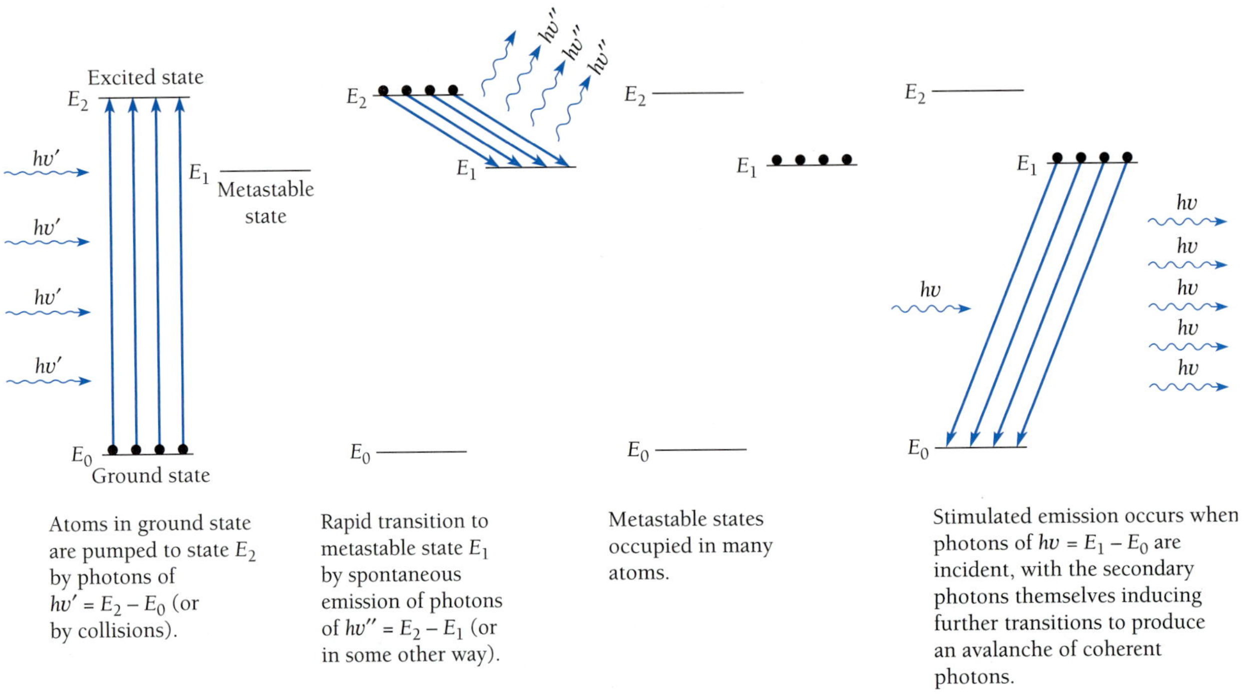

Figure 4.26 The principle of the laser.

of atoms, the more upward transitions there will be from the ground state to the metastable state. However, at the same time the pumping will stimulate downward transitions from the metastable state to the ground state. When half the atoms are in each state, the rate of stimulated emissions will equal the rate of stimulated absorptions, so the assembly cannot ever have more than half its atoms in the metastable state. In this situation laser amplification cannot occur. A population inversion is only possible when the stimulated absorptions are to a higher energy level than the metastable one from which the stimulated emission takes place, which prevents the pumping from depopulating the metastable state.

In a three-level laser, more than half the atoms must be in the metastable state for stimulated induced emission to predominate. This is not the case for a **four-level laser.**

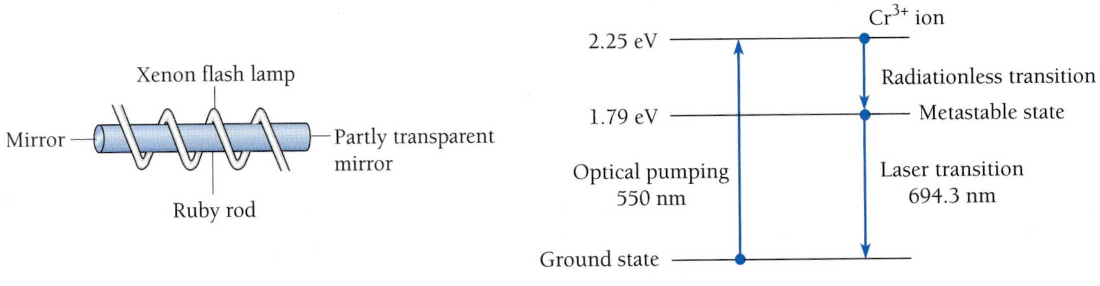

Figure 4.27 The ruby laser. In order for stimulated emission to exceed stimulated absorption, more than half the Cr^{3+} ions in the ruby rod must be in the metastable state. This laser produces a pulse of red light after each flash of the lamp.

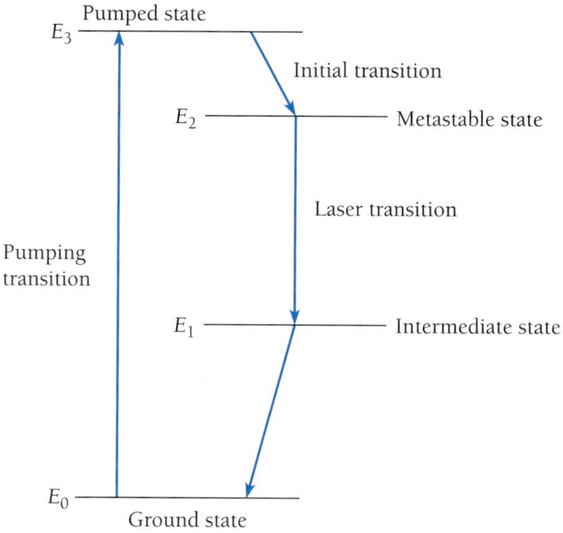

Figure 4.28 A four-level laser.

As in Fig. 4.28, the laser transition from the metastable state ends at an unstable intermediate state rather than at the ground state. Because the intermediate state decays rapidly to the ground state, very few atoms are in the intermediate state. Hence even a modest amount of pumping is enough to populate the metastable state to a greater extent than the intermediate state, as required for laser amplification.

Practical Lasers

The first successful laser, the **ruby laser**, is based on the three energy levels in the chromium ion Cr^{3+} shown in Fig. 4.27. A ruby is a crystal of aluminum oxide, Al_2O_3,

Precision laser cutting of a metal plate. (Courtesy of AEI Fabrication)

in which some of the Al^{3+} ions are replaced by Cr^{3+} ions, which are responsible for the red color. A Cr^{3+} ion has a metastable level whose lifetime is about 0.003 s. In the ruby laser, a xenon flash lamp excites the Cr^{3+} ions to a level of higher energy from which they fall to the metastable level by losing energy to other ions in the crystal. Photons from the spontaneous decay of some Cr^{3+} ions are reflected back and forth between the mirrored ends of the ruby rod, stimulating other excited Cr^{3+} ions to radiate. After a few microseconds the result is a large pulse of monochromatic, coherent red light from the partly transparent end of the rod.

The rod's length is made precisely an integral number of half-wavelengths long, so the radiation trapped in it forms an optical standing wave. Since the stimulated emissions are induced by the standing wave, their waves are all in step with it.

The common **helium-neon gas laser** achieves a population inversion in a different way. A mixture of about 10 parts of helium and 1 part of neon at a low pressure ($\sim$1 torr) is placed in a glass tube that has parallel mirrors, one of them partly transparent, at both ends. The spacing of the mirrors is again (as in all lasers) equal to an integral number of half-wavelengths of the laser light. An electric discharge is produced in the gas by means of electrodes outside the tube connected to a source of high-frequency alternating current, and collisions with electrons from the discharge excite He and Ne atoms to metastable states respectively 20.61 and 20.66 eV above their ground states (Fig. 4.29). Some of the excited He atoms transfer their energy to ground-state Ne atoms in collisions, with the 0.05 eV of additional energy being provided by the kinetic energy of the atoms. The purpose of the He atoms is thus to help achieve a population inversion in the Ne atoms.

The laser transition in Ne is from the metastable state at 20.66 eV to an excited state at 18.70 eV, with the emission of a 632.8-nm photon. Then another photon is spontaneously emitted in a transition to a lower metastable state; this transition yields only incoherent light. The remaining excitation energy is lost in collisions with the tube walls. Because the electron impacts that excite the He and Ne atoms occur all the time, unlike the pulsed excitation from the xenon flash lamp in a ruby laser, a He-Ne laser operates continuously. This is the laser whose narrow red beam is used in supermarkets to read bar codes. In a He-Ne laser, only a tiny

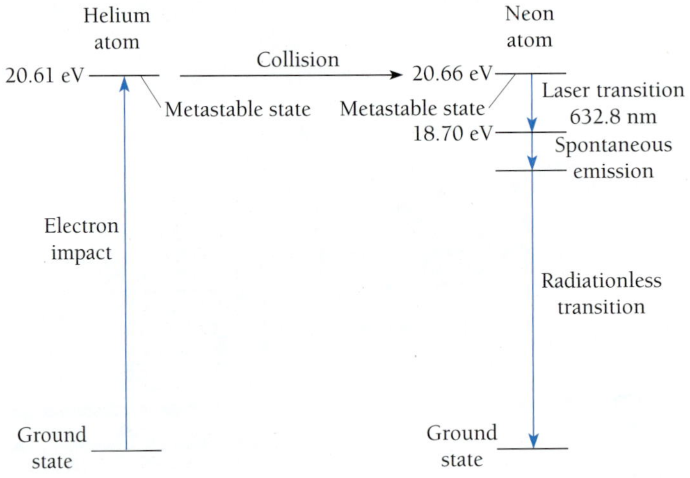

Figure 4.29 The helium-neon laser. In a four-level laser such as this, continuous operation is possible. Helium-neon lasers are commonly used to read bar codes.

Chirped Pulse Amplification

The most powerful lasers are pulsed, producing phenomenal outputs for very short periods. The petawatt (10^{15} W) threshold was crossed in 1996 with pulses less than a trillionth of a second long—not all that much energy per pulse, but at a rate of delivery over 1000 times that of the entire electrical grid of the United States. An ingenious method called chirped pulse amplification made this possible without the laser apparatus itself being destroyed in the process. What was done was to start with a low-power laser pulse that was quite short, only 0.1 picosecond (10^{-13} s). Because the pulse was short, it consisted of a large span of wavelengths, as discussed in Sec. 3.7 (see Figs. 3.13 and 3.14). A diffraction grating then spread out the light into different paths according to wavelength, which stretched the pulse to 3 nanoseconds (3×10^{-9} s), 30,000 times longer. The result was to decrease the peak power so that laser amplifiers could boost the energy of each beam. Finally the amplified beams, each of slightly different wavelength, were recombined by another grating to produce a pulse less than 0.5 picoseconds long whose power was 1.3 petawatts.

fraction (one in millions) of the atoms present participates in the laser process at any moment.

Many other types of lasers have been devised. A number of them employ molecules rather than atoms. **Chemical lasers** are based on the production by chemical reactions of molecules in metastable excited states. Such lasers are efficient and can be very powerful: one chemical laser, in which hydrogen and fluorine combine to form hydrogen fluoride, has generated an infrared beam of over 2 MW. **Dye lasers** use dye molecules whose energy levels are so close together that they can "lase" over a virtually continuous range of wavelengths (see Sec. 8.7). A dye laser can be tuned to any desired wavelength in its range. **Nd:YAG lasers,** which use the glassy solid yttrium aluminum garnet with neodymium as an impurity, are helpful in surgery because they seal small blood vessels while cutting through tissue by vaporizing water in the path of their beams. Powerful **carbon dioxide gas lasers** with outputs up to many kilowatts are used industrially for the precise cutting of almost any material, including steel, and for welding.

Tiny **semiconductor lasers** by the million process and transmit information today. (How such lasers work is described in Chap. 10.) In a compact disk player, a semiconductor laser beam is focused to a spot a micrometer (10^{-6} m) across to read data coded as pits that appear as dark spots on a reflective disk 12 cm in diameter. A compact disk can store over 600 megabytes of digital data, about 1000 times as much as the floppy disks used in personal computers. If the stored data is digitized music, the playing time can be over an hour.

Semiconductor lasers are ideal for fiber-optic transmission lines in which the electric signals that would normally be sent along copper wires are first converted into a series of pulses according to a standard code. Lasers then turn the pulses into flashes of infrared light that travel along thin (5–50 μm diameter) glass fibers and at the other end are changed back into electric signals. Over a million telephone conversations can be carried by a single fiber; by contrast, no more than 32 conversations can be carried at the same time by a pair of wires. Telephone fiber-optic systems today link many cities and exchanges within cities everywhere, and fiber-optic cables span the world's seas and oceans.

Appendix to Chapter 4

Rutherford Scattering

Rutherford's model of the atom was accepted because he was able to arrive at a formula to describe the scattering of alpha particles by thin foils on the basis of this model that agreed with the experimental results. He began by assuming that the alpha particle and the nucleus it interacts with are both small enough to be considered as point masses and charges; that the repulsive electric force between alpha particle and nucleus (which are both positively charged) is the only one acting; and that the nucleus is so massive compared with the alpha particle that it does not move during their interaction. Let us see how these assumptions lead to Eq. (4.1).

Scattering Angle

Owing to the variation of the electric force with $1/r^2$, where r is the instantaneous separation between alpha particle and nucleus, the alpha particle's path is a hyperbola with the nucleus at the outer focus (Fig. 4.30). The **impact parameter** b is the minimum distance to which the alpha particle would approach the nucleus if there were no force between them, and the **scattering angle** θ is the angle between the asymptotic direction of approach of the alpha particle and the asymptotic direction in which it recedes. Our first task is to find a relationship between b and θ.

As a result of the impulse $\int \mathbf{F}\, dt$ given it by the nucleus, the momentum of the alpha particle changes by $\Delta\mathbf{p}$ from the initial value $\mathbf{p}_1$ to the final value $\mathbf{p}_2$. That is,

$$\Delta \mathbf{p} = \mathbf{p}_2 - \mathbf{p}_1 = \int \mathbf{F}\, dt \tag{4.24}$$

Because the nucleus remains stationary during the passage of the alpha particle, by hypothesis, the alpha-particle kinetic energy is the same before and after the scattering. Hence the *magnitude* of its momentum is also the same before and after, and

$$p_1 = p_2 = mv$$

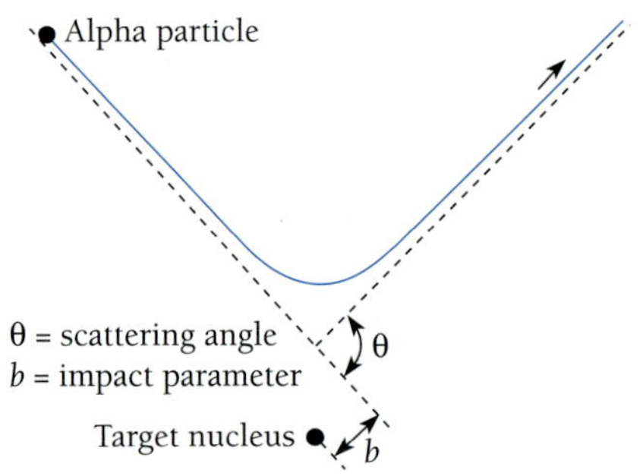

Figure 4.30 Rutherford scattering.

Here v is the alpha-particle velocity far from the nucleus.

From Fig. 4.31 we see that according to the law of sines,

$$\frac{\Delta p}{\sin\theta} = \frac{mv}{\sin\dfrac{\pi-\theta}{2}}$$

Since $$\sin\frac{1}{2}(\pi-\theta) = \cos\frac{\theta}{2}$$

and $$\sin\theta = 2\sin\frac{\theta}{2}\cos\frac{\theta}{2}$$

we have for the magnitude of the momentum change

$$\Delta p = 2mv\sin\frac{\theta}{2} \tag{4.25}$$

Because the impulse $\int \mathbf{F}\,dt$ is in the same direction as the momentum change $\Delta\mathbf{p}$, its magnitude is

$$\left|\int \mathbf{F}\,dt\right| = \int F\cos\phi\,dt \tag{4.26}$$

where ϕ is the instantaneous angle between $\mathbf{F}$ and Δp along the path of the alpha particle. Inserting Eqs. (4.25) and (4.26) in Eq. (4.24),

$$2mv\sin\frac{\theta}{2} = \int_{-\infty}^{\infty} F\cos\phi\,dt$$

To change the variable on the right-hand side from t to ϕ, we note that the limits of integration will change to $-\frac{1}{2}(\pi-\theta)$ and $+\frac{1}{2}(\pi-\theta)$, corresponding to ϕ at $t=-\infty$ and $t=\infty$ respectively, and so

$$2mv\sin\frac{\theta}{2} = \int_{-(\pi-\theta)/2}^{+(\pi-\theta)/2} F\cos\phi\,\frac{dt}{d\phi}\,d\phi \tag{4.27}$$

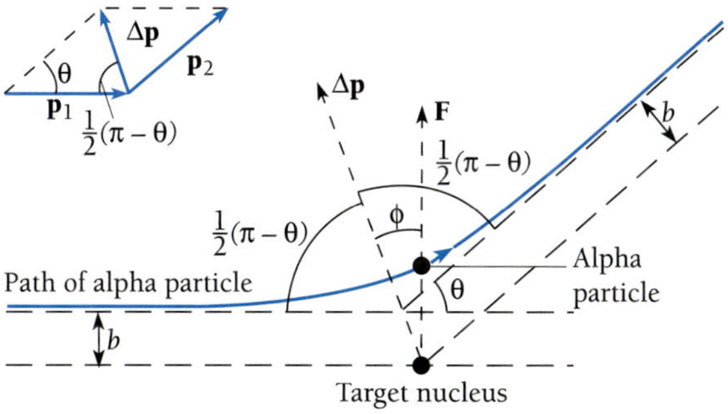

Figure 4.31 Geometrical relationships in Rutherford scattering.

The quantity $d\phi/dt$ is just the angular velocity ω of the alpha particle about the nucleus (this is evident from Fig. 4.31).

The electric force exerted by the nucleus on the alpha particle acts along the radius vector joining them, so there is no torque on the alpha particle and its angular momentum $m\omega r^2$ is constant. Hence

$$m\omega r^2 = \text{constant} = mr^2 \frac{d\phi}{dt} = mvb$$

from which we obtain

$$\frac{dt}{d\phi} = \frac{r^2}{vb}$$

Substituting this expression for $dt/d\phi$ in Eq. (4.27) gives

$$2mv^2b \sin\frac{\theta}{2} = \int_{-(\pi-\theta)/2}^{+(\pi-\theta)/2} Fr^2 \cos\phi \, d\phi \tag{4.28}$$

As we recall, F is the electric force exerted by the nucleus on the alpha particle. The charge on the nucleus is Ze, corresponding to the atomic number Z, and that on the alpha particle is $2e$. Therefore

$$F = \frac{1}{4\pi\epsilon_0} \frac{2Ze^2}{r^2}$$

and

$$\frac{4\pi\epsilon_0 mv^2b}{Ze^2} \sin\frac{\theta}{2} = \int_{-(\pi-\theta)/2}^{+(\pi-\theta)/2} \cos\phi \, d\phi = 2\cos\frac{\theta}{2}$$

The scattering angle θ is related to the impact parameter b by the equation

$$\cot\frac{\theta}{2} = \frac{2\pi\epsilon_0 mv^2}{Ze^2} b$$

It is more convenient to specify the alpha-particle energy KE instead of its mass and velocity separately; with this substitution,

Scattering angle

$$\cot\frac{\theta}{2} = \frac{4\pi\epsilon_0 \text{KE}}{Ze^2} b \tag{4.29}$$

Figure 4.32 is a schematic representation of Eq. (4.29); the rapid decrease in θ as b increases is evident. A very near miss is required for a substantial deflection.

Rutherford Scattering Formula

Equation (4.29) cannot be directly confronted with an experiment because there is no way of measuring the impact parameter corresponding to a particular observed scattering angle. An indirect strategy is required.

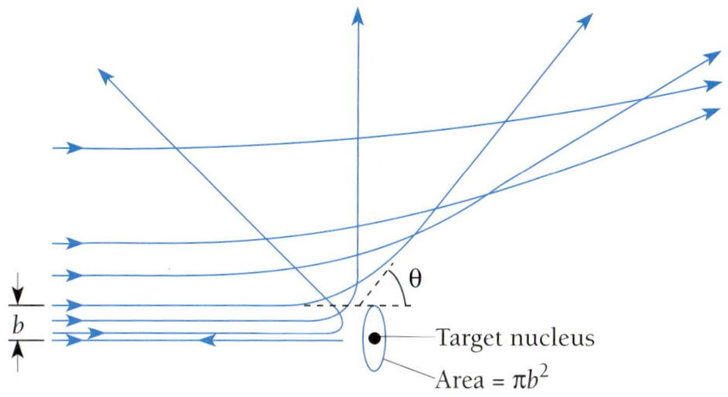

Figure 4.32 The scattering angle decreases with increasing impact parameter.

Our first step is to note that all alpha particles approaching a target nucleus with an impact parameter from 0 to b will be scattered through an angle of θ or more, where θ is given in terms of b by Eq. (4.29). This means that an alpha particle that is initially directed anywhere within the area πb^2 around a nucleus will be scattered through θ or more (Fig. 4.32). The area πb^2 is accordingly called the **cross section** for the interaction. The general symbol for the cross section is σ, and so here

Cross section
$$\sigma = \pi b^2 \tag{4.30}$$

Of course, the incident alpha particle is actually scattered before it reaches the immediate vicinity of the nucleus and hence does not necessarily pass within a distance b of it.

Now we consider a foil of thickness t that contains n atoms per unit volume. The number of target nuclei per unit area is nt, and an alpha-particle beam incident upon an area A therefore encounters ntA nuclei. The aggregate cross section for scatterings of θ or more is the number of target nuclei ntA multiplied by the cross section σ for such scattering per nucleus, or $ntA\sigma$. Hence the fraction f of incident alpha particles scattered by θ or more is the ratio between the aggregate cross section $ntA\sigma$ for such scattering and the total target area A. That is,

$$\begin{aligned} f &= \frac{\text{alpha particles scattered by } \theta \text{ or more}}{\text{incident alpha particles}} \\ &= \frac{\text{aggregate cross section}}{\text{target area}} = \frac{ntA\sigma}{A} \\ &= nt\pi b^2 \end{aligned}$$

Substituting for b from Eq. (4.30),

$$f = \pi nt \left(\frac{Ze^2}{4\pi\epsilon_0 \text{KE}} \right)^2 \cot^2 \frac{\theta}{2} \tag{4.31}$$

In this calculation it was assumed that the foil is sufficiently thin so that the cross sections of adjacent nuclei do not overlap and that a scattered alpha particle receives its entire deflection from an encounter with a single nucleus.

Example 4.8

Find the fraction of a beam of 7.7-MeV alpha particles that is scattered through angles of more than 45° when incident upon a gold foil 3×10^{-7} m thick. These values are typical of the alpha-particle energies and foil thicknesses used by Geiger and Marsden. For comparison, a human hair is about 10^{-4} m in diameter.

Solution

We begin by finding n, the number of gold atoms per unit volume in the foil, from the relationship

$$n = \frac{\text{atoms}}{\text{m}^3} = \frac{\text{mass/m}^3}{\text{mass/atom}}$$

Since the density of gold is 1.93×10^4 kg/m^3, its atomic mass is 197 u, and 1 u = 1.66×10^{-27} kg, we have

$$n = \frac{1.93 \times 10^4 \text{ kg/m}^3}{(197 \text{ u/atom})(1.66 \times 10^{-27} \text{ kg/u})}$$
$$= 5.90 \times 10^{28} \text{ atoms/m}^3$$

The atomic number Z of gold is 79, a kinetic energy of 7.7 MeV is equal to 1.23×10^{-12} J, and $\theta = 45°$; from these figures we find that

$$f = 7 \times 10^{-5}$$

of the incident alpha particles are scattered through 45° or more—only 0.007 percent! A foil this thin is quite transparent to alpha particles.

In an actual experiment, a detector measures alpha particles scattered between θ and $\theta + d\theta$, as in Fig. 4.33. The fraction of incident alpha particles so scattered is found by differentiating Eq. (4.31) with respect to θ, which gives

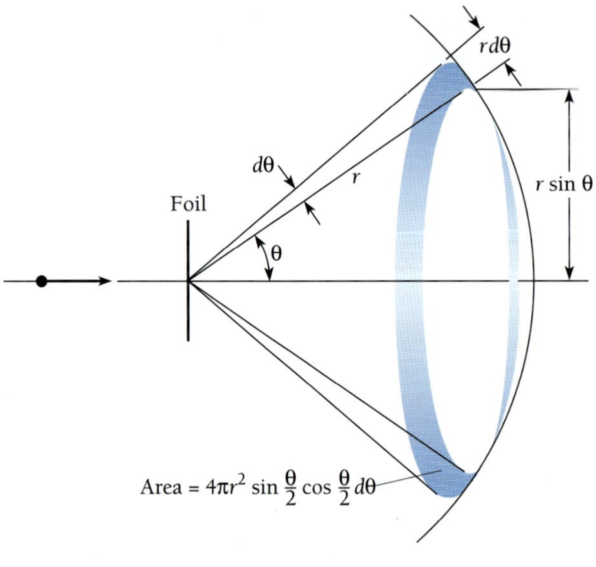

Figure 4.33 In the Rutherford experiment, particles are detected that have been scattered between θ and $\theta + d\theta$.

$$df = -\pi nt\left(\frac{Ze^2}{4\pi\epsilon_0 \text{KE}}\right)^2 \cot\frac{\theta}{2}\csc^2\frac{\theta}{2}\,d\theta \tag{4.32}$$

The minus sign expresses the fact that f decreases with increasing θ.

As we saw in Fig. 4.2, Geiger and Marsden placed a fluorescent screen a distance r from the foil and the scattered alpha particles were detected by means of the scintillations they caused. Those alpha particles scattered between θ and $\theta + d\theta$ reached a zone of a sphere of radius r whose width is $r\,d\theta$. The zone radius itself is $r \sin\theta$, and so the area dS of the screen struck by these particles is

$$dS = (2\pi r \sin\theta)(r\,d\theta) = 2\pi r^2 \sin\theta\,d\theta$$

$$= 4\pi r^2 \sin\frac{\theta}{2}\cos\frac{\theta}{2}\,d\theta$$

If a total of N_i alpha particles strike the foil during the course of the experiment, the number scattered into $d\theta$ at θ is $N_i\,df$. The number $N(\theta)$ per unit area striking the screen at θ, which is the quantity actually measured, is

$$N(\theta) = \frac{N_i|df|}{dS} = \frac{N_i\pi nt\left(\dfrac{Ze^2}{4\pi\epsilon_0 \text{KE}}\right)^2 \cot\dfrac{\theta}{2}\csc^2\dfrac{\theta}{2}\,d\theta}{4\pi r^2 \sin\dfrac{\theta}{2}\cos\dfrac{\theta}{2}\,d\theta}$$

Rutherford scattering formula

$$N(\theta) = \frac{N_i nt Z^2 e^4}{(8\pi\epsilon_0)^2 r^2\, \text{KE}^2 \sin^4(\theta/2)} \tag{4.1}$$

Equation (4.1) is the Rutherford scattering formula. Figure 4.4 shows how $N(\theta)$ varies with θ.

EXERCISES

It isn't that they can't see the solution. It is that they can't see the problem. —Gilbert Chesterton

4.1 The Nuclear Atom

1. The great majority of alpha particles passes through gases and thin metal foils with no deflections. To what conclusion about atomic structure does this observation lead?

2. The electric field intensity at a distance r from the center of a uniformly charged sphere of radius R and total charge Q is $Qr/4\pi\epsilon_0 R^3$ when $r < R$. Such a sphere corresponds to the Thomson model of the atom. Show that an electron in this sphere executes simple harmonic motion about its center and derive a formula for the frequency of this motion. Evaluate the frequency of the electron oscillations for the case of the hydrogen atom and compare it with the frequencies of the spectral lines of hydrogen.

3. Determine the distance of closest approach of 1.00-MeV protons incident on gold nuclei.

4.2 Electron Orbits

4. Find the frequency of revolution of the electron in the classical model of the hydrogen atom. In what region of the spectrum are electromagnetic waves of this frequency?

4.3 Atomic Spectra

5. What is the shortest wavelength present in the Brackett series of spectral lines?

6. What is the shortest wavelength present in the Paschen series of spectral lines?

4.4 The Bohr Atom

7. In the Bohr model, the electron is in constant motion. How can such an electron have a negative amount of energy?

8. Lacking de Broglie's hypothesis to guide his thinking, Bohr arrived at his model by postulating that the angular momentum of an orbital electron must be an integral multiple of $\hbar$. Show that this postulate leads to Eq. (4.13).

9. The **fine structure constant** is defined as $\alpha = e^2/2\epsilon_0 hc$. This quantity got its name because it first appeared in a theory by the German physicist Arnold Sommerfeld that tried to explain the fine structure in spectral lines (multiple lines close together instead of single lines) by assuming that elliptical as well as circular orbits are possible in the Bohr model. Sommerfeld's approach was on the wrong track, but α has nevertheless turned out to be a useful quantity in atomic physics. (*a*) Show that $\alpha = v_1/c$, where v_1 is the velocity of the electron in the ground state of the Bohr atom. (*b*) Show that the value of α is very close to 1/137 and is a pure number with no dimensions. Because the magnetic behavior of a moving charge depends on its velocity, the small value of α is representative of the relative magnitudes of the magnetic and electric aspects of electron behavior in an atom. (*c*) Show that $\alpha a_0 = \lambda_C/2\pi$, where a_0 is the radius of the ground-state Bohr orbit and λ_C is the Compton wavelength of the electron.

10. An electron at rest is released far away from a proton, toward which it moves. (*a*) Show that the de Broglie wavelength of the electron is proportional to $\sqrt{r}$, where r is the distance of the electron from the proton. (*b*) Find the wavelength of the electron when it is a_0 from the proton. How does this compare with the wavelength of an electron in a ground-state Bohr orbit? (*c*) In order for the electron to be captured by the proton to form a ground-state hydrogen atom, energy must be lost by the system. How much energy?

11. Find the quantum number that characterizes the earth's orbit around the sun. The earth's mass is 6.0×10^{24} kg, its orbital radius is 1.5×10^{11} m, and its orbital speed is 3.0×10^4 m/s.

12. Suppose a proton and an electron were held together in a hydrogen atom by gravitational forces only. Find the formula for the energy levels of such an atom, the radius of its ground-state Bohr orbit, and its ionization energy in eV.

13. Compare the uncertainty in the momentum of an electron confined to a region of linear dimension a_0 with the momentum of an electron in a ground-state Bohr orbit.

4.5 Energy Levels and Spectra

14. When radiation with a continuous spectrum is passed through a volume of hydrogen gas whose atoms are all in the ground state, which spectral series will be present in the resulting absorption spectrum?

15. What effect would you expect the rapid random motion of the atoms of an excited gas to have on the spectral lines they produce?

16. A beam of 13.0-eV electrons is used to bombard gaseous hydrogen. What series of wavelengths will be emitted?

17. A proton and an electron, both at rest initially, combine to form a hydrogen atom in the ground state. A single photon is emitted in this process. What is its wavelength?

18. How many different wavelengths would appear in the spectrum of hydrogen atoms initially in the $n = 5$ state?

19. Find the wavelength of the spectral line that corresponds to a transition in hydrogen from the $n = 10$ state to the ground state. In what part of the spectrum is this?

20. Find the wavelength of the spectral line that corresponds to a transition in hydrogen from the $n = 6$ state to the $n = 3$ state. In what part of the spectrum is this?

21. A beam of electrons bombards a sample of hydrogen. Through what potential difference must the electrons have been accelerated if the first line of the Balmer series is to be emitted?

22. How much energy is required to remove an electron in the $n = 2$ state from a hydrogen atom?

23. The longest wavelength in the Lyman series is 121.5 nm and the shortest wavelength in the Balmer series is 364.6 nm. Use the figures to find the longest wavelength of light that could ionize hydrogen.

24. The longest wavelength in the Lyman series is 121.5 nm. Use this wavelength together with the values of c and h to find the ionization energy of hydrogen.

25. An excited hydrogen atom emits a photon of wavelength λ in returning to the ground state. (*a*) Derive a formula that gives the quantum number of the initial excited state in terms of λ and R. (*b*) Use this formula to find n_i for a 102.55-nm photon.

26. An excited atom of mass m and initial speed v emits a photon in its direction of motion. If $v \ll c$, use the requirement that linear momentum and energy must both be conserved to show that the frequency of the photon is higher by $\Delta\nu/\nu \approx v/c$ than it would have been if the atom had been at rest. (See also Exercise 56 of Chap. 1.)

27. When an excited atom emits a photon, the linear momentum of the photon must be balanced by the recoil momentum of the atom. As a result, some of the excitation energy of the atom goes into the kinetic energy of its recoil. (*a*) Modify Eq. (4.16) to include this effect. (*b*) Find the ratio between the recoil energy and the photon energy for the $n = 3 \longrightarrow n = 2$ transition in hydrogen, for which $E_i - E_f = 1.9$ eV. Is the effect a major one? A nonrelativistic calculation is sufficient here.

4.6 Correspondence Principle

28. Of the following quantities, which increase and which decrease in the Bohr model as n increases? Frequency of revolution, electron speed, electron wavelength, angular momentum, potential energy, kinetic energy, total energy.

29. Show that the frequency of the photon emitted by a hydrogen atom in going from the level $n + 1$ to the level n is always intermediate between the frequencies of revolution of the electron in the respective orbits.

4.7 Nuclear Motion

30. An antiproton has the mass of a proton but a charge of $-e$. If a proton and an antiproton orbited each other, how far apart would they be in the ground state of such a system? Why might you think such a system could not occur?

31. A μ^- muon is in the $n = 2$ state of a muonic atom whose nucleus is a proton. Find the wavelength of the photon emitted when the muonic atom drops to its ground state. In what part of the spectrum is this wavelength?

32. Compare the ionization energy in positronium with that in hydrogen.

33. A mixture of ordinary hydrogen and tritium, a hydrogen isotope whose nucleus is approximately 3 times more massive than ordinary hydrogen, is excited and its spectrum observed. How far apart in wavelength will the H_α lines of the two kinds of hydrogen be?

34. Find the radius and speed of an electron in the ground state of doubly ionized lithium and compare them with the radius and speed of the electron in the ground state of the hydrogen atom. (Li^{++} has a nuclear charge of $3e$.)

35. (*a*) Derive a formula for the energy levels of a **hydrogenic atom,** which is an ion such as He^+ or Li^{2+} whose nuclear charge is $+Ze$ and which contains a single electron. (*b*) Sketch the energy levels of the He^+ ion and compare them with the energy levels of the H atom. (*c*) An electron joins a bare helium nucleus to form a He^+ ion. Find the wavelength of the photon emitted in this process if the electron is assumed to have had no kinetic energy when it combined with the nucleus.

4.9 The Laser

36. For laser action to occur, the medium used must have at least three energy levels. What must be the nature of each of these levels? Why is three the minimum number?

37. A certain ruby laser emits 1.00-J pulses of light whose wavelength is 694 nm. What is the minimum number of Cr^{3+} ions in the ruby?

38. Steam at 100°C can be thought of as an excited state of water at 100°C. Suppose that a laser could be built based upon the transition from steam to water, with the energy lost per molecule of steam appearing as a photon. What would the frequency of such a photon be? To what region of the spectrum does this correspond? The heat of vaporization of water is 2260 kJ/kg and its molar mass is 18.02 kg/kmol.

Appendix: Rutherford Scattering

39. The Rutherford scattering formula fails to agree with the data at very small scattering angles. Can you think of a reason?

40. Show that the probability for a 2.0-MeV proton to be scattered by more than a given angle when it passes through a thin foil is the same as that for a 4.0-MeV alpha particle.

41. A 5.0-MeV alpha particle approaches a gold nucleus with an impact parameter of 2.6×10^{-13} m. Through what angle will it be scattered?

42. What is the impact parameter of a 5.0-MeV alpha particle scattered by 10° when it approaches a gold nucleus?

43. What fraction of a beam of 7.7-MeV alpha particles incident upon a gold foil 3.0×10^{-7} m thick is scattered by less than 1°?

44. What fraction of a beam of 7.7-MeV alpha particles incident upon a gold foil 3.0×10^{-7} m thick is scattered by 90° or more?

45. Show that twice as many alpha particles are scattered by a foil through angles between 60° and 90° as are scattered through angles of 90° or more.

46. A beam of 8.3-MeV alpha particles is directed at an aluminum foil. It is found that the Rutherford scattering formula ceases to be obeyed at scattering angles exceeding about 60°. If the alpha-particle radius is assumed small enough to neglect here, find the radius of the aluminum nucleus.

47. In special relativity, a photon can be thought of as having a "mass" of $m = E_\nu/c^2$. This suggests that we can treat a photon that passes near the sun in the same way as Rutherford treated an alpha particle that passes near a nucleus, with an attractive gravitational force replacing the repulsive electrical force. Adapt Eq. (4.29) to this situation and find the angle of deflection θ for a photon that passes $b = R_{sun}$ from the center of the sun. The mass and radius of the sun are respectively 2.0×10^{30} kg and 7.0×10^8 m. In fact, general relativity shows that this result is exactly half the actual deflection, a conclusion supported by observations made during solar eclipses as mentioned in Sec. 1.10.

CHAPTER 5

Quantum Mechanics

A scanning tunneling microscope(STM) image of a silicon chip surface. Each protrusion is a single silicon atom and the distance between atoms is about 0.7nm. Quantum mechanical tunneling effect is used in STM.

Although the Bohr theory of the atom, which can be extended further than was done in Chap. 4, is able to account for many aspects of atomic phenomena, it has a number of severe limitations as well. First of all, it applies only to hydrogen and one-electron ions such as He^+ and Li^{2+}—it does not even work for ordinary helium. The Bohr theory cannot explain why certain spectral lines are more intense than others (that is, why certain transitions between energy levels have greater probabilities of occurrence than others). It cannot account for the observation that many spectral lines actually consist of several separate lines whose wavelengths differ slightly. And perhaps most important, it does not permit us to obtain what a really successful theory of the atom should make possible: an understanding of how individual atoms interact with one another to endow macroscopic aggregates of matter with the physical and chemical properties we observe.

The preceding objections to the Bohr theory are not put forward in an unfriendly way, for the theory was one of those seminal achievements that transformed scientific thought, but rather to emphasize that a more general approach to atomic phenomena is required. Such an approach was developed in 1925 and 1926 by Erwin Schrödinger, Werner Heisenberg, Max Born, Paul Dirac, and others under the apt name of **quantum mechanics.** "The discovery of quantum mechanics was nearly a total surprise. It described the physical world in a way that was fundamentally new. It seemed to many of us a miracle," noted Eugene Wigner, one of the early workers in the field. By the early 1930s the application of quantum mechanics to problems involving nuclei, atoms, molecules, and matter in the solid state made it possible to understand a vast body of data ("a large part of physics and the whole of chemistry," according to Dirac) and—vital for any theory—led to predictions of remarkable accuracy. Quantum mechanics has survived every experimental test thus far of even its most unexpected conclusions.

5.1 QUANTUM MECHANICS

Classical mechanics is an approximation of quantum mechanics

The fundamental difference between classical (or Newtonian) mechanics and quantum mechanics lies in what they describe. In classical mechanics, the future history of a particle is completely determined by its initial position and momentum together with the forces that act upon it. In the everyday world these quantities can all be determined well enough for the predictions of Newtonian mechanics to agree with what we find.

Quantum mechanics also arrives at relationships between observable quantities, but the uncertainty principle suggests that the nature of an observable quantity is different in the atomic realm. Cause and effect are still related in quantum mechanics, but what they concern needs careful interpretation. In quantum mechanics the kind of certainty about the future characteristic of classical mechanics is impossible because the initial state of a particle cannot be established with sufficient accuracy. As we saw in Sec. 3.7, the more we know about the position of a particle now, the less we know about its momentum and hence about its position later.

The quantities whose relationships quantum mechanics explores are *probabilities.* Instead of asserting, for example, that the radius of the electron's orbit in a ground-state hydrogen atom is always exactly 5.3×10^{-11} m, as the Bohr theory does, quantum mechanics states that this is the *most probable* radius. In a suitable experiment most trials will yield a different value, either larger or smaller, but the value most likely to be found will be 5.3×10^{-11} m.

Quantum mechanics might seem a poor substitute for classical mechanics. However, classical mechanics turns out to be just an approximate version of quantum mechanics. The certainties of classical mechanics are illusory, and their apparent agreement with experiment occurs because ordinary objects consist of so many individual atoms that departures from average behavior are unnoticeable. Instead of two sets of physical principles, one for the macroworld and one for the microworld, there is only the single set included in quantum mechanics.

Wave Function

As mentioned in Chap. 3, the quantity with which quantum mechanics is concerned is the **wave function** Ψ of a body. While Ψ itself has no physical interpretation, the square of its absolute magnitude $|\Psi|^2$ evaluated at a particular place at a particular time is proportional to the probability of finding the body there at that time. The linear momentum, angular momentum, and energy of the body are other quantities that can be established from Ψ. The problem of quantum mechanics is to determine Ψ for a body when its freedom of motion is limited by the action of external forces.

Wave functions are usually complex with both real and imaginary parts. A probability, however, must be a positive real quantity. The probability density $|\Psi|^2$ for a complex Ψ is therefore taken as the product $\Psi^*\Psi$ of Ψ and its **complex conjugate** Ψ^*. The complex conjugate of any function is obtained by replacing $i(=\sqrt{-1})$ by $-i$ wherever it appears in the function. Every complex function Ψ can be written in the form

Wave function $$\Psi = A + iB$$

where A and B are real functions. The complex conjugate Ψ^* of Ψ is

Complex conjugate $$\Psi^* = A - iB$$

and so $$|\Psi|^2 = \Psi^*\Psi = A^2 - i^2B^2 = A^2 + B^2$$

since $i^2 = -1$. Hence $|\Psi|^2 = \Psi^*\Psi$ is always a positive real quantity, as required.

Normalization

Even before we consider the actual calculation of Ψ, we can establish certain requirements it must always fulfill. For one thing, since $|\Psi|^2$ is proportional to the probability density P of finding the body described by Ψ, the integral of $|\Psi|^2$ over all space must be finite—the body is *somewhere*, after all. If

$$\int_{-\infty}^{\infty} |\Psi|^2 \, dV = 0$$

the particle does not exist, and the integral obviously cannot be ∞ and still mean anything. Furthermore, $|\Psi|^2$ cannot be negative or complex because of the way it is defined. The only possibility left is that the integral be a finite quantity if Ψ is to describe properly a real body.

It is usually convenient to have $|\Psi|^2$ be *equal* to the probability density P of finding the particle described by Ψ, rather than merely be proportional to P. If $|\Psi|^2$ is to equal P, then it must be true that

Normalization

$$\int_{-\infty}^{\infty} |\Psi|^2 \, dV = 1 \tag{5.1}$$

since if the particle exists somewhere at all times,

$$\int_{-\infty}^{\infty} P \, dV = 1$$

A wave function that obeys Eq. (5.1) is said to be **normalized.** Every acceptable wave function can be normalized by multiplying it by an appropriate constant; we shall shortly see how this is done.

Well-Behaved Wave Functions

Besides being normalizable, Ψ must be single-valued, since P can have only one value at a particular place and time, and continuous. Momentum considerations (see Sec. 5.6) require that the partial derivatives $\partial\Psi/\partial x$, $\partial\Psi/\partial y$, $\partial\Psi/\partial z$ be finite, continuous, and single-valued. Only wave functions with all these properties can yield physically meaningful results when used in calculations, so only such "well-behaved" wave functions are admissible as mathematical representations of real bodies. To summarize:

1 Ψ must be continuous and single-valued everywhere.
2 $\partial\Psi/\partial x$, $\partial\Psi/\partial y$, $\partial\Psi/\partial z$ must be continuous and single-valued everywhere.
3 $\partial\Psi$ must be normalizable, which means that Ψ must go to 0 as $x \to \pm\infty$, $y \to \pm\infty$, $z \to \pm\infty$ in order that $\int |\Psi|^2 \, dV$ over all space will be a finite constant.

These rules are not always obeyed by the wave functions of particles in model situations that only approximate actual ones. For instance, the wave functions of a particle in a box with infinitely hard walls do not have continuous derivatives at the walls, since $\Psi = 0$ outside the box (see Fig. 5.4). But in the real world, where walls are never infinitely hard, there is no sharp change in Ψ at the walls (see Fig. 5.7) and the derivatives are continuous. Exercise 7 gives another example of a wave function that is not well-behaved.

Given a normalized and otherwise acceptable wave function Ψ, the probability that the particle it describes will be found in a certain region is simply the integral of the probability density $|\Psi|^2$ over that region. Thus for a particle restricted to motion in the x direction, the probability of finding it between x_1 and x_2 is given by

Probability

$$P_{x_1x_2} = \int_{x_1}^{x_2} |\Psi|^2 \, dx \tag{5.2}$$

We will see examples of such calculations later in this chapter and in Chap. 6.

5.2 THE WAVE EQUATION

It can have a variety of solutions, including complex ones

Schrödinger's equation, which is the fundamental equation of quantum mechanics in the same sense that the second law of motion is the fundamental equation of Newtonian mechanics, is a wave equation in the variable Ψ.

Before we tackle Schrödinger's equation, let us review the wave equation

Wave equation
$$\frac{\partial^2 y}{\partial x^2} = \frac{1}{v^2}\frac{\partial^2 y}{\partial t^2} \tag{5.3}$$

which governs a wave whose variable quantity is y that propagates in the x direction with the speed $\mathbf{v}$. In the case of a wave in a stretched string, y is the displacement of the string from the x axis; in the case of a sound wave, y is the pressure difference; in the case of a light wave, y is either the electric or the magnetic field magnitude. Equation (5.3) can be derived from the second law of motion for mechanical waves and from Maxwell's equations for electromagnetic waves.

Partial Derivatives

Suppose we have a function $f(x, y)$ of two variables, x and y, and we want to know how f varies with only one of them, say x. To find out, we differentiate f with respect to x while treating the other variable y as a constant. The result is the **partial derivative** of f with respect to x, which is written $\partial f/\partial x$

$$\frac{\partial f}{\partial x} = \left(\frac{df}{dx}\right)_{y=\text{constant}}$$

The rules for ordinary differentiation hold for partial differentiation as well. For instance, if $f = cx^2$,

$$\frac{df}{dx} = 2cx$$

and so, if $f = yx^2$,

$$\frac{\partial f}{\partial x} = \left(\frac{df}{dx}\right)_{y=\text{constant}} = 2yx$$

The partial derivative of $f = yx^2$ with respect to the other variable, y, is

$$\frac{\partial f}{\partial y} = \left(\frac{df}{dy}\right)_{x=\text{constant}} = x^2$$

Second order partial derivatives occur often in physics, as in the wave equation. To find $\partial^2 f/\partial x^2$, we first calculate $\partial f/\partial x$ and then differentiate again, still keeping y constant:

$$\frac{\partial^2 f}{\partial x^2} = \frac{\partial}{\partial x}\left(\frac{\partial f}{\partial x}\right)$$

For $f = yx^2$,

$$\frac{\partial^2 f}{\partial x^2} = \frac{\partial}{\partial x}(2yx) = 2y$$

Similarly
$$\frac{\partial^2 f}{\partial y^2} = \frac{\partial}{\partial y}(x^2) = 0$$

Solutions of the wave equation may be of many kinds, reflecting the variety of waves that can occur—a single traveling pulse, a train of waves of constant amplitude and wavelength, a train of superposed waves of the same amplitudes and wavelengths, a train of superposed waves of different amplitudes and wavelengths,

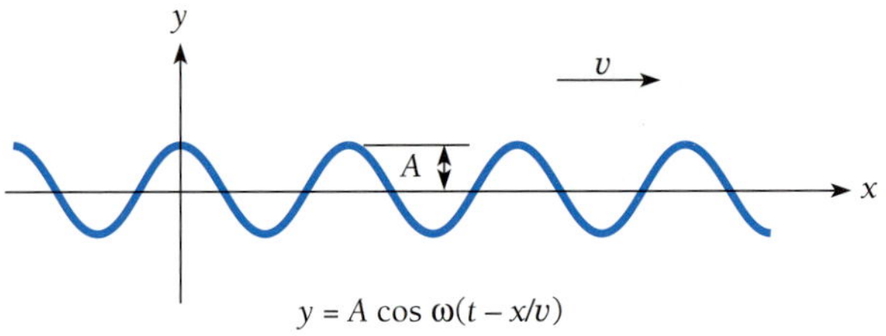

Figure 5.1 Waves in the xy plane traveling in the $+x$ direction along a stretched string lying on the x axis.

a standing wave in a string fastened at both ends, and so on. All solutions must be of the form

$$y = F\left(t \pm \frac{x}{v}\right) \tag{5.4}$$

where F is any function that can be differentiated. The solutions $F(t - x/v)$ represent waves traveling in the $+x$ direction, and the solutions $F(t + x/v)$ represent waves traveling in the $-x$ direction.

Let us consider the wave equivalent of a "free particle," which is a particle that is not under the influence of any forces and therefore pursues a straight path at constant speed. This wave is described by the general solution of Eq. (5.3) for undamped (that is, constant amplitude A), monochromatic (constant angular frequency ω) harmonic waves in the $+x$ direction, namely

$$y = Ae^{-i\omega(t-x/v)} \tag{5.5}$$

In this formula y is a complex quantity, with both real and imaginary parts.

Because

$$e^{-i\theta} = \cos\theta - i\sin\theta$$

Eq. (5.5) can be written in the form

$$y = A\cos\omega\left(t - \frac{x}{v}\right) - iA\sin\omega\left(t - \frac{x}{v}\right) \tag{5.6}$$

Only the real part of Eq. (5.6) [which is the same as Eq. (3.5)] has significance in the case of waves in a stretched string. There y represents the displacement of the string from its normal position (Fig. 5.1), and the imaginary part of Eq. (5.6) is discarded as irrelevant.

Example 5.1

Verify that Eq. (5.5) is a solution of the wave equation.

Solution

The derivative of an exponential function e^u is

$$\frac{d}{dx}(e^u) = e^u\frac{du}{dx}$$

The partial derivative of y with respect to x (which means t is treated as a constant) from Eq. (5.5) is therefore

$$\frac{\partial y}{\partial x} = \frac{i\omega}{v}y$$

and the second partial derivative is

$$\frac{\partial^2 y}{\partial x^2} = \frac{i^2\omega^2}{v^2}y = -\frac{\omega^2}{v^2}y$$

since $i^2 = -1$. The partial derivative of y with respect to t (now holding x constant) is

$$\frac{\partial y}{\partial t} = -i\omega y$$

and the second partial derivative is

$$\frac{\partial^2 y}{\partial t^2} = i^2\omega^2 y = -\omega^2 y$$

Combining these results gives

$$\frac{\partial^2 y}{\partial x^2} = \frac{1}{v^2}\frac{\partial^2 y}{\partial t^2}$$

which is Eq. (5.3). Hence Eq. (5.5) is a solution of the wave equation.

5.3 SCHRÖDINGER'S EQUATION: TIME-DEPENDENT FORM

A basic physical principle that cannot be derived from anything else

In quantum mechanics the wave function Ψ corresponds to the wave variable y of wave motion in general. However, Ψ, unlike y, is not itself a measurable quantity and may therefore be complex. For this reason we assume that Ψ for a particle moving freely in the $+x$ direction is specified by

$$\Psi = Ae^{-i\omega(t-x/v)} \tag{5.7}$$

Replacing ω in the above formula by $2\pi\nu$ and v by $\lambda\nu$ gives

$$\Psi = Ae^{-2\pi i(\nu t-x/\lambda)} \tag{5.8}$$

This is convenient since we already know what ν and λ are in terms of the total energy E and momentum p of the particle being described by Ψ. Because

$$E = h\nu = 2\pi\hbar\nu \qquad \text{and} \qquad \lambda = \frac{h}{p} = \frac{2\pi\hbar}{p}$$

we have

Free particle
$$\Psi = Ae^{-(i/\hbar)(Et-px)} \tag{5.9}$$

Equation (5.9) describes the wave equivalent of an unrestricted particle of total energy E and momentum p moving in the $+x$ direction, just as Eq. (5.5) describes, for example, a harmonic displacement wave moving freely along a stretched string.

The expression for the wave function Ψ given by Eq. (5.9) is correct only for freely moving particles. However, we are most interested in situations where the motion of a particle is subject to various restrictions. An important concern, for example, is an electron bound to an atom by the electric field of its nucleus. What we must now do is obtain the fundamental differential equation for Ψ, which we can then solve for Ψ in a specific situation. This equation, which is Schrödinger's equation, can be arrived at in various ways, but it *cannot* be rigorously derived from existing physical principles:

the equation represents something new. What will be done here is to show one route to the wave equation for Ψ and then to discuss the significance of the result.

We begin by differentiating Eq. (5.9) for Ψ twice with respect to x, which gives

$$\frac{\partial^2\Psi}{\partial x^2} = -\frac{p^2}{\hbar^2}\Psi$$

$$p^2\Psi = -\hbar^2\frac{\partial^2\Psi}{\partial x^2} \tag{5.10}$$

Differentiating Eq. (5.9) once with respect to t gives

$$\frac{\partial\Psi}{\partial t} = -\frac{iE}{\hbar}\Psi$$

$$E\Psi = -\frac{\hbar}{i}\frac{\partial\Psi}{\partial t} \tag{5.11}$$

At speeds small compared with that of light, the total energy E of a particle is the sum of its kinetic energy $p^2/2m$ and its potential energy U, where U is in general a function of position x and time t:

$$E = \frac{p^2}{2m} + U(x, t) \tag{5.12}$$

The function U represents the influence of the rest of the universe on the particle. Of course, only a small part of the universe interacts with the particle to any extent; for

About Scientists

Erwin Schrödinger (1887–1961) was born in Vienna to an Austrian father and a half-English mother and received his doctorate at the university there. After World War I, during which he served as an artillery officer, Schrödinger had appointments at several German universities before becoming professor of physics in Zurich, Switzerland. Late in November, 1925, Schrödinger gave a talk on de Broglie's notion that a moving particle has a wave character. A colleague remarked to him afterward that to deal properly with a wave, one needs a wave equation. Schrödinger took this to heart, and a few weeks later he was "struggling with a new atomic theory. If only I knew more mathematics! I am very optimistic about this thing and expect that if I can only . . . solve it, it will be *very* beautiful." (Schrödinger was not the only physicist to find the mathematics he needed difficult; the eminent mathematician David Hilbert said at about this time, "Physics is much too hard for physicists.")

The struggle was successful, and in January 1926 the first of four papers on "Quantization as an Eigenvalue Problem" was completed. In this epochal paper Schrödinger introduced the equation that bears his name and solved it for the hydrogen atom, thereby opening wide the door to the modern view of the atom which others had only pushed ajar. By June, Schrödinger had applied wave mechanics to the harmonic oscillator, the diatomic molecule, the hydrogen atom in an electric field, the absorption and emission of radiation, and the scattering of radiation by atoms and molecules. He had also shown that his wave mechanics was mathematically equivalent to the more abstract Heisenberg-Born-Jordan matrix mechanics.

The significance of Schrödinger's work was at once realized. In 1927 he succeeded Planck at the University of Berlin but left Germany in 1933, the year he received the Nobel Prize, when the Nazis came to power. He was at Dublin's Institute for Advanced Study from 1939 until his return to Austria in 1956. In Dublin, Schrödinger became interested in biology, in particular the mechanism of heredity. He seems to have been the first to make definite the idea of a genetic code and to identify genes as long molecules that carry the code in the form of variations in how their atoms are arranged. Schrödinger's 1944 book *What Is Life?* was enormously influential, not only by what it said but also by introducing biologists to a new way of thinking—that of the physicist—about their subject. *What Is Life?* started James Watson on his search for "the secret of the gene," which he and Francis Crick (a physicist) discovered in 1953 to be the structure of the DNA molecule.

instance, in the case of the electron in a hydrogen atom, only the electric field of the nucleus must be taken into account.

Multiplying both sides of Eq. (5.12) by the wave function Ψ gives

$$E\Psi = \frac{p^2\Psi}{2m} + U\Psi \tag{5.13}$$

Now we substitute for $E\Psi$ and $p^2\Psi$ from Eqs. (5.10) and (5.11) to obtain the **time-dependent form of Schrödinger's equation:**

Time-dependent Schrödinger equation in one dimension

$$i\hbar\frac{\partial\Psi}{\partial t} = -\frac{\hbar^2}{2m}\frac{\partial^2\Psi}{\partial x^2} + U\Psi \tag{5.14}$$

In three dimensions the time-dependent form of Schrödinger's equation is

$$i\hbar\frac{\partial\Psi}{\partial t} = -\frac{\hbar^2}{2m}\left(\frac{\partial^2\Psi}{\partial x^2} + \frac{\partial^2\Psi}{\partial y^2} + \frac{\partial^2\Psi}{\partial z^2}\right) + U\Psi \tag{5.15}$$

where the particle's potential energy U is some function of x, y, z, and t.

Any restrictions that may be present on the particle's motion will affect the potential-energy function U. Once U is known, Schrödinger's equation may be solved for the wave function Ψ of the particle, from which its probability density $|\Psi|^2$ may be determined for a specified x, y, z, t.

Validity of Schrödinger's Equation

Schrödinger's equation was obtained here using the wave function of a freely moving particle (potential energy U = constant). How can we be sure it applies to the general case of a particle subject to arbitrary forces that vary in space and time [$U = U(x, y, z, t)$]? Substituting Eqs. (5.10) and (5.11) into Eq. (5.13) is really a wild leap with no formal justification; this is true for all other ways in which Schrödinger's equation can be arrived at, including Schrödinger's own approach.

What we must do is postulate Schrödinger's equation, solve it for a variety of physical situations, and compare the results of the calculations with the results of experiments. If both sets of results agree, the postulate embodied in Schrödinger's equation is valid. If they disagree, the postulate must be discarded and some other approach would then have to be explored. In other words,

> Schrödinger's equation cannot be derived from other basic principles of physics; it is a basic principle in itself.

What has happened is that Schrödinger's equation has turned out to be remarkably accurate in predicting the results of experiments. To be sure, Eq. (5.15) can be used only for nonrelativistic problems, and a more elaborate formulation is needed when particle speeds near that of light are involved. But because it is in accord with experience within its range of applicability, we must consider Schrödinger's equation as a valid statement concerning certain aspects of the physical world.

It is worth noting that Schrödinger's equation does not increase the number of principles needed to describe the workings of the physical world. Newton's second law of motion $F = ma$, the basic principle of classical mechanics, can be derived from Schrödinger's equation provided the quantities it relates are understood to be averages rather than precise values. (Newton's laws of motion were also not derived from any other principles. Like Schrödinger's equation, these laws are considered valid in their range of applicability because of their agreement with experiment.)

5.4 LINEARITY AND SUPERPOSITION

Wave functions add, not probabilities

An important property of Schrödinger's equation is that it is linear in the wave function Ψ. By this is meant that the equation has terms that contain Ψ and its derivatives but no terms independent of Ψ or that involve higher powers of Ψ or its derivatives. As a result, a linear combination of solutions of Schrödinger's equation for a given system is also itself a solution. If Ψ_1 and Ψ_2 are two solutions (that is, two wave functions that satisfy the equation), then

$$\Psi = a_1\Psi_1 + a_2\Psi_2$$

is also a solution, where a_1 and a_2 are constants (see Exercise 8). Thus the wave functions Ψ_1 and Ψ_2 obey the superposition principle that other waves do (see Sec. 2.1) and we conclude that interference effects can occur for wave functions just as they can for light, sound, water, and electromagnetic waves. In fact, the discussions of Secs. 3.4 and 3.7 assumed that de Broglie waves are subject to the superposition principle.

Let us apply the superposition principle to the diffraction of an electron beam. Figure 5.2*a* shows a pair of slits through which a parallel beam of monoenergetic electrons pass on their way to a viewing screen. If slit 1 only is open, the result is the intensity variation shown in Fig. 5.2*b* that corresponds to the probability density

$$P_1 = |\Psi_1|^2 = \Psi_1^*\Psi_1$$

If slit 2 only is open, as in Fig. 5.2*c*, the corresponding probability density is

$$P_2 = |\Psi_2|^2 = \Psi_2^*\Psi_2$$

We might suppose that opening both slits would give an electron intensity variation described by $P_1 + P_2$, as in Fig. 5.2*d*. However, this is not the case because in quantum

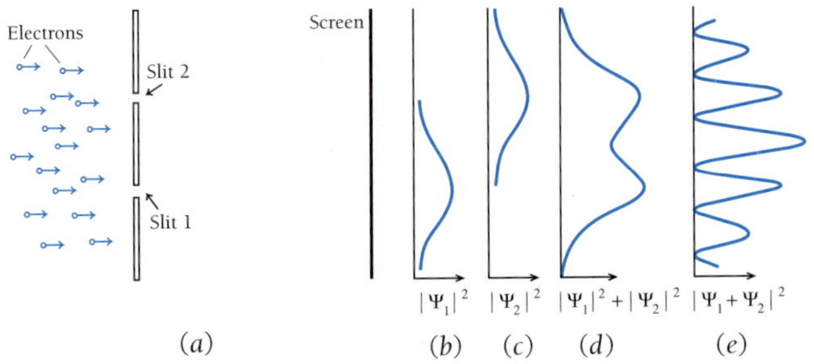

Figure 5.2 (*a*) Arrangement of double-slit experiment. (*b*) The electron intensity at the screen with only slit 1 open. (*c*) The electron intensity at the screen with only slit 2 open. (*d*) The sum of the intensities of (*b*) and (*c*). (*e*) The actual intensity at the screen with slits 1 and 2 both open. The wave functions Ψ_1 and Ψ_2 add to produce the intensity at the screen, not the probability densities $|\Psi_1|^2$ and $|\Psi_2|^2$.

mechanics wave functions add, *not* probabilities. Instead the result with both slits open is as shown in Fig. 5.2*e*, the same pattern of alternating maxima and minima that occurs when a beam of monochromatic light passes through the double slit of Fig. 2.4.

The diffraction pattern of Fig. 5.2*e* arises from the superposition Ψ of the wave functions Ψ_1 and Ψ_2 of the electrons that have passed through slits 1 and 2:

$$\Psi = \Psi_1 + \Psi_2$$

The probability density at the screen is therefore

$$\begin{aligned} P = |\Psi|^2 &= |\Psi_1 + \Psi_2|^2 = (\Psi_1^* + \Psi_2^*)(\Psi_1 + \Psi_2) \\ &= \Psi_1^*\Psi_1 + \Psi_2^*\Psi_2 + \Psi_1^*\Psi_2 + \Psi_2^*\Psi_1 \\ &= P_1 + P_2 + \Psi_1^*\Psi_2 + \Psi_2^*\Psi_1 \end{aligned}$$

The two terms at the right of this equation represent the difference between Fig. 5.2*d* and *e* and are responsible for the oscillations of the electron intensity at the screen. In Sec. 6.8 a similar calculation will be used to investigate why a hydrogen atom emits radiation when it undergoes a transition from one quantum state to another of lower energy.

5.5 EXPECTATION VALUES

How to extract information from a wave function

Once Schrödinger's equation has been solved for a particle in a given physical situation, the resulting wave function $\Psi(x, y, z, t)$ contains all the information about the particle that is permitted by the uncertainty principle. Except for those variables that are quantized this information is in the form of probabilities and not specific numbers.

As an example, let us calculate the **expectation value** $\langle x \rangle$ of the position of a particle confined to the x axis that is described by the wave function $\Psi(x, t)$. This is the value of x we would obtain if we measured the positions of a great many particles described by the same wave function at some instant t and then averaged the results.

To make the procedure clear, we first answer a slightly different question: What is the average position $\bar{x}$ of a number of identical particles distributed along the x axis in such a way that there are N_1 particles at x_1, N_2 particles at x_2, and so on? The average position in this case is the same as the center of mass of the distribution, and so

$$\bar{x} = \frac{N_1x_1 + N_2x_2 + N_3x_3 + \cdots}{N_1 + N_2 + N_3 + \cdots} = \frac{\Sigma N_i x_i}{\Sigma N_i} \tag{5.16}$$

When we are dealing with a single particle, we must replace the number N_i of particles at x_i by the probability P_i that the particle be found in an interval dx at x_i. This probability is

$$P_i = |\Psi_i|^2\, dx \tag{5.17}$$

where Ψ_i is the particle wave function evaluated at $x = x_i$. Making this substitution and changing the summations to integrals, we see that the expectation value of the position of the single particle is

$$\langle x \rangle = \frac{\int_{-\infty}^{\infty} x|\Psi|^2 \, dx}{\int_{-\infty}^{\infty} |\Psi|^2 \, dx} \tag{5.18}$$

If Ψ is a normalized wave function, the denominator of Eq. (5.18) equals the probability that the particle exists somewhere between $x = -\infty$ and $x = \infty$ and therefore has the value 1. In this case

Expectation value for position

$$\langle x \rangle = \int_{-\infty}^{\infty} x|\Psi|^2 \, dx \tag{5.19}$$

Example 5.2

A particle limited to the x axis has the wave function $\Psi = ax$ between $x = 0$ and $x = 1$; $\Psi = 0$ elsewhere. (*a*) Find the probability that the particle can be found between $x = 0.45$ and $x = 0.55$. (*b*) Find the expectation value $\langle x \rangle$ of the particle's position.

Solution

(*a*) The probability is

$$\int_{x_1}^{x_2} |\Psi|^2 \, dx = a^2 \int_{0.45}^{0.55} x^2 dx = a^2 \left[\frac{x^3}{3} \right]_{0.45}^{0.55} = 0.0251a^2$$

(*b*) The expectation value is

$$\langle x \rangle = \int_0^1 x|\Psi|^2 \, dx = a^2 \int_0^1 x^3 dx = a^2 \left[\frac{x^4}{4} \right]_0^1 = \frac{a^2}{4}$$

The same procedure as that followed above can be used to obtain the expectation value $\langle G(x) \rangle$ of any quantity—for instance, potential energy $U(x)$—that is a function of the position x of a particle described by a wave function Ψ. The result is

Expectation value

$$\langle G(x) \rangle = \int_{-\infty}^{\infty} G(x)|\Psi|^2 \, dx \tag{5.20}$$

The expectation value $\langle p \rangle$ for momentum cannot be calculated this way because, according to the uncertainty principles, no such function as $p(x)$ can exist. If we specify x, so that $\Delta x = 0$, we cannot specify a corresponding p since $\Delta x \, \Delta p \geq \hbar/2$. The same problem occurs for the expectation value $\langle E \rangle$ for energy because $\Delta E \Delta t \geq \hbar/2$ means that, if we specify t, the function $E(t)$ is impossible. In Sec. 5.6 we will see how $\langle p \rangle$ and $\langle E \rangle$ can be determined.

In classical physics no such limitation occurs, because the uncertainty principle can be neglected in the macroworld. When we apply the second law of motion to the motion of a body subject to various forces, we expect to get $p(x, t)$ and $E(x, t)$ from the solution as well as $x(t)$. Solving a problem in classical mechanics gives us the entire future course of the body's motion. In quantum physics, on the other hand, all we get directly by applying Schrödinger's equation to the motion of a particle is the wave function Ψ, and the future course of the particle's motion—like its initial state—is a matter of probabilities instead of certainties.

5.6 OPERATORS

Another way to find expectation values

A hint as to the proper way to evaluate $\langle p \rangle$ and $\langle E \rangle$ comes from differentiating the free-particle wave function $\Psi = Ae^{-(i/\hbar)(Et-px)}$ with respect to x and to t. We find that

$$\frac{\partial \Psi}{\partial x} = \frac{i}{\hbar} p\Psi$$

$$\frac{\partial \Psi}{\partial t} = -\frac{i}{\hbar} E\Psi$$

which can be written in the suggestive forms

$$p\Psi = \frac{\hbar}{i}\frac{\partial}{\partial x}\Psi \tag{5.21}$$

$$E\Psi = i\hbar \frac{\partial}{\partial t}\Psi \tag{5.22}$$

Evidently the dynamical quantity p in some sense corresponds to the differential operator $(\hbar/i)\ \partial/\partial x$ and the dynamical quantity E similarly corresponds to the differential operator $i\hbar\ \partial/\partial t$.

An **operator** tells us what operation to carry out on the quantity that follows it. Thus the operator $i\hbar\ \partial/\partial t$ instructs us to take the partial derivative of what comes after it with respect to t and multiply the result by $i\hbar$. Equation (5.22) was on the postmark used to cancel the Austrian postage stamp issued to commemorate the 100th anniversary of Schrödinger's birth.

It is customary to denote operators by using a caret, so that $\hat{p}$ is the operator that corresponds to momentum p and $\hat{E}$ is the operator that corresponds to total energy E. From Eqs. (5.21) and (5.22) these operators are

Momentum operator

$$\hat{p} = \frac{\hbar}{i}\frac{\partial}{\partial x} \tag{5.23}$$

Total-energy operator

$$\hat{E} = i\hbar \frac{\partial}{\partial t} \tag{5.24}$$

Though we have only shown that the correspondences expressed in Eqs. (5.23) and (5.24) hold for free particles, they are entirely general results whose validity is the same as that of Schrödinger's equation. To support this statement, we can replace the equation $E = \mathrm{KE} + U$ for the total energy of a particle with the operator equation

$$\hat{E} = \hat{\mathrm{KE}} + \hat{U} \tag{5.25}$$

The operator $\hat{U}$ is just $U(\Psi)$. The kinetic energy KE is given in terms of momentum p by

$$\mathrm{KE} = \frac{p^2}{2m}$$

and so we have

Kinetic-energy operator

$$\hat{KE} = \frac{\hat{p}^2}{2m} = \frac{1}{2m}\left(\frac{\hbar}{i}\frac{\partial}{\partial x}\right)^2 = -\frac{\hbar^2}{2m}\frac{\partial^2}{\partial x^2} \tag{5.26}$$

Equation (5.25) therefore reads

$$i\hbar\frac{\partial}{\partial t} = -\frac{\hbar^2}{2m}\frac{\partial^2}{\partial x^2} + U \tag{5.27}$$

Now we multiply the identity $\Psi = \Psi$ by Eq. (5.27) and obtain

$$i\hbar\frac{\partial\Psi}{\partial t} = -\frac{\hbar^2}{2m}\frac{\partial^2\Psi}{\partial x^2} + U\Psi$$

which is Schrödinger's equation. Postulating Eqs. (5.23) and (5.24) is equivalent to postulating Schrödinger's equation.

Operators and Expectation Values

Because p and E can be replaced by their corresponding operators in an equation, we can use these operators to obtain expectation values for p and E. Thus the expectation value for p is

$$\langle p\rangle = \int_{-\infty}^{\infty}\Psi^*\hat{p}\Psi\,dx = \int_{-\infty}^{\infty}\Psi^*\left(\frac{\hbar}{i}\frac{\partial}{\partial x}\right)\Psi\,dx = \frac{\hbar}{i}\int_{-\infty}^{\infty}\Psi^*\frac{\partial\Psi}{\partial x}\,dx \tag{5.28}$$

and the expectation value for E is

$$\langle E\rangle = \int_{-\infty}^{\infty}\Psi^*\hat{E}\Psi\,dx = \int_{-\infty}^{\infty}\Psi^*\left(i\hbar\frac{\partial}{\partial t}\right)\Psi\,dx = i\hbar\int_{-\infty}^{\infty}\Psi^*\frac{\partial\Psi}{\partial t}\,dx \tag{5.29}$$

Both Eqs. (5.28) and (5.29) can be evaluated for any acceptable wave function $\Psi(x, t)$.

Let us see why expectation values involving operators have to be expressed in the form

$$\langle p\rangle = \int_{-\infty}^{\infty}\Psi^*\hat{p}\Psi\,dx$$

The other alternatives are

$$\int_{-\infty}^{\infty}\hat{p}\Psi^*\Psi\,dx = \frac{\hbar}{i}\int_{-\infty}^{\infty}\frac{\partial}{\partial x}(\Psi^*\Psi)\,dx = \frac{\hbar}{i}\Big[\Psi^*\Psi\Big]_{-\infty}^{\infty} = 0$$

since Ψ^* and Ψ must be 0 at $x = \pm\infty$, and

$$\int_{-\infty}^{\infty}\Psi^*\Psi\hat{p}\,dx = \frac{\hbar}{i}\int_{-\infty}^{\infty}\Psi^*\Psi\frac{\partial}{\partial x}\,dx$$

which makes no sense. In the case of algebraic quantities such as x and $V(x)$, the order of factors in the integrand is unimportant, but when differential operators are involved, the correct order of factors must be observed.

Every observable quantity G characteristic of a physical system may be represented by a suitable quantum-mechanical operator $\hat{G}$. To obtain this operator, we express G in terms of x and p and then replace p by $(\hbar/i)\partial/\partial x$. If the wave function Ψ of the system is known, the expectation value of $G(x, p)$ is

Expectation value of an operator

$$\langle G(x, p)\rangle = \int_{-\infty}^{\infty} \Psi^* \hat{G} \Psi \, dx \tag{5.30}$$

In this way all the information about a system that is permitted by the uncertainty principle can be obtained from its wave function Ψ.

5.7 SCHRÖDINGER'S EQUATION: STEADY-STATE FORM

Eigenvalues and eigenfunctions

In a great many situations the potential energy of a particle does not depend on time explicitly; the forces that act on it, and hence U, vary with the position of the particle only. When this is true, Schrödinger's equation may be simplified by removing all reference to t.

We begin by noting that the one-dimensional wave function Ψ of an unrestricted particle may be written

$$\Psi = Ae^{-(i/\hbar)(Et-px)} = Ae^{-(iE/\hbar)t}e^{+(ip/\hbar)x} = \psi e^{-(iE/\hbar)t} \tag{5.31}$$

Evidently Ψ is the product of a time-dependent function $e^{-(iE/\hbar)t}$ and a position-dependent function ψ. As it happens, the time variations of *all* wave functions of particles acted on by forces independent of time have the same form as that of an unrestricted particle. Substituting the Ψ of Eq. (5.31) into the time-dependent form of Schrödinger's equation, we find that

$$E\psi e^{-(iE/\hbar)t} = -\frac{\hbar^2}{2m} e^{-(iE/\hbar)t} \frac{\partial^2 \psi}{\partial x^2} + U\psi e^{-(iE/\hbar)t}$$

Dividing through by the common exponential factor gives

Steady-state Schrödinger equation in one dimension

$$\frac{\partial^2 \psi}{\partial x^2} + \frac{2m}{\hbar^2}(E - U)\psi = 0 \tag{5.32}$$

Equation (5.32) is the **steady-state form of Schrödinger's equation.** In three dimensions it is

Steady-state Schrödinger equation in three dimensions

$$\frac{\partial^2 \psi}{\partial x^2} + \frac{\partial^2 \psi}{\partial y^2} + \frac{\partial^2 \psi}{\partial z^2} + \frac{2m}{\hbar^2}(E - U)\psi = 0 \tag{5.33}$$

An important property of Schrödinger's steady-state equation is that, if it has one or more solutions for a given system, each of these wave functions corresponds to a specific value of the energy E. Thus energy quantization appears in wave mechanics as a natural element of the theory, and energy quantization in the physical world is revealed as a universal phenomenon characteristic of *all* stable systems.

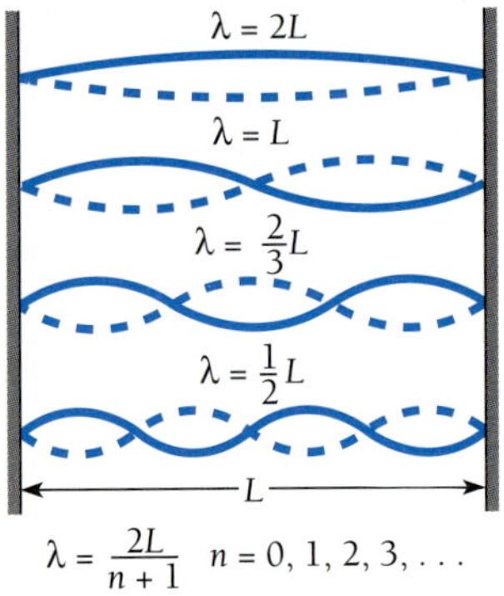

Figure 5.3 Standing waves in a stretched string fastened at both ends.

A familiar and quite close analogy to the manner in which energy quantization occurs in solutions of Schrödinger's equation is with standing waves in a stretched string of length L that is fixed at both ends. Here, instead of a single wave propagating indefinitely in one direction, waves are traveling in both the $+x$ and $-x$ directions simultaneously. These waves are subject to the condition (called a **boundary condition**) that the displacement y always be zero at both ends of the string. An acceptable function $y(x, t)$ for the displacement must, with its derivatives (except at the ends), be as well-behaved as ψ and its derivatives—that is, be continuous, finite, and single-valued. In this case y must be real, not complex, as it represents a directly measurable quantity. The only solutions of the wave equation, Eq. (5.3), that are in accord with these various limitations are those in which the wavelengths are given by

$$\lambda_n = \frac{2L}{n+1} \qquad n = 0, 1, 2, 3, \ldots$$

as shown in Fig. 5.3. It is the *combination* of the wave equation and the restrictions placed on the nature of its solution that leads us to conclude that $y(x, t)$ can exist only for certain wavelengths λ_n.

Eigenvalues and Eigenfunctions

The values of energy E_n for which Schrödinger's steady-state equation can be solved are called **eigenvalues** and the corresponding wave functions ψ_n are called **eigenfunctions.** (These terms come from the German *Eigenwert,* meaning "proper or characteristic value," and *Eigenfunktion,* "proper or characteristic function.") The discrete energy levels of the hydrogen atom

$$E_n = -\frac{me^4}{32\pi^2\epsilon_0^2\hbar^2}\left(\frac{1}{n^2}\right) \qquad n = 1, 2, 3, \ldots$$

are an example of a set of eigenvalues. We shall see in Chap. 6 why these particular values of E are the only ones that yield acceptable wave functions for the electron in the hydrogen atom.

An important example of a dynamical variable other than total energy that is found to be quantized in stable systems is angular momentum **L**. In the case of the hydrogen atom, we shall find that the eigenvalues of the magnitude of the total angular momentum are specified by

$$L = \sqrt{l(l+1)}\,\hbar \qquad l = 0, 1, 2, \ldots, (n-1)$$

Of course, a dynamical variable G may not be quantized. In this case, measurements of G made on a number of identical systems will not yield a unique result but instead a spread of values whose average is the expectation value

$$\langle G \rangle = \int_{-\infty}^{\infty} G|\psi|^2 \, dx$$

In the hydrogen atom, the electron's position is not quantized, for instance, so that we must think of the electron as being present in the vicinity of the nucleus with a certain probability $|\psi|^2$ per unit volume but with no predictable position or even orbit in the classical sense. This probabilistic statement does not conflict with the fact that experiments performed on hydrogen atoms always show that each one contains a whole electron, not 27 percent of an electron in a certain region and 73 percent elsewhere. The probability is one of *finding* the electron, and although this probability is smeared out in space, the electron itself is not.

Operators and Eigenvalues

The condition that a certain dynamical variable G be restricted to the discrete values G_n—in other words, that G be quantized—is that the wave functions ψ_n of the system be such that

Eigenvalue equation
$$\hat{G}\psi_n = G_n\psi_n \tag{5.34}$$

where $\hat{G}$ is the operator that corresponds to G and each G_n is a real number. When Eq. (5.34) holds for the wave functions of a system, it is a fundamental postulate of quantum mechanics that any measurement of G can only yield one of the values G_n. If measurements of G are made on a number of identical systems all in states described by the particular eigenfunction ψ_k, each measurement will yield the single value G_k.

Example 5.3

An eigenfunction of the operator d^2/dx^2 is $\psi = e^{2x}$. Find the corresponding eigenvalue.

Solution

Here $\hat{G} = d^2/dx^2$, so

$$\hat{G}\psi = \frac{d^2}{dx^2}(e^{2x}) = \frac{d}{dx}\left[\frac{d}{dx}(e^{2x})\right] = \frac{d}{dx}(2e^{2x}) = 4e^{2x}$$

But $e^{2x} = \psi$, so

$$\hat{G}\psi = 4\psi$$

From Eq. (5.34) we see that the eigenvalue G here is just $G = 4$.

In view of Eqs. (5.25) and (5.26) the total-energy operator $\hat{E}$ of Eq. (5.24) can also be written as

Hamiltonian operator
$$\hat{H} = -\frac{\hbar^2}{2m}\frac{\partial^2}{\partial x^2} + U \tag{5.35}$$

and is called the **Hamiltonian operator** because it is reminiscent of the Hamiltonian function in advanced classical mechanics, which is an expression for the total energy of a system in terms of coordinates and momenta only. Evidently the steady-state Schrödinger equation can be written simply as

Schrödinger's equation
$$\hat{H}\psi_n = E_n\psi_n \tag{5.36}$$

Table 5.1 Operators Associated with Various Observable Quantities

Quantity	Operator
Position, x	x
Linear momentum, p	$\frac{\hbar}{i}\frac{\partial}{\partial x}$
Potential energy, $U(x)$	$U(x)$
Kinetic energy, $KE = \frac{p^2}{2m}$	$-\frac{\hbar^2}{2m}\frac{\partial^2}{\partial x^2}$
Total energy, E	$i\hbar\frac{\partial}{\partial t}$
Total energy (Hamiltonian form), H	$-\frac{\hbar^2}{2m}\frac{\partial^2}{\partial x^2} + U(x)$

so we can say that the various E_n are the eigenvalues of the Hamiltonian operator $\hat{H}$. This kind of association between eigenvalues and quantum-mechanical operators is quite general. Table 5.1 lists the operators that correspond to various observable quantities.

5.8 PARTICLE IN A BOX

How boundary conditions and normalization determine wave functions

To solve Schrödinger's equation, even in its simpler steady-state form, usually requires elaborate mathematical techniques. For this reason the study of quantum mechanics has traditionally been reserved for advanced students who have the required proficiency in mathematics. However, since quantum mechanics is the theoretical structure whose results are closest to experimental reality, we must explore its methods and applications to understand modern physics. As we shall see, even a modest mathematical background is enough for us to follow the trains of thought that have led quantum mechanics to its greatest achievements.

The simplest quantum-mechanical problem is that of a particle trapped in a box with infinitely hard walls. In Sec. 3.6 we saw how a quite simple argument yields the energy levels of the system. Let us now tackle the same problem in a more formal way, which will give us the wave function ψ_n that corresponds to each energy level.

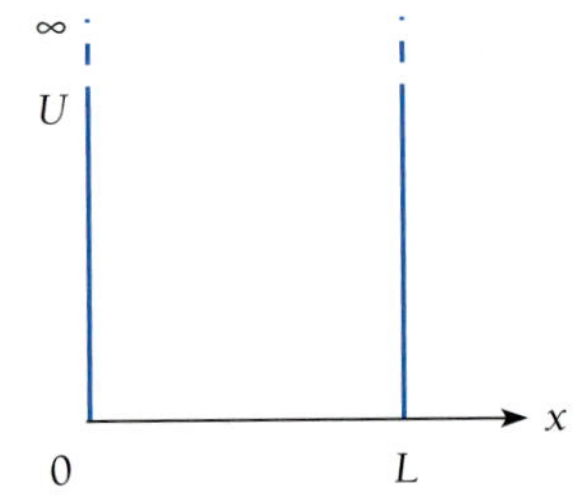

Figure 5.4 A square potential well with infinitely high barriers at each end corresponds to a box with infinitely hard walls.

We may specify the particle's motion by saying that it is restricted to traveling along the x axis between $x = 0$ and $x = L$ by infintely hard walls. A particle does not lose energy when it collides with such walls, so that its total energy stays constant. From a formal point of view the potential energy U of the particle is infinite on both sides of the box, while U is a constant—say 0 for convenience—on the inside (Fig. 5.4). Because the particle cannot have an infinite amount of energy, it cannot exist outside the box, and so its wave function ψ is 0 for $x \leq 0$ and $x \geq L$. Our task is to find what ψ is within the box, namely, between $x = 0$ and $x = L$.

Within the box Schrödinger's equation becomes

$$\frac{d^2\psi}{dx^2} + \frac{2m}{\hbar^2}E\psi = 0 \tag{5.37}$$

since $U = 0$ there. (The total derivative $d^2\psi/dx^2$ is the same as the partial derivative $\partial^2\psi/\partial x^2$ because ψ is a function only of x in this problem.) Equation (5.37) has the solution

$$\psi = A \sin \frac{\sqrt{2mE}}{\hbar} x + B \cos \frac{\sqrt{2mE}}{\hbar} x \tag{5.38}$$

which we can verify by substitution back into Eq. (5.37). A and B are constants to be evaluated.

This solution is subject to the boundary conditions that $\psi = 0$ for $x = 0$ and for $x = L$. Since $\cos 0 = 1$, the second term cannot describe the particle because it does not vanish at $x = 0$. Hence we conclude that $B = 0$. Since $\sin 0 = 0$, the sine term always yields $\psi = 0$ at $x = 0$, as required, but ψ will be 0 at $x = L$ only when

$$\frac{\sqrt{2mE}}{\hbar} L = n\pi \qquad n = 1, 2, 3, \ldots \tag{5.39}$$

This result comes about because the sines of the angles π, 2π, 3π, . . . are all 0.

From Eq. (5.39) it is clear that the energy of the particle can have only certain values, which are the eigenvalues mentioned in the previous section. These eigenvalues, constituting the **energy levels** of the system, are found by solving Eq. (5.39) for E_n, which gives

Particle in a box

$$E_n = \frac{n^2\pi^2\hbar^2}{2mL^2} \qquad n = 1, 2, 3, \ldots \tag{5.40}$$

Equation (5.40) is the same as Eq. (3.18) and has the same interpretation [see the discussion that follows Eq. (3.18) in Sec. 3.6].

Wave Functions

The wave functions of a particle in a box whose energies are E_n are, from Eq. (5.38) with $B = 0$,

$$\psi_n = A \sin \frac{\sqrt{2mE_n}}{\hbar} x \tag{5.41}$$

Substituting Eq. (5.40) for E_n gives

$$\psi_n = A \sin \frac{n\pi x}{L} \tag{5.42}$$

for the eigenfunctions corresponding to the energy eigenvalues E_n.

It is easy to verify that these eigenfunctions meet all the requirements discussed in Sec. 5.1: for each quantum number n, ψ_n is a finite, single-valued function of x, and ψ_n and $\partial\psi_n/\partial x$ are continuous (except at the ends of the box). Furthermore, the integral of $|\psi_n|^2$ over all space is finite, as we can see by integrating $|\psi_n|^2\,dx$ from $x = 0$ to $x = L$ (since the particle is confined within these limits). With the help of the trigonometric identity $\sin^2\theta = \frac{1}{2}(1 - \cos 2\theta)$ we find that

$$\int_{-\infty}^{\infty} |\psi_n|^2\,dx = \int_0^L |\psi_n|^2\,dx = A^2 \int_0^L \sin^2\left(\frac{n\pi x}{L}\right) dx$$

$$= \frac{A^2}{2}\left[\int_0^L dx - \int_0^L \cos\left(\frac{2n\pi x}{L}\right) dx\right]$$

$$= \frac{A^2}{2}\left[x - \left(\frac{L}{2n\pi}\right)\sin\frac{2n\pi x}{L}\right]_0^L = A^2\left(\frac{L}{2}\right) \tag{5.43}$$

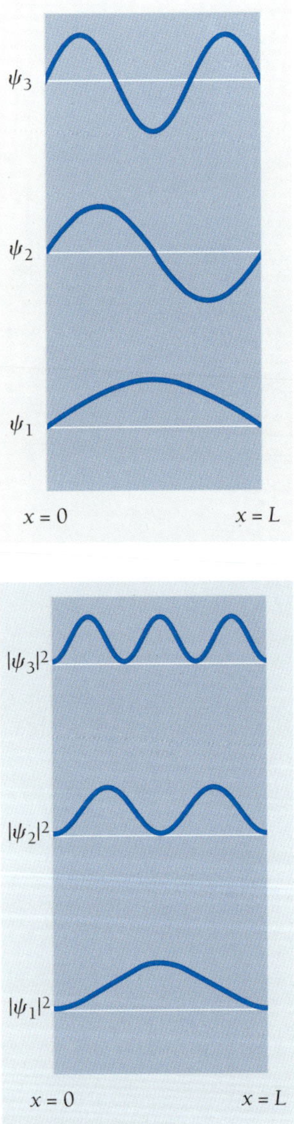

Figure 5.5 Wave functions and probability densities of a particle confined to a box with rigid walls.

To normalize ψ we must assign a value to A such that $|\psi_n|^2\,dx$ is *equal* to the probability $P\,dx$ of finding the particle between x and $x + dx$, rather than merely proportional to $P\,dx$. If $|\psi_n|^2\,dx$ is to equal $P\,dx$, then it must be true that

$$\int_{-\infty}^{\infty} |\psi_n|^2\,dx = 1 \tag{5.44}$$

Comparing Eqs. (5.43) and (5.44), we see that the wave functions of a particle in a box are normalized if

$$A = \sqrt{\frac{2}{L}} \tag{5.45}$$

The normalized wave functions of the particle are therefore

Particle in a box

$$\psi_n = \sqrt{\frac{2}{L}}\sin\frac{n\pi x}{L} \qquad n = 1, 2, 3, \ldots \tag{5.46}$$

The normalized wave functions ψ_1, ψ_2, and ψ_3 together with the probability densities $|\psi_1|^2$, $|\psi_2|^2$, and $|\psi_3|^2$ are plotted in Fig. 5.5. Although ψ_n may be negative as well as positive, $|\psi_n|^2$ is never negative and, since ψ_n is normalized, its value at a given x is equal to the probability density of finding the particle there. In every case $|\psi_n|^2 = 0$ at $x = 0$ and $x = L$, the boundaries of the box.

At a particular place in the box the probability of the particle being present may be very different for different quantum numbers. For instance, $|\psi_1|^2$ has its maximum value of $2/L$ in the middle of the box, while $|\psi_2|^2 = 0$ there. A particle in the lowest energy level of $n = 1$ is most likely to be in the middle of the box, while a particle in the next higher state of $n = 2$ is *never* there! Classical physics, of course, suggests the same probability for the particle being anywhere in the box.

The wave functions shown in Fig. 5.5 resemble the possible vibrations of a string fixed at both ends, such as those of the stretched string of Fig. 5.2. This follows from the fact that waves in a stretched string and the wave representing a moving particle are described by equations of the same form, so that when identical restrictions are placed upon each kind of wave, the formal results are identical.

Example 5.4

Find the probability that a particle trapped in a box L wide can be found between $0.45L$ and $0.55L$ for the ground and first excited states.

Solution

This part of the box is one-tenth of the box's width and is centered on the middle of the box (Fig. 5.6). Classically we would expect the particle to be in this region 10 percent of the time. Quantum mechanics gives quite different predictions that depend on the quantum number of the particle's state. From Eqs. (5.2) and (5.46) the probability of finding the particle between x_1 and x_2 when it is in the nth state is

$$P_{x_1,x_2} = \int_{x_1}^{x_2} |\psi_n|^2\,dx = \frac{2}{L}\int_{x_1}^{x_2} \sin^2\frac{n\pi x}{L}\,dx$$

$$= \left[\frac{x}{L} - \frac{1}{2n\pi}\sin\frac{2n\pi x}{L}\right]_{x_1}^{x_2}$$

Here $x_1 = 0.45L$ and $x_2 = 0.55L$. For the ground state, which corresponds to $n = 1$, we have

$$P_{x_1,x_2} = 0.198 = 19.8 \text{ percent}$$

This is about twice the classical probability. For the first excited state, which corresponds to $n = 2$, we have

$$P_{x_1,x_2} = 0.0065 = 0.65 \text{ percent}$$

This low figure is consistent with the probability density of $|\psi_n|^2 = 0$ at $x = 0.5L$.

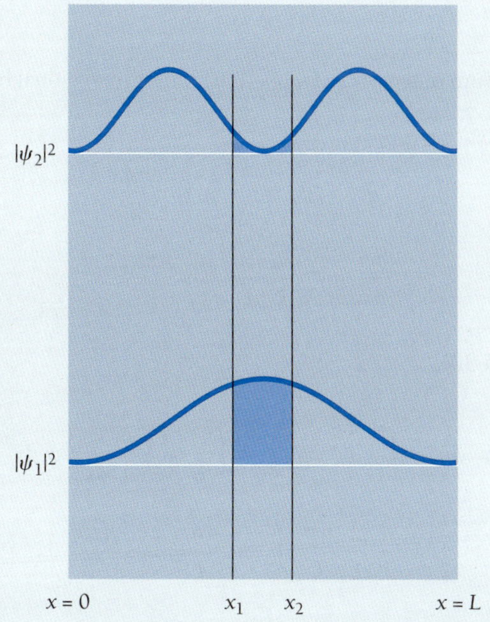

Figure 5.6 The probability P_{x_1,x_2} of finding a particle in the box of Fig. 5.5 between $x_1 = 0.45L$ and $x_2 = 0.55L$ is equal to the area under the $|\psi|^2$ curves between these limits.

Example 5.5

Find the expectation value $\langle x \rangle$ of the position of a particle trapped in a box L wide.

Solution

From Eqs. (5.19) and (5.46) we have

$$\langle x \rangle = \int_{-\infty}^{\infty} x|\psi|^2\, dx = \frac{2}{L}\int_0^L x \sin^2 \frac{n\pi x}{L}\, dx$$

$$= \frac{2}{L}\left[\frac{x^2}{4} - \frac{x\sin(2n\pi x/L)}{4n\pi/L} - \frac{\cos(2n\pi x/L)}{8(n\pi/L)^2}\right]_0^L$$

Since $\sin n\pi = 0$, $\cos 2n\pi = 1$, and $\cos 0 = 1$, for all the values of n the expectation value of x is

$$\langle x \rangle = \frac{2}{L}\left(\frac{L^2}{4}\right) = \frac{L}{2}$$

This result means that the average position of the particle is the middle of the box in all quantum states. There is no conflict with the fact that $|\psi|^2 = 0$ at $L/2$ in the $n = 2, 4, 6, \ldots$ states because $\langle x \rangle$ is an *average*, not a probability, and it reflects the symmetry *of* $|\psi|^2$ about the middle of the box.

Momentum

Finding the momentum of a particle trapped in a one-dimensional box is not as straightforward as finding $\langle x \rangle$. Here

$$\psi^* = \psi_n = \sqrt{\frac{2}{L}}\sin\frac{n\pi x}{L}$$

$$\frac{d\psi}{dx} = \sqrt{\frac{2}{L}}\frac{n\pi}{L}\cos\frac{n\pi x}{L}$$

and so, from Eq. (5.30),

$$\langle p \rangle = \int_{-\infty}^{\infty} \psi^* \hat{p} \psi\, dx = \int_{-\infty}^{\infty} \psi^* \left(\frac{\hbar}{i}\frac{d}{dx}\right)\psi\, dx$$

$$= \frac{\hbar}{i}\frac{2}{L}\frac{n\pi}{L}\int_0^L \sin\frac{n\pi x}{L}\cos\frac{n\pi x}{L}\, dx$$

We note that

$$\int \sin ax \cos ax\, dx = \frac{1}{2a}\sin^2 ax$$

With $a = n\pi/L$ we have

$$\langle p \rangle = \frac{\hbar}{iL}\left[\sin^2 \frac{n\pi x}{L}\right]_0^L = 0$$

since $$\sin^2 0 = \sin^2 n\pi = 0 \qquad n = 1, 2, 3, \ldots$$

The expectation value $\langle p \rangle$ of the particle's momentum is 0.

At first glance this conclusion seems strange. After all, $E = p^2/2m$, and so we would anticipate that

Momentum eigenvalues for trapped particle

$$p_n = \pm\sqrt{2mE_n} = \pm\frac{n\pi\hbar}{L} \tag{5.47}$$

The $\pm$ sign provides the explanation: The particle is moving back and forth, and so its *average* momentum for any value of n is

$$p_{\text{av}} = \frac{(+n\pi\hbar/L) + (-n\pi\hbar/L)}{2} = 0$$

which is the expectation value.

According to Eq. (5.47) there should be two momentum eigenfunctions for every energy eigenfunction, corresponding to the two possible directions of motion. The general procedure for finding the eigenvalues of a quantum-mechanical operator, here $\hat{p}$, is to start from the eigenvalue equation

$$\hat{p}\psi_n = p_n\psi_n \tag{5.48}$$

where each p_n is a real number. This equation holds only when the wave functions ψ_n are eigenfunctions of the momentum operator $\hat{p}$, which here is

$$\hat{p} = \frac{\hbar}{i}\frac{d}{dx}$$

We can see at once that the energy eigenfunctions

$$\psi_n = \sqrt{\frac{2}{L}}\sin\frac{n\pi x}{L}$$

are not also momentum eigenfunctions, because

$$\frac{\hbar}{i}\frac{d}{dx}\left(\sqrt{\frac{2}{L}}\sin\frac{n\pi x}{L}\right) = \frac{\hbar}{i}\frac{n\pi}{L}\sqrt{\frac{2}{L}}\cos\frac{n\pi x}{L} \neq p_n\psi_n$$

To find the correct momentum eigenfunctions, we note that

$$\sin\theta = \frac{e^{i\theta} - e^{-i\theta}}{2i} = \frac{1}{2i}e^{i\theta} - \frac{1}{2i}e^{-i\theta}$$

Hence each energy eigenfunction can be expressed as a linear combination of the two wave functions

Momentum eigenfunctions for trapped particle

$$\psi_n^+ = \frac{1}{2i}\sqrt{\frac{2}{L}}\ e^{in\pi x/L} \tag{5.49}$$

$$\psi_n^- = \frac{1}{2i}\sqrt{\frac{2}{L}}\ e^{-in\pi x/L} \tag{5.50}$$

Inserting the first of these wave functions in the eigenvalue equation, Eq. (5.48), we have

$$\hat{p}\psi_n^+ = p_n^+\psi_n^+$$

$$\frac{\hbar}{i}\frac{d}{dx}\psi_n^+ = \frac{\hbar}{i}\frac{1}{2i}\sqrt{\frac{2}{L}}\frac{in\pi}{L}e^{in\pi x/L} = \frac{n\pi\hbar}{L}\psi_n^+ = p_n^+\psi_n^+$$

so that

$$p_n^+ = +\frac{n\pi\hbar}{L} \tag{5.51}$$

Similarly the wave function ψ_n^- leads to the momentum eigenvalues

$$p_n^- = -\frac{n\pi\hbar}{L} \tag{5.52}$$

We conclude that ψ_n^+ and ψ_n^- are indeed the momentum eigenfunctions for a particle in a box, and that Eq. (5.47) correctly states the corresponding momentum eigenvalues.

5.9 FINITE POTENTIAL WELL

The wave function penetrates the walls, which lowers the energy levels

Potential energies are never infinite in the real world, and the box with infinitely hard walls of the previous section has no physical counterpart. However, potential wells with barriers of finite height certainly do exist. Let us see what the wave functions and energy levels of a particle in such a well are.

Figure 5.7 shows a potential well with square corners that is U high and L wide and contains a particle whose energy E is less than U. According to classical mechanics, when the particle strikes the sides of the well, it bounces off without entering regions I and III. In quantum mechanics, the particle also bounces back and forth, but now it has a certain probability of penetrating into regions I and III even though $E < U$.

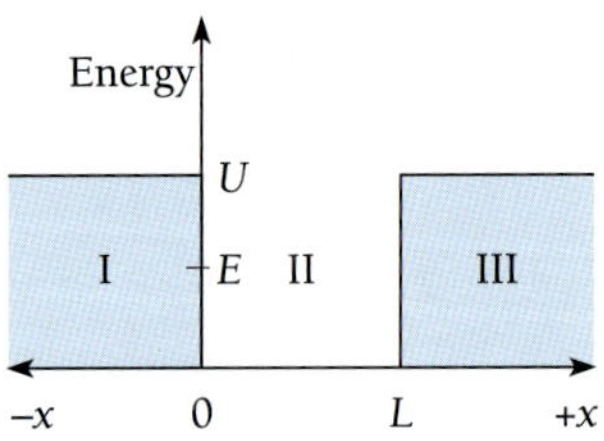

Figure 5.7 A square potential well with finite barriers. The energy E of the trapped particle is less than the height U of the barriers.

In regions I and III Schrödinger's steady-state equation is

$$\frac{d^2\psi}{dx^2} + \frac{2m}{\hbar^2}(E - U)\psi = 0$$

which we can rewrite in the more convenient form

$$\frac{d^2\psi}{dx^2} - a^2\psi = 0 \qquad \begin{matrix} x < 0 \\ x > L \end{matrix} \tag{5.53}$$

where

$$a = \frac{\sqrt{2m(U - E)}}{\hbar} \tag{5.54}$$

The solutions to Eq. (5.53) are real exponentials:

$$\psi_{\text{I}} = Ce^{ax} + De^{-ax} \tag{5.55}$$

$$\psi_{\text{III}} = Fe^{ax} + Ge^{-ax} \tag{5.56}$$

Both ψ_{I} and ψ_{III} must be finite everywhere. Since $e^{-ax} \to \infty$ as $x \to -\infty$ and $e^{ax} \to \infty$ as $x \to \infty$, the coefficients D and F must therefore be 0. Hence we have

$$\psi_{\text{I}} = Ce^{ax} \tag{5.57}$$

$$\psi_{\text{III}} = Ge^{-ax} \tag{5.58}$$

These wave functions decrease exponentially inside the barriers at the sides of the well.

Within the well Schrödinger's equation is the same as Eq. (5.37) and its solution is again

$$\psi_{\text{II}} = A \sin\frac{\sqrt{2mE}}{\hbar}x + B \cos\frac{\sqrt{2mE}}{\hbar}x \tag{5.59}$$

In the case of a well with infinitely high barriers, we found that $B = 0$ in order that $\psi = 0$ at $x = 0$ and $x = L$. Here, however, $\psi_{\text{II}} = C$ at $x = 0$ and $\psi_{\text{II}} = G$ at $x = L$, so both the sine and cosine solutions of Eq. (5.59) are possible.

For either solution, both ψ and $d\psi/dx$ must be continuous at $x = 0$ and $x = L$: the wave functions inside and outside each side of the well must not only have the same value where they join but also the same slopes, so they match up perfectly. When these boundary conditions are taken into account, the result is that exact matching only occurs for certain specific values E_n of the particle energy. The complete wave functions and their probability densities are shown in Fig. 5.8.

Because the wavelengths that fit into the well are longer than for an infinite well of the same width (see Fig. 5.5), the corresponding particle momenta are lower (we recall that $\lambda = h/p$). Hence the energy levels E_n are lower for each n than they are for a particle in an infinite well.

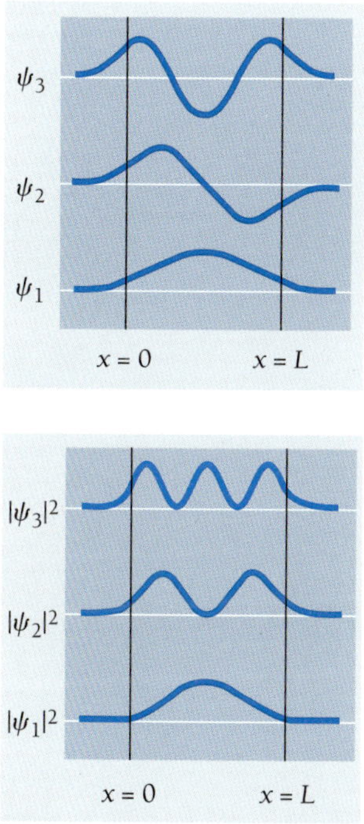

Figure 5.8 Wave functions and probability densities of a particle in a finite potential well. The particle has a certain probability of being found outside the wall.

5.10 TUNNEL EFFECT

A particle without the energy to pass over a potential barrier may still tunnel through it

Although the walls of the potential well of Fig. 5.7 were of finite height, they were assumed to be infinitely thick. As a result the particle was trapped forever even though it could penetrate the walls. We next look at the situation of a particle that strikes a potential barrier of height U, again with $E < U$, but here the barrier has a finite width (Fig. 5.9). What we will find is that the particle has a certain probability—not

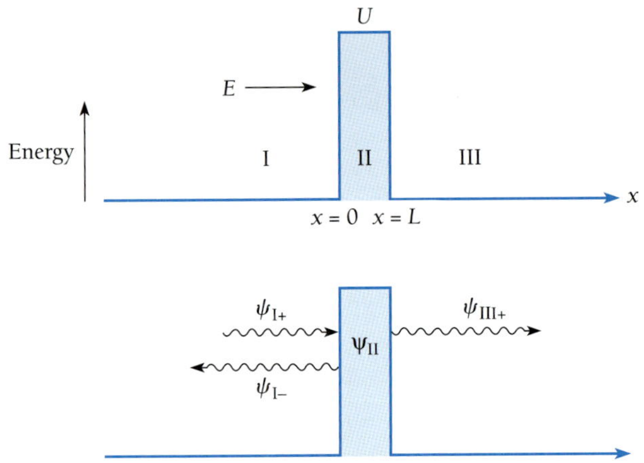

Figure 5.9 When a particle of energy $E < U$ approaches a potential barrier, according to classical mechanics the particle must be reflected. In quantum mechanics, the de Broglie waves that correspond to the particle are partly reflected and partly transmitted, which means that the particle has a finite chance of penetrating the barrier.

necessarily great, but not zero either—of passing through the barrier and emerging on the other side. The particle lacks the energy to go over the top of the barrier, but it can nevertheless tunnel through it, so to speak. Not surprisingly, the higher the barrier and the wider it is, the less the chance that the particle can get through.

The **tunnel effect** actually occurs, notably in the case of the alpha particles emitted by certain radioactive nuclei. As we shall learn in Chap. 12, an alpha particle whose kinetic energy is only a few MeV is able to escape from a nucleus whose potential wall is perhaps 25 MeV high. The probability of escape is so small that the alpha particle might have to strike the wall 10^{38} or more times before it emerges, but sooner or later it does get out. Tunneling also occurs in the operation of certain semiconductor diodes (Sec. 10.7) in which electrons pass through potential barriers even though their kinetic energies are smaller than the barrier heights.

Let us consider a beam of identical particles all of which have the kinetic energy E. The beam is incident from the left on a potential barrier of height U and width L, as in Fig. 5.9. On both sides of the barrier $U = 0$, which means that no forces act on the particles there. The wave function ψ_{I+} represents the incoming particles moving to the right and ψ_{I-} represents the reflected particles moving to the left; ψ_{III} represents the transmitted particles moving to the right. The wave function ψ_{II} represents the particles inside the barrier, some of which end up in region III while the others return to region I. The transmission probability T for a particle to pass through the barrier is equal to the fraction of the incident beam that gets through the barrier. This probability is calculated in the Appendix to this chapter. Its approximate value is given by

Approximate transmission probability

$$T = e^{-2k_2L} \tag{5.60}$$

where

$$k_2 = \frac{\sqrt{2m(U - E)}}{\hbar} \tag{5.61}$$

and L is the width of the barrier.

Example 5.6

Electrons with energies of 1.0 eV and 2.0 eV are incident on a barrier 10.0 eV high and 0.50 nm wide. (*a*) Find their respective transmission probabilities. (*b*) How are these affected if the barrier is doubled in width?

Solution

(*a*) For the 1.0-eV electrons

$$\begin{aligned} k_2 &= \frac{\sqrt{2m(U-E)}}{\hbar} \\ &= \frac{\sqrt{(2)(9.1 \times 10^{-31}\ \text{kg})[(10.0-1.0)\ \text{eV}](1.6 \times 10^{-19}\ \text{J/eV})}}{1.054 \times 10^{-34}\ \text{J} \cdot \text{s}} \\ &= 1.6 \times 10^{10}\ \text{m}^{-1} \end{aligned}$$

Since $L = 0.50$ nm $= 5.0 \times 10^{-10}$ m, $2k_2L = (2)(1.6 \times 10^{10}\ \text{m}^{-1})(5.0 \times 10^{-10}\ \text{m}) = 16$, and the approximate transmission probability is

$$T_1 = e^{-2k_2L} = e^{-16} = 1.1 \times 10^{-7}$$

One 1.0-eV electron out of 8.9 million can tunnel through the 10-eV barrier on the average. For the 2.0-eV electrons a similar calculation gives $T_2 = 2.4 \times 10^{-7}$. These electrons are over twice as likely to tunnel through the barrier.

(*b*) If the barrier is doubled in width to 1.0 nm, the transmission probabilities become

$$T_1' = 1.3 \times 10^{-14} \qquad T_2' = 5.1 \times 10^{-14}$$

Evidently T is more sensitive to the width of the barrier than to the particle energy here.

Scanning Tunneling Microscope

The ability of electrons to tunnel through a potential barner is used in an ingenious way in the **scanning tunneling microscope** (STM) to study surfaces on an atomic scale of size. The STM was invented in 1981 by Gert Binning and Heinrich Rohrer, who shared the 1986 Nobel Prize in physics with Ernst Ruska, the inventor of the electron microscope. In an STM, a metal probe with a point so fine that its tip is a single atom is brought close to the surface of a conducting or semiconducting material. Normally even the most loosely bound electrons in an atom on a surface need several electron-volts of energy to escape—this is the work function discussed in Chap. 2 in connection with the photoelectric effect. However, when a voltage of only 10 mV or so is applied between the probe and the surface, electrons can tunnel across the gap between them if the gap is small enough, a nanometer or two.

According to Eq. (5.60) the electron transmission probability is proportional to e^{-L}, where L is the gap width, so even a small change in L (as little as 0.01 nm, less than a twentieth the diameter of most atoms) means a detectable change in the tunneling current. What is done is to move the probe across the surface in a series of closely spaced back-and-forth scans in about the same way an electron beam traces out an image on the screen of a television picture tube. The height of the probe is continually adjusted to give a constant tunneling current, and the adjustments are recorded so that a map of surface height versus position is built up. Such a map is able to resolve individual atoms on a surface.

The tungsten probe of a scanning tunneling microscope.

How can the position of the probe be controlled precisely enough to reveal the outlines of individual atoms? The thickness of certain ceramics changes when a voltage is applied across them, a property called **piezoelectricity.** The changes might be several tenths of a nanometer per volt. In an STM, piezoelectric controls move the probe in x and y directions across a surface and in the z direction perpendicular to the surface.

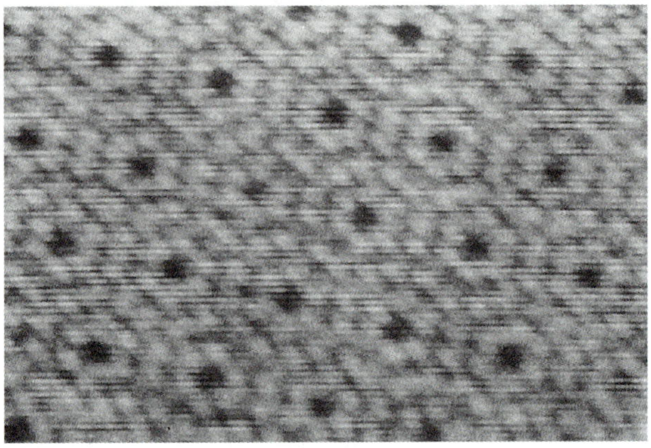

Silicon atoms on the surface of a silicon crystal form a regular, repeated pattern in this image produced by an STM.

Actually, the result of an STM scan is not a true topographical map of surface height but a contour map of constant electron density on the surface. This means that atoms of different elements appear differently, which greatly increases the value of the STM as a research tool.

Although many biological materials conduct electricity, they do so by the flow of ions rather than of electrons and so cannot be studied with STMs. A more recent development, the **atomic force microscope** (AFM) can be used on any surface, although with somewhat less resolution than an STM. In an AFM, the sharp tip of a fractured diamond presses gently against the atoms on a surface. A spring keeps the pressure of the tip constant, and a record is made of the deflections of the tip as it moves across the surface. The result is a map showing contours of constant repulsive force between the electrons of the probe and the electrons of the surface atoms. Even relatively soft biological materials can be examined with an AFM and changes in them monitored. For example, the linking together of molecules of the blood protein fibrin, which occurs when blood clots, has been watched with an AFM.

5.11 HARMONIC OSCILLATOR

Its energy levels are evenly spaced

Harmonic motion takes place when a system of some kind vibrates about an equilibrium configuration. The system may be an object supported by a spring or floating in a liquid, a diatomic molecule, an atom in a crystal lattice—there are countless examples on all scales of size. The condition for harmonic motion is the presence of a restoring force that acts to return the system to its equilibrium configuration when it is disturbed. The inertia of the masses involved causes them to overshoot equilibrium, and the system oscillates indefinitely if no energy is lost.

In the special case of simple harmonic motion, the restoring force F on a particle of mass m is linear; that is, F is proportional to the particle's displacement x from its equilibrium position and in the opposite direction. Thus

Hooke's law
$$F = -kx$$

This relationship is customarily called Hooke's law. From the second law of motion, $\mathbf{F} = m\mathbf{a}$, we have

$$-kx = m\frac{d^2x}{dt^2}$$

Harmonic oscillator

$$\frac{d^2x}{dt^2} + \frac{k}{m}x = 0 \tag{5.62}$$

There are various ways to write the solution to Eq. (5.62). A common one is

$$x = A \cos (2\pi\nu t + \phi) \tag{5.63}$$

where

Frequency of harmonic oscillator

$$\nu = \frac{1}{2\pi}\sqrt{\frac{k}{m}} \tag{5.64}$$

is the frequency of the oscillations and A is their amplitude. The value of ϕ, the phase angle, depends upon what x is at the time $t = 0$ and on the direction of motion then.

The importance of the simple harmonic oscillator in both classical and modern physics lies not in the strict adherence of actual restoring forces to Hooke's law, which is seldom true, but in the fact that these restoring forces reduce to Hooke's law for small displacements x. As a result, any system in which something executes small vibrations about an equilibrium position behaves very much like a simple harmonic oscillator.

To verify this important point, we note that any restoring force which is a function of x can be expressed in a Maclaurin's series about the equilibrium position $x = 0$ as

$$F(x) = F_{x=0} + \left(\frac{dF}{dx}\right)_{x=0} x + \frac{1}{2}\left(\frac{d^2F}{dx^2}\right)_{x=0} x^2 + \frac{1}{6}\left(\frac{d^3F}{dx^3}\right)_{x=0} x^3 + \cdots$$

Since $x = 0$ is the equilibrium position, $F_{x=0} = 0$. For small x the values of $x^2, x^3, \ldots$ are very small compared with x, so the third and higher terms of the series can be neglected. The only term of significance when x is small is therefore the second one. Hence

$$F(x) = \left(\frac{dF}{dx}\right)_{x=0} x$$

which is Hooke's law when $(dF/dx)_{x=0}$ is negative, as of course it is for any restoring force. The conclusion, then, is that *all* oscillations are simple harmonic in character when their amplitudes are sufficiently small.

The potential-energy function $U(x)$ that corresponds to a Hooke's law force may be found by calculating the work needed to bring a particle from $x = 0$ to $x = x$ against such a force. The result is

$$U(x) = -\int_0^x F(x)\, dx = k\int_0^x x\, dx = \frac{1}{2}kx^2 \tag{5.65}$$

which is plotted in Fig. 5.10. The curve of $U(x)$ versus x is a parabola. If the energy of the oscillator is E, the particle vibrates back and forth between $x = -A$ and $x = +A$, where E and A are related by $E = \frac{1}{2}kA^2$. Figure 8.18 shows how a nonparabolic potential energy curve can be approximated by a parabola for small displacements.

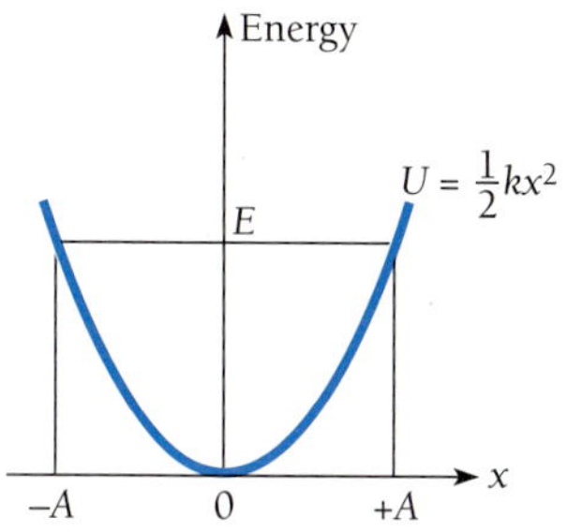

Figure 5.10 The potential energy of a harmonic oscillator is proportional to x^2, where x is the displacement from the equilibrium position. The amplitude A of the motion is determined by the total energy E of the oscillator, which classically can have any value.

Even before we make a detailed calculation we can anticipate three quantum-mechanical modifications to this classical picture:

1 The allowed energies will not form a continuous spectrum but instead a discrete spectrum of certain specific values only.
2 The lowest allowed energy will not be $E = 0$ but will be some definite minimum $E = E_0$.
3 There will be a certain probability that the particle can penetrate the potential well it is in and go beyond the limits of $-A$ and $+A$.

Energy Levels

Schrödinger's equation for the harmonic oscillator is, with $U = \frac{1}{2}kx^2$,

$$\frac{d^2\psi}{dx^2} + \frac{2m}{\hbar^2}\left(E - \frac{1}{2}kx^2\right)\psi = 0 \tag{5.66}$$

It is convenient to simplify Eq. (5.75) by introducing the dimensionless quantities

$$y = \left(\frac{1}{\hbar}\sqrt{km}\right)^{1/2} x = \sqrt{\frac{2\pi m\nu}{\hbar}}\, x \tag{5.67}$$

and

$$\alpha = \frac{2E}{\hbar}\sqrt{\frac{m}{k}} = \frac{2E}{h\nu} \tag{5.68}$$

where ν is the classical frequency of the oscillation given by Eq. (5.64). In making these substitutions, what we have done is change the units in which x and E are expressed from meters and joules, respectively, to dimensionless units.

In terms of y and α Schrödinger's equation becomes

$$\frac{d^2\psi}{dy^2} + (\alpha - y^2)\psi = 0 \tag{5.69}$$

The solutions to this equation that are acceptable here are limited by the condition that $\psi \to 0$ as $y \to \infty$ in order that

$$\int_{-\infty}^{\infty} |\psi|^2 \, dy = 1$$

Otherwise the wave function cannot represent an actual particle. The mathematical properties of Eq. (5.69) are such that this condition will be fulfilled only when

$$\alpha = 2n + 1 \qquad n = 0, 1, 2, 3, \ldots$$

Since $\alpha = 2E/h\nu$ according to Eq. (5.68), the energy levels of a harmonic oscillator whose classical frequency of oscillation is ν are given by the formula

Energy levels of harmonic oscillator

$$E_n = (n + \tfrac{1}{2})h\nu \qquad n = 0, 1, 2, 3, \ldots \tag{5.70}$$

The energy of a harmonic oscillator is thus quantized in steps of $h\nu$.

We note that when $n = 0$,

Zero-point energy

$$E_0 = \tfrac{1}{2}h\nu \tag{5.71}$$

which is the lowest value the energy of the oscillator can have. This value is called the **zero-point energy** because a harmonic oscillator in equilibrium with its surroundings would approach an energy of $E = E_0$ and not $E = 0$ as the temperature approaches 0 K.

Figure 5.11 is a comparison of the energy levels of a harmonic oscillator with those of a hydrogen atom and of a particle in a box with infinitely hard walls. The shapes of the respective potential-energy curves are also shown. The spacing of the energy levels is constant only for the harmonic oscillator.

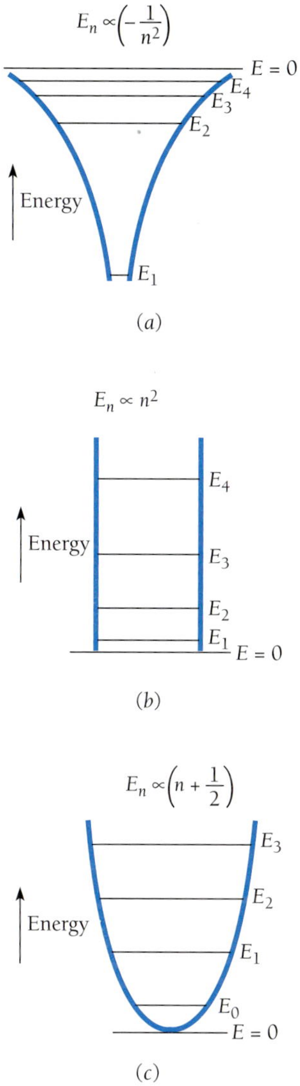

Figure 5.11 Potential wells and energy levels of (a) a hydrogen atom, (b) a particle in a box, and (c) a harmonic oscillator. In each case the energy levels depend in a different way on the quantum number n. Only for the harmonic oscillator are the levels equally spaced. The symbol $\propto$ means "is proportional to."

Wave Functions

For each choice of the parameter α_n there is a different wave function ψ_n. Each function consists of a polynomial $H_n(y)$ (called a **Hermite polynomial**) in either odd or even powers of y, the exponential factor $e^{-y^2/2}$, and a numerical coefficient which is needed for ψ_n to meet the normalization condition

$$\int_{-\infty}^{\infty} |\psi_n|^2 \, dy = 1 \qquad n = 0, 1, 2 \ldots$$

The general formula for the nth wave function is

Harmonic oscillator

$$\psi_n = \left(\frac{2m\nu}{\hbar}\right)^{1/4} (2^n n!)^{-1/2} H_n(y) e^{-y^2/2} \tag{5.72}$$

The first six Hermite polynomials $H_n(y)$ are listed in Table 5.2.

The wave functions that correspond to the first six energy levels of a harmonic oscillator are shown in Fig. 5.12. In each case the range to which a particle oscillating classically with the same total energy E_n would be confined is indicated. Evidently the particle is able to penetrate into classically forbidden regions—in other words, to exceed the amplitude A determined by the energy—with an exponentially decreasing probability, just as in the case of a particle in a finite square potential well.

It is interesting and instructive to compare the probability densities of a classical harmonic oscillator and a quantum-mechanical harmonic oscillator of the same energy. The upper curves in Fig. 5.13 show this density for the classical oscillator. The probability P of finding the particle at a given position is greatest at the endpoints of its motion,

Table 5.2 Some Hermite Polynomials

n	$H_n(y)$	α_n	E_n
0	1	1	$\frac{1}{2}h\nu$
1	$2y$	3	$\frac{3}{2}h\nu$
2	$4y^2 - 2$	5	$\frac{5}{2}h\nu$
3	$8y^3 - 12y$	7	$\frac{7}{2}h\nu$
4	$16y^4 - 48y^2 + 12$	9	$\frac{9}{2}h\nu$
5	$32y^5 - 160y^3 + 120y$	11	$\frac{11}{2}h\nu$

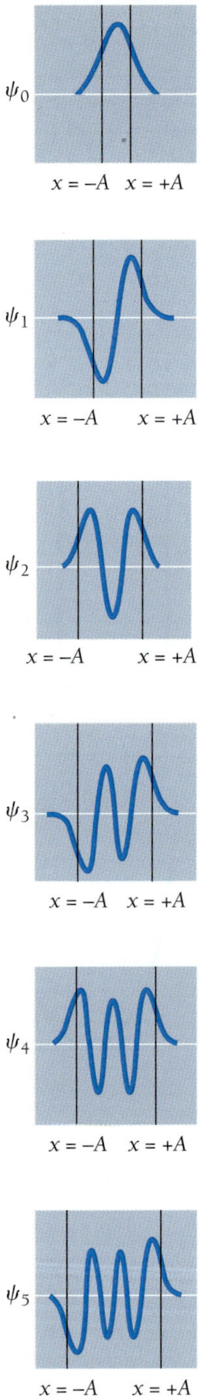

Figure 5.12 The first six harmonic-oscillator wave functions. The vertical lines show the limits $-A$ and $+A$ between which a classical oscillator with the same energy would vibrate.

where it moves slowly, and least near the equilibrium position ($x = 0$), where it moves rapidly.

Exactly the opposite behavior occurs when a quantum-mechanical oscillator is in its lowest energy state of $n = 0$. As shown, the probability density $|\psi_0|^2$ has its maximum value at $x = 0$ and drops off on either side of this position. However, this disagreement becomes less and less marked with increasing n. The lower graph of Fig. 5.13 corresponds to $n = 10$, and it is clear that $|\psi_{10}|^2$ when averaged over x has approximately the general character of the classical probability P. This is another example of the correspondence principle mentioned in Chap. 4: In the limit of large quantum numbers, quantum physics yields the same results as classical physics.

It might be objected that although $|\psi_{10}|^2$ does indeed approach P when smoothed out, nevertheless $|\psi_{10}|^2$ fluctuates rapidly with x whereas P does not. However, this objection has meaning only if the fluctuations are observable, and the smaller the spacing of the peaks and hollows, the more difficult it is to detect them experimentally. The exponential "tails" of $|\psi_{10}|^2$ beyond $x = \pm A$ also decrease in magnitude with increasing n. Thus the classical and quantum pictures begin to resemble each other more and more the larger the value of n, in agreement with the correspondence principle, although they are very different for small n.

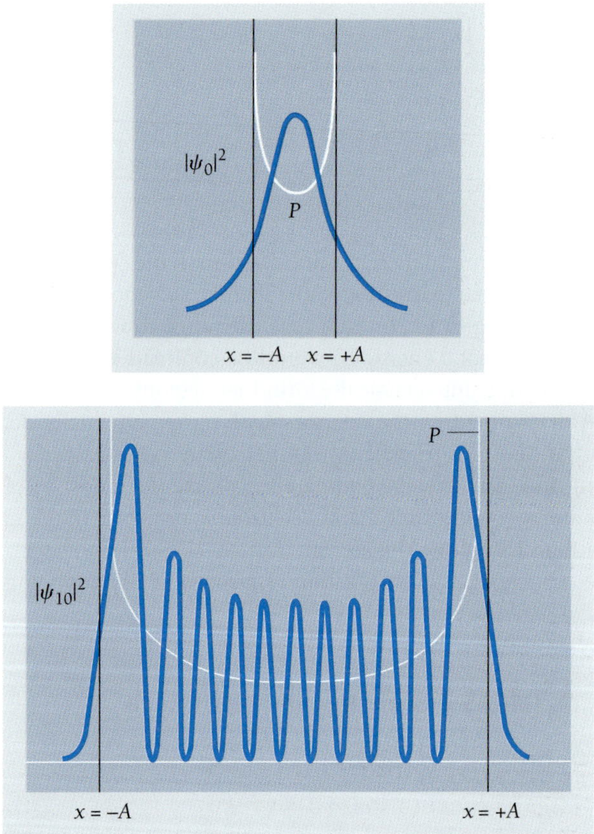

Figure 5.13 Probability densities for the $n = 0$ and $n = 10$ states of a quantum-mechanical harmonic oscillator. The probability densities for classical harmonic oscillators with the same energies are shown in white. In the $n = 10$ state, the wavelength is shortest at $x = 0$ and longest at $x = -A$.

Example 5.7

Find the expectation value $\langle x \rangle$ for the first two states of a harmonic oscillator.

Solution

The general formula for $\langle x \rangle$ is

$$\langle x \rangle = \int_{-\infty}^{\infty} x\,|\psi|^2\, dx$$

In calculations such as this it is easier to begin with y in place of x and afterward use Eq. (5.67) to change to x. From Eq. (5.72) and Table 5.2,

$$\psi_0 = \left(\frac{2m\nu}{\hbar}\right)^{1/4} e^{-y^2/2}$$

$$\psi_1 = \left(\frac{2m\nu}{\hbar}\right)^{1/4} \left(\frac{1}{2}\right)^{1/2} (2y)\, e^{-y^2/2}$$

The values of $\langle x \rangle$ for $n = 0$ and $n = 1$ will respectively be proportional to the integrals

$$n = 0: \int_{-\infty}^{\infty} y|\psi_0|^2\, dy = \int_{-\infty}^{\infty} y e^{-y^2}\, dy = -\left[\frac{1}{2} e^{-y^2}\right]_{-\infty}^{\infty} = 0$$

$$n = 1: \int_{-\infty}^{\infty} y|\psi_1|^2\, dy = \int_{-\infty}^{\infty} y^3 e^{-y^2}\, dy = -\left[\left(\frac{1}{4} + \frac{y^2}{2}\right) e^{-y^2}\right]_{-\infty}^{\infty} = 0$$

The expectation value $\langle x \rangle$ is therefore 0 in both cases. In fact, $\langle x \rangle = 0$ for *all* states of a harmonic oscillator, which could be predicted since $x = 0$ is the equilibrium position of the oscillator where its potential energy is a minimum.

Appendix to Chapter 5

The Tunnel Effect

We consider the situation that was shown in Fig. 5.9 of a particle of energy $E < U$ that approaches a potential barrier U high and L wide. Outside the barrier in regions I and III Schrödinger's equation for the particle takes the forms

$$\frac{d^2\psi_{\text{I}}}{dx^2} + \frac{2m}{\hbar^2}E\psi_{\text{I}} = 0 \tag{5.73}$$

$$\frac{d^2\psi_{\text{III}}}{dx^2} + \frac{2m}{\hbar^2}E\psi_{\text{III}} = 0 \tag{5.74}$$

The solutions to these equations that are appropriate here are

$$\psi_{\text{I}} = Ae^{ik_1x} + Be^{-ik_1x} \tag{5.75}$$

$$\psi_{\text{III}} = Fe^{ik_1x} + Ge^{-ik_1x} \tag{5.76}$$

where

Wave number outside barrier

$$k_1 = \frac{\sqrt{2mE}}{\hbar} = \frac{p}{\hbar} = \frac{2\pi}{\lambda} \tag{5.77}$$

is the wave number of the de Broglie waves that represent the particles outside the barrier.

Because

$$e^{i\theta} = \cos\theta + i\sin\theta$$

$$e^{-i\theta} = \cos\theta - i\sin\theta$$

these solutions are equivalent to Eq. (5.38)—the values of the coefficients are different in each case, of course—but are in a more suitable form to describe particles that are not trapped.

The various terms in Eqs. (5.75) and (5.76) are not hard to interpret. As was shown schematically in Fig. 5.9, Ae^{ik_1x} is a wave of amplitude A incident from the left on the barrier. Hence we can write

Incoming wave

$$\psi_{\text{I}+} = Ae^{ik_1x} \tag{5.78}$$

This wave corresponds to the incident beam of particles in the sense that $|\psi_{\text{I}+}|^2$ is their probability density. If $v_{\text{I}+}$ is the group velocity of the incoming wave, which equals the velocity of the particles, then

$$S = |\psi_{\text{I}+}|^2 v_{\text{I}+}$$

is the flux of particles that arrive at the barrier. That is, S is the number of particles per second that arrives there.

At $x = 0$ the incident wave strikes the barrier and is partially reflected, with

Reflected wave
$$\psi_{I-} = Be^{-ik_1x} \tag{5.79}$$

representing the reflected wave. Hence

$$\psi_I = \psi_{I+} + \psi_{I-} \tag{5.80}$$

On the far side of the barrier ($x > L$) there can only be a wave

Transmitted wave
$$\psi_{III+} = Fe^{ik_1x} \tag{5.81}$$

traveling in the $+x$ direction at the velocity v_{III+} since region III contains nothing that could reflect the wave. Hence $G = 0$ and

$$\psi_{III} = \psi_{III+} = Fe^{ik_1x} \tag{5.82}$$

The transmission probability T for a particle to pass through the barrier is the ratio

Transmission probability
$$T = \frac{|\psi_{III+}|^2 v_{III+}}{|\psi_{I+}|^2 v_{I+}} = \frac{FF^* v_{III+}}{AA^* v_{I+}} \tag{5.83}$$

between the flux of particles that emerges from the barrier and the flux that arrives at it. In other words, T is the fraction of incident particles that succeed in tunneling through the barrier. Classically $T = 0$ because a particle with $E < U$ cannot exist inside the barrier; let us see what the quantum-mechanical result is.

In region II Schrödinger's equation for the particles is

$$\frac{d^2\psi_{II}}{dx^2} + \frac{2m}{\hbar^2}(E - U)\psi_{II} = \frac{d^2\psi_{II}}{dx^2} - \frac{2m}{\hbar^2}(U - E)\psi_{II} = 0 \tag{5.84}$$

Since $U > E$ the solution is

Wave function inside barrier
$$\psi_{II} = Ce^{-k_2x} + De^{k_2x} \tag{5.85}$$

where the wave number inside the barrier is

Wave number inside barrier
$$k_2 = \frac{\sqrt{2m(U - E)}}{\hbar} \tag{5.86}$$

Since the exponents are real quantities, ψ_{II} does not oscillate and therefore does not represent a moving particle. However, the probability density $|\psi_{II}|^2$ is not zero, so there is a finite probability of finding a particle within the barrier. Such a particle may emerge into region III or it may return to region I.

Applying the Boundary Conditions

In order to calculate the transmission probability T we have to apply the appropriate boundary conditions to ψ_{I}, ψ_{II}, and ψ_{III}. Fig. 5.14 shows the wave functions in regions I, II, and III. As discussed earlier, both ψ and its derivative $\partial\psi/\partial x$ must be continuous everywhere. With reference to Fig. 5.14, these conditions mean that for a perfect fit at each side of the barrier, the wave functions inside and outside must have the same value and the same slope. Hence at the left-hand side of the barrier

Boundary conditions at $x = 0$

$$\left.\begin{aligned} \psi_{\text{I}} &= \psi_{\text{II}} \quad &(5.87)\\ \frac{d\psi_{\text{I}}}{dx} &= \frac{d\psi_{\text{II}}}{dx} \quad &(5.88)\end{aligned}\right\} x = 0$$

and at the right-hand side

Boundary conditions at $x = L$

$$\left.\begin{aligned} \psi_{\text{II}} &= \psi_{\text{III}} \quad &(5.89)\\ \frac{d\psi_{\text{II}}}{dx} &= \frac{d\psi_{\text{III}}}{dx} \quad &(5.90)\end{aligned}\right\} x = L$$

Now we substitute ψ_{I}, ψ_{II}, and ψ_{III} from Eqs. (5.75), (5.81), and (5.85) into the above equations. This yields in the same order

$$A + B = C + D \tag{5.91}$$

$$ik_1A - ik_1B = -k_2C + k_2D \tag{5.92}$$

$$Ce^{-k_2L} + De^{k_2L} = Fe^{ik_1L} \tag{5.93}$$

$$-k_2Ce^{-k_2L} + k_2De^{k_2L} = ik_1Fe^{ik_1L} \tag{5.94}$$

Equations (5.91) to (5.94) may be solved for (A/F) to give

$$\left(\frac{A}{F}\right) = \left[\frac{1}{2} + \frac{i}{4}\left(\frac{k_2}{k_1} - \frac{k_1}{k_2}\right)\right]e^{(ik_1+k_2)L} + \left[\frac{1}{2} - \frac{i}{4}\left(\frac{k_2}{k_1} - \frac{k_1}{k_2}\right)\right]e^{(ik_1-k_2)L} \tag{5.95}$$

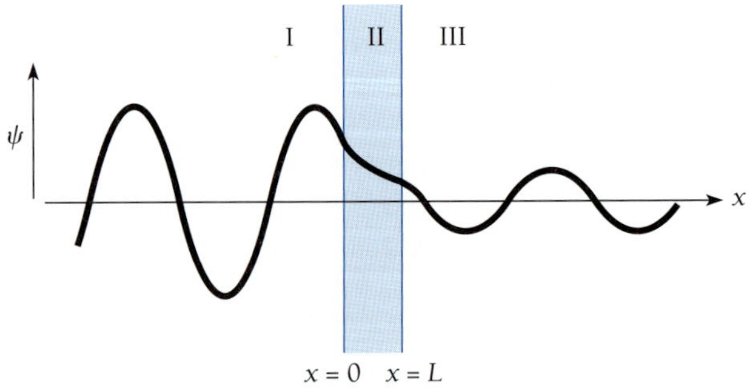

Figure 5.14 At each wall of the barrier, the wave functions inside and outside it must match up perfectly, which means that they must have the same values and slopes there.

Let us assume that the potential barrier U is high relative to the energy E of the incident particles. If this is the case, then $k_2/k_1 > k_1/k_2$ and

$$\frac{k_2}{k_1} - \frac{k_1}{k_2} \approx \frac{k_2}{k_1} \tag{5.96}$$

Let us also assume that the barrier is wide enough for ψ_{II} to be severely weakened between $x = 0$ and $x = L$. This means that $k_2 L \gg 1$ and

$$e^{k_2 L} \gg e^{-k_2 L}$$

Hence Eq. (5.95) can be approximated by

$$\left(\frac{A}{F}\right) = \left(\frac{1}{2} + \frac{ik_2}{4k_1}\right) e^{(ik_1 + k_2)L} \tag{5.97}$$

The complex conjugate of (A/F), which we need to compute the transmission probability T, is found by replacing i by $-i$ wherever it occurs in (A/F):

$$\left(\frac{A}{F}\right)^* = \left(\frac{1}{2} - \frac{ik_2}{4k_1}\right) e^{(-ik_1 + k_2)L} \tag{5.98}$$

Now we multiply (A/F) and $(A/F)^*$ to give

$$\frac{AA^*}{FF^*} = \left(\frac{1}{4} + \frac{k_2^2}{16k_1^2}\right) e^{2k_2 L}$$

Here $v_{\text{III}+} = v_{\text{I}+}$ so $v_{\text{III}+}/v_{\text{I}+} = 1$ in Eq. (5.83), which means that the transmission probability is

Transmission probability

$$T = \frac{FF^* v_{\text{III}+}}{AA^* v_{\text{I}+}} = \left(\frac{AA^*}{FF^*}\right)^{-1} = \left[\frac{16}{4 + (k_2/k_1)^2}\right] e^{-2k_2 L} \tag{5.99}$$

From the definitions of k_1, Eq. (5.77), and of k_2, Eq. (5.86), we see that

$$\left(\frac{k_2}{k_1}\right)^2 = \frac{2m(U - E)/\hbar^2}{2mE/\hbar^2} = \frac{U}{E} - 1 \tag{5.100}$$

This formula means that the quantity in brackets in Eq. (5.99) varies much less with E and U than does the exponential. The bracketed quantity, furthermore, always is of the order of magnitude of 1 in value. A reasonable approximation of the transmission probability is therefore

Approximate transmission probability

$$T = e^{-2k_2 L} \tag{5.101}$$

as stated in Sec. 5.10.

EXERCISES

Press on, and faith will catch up with you. —Jean D'Alembert

5.1 Quantum Mechanics

1. Which of the wave functions in Fig. 5.15 cannot have physical significance in the interval shown? Why not?

2. Which of the wave functions in Fig. 5.16 cannot have physical significance in the interval shown? Why not?

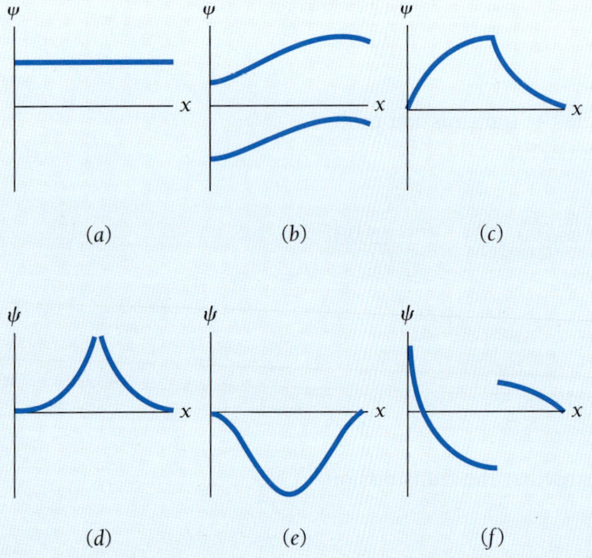

Figure 5.15

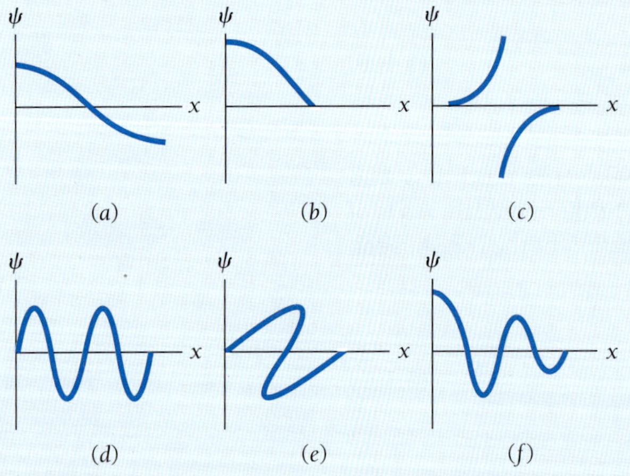

Figure 5.16

3. Which of the following wave functions cannot be solutions of Schrödinger's equation for all values of x? Why not? (*a*) $\psi = A \sec x$; (*b*) $\psi = A \tan x$; (*c*) $\psi = Ae^{x^2}$; (*d*) $\psi = Ae^{-x^2}$.

4. Find the value of the normalization constant A for the wave function $\psi = Axe^{-x^2/2}$.

5. The wave function of a certain particle is $\psi = A \cos^2 x$ for $-\pi/2 < x < \pi/2$. (*a*) Find the value of A. (*b*) Find the probability that the particle be found between $x = 0$ and $x = \pi/4$.

5.2 The Wave Equation

6. The formula $y = A \cos \omega(t - x/v)$, as we saw in Sec. 3.3, describes a wave that moves in the $+x$ direction along a stretched string. Show that this formula is a solution of the wave equation, Eq.(5.3).

7. As mentioned in Sec. 5.1, in order to give physically meaningful results in calculations a wave function and its partial derivatives must be finite, continuous, and single-valued, and in addition must be normalizable. Equation (5.9) gives the wave function of a particle moving freely (that is, with no forces acting on it) in the $+x$ direction as

$$\Psi = Ae^{-(i/\hbar)(Et-px)}$$

where E is the particle's total energy and p is its momentum. Does this wave function meet all the above requirements? If not, could a linear superposition of such wave functions meet these requirements? What is the significance of such a superposition of wave functions?

5.4 Linearity and Superposition

8. Prove that Schrödinger's equation is linear by showing that

$$\Psi = a_1\Psi_1(x, t) + a_2\Psi_2(x, t)$$

is also a solution of Eq. (5.14) if Ψ_1 and Ψ_2 are themselves solutions.

5.6 Operators

9. Show that the expectation values $\langle px \rangle$ and $\langle xp \rangle$ are related by

$$\langle px \rangle - \langle xp \rangle = \frac{\hbar}{i}$$

This result is described by saying that p and x do not **commute** and it is intimately related to the uncertainty principle.

10. An eigenfunction of the operator d^2/dx^2 is $\sin nx$, where $n = 1, 2, 3, \ldots$. Find the corresponding eigenvalues.

5.7 Schrödinger's Equation: Steady-State Form

11. Obtain Schrödinger's steady-state equation from Eq. (3.5) with the help of de Broglie's relationship $\lambda = h/mv$ by letting $y = \psi$ and finding $\partial^2\psi/\partial x^2$.

5.8 Particle in a Box

12. According to the correspondence principle, quantum theory should give the same results as classical physics in the limit of large quantum numbers. Show that as $n \to \infty$, the probability of finding the trapped particle of Sec. 5.8 between x and $x + \Delta x$ is $\Delta x/L$ and so is independent of x, which is the classical expectation.

13. One of the possible wave functions of a particle in the potential well of Fig. 5.17 is sketched there. Explain why the wavelength and amplitude of ψ vary as they do.

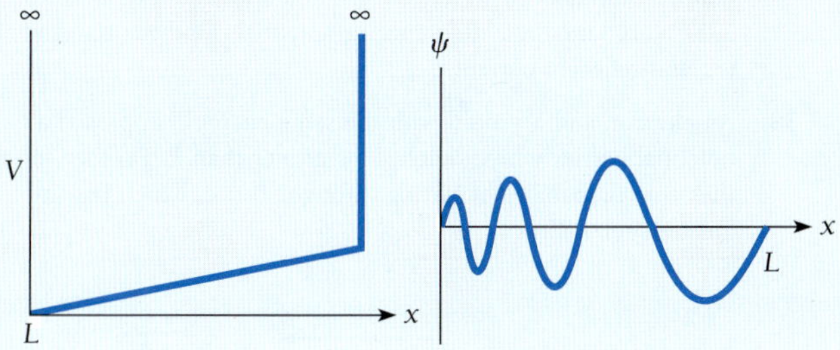

Figure 5.17

14. In Sec. 5.8 a box was considered that extends from $x = 0$ to $x = L$. Suppose the box instead extends from $x = x_0$ to $x = x_0 + L$, where $x_0 \neq 0$. Would the expression for the wave functions of a particle in this box be any different from those in the box that extends from $x = 0$ to $x = L$? Would the energy levels be different?

15. An important property of the eigenfunctions of a system is that they are **orthogonal** to one another, which means that

$$\int_{-\infty}^{\infty} \psi_n \psi_m \, dV = 0 \qquad n \neq m$$

Verify this relationship for the eigenfunctions of a particle in a one-dimensional box given by Eq. (5.46).

16. A rigid-walled box that extends from $-L$ to L is divided into three sections by rigid interior walls at $-x$ and x, where $x < L$. Each section contains one particle in its ground state. (*a*) What is the total energy of the system as a function of x? (*b*) Sketch $E(x)$ versus x. (*c*) At what value of x is $E(x)$ a minimum?

17. As shown in the text, the expectation value $\langle x \rangle$ of a particle trapped in a box L wide is $L/2$, which means that its average position is the middle of the box. Find the expectation value $\langle x^2 \rangle$.

18. As noted in Exercise 8, a linear combination of two wave functions for the same system is also a valid wave function. Find the normalization constant B for the combination

$$\psi = B\left(\sin\frac{\pi x}{L} + \sin\frac{2\pi x}{L}\right)$$

of the wave functions for the $n = 1$ and $n = 2$ states of a particle in a box L wide.

19. Find the probability that a particle in a box L wide can be found between $x = 0$ and $x = L/n$ when it is in the nth state.

20. In Sec. 3.7 the standard deviation σ of a set of N measurements of some quantity x was defined as

$$\sigma = \sqrt{\frac{1}{N}\sum_{i=1}^{N}(x_i - x_0)^2}$$

(*a*) Show that, in terms of expectation values, this formula can be written as

$$\sigma = \sqrt{\langle (x - \langle x \rangle)^2 \rangle} = \sqrt{\langle x^2 \rangle - \langle x \rangle^2}$$

(*b*) If the uncertainty in position of a particle in a box is taken as the standard deviation, find the uncertainty in the expectation value $\langle x \rangle = L/2$ for $n = 1$. (*c*) What is the limit of Δx as n increases?

21. A particle is in a cubic box with infinitely hard walls whose edges are L long (Fig. 5.18). The wave functions of the particle are given by

$$\psi = A \sin\frac{n_x \pi x}{L} \sin\frac{n_y \pi y}{L} \sin\frac{n_z \pi z}{L} \qquad \begin{aligned} n_x &= 1, 2, 3, \ldots \\ n_y &= 1, 2, 3, \ldots \\ n_z &= 1, 2, 3, \ldots \end{aligned}$$

Find the value of the normalization constant A.

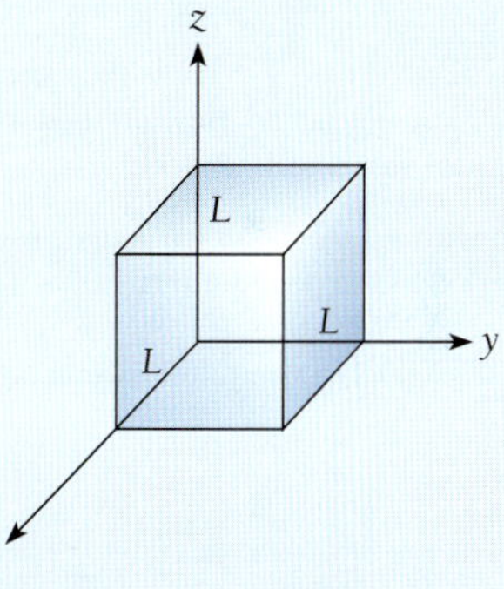

Figure 5.18 A cubic box.

22. The particle in the box of Exercise 21 is in its ground state of $n_x = n_y = n_z = 1$. (*a*) Find the probability that the particle will be found in the volume defined by $0 \le x \le L/4$, $0 \le y \le L/4$, $0 \le z \le L/4$. (*b*) Do the same for $L/2$ instead of $L/4$.

23. (*a*) Find the possible energies of the particle in the box of Exercise 21 by substituting its wave function ψ in Schrödinger's equation and solving for E. (*Hint*: Inside the box $U = 0$.) (*b*) Compare the ground-state energy of a particle in a one-dimensional box of length L with that of a particle in the three-dimensional box.

5.10 Tunnel Effect

24. Electrons with energies of 0.400 eV are incident on a barrier 3.00 eV high and 0.100 nm wide. Find the approximate probability for these electrons to penetrate the barrier.

25. A beam of electrons is incident on a barrier 6.00 eV high and 0.200 nm wide. Use Eq. (5.60) to find the energy they should have if 1.00 percent of them are to get through the barrier.

5.11 Harmonic Oscillator

26. Show that the energy-level spacing of a harmonic oscillator is in accord with the correspondence principle by finding the ratio $\Delta E_n/E_n$ between adjacent energy levels and seeing what happens to this ratio as $n \to \infty$.

27. What bearing would you think the uncertainty principle has on the existence of the zero-point energy of a harmonic oscillator?

28. In a harmonic oscillator, the particle varies in position from $-A$ to $+A$ and in momentum from $-p_0$ to $+p_0$. In such an oscillator, the standard deviations of x and p are $\Delta x = A/\sqrt{2}$ and $\Delta p = p_0/\sqrt{2}$. Use this observation to show that the minimum energy of a harmonic oscillator is $\frac{1}{2}h\nu$.

29. Show that for the $n = 0$ state of a harmonic oscillator whose classical amplitude of motion is A, $y = 1$ at $x = A$, where y is the quantity defined by Eq. (5.67).

30. Find the probability density $|\psi_0|^2\,dx$ at $x = 0$ and at $x = \pm A$ of a harmonic oscillator in its $n = 0$ state (see Fig. 5.13).

31. Find the expectation values $\langle x \rangle$ and $\langle x^2 \rangle$ for the first two states of a harmonic oscillator.

32. The potential energy of a harmonic oscillator is $U = \frac{1}{2}kx^2$. Show that the expectation value $\langle U \rangle$ of U is $E_0/2$ when the oscillator is in the $n = 0$ state. (This is true of all states of the harmonic oscillator, in fact.) What is the expectation value of the oscillator's kinetic energy? How do these results compare with the classical values of $\overline{U}$ and $\overline{KE}$?

33. A pendulum with a 1.00-g bob has a massless string 250 mm long. The period of the pendulum is 1.00 s. (*a*) What is its zero-point energy? Would you expect the zero-point oscillations to be detectable? (*b*) The pendulum swings with a very small amplitude such that its bob rises a maximum of 1.00 mm above its equilibrium position. What is the corresponding quantum number?

34. Show that the harmonic-oscillator wave function ψ_1 is a solution of Schrödinger's equation.

35. Repeat Exercise 34 for ψ_2.

36. Repeat Exercise 34 for ψ_3.

Appendix: The Tunnel Effect

37. Consider a beam of particles of kinetic energy E incident on a potential step at $x = 0$ that is U high, where $E > U$ (Fig. 5.19). (*a*) Explain why the solution $De^{-ik'x}$ (in the notation of appendix) has no physical meaning in this situation, so that $D = 0$. (*b*) Show that the transmission probability here is $T = CC^*v'/AA^*v_1 = 4k_1^2/(k_1 + k')^2$. (*c*) A 1.00-mA beam of electrons moving at 2.00×10^6 m/s enters a region with a sharply defined boundary in which the electron speeds are reduced to 1.00×10^6 m/s by a difference in potential. Find the transmitted and reflected currents.

38. An electron and a proton with the same energy E approach a potential barrier whose height U is greater than E. Do they have the same probability of getting through? If not, which has the greater probability?

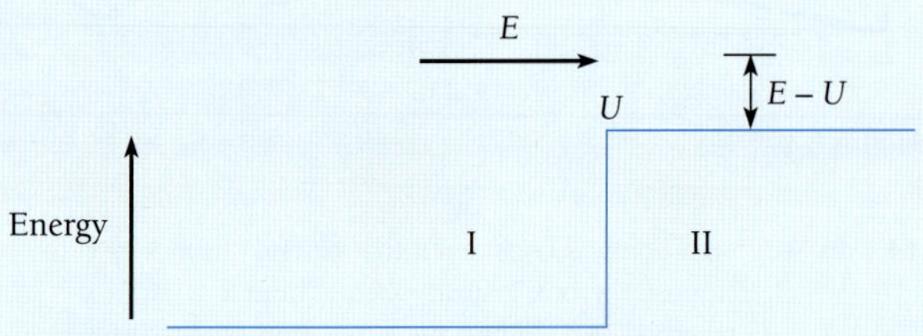

Figure 5.19

CHAPTER 6

Quantum Theory of the Hydrogen Atom

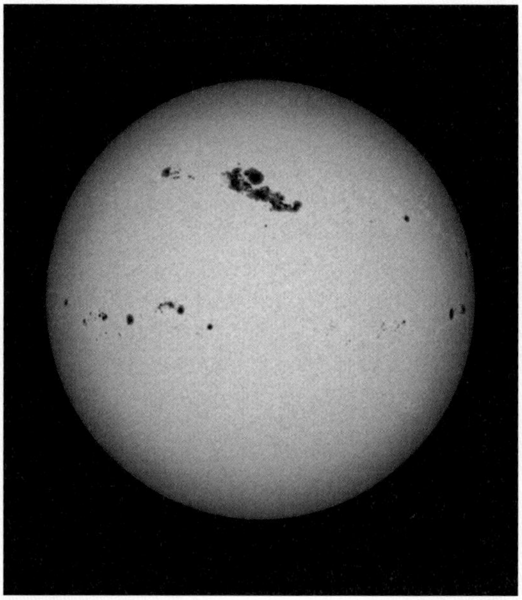

The strong magnetic fields associated with sunspots were detected by means of the Zeeman effect. Sunspots appear dark because they are cooler than the rest of the solar surface, although quite hot themselves. The number of spots varies in an 11-year cycle, and a number of terrestrial phenomena follow this cycle.

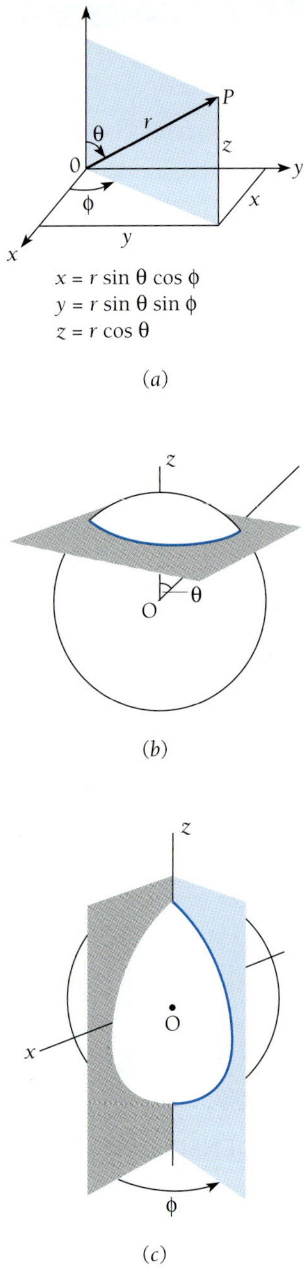

Figure 6.1 (*a*) Spherical polar coordinates. (*b*) A line of constant zenith angle θ on a sphere is a circle whose plane is perpendicular to the z axis. (*c*) A line of constant azimuth angle ϕ is a circle whose plane includes the z axis.

The first problem that Schrödinger tackled with his new wave equation was that of the hydrogen atom. He found the mathematics heavy going, but was rewarded by the discovery of how naturally quantization occurs in wave mechanics: "It has its basis in the requirement that a certain spatial function be finite and single-valued." In this chapter we shall see how Schrödinger's quantum theory of the hydrogen atom achieves its results, and how these results can be interpreted in terms of familiar concepts.

6.1 SCHRÖDINGER'S EQUATION FOR THE HYDROGEN ATOM

Symmetry suggests spherical polar coordinates

A hydrogen atom consists of a proton, a particle of electric charge $+e$, and an electron, a particle of charge $-e$ which is 1836 times lighter than the proton. For the sake of convenience we shall consider the proton to be stationary, with the electron moving about in its vicinity but prevented from escaping by the proton's electric field. As in the Bohr theory, the correction for proton motion is simply a matter of replacing the electron mass m by the reduced mass m' given by Eq. (4.22).

Schrödinger's equation for the electron in three dimensions, which is what we must use for the hydrogen atom, is

$$\frac{\partial^2 \psi}{\partial x^2} + \frac{\partial^2 \psi}{\partial y^2} + \frac{\partial^2 \psi}{\partial z^2} + \frac{2m}{\hbar^2}(E - U)\psi = 0 \tag{6.1}$$

The potential energy U here is the electric potential energy

Electric potential energy

$$U = -\frac{e^2}{4\pi\epsilon_0 r} \tag{6.2}$$

of a charge $-e$ when it is the distance r from another charge $+e$.

Since U is a function of r rather than of x, y, z, we cannot substitute Eq. (6.2) directly into Eq. (6.1). There are two alternatives. One is to express U in terms of the cartesian coordinates x, y, z by replacing r by $\sqrt{x^2 + y^2 + z^2}$. The other is to express Schrödinger's equation in terms of the spherical polar coordinates r, θ, ϕ defined in Fig. 6.1. Owing to the symmetry of the physical situation, doing the latter is appropriate here, as we shall see in Sec. 6.2.

The spherical polar coordinates r, θ, ϕ of the point P shown in Fig. 6.1 have the following interpretations:

Spherical polar coordinates

$$r = \text{length of radius vector from origin } O \text{ to point } P$$
$$= \sqrt{x^2 + y^2 + z^2}$$

$$\theta = \text{angle between radius vector and } +z \text{ axis}$$
$$= \text{zenith angle}$$
$$= \cos^{-1}\frac{z}{\sqrt{x^2 + y^2 + z^2}}$$
$$= \cos^{-1}\frac{z}{r}$$

ϕ = angle between the projection of the radius vector in the xy plane and the $+x$ axis, measured in the direction shown

= azimuth angle

$= \tan^{-1}\dfrac{y}{x}$

On the surface of a sphere whose center is at O, lines of constant zenith angle θ are like parallels of latitude on a globe (but we note that the value of θ of a point is *not* the same as its latitude; $\theta = 90°$ at the equator, for instance, but the latitude of the equator is 0°). Lines of constant azimuth angle ϕ are like meridians of longitude (here the definitions coincide if the axis of the globe is taken as the $+z$ axis and the $+x$ axis is at $\phi = 0°$).

In spherical polar coordinates Schrödinger's equation is written

$$\frac{1}{r^2}\frac{\partial}{\partial r}\left(r^2\frac{\partial\psi}{\partial r}\right) + \frac{1}{r^2\sin\theta}\frac{\partial}{\partial\theta}\left(\sin\theta\frac{\partial\psi}{\partial\theta}\right) + \frac{1}{r^2\sin^2\theta}\frac{\partial^2\psi}{\partial\phi^2} + \frac{2m}{\hbar^2}(E - U)\psi = 0 \tag{6.3}$$

Substituting Eq. (6.2) for the potential energy U and multiplying the entire equation by $r^2 \sin^2\theta$, we obtain

Hydrogen atom

$$\sin^2\theta\frac{\partial}{\partial r}\left(r^2\frac{\partial\psi}{\partial r}\right) + \sin\theta\frac{\partial}{\partial\theta}\left(\sin\theta\frac{\partial\psi}{\partial\theta}\right) + \frac{\partial^2\psi}{\partial\phi^2} + \frac{2mr^2\sin^2\theta}{\hbar^2}\left(\frac{e^2}{4\pi\epsilon_0 r} + E\right)\psi = 0 \tag{6.4}$$

Equation (6.4) is the partial differential equation for the wave function ψ of the electron in a hydrogen atom. Together with the various conditions ψ must obey, namely that ψ be normalizable and that ψ and its derivatives be continuous and single-valued at each point r, θ, ϕ, this equation completely specifies the behavior of the electron. In order to see exactly what this behavior is, we must solve Eq. (6.4) for ψ.

When Eq. (6.4) is solved, it turns out that three quantum numbers are required to describe the electron in a hydrogen atom, in place of the single quantum number of the Bohr theory. (In Chap. 7 we shall find that a fourth quantum number is needed to describe the spin of the electron.) In the Bohr model, the electron's motion is basically one-dimensional, since the only quantity that varies as it moves is its position in a definite orbit. One quantum number is enough to specify the state of such an electron, just as one quantum number is enough to specify the state of a particle in a one-dimensional box.

A particle in a three-dimensional box needs three quantum numbers for its description, since there are now three sets of boundary conditions that the particle's wave function ψ must obey: ψ must be 0 at the walls of the box in the x, y, and z directions independently. In a hydrogen atom the electron's motion is restricted by the inverse-square electric field of the nucleus instead of by the walls of a box, but the electron is nevertheless free to move in three dimensions, and it is accordingly not surprising that three quantum numbers govern its wave function also.

6.2 SEPARATION OF VARIABLES

A differential equation for each variable

The advantage of writing Schrödinger's equation in spherical polar coordinates for the problem of the hydrogen atom is that in this form it may be separated into three independent equations, each involving only a single coordinate. Such a separation is possible here because the wave function $\psi(r, \theta, \phi)$ has the form of a product of three different functions: $R(r)$, which depends on r alone; $\Theta(\theta)$ which depends on θ alone; and $\Phi(\phi)$, which depends on ϕ alone. Of course, we do not really know that this separation is possible yet, but we can proceed by assuming that

Hydrogen-atom wave function

$$\psi(r, \theta, \phi) = R(r)\Theta(\theta)\Phi(\phi) \tag{6.5}$$

and then seeing if it leads to the desired separation. The function $R(r)$ describes how the wave function ψ of the electron varies along a radius vector from the nucleus, with θ and ϕ constant. The function $\Theta(\theta)$ describes how ψ varies with zenith angle θ along a meridian on a sphere centered at the nucleus, with r and ϕ constant (Fig. 6.1*c*). The function $\Phi(\phi)$ describes how ψ varies with azimuth angle ϕ along a parallel on a sphere centered at the nucleus, with r and θ constant (Fig. 6.1*b*).

From Eq. (6.5), which we may write more simply as

$$\psi = R\Theta\Phi$$

we see that

$$\frac{\partial\psi}{\partial r} = \Theta\Phi\frac{\partial R}{\partial r} = \Theta\Phi\frac{dR}{dr}$$

$$\frac{\partial\psi}{\partial\theta} = R\Phi\frac{\partial\Theta}{\partial\theta} = R\Phi\frac{d\Theta}{d\theta}$$

$$\frac{\partial^2\psi}{\partial\phi^2} = R\Theta\frac{\partial^2\Phi}{\partial\phi^2} = R\Theta\frac{d^2\Phi}{d\phi^2}$$

The change from partial derivatives to ordinary derivatives can be made because we are assuming that each of the functions R, Θ, and Φ depends only on the respective variables r, θ, and ϕ.

When we substitute $R\Theta\Phi$ for ψ in Schrödinger's equation for the hydrogen atom and divide the entire equation by $R\Theta\Phi$, we find that

$$\frac{\sin^2\theta}{R}\frac{d}{dr}\left(r^2\frac{dR}{dr}\right) + \frac{\sin\theta}{\Theta}\frac{d}{d\theta}\left(\sin\theta\frac{d\Theta}{d\theta}\right) + \frac{1}{\Phi}\frac{d^2\Phi}{d\phi^2} + \frac{2mr^2\sin^2\theta}{\hbar^2}\left(\frac{e^2}{4\pi\epsilon_0 r} + E\right) = 0 \tag{6.6}$$

The third term of Eq. (6.6) is a function of azimuth angle ϕ only, whereas the other terms are functions of r and θ only.

Let us rearrange Eq. (6.6) to read

$$\frac{\sin^2\theta}{R}\frac{d}{dr}\left(r^2\frac{dR}{dr}\right)+\frac{\sin\theta}{\Theta}\frac{d}{d\theta}\left(\sin\theta\frac{d\Theta}{d\theta}\right)+\frac{2mr^2\sin^2\theta}{\hbar^2}\left(\frac{e^2}{4\pi\epsilon_0 r}+E\right)=-\frac{1}{\Phi}\frac{d^2\Phi}{d\phi^2} \tag{6.7}$$

This equation can be correct only if both sides of it are equal to the same constant, since they are functions of *different* variables. As we shall see, it is convenient to call this constant m_l^2. The differential equation for the function ϕ is therefore

$$-\frac{1}{\Phi}\frac{d^2\Phi}{d\phi^2}=m_l^2 \tag{6.8}$$

Next we substitute m_l^2 for the right-hand side of Eq. (6.7), divide the entire equation by $\sin^2\theta$, and rearrange the various terms, which yields

$$\frac{1}{R}\frac{d}{dr}\left(r^2\frac{dR}{dr}\right)+\frac{2mr^2}{\hbar^2}\left(\frac{e^2}{4\pi\epsilon_0 r}+E\right)=\frac{m_l^2}{\sin^2\theta}-\frac{1}{\Theta\sin\theta}\frac{d}{d\theta}\left(\sin\theta\frac{d\Theta}{d\theta}\right) \tag{6.9}$$

Again we have an equation in which different variables appear on each side, requiring that both sides be equal to the same constant. This constant is called $l(l+1)$, once more for reasons that will be apparent later. The equations for the functions Θ and R are therefore

$$\frac{m_l^2}{\sin^2\theta}-\frac{1}{\Theta\sin\theta}\frac{d}{d\theta}\left(\sin\theta\frac{d\Theta}{d\theta}\right)=l(l+1) \tag{6.10}$$

$$\frac{1}{R}\frac{d}{dr}\left(r^2\frac{dR}{dr}\right)+\frac{2mr^2}{\hbar^2}\left(\frac{e^2}{4\pi\epsilon_0 r}+E\right)=l(l+1) \tag{6.11}$$

Equations (6.8), (6.10), and (6.11) are usually written

Equation for Φ

$$\frac{d^2\Phi}{d\phi^2}+m_l^2\Phi=0 \tag{6.12}$$

Equation for Θ

$$\frac{1}{\sin\theta}\frac{d}{d\theta}\left(\sin\theta\frac{d\Theta}{d\theta}\right)+\left[l(l+1)-\frac{m_l^2}{\sin^2\theta}\right]\Theta=0 \tag{6.13}$$

Equation for R

$$\frac{1}{r^2}\frac{d}{dr}\left(r^2\frac{dR}{dr}\right)+\left[\frac{2m}{\hbar^2}\left(\frac{e^2}{4\pi\epsilon_0 r}+E\right)-\frac{l(l+1)}{r^2}\right]R=0 \tag{6.14}$$

Each of these is an ordinary differential equation for a single function of a single variable. Only the equation for R depends on the potential energy $U(r)$.

We have therefore accomplished our task of simplifying Schrödinger's equation for the hydrogen atom, which began as a partial differential equation for a function ψ of three variables. The assumption embodied in Eq. (6.5) is evidently valid.

6.3 QUANTUM NUMBERS

Three dimensions, three quantum numbers

The first of these equations, Eq. (6.12), is readily solved. The result is

$$\Phi(\phi) = Ae^{im_l\phi} \tag{6.15}$$

As we know, one of the conditions that a wave function—and hence Φ, which is a component of the complete wave function ψ—must obey is that it have a single value at a given point in space. From Fig. 6.2 it is clear that ϕ and $\phi + 2\pi$ both identify the same meridian plane. Hence it must be true that $\Phi(\phi) = \Phi(\phi + 2\pi)$, or

$$Ae^{im_l\phi} = Ae^{im_l(\phi+2\pi)}$$

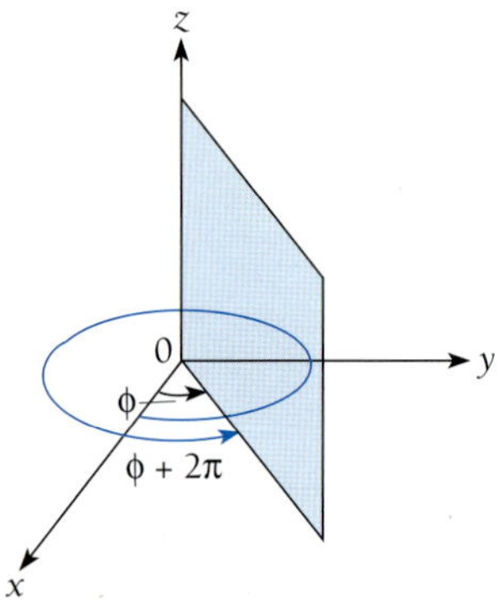

Figure 6.2 The angles ϕ and $\phi + 2\pi$ both indentify the same meridian plane.

which can happen only when m_l is 0 or a positive or negative integer (± 1, ± 2, ± 3, . . .). The constant m_l is known as the **magnetic quantum number** of the hydrogen atom.

The differential equation for $\Theta(\theta)$, Eq. (6.13), has a solution provided that the constant l is an integer equal to or greater than $|m_l|$, the absolute value of m_l. This requirement can be expressed as a condition on m_l in the form

$$m_l = 0, \pm 1, \pm 2, \ldots, \pm l$$

The constant l is known as the **orbital quantum number.**

The solution of the final equation, Eq. (6.14), for the radial part $R(r)$ of the hydrogen-atom wave function ψ also requires that a certain condition be fulfilled. This condition is that E be positive or have one of the negative values E_n (signifying that the electron is bound to the atom) specified by

$$E_n = -\frac{me^4}{32\pi^2\epsilon_0^2\hbar^2}\left(\frac{1}{n^2}\right) = \frac{E_1}{n^2} \qquad n = 1, 2, 3, \ldots \tag{6.16}$$

We recognize that this is precisely the same formula for the energy levels of the hydrogen atom that Bohr obtained.

Another condition that must be obeyed in order to solve Eq. (6.14) is that n, known as the **principal quantum number,** must be equal to or greater than $l + 1$. This requirement may be expressed as a condition on l in the form

$$l = 0, 1, 2, \ldots, (n - 1)$$

Hence we may tabulate the three quantum numbers n, l, and m together with their permissible values as follows:

Principal quantum number $\quad n = 1, 2, 3, \ldots$

Orbital quantum number $\quad l = 0, 1, 2, \ldots, (n-1) \qquad (6.17)$

Magnetic quantum number $\quad m_l = 0, \pm 1, \pm 2, \ldots, \pm l$

It is worth noting again the natural way in which quantum numbers appear in quantum-mechanical theories of particles trapped in a particular region of space.

To exhibit the dependence of R, Θ, and Φ upon the quantum numbers n, l, m, we may write for the electron wave functions of the hydrogen atom

$$\psi = R_{nl}\Theta_{lm_l}\Phi_{m_l} \tag{6.18}$$

The wave functions R, Θ, and Φ together with ψ are given in Table 6.1 for $n = 1$, 2, and 3.

Table 6.1 Normalized Wave Functions of the Hydrogen Atom for $n = 1$, 2, and 3*

n	l	m_l	$\Phi(\phi)$	$\Theta(\theta)$	$R(r)$	$\psi(r, \theta, \phi)$
1	0	0	$\frac{1}{\sqrt{2\pi}}$	$\frac{1}{\sqrt{2}}$	$\frac{2}{a_0^{3/2}} e^{-r/a_0}$	$\frac{1}{\sqrt{\pi}\, a_0^{3/2}} e^{-r/a_0}$
2	0	0	$\frac{1}{\sqrt{2\pi}}$	$\frac{1}{\sqrt{2}}$	$\frac{1}{2\sqrt{2}\, a_0^{3/2}}\left(2 - \frac{r}{a_0}\right)e^{-r/2a_0}$	$\frac{1}{4\sqrt{2\pi}\, a_0^{3/2}}\left(2 - \frac{r}{a_0}\right)e^{-r/2a_0}$
2	1	0	$\frac{1}{\sqrt{2\pi}}$	$\frac{\sqrt{6}}{2}\cos\theta$	$\frac{1}{2\sqrt{6}\, a_0^{3/2}}\frac{r}{a_0}e^{-r/2a_0}$	$\frac{1}{4\sqrt{2\pi}\, a_0^{3/2}}\frac{r}{a_0}e^{-r/2a_0}\cos\theta$
2	1	±1	$\frac{1}{\sqrt{2\pi}}e^{\pm i\phi}$	$\frac{\sqrt{3}}{2}\sin\theta$	$\frac{1}{2\sqrt{6}\, a_0^{3/2}}\frac{r}{a_0}e^{-r/2a_0}$	$\frac{1}{8\sqrt{\pi}\, a_0^{3/2}}\frac{r}{a_0}e^{-r/2a_0}\sin\theta\, e^{\pm i\phi}$
3	0	0	$\frac{1}{\sqrt{2\pi}}$	$\frac{1}{\sqrt{2}}$	$\frac{2}{81\sqrt{3}\, a_0^{3/2}}\left(27 - 18\frac{r}{a_0} + 2\frac{r^2}{a_0^2}\right)e^{-r/3a_0}$	$\frac{1}{81\sqrt{3\pi}\, a_0^{3/2}}\left(27 - 18\frac{r}{a_0} + 2\frac{r^2}{a_0^2}\right)e^{-r/3a_0}$
3	1	0	$\frac{1}{\sqrt{2\pi}}$	$\frac{\sqrt{6}}{2}\cos\theta$	$\frac{4}{81\sqrt{6}\, a_0^{3/2}}\left(6 - \frac{r}{a_0}\right)\frac{r}{a_0}e^{-r/3a_0}$	$\frac{\sqrt{2}}{81\sqrt{\pi}\, a_0^{3/2}}\left(6 - \frac{r}{a_0}\right)\frac{r}{a_0}e^{-r/3a_0}\cos\theta$
3	1	±1	$\frac{1}{\sqrt{2\pi}}e^{\pm i\phi}$	$\frac{\sqrt{3}}{2}\sin\theta$	$\frac{4}{81\sqrt{6}\, a_0^{3/2}}\left(6 - \frac{r}{a_0}\right)\frac{r}{a_0}e^{-r/3a_0}$	$\frac{1}{81\sqrt{\pi}\, a_0^{3/2}}\left(6 - \frac{r}{a_0}\right)\frac{r}{a_0}e^{-r/3a_0}\sin\theta\, e^{\pm i\phi}$
3	2	0	$\frac{1}{\sqrt{2\pi}}$	$\frac{\sqrt{10}}{4}(3\cos^2\theta - 1)$	$\frac{4}{81\sqrt{30}\, a_0^{3/2}}\frac{r^2}{a_0^2}e^{-r/3a_0}$	$\frac{1}{81\sqrt{6\pi}\, a_0^{3/2}}\frac{r^2}{a_0^2}e^{-r/3a_0}(3\cos^2\theta - 1)$
3	2	±1	$\frac{1}{\sqrt{2\pi}}e^{\pm i\phi}$	$\frac{\sqrt{15}}{2}\sin\theta\cos\theta$	$\frac{4}{81\sqrt{30}\, a_0^{3/2}}\frac{r^2}{a_0^2}e^{-r/3a_0}$	$\frac{1}{81\sqrt{\pi}\, a_0^{3/2}}\frac{r^2}{a_0^2}e^{-r/3a_0}\sin\theta\cos\theta\, e^{\pm i\phi}$
3	2	±2	$\frac{1}{\sqrt{2\pi}}e^{\pm 2i\phi}$	$\frac{\sqrt{15}}{4}\sin^2\theta$	$\frac{4}{81\sqrt{30}\, a_0^{3/2}}\frac{r^2}{a_0^2}e^{-r/3a_0}$	$\frac{1}{162\sqrt{\pi}\, a_0^{3/2}}\frac{r^2}{a_0^2}e^{-r/3a_0}\sin^2\theta\, e^{\pm 2i\phi}$

*The quantity $a_0 = 4\pi\epsilon_0\hbar^2/me^2 = 5.292 \times 10^{-11}$ m is equal to the radius of the innermost Bohr orbit.

Example 6.1

Find the ground-state electron energy E_1 by substituting the radial wave function R that corresponds to $n = 1, l = 0$ into Eq. (6.14).

Solution

From Table 6.1 we see that $R = (2/a_0^{3/2})e^{-r/a_0}$. Hence

$$\frac{dR}{dr} = \left(\frac{2}{a_0^{5/2}}\right)e^{-r/a_0}$$

and

$$\frac{1}{r^2}\frac{d}{dr}\left(r^2\frac{dR}{dr}\right) = \left(\frac{2}{a_0^{7/2}} - \frac{4}{a_0^{5/2}r}\right)e^{-r/a_0}$$

Substituting in Eq. (6.14) with $E = E_1$ and $l = 0$ gives

$$\left[\left(\frac{2}{a_0^{7/2}} + \frac{4mE_1}{\hbar^2 a_0^{3/2}}\right) + \left(\frac{me^2}{\pi\epsilon_0\hbar a_0^{3/2}} - \frac{4}{a_0^{5/2}}\right)\frac{1}{r}\right]e^{-r/a_0} = 0$$

Each parenthesis must equal 0 for the entire equation to equal 0. For the second parenthesis this gives

$$\frac{me^2}{\pi\epsilon_0\hbar^2 a^{3/2}} - \frac{4}{a_0^{5/2}} = 0$$

$$a_0 = \frac{4\pi\epsilon_0\hbar^2}{me^2}$$

which is the Bohr radius $a_0 = r_1$ given by Eq. (4.13)—we recall that $\hbar = h/2\pi$. For the first parenthesis,

$$\frac{2}{a_0^{7/2}} + \frac{4mE_1}{\hbar^2 a_0^{3/2}} = 0$$

$$E_1 = -\frac{\hbar^2}{2ma_0^2} = -\frac{me^4}{32\pi^2\epsilon_0^2\hbar^2}$$

which agrees with Eq. (6.16).

6.4 PRINCIPAL QUANTUM NUMBER

Quantization of energy

It is interesting to consider what the hydrogen-atom quantum numbers signify in terms of the classical model of the atom. This model, as we saw in Chap. 4, corresponds exactly to planetary motion in the solar system except that the inverse-square force holding the electron to the nucleus is electrical rather than gravitational. Two quantities are conserved—that is, maintain a constant value at all times—in planetary motion: the scalar total energy and the vector angular momentum of each planet.

Classically the total energy can have any value whatever, but it must, of course, be negative if the planet is to be trapped permanently in the solar system. In the quantum theory of the hydrogen atom the electron energy is also a constant, but while it may have any positive value (corresponding to an ionized atom), the *only* negative values the electron can have are specified by the formula $E_n = E_1/n^2$. The quantization of electron energy in the hydrogen atom is therefore described by the principal quantum number n.

The theory of planetary motion can also be worked out from Schrödinger's equation, and it yields a similar energy restriction. However, the total quantum number n for any of the planets turns out to be so immense (see Exercise 11 of Chap. 4) that the separation of permitted levels is far too small to be observable. For this reason classical physics provides an adequate description of planetary motion but fails within the atom.

6.5 ORBITAL QUANTUM NUMBER

Quantization of angular-momentum magnitude

The interpretation of the orbital quantum number l is less obvious. Let us look at the differential equation for the radial part $R(r)$ of the wave function ψ:

$$\frac{1}{r^2}\frac{d}{dr}\left(r^2\frac{dR}{dr}\right)+\left[\frac{2m}{\hbar^2}\left(\frac{e^2}{4\pi\epsilon_0 r}+E\right)-\frac{l(l+1)}{r^2}\right]R=0 \tag{6.14}$$

This equation is solely concerned with the radial aspect of the electron's motion, that is, its motion toward or away from the nucleus. However, we notice the presence of E, the total electron energy, in the equation. The total energy E includes the electron's kinetic energy of orbital motion, which should have nothing to do with its radial motion.

This contradiction may be removed by the following argument. The kinetic energy KE of the electron has two parts, KE_{radial} due to its motion toward or away from the nucleus, and $KE_{orbital}$ due to its motion around the nucleus. The potential energy U of the electron is the electric energy

$$U=-\frac{e^2}{4\pi\epsilon_0 r} \tag{6.2}$$

Hence the total energy of the electron is

$$E=KE_{radial}+KE_{orbital}+U=KE_{radial}+KE_{orbital}-\frac{e^2}{4\pi\epsilon_0 r}$$

Inserting this expression for E in Eq. (6.14) we obtain, after a slight rearrangement,

$$\frac{1}{r^2}\frac{d}{dr}\left(r^2\frac{dR}{dr}\right)+\frac{2m}{\hbar^2}\left[KE_{radial}+KE_{orbital}-\frac{\hbar^2 l(l+1)}{2mr^2}\right]R=0 \tag{6.19}$$

If the last two terms in the square brackets of this equation cancel each other out, we shall have what we want: a differential equation for $R(r)$ that involves functions of the radius vector r exclusively.

We therefore require that

$$KE_{orbital}=\frac{\hbar^2 l(l+1)}{2mr^2} \tag{6.20}$$

Since the orbital kinetic energy of the electron and the magnitude of its angular momentum are respectively

$$\mathrm{KE}_{\text{orbital}} = \frac{1}{2} m v_{\text{orbital}}^2 \qquad L = m v_{\text{orbital}} r$$

we may write for the orbital kinetic energy

$$\mathrm{KE}_{\text{orbital}} = \frac{L^2}{2mr^2}$$

Hence, from Eq. (6.20),

$$\frac{L^2}{2mr^2} = \frac{\hbar^2 l(l+1)}{2mr^2}$$

Electron angular momentum

$$L = \sqrt{l(l+1)}\,\hbar \tag{6.21}$$

With the orbital quantum number l restricted to the values

$$l = 0, 1, 2, \ldots, (n-1)$$

The electron can have only the angular momenta L specified by Eq. (6.21). Like total energy E, *angular momentum is both conserved and quantized.* The quantity

$$\hbar = \frac{h}{2\pi} = 1.054 \times 10^{-34}\ \mathrm{J \cdot s}$$

is thus the natural unit of angular momentum.

In macroscopic planetary motion, as in the case of energy, the quantum number describing angular momentum is so large that the separation into discrete angular momentum states cannot be experimentally observed. For example, an electron (or, for that matter, any other body) whose orbital quantum number is 2 has the angular momentum

$$L = \sqrt{2(2+1)}\,\hbar = \sqrt{6}\,\hbar$$
$$= 2.6 \times 10^{-34}\ \mathrm{J \cdot s}$$

By contrast the orbital angular momentum of the earth is $2.7 \times 10^{40}\ \mathrm{J \cdot s}$!

Designation of Angular-Momentum States

It is customary to specify electron angular-momentum states by a letter, with s corresponding to $l = 0$, p to $l = 1$, and so on, according to the following scheme:

Angular-momentum states

$l =$	0	1	2	3	4	5	6	...
	s	p	d	f	g	h	i	...

Table 6.2 Atomic Electron States

	$l = 0$	$l = 1$	$l = 2$	$l = 3$	$l = 4$	$l = 5$
$n = 1$	1*s*					
$n = 2$	2*s*	2*p*				
$n = 3$	3*s*	3*p*	3*d*			
$n = 4$	4*s*	4*p*	4*d*	4*f*		
$n = 5$	5*s*	5*p*	5*d*	5*f*	5*g*	
$n = 6$	6*s*	6*p*	6*d*	6*f*	6*g*	6*h*

This peculiar code originated in the empirical classification of spectra into series called sharp, principal, diffuse, and fundamental which occurred before the theory of the atom was developed. Thus an *s* state is one with no angular momentum, a *p* state has the angular moment $\sqrt{2}\hbar$, and so forth.

The combination of the total quantum number with the letter that represents orbital angular momentum provides a convenient and widely used notation for atomic electron states. In this notation a state in which $n = 2$, $l = 0$ is a 2*s* state, for example, and one in which $n = 4$, $l = 2$ is a 4*d* state. Table 6.2 gives the designations of electron states in an atom through $n = 6$, $l = 5$.

6.6 MAGNETIC QUANTUM NUMBER

Quantization of angular-momentum direction

The orbital quantum number l determines the *magnitude* L of the electron's angular momentum **L**. However, angular momentum, like linear momentum, is a vector quantity, and to describe it completely means that its *direction* be specified as well as its magnitude. (The vector **L**, we recall, is perpendicular to the plane in which the rotational motion takes place, and its sense is given by the right-hand rule: When the fingers of the right hand point in the direction of the motion, the thumb is in the direction of **L**. This rule is illustrated in Fig. 6.3.)

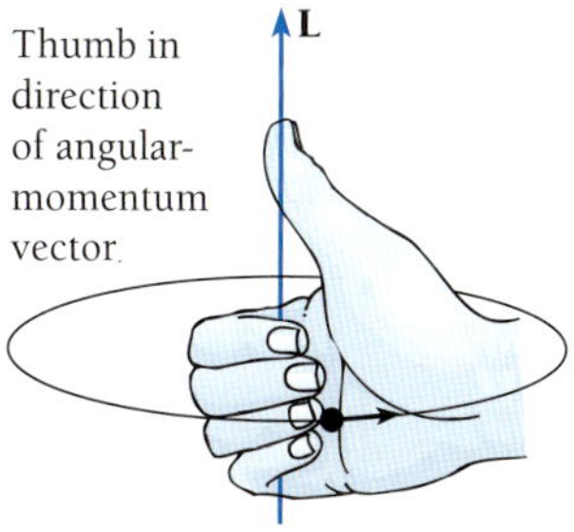

Figure 6.3 The right-hand rule for angular momentum.

What possible significance can a direction in space have for a hydrogen atom? The answer becomes clear when we reflect that an electron revolving about a nucleus is a minute current loop and has a magnetic field like that of a magnetic dipole. Hence an atomic electron that possesses angular momentum interacts with an external magnetic field **B**. The magnetic quantum number m_l specifies the direction of **L** by determining the component of **L** in the field direction. This phenomenon is often referred to as **space quantization.**

If we let the magnetic-field direction be parallel to the z axis, the component of **L** in this direction is

Space quantization $$L_z = m_l\hbar \qquad m_l = 0, \pm 1, \pm 2, \ldots, \pm l \tag{6.22}$$

The possible values of m_l for a given value of l range from $+l$ through 0 to $-l$, so that the number of possible orientations of the angular-momentum vector **L** in a magnetic field is $2l + 1$. When $l = 0$, L_z can have only the single value of 0; when $l = 1$, L_z may be $\hbar$, 0, or $-\hbar$; when $l = 2$, L_z may be $2\hbar$, $\hbar$, 0, $-\hbar$, or $-2\hbar$; and so on.

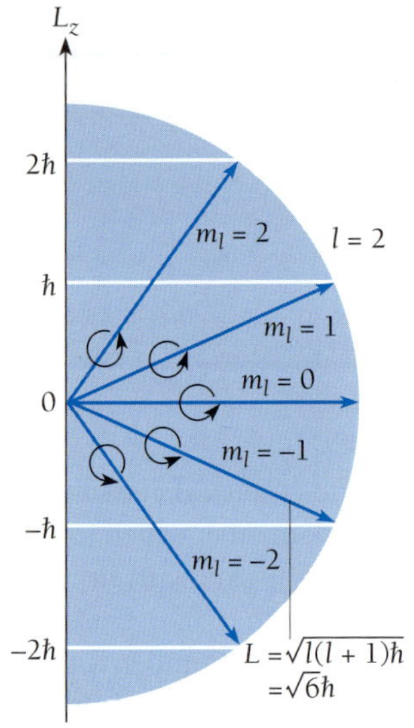

Figure 6.4 Space quantization of orbital angular momentum. Here the orbital quantum number is $l = 2$ and there are accordingly $2l + 1 = 5$ possible values of the magnetic quantum number m_l, with each value corresponding to a different orientation relative to the z axis.

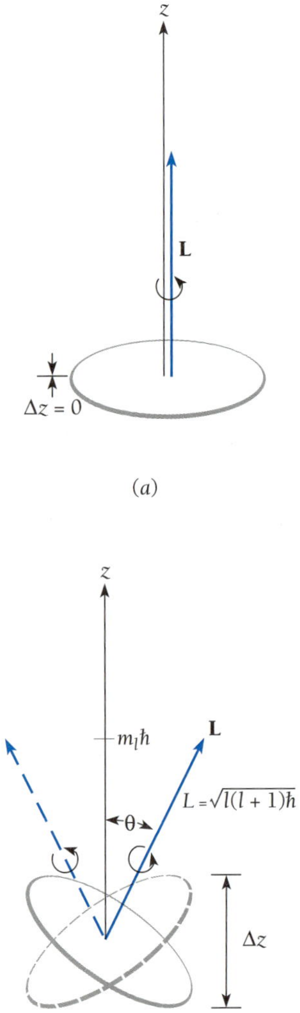

Figure 6.5 The uncertainty principle prohibits the angular momentum vector **L** from having a definite direction in space.

The space quantization of the orbital angular momentum of the hydrogen atom is show in Fig. 6.4. An atom with a certain value of m_l will assume the corresponding orientation of its angular momentum **L** relative to an external magnetic field if it finds itself in such a field. We note that **L** can never be aligned exactly parallel or antiparallel to **B** because L_z is always smaller than the magnitude $\sqrt{l(l+1)}\hbar$ of the total angular momentum.

In the absence of an external magnetic field, the direction of the z axis is arbitrary. What must be true is that the component of **L** in *any* direction we choose is $m_l\hbar$. What an external magnetic field does is to provide an experimentally meaningful reference direction. A magnetic field is not the only such reference direction possible. For example, the line between the two H atoms in the hydrogen molecule H_2 is just as experimentally meaningful as the direction of a magnetic field, and along this line the components of the angular momenta of the H atoms are determined by their m_l values.

The Uncertainty Principle and Space Quantization

Why is only one component of **L** quantized? The answer is related to the fact that **L** can never point in any specific direction but instead is somewhere on a cone in space such that its projection L_z is $m_l\hbar$. Were this not so, the uncertainty principle would be violated. If **L** were fixed in space, so that L_x and L_y as well as L_z had definite values, the electron would be confined to a definite plane. For instance, if L were in the z direction, the electron would have to be in the xy plane at all times (Fig. 6.5*a*). This can occur only if the electron's momentum component p_z in the z direction is infinitely uncertain, which of course is impossible if it is to be part of a hydrogen atom.

However, since in reality only *one* component L_z of **L** together with its magnitude L have definite values and $|L| > |L_z|$, the electron is not limited to a single plane (Fig.6.5*b*). Thus there is a built-in uncertainty in the electron's z coordinate. The direction of **L** is not fixed, as in Fig. 6.6, and so the average values of L_x and L_y are 0, although L_z always has the specific value $m_l\hbar$.

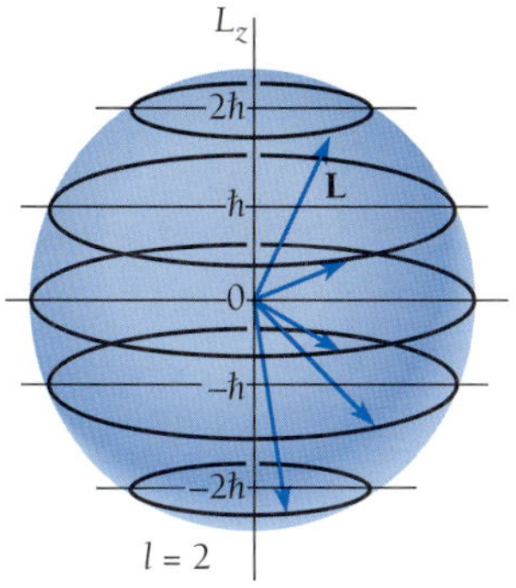

Figure 6.6 The angular momentum vector **L** precesses constantly about the z axis.

6.7 ELECTRON PROBABILITY DENSITY

No definite orbits

In Bohr's model of the hydrogen atom the electron is visualized as revolving around the nucleus in a circular path. This model is pictured in a spherical polar coordinate system in Fig. 6.7. It implies that if a suitable experiment were performed, the electron would always be found a distance of $r = n^2a_0$ (where n is the quantum number of the orbit and a_0 is the radius of the innermost orbit) from the nucleus and in the equatorial plane $\theta = 90°$, while its azimuth angle ϕ changes with time.

The quantum theory of the hydrogen atom modifies the Bohr model in two ways:

1 No definite values for r, θ, or ϕ can be given, but only the relative probabilities for finding the electron at various locations. This imprecision is, of course, a consequence of the wave nature of the electron.
2 We cannot even think of the electron as moving around the nucleus in any conventional sense since the probability density $|\psi|^2$ is independent of time and varies from place to place.

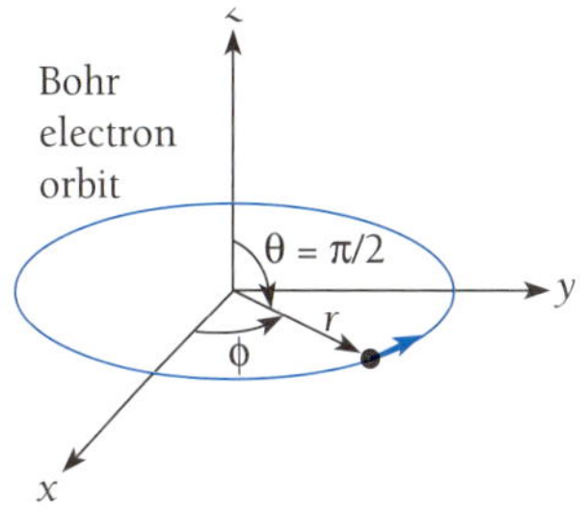

Figure 6.7 The Bohr model of the hydrogen atom in a spherical polar coordinate system.

The probability density $|\psi|^2$ that corresponds to the electron wave function $\psi = R\Theta\Phi$ in the hydrogen atom is

$$|\psi|^2 = |R|^2|\Theta|^2|\Phi|^2 \tag{6.23}$$

As usual the square of any function that is complex is to be replaced by the product of the function and its complex conjugate. (We recall that the complex conjugate of a function is formed by changing i to $-i$ whenever it appears.)

From Eq. (6.15) we see that the azimuthal wave function is given by

$$\Phi(\phi) = Ae^{im_l\phi}$$

The azimuthal probability density $|\Phi|^2$ is therefore

$$|\Phi|^2 = \Phi^*\Phi = A^2e^{-im_l\phi}e^{im_l\phi} = A^2e^0 = A^2$$

The likelihood of finding the electron at a particular azimuth angle ϕ is a constant that does not depend upon ϕ at all. The electron's probability density is symmetrical about the z axis regardless of the quantum state it is in, and the electron has the same chance of being found at one angle ϕ as at another.

The radial part R of the wave function, in contrast to Φ, not only varies with r but does so in a different way for each combination of quantum numbers n and l. Figure 6.8 contains graphs of R versus r for 1s, 2s, 2p, 3s, 3p, and 3d states of the hydrogen atom. Evidently R is a maximum at $r = 0$—that is, at the nucleus itself—for all s states, which correspond to $L = 0$ since $l = 0$ for such states. The value of R is zero at $r = 0$ for states that possess angular momentum.

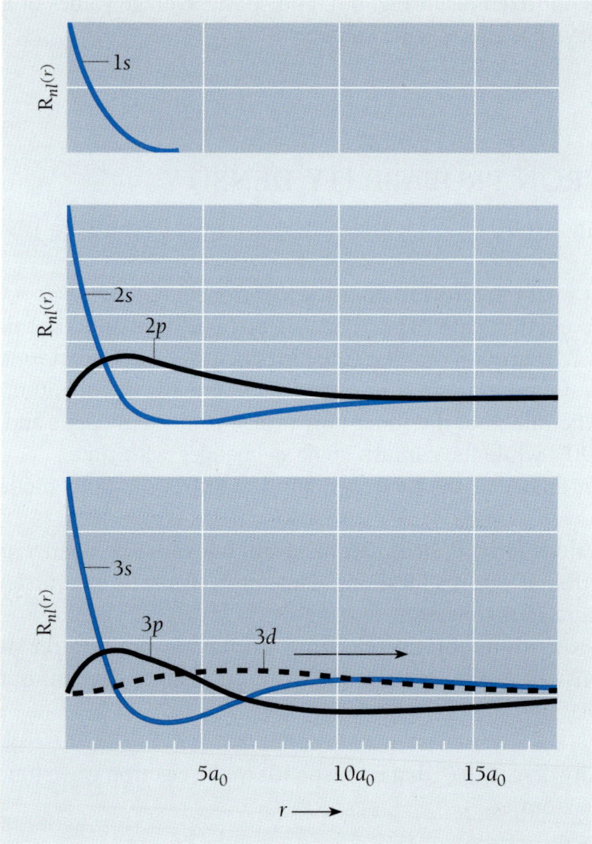

Figure 6.8 The variation with distance from the nucleus of the radial part of the electron wave function in hydrogen for various quantum states. The quantity $a_0 = 4\pi\epsilon_0\hbar^2/me^2 = 0.053$ nm is the radius of the first Bohr orbit.

Probability of Finding the Electron

The *probability density* of the electron at the point r, θ, ϕ is proportional to $|\psi|^2$, but the *actual probability* of finding it in the infinitesimal volume element dV there is $|\psi|^2\, dV$. In spherical polar coordinates (Fig. 6.9),

Volume element

$$\begin{aligned} dV &= (dr)\,(r\,d\theta)\,(r \sin\theta\, d\phi) \\ &= r^2 \sin\theta\, dr\, d\theta\, d\phi \end{aligned} \tag{6.24}$$

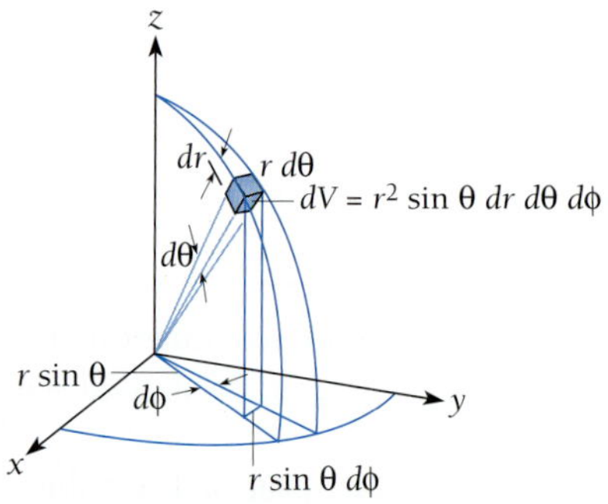

Figure 6.9 Volume element dV in spherical polar coordinates.

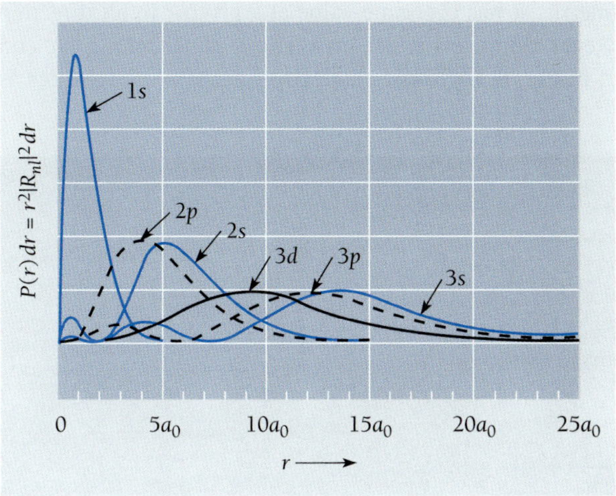

Figure 6.11 The probability of finding the electron in a hydrogen atom at a distance between r and $r + dr$ from the nucleus for the quantum states of Fig. 6.8.

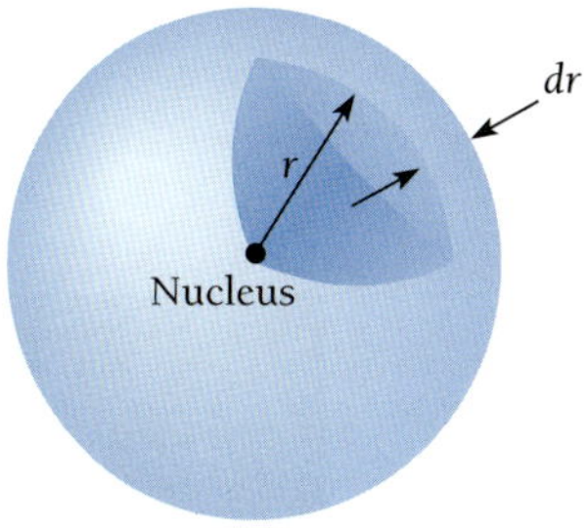

Figure 6.10 The probability of finding the electron in a hydrogen atom in the spherical shell between r and $r + dr$ from the nucleus is $P(r)\,dr$.

As Θ and Φ are normalized functions, the actual probability $P(r)\,dr$ of finding the electron in a hydrogen atom somewhere in the spherical shell between r and $r + dr$ from the nucleus (Fig. 6.10) is

$$\begin{aligned} P(r)\,dr &= r^2|R|^2\,dr \int_0^{\pi} |\Theta|^2 \sin\theta\,d\theta \int_0^{2\pi} |\Phi|^2\,d\phi \\ &= r^2|R|^2\,dr \qquad (6.25) \end{aligned}$$

Equation (6.25) is plotted in Fig. 6.11 for the same states whose radial functions R were shown in Fig. 6.8. The curves are quite different as a rule. We note immediately that P is not a maximum at the nucleus for s states, as R itself is, but has its maximum a definite distance from it.

The most probable value of r for a 1s electron turns out to be exactly a_0, the orbital radius of a ground-state electron in the Bohr model. However, the *average* value of r for a 1s electron is $1.5a_0$, which is puzzling at first sight because the energy levels are the same in both the quantum-mechanical and Bohr atomic models. This apparent discrepancy is removed when we recall that the electron energy depends upon $1/r$ rather than upon r directly, and the average value of $1/r$ for a 1s electron is exactly $1/a_0$.

Angular Variation of Probability Density

The function Θ varies with zenith angle θ for all quantum numbers l and m_l except $l = m_l = 0$, which are s states. The value of $|\Theta|^2$ for an s state is a constant; $\frac{1}{2}$, in fact. This means that since $|\Phi|^2$ is also a constant, the electron probability density $|\psi|^2$ is spherically symmetric: it has the same value at a given r in all directions. Electrons in other states, however, do have angular preferences, sometimes quite complicated ones. This can be seen in Fig. 6.12, in which electron probability densities as functions of r

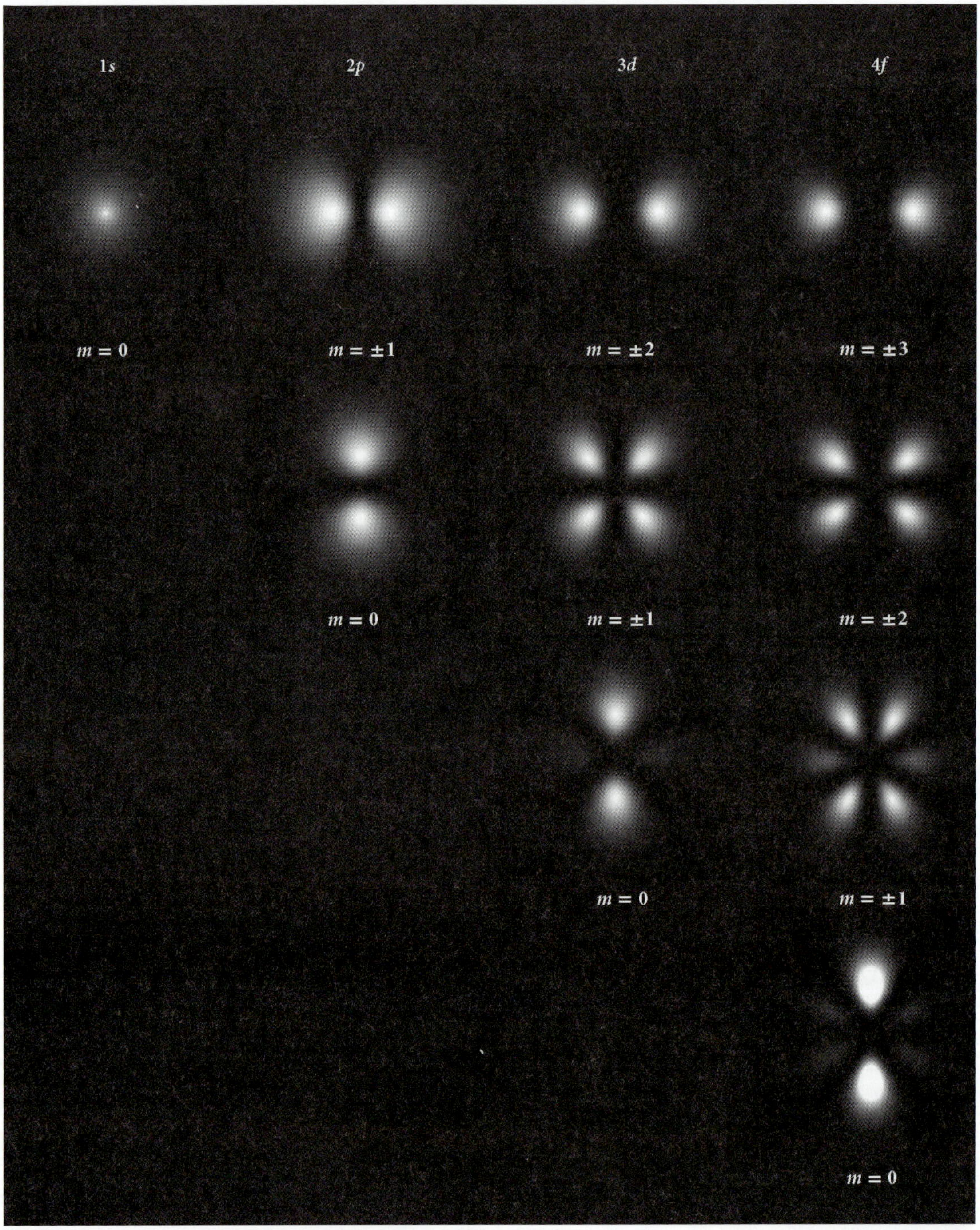

Figure 6.12 Photographic representation of the electron probability-density distribution $|\psi|^2$ for several energy states. These may be regarded as sectional views of the distribution in a plane containing the polar axis, which is vertical and in the plane of the paper. The scale varies from figure to figure.

and θ are shown for several atomic states. (The quantity plotted is $|\psi|^2$, not $|\psi|^2 \, dV$.) Since $|\psi|^2$ is independent of ϕ, we can obtain a three-dimensional picture of $|\psi|^2$ by rotating a particular representation about a vertical axis. When this is done, we see that the probability densities for *s* states are spherically symmetric whereas those for other

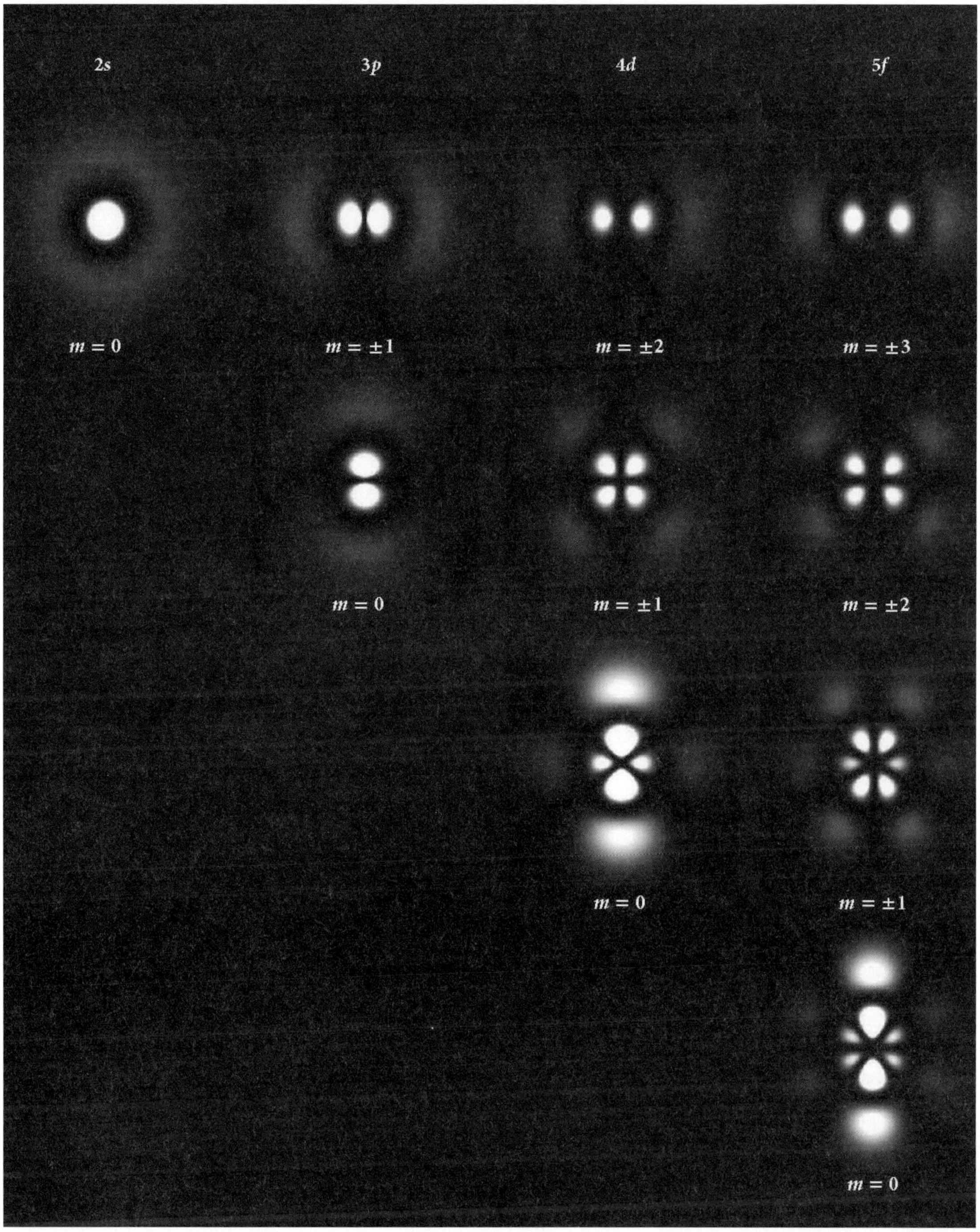

Example 6.2

Verify that the average value of $1/r$ for a 1s electron in the hydrogen atom is $1/a_0$.

Solution

The wave function of a 1s electron is, from Table 6.1,

$$\psi = \frac{e^{-r/a_0}}{\sqrt{\pi}a_0^{3/2}}$$

Since $dV = r^2 \sin\theta\, dr\, d\theta\, d\phi$ we have for the expectation value of $1/r$

$$\left\langle \frac{1}{r} \right\rangle = \int_0^\infty \left(\frac{1}{r}\right) |\psi|^2\, dV$$

$$= \frac{1}{\pi a_0^3} \int_0^\infty r e^{-2r/a_0}\, dr \int_0^\pi \sin\theta\, d\theta \int_0^{2\pi} d\phi$$

The integrals have the respective values

$$\int_0^\infty r e^{-2r/a_0}\, dr = \left[\frac{a_0^2}{4} e^{-2r/a_0} - \frac{r}{2} e^{-2r/a_0}\right]_0^\infty = \frac{a_0^2}{4}$$

$$\int_0^\pi \sin\,\theta\, d\theta = [-\cos\,\theta]_0^\pi = 2$$

$$\int_0^{2\pi} d\phi = [\phi]_0^{2\pi} = 2\pi$$

Hence
$$\left\langle \frac{1}{r} \right\rangle = \left(\frac{1}{\pi a_0^3}\right)\left(\frac{a_0^2}{4}\right)(2)(2\pi) = \frac{1}{a_0}$$

Example 6.3

How much more likely is a 1s electron in a hydrogen atom to be at the distance a_0 from the nucleus than at the distance $a_0/2$?

Solution

According to Table 6.1 the radial wave function for a 1s electron is

$$R = \frac{2}{a_0^{3/2}} e^{-r/a_0}$$

From Eq. (6.25) we have for the ratio of the probabilities that an electron in a hydrogen atom will be at the distances r_1 and r_2 from the nucleus

$$\frac{P_1}{P_2} = \frac{r_1^2|R_1|^2}{r_2^2|R_2|^2} = \frac{r_1^2\, e^{-2r_1/a_0}}{r_2^2\, e^{-2r_2/a_0}}$$

Here $r_1 = a_0$ and $r_2 = a_0/2$, so

$$\frac{P_{a_0}}{P_{a_0/2}} = \frac{(a_0)^2 e^{-2}}{(a_0/2)^2 e^{-1}} = 4e^{-1} = 1.47$$

The electron is 47 percent more likely to be a_0 from the nucleus than half that distance (see Fig. 6.11).

states are not. The pronounced lobe patterns characteristic of many of the states turn out to be significant in chemistry since these patterns help determine the manner in which adjacent atoms in a molecule interact.

A look at Figure 6.12 also reveals quantum-mechanical states that resemble those of the Bohr model. The electron probability-density distribution for a $2p$ state with $m_l = \pm 1$, for instance, is like a doughnut in the equatorial plane centered at the nucleus. Calculation shows the most probable distance of such an electron from the nucleus to be $4a_0$—precisely the radius of the Bohr orbit for the same principal quantum number $n = 2$. Similar correspondences exist for $3d$ states with $m_l = \pm 2$, $4f$ states with $m_l = \pm 3$, and so on. In each of these cases the angular momentum is the highest possible for that energy level, and the angular-momentum vector is as near the z axis as possible so that the probability density is close to the equatorial plane. Thus the Bohr model predicts the most probable location of the electron in *one* of the several possible states in each energy level.

6.8 RADIATIVE TRANSITIONS

What happens when an electron goes from one state to another

In formulating his theory of the hydrogen atom, Bohr was obliged to postulate that the frequency ν of the radiation emitted by an atom dropping from an energy level E_m to a lower level E_n is

$$\nu = \frac{E_m - E_n}{h}$$

It is not hard to show that this relationship arises naturally in quantum mechanics. For simplicity we shall consider a system in which an electron moves only in the x direction.

From Sec. 5.7 we know that the time-dependent wave function Ψ_n of an electron in a state of quantum number n and energy E_n is the product of a time-independent wave function ψ_n and a time-varying function whose frequency is

$$\nu_n = \frac{E_n}{h}$$

Hence

$$\Psi_n = \psi_n e^{-(iE_n/h)t} \qquad \Psi_n^* = \psi_n^* e^{+(iE_n/\hbar)t} \tag{6.26}$$

The expectation value $\langle x \rangle$ of the position of such an electron is

$$\langle x \rangle = \int_{-\infty}^{\infty} x\Psi_n^*\Psi_n \, dx = \int_{-\infty}^{\infty} x\psi_n^*\psi_n e^{[(iE_n/\hbar)-(iE_n/\hbar)]t} \, dx$$

$$= \int_{-\infty}^{\infty} x\psi_n^*\psi_n \, dx \tag{6.27}$$

The expectation value $\langle x \rangle$ is constant in time since ψ_n and ψ_n^* are, by definition, functions of position only. The electron does not oscillate, and no radiation occurs. Thus quantum mechanics predicts that a system in a specific quantum state does not radiate, as observed.

We next consider an electron that shifts from one energy state to another. A system might be in its ground state n when an excitation process of some kind (a beam of radiation, say, or collisions with other particles) begins to act upon it. Subsequently we find that the system emits radiation corresponding to a transition from an excited state of energy E_m to the ground state. We conclude that at some time during the intervening period the system existed in the state m. What is the frequency of the radiation?

The wave function Ψ of an electron that can exist in both states n and m is

$$\Psi = a\Psi_n + b\Psi_m \tag{6.28}$$

where a^*a is the probability that the electron is in state n and b^*b the probability that it is in state m. Of course, it must always be true that $a^*a + b^*b = 1$. Initially $a = 1$ and $b = 0$; when the electron is in the excited state, $a = 0$ and $b = 1$; and ultimately $a = 1$ and $b = 0$ once more. While the electron is in either state, there is no radiation, but when it is in the midst of the transition from m to n (that is, when both a and b have nonvanishing values), electromagnetic waves are produced.

The expectation value $\langle x\rangle$ that corresponds to the composite wave function of Eq. (6.28) is

$$\begin{aligned}\langle x\rangle &= \int_{-\infty}^{\infty} x(a^*\Psi_n^* + b^*\Psi_m^*)(a\Psi_n + b\Psi_m)\,dx \\ &= \int_{-\infty}^{\infty} x(a^2\Psi_n^*\Psi_n + b^*a\Psi_m^*\Psi_n + a^*b\Psi_n^*\Psi_m + b^2\Psi_m^*\Psi_m)\,dx\end{aligned} \tag{6.29}$$

Here, as before, we let $a^*a = a^2$ and $b^*b = b^2$. The first and last integrals do not vary with time, so the second and third integrals are the only ones able to contribute to a time variation in $\langle x\rangle$.

With the help of Eq. (6.26) we expand Eq. (6.29) to give

$$\begin{aligned}\langle x\rangle = a^2\int_{-\infty}^{\infty} x\psi_n^*\psi_n\,dx &+ b^*a\int_{-\infty}^{\infty} x\psi_m^*e^{+(iE_m/\hbar)t}\,\psi_n e^{-(iE_n/\hbar)t}\,dx \\ &+ a^*b\int_{-\infty}^{\infty} x\psi_n^*e^{+(iE_n/\hbar)t}\,\psi_m e^{-(iE_m/\hbar)t}\,dx + b^2\int_{-\infty}^{\infty} x\psi_m^*\psi_m\,dx\end{aligned} \tag{6.30}$$

Because

$$e^{i\theta} = \cos\theta + i\sin\theta \qquad \text{and} \qquad e^{-i\theta} = \cos\theta - i\sin\theta$$

the two middle terms of Eq. (6.30), which are functions of time, become

$$\begin{aligned}&\cos\left(\frac{E_m - E_n}{\hbar}\right)t\int_{-\infty}^{\infty} x[b^*a\psi_m^*\psi_n + a^*b\psi_n^*\psi_m]\,dx \\ &\qquad + i\sin\left(\frac{E_m - E_n}{\hbar}\right)t\int_{-\infty}^{\infty} x[b^*a\psi_m^*\psi_n - a^*b\psi_n^*\psi_m)\,dx\end{aligned} \tag{6.31}$$

The real part of this result varies with time as

$$\cos\left(\frac{E_m - E_n}{\hbar}\right)t = \cos 2\pi\left(\frac{E_m - E_n}{h}\right)t = \cos 2\pi\nu t \tag{6.32}$$

The electron's position therefore oscillates sinusoidally at the frequency

$$\nu = \frac{E_m - E_n}{h} \tag{6.33}$$

When the electron is in state n or state m the expectation value of the electron's position is constant. When the electron is undergoing a transition between these states, its position oscillates with the frequency ν. Such an electron, of course, is like an electric dipole and radiates electromagnetic waves of the same frequency ν. This result is the same as that postulated by Bohr and verified by experiment. As we have seen, quantum mechanics gives Eq. (6.33) without the need for any special assumptions.

6.9 SELECTION RULES

Some transitions are more likely to occur than others

We did not have to know the values of the probabilities a and b as functions of time, nor the electron wave functions ψ_n and ψ_m, in order to find the frequency ν. We need these quantities, however, to calculate the chance a given transition will occur. The general condition necessary for an atom in an excited state to radiate is that the integral

$$\int_{-\infty}^{\infty} x\psi_n\psi_m^* \, dx \tag{6.34}$$

not be zero, since the intensity of the radiation is proportional to it. Transitions for which this integral is finite are called **allowed transitions,** while those for which it is zero are called **forbidden transitions.**

In the case of the hydrogen atom, three quantum numbers are needed to specify the initial and final states involved in a radiative transition. If the principal, orbital, and magnetic quantum numbers of the initial state are n', l', m_l', respectively, and those of the final state are n, l, m_l, and u represents either the x, y, or z coordinate, the condition for an allowed transition is

Allowed transitions

$$\int_{-\infty}^{\infty} u\psi_{n,l,m_l}\psi_{n',l',m_l'}^* \, dV \neq 0 \tag{6.35}$$

where the integral is now over all space. When u is taken as x, for example, the radiation would be that produced by a dipole antenna lying on the x axis.

Since the wave functions ψ_{n,l,m_l} for the hydrogen atom are known, Eq. (6.35) can be evaluated for $u = x$, $u = y$, and $u = z$ for all pairs of states differing in one or more quantum numbers. When this is done, it is found that the only transitions between states of different n that can occur are those in which the orbital quantum number l changes by $+1$ or -1 and the magnetic quantum number m_l does not change or changes by $+1$ or -1. That is, the condition for an allowed transition is that

Selection rules

$$\Delta l = \pm 1 \tag{6.36}$$

$$\Delta m_l = 0, \pm 1 \tag{6.37}$$

The change in total quantum number n is not restricted. Equations (6.36) and (6.37) are known as the **selection rules** for allowed transitions (Fig. 6.13).

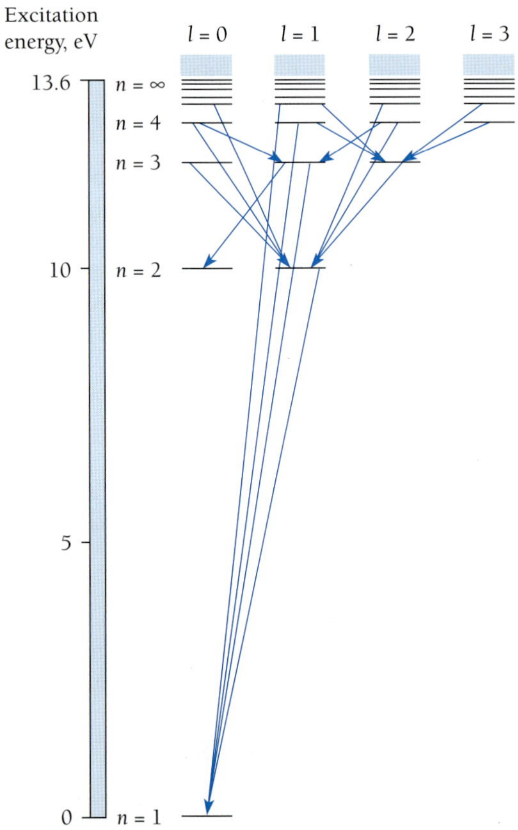

Figure 6.13 Energy-level diagram for hydrogen showing transitions allowed by the selection rule $\Delta l = \pm 1$. In this diagram the vertical axis represents excitation energy above the ground state.

The selection rule requiring that l change by ± 1 if an atom is to radiate means that an emitted photon carries off the angular momentum $\pm\hbar$ equal to the difference between the angular momenta of the atom's initial and final states. The classical analog of a photon with angular momentum $\pm\hbar$ is a left or right circularly polarized electromagnetic wave, so this notion is not unique with quantum theory.

Quantum Electrodynamics

The preceding analysis of radiative transitions in an atom is based on a mixture of classical and quantum concepts. As we have seen, the expectation value of the position of an atomic electron oscillates at the frequency ν of Eq. (6.33) while passing from an initial eigenstate to another one of lower energy. Classically, such an oscillating charge gives rise to electromagnetic waves of the same frequency ν, and indeed the observed radiation has this frequency. However, classical concepts are not always reliable guides to atomic processes, and a deeper treatment is required. Such a treatment, called **quantum electrodynamics**, shows that the radiation emitted during a transition from state m to state n is in the form of a single photon.

In addition, quantum electrodynamics provides an explanation for the mechanism that causes the "spontaneous" transition of an atom from one energy state to a lower one. All electric and

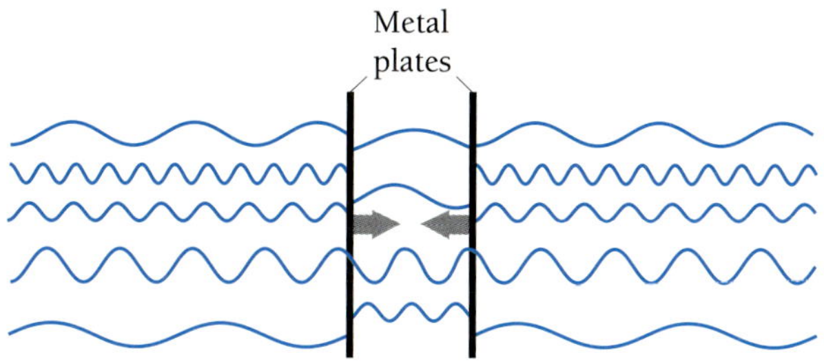

Figure 6.14 Two parallel metal plates exhibit the Casimir effect even in empty space. Virtual photons of any wavelength can strike the plates from the outside, but photons trapped between the plates can have only certain wavelengths. The resulting imbalance produces inward forces on the plates.

magnetic fields turn out to fluctuate constantly about the **E** and **B** that would be expected on purely classical grounds. Such fluctuations occur even when electromagnetic waves are absent and when, classically, $\mathbf{E} = \mathbf{B} = 0$. It is these fluctuations (often called "vacuum fluctuations" and analogous to the zero-point vibrations of a harmonic oscillator) that induce the apparently spontaneous emission of photons by atoms in excited states.

The vacuum fluctuations can be regarded as a sea of "virtual" photons so short-lived that they do not violate energy conservation because of the uncertainty principle in the form $\Delta E \, \Delta t \geq \hbar/2$. These photons, among other things, give rise to the **Casimir effect** (Fig. 6.14), which was proposed by the Dutch physicist Hendrik Casimir in 1948. Only virtual photons with certain specific wavelengths can be reflected back-and-forth between two parallel metal plates, whereas outside the plates virtual photons of all wavelengths can be reflected by them. The result is a very small but detectable force that tends to push the plates together.

Can the Casimir effect be used as a source of energy? If the parallel plates are released, they would fly together and thereby pick up kinetic energy from the vacuum fluctuations that would become heat if the plates were allowed to collide. Unfortunately not much energy is available in this way: about half a nanojoule (0.5×10^{-9} J) per square meter of plate area.

About Scientists

Richard P. Feynman (1918–1988) was born in Far Rockaway, a suburb of New York City, and studied at the Massachusetts Institute of Technology and Princeton. After receiving his Ph.D. in 1942, he helped develop the atomic bomb at Los Alamos, New Mexico, along with many other young physicists. When the war was over, he went first to Cornell and, in 1951, to the California Institute of Technology.

In the late 1940s Feynman made important contributions to quantum electrodynamics, the relativistic quantum theory that describes the electromagnetic interaction between charged particles. A serious problem in this theory is the presence of infinite quantities in its results, which in the procedure called renormalization are removed by subtracting other infinite quantities. Although this step is mathematically dubious and still leaves some physicists uneasy, the final theory has proven extraordinarily accurate in all its predictions. An unrepentant Feynman remarked, "It is not philosophy we are after, but the behavior of real things," and compared the agreement between quantum electrodynamics and experiment to finding the distance from New York to Los Angeles to within the thickness of a single hair.

Feynman articulated the feelings of many physicists when he wrote: "We have always had a great deal of difficulty understanding the world view that quantum mechanics represents . . . I cannot define the real problem, therefore I suspect there's no real problem, but I'm not sure there's no real problem."

In 1965 Feynman received the Nobel Prize together with two other pioneers in quantum electrodynamics, Julian Schwinger, also an American, and Sin-Itiro Tomonaga, a Japanese. Feynman made other major contributions to physics, notably in explaining the behavior of liquid helium near absolute zero and in elementary particle theory. His three-volume *Lectures on Physics* has stimulated and enlightened both students and teachers since its publication in 1963.

6.10 ZEEMAN EFFECT

How atoms interact with a magnetic field

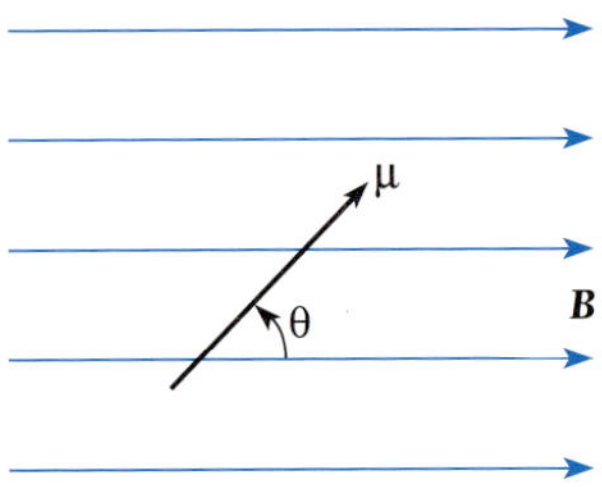

Figure 6.15 A magnetic dipole of moment $\boldsymbol{\mu}$ at the angle θ relative to a magnetic field $\mathbf{B}$.

In an external magnetic field **B**, a magnetic dipole has an amount of potential energy U_m that depends upon both the magnitude μ of its magnetic moment and the orientation of this moment with respect to the field (Fig. 6.15).

The torque τ on a magnetic dipole in a magnetic field of flux density **B** is

$$\tau = \mu B \sin \theta$$

where θ is the angle between $\boldsymbol{\mu}$ and **B**. The torque is a maximum when the dipole is perpendicular to the field, and zero when it is parallel or antiparallel to it. To calculate the potential energy U_m we must first establish a reference configuration in which U_m is zero by definition. (Since only *changes* in potential energy are ever experimentally observed, the choice of a reference configuration is arbitrary.) It is convenient to set $U_m = 0$ when $\theta = \pi/2 = 90°$, that is, when $\boldsymbol{\mu}$ is perpendicular to **B**. The potential energy at any other orientation of $\boldsymbol{\mu}$ is equal to the external work that must be done to rotate the dipole from $\theta_0 = \pi/2$ to the angle θ that corresponds to that orientation. Hence

$$\begin{aligned} U_m &= \int_{\pi/2}^{\theta} \tau \, d\theta = \mu B \int_{\pi/2}^{\theta} \sin \theta \, d\theta \\ &= -\mu B \cos \theta \end{aligned} \tag{6.38}$$

When $\boldsymbol{\mu}$ points in the same direction as **B**, then $U_m = -\mu B$, its minimum value. This follows from the fact that a magnetic dipole tends to align itself with an external magnetic field.

The magnetic moment of the orbital electron in a hydrogen atom depends on its angular momentum **L**. Hence both the magnitude of **L** and its orientation with respect to the field determine the extent of the magnetic contribution to the total energy of the atom when it is in a magnetic field. The magnetic moment of a current loop has the magnitude

$$\mu = IA$$

where I is the current and A the area it encloses. An electron that makes f rev/s in a circular orbit of radius r is equivalent to a current of $-ef$ (since the electronic charge is $-e$), and its magnetic moment is therefore

$$\mu = -ef\pi r^2$$

Because the linear speed v of the electron is $2\pi fr$ its angular momentum is

$$L = mvr = 2\pi mfr^2$$

Comparing the formulas for magnetic moment μ and angular momentum L shows that

Electron magnetic moment

$$\boldsymbol{\mu} = -\left(\frac{e}{2m}\right)\mathbf{L} \tag{6.39}$$

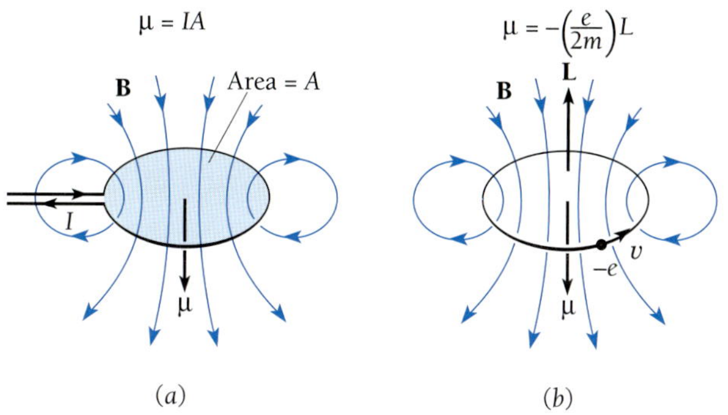

Figure 6.16 (*a*) Magnetic moment of a current loop enclosing area *A*. (*b*) Magnetic moment of an orbiting electron of angular momentum **L**.

for an orbital electron (Fig. 6.16). The quantity $(-e/2m)$, which involves only the charge and mass of the electron, is called its **gyromagnetic ratio.** The minus sign means that $\boldsymbol{\mu}$ is in the opposite direction to **L** and is a consequence of the negative charge of the electron. While the above expression for the magnetic moment of an orbital electron has been obtained by a classical calculation, quantum mechanics yields the same result. The magnetic potential energy of an atom in a magnetic field is therefore

$$U_m = \left(\frac{e}{2m}\right)LB\cos\theta \tag{6.40}$$

which depends on both B and θ.

Magnetic Energy

From Fig. 6.4 we see that the angle θ between **L** and the z direction can have only the values specified by

$$\cos\theta = \frac{m_l}{\sqrt{l(l+1)}}$$

with the permitted values of L specified by

$$L = \sqrt{l(l+1)}\hbar$$

To find the magnetic energy that an atom of magnetic quantum number m_l has when it is in a magnetic field **B**, we put the above expressions for cos θ and L in Eq. (6.40) to give

Magnetic energy
$$U_m = m_l\left(\frac{e\hbar}{2m}\right)B \tag{6.41}$$

The quantity $e\hbar/2m$ is called the **Bohr magneton:**

Bohr magneton
$$\mu_B = \frac{e\hbar}{2m} = 9.274 \times 10^{-24}\ \text{J/T} = 5.788 \times 10^{-5}\ \text{eV/T} \tag{6.42}$$

In a magnetic field, then, the energy of a particular atomic state depends on the value of m_l as well as on that of n. A state of total quantum number n breaks up into several substates when the atom is in a magnetic field, and their energies are slightly more or slightly less than the energy of the state in the absence of the field. This phenomenon leads to a "splitting" of individual spectral lines into separate lines when atoms radiate in a magnetic field. The spacing of the lines depends on the magnitude of the field.

The splitting of spectral lines by a magnetic field is called the **Zeeman effect** after the Dutch physicist Pieter Zeeman, who first observed it in 1896. The Zeeman effect is a vivid confirmation of space quantization.

Because m_l can have the $2l + 1$ values of $+l$ through 0 to $-l$, a state of given orbital quantum number l is split into $2l + 1$ substates that differ in energy by $\mu_B B$ when the atom is in a magnetic field. However, because changes in m_l are restricted to $\Delta m_l = 0, \pm 1$, we expect a spectral line from a transition between two states of different l to be split into only three components, as shown in Fig. 6.17. The **normal Zeeman effect** consists of the splitting of a spectral line of frequency ν_0 into three components whose frequencies are

Normal Zeeman effect

$$\begin{aligned} \nu_1 &= \nu_0 - \mu_B \frac{B}{h} = \nu_0 - \frac{e}{4\pi m} B \\ \nu_2 &= \nu_0 \\ \nu_3 &= \nu_0 + \mu_B \frac{B}{h} = \nu_0 + \frac{e}{4\pi m} B \end{aligned} \tag{6.43}$$

In Chap. 7 we will see that this is not the whole story of the Zeeman effect.

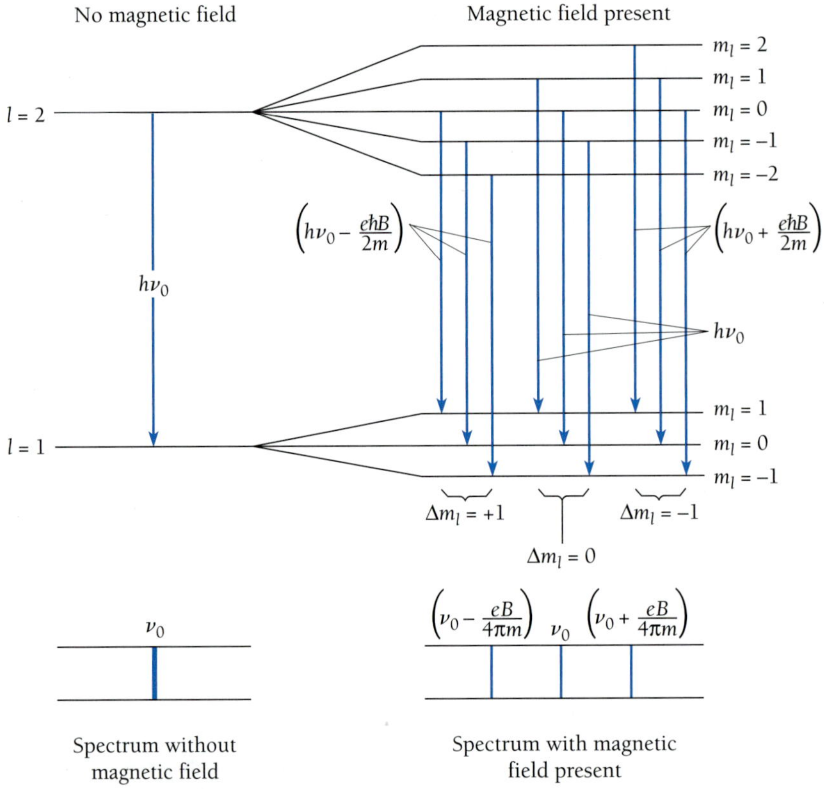

Figure 6.17 In the normal Zeeman effect a spectral line of frequency ν_0 is split into three components when the radiating atoms are in a magnetic field of magnitude B. One component is ν_0 and the others are less than or greater than ν_0 by $eB/4\pi m$. There are only three components because of the selection rule $\Delta m_l = 0, \pm 1$.

Example 6.4

A sample of a certain element is placed in a 0.300-T magnetic field and suitably excited. How far apart are the Zeeman components of the 450-nm spectral line of this element?

Solution

The separation of the Zeeman components is

$$\Delta\nu = \frac{eB}{4\pi m}$$

Since $\nu = c/\lambda$, $d\nu = -c\, d\lambda/\lambda^2$, and so, disregarding the minus sign,

$$\Delta\lambda = \frac{\lambda^2\,\Delta\nu}{c} = \frac{eB\lambda^2}{4\pi mc}$$

$$= \frac{(1.60 \times 10^{-19}\text{ C})(0.300\text{ T})(4.50 \times 10^{-7}\text{ m})^2}{(4\pi)(9.11 \times 10^{-31}\text{ kg})(3.00 \times 10^{8}\text{ m/s})}$$

$$= 2.83 \times 10^{-12}\text{ m} = 0.00283\text{ nm}$$

EXERCISES

To strive, to seek, to find, and not to yield. —Alfred, Lord Tennyson

6.3 Quantum Numbers

1. Why is it natural that three quantum numbers are needed to describe an atomic electron (apart from electron spin)?

2. Show that

$$\Theta_{20}(\theta) = \frac{\sqrt{10}}{4}(3\cos^2\theta - 1)$$

is a solution of Eq. (6.13) and that it is normalized.

3. Show that

$$R_{10}(r) = \frac{2}{a_0^{3/2}} e^{-r/a_0}$$

is a solution of Eq. (6.14) and that it is normalized.

4. Show that

$$R_{21}(r) = \frac{1}{2\sqrt{6}a_0^{3/2}} \frac{r}{a_0} e^{-r/2a_0}$$

is a solution of Eq. (6.14) and that it is normalized.

5. In Exercise 12 of Chap. 5 it was stated that an important property of the eigenfunctions of a system is that they are orthogonal to one another, which means that

$$\int_{-\infty}^{\infty} \psi_n^*\psi_m\, dV = 0 \qquad n \neq m$$

Verify that this is true for the azimuthal wave functions Φ_{m_l} of the hydrogen atom by calculating

$$\int_0^{2\pi} \Phi_{m_l}^*\Phi_{m_l'}\, d\phi$$

for $m_l \neq m_l'$.

6. The azimuthal wave function for the hydrogen atom is

$$\Phi(\phi) = Ae^{im_l\phi}$$

Show that the value of the normalization constant A is $1/\sqrt{2\pi}$ by integrating $|\Phi|^2$ over all angles from 0 to 2π.

6.5 Orbital Quantum Number

7. Compare the angular momentum of a ground-state electron in the Bohr model of the hydrogen atom with its value in the quantum theory.

8. (*a*) What is Schrödinger's equation for a particle of mass m that is constrained to move in a circle of radius R, so that ψ depends only on ϕ? (*b*) Solve this equation for ψ and evaluate the normalization constant. (*Hint*: Review the solution of Schrödinger's equation for the hydrogen atom.) (*c*) Find the possible energies of the particle. (*d*) Find the possible angular momenta of the particle.

6.6 Magnetic Quantum Number

9. Under what circumstances, if any, is L_z equal to L?

10. What are the angles between **L** and the z axis for $l = 1$? For $l = 2$?

11. What are the possible values of the magnetic quantum number m_l of an atomic electron whose orbital quantum number is $l = 4$?

12. List the sets of quantum numbers possible for an $n = 4$ hydrogen atom.

13. Find the percentage difference between L and the maximum value of L_z for an atomic electron in p, d, and f states.

6.7 Electron Probability Density

14. Under what circumstances is an atomic electron's probability-density distribution spherically symmetric? Why?

15. In Sec. 6.7 it is stated that the most probable value of r for a 1s electron in a hydrogen atom is the Bohr radius a_0. Verify this.

16. At the end of Sec. 6.7 it is stated that the most probable value of r for a 2p electron in a hydrogen atom is $4a_0$, which is the same as the radius of the $n = 2$ Bohr orbit. Verify this.

17. Find the most probable value of r for a 3d electron in a hydrogen atom.

18. According to Fig. 6.11, $P\,dr$ has *two* maxima for a 2s electron. Find the values of r at which these maxima occur.

19. How much more likely is the electron in a ground-state hydrogen atom to be at the distance a_0 from the nucleus than at the distance $2a_0$?

20. In Section 6.7 it is stated that the average value of r for a 1s electron in a hydrogen atom is $1.5a_0$. Verify this statement by calculating the expectation value $\langle r \rangle = \int r|\psi|^2\, dV$.

21. The probability of finding an atomic electron whose radial wave function is $R(r)$ outside a sphere of radius r_0 centered on the nucleus is

$$\int_{r_0}^{\infty} |R(r)|^2 r^2\, dr$$

(*a*) Calculate the probability of finding a 1s electron in a hydrogen atom at a distance greater than a_0 from the nucleus.
(*b*) When a 1s electron in a hydrogen atom is $2a_0$ from the nucleus, all its energy is potential energy. According to classical physics, the electron therefore cannot ever exceed the distance $2a_0$ from the nucleus. Find the probability $r > 2a_0$ for a 1s electron in a hydrogen atom.

22. According to Fig. 6.11, a 2s electron in a hydrogen atom is more likely than a 2p electron to be closer to the nucleus than $r = a_0$ (that is, to be between $r = 0$ and $r = a_0$). Verify this by calculating the relevant probabilities.

23. **Unsöld's theorem** states that for any value of the orbital quantum number l, the probability densities summed over all possible states from $m_l = -l$ to $m_l = +l$ yield a constant independent of angles θ or ϕ; that is,

$$\sum_{m_l=-l}^{+l} |\Theta|^2|\Phi|^2 = \text{constant}$$

This theorem means that every closed subshell atom or ion (Sec. 7.6) has a spherically symmetric distribution of electric charge. Verify Unsöld's theorem for $l = 0$, $l = 1$, and $l = 2$ with the help of Table 6.1.

6.9 Selection Rules

24. A hydrogen atom is in the 4p state. To what state or states can it go by radiating a photon in an allowed transition?

25. With the help of the wave functions listed in Table 6.1 verify that $\Delta l = \pm 1$ for $n = 2 \rightarrow n = 1$ transitions in the hydrogen atom.

26. The selection rule for transitions between states in a harmonic oscillator is $\Delta n = \pm 1$. (*a*) Justify this rule on classical grounds. (*b*) Verify from the relevant wave functions that the $n = 1 \rightarrow n = 3$ transition in a harmonic oscillator is forbidden whereas the $n = 1 \rightarrow n = 0$ and $n = 1 \rightarrow n = 2$ transitions are allowed.

27. Verify that the $n = 3 \rightarrow n = 1$ transition for the particle in a box of Sec. 5.8 is forbidden whereas the $n = 3 \rightarrow n = 2$ and $n = 2 \rightarrow n = 1$ transitions are allowed.

6.10 Zeeman Effect

28. In the Bohr model of the hydrogen atom, what is the magnitude of the orbital magnetic moment of an electron in the nth energy level?

29. Show that the magnetic moment of an electron in a Bohr orbit of radius r_n is proportional to $\sqrt{r_n}$.

30. Example 4.7 considered a muonic atom in which a negative muon ($m = 207m_e$) replaces the electron in a hydrogen atom. What difference, if any, would you expect between the Zeeman effect in such atoms and in ordinary hydrogen atoms?

31. Find the minimum magnetic field needed for the Zeeman effect to be observed in a spectral line of 400-nm wavelength when a spectrometer whose resolution is 0.010 nm is used.

32. The Zeeman components of a 500-nm spectral line are 0.0116 nm apart when the magnetic field is 1.00 T. Find the ratio e/m for the electron from these data.

Many-Electron Atoms

Helium, whose atoms have only closed electron shells, is inert chemically and cannot burn or explode. Because it is also less dense than air, it is used in airships.

Quantum mechanics explains certain properties of the hydrogen atom in an accurate, straightforward, and beautiful way. However, it cannot approach a complete description of this atom or of any other without taking into account electron spin and the exclusion principle. In this chapter we will look into the role of electron spin in atomic phenomena and into why the exclusion principle is the key to understanding the structures of atoms with more than one electron.

7.1 ELECTRON SPIN

Round and round it goes forever

The theory of the atom developed in the previous chapter cannot account for a number of well-known experimental observations. One is the fact that many spectral lines actually consist of two separate lines that are very close together. An example of this **fine structure** is the first line of the Balmer series of hydrogen, which arises from transitions between the $n = 3$ and $n = 2$ levels in hydrogen atoms. Here the theoretical prediction is for a single line of wavelength 656.3 nm while in reality there are two lines 0.14 nm apart—a small effect, but a conspicuous failure for the theory.

Another failure of the simple quantum-mechanical theory of the atom occurs in the Zeeman effect, which was discussed in Sec. 6.10. There we saw that the spectral lines of an atom in a magnetic field should each be split into the three components specified by Eq. (6.43). While the normal Zeeman effect is indeed observed in the spectra of a few elements under certain circumstances, more often it is not. Four, six, or even more components may appear, and even when three components are present their spacing may not agree with Eq. (6.43). Several anomalous Zeeman patterns are shown in Fig. 7.1 together with the predictions of Eq. (6.43). (When reproached in 1923 for looking sad, the physicist Wolfgang Pauli replied, "How can one look happy when he is thinking about the anomalous Zeeman effect?")

In order to account for both fine structure in spectral lines and the anomalous Zeeman effect, two Dutch graduate students, Samuel Goudsmit and George Uhlenbeck, proposed in 1925 that

> Every electron has an intrinsic angular momentum, called spin, whose magnitude is the same for all electrons. Associated with this angular momentum is a magnetic moment.

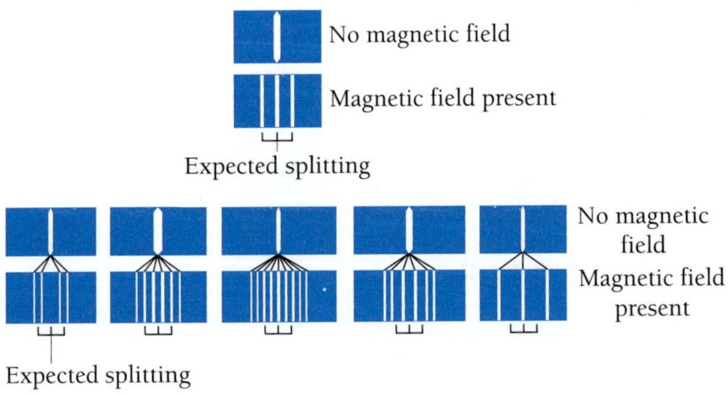

Figure 7.1 The normal and anomalous Zeeman effects in various spectral lines.

What Goudsmit and Uhlenbeck had in mind was a classical picture of an electron as a charged sphere spinning on its axis. The rotation involves angular momentum, and because the electron is negatively charged, it has a magnetic moment $\boldsymbol{\mu}_s$ opposite in direction to its angular momentum vector **S**. The notion of electron spin proved to be successful in explaining not only fine structure and the anomalous Zeeman effect but a wide variety of other atomic effects as well.

To be sure, the picture of an electron as a spinning charged sphere is open to serious objections. For one thing, observations of the scattering of electrons by other electrons at high energy indicate that the electron must be less than 10^{-16} m across, and quite possibly is a point particle. In order to have the observed angular momentum associated with electron spin, so small an object would have to rotate with an equatorial velocity many times greater than the velocity of light.

But the failure of a model taken from everyday life does not invalidate the idea of electron spin. We have already found plenty of ideas in relativity and quantum physics that are mandated by experiment although at odds with classical concepts. In 1929 the fundamental nature of electron spin was confirmed by Paul Dirac's development of relativistic quantum mechanics. He found that a particle with the mass and charge of the electron *must* have the intrinsic angular momentum and magnetic moment proposed for the electron by Goudsmit and Uhlenbeck.

The quantum number s describes the spin angular momentum of the electron. The only value s can have is $s = \frac{1}{2}$, which follows both from Dirac's theory and from spectral data. The magnitude S of the angular momentum due to electron spin is given in terms of the spin quantum number s by

Spin angular momentum

$$S = \sqrt{s(s+1)}\hbar = \frac{\sqrt{3}}{2}\hbar \qquad (7.1)$$

This is the same formula as that giving the magnitude L of the orbital angular momentum in terms of the orbital quantum number l, $L = \sqrt{l(l+1)}\hbar$.

Example 7.1

Find the equatorial velocity v of an electron under the assumption that it is a uniform sphere of radius $r = 5.00 \times 10^{-17}$ m that is rotating about an axis through its center.

Solution

The angular momentum of a spinning sphere is $I\omega$, where $I = \frac{2}{5}mr^2$ is its moment of inertia and $\omega = v/r$ is its angular velocity. From Eq. (7.1) the spin angular momentum of an electron is $S = (\sqrt{3}/2)\hbar$, so

$$S = \frac{\sqrt{3}}{2}\hbar = I\omega = \left(\frac{2}{5}mr^2\right)\left(\frac{v}{r}\right) = \frac{2}{5}mvr$$

$$v = \left(\frac{5\sqrt{3}}{4}\right)\frac{\hbar}{mr} = \frac{(5\sqrt{3})(1.055 \times 10^{-34}\ \text{J}\cdot\text{s})}{(4)(9.11 \times 10^{-31}\ \text{kg})(5.00 \times 10^{-17}\ \text{m})} = 5.01 \times 10^{12}\ \text{m/s} = 1.67 \times 10^4\ c$$

The equatorial velocity of an electron on the basis of this model must be over 10,000 times the velocity of light, which is impossible. No classical model of the electron can overcome this difficulty.

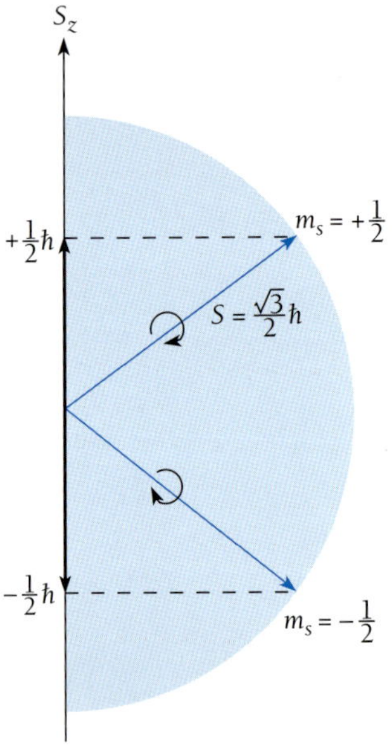

Figure 7.2 The two possible orientations of the spin angular-momentum vector are "spin up" ($m_s = +\frac{1}{2}$) and "spin down" ($m_s = -\frac{1}{2}$).

Table 7.1 Quantum Numbers of an Atomic Electron

Name	Symbol	Possible Values	Quantity Determined
Principal	n	$1, 2, 3, \ldots$	Electron energy
Orbital	l	$0, 1, 2, \ldots, n-1$	Orbital angular-momentum magnitude
Magnetic	m_l	$-l, \ldots, 0, \ldots, +l$	Orbital angular-momentum direction
Spin magnetic	m_s	$-\frac{1}{2}, +\frac{1}{2}$	Electron spin direction

The space quantization of electron spin is described by the spin magnetic quantum number m_s. We recall that the orbital angular-momentum vector can have the $2l + 1$ orientations in a magnetic field from $+l$ to $-l$. Similarly the spin angular-momentum vector can have the $2s + 1 = 2$ orientations specified by $m_s = +\frac{1}{2}$ ("spin up") and $m_s = -\frac{1}{2}$ ("spin down"), as in Fig. 7.2. The component S_z of the spin angular momentum of an electron along a magnetic field in the z direction is determined by the spin magnetic quantum number, so that

z component of spin angular momentum

$$S_z = m_s\hbar = \pm\frac{1}{2}\hbar \tag{7.2}$$

We recall from Sec. 6.10 that gyromagnetic ratio is the ratio between magnetic moment and angular momentum. The gyromagnetic ratio for electron orbital motion is $-e/2m$. The gyromagnetic ratio characteristic of electron spin is almost exactly twice that characteristic of electron orbital motion. Taking this ratio as equal to 2, the spin magnetic moment $\boldsymbol{\mu}_s$ of an electron is related to its spin angular momentum **S** by

Spin magnetic moment

$$\boldsymbol{\mu}_s = -\frac{e}{m}\mathbf{S} \tag{7.3}$$

The possible components of $\boldsymbol{\mu}_s$ along any axis, say the z axis, are therefore limited to

z component of spin magnetic moment

$$\mu_{sz} = \pm\frac{e\hbar}{2m} = \pm\mu_B \tag{7.4}$$

where μ_B is the Bohr magneton ($= 9.274 \times 10^{-24}$ J/T $= 5.788 \times 10^{-5}$ eV/T).

The introduction of electron spin into the theory of the atom means that a total of four quantum numbers, n, l, m_l, and m_s, is needed to describe each possible state of an atomic electron. These are listed in Table 7.1.

7.2 EXCLUSION PRINCIPLE

A different set of quantum numbers for each electron in an atom

In a normal hydrogen atom, the electron is in its quantum state of lowest energy. What about more complex atoms? Are all 92 electrons of a uranium atom in the same quantum state, jammed into a single probability cloud? Many lines of evidence make this idea unlikely.

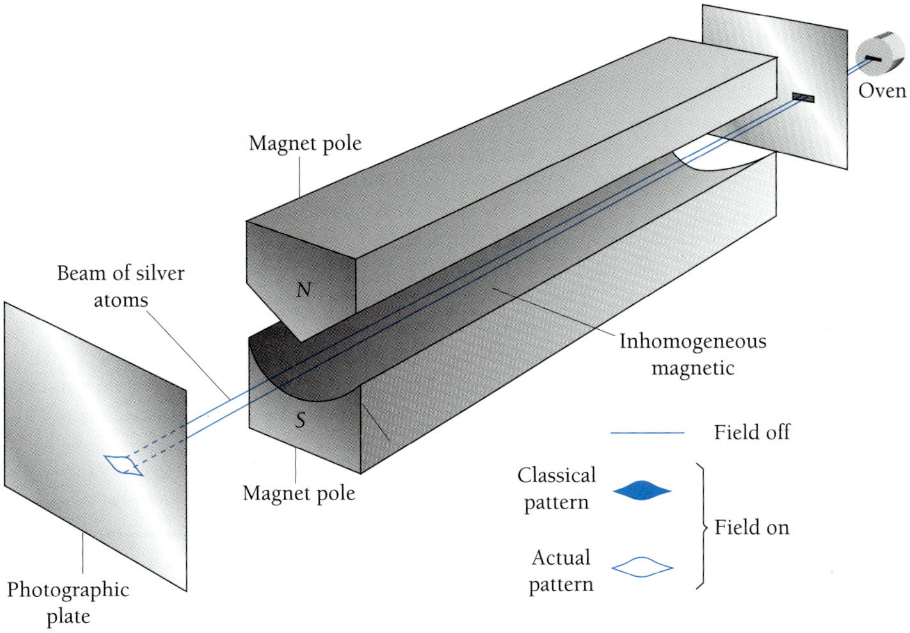

Figure 7.3 The Stern-Gerlach experiment.

The Stern-Gerlach Experiment

Space quantization was first explictly demonstrated in 1921 by Otto Stern and Walter Gerlach. They directed a beam of neutral silver atoms from an oven through a set of collimating slits into an inhomogeneous magnetic field as in Fig. 7.3. A photographic plate recorded the shape of the beam after it had passed through the field.

In its normal state the entire magnetic moment of a silver atom is due to the spin of only one of its electrons. In a uniform magnetic field, such a dipole would merely experience a torque tending to align it with the field. In an inhomogeneous field, however, each "pole" of the dipole is subject to a force of different magnitude and therefore there is a resultant force on the dipole that varies with its orientation relative to the field.

Classically, all orientations should be present in a beam of atoms. The result would merely be a broad trace on the photographic plate instead of the thin line formed without any magnetic field. Stern and Gerlach found, however, that the initial beam split into two distinct parts that correspond to the two opposite spin orientations in the magnetic field permitted by space quantization.

An example is the great difference in chemical behavior shown by certain elements whose atomic structures differ by only one electron. Thus the elements that have the atomic numbers 9, 10, and 11 are respectively the chemically active halogen gas fluorine, the inert gas neon, and the alkali metal sodium. Since the electron structure of an atom controls how it interacts with other atoms, it makes no sense that the chemical properties of the elements should change so sharply with a small change in atomic number if all the electrons in an atom were in the same quantum state.

About Scientists

Wolfgang Pauli (1900–1958) was born in Vienna and at nineteen had prepared a detailed account of special and general relativity that impressed Einstein and remained the standard work on the subject for many years. Pauli received his doctorate from the University of Munich in 1922 and then spent short periods in Göttingen, Copenhagen, and Hamburg before becoming professor of physics at the Institute of Technology in Zurich, Switzerland, in 1928. In 1925 he proposed that four quantum numbers (what one of them governed was then unknown) are needed to characterize each atomic electron and that no two electrons in an atom have the same set of quantum numbers. This exclusion principle turned out to be the missing link in understanding the arrangement of electrons in an atom.

Late in 1925 Goudsmit and Uhlenbeck, two young Dutch physicists, showed that the electron possesses intrinsic angular momentum, so it must be thought of as spinning, and that Pauli's fourth quantum number described the direction of the spin. The American physicist Ralph Kronig had conceived of electron spin a few months earlier and had told Pauli about it. However, because Pauli had "ridiculed the idea" Kronig did not publish his work.

In 1931 Pauli resolved the problem of the apparently missing energy in the beta decay of a nucleus by proposing that a neutral, massless particle leaves the nucleus together with the electron emitted. Two years later Fermi developed the theory of beta decay with the help of this particle (today believed to have a small mass), which he called the neutrino ("small neutral one" in Italian). Pauli spent the war years in the United States, and received the Nobel Prize in 1945.

In 1925 Wolfgang Pauli discovered the fundamental principle that governs the electronic configurations of atoms having more than one electron. His **exclusion principle** states that

> No two electrons in an atom can exist in the same quantum state. Each electron must have a different set of quantum numbers n, l, m_l, m_s.

Pauli was led to the exclusion principle by a study of atomic spectra. The various states of an atom can be determined from its spectrum, and the quantum numbers of these states can be inferred. In the spectra of every element but hydrogen a number of lines are *missing* that correspond to transitions to and from states having certain combinations of quantum numbers. For instance, no transitions are observed in helium to or from the ground-state configuration in which the spins of both electrons are in the same direction. However, transitions *are* observed to and from the other ground-state configuration, in which the spins are in opposite directions.

In the absent state in helium the quantum numbers of *both* electrons would be $n = 1, l = 0, m_l = 0, m_s = \frac{1}{2}$. On the other hand, in the state known to exist one of the electrons has $m_s = \frac{1}{2}$ and the other $m_s = -\frac{1}{2}$. Pauli showed that every unobserved atomic state involves two or more electrons with identical quantum numbers, and the exclusion principle is a statement of this finding.

7.3 SYMMETRIC AND ANTISYMMETRIC WAVE FUNCTIONS

Fermions and bosons

Before we explore the role of the exclusion principle in determining atomic structures, it is interesting to look into its quantum-mechanical implications.

The complete wave function $\psi(1, 2, 3, \ldots, n)$ of a system of n noninteracting particles can be expressed as the product of the wave functions $\psi(1)$, $\psi(2)$, $\psi(3)$, . . . , $\psi(n)$ of the individual particles. That is,

$$\psi(1, 2, 3, \ldots, n) = \psi(1)\,\psi(2)\,\psi(3) \ldots \psi(n) \tag{7.5}$$

Let us use Eq. (7.5) to look into the kinds of wave functions that can be used to describe a system of two identical particles.

Suppose one of the particles is in quantum state a and the other in state b. Because the particles are identical, it should make no difference in the probability density $|\psi|^2$ of the system if the particles are exchanged, with the one in state a replacing the one in state b, and vice versa. Symbolically, we require that

$$|\psi|^2(1, 2) = |\psi|^2(2, 1) \tag{7.6}$$

The wave function $\psi(2, 1)$ that represents the exchanged particles can be either

Symmetric $$\psi(2, 1) = \psi(1, 2) \tag{7.7}$$

or

Antisymmetric $$\psi(2, 1) = -\psi(1, 2) \tag{7.8}$$

and still fulfill Eq. (7.6). The wave function of the system is not itself a measurable quantity, and so it can be altered in sign by the exchange of the particles. Wave functions that are unaffected by an exchange of particles are said to be **symmetric**, while those that reverse sign upon such an exchange are said to be **antisymmetric.**

If particle 1 is in state a and particle 2 is in state b, the wave function of the system is, according to Eq. (7.5),

$$\psi_{\text{I}} = \psi_a(1)\psi_b(2) \tag{7.9}$$

If particle 2 is in state a and particle 1 is in state b, the wave function is

$$\psi_{\text{II}} = \psi_a(2)\psi_b(1) \tag{7.10}$$

Because the two particles are indistinguishable, we have no way to know at any moment whether ψ_{I} or ψ_{II} describes the system. The likelihood that ψ_{I} is correct at any moment is the same as the likelihood that ψ_{II} is correct.

Equivalently, we can say that the system spends half the time in the configuration whose wave function is ψ_{I} and the other half in the configuration whose wave function is ψ_{II}. Therefore a linear combination of ψ_{I} and ψ_{II} is the proper description of the system. Two such combinations, symmetric and antisymmetric, are possible:

Symmetric $$\psi_S = \frac{1}{\sqrt{2}}\,[\psi_a(1)\psi_b(2) + \psi_a(2)\psi_b(1)] \tag{7.11}$$

Antisymmetric $$\psi_A = \frac{1}{\sqrt{2}}\,[\psi_a(1)\psi_b(2) - \psi_a(2)\psi_b(1)] \tag{7.12}$$

The factor $1/\sqrt{2}$ is needed to normalize ψ_S and ψ_A. Exchanging particles 1 and 2 leaves ψ_S unaffected, while it reverses the sign of ψ_A. Both ψ_S and ψ_A obey Eq. (7.6).

There are a number of important distinctions between the behavior of particles in systems whose wave functions are symmetric and that of particles in systems whose wave functions are antisymmetric. The most obvious is that in the symmetric case, both particles 1 and 2 can simultaneously exist in the same state, with $a = b$. In the antisymmetric case, if we set $a = b$, we find that

$$\psi_A = \frac{1}{\sqrt{2}}[\psi_a(1)\psi_a(2) - \psi_a(2)\psi_a(1)] = 0$$

Hence the two particles *cannot* be in the same quantum state. Pauli found that no two electrons in an atom can be in the same quantum state, so we conclude that systems of electrons are described by wave functions that reverse sign upon the exchange of any pair of them.

Fermions and Bosons

The results of various experiments show that *all* particles which have odd half-integral spins ($\frac{1}{2}$, $\frac{3}{2}$, . . .) have wave functions that are antisymmetric to an exchange of any pair of them. Such particles, which include protons and neutrons as well as electrons, obey the exclusion principle when they are in the same system. That is, when they move in a common force field, each member of the system must be in a different quantum state. Particles of odd half-integral spin are often referred to as **fermions** because, as we shall learn in Chap. 9, the behavior of systems of them (such as free electrons in a metal) is governed by a statistical distribution law discovered by Fermi and Dirac.

Particles whose spins are 0 or an integer have wave functions that are symmetric to an exchange of any pair of them. These particles, which include photons, alpha particles, and helium atoms, do not obey the exclusion principle. Particles of 0 or integral spin are often referred to as **bosons** because the behavior of systems of them (such as photons in a cavity) is governed by a statistical distribution law discovered by Bose and Einstein.

There are other consequences of the symmetry or antisymmetry of particle wave functions besides that expressed in the exclusion principle. It is these consequences that make it useful to classify particles according to the natures of their wave functions rather than merely according to whether or not they obey the exclusion principle.

7.4 PERIODIC TABLE

Organizing the elements

In 1869 the Russian chemist Dmitri Mendeleev formulated the **periodic law** whose modern statement is

> When the elements are listed in order of atomic number, elements with similar chemical and physical properties recur at regular intervals.

Although the modern quantum theory of the atom was many years in the future, Mendeleev was fully aware of the significance his work would turn out to have. As he

About Scientists

Dmitri Mendeleev (1834–1907) was born in Siberia and grew up there, going on to Moscow and later France and Germany to study chemistry. In 1866 he became professor of chemistry at the University of St. Petersburg and three years later published the first version of the periodic table. The notion of atomic number was then unknown and Mendeleev had to deviate from the strict sequence of atomic masses for some elements and leave gaps in the table in order that the known elements (only 63 at that time) occupy places appropriate to their properties. Other chemists of the time were thinking along the same lines, but Mendeleev went further in 1871 by proposing that the gaps correspond to then-unknown elements. When his detailed predictions of the properties of these elements were fulfilled upon their discovery, Mendeleev became world famous. A further triumph for the periodic table came at the end of the nineteenth century, when the inert gases were discovered. Here were six elements of whose existence Mendeleev had been unaware, but they fit perfectly as a new group in the table. The element of atomic number 101 is called mendelevium in his honor.

remarked, "The periodic law, together with the revelations of spectrum analysis, have contributed to again revive an old but remarkably long-lived hope—that of discovering, if not by experiment, at least by mental effort, the *primary matter*."

A **periodic table** is an arrangement of the elements according to atomic number in a series of rows such that elements with similar properties form vertical columns. Table 7.2 is a simple form of periodic table.

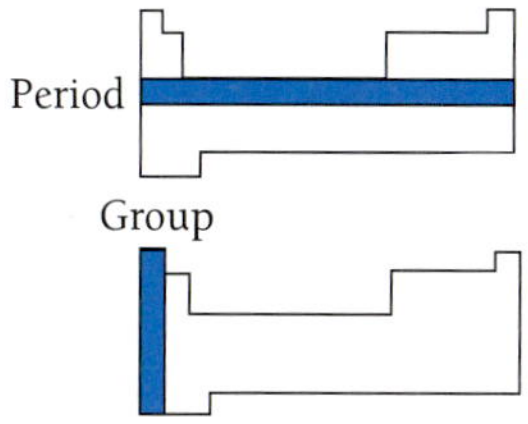

Figure 7.4 The elements in a group of the periodic table have similar properties, while those in a period have different properties.

Elements with similar properties form the **groups** shown as vertical columns in Table 7.2 (Fig. 7.4). Thus group 1 consists of hydrogen plus the alkali metals, which are all soft, have low melting points, and are very active chemically. Lithium, sodium, and potassium are examples. Hydrogen, although physically a nonmetal, behaves chemically much like an active metal. Group 7 consists of the halogens, volatile nonmetals that form diatomic molecules in the gaseous state. Like the alkali metals, the halogens are chemically active, but as oxidizing agents rather than as reducing agents. Fluorine, chlorine, bromine, and iodine are examples; fluorine is so active it can corrode platinum. Group 8 consists of the inert gases, of which helium, neon, and argon are examples. As their names suggest, they are inactive chemically: they form virtually no compounds with other elements, and their atoms do not join together into molecules.

The horizontal rows in Table 7.2 are called **periods.** The first three periods are broken in order to keep their members aligned with the most closely related elements of the long periods below. Most of the elements are metals (Fig. 7.5). Across each period is a more or less steady transition from an active metal through less active metals and weakly active nonmetals to highly active nonmetals and finally to an inert gas (Fig. 7.6). Within each column there are also regular changes in properties, but they are far less conspicuous than those in each period. For example, increasing atomic number in the alkali metals is accompanied by greater chemical activity, while the reverse is true in the halogens.

A series of **transition elements** appears in each period after the third between the group 2 and group 3 elements (Fig. 7.7). The transition elements are metals, in general hard and brittle with high melting points, that have similar chemical behavior. Fifteen of the transition elements in period 6 are virtually indistinguishable in their properties and are known as the **lanthanide** elements (or **rare earths**). Another group of closely related metals, the **actinide** elements, is found in period 7.

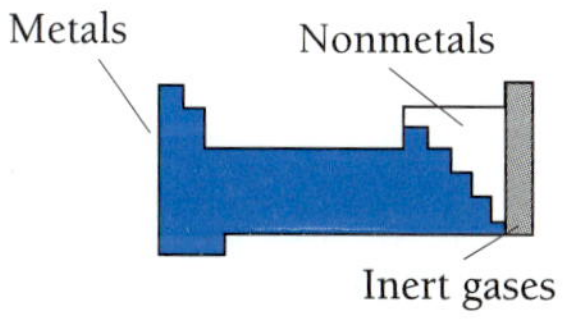

Figure 7.5 The majority of the elements are metals.

For over a century the periodic law has been indispensable to chemists because it provides a framework for organizing their knowledge of the elements. It is one of the

Table 7.2

The Periodic Table of the Elements

The number above the symbol of each element is its atomic number, and the number below its name is its average atomic mass. The elements whose atomic masses are given in parentheses do not occur in nature but have been created in nuclear reactions. The atomic mass in such a case is the mass number of the most long-lived radioisotope of the element.

Elements with atomic numbers 110, 111, 112, 114, and 116 have also been created but not yet named.

Group → Period	1	2											3	4	5	6	7	8
1	1 H Hydrogen 1.008																	2 He Helium 4.003
2	3 Li Lithium 6.941	4 Be Beryllium 9.012											5 B Boron 10.81	6 C Carbon 12.01	7 N Nitrogen 14.01	8 O Oxygen 16.00	9 F Fluorine 19.00	10 Ne Neon 20.18
3	11 Na Sodium 22.99	12 Mg Magnesium 24.31				Transition metals							13 Al Aluminium 26.98	14 Si Silicon 28.09	15 P Phosphorus 30.97	16 S Sulfur 32.07	17 Cl Chlorine 35.45	18 Ar Argon 39.95
4	19 K Potassium 39.10	20 Ca Calcium 40.08	21 Sc Scandium 44.96	22 Ti Titanium 47.88	23 V Vanadium 50.94	24 Cr Chromium 52.00	25 Mn Manganese 54.94	26 Fe Iron 55.8	27 Co Cobalt 58.93	28 Ni Nickel 58.69	29 Cu Copper 63.55	30 Zn Zinc 65.39	31 Ga Gallium 69.72	32 Ge Germanium 72.59	33 As Arsenic 74.92	34 Se Selenium 78.96	35 Br Bromine 79.90	36 Kr Krypton 83.80
5	37 Rb Rubidium 85.47	38 Sr Strontium 87.62	39 Y Yttrium 88.91	40 Zr Zirconium 91.22	41 Nb Niobium 92.91	42 Mo Molybdenum 95.94	43 Tc Technetium (98)	44 Ru Ruthenium 101.1	45 Rh Rhodium 102.9	46 Pd Palladium 106.4	47 Ag Silver 107.9	48 Cd Cadmium 112.4	49 In Indium 114.8	50 Sn Tin 118.7	51 Sb Antimony 121.9	52 Te Tellurium 127.6	53 I Iodine 126.9	54 Xe Xenon 131.8
6	55 Cs Cesium 132.9	56 Ba Barium 137.3		72 Hf Hafnium 178.5	73 Ta Tantalum 180.9	74 W Tungsten 183.9	75 Re Rhenium 186.2	76 Os Osmium 190.2	77 Ir Iridium 192.2	78 Pt Platinum 195.1	79 Au Gold 197.0	80 Hg Mercury 200.6	81 Tl Thallium 204.4	82 Pb Lead 207.2	83 Bi Bismuth 209.0	84 Po Polonium (209)	85 At Astatine (210)	86 Rn Radon (222)
7	87 Fr Francium (223)	88 Ra Radium 226.0		104 Rf Rutherfordium (261)	105 Db Dubnium (262)	106 Sg Seaborgium (263)	107 Bh Nielsbohrium (262)	108 Hs Hassium (264)	109 Mt Meitnerium (266)								Halogens	Inert gases
	Alkali metals																	

Lanthanides (rare earths)

57 La Lanthanum 138.9	58 Ce Cerium 140.1	59 Pr Praseodymium 140.9	60 Nd Neodymium 144.2	61 Pm Promethium (145)	62 Sm Sarnarium 150.4	63 Eu Europium 152.0	64 Gd Gadolinium 157.3	65 Tb Terbium 158.9	66 Dy Dysprosium 162.5	67 Ho Holmium 184.9	68 Er Erbium 167.3	69 Tm Thulium 168.9	70 Yb Ytterbium 173.0	71 Lu Lutetium 175.0
89 Ac Actinium (227)	90 Th Thorium 232.0	91 Pa Protactinium 231.0	92 U Uranium 238.0	93 Np Neptunium (237)	94 Pu Plutonium (244)	95 Am Americium (243)	96 Cm Curium (247)	97 Bk Berkelium (247)	98 Cf Californium (251)	99 Es Einsteinium (252)	100 Fm Fermium (257)	101 Md Mendelevium (260)	102 No Nobelium (259)	103 Lr Lawrencium (262)

Actinides

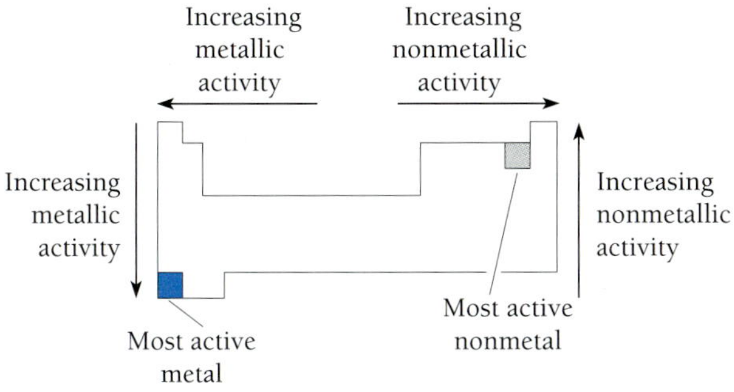

Figure 7.6 How chemical activity varies in the periodic table.

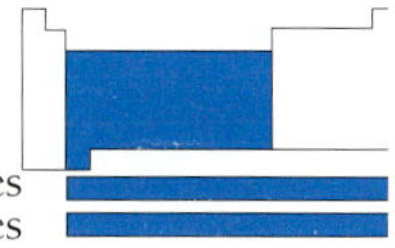

Figure 7.7 The transition elements are metals.

triumphs of the quantum theory of the atom that it enables us to account in a natural way for the periodic law without invoking any new assumptions.

7.5 ATOMIC STRUCTURES

Shells and subshells of electrons

Two basic principles determine the structures of atoms with more than one electron:

1 A system of particles is stable when its total energy is a minimum.
2 Only one electron can exist in any particular quantum state in an atom.

Before we apply these rules to actual atoms, let us examine the variation of electron energy with quantum state.

While the various electrons in a complex atom certainly interact directly with one another, much about atomic structure can be understood by simply considering each electron as though it exists in a constant mean electric field. For a given electron this effective field is approximately that of the nuclear charge Ze decreased by the partial shielding of those other electrons that are closer to the nucleus (see Fig. 7.9 in Sec. 7.6).

Electrons that have the same principal quantum number n usually (though not always) average roughly the same distance from the nucleus. These electrons therefore interact with roughly the same electric field and have similar energies. It is conventional to speak of such electrons as occupying the same atomic **shell.** Shells are denoted by capital letters according to the following scheme:

Atomic shells

$$\begin{array}{llllll} n = 1 & 2 & 3 & 4 & 5 & \ldots \\ \quad K & L & M & N & O & \ldots \end{array} \tag{7.13}$$

The energy of an electron in a particular shell also depends to a certain extent on its orbital quantum number l, though not as much as on n. In a complex atom the degree to which the full nuclear charge is shielded from a given electron by intervening shells of other electrons varies with its probability-density distribution. An electron of small l is more likely to be found near the nucleus where it is poorly shielded by the other electrons than is one of higher l (see Fig. 6.11). The result is a lower total energy (that is, higher binding energy) for the electron. The electrons in each shell

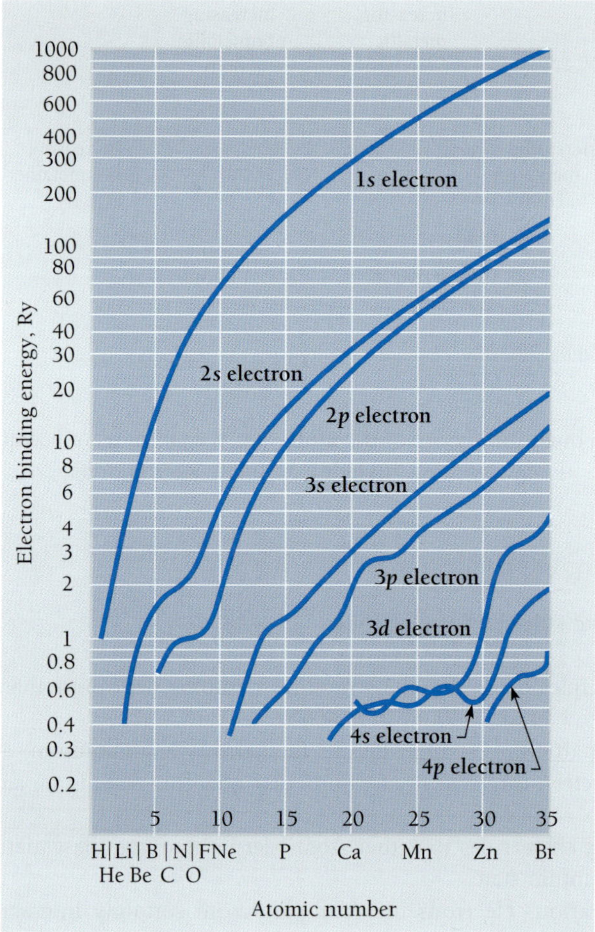

Figure 7.8 The binding energies of atomic electrons in rydbergs. (1 Ry = 13.6 eV = ground-state energy of H atom.)

accordingly increase in energy with increasing l. This effect is illustrated in Fig. 7.8, which is a plot of the binding energies of various atomic electrons as a function of atomic number for the lighter elements.

Electrons that share a certain value of l in a shell are said to occupy the same **subshell.** All the electrons in a subshell have almost identical energies, since the dependence of electron energy upon m_l and m_s is comparatively minor.

The occupancy of the various subshells in an atom is usually expressed with the help of the notation introduced in the previous chapter for the various quantum states of the hydrogen atom. As indicated in Table 6.2, each subshell is identified by its principal quantum number n followed by the letter corresponding to its orbital quantum number l. A superscript after the letter indicates the number of electrons in that subshell. For example, the electron configuration of sodium is written

$$1s^2 2s^2 2p^6 3s^1$$

which means that the 1s ($n = 1$, $l = 0$) and 2s ($n = 2$, $l = 0$) subshells contain two electrons each, the 2p ($n = 2$, $l = 1$) subshell contains six electrons, and the 3s ($n = 3$, $l = 0$) subshell contains one electron.

Table 7.3 Subshell Capacities in the M ($n = 3$) Shell of an Atom

	$m_l = 0$	$m_l = -1$	$m_l = +1$	$m_l = -2$	$m_l = +2$	
$l = 0$:	↓↑					↑$m_s = +\frac{1}{2}$
$l = 1$:	↓↑	↓↑	↓↑			↓$m_s = -\frac{1}{2}$
$l = 2$:	↓↑	↓↑	↓↑	↓↑	↓↑	

Shell and Subshell Capacities

The exclusion principle limits the number of electrons that can occupy a given subshell. A subshell is characterized by a certain principal quantum number n and orbital quantum number l, where l can have the values $0, 1, 2, \ldots, (n-1)$. There are $2l+1$ different values of the magnetic quantum number m_l for any l, since $m_l = 0, \pm 1, \pm 2, \ldots, \pm l$. Finally, the spin magnetic quantum number m_s has the two possible values of $+\frac{1}{2}$ and $-\frac{1}{2}$ for any m_l. The result is that each subshell can contain a maximum of $2(2l+1)$ electrons (Table 7.3).

The maximum number of electrons a shell can hold is the sum of the electrons in its filled subshells. This number is

$$N_{\max} = \sum_{l=0}^{l=n-1} 2(2l+1) = 2[1+3+5+\cdots+(2(n-1)+1)]$$
$$= 2[1+3+5+\cdots+2n-1]$$

The quantity in brackets has n terms whose average value is $\frac{1}{2}[1+(2n-1)]$. The number of electrons in a filled shell is therefore

$$N_{\max} = (n)(2)(\tfrac{1}{2})[1+(2n-1)] = 2n^2 \tag{7.14}$$

Thus a closed K shell holds 2 electrons, a closed L shell holds 8 electrons, a closed M shell holds 18 electrons, and so on.

7.6 EXPLAINING THE PERIODIC TABLE

How an atom's electron structure determines its chemical behavior

The notion of electron shells and subshells fits perfectly into the pattern of the periodic table, which mirrors the atomic structures of the elements. Let us see how this pattern arises.

An atomic shell or subshell that contains its full quota of electrons is said to be **closed.** A closed s subshell ($l = 0$) holds two electrons, a closed p subshell ($l = 1$) six electrons, a closed d subshell ($l = 2$) ten electrons, and so on.

The total orbital and spin angular momenta of the electrons in a closed subshell are zero, and their effective charge distributions are perfectly symmetrical (see Exercise 23 of Chap. 6). The electrons in a closed shell are all very tightly bound, since the positive nuclear charge is large relative to the negative charge of the inner shielding electrons (Fig. 7.9). Because an atom with only closed shells has no dipole moment, it does not attract other electrons, and its electrons cannot be easily

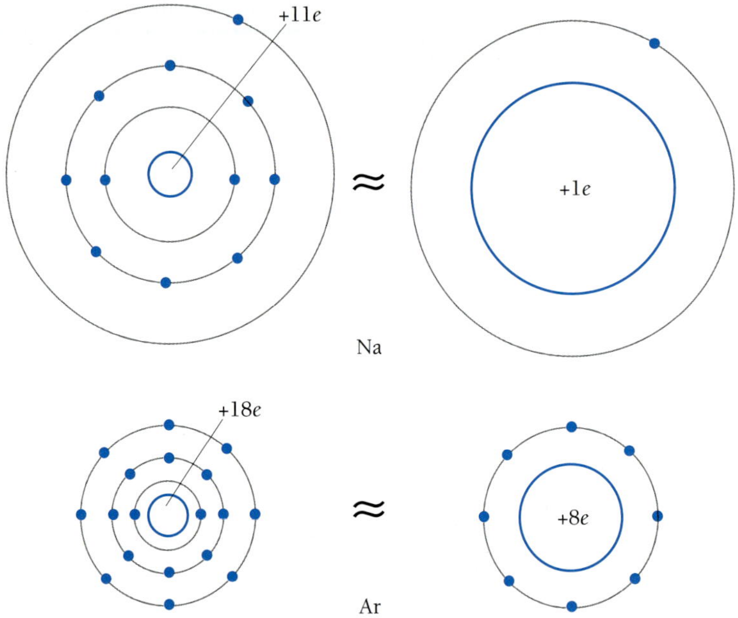

Figure 7.9 Schematic representation of electron shielding in the sodium and argon atoms. In this crude model, each outer electron in an Ar atom is acted upon by an effective nuclear charge 8 times greater than that acting upon the outer electron in a Na atom. The Ar atom is accordingly smaller in size and has higher ionization energy. In the actual atoms, the probability-density distributions of the various electrons overlap in complex ways and thus alter the amount of shielding, but the basic effect remains the same.

detached. We expect such atoms to be passive chemically, like the inert gases—and the inert gases all turn out to have closed-shell electron configurations or their equivalents. This is evident from Table 7.4, which shows the electron configurations of the elements.

An atom of any of the alkali metals of group 1 has a single *s* electron in its outer shell. Such an electron is relatively far from the nucleus. It is also shielded by the inner electrons from all but an effective nuclear charge of approximately $+e$ rather than $+Ze$. Relatively little work is needed to detach an electron from such an atom, and the alkali metals accordingly form positive ions of charge $+e$ readily.

Example 7.2

The ionization energy of lithium is 5.39 eV. Use this figure to find the effective charge that acts on the outer (2*s*) electron of the lithium atom.

Solution

If the effective nuclear charge is Ze instead of e, Eq. (4.15) becomes

$$E_n = \frac{Z^2 E_1}{n^2}$$

Here $n = 2$ for the $2s$ electron, its ionization energy is $E_2 = -5.39$ eV, and $E_1 = -13.6$ eV is the ionization energy of the hydrogen atom. Hence

$$Z = n\sqrt{\frac{E_2}{E_1}} = 2\sqrt{\frac{5.39 \text{ eV}}{13.6 \text{ eV}}} = 1.26$$

The effective charge is $1.26e$ and not e because the shielding of $2e$ of the nuclear charge of $3e$ by the two $1s$ electrons is not complete: as we can see in Fig. 6.11, the $2s$ electron has a certain probability of being found inside the $1s$ electrons.

Ionization Energy

Figure 7.10 shows how the ionization energies of the elements vary with atomic number. As we expect, the inert gases have the highest ionization energies and the alkali metals the lowest. The larger an atom, the farther the outer electron is from the nucleus and the weaker the force that holds it to the atom. This is why the ionization energy generally decreases as we go down a group in the periodic table. The increase in ionization energy from left to right across any period is accounted for by the increase in nuclear charge while the number of inner shielding electrons stays constant. In period 2, for instance, the outer electron in a lithium atom is held by an effective charge of about $+e$, while each outer electron in beryllium, boron, carbon, and so on, is held by effective charges of about $+2e$, $+3e$, $+4e$, and so on. The ionization energy of lithium is 5.4 eV whereas that of neon, which ends the period, is 21.6 eV.

At the other extreme from alkali metal atoms, which tend to lose their outermost electrons, are halogen atoms, whose imperfectly shielded nuclear charges tend to

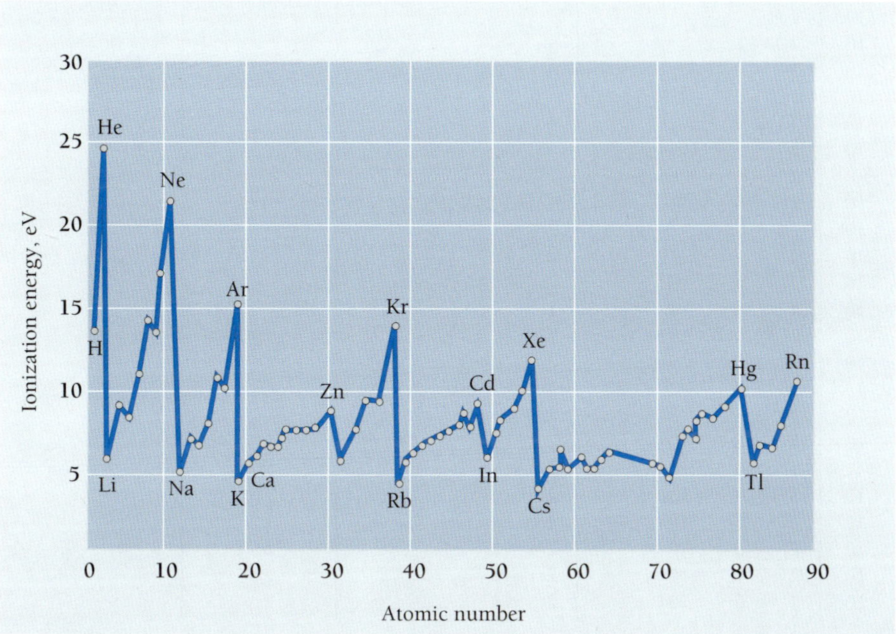

Figure 7.10 The variation of ionization energy with atomic number.

Table 7.4 Electron Configurations of the Elements

	K	*L*		*M*			*N*				*O*				*P*			*Q*	
	1*s*	2*s*	2*p*	3*s*	3*p*	3*d*	4*s*	4*p*	4*d*	4*f*	5*s*	5*p*	5*d*	5*f*	6*s*	6*p*	6*d*	7*s*	
1 H	1																		
2 He	2																		← Inert gas
3 Li	2	1																	← Alkali metal
4 Be	2	2																	
5 B	2	2	1																
6 C	2	2	2																
7 N	2	2	3																
8 O	2	2	4																
9 F	2	2	5																← Halogen
10 Ne	2	2	6																← Inert gas
11 Na	2	2	6	1															← Alkali metal
12 Mg	2	2	6	2															
13 Al	2	2	6	2	1														
14 Si	2	2	6	2	2														
15 P	2	2	6	2	3														
16 S	2	2	6	2	4														
17 Cl	2	2	6	2	5														← Halogen
18 Ar	2	2	6	2	6														← Inert gas
19 K	2	2	6	2	6		1												← Alkali metal
20 Ca	2	2	6	2	6		2												
21 Sc	2	2	6	2	6	1	2												
22 Ti	2	2	6	2	6	2	2												
23 V	2	2	6	2	6	3	2												
24 Cr	2	2	6	2	6	5	1												
25 Mn	2	2	6	2	6	5	2												Transition elements (21–30)
26 Fe	2	2	6	2	6	6	2												
27 Co	2	2	6	2	6	7	2												
28 Ni	2	2	6	2	6	8	2												
29 Cu	2	2	6	2	6	10	1												
30 Zn	2	2	6	2	6	10	2												
31 Ga	2	2	6	2	6	10	2	1											
32 Ge	2	2	6	2	6	10	2	2											
33 As	2	2	6	2	6	10	2	3											
34 Se	2	2	6	2	6	10	2	4											
35 Br	2	2	6	2	6	10	2	5											← Halogen
36 Kr	2	2	6	2	6	10	2	6											← Inert gas
37 Rb	2	2	6	2	6	10	2	6			1								← Alkali metal
38 Sr	2	2	6	2	6	10	2	6			2								
39 Y	2	2	6	2	6	10	2	6	1		2								
40 Zr	2	2	6	2	6	10	2	6	2		2								
41 Nb	2	2	6	2	6	10	2	6	4		1								
42 Mo	2	2	6	2	6	10	2	6	5		1								
43 Tc	2	2	6	2	6	10	2	6	5		2								Transition elements (39–48)
44 Ru	2	2	6	2	6	10	2	6	7		1								
45 Rh	2	2	6	2	6	10	2	6	8		1								
46 Pd	2	2	6	2	6	10	2	6	10										
47 Ag	2	2	6	2	6	10	2	6	10		1								
48 Cd	2	2	6	2	6	10	2	6	10		2								
49 In	2	2	6	2	6	10	2	6	10		2	1							
50 Sn	2	2	6	2	6	10	2	6	10		2	2							
51 Sb	2	2	6	2	6	10	2	6	10		2	3							
52 Te	2	2	6	2	6	10	2	6	10		2	4							

Table 7.4 (*Cont.*)

	K	*L*		*M*			*N*				*O*				*P*			*Q*	
	1*s*	2*s*	2*p*	3*s*	3*p*	3*d*	4*s*	4*p*	4*d*	4*f*	5*s*	5*p*	5*d*	5*f*	6*s*	6*p*	6*d*	7*s*	
53 I	2	2	6	2	6	10	2	6	10		2	5 ← Halogen							
54 Xe	2	2	6	2	6	10	2	6	10		2	6 ← Inert gas							
55 Cs	2	2	6	2	6	10	2	6	10		2	6			1 ← Alkali metal				
56 Ba	2	2	6	2	6	10	2	6	10		2	6			2				
57 La	2	2	6	2	6	10	2	6	10		2	6	1		2				
58 Ce	2	2	6	2	6	10	2	6	10	2	2	6			2				
59 Pr	2	2	6	2	6	10	2	6	10	3	2	6			2				
60 Nd	2	2	6	2	6	10	2	6	10	4	2	6			2				
61 Pm	2	2	6	2	6	10	2	6	10	5	2	6			2				
62 Sm	2	2	6	2	6	10	2	6	10	6	2	6			2				
63 Eu	2	2	6	2	6	10	2	6	10	7	2	6			2				
64 Gd	2	2	6	2	6	10	2	6	10	7	2	6	1		2				
65 Tb	2	2	6	2	6	10	2	6	10	9	2	6			2				Lanthanides
66 Dy	2	2	6	2	6	10	2	6	10	10	2	6			2				
67 Ho	2	2	6	2	6	10	2	6	10	11	2	6			2				
68 Er	2	2	6	2	6	10	2	6	10	12	2	6			2				Transition elements
69 Tm	2	2	6	2	6	10	2	6	10	13	2	6			2				
70 Yb	2	2	6	2	6	10	2	6	10	14	2	6			2				
71 Lu	2	2	6	2	6	10	2	6	10	14	2	6	1		2				
72 Hf	2	2	6	2	6	10	2	6	10	14	2	6	2		2				
73 Ta	2	2	6	2	6	10	2	6	10	14	2	6	3		2				
74 W	2	2	6	2	6	10	2	6	10	14	2	6	4		2				
75 Re	2	2	6	2	6	10	2	6	10	14	2	6	5		2				
76 Os	2	2	6	2	6	10	2	6	10	14	2	6	6		2				
77 Ir	2	2	6	2	6	10	2	6	10	14	2	6	7		2				
78 Pt	2	2	6	2	6	10	2	6	10	14	2	6	9		1				
79 Au	2	2	6	2	6	10	2	6	10	14	2	6	10		1				
80 Hg	2	2	6	2	6	10	2	6	10	14	2	6	10		2				
81 Tl	2	2	6	2	6	10	2	6	10	14	2	6	10		2	1			
82 Pb	2	2	6	2	6	10	2	6	10	14	2	6	10		2	2			
83 Bi	2	2	6	2	6	10	2	6	10	14	2	6	10		2	3			
84 Po	2	2	6	2	6	10	2	6	10	14	2	6	10		2	4			
85 At	2	2	6	2	6	10	2	6	10	14	2	6	10		2	5 ← Halogen			
86 Rn	2	2	6	2	6	10	2	6	10	14	2	6	10		2	6 ← Inert gas			
87 Fr	2	2	6	2	6	10	2	6	10	14	2	6	10		2	6		1 ← Alkali metal	
88 Ra	2	2	6	2	6	10	2	6	10	14	2	6	10		2	6		2	
89 Ac	2	2	6	2	6	10	2	6	10	14	2	6	10		2	6	1	2	
90 Th	2	2	6	2	6	10	2	6	10	14	2	6	10		2	6	2	2	
91 Pa	2	2	6	2	6	10	2	6	10	14	2	6	10	2	2	6	1	2	
92 U	2	2	6	2	6	10	2	6	10	14	2	6	10	3	2	6	1	2	
93 Np	2	2	6	2	6	10	2	6	10	14	2	6	10	4	2	6	1	2	
94 Pu	2	2	6	2	6	10	2	6	10	14	2	6	10	5	2	6	1	2	
95 Am	2	2	6	2	6	10	2	6	10	14	2	6	10	6	2	6	1	2	Actinides
96 Cm	2	2	6	2	6	10	2	6	10	14	2	6	10	7	2	6	1	2	
97 Bk	2	2	6	2	6	10	2	6	10	14	2	6	10	8	2	6	1	2	
98 Cf	2	2	6	2	6	10	2	6	10	14	2	6	10	10	2	6		2	
99 Es	2	2	6	2	6	10	2	6	10	14	2	6	10	11	2	6		2	
100 Fm	2	2	6	2	6	10	2	6	10	14	2	6	10	12	2	6		2	
101 Md	2	2	6	2	6	10	2	6	10	14	2	6	10	13	2	6		2	
102 No	2	2	6	2	6	10	2	6	10	14	2	6	10	14	2	6		2	
103 Lr	2	2	6	2	6	10	2	6	10	14	2	6	10	14	2	6	1	2	

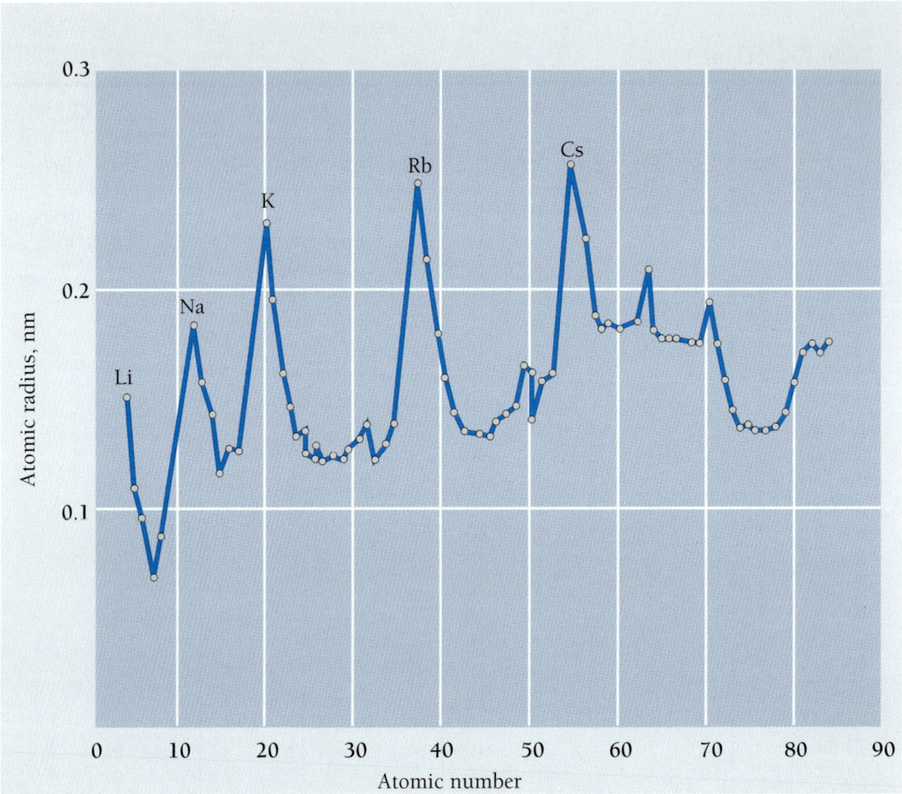

Figure 7.11 Atomic radii of the elements.

complete their outer subshells by picking up an additional electron each. Halogen atoms accordingly form negative ions of charge $-e$ readily. Reasoning of this kind accounts for the similarities of the members of the various groups of the periodic table.

Size

Although, strictly speaking, an atom of a certain kind cannot be said to have a definite size, from a practical point of view a fairly definite size can usually be attributed to it on the basis of the observed interatomic spacings in closely packed crystal lattices. Figure 7.11 shows how the resulting radii vary with atomic number. The periodicity here is as conspicuous as in the case of ionization energy and has a similar origin in the partial shielding by inner electrons of the full nuclear charge. The greater the shielding, the lower the binding energy of an outer electron and the farther it is on the average from the nucleus.

The relatively small range of atomic radii is not surprising in view of the binding-energy curves of Fig. 7.8. There we see that in contrast to the enormous increase in the binding energies of the unshielded 1*s* electrons with Z, the binding energies of the outermost electrons (whose probability-density distributions are what determine atomic size) vary through a narrow range. The heaviest atoms, with over 90 electrons, have radii only about 3 times that of the hydrogen atom, and even the cesium atom, the largest in size, has a radius only 4.4 times that of the hydrogen atom.

Transition Elements

The origin of the transition elements lies in the tighter binding of *s* electrons than *d* or *f* electrons in complex atoms, discussed in the previous section (see Fig. 7.8). The first element to exhibit this effect is potassium, whose outermost electron is in a 4*s* instead of a 3*d* substate. The difference in binding energy between 3*d* and 4*s* electrons is not very great, as the configurations of chromium and copper show. In both these elements an additional 3*d* electron is present at the expense of a vacancy in the 4*s* subshell.

The order in which electron subshells tend to be filled, together with the maximum occupancy of each subshell, is usually as follows:

$$1s^2 \quad 2s^2 \quad 2p^6 \quad 3s^2 \quad 3p^6 \quad 4s^2 \quad 3d^{10} \quad 4p^6 \quad 5s^2$$
$$4d^{10} \quad 5p^6 \quad 6s^2 \quad 4f^{14} \quad 5d^{10} \quad 6p^6 \quad 7s^2 \quad 6d^{10} \quad 5f^{14}$$

Figure 7.12 illustrates this sequence. The remarkable similarities in chemical behavior among the lanthanides and actinides are easy to understand on the basis of this sequence. All the lanthanides have the same $5s^25p^66s^2$ configurations but have incomplete 4*f* subshells. The addition of 4*f* electrons has almost no effect on the chemical properties of the lanthanide elements, which are determined by the outer electrons. Similarly, all the actinides have $6s^26p^67s^2$ configurations and differ only in the numbers of their 5*f* and 6*d* electrons.

These irregularities in the binding energies of atomic electrons are also responsible for the lack of completely full outer shells in the heavier inert gases. Helium ($Z = 2$) and neon ($Z = 10$) contain closed *K* and *L* shells, respectively, but argon ($Z = 18$) has only 8 electrons in its *M* shell, corresponding to closed 3*s* and 3*p* subshells. The reason the 3*d* subshell is not filled next is that 4*s* electrons have higher binding energies

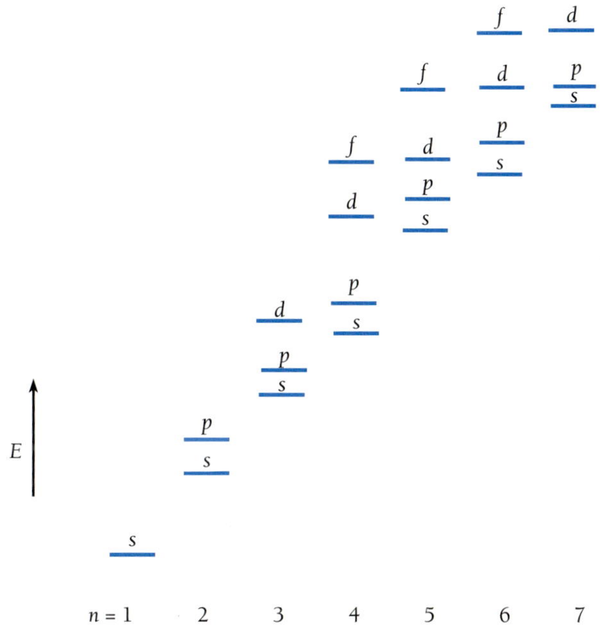

Figure 7.12 The sequence of quantum states in an atom. Not to scale.

Table 7.5 Electron Configurations of Elements from $Z = 5$ to $Z = 10$. The p electrons have parallel spins whenever possible, in accord with Hund's rule.

Element	Atomic Number	Configuration	Spins of p Electrons		
Boron	5	$1s^22s^22p^1$	↑		
Carbon	6	$1s^22s^22p^2$	↑	↑	
Nitrogen	7	$1s^22s^22p^3$	↑	↑	↑
Oxygen	8	$1s^22s^22p^4$	↑↓	↑	↑
Fluorine	9	$1s^22s^22p^5$	↑↓	↑↓	↑
Neon	10	$1s^22s^22p^6$	↑↓	↑↓	↑↓

than do $3d$ electrons. Hence the $4s$ subshell is filled first in potassium and calcium. As the $3d$ subshell is filled in successively heavier transition elements, there are still one or two outer $4s$ electrons that make possible chemical activity. Not until krypton ($Z = 36$) is another inert gas reached, and here a similarly incomplete outer shell occurs with only the $4s$ and $4p$ subshells filled. Following krypton is rubidium ($Z = 37$), which skips both the $4d$ and $4f$ subshells to have a $5s$ electron. The next inert gas is xenon ($Z = 54$), which has filled $4d$, $5s$, and $5p$ subshells, but now even the inner $4f$ subshell is empty as well as the $5d$ and $5f$ subshells. The same pattern recurs with the last inert gas, radon.

Hund's Rule

In general, the electrons in a subshell remain unpaired—that is, have parallel spins—whenever possible (**Table 7.5**). This principal is called *Hund's rule.* The ferromagnetism of iron, cobalt, and nickle (Z 5 26, 27, 28) is in part a consequence of Hund's rule. The $3d$ subshells of their atoms are only partially occupied, and the electrons in these subshells do not pair off to permit their spin magnetic moments to cancel out. In iron, for instance, five of the six $3d$ electrons have parallel spins, so that each iron atom has a large resultant magnetic moment.

The origin of Hund's rule lies in the mutual repulsion of atomic electrons. Because of this repulsion, the farther apart the electrons in an atom are, the lower the energy of the atom. Electrons in the same subshell with the same spin must have different m_l values and accordingly are described by wave functions whose spatial distributions are different. Electrons with parallel spins are therefore more separated in space than they would be if they paired off. This arrangement, having less energy, is the more stable one.

7.7 SPIN-ORBIT COUPLING

Angular momenta linked magnetically

The fine-structure doubling of spectral lines arises from a magnetic interaction between the spin and orbital angular momenta of an atomic electron called **spin-orbit coupling.**

Spin-orbit coupling can be understood in terms of a straightforward classical model. An electron revolving about a nucleus finds itself in a magnetic field because in its own frame of reference, the nucleus is circling about *it* Fig. 7.13. This magnetic field then acts upon the electron's own spin magnetic moment to produce a kind of internal Zeeman effect.

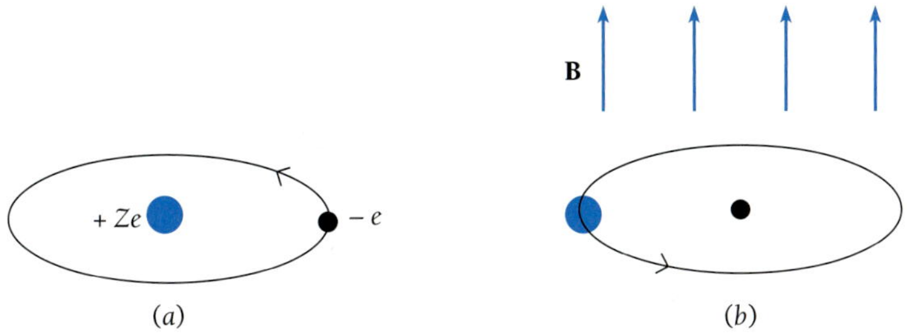

Figure 7.13 (a) An electron circles an atomic nucleus, as viewed from the frame of reference of the nucleus. (b) From the electron's frame of reference, the nucleus is circling it. The magnetic field the electron experiences as a result is directed upward from the plane of the orbit. The interaction between the electron's spin magnetic moment and this magnetic field leads to the phenomenon of spin-orbit coupling.

The potential energy U_m of a magnetic dipole of moment $\boldsymbol{\mu}$ in a magnetic field **B** is, as we know,

$$U_m = -\mu B \cos \theta \tag{6.38}$$

where θ is the angle between $\boldsymbol{\mu}$ and **B**. The quantity $\mu \cos \theta$ is the component of $\boldsymbol{\mu}$ parallel to **B**. In the case of the spin magnetic moment of the electron this component is $\mu_{sz} = \pm\mu_B$. Hence

$$\mu \cos \theta = \pm\mu_B$$

and so

Spin-orbit coupling

$$U_m = \pm\mu_B B \tag{7.15}$$

Depending on the orientation of its spin vector **S**, the energy of an atomic electron will be higher or lower by $\mu_B B$ than its energy without spin-orbit coupling. The result is that every quantum state (except s states in which there is no orbital angular momentum) is split into two substates.

The assignment of $s = \frac{1}{2}$ is the only one that agrees with the observed fine-structure doubling. Because what would be single states without spin are in fact twin states, the $2s + 1$ possible orientations of the spin vector **S** must total 2. With $2s + 1 = 2$, the result is $s = \frac{1}{2}$.

Example 7.3

Estimate the magnetic energy U_m for an electron in the $2p$ state of a hydrogen atom using the Bohr model, whose $n = 2$ state corresponds to the $2p$ state.

Solution

A circular wire loop of radius r that carries the current I has a magnetic field at its center of magnitude

$$B = \frac{\mu_0 I}{2r}$$

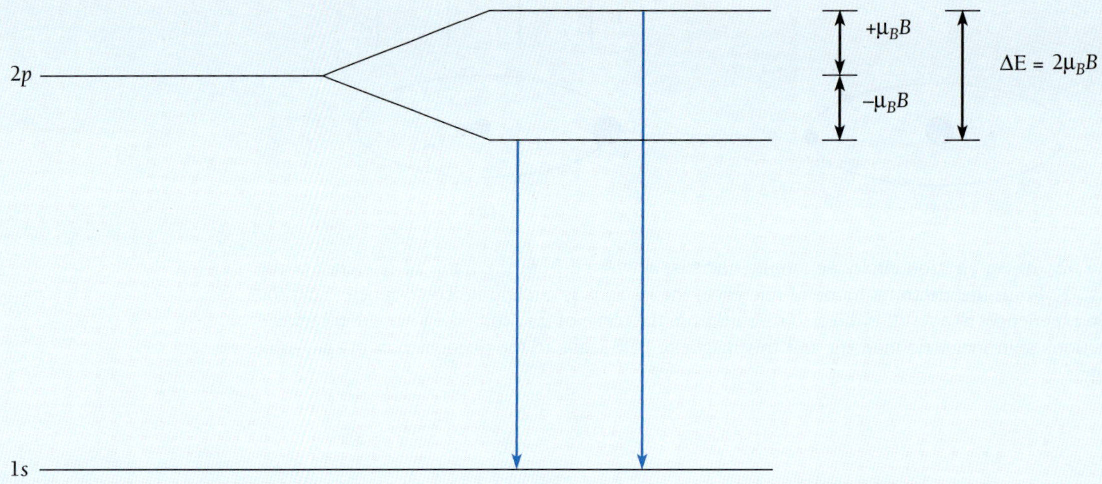

Figure 7.14 Spin-orbit coupling splits the $2p$ state in the hydrogen atom into two substates ΔE apart. The result is a doublet (two closely spaced lines) instead of a single spectral line for the $2p \rightarrow 1s$ transition.

The orbiting electron "sees" itself circled f times per second by the proton of charge $+e$ that is the nucleus, for a resulting magnetic field of

$$B = \frac{\mu_0 fe}{2r}$$

The frequency of revolution and orbital radius for $n = 2$ are, from Eqs. (4.4) and (4.14),

$$f = \frac{v}{2\pi r} = 8.4 \times 10^{14}\ \text{s}^{-1}$$

$$r = n^2 a_0 = 4a_0 = 2.1 \times 10^{-10}\ \text{m}$$

Hence the magnetic field experienced by the electron is

$$B = \frac{(4\pi \times 10^{-7}\ \text{T} \cdot \text{m/A})(8.4 \times 10^{14}\ \text{s}^{-1})(1.6 \times 10^{-19}\ \text{C})}{(2)(2.1 \times 10^{-10}\ \text{m})} = 0.40\ \text{T}$$

which is a fairly strong field. Since the value of the Bohr magneton is $\mu_B = e\hbar/2m = 9.27 \times 10^{-24}$ J/T, the magnetic energy of the electron is

$$U_m = \mu_B B = 3.7 \times 10^{-24}\ \text{J} = 2.3 \times 10^{-5}\ \text{eV}$$

The energy difference between the upper and lower substates is twice this, 4.6×10^{-5} eV, which is not far from what is observed (Fig. 7.14).

7.8 TOTAL ANGULAR MOMENTUM

Both magnitude and direction are quantized

Each electron in an atom has a certain orbital angular momentum **L** and a certain spin angular momentum **S**, both of which contribute to the total angular momentum **J** of the atom. Let us first consider an atom whose total angular momentum is provided by a single electron. Atoms of the elements in group 1 of the periodic table—hydrogen, lithium, sodium, and so on—are of this kind. They have single electrons outside closed inner shells (except for hydrogen, which has no inner electrons) and the exclusion principle ensures that the total angular momentum and magnetic moment of a closed shell are zero. Also in this category are the ions He^+, Be^+, Mg^+, B^{2+}, Al^{2+}, and so on.

In these atoms and ions, the outer electron's total angular momentum **J** is the vector sum of **L** and **S**:

Total atomic angular momentum

$$\mathbf{J} = \mathbf{L} + \mathbf{S} \tag{7.16}$$

Like all angular momenta, **J** is quantized in both magnitude and direction. The magnitude of **J** is given by

$$J = \sqrt{j(j+1)}\hbar \quad j = l + s = l \pm \tfrac{1}{2} \tag{7.17}$$

If $l = 0$, j has the single value $j = \frac{1}{2}$. The component J_z of **J** in the z direction is given by

$$J_z = m_j\hbar \quad m_j = -j, -j+1, \ldots, j-1, j \tag{7.18}$$

Because of the simultaneous quantization of **J**, **L**, and **S** they can have only certain specific relative orientations. This is a general conclusion; in the case of a one-electron atom, there are only two relative orientations possible. One relative orientation corresponds to $j = l + s$, so that $J > L$, and the other to $j = l - s$, so that $J < L$. Figure 7.15 shows the two ways in which **L** and **S** can combine to form **J** when $l = 1$. Evidently the orbital and spin angular-momentum vectors can never be exactly parallel or antiparallel to each other or to the total angular-momentum vector.

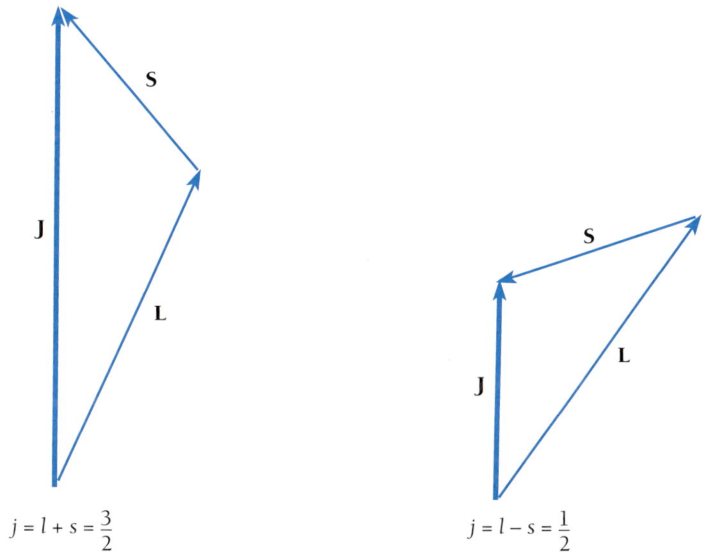

Figure 7.15 The two ways in which **L** and **S** can be added to form **J** when $l = 1$, $s = \frac{1}{2}$.

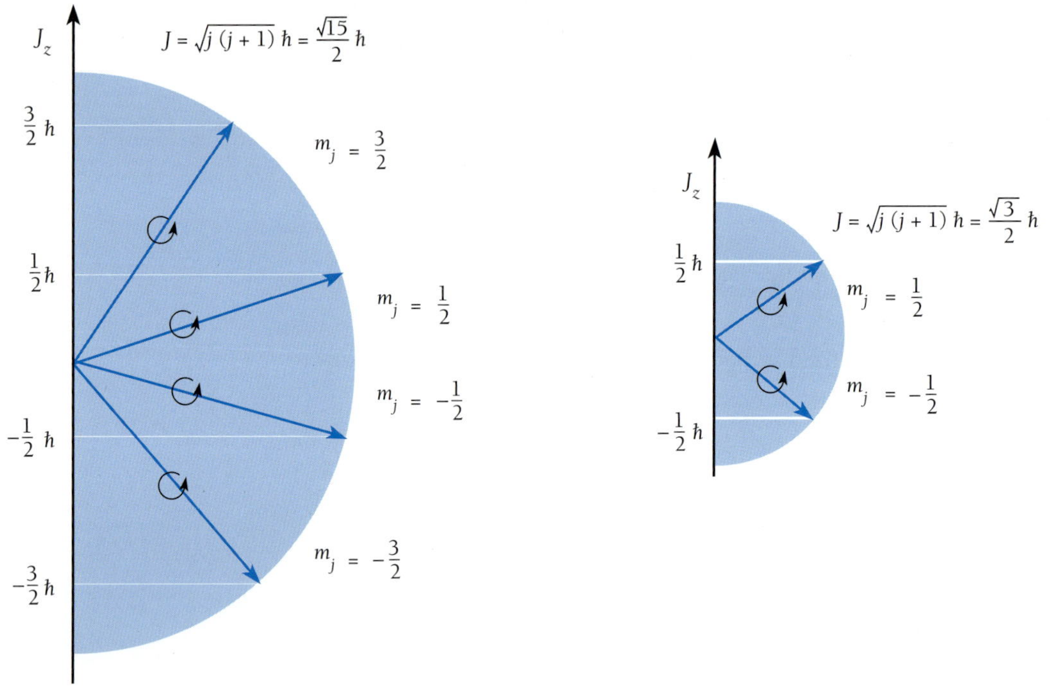

Figure 7.16 Space quantization of total angular momentum when the orbital angular momentum is $l = 1$.

Example 7.4

What are the possible orientations of **J** for the $j = \frac{3}{2}$ and $j = \frac{1}{2}$ states that correspond to $l = 1$?

Solution

For the $j = \frac{3}{2}$ state, Eq. (7.18) gives $m_j = -\frac{3}{2}, -\frac{1}{2}, \frac{1}{2}, \frac{3}{2}$. For the $j = \frac{1}{2}$ state, $m_j = -\frac{1}{2}, \frac{1}{2}$. Figure 7.16 shows the orientations of **J** relative to the z axis for these values of j.

The angular momenta **L** and **S** interact magnetically, as we saw in Sec. 7.7. If there is no external magnetic field, the total angular momentum **J** is conserved in magnitude and direction, and the effect of the internal torques is the precession of **L** and **S** around the direction of their resultant **J** (Fig. 7.17). However, if there is an external magnetic field **B** present, then **J** precesses about the direction of **B** while **L** and **S** continue precessing about **J**, as in Fig. 7.18. The precession of **J** about **B** is what gives rise to the anomalous Zeeman effect, since different orientations of **J** involve slightly different energies in the presence of **B**.

LS Coupling

When more than one electron contributes orbital and spin angular momenta to the total angular momentum **J** of an atom, **J** is still the vector sum of these individual momenta. The usual pattern for all but the heaviest atoms is that the orbital angular momenta $\mathbf{L}_i$ of

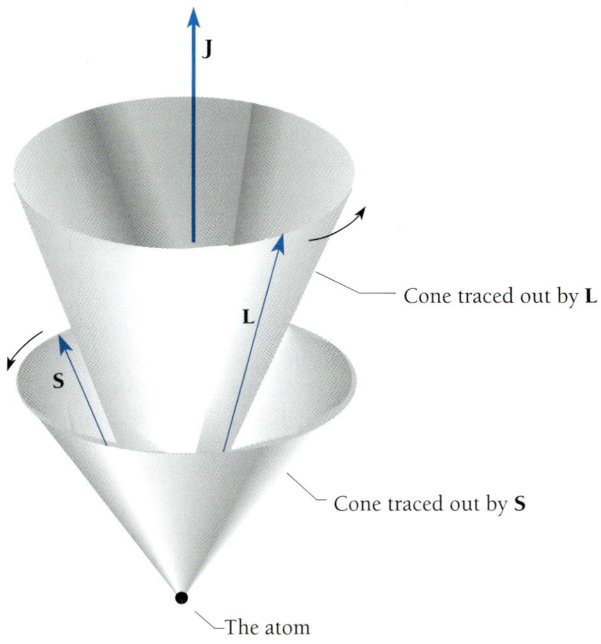

Figure 7.17 The orbital and spin angular-momentum vectors **L** and **S** precess about **J**.

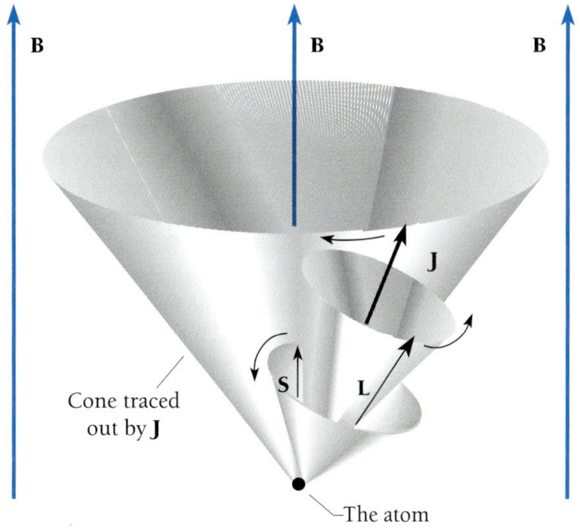

Figure 7.18 In the presence of an external magnetic field **B**, the total angular-momentum vector **J** precesses about **B**.

the various electrons are coupled together into a single resultant **L**. The spin angular momenta $\mathbf{S}_i$ are also coupled together into another single resultant **S**. The momenta **L** and **S** then interact via the spin-orbit effect to form a total angular momentum **J**. This scheme, called ***LS* coupling**, can be summarized as follows:

LS coupling

$$\mathbf{L} = \sum \mathbf{L}_i$$
$$\mathbf{S} = \sum \mathbf{S}_i \tag{7.19}$$
$$\mathbf{J} = \mathbf{L} + \mathbf{S}$$

The angular momentum magnitudes L, S, J and their z components L_z, S_z, and J_z are all quantized in the usual ways, with the respective quantum numbers L, S, J, $\mathsf{M_L}$, $\mathsf{M_S}$, and $\mathsf{M_J}$. Hence

$$L = \sqrt{\mathsf{L}(\mathsf{L}+1)}\hbar$$
$$L_z = \mathsf{M_L}\hbar$$
$$S = \sqrt{\mathsf{S}(\mathsf{S}+1)}\hbar$$
$$S_z = \mathsf{M_S}\hbar$$
$$J = \sqrt{\mathsf{J}(\mathsf{J}+1)}\hbar$$
$$J_z = \mathsf{M_J}\hbar \tag{7.20}$$

Both L and $\mathsf{M_L}$ are always integers or 0, while the other quantum numbers are half-integral if an odd number of electrons is involved and integral or 0 if an even number of electrons is involved. When $\mathsf{L} > \mathsf{S}$, J can have $2\mathsf{S} + 1$ values; when $\mathsf{L} < \mathsf{S}$, J can have $2\mathsf{L} + 1$ values.

Example 7.5

Find the possible values of the total angular-momentum quantum number J under *LS* coupling of two atomic electrons whose orbital quantum numbers are $l_1 = 1$ and $l_2 = 2$.

Solution

As in Fig. 7.19*a*, the vectors $\mathbf{L}_1$ and $\mathbf{L}_2$ can be combined in three ways into a single vector $\mathbf{L}$ that is quantized according to Eq. (7.20). These correspond to $L = 1$, 2, and 3 since all values of L are possible from $|l_1 - l_2|$ (= 1 here) to $l_1 + l_2$. The spin quantum number s is always $\frac{1}{2}$, which gives the two possibilities for $\mathbf{S}_1 + \mathbf{S}_2$ shown in Fig. 7.19*b*, corresponding to $S = 0$ and $S = 1$.

We note that if the vector sums are not 0, $\mathbf{L}_1$ and $\mathbf{L}_2$ can never be exactly parallel to $\mathbf{L}$, nor can $\mathbf{S}_1$ and $\mathbf{S}_2$ be parallel to $\mathbf{S}$. Because J can have any value between $|L - S|$ and $L + S$, the five possible values here are $J = 0$, 1, 2, 3, and 4.

Atomic nuclei also have intrinsic angular momenta and magnetic moments, and these contribute to the total atomic angular momenta and magnetic moments. Such contributions are small because nuclear magnetic moments are $\sim 10^{-3}$ the magnitude of electronic moments. They lead to the **hyperfine structure** of spectral lines with typical spacings between components of $\sim 10^{-3}$ nm as compared with typical fine-structure spacings a hundred times greater.

Term Symbols

In Sec. 6.5 we saw that individual orbital angular-momentum states are customarily described by a lowercase letter, with s corresponding to $l = 0$, p to $l = 1$, d to $l = 2$, and so on. A similar scheme using capital letters is used to designate the entire electronic state of an atom according to its total orbital angular-momentum quantum number L as follows:

$$L = 0 \quad 1 \quad 2 \quad 3 \quad 4 \quad 5 \quad 6 \ldots$$
$$S \quad P \quad D \quad F \quad G \quad H \quad I \ldots$$

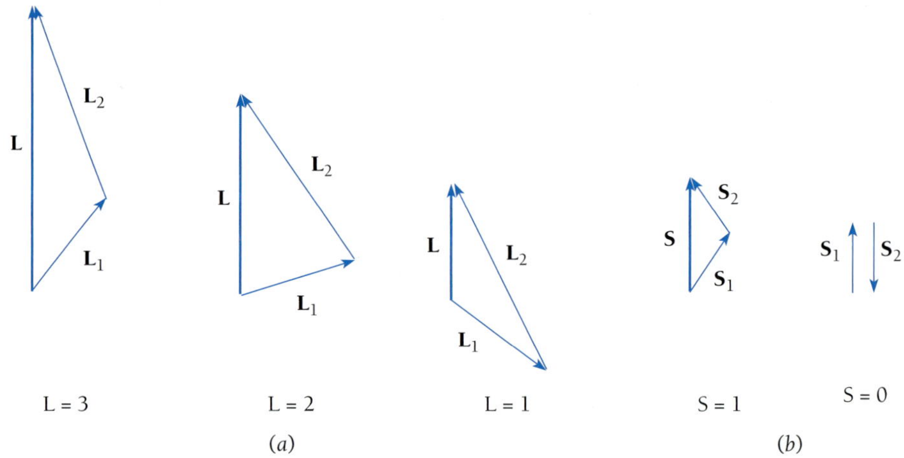

Figure 7.19 When $l_1 = 1$, $s_1 = \frac{1}{2}$, and $l_2 = 2$, $s_2 = \frac{1}{2}$, there are three ways in which $\mathbf{L}_1$ and $\mathbf{L}_2$ can combine to form $\mathbf{L}$ and two ways in which $\mathbf{S}_1$ and $\mathbf{S}_2$ can combine to form $\mathbf{S}$.

A superscript number before the letter (^2P, for instance) is used to indicate the **multiplicity** of the state, which is the number of different possible orientations of **L** and **S** and hence the number of different possible values of J. The multiplicity is equal to $2S + 1$ in the usual situation where $L > S$, since J ranges from $L + S$ to $L - S$. Thus when $S = 0$, the multiplicity is 1 (a **singlet** state) and $J = L$; when $S = \frac{1}{2}$, the multiplicity is 2 (a **doublet** state) and $J = L \pm \frac{1}{2}$; when $S = 1$, the multiplicity is 3 (a **triplet** state) and $J = L + 1$, L, or $L - 1$; and so on. (In a configuration in which $S > L$, the multiplicity is given by $2L + 1$.) The total angular-momentum quantum number J is used as a subscript after the letter, so that a $^2P_{3/2}$ state (read as "doublet P three-halves") refers to an electronic configuration in which $S = \frac{1}{2}$, $L = 1$, and $J = \frac{3}{2}$. For historical reasons, these designations are called **term symbols.**

In the event that the angular momentum of the atom arises from a single outer electron, the principal quantum number n of this electron is used as a prefix. Thus the ground state of the sodium atom is described by $3^2S_{1/2}$, since its electronic configuration has an electron with $n = 3$, $l = 0$, and $s = \frac{1}{2}$ (and hence $j = \frac{1}{2}$) outside closed $n = 1$ and $n = 2$ shells. For consistency it is conventional to denote the above state by $3^2S_{1/2}$ with the superscript 2 indicating a doublet, even though there is only a single possibility for J since $L = 0$.

Example 7.6

The term symbol of the ground state of sodium is $3^2S_{1/2}$ and that of its first excited state is $3^2P_{1/2}$. List the possible quantum numbers n, l, j, and m_j of the outer electron in each case.

Solution

$$3^2S_{1/2}\text{: } n = 3, l = 0, j = \tfrac{1}{2}, m_j = \pm\tfrac{1}{2}$$
$$3^2P_{1/2}\text{: } n = 3, l = 1, j = \tfrac{3}{2}, m_j = \pm\tfrac{1}{2}, \pm\tfrac{3}{2}$$
$$n = 3, l = 1, j = \tfrac{1}{2}, m_j = \pm\tfrac{1}{2}$$

Example 7.7

Why is it impossible for a $2^2P_{5/2}$ state to exist?

Solution

A P state has $L = 1$ and $J = L \pm \frac{1}{2}$, so $J = \frac{5}{2}$ is impossible.

7.9 X-RAY SPECTRA

They arise from transitions to inner shells

In Chap. 2 we learned that the x-ray spectra of targets bombarded by fast electrons show narrow spikes at wavelengths characteristic of the target material. These are besides a continuous distribution of wavelengths down to a minimum wavelength inversely proportional to the electron energy (see Fig. 2.17). The continuous x-ray spectrum is the result of the inverse photoelectric effect, with electron kinetic energy being transformed into photon energy $h\nu$. The line spectrum, on the other hand, comes from electronic transitions within atoms that have been disturbed by the incident electrons.

The transitions of the outer electrons of an atom usually involve only a few electronvolts of energy, and even removing an outer electron requires at most 24.6 eV (for helium). Such transitions accordingly are associated with photons whose wavelengths lie in or near the visible part of the electromagnetic spectrum. The inner electrons of heavier elements are a quite different matter, because these electrons are not well shielded from the full nuclear charge by intervening electron shells and so are very tightly bound.

In sodium, for example, only 5.13 eV is needed to remove the outermost 3*s* electron, whereas the corresponding figures for the inner ones are 31 eV for each 2*p* electron, 63 eV for each 2*s* electron, and 1041 eV for each 1*s* electron. Transitions that involve the inner electrons in an atom are what give rise to x-ray line spectra because of the high photon energies involved.

Figure 7.20 shows the energy levels (not to scale) of a heavy atom. The energy differences between angular momentum states within a shell are minor compared with

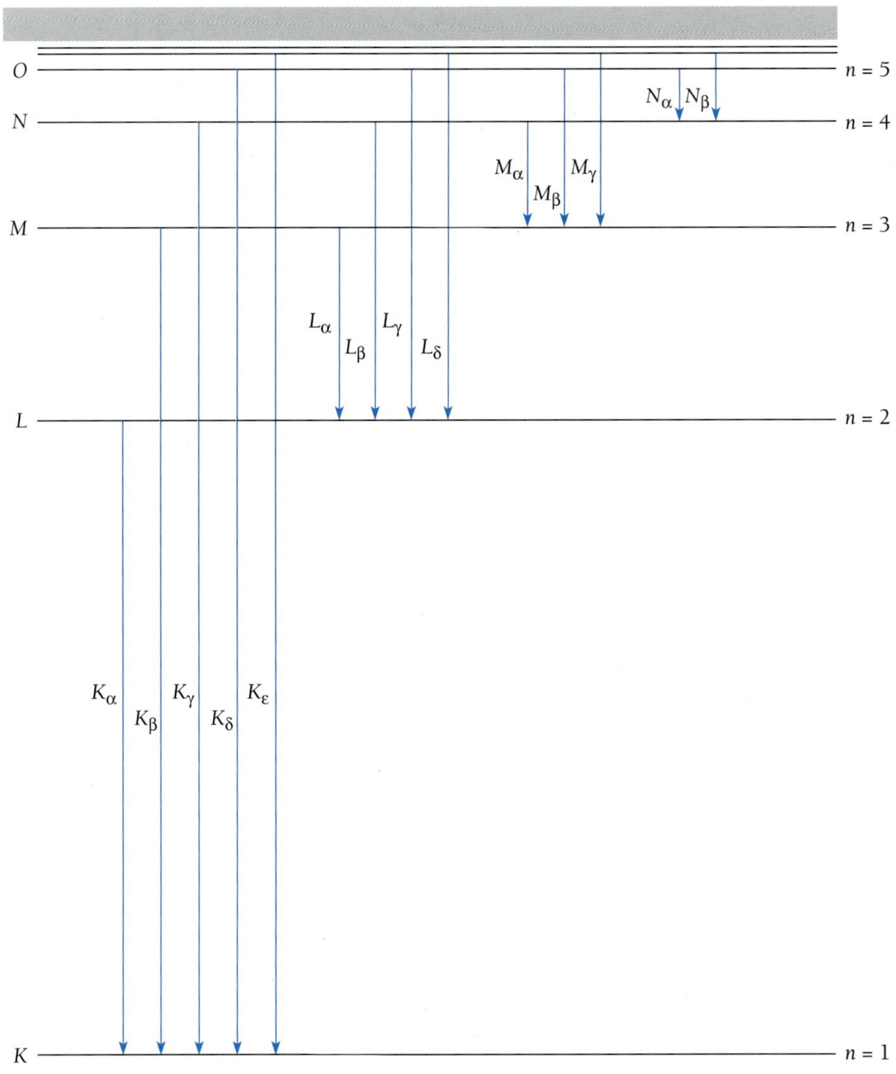

Figure 7.20 The origin of x-ray spectra.

About Scientists

Henry G. J. Moseley (1887–1915) was born in Weymouth, on England's south coast. He studied physics at Oxford, where his father had been professor of anatomy. After graduating in 1910, Moseley joined Rutherford at Manchester, where he began a systematic study of x-ray spectra that he later continued at Oxford. From the data he was able to infer a relationship between the x-ray wavelengths of an element and its atomic number, a relationship that permitted him to correct ambiguities in then-current atomic number assignments and to predict the existence of several then-unknown elements. Moseley soon recognized the important link between his discovery and Bohr's atomic model. By then World War I had broken out and Moseley enlisted in the British Army. Rutherford unsuccessfully tried to have him assigned to scientific work, but in 1915 Moseley was sent to Turkey on the ill-conceived and disastrous Dardanelles campaign and was killed at the age of twenty-seven.

the energy differences between shells. Let us look at what happens when an energetic electron strikes the atom and knocks out one of the K-shell electrons. The K electron could also be raised to one of the unfilled upper states of the atom, but the difference between the energy needed to do this and that needed to remove the electron completely is insignificant, only 0.2 percent in sodium and still less in heavier atoms.

An atom with a missing K electron gives up most of its considerable excitation energy in the form of an x-ray photon when an electron from an outer shell drops into the "hole" in the K shell. As indicated in Fig. 7.20, the **K series** of lines in the x-ray spectrum of an element consists of wavelengths arising in transitions from the L, M, N, . . . levels to the K level. Similarly the longer-wavelength **L series** originates when an L electron is knocked out of the atom, the **M series** when an M electron is knocked out, and so on. The two spikes in the x-ray spectrum of molybdenum in Fig. 2.17 are the K_α and K_β lines of its K series.

It is easy to find an approximate relationship between the frequency of the K_α x-ray line of an element and its atomic number Z. A K_α photon is emitted when an L ($n = 2$) electron undergoes a transition to a vacant $K(n = 1)$ state. The L electron experiences a nuclear charge of Ze that is reduced to an effective charge in the neighborhood of $(Z - 1)e$ by the shielding effect of the remaining K electron. Thus we can use Eqs. (4.15) and (4.16) to find the K_α photon frequency by letting $n_i = 2$ and $n_f = 1$, and replacing e^4 by $(Z - 1)^2e^4$. This gives

$$\nu = \frac{m(Z-1)^2e^4}{8\epsilon_0^2h^3}\left(\frac{1}{n_f^2} - \frac{1}{n_i^2}\right) = cR(Z-1)^2\left(\frac{1}{1^2} - \frac{1}{2^2}\right)$$

K_α x-rays

$$\nu = \frac{3cR(Z-1)^2}{4} \tag{7.21}$$

where $R = me^4/8\epsilon_0^2\,ch^3 = 1.097 \times 10^7\ \text{m}^{-1}$ is the Rydberg constant. The energy of a K_α x-ray photon is given in electronvolts in terms of $(Z - 1)$ by the formula

$$E(K_\alpha) = (10.2\ \text{eV})(Z-1)^2 \tag{7.22}$$

In 1913 and 1914 the young British physicist H. G. J. Moseley confirmed Eq. (7.21) by measuring the K_α frequencies of most of the then-known elements using the diffraction method described in Sec. 2.6. Besides supporting Bohr's newly formulated atomic model, Moseley's work provided for the first time a way to determine experimentally the atomic number Z of an element. As a result, the correct sequence of elements in the

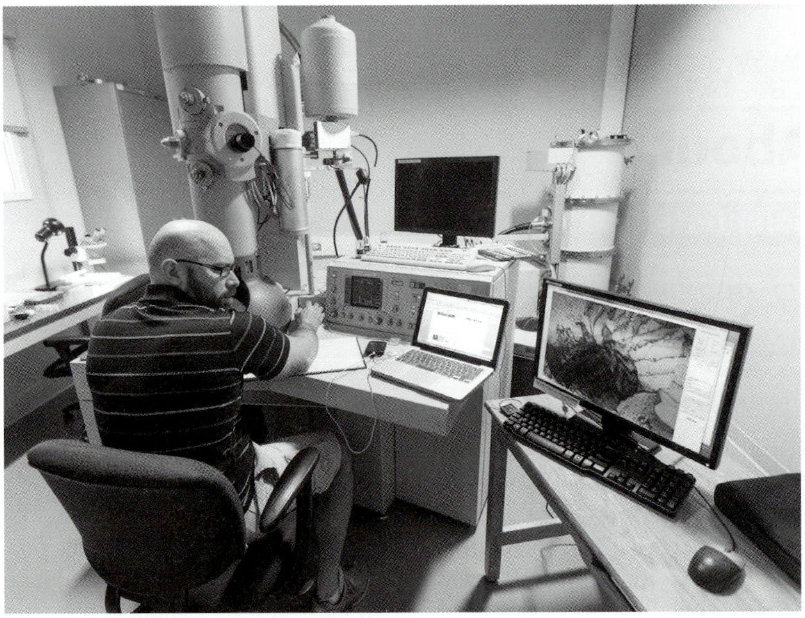

The energy-dispersive x-ray spectroscopy(EDX, EDS) system (the cylinder at the top center position) is commonly attached to the transmission electron microscope(TEM, long column at the left). In the operation of TEM, a stream of fast electrons is directed at a sample and the image is obtained. Some of the electrons knock out inner electrons in the target atoms, and when outer electrons replace them, x-ray are emitted whose wavelengths are characteristic of the elements present. EDX system takes the x-ray spectrum. The identity and relative amounts of the elements in the sample can be found in this way.

periodic table could be established. The ordering of the elements by atomic number (which is what matters) is not always the same as their ordering by atomic mass, which until then was the method used. Atomic number was originally just the number of an element in the list of atomic masses. For instance, $Z = 27$ for cobalt and $Z = 28$ for nickel, but their respective atomic masses are 58.93 and 58.71. The order dictated by atomic mass could not be understood on the basis of the chemical properties of cobalt and nickel.

In addition, Moseley found gaps in his data that corresponded to $Z = 43$, 61, 72, and 75, which suggested the existence of hitherto unknown elements that were later discovered. The first two, technetium and promethium, have no stable isotopes and were first produced in the laboratory many years later. The last two, hafnium and rhenium, were isolated in the 1920s.

Example 7.8

Which element has a K_α x-ray line whose wavelength is 0.180 nm?

Solution

The frequency corresponding to a wavelength of 0.180 nm = 1.80×10^{-10} m is

$$\nu = \frac{c}{\lambda} = \frac{3.00 \times 10^{8}\ \text{m/s}}{1.80 \times 10^{-10}\ \text{m}} = 1.67 \times 10^{18}\ \text{Hz}$$

From Eq. (7.21) we have

$$Z - 1 = \sqrt{\frac{4}{3cR}} = \sqrt{\frac{(4)(1.67 \times 10^{18}\ \text{Hz})}{(3)(3.00 \times 10^{8}\ \text{m/s})(1.097 \times 10^{7}\ \text{m}^{-1})}} = 26$$

$$Z = 27$$

The element with atomic number 27 is cobalt.

Auger Effect

An atom with a missing inner electron can also lose excitation energy by the **Auger effect** without emitting an x-ray photon. In this effect, which was discovered by the French physicist Pierre Auger, an outer-shell electron is ejected from the atom at the same time that another outer-shell electron drops to the incomplete inner shell. Thus the ejected electron carries off the atom's excitation energy instead of a photon doing this (Fig. 7.21). In a sense the Auger effect represents an internal photoelectric effect, although the photon never actually comes into being within the atom.

The Auger process is competitive with x-ray emission in most atoms, but the resulting electrons are usually absorbed in the target material while the x-rays emerge to be detected. Those Auger electrons that do emerge come either from atoms on the surface of the material or just below the surface. Because the energy levels of an atom are affected by its participation in a chemical bond, the energies of Auger electrons provide insight into the chemical environment of the atoms involved. Auger spectroscopy has turned out to be a valuable method for studying the properties of surfaces, information especially needed by manufacturers of semiconductor devices that consist of thin layers of different materials deposited on one another.

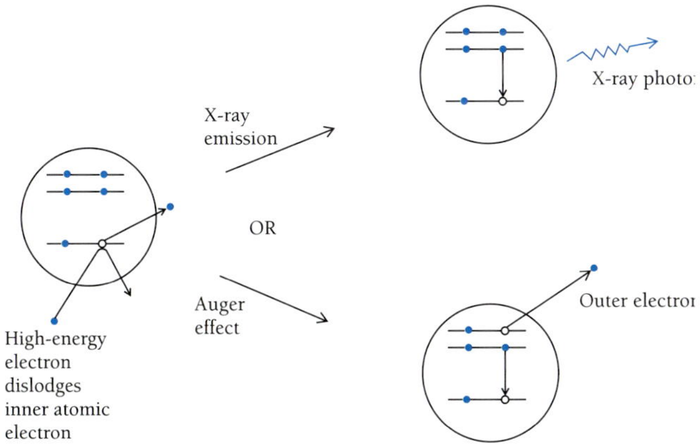

Figure 7.21 When an electron from an outer shell of an atom with a missing inner electron drops to fill the vacant state, the excitation energy can be carried off by an x-ray photon or by another outer electron. The latter process is called the Auger effect.

Appendix to Chapter 7

Atomic Spectra

We are now in a position to understand the chief features of the spectra of the various elements. Before we examine some representative examples, it should be mentioned that further complications exist which have not been considered here, for instance those that originate in relativistic effects and in the coupling between electrons and vacuum fluctuations in the electromagnetic field (see Sec. 6.9). These additional factors split certain energy states into closely spaced substates and therefore represent other sources of fine structure in spectral lines.

Hydrogen

Figure 7.22 shows the various states of the hydrogen atom classified by their total quantum number n and orbital angular-momentum quantum number l. The selection rule

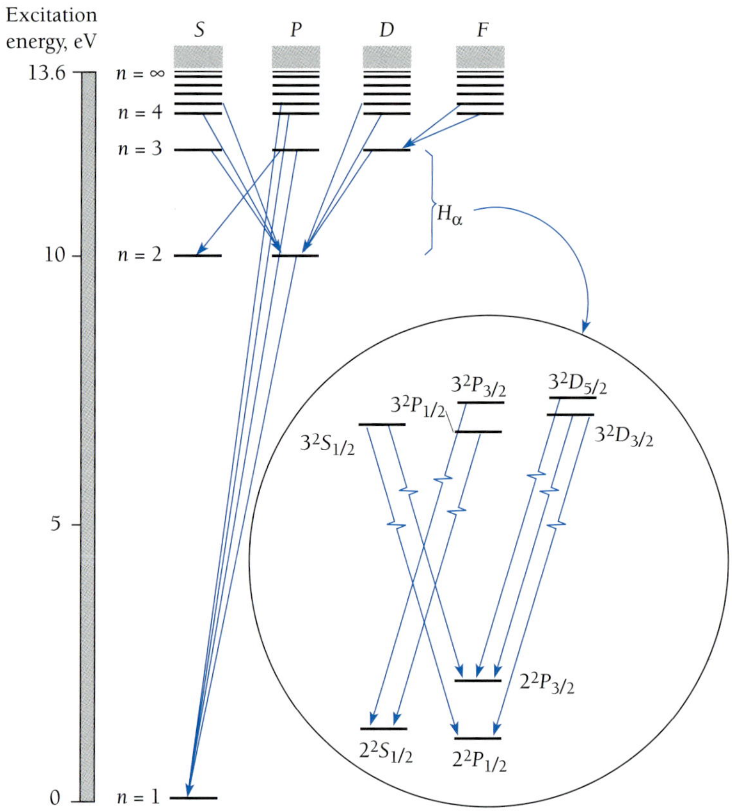

Figure 7.22 Energy-level diagram for hydrogen showing the origins of some of the more prominent spectral lines. The detailed structures of the $n = 2$ and $n = 3$ levels and the transitions that lead to the various components of the H_α line are pictured in the inset.

for allowed transitions here is

Selection rule $$\Delta l = \pm 1$$

which is illustrated by the transitions shown. The principal quantum number n can change by any amount.

To indicate some of the detail that is omitted in a simple diagram of this kind, the detailed structures of the $n = 2$ and $n = 3$ levels are pictured. Not only are all substates of the same n and different j separated in energy, but the same is true of states of the same n and j but with different l. The latter effect is most marked for states of small n and l, and was first established in 1947 in the "Lamb shift" of the $2^2S_{1/2}$ state relative to the $2^2P_{1/2}$ state. The various separations conspire to split the H_α spectral line ($n = 3 \rightarrow n = 2$) into seven closely spaced components.

Sodium

The sodium atom has a single 3s electron outside closed inner shells, and so if we assume that the 10 electrons in its inner core completely shield $+10e$ of nuclear charge (which is not quite true), the outer electron is acted upon by an effective nuclear charge of $+e$ just as in the hydrogen atom. Hence we expect, as a first approximation, that the energy levels of sodium will be the same as those of hydrogen except that the lowest one will correspond to $n = 3$ instead of $n = 1$ because of the exclusion principle. Figure 7.23 is the energy-level diagram for sodium. By comparison with the hydrogen levels also shown, there is indeed agreement for the states of highest l, that is, for the states of highest angular momentum.

To understand the reason for the discrepancies at lower values of l, we need only refer to Fig. 6.11 to see how the probability for finding the electron in a hydrogen atom varies with distance from the nucleus. The smaller the value of l for a given n, the closer the electron gets to the nucleus on occasion. Although the sodium wave functions are not identical with those of hydrogen, their general behavior is similar. Accordingly we expect the outer electron in a sodium atom to penetrate the core of inner electrons most often when it is in an s state, less often when it is in a p state, still less often when it is in a d state, and so on. The less shielded an outer electron is from the full nuclear charge, the greater the average force acting on it, and the smaller (that is, the more negative) its total energy. For this reason the states of small l in sodium are displaced downward from their equivalents in hydrogen, as in Fig. 7.23, and there are pronounced differences in energy between states of the same n but different l.

Helium

A single electron is responsible for the energy levels of both hydrogen and sodium. However, there are two 1s electrons in the ground state of helium, and coupling affects the properties and behavior of the helium atom. These are the selection rules for allowed transitions under LS coupling:

LS selection rules
$$\Delta L = 0, \pm 1$$
$$\Delta J = 0, \pm 1$$
$$\Delta S = 0$$

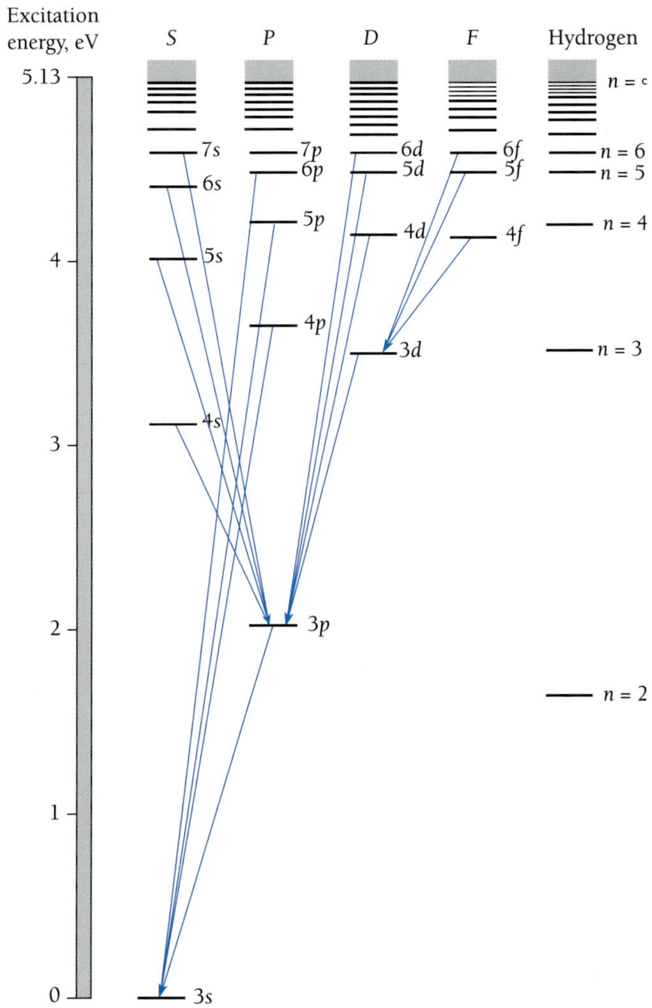

Figure 7.23 Energy-level diagram for sodium. The energy levels of hydrogen are included for comparison.

When only a single electron is involved, $\Delta L = 0$ is prohibited and $\Delta L = \Delta l = \pm 1$ is the only possibility. Furthermore, J must change when the initial state has $J = 0$, so that $J = 0 \rightarrow J = 0$ is prohibited.

The helium energy-level diagram is shown in Fig. 7.24. The various levels represent configurations in which one electron is in its ground state and the other is in an excited state. Because the angular momenta of the two electrons are coupled, the levels are characteristic of the entire atom. Three differences between this diagram and the corresponding ones for hydrogen and sodium are conspicuous:

1 There is a division into singlet and triplet states. These are, respectively, states in which the spins of the two electrons are antiparallel (to give $S = 0$) and parallel (to give $S = 1$). Because of the selection rule $\Delta S = 0$, no allowed transitions can occur between singlet states and triplet states, and the helium spectrum arises from transitions in one set or the other.

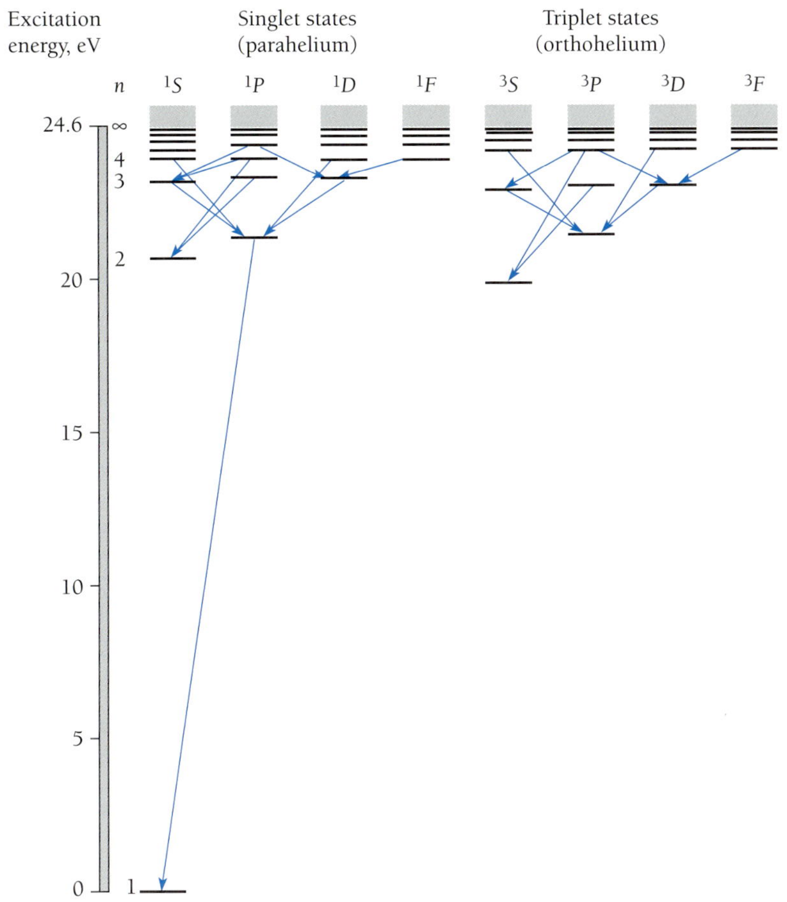

Figure 7.24 Energy-level diagram for helium showing the division into singlet (parahelium) and triplet (orthohelium) states. There is no 1^3S state because the exclusion principle prohibits two electrons with parallel spins in the same state.

Helium atoms in singlet states (antiparallel spins) constitute **parahelium** and those in triplet states (parallel spins) constitute **orthohelium.** An orthohelium atom can lose excitation energy in a collision and become one of parahelium, while a parahelium atom can gain excitation energy in a collision and become one of orthohelium. Ordinary liquid or gaseous helium is therefore a mixture of both. The lowest triplet states are metastable because, in the absence of collisions, an atom in one of them can retain its excitation energy for a relatively long time (a second or more) before radiating.

2 Another obvious peculiarity in Fig. 7.24 is the absence of the 1^3S state in helium. The lowest triplet state is 2^3S, although the lowest singlet state is 1^1S. The 1^3S state is missing because of the exclusion principle, since in this state the two electrons would have parallel spins and therefore identical sets of quantum numbers.

3 The energy difference between the ground state and the lowest excited state in helium is relatively large. This reflects the tight binding of closed-shell electrons discussed earlier in this chapter. The ionization energy of helium—the work that must be done to remove an electron from a helium atom—is 24.6 eV, the highest of any element.

Mercury

The last energy-level diagram we consider is that of mercury, which has two electrons outside an inner core of 78 electrons in closed shells or subshells (Table 7.4). We expect a division into singlet and triplet states as in helium. Because the atom is so heavy we might also expect signs of a breakdown in the *LS* coupling of angular momenta.

As Fig. 7.25 reveals, both of these expectations are realized, and several prominent lines in the mercury spectrum arise from transitions that violate the $\Delta \mathbf{S} = 0$ selection rule. The transition $^3P_1 \rightarrow {}^1S_0$ is an example, and is responsible for the strong 253.7-nm line in the ultraviolet. To be sure, this does not mean that the transition probability is necessarily very high, since the three $^3P_1$ states tend to be highly populated in excited mercury vapor. The $^3P_0 \rightarrow {}^1S_0$ and $^3P_2 \rightarrow {}^1S_0$ transitions, respectively, violate the rules that forbid transitions from $\mathbf{J} = 0$ to $\mathbf{J} = 0$ and that limit $\Delta \mathbf{J}$ to 0 or ± 1, as well as violating $\Delta \mathbf{S} = 0$, and hence are considerably less likely to occur than the $^3P_1 \rightarrow {}^1S_0$ transition. The $^3P_0$ and $^3P_2$ states are therefore metastable, and in the absence of collisions, an atom can persist in either of them for a relatively long time. The strong spin-orbit interaction in mercury that leads to the partial failure of *LS* coupling is also responsible for the wide spacing of the elements of the ^3P triplet.

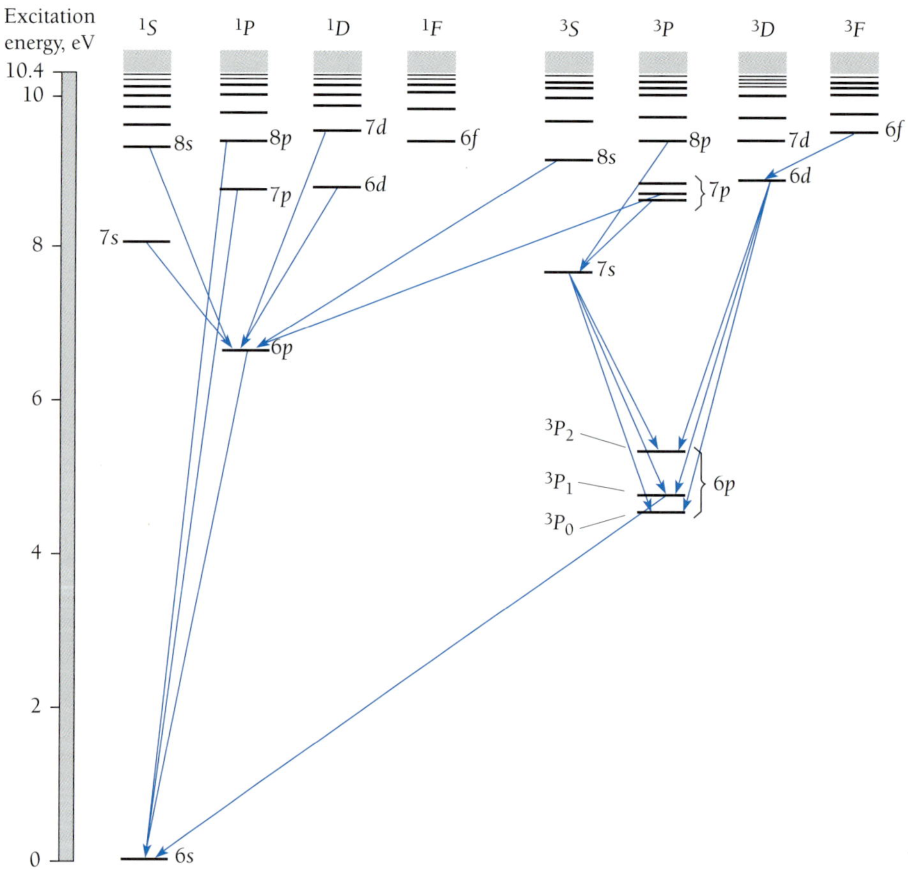

Figure 7.25 Energy-level diagram for mercury. In each excited level one outer electron is in the ground state, and the designation of the levels in the diagram corresponds to the state of the other electron.

EXERCISES

No plan survives contact with the enemy. —Field Marshal von Moltke

7.1 Electron Spin

1. A beam of electrons enters a uniform 1.20-T magnetic field. (*a*) Find the energy difference between electrons whose spins are parallel and antiparallel to the field. (*b*) Find the wavelength of the radiation that can cause the electrons whose spins are parallel to the field to flip so that their spins are antiparallel.

2. Radio astronomers can detect clouds of hydrogen in our galaxy too cool to radiate in the optical part of the spectrum by means of the 21-cm spectral line that corresponds to the flipping of the electron in a hydrogen atom from having its spin parallel to the spin of the proton to having it antiparallel. Find the magnetic field experienced by the electron in a hydrogen atom.

3. Find the possible angles between the z axis and the direction of the spin angular-momentum vector **S**.

7.3 Symmetric and Antisymmetric Wave Functions

4. In superconductivity, which occurs in certain materials at very low temperatures, electrons are linked together in "Cooper pairs" by their interaction with the crystal lattices of the materials. Cooper pairs do not obey the exclusion principle. What aspect of these pairs do you think permits this?

5. Protons and neutrons, like electrons, are spin-$\frac{1}{2}$ particles. The nuclei of ordinary helium atoms, $^4_2$He, contain two protons and two neutrons each; the nuclei of another type of helium atom, $^3_2$He, contain two protons and one neutron each. The properties of liquid $^4_2$He and liquid $^3_2$He are different because one type of helium atom obeys the exclusion principle but the other does not. Which is which, and why?

6. A one-dimensional potential well like those of Secs. 3.6 and 5.8 has a width of 1.00 nm and contains 10 electrons. The system of electrons has the minimum total energy possible. What is the least energy, in eV, a photon must have in order to excite a ground-state ($n = 1$) electron in this system to the lowest higher state it can occupy?

7.6 Explaining the Periodic Table

7. In what way does the electron structure of an alkali metal atom differ from that of a halogen atom? From that of an inert gas atom?

8. What is true in general of the properties of elements in the same period of the periodic table? Of elements in the same group?

9. How many electrons can occupy an f subshell?

10. (*a*) How would the periodic table be modified if the electron had a spin of 1, so it could have spin states of -1, 0, and $+1$? Assume (wrongly) that such electrons are fermions and so obey the exclusion principle. Which elements would then be inert gases? (*b*) Such electrons would in fact be bosons. Which elements in this case would be inert gases?

11. If atoms could contain electrons with principal quantum numbers up to and including $n = 6$, how many elements would there be?

12. Verify that atomic subshells are filled in order of increasing $n + l$, and within a group of given $n + l$ in order of increasing n.

13. The ionization energies of Li, Na, K, Rb, and Cs are, respectively, 5.4, 5.1, 4.3, 4.2, and 3.9 eV. All are in group 1 of the periodic table. Account for the decrease in ionization energy with increasing atomic number.

14. The ionization energies of the elements of atomic numbers 20 through 29 are very nearly equal. Why should this be so when considerable variations exist in the ionization energies of other consecutive sequences of elements?

15. (*a*) Make a rough estimate of the effective nuclear charge that acts on each electron in the outer shell of the calcium ($Z = 20$) atom. Would you think that such an electron is relatively easy or relatively hard to detach from the atom? (*b*) Do the same for the sulfur ($Z = 16$) atom.

16. The effective nuclear charge that acts on the outer electron in the sodium atom is 1.84e. Use this figure to calculate the ionization energy of sodium.

17. Why are Cl atoms more chemically active than Cl^- ions? Why are Na atoms more chemically active than Na^+ ions?

18. Account for the general trends of the variation of atomic radius with atomic number shown in Fig. 7.11.

19. In each of the following pairs of atoms, which would you expect to be larger in size? Why? Li and F; Li and Na; F and Cl; Na and Si.

20. The nucleus of a helium atom consists of two protons and two neutrons. The Bohr model of this atom has two electrons in the same orbit around the nucleus. Estimate the average separation of the electrons in a helium atom in the following way. (1) Assume that each electron moves independently of the other in a ground-state Bohr orbit and calculate its ionization energy on this basis. (2) Use the difference between the calculated ionization energy and the measured one of 24.6 eV to find the interaction energy between the two electrons. (3) On the assumption that the interaction energy results from the repulsion between the electrons, find their separation. How does this compare with the radius of the orbit?

21. Why is the normal Zeeman effect observed only in atoms with an even number of electrons?

7.7 Spin-Orbit Coupling

22. Why is the ground state of the hydrogen atom not split into two sublevels by spin-orbit coupling?

23. The spin-orbit effect splits the $3P \rightarrow 3S$ transition in sodium (which gives rise to the yellow light of sodium-vapor highway lamps) into two lines, 589.0 nm corresponding to $3P_{3/2} \rightarrow 3S_{1/2}$ and 589.6 nm corresponding to $3P_{1/2} \rightarrow 3S_{1/2}$. Use these wavelengths to calculate the effective magnetic field experienced by the outer electron in the sodium atom as a result of its orbital motion.

7.8 Total Angular Momentum

24. An atom has a single electron outside closed inner shells. What total angular momentum J can the atom have if it is in a P state? In a D state?

25. If $j = \frac{5}{2}$, what values of l are possible?

26. (*a*) What are the possible values of **L** for a system of two electrons whose orbital quantum numbers are $l_1 = 1$ and $l_2 = 3$? (*b*) What are the possible values of **S**? (*c*) What are the possible values of **J**?

27. What must be true of the subshells of an atom which has a $^1S_0$ ground state?

28. Find the **S**, **L**, and **J** values that correspond to each of the following states: $^1S_0$, $^3P_2$, $^2D_{3/2}$, $^5F_5$, $^6H_{5/2}$.

29. The lithium atom has one 2*s* electron outside a filled inner shell. Its ground state is $^2S_{1/2}$. (*a*) What are the term symbols of the other allowed states, if any? (*b*) Why would you think the $^2S_{1/2}$ state is the ground state?

30. The magnesium atom has two 3*s* electrons outside filled inner shells. Find the term symbol of its ground state.

31. The aluminum atom has two 3*s* electrons and one 3*p* electron outside filled inner shells. Find the term symbol of its ground state.

32. In a carbon atom, only the two 2*p* electrons contribute to its angular momentum. The ground state of this atom is $^3P_0$, and the first four excited states, in order of increasing energy, are $^3P_1$, $^3P_2$, $^1D_2$, and $^1S_0$. (*a*) Give the **L**, **S**, and **J** values for each of these five states. (*b*) Why do you think the $^3P_0$ state is the ground state?

33. Why is it impossible for a $2^2D_{3/2}$ state to exist?

34. (*a*) What values can the quantum number j have for a d electron in an atom whose total angular momentum is provided by this electron? (*b*) What are the magnitudes of the corresponding angular momenta of the electron? (*c*) what are the angles between the directions of **L** and **S** in each case? (*d*) What are the term symbols for this atom?

35. Answer the questions of Exercise 34 for an f electron in an atom whose total angular momentum is provided by this electron.

36. Show that if the angle between the directions of **L** and **S** in Fig. 7.15 is θ,

$$\cos\theta = \frac{j(j+1) - l(l+1) - s(s+1)}{2\sqrt{l(l+1)\, s(s+1)}}$$

37. The magnetic moment $\boldsymbol{\mu}_J$ of an atom in which LS coupling holds has the magnitude

$$\mu_J = \sqrt{J(J+1)}g_J\mu_B$$

where $\mu_B = e\hbar/2m$ is the Bohr magneton and

$$g_J = 1 + \frac{\mathrm{J(J+1) - L(L+1) + S(S+1)}}{\mathrm{2J(J+1)}}$$

is the **Landé g factor.** (*a*) Derive this result with the help of the law of cosines starting from the fact that averaged over time, only the components of $\boldsymbol{\mu}_L$ and $\boldsymbol{\mu}_S$ parallel to **J** contribute to $\boldsymbol{\mu}_J$. (*b*) Consider an atom that obeys LS coupling that is in a weak magnetic field **B** in which the coupling is preserved. How many substates are there for a given value of **J**? What is the energy difference between different substates?

38. The ground state of chlorine is $^2P_{3/2}$. Find its magnetic moment (see previous exercise). Into how many substates will the ground state split in a weak magnetic field?

7.9 X-Ray Spectra

39. Explain why the x-ray spectra of elements of nearby atomic numbers are qualitatively very similar, although the optical spectra of these elements may differ considerably.

40. What element has a K_α x-ray line of wavelength 0.144 nm?

41. Find the energy and the wavelength of the K_α x-rays of aluminum.

42. The effective charge experienced by an M ($n = 3$) electron in an atom of atomic number Z is about $(Z - 7.4)e$. Show that the frequency of the L_α x-rays of such an element is given by $5cR(Z - 7.4)^2/36$.

Appendix: Atomic Spectra

43. Distinguish between singlet and triplet states in atoms with two outer electrons.

44. Which of the following elements would you expect to have energy levels divided into singlet and triplet states: Ne, Mg, Cl, Ca, Cu, Ag, Ba?

CHAPTER 8

Molecules

KRAKOW, POLAND, CIRCA MAY 2016. NANOIR2 system, IR microscope with nanometric spatial resolution in a Laboratory of Spectroscopic Imaging at Institute of Nuclear Physics, Polish Academy of Science.

Individual atoms are rare on the earth and in the lower part of its atmosphere. Only inert gas atoms occur by themselves. All other atoms are found joined together in small groups called molecules and in large groups as liquids and solids. Some molecules, liquids, and solids are composed entirely of atoms of the same element; others are composed of atoms of different elements.

What holds atoms together? This question, of fundamental importance to the chemist, is no less important to the physicist, whose quantum theory of the atom cannot be correct unless it provides a satisfactory answer. The ability of the quantum theory to explain chemical bonding with no special assumptions is further testimony to the power of this approach.

8.1 THE MOLECULAR BOND

Electric forces hold atoms together to form molecules

A molecule is an electrically neutral group of atoms held together strongly enough to behave as a single particle.

A molecule of a given kind always has a certain definite composition and structure. Hydrogen molecules, for instance, always consist of two hydrogen atoms each, and water molecules always consist of one oxygen atom and two hydrogen atoms each. If one of the atoms of a molecule is somehow removed or another atom becomes attached, the result is a molecule of a different kind with different properties.

A molecule exists because its energy is less than that of the system of separate noninteracting atoms. If the interactions among a certain group of atoms reduce their total energy, a molecule can be formed. If the interactions increase their total energy, the atoms repel one another.

Let us see what happens when two atoms are brought closer and closer together. Three extreme situations can occur:

1 *A covalent bond is formed.* One or more pairs of electrons are shared by the two atoms. As these electrons circulate between the atoms, they spend more time between the atoms than elsewhere, which produces an attractive force. An example is H_2, the hydrogen molecule, whose electrons belong to both protons (Fig. 8.1). The attractive force the electrons exert on the protons is more than enough to counterbalance the direct repulsion between them. If the protons are too close together, however, their repulsion becomes dominant and the molecule is not stable.

The balance between attractive and repulsive forces occurs at a separation of 7.42×10^{-11} m, where the total energy of the H_2 molecule is -4.5 eV. Hence 4.5 eV of work must be done to break a H_2 molecule into two H atoms:

$$H_2 + 4.5 \text{ eV} \rightarrow H + H$$

By comparison, the binding energy of the hydrogen atom is 13.6 eV:

$$H + 13.6 \text{ eV} \rightarrow p^+ + e^-$$

This is an example of the general rule that it is easier to break up a molecule than to break up an atom.

2 *An ionic bond is formed.* One or more electrons from one atom may transfer to the other and the resulting positive and negative ions attract each other. An example is

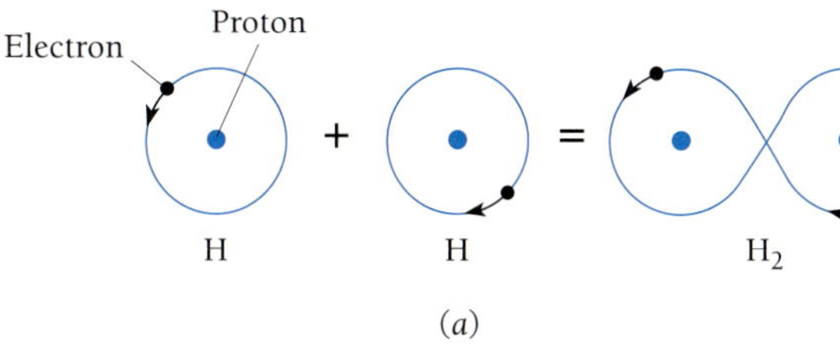

Figure 8.1 (*a*) Orbit model of the hydrogen molecule. (*b*) Quantum-mechanical model of the hydrogen molecule. In both models the shared electrons spend more time on the average between the nuclei, which leads to an attractive force. Such a bond is said to be covalent.

rock salt, NaCl, where the bond exists between Na^+ and Cl^- ions and not between Na and Cl atoms (Fig. 8.2). Ionic bonds usually do not result in the formation of molecules. The crystals of rock salt are aggregates of sodium and chlorine ions which, although always arranged in a certain definite structure (Fig. 8.3), do not pair off into molecules consisting of one Na^+ ion and one Cl^- ion. Rock salt crystals may have any size and shape. There are always equal numbers of Na^+ and Cl^- ions in rock salt, so that the formula NaCl correctly represents its composition. Molten NaCl also consists of Na^+ and Cl^- ions: these ions form molecules rather than crystals only in the gaseous state. Ionic bonding is further discussed in Chap. 10.

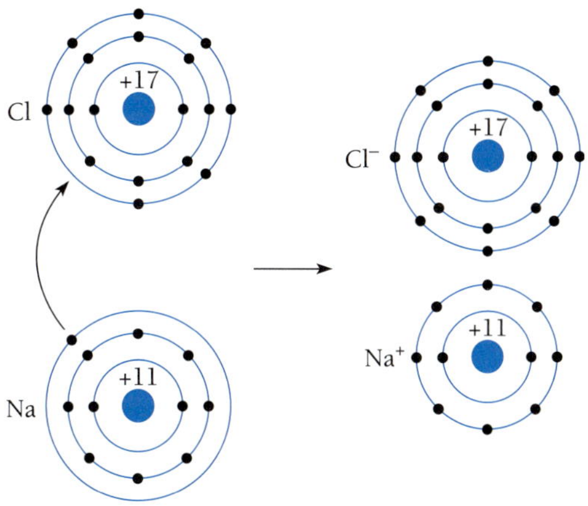

Figure 8.2 An example of ionic bonding. Sodium and chlorine combine chemically by the transfer of electrons from sodium atoms to chlorine atoms; the resulting ions attract each other electrically.

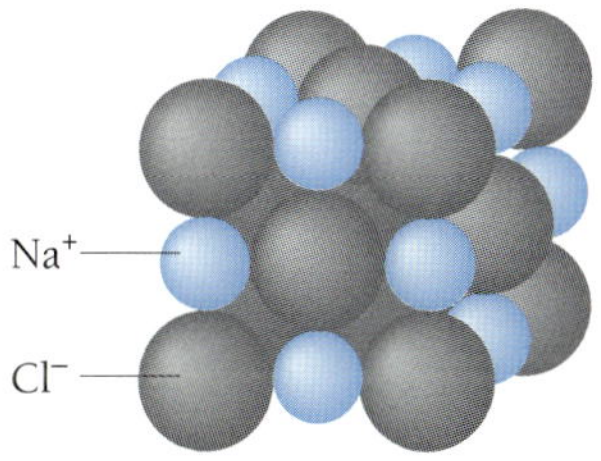

Figure 8.3 Scale model of an NaCl crystal.

In H_2 the bond is purely covalent and in NaCl it is purely ionic. In many molecules an intermediate type of bond occurs in which the atoms share electrons to an unequal extent. An example is the HCl molecule, where the Cl atom attracts the shared electrons more strongly than the H atom. We can think of the ionic bond as an extreme case of the covalent bond.

3 *No bond is formed.* When the electron structures of two atoms overlap, they constitute a single system. According to the exclusion principle, no two electrons in such a system can exist in the same quantum state. If some of the interacting electrons are forced into higher energy states than they occupied in the separate atoms, the system may have more energy than before and be unstable. Even when the exclusion principle can be obeyed with no increase in energy, there will be an electric repulsive force between the various electrons. This is a much less significant factor than the exclusion principle in influencing bond formation, however.

8.2 ELECTRON SHARING

The mechanism of the covalent bond

The simplest possible molecular system is H_2^+, the hydrogen molecular ion, in which a single electron bonds two protons. Before we consider the bond in H_2^+ in detail, let us look in a general way into how it is possible for two protons to share an electron and why such sharing should lead to a lower total energy and hence to a stable system.

In Chap. 5 the phenomenon of quantum-mechanical barrier penetration was examined. There we saw that a particle can "leak" out of a box even without enough energy to break through the wall because the particle's wave function extends beyond it. Only if the wall is infinitely strong is the wave function wholly inside the box.

The electric field around a proton is in effect a box for an electron, and two nearby protons correspond to a pair of boxes with a wall between them (Fig. 8.4). No mechanism in classical physics permits the electron in a hydrogen atom to jump spontaneously to a neighboring proton more distant than its parent proton. In quantum physics, however, such a mechanism does exist. There is a certain probability that an electron trapped in one box will tunnel through the wall and get into the other box, and once there it has the same probability for tunneling back. This situation can be described by saying the electron is shared by the protons.

To be sure, the likelihood that an electron will pass through the region of high potential energy—the "wall"—between two protons depends strongly on how far apart the protons are. If the proton-proton distance is 0.1 nm, the electron may be regarded as going from one proton to the other about every 10^{-15} s. We can legitimately consider such an electron as being shared by both. If the proton-proton distance is 1 nm, however, the electron shifts across an average of only about once per second, which is practically an infinite time on an atomic scale. Since the effective radius of the 1*s* wave function in hydrogen is 0.053 nm, we conclude that electron sharing can take place only between atoms whose wave functions overlap appreciably.

Granting that two protons can share an electron, a simple argument shows why the energy of such a system could be less than that of a separate hydrogen atom and proton. According to the uncertainty principle, the smaller the region to which we restrict a particle, the greater must be its momentum and hence kinetic energy. An electron

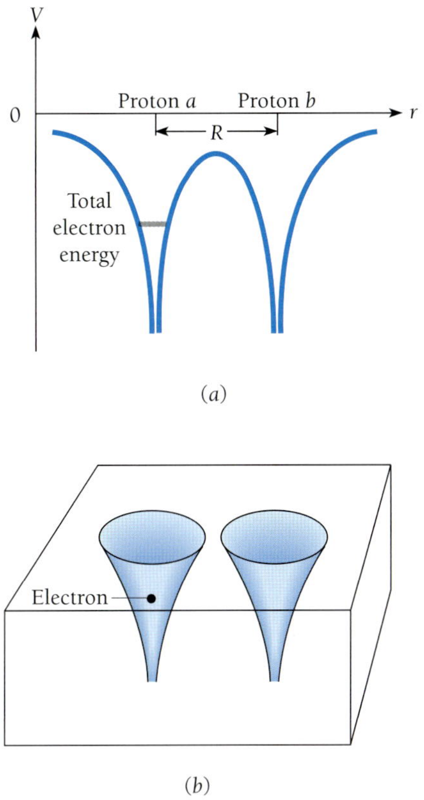

Figure 8.4 (*a*) Potential energy of an electron in the electric field of two nearby protons. The total energy of a ground-state electron in the hydrogen atom is indicated. (*b*) Two nearby protons correspond quantum-mechanically to a pair of boxes separated by a barrier.

shared by two protons is less confined than one belonging to a single proton, which means that it has less kinetic energy. The total energy of the electron in H_2^+ is therefore less than that of the electron in $H + H^+$. Provided the magnitude of the proton-proton repulsion in H_2^+ is not too great, then, H_2^+ ought to be stable.

8.3 THE H_2^+ MOLECULAR ION

Bonding requires a symmetric wave function

What we would like to know is the wave function ψ of the electron in H_2^+, since from ψ we can calculate the energy of the system as a function of the separation R of the protons. If $E(R)$ has a minimum, we will know that a bond can exist, and we can also determine the bond energy and the equilibrium spacing of the protons.

Solving Schrödinger's equation for ψ is a long and complicated procedure. An intuitive approach that brings out the physics of the situation is more appropriate here. Let us begin by trying to predict what ψ is when R, the distance between the protons, is large compared with a_0, the radius of the smallest Bohr orbit in the hydrogen atom. In this event ψ near each proton must closely resemble the 1s wave function of the hydrogen atom, as pictured in Fig. 8.5. The 1s wave function around proton a is called ψ_a and that around proton b is called ψ_b.

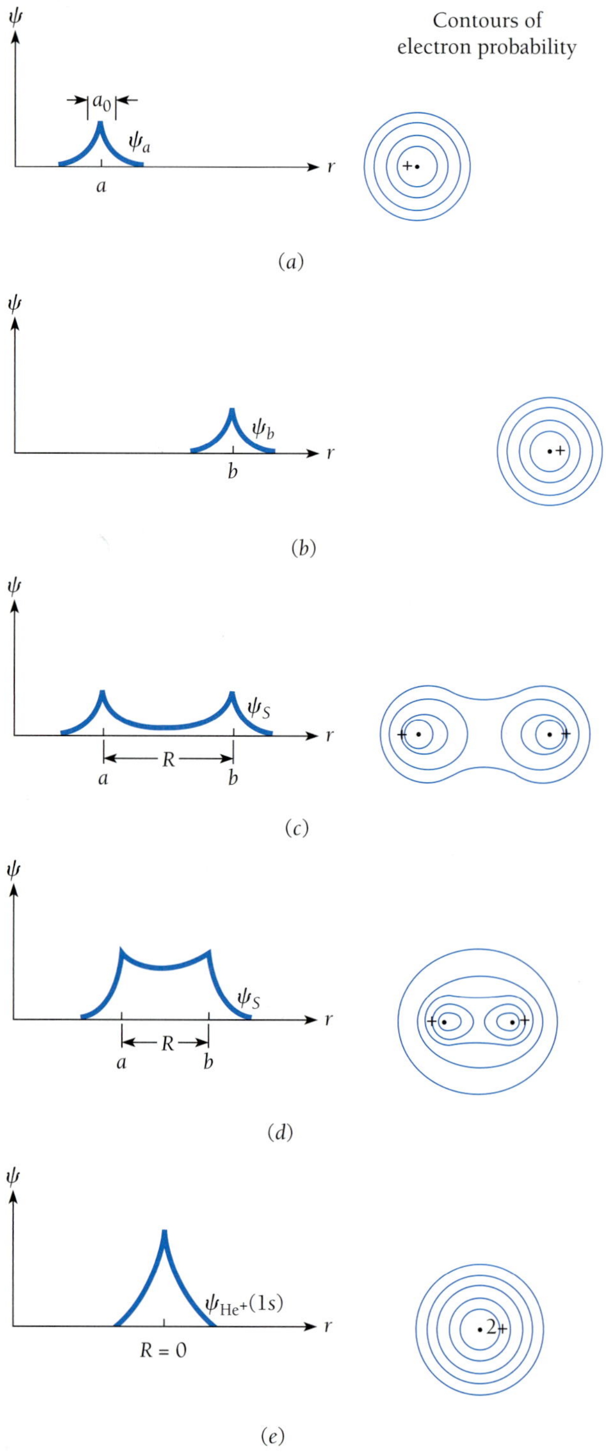

Figure 8.5 (*a*)–(*d*) The combination of two hydrogen-atom 1*s* wave functions to form the symmetric H_2^+ wave function ψ_S. The result is a stable H_2^+ molecular ion because the electron has a greater probability of being between the protons than outside them. (*e*) If the protons could join together, the resulting wave function would be the same as the 1*s* wave function of a He^+ ion.

We also know what ψ looks like when R is 0, that is, when the protons are imagined to be fused together. Here the situation is that of the He^+ ion, since the electron is now near a single nucleus whose charge is $+2e$. The 1s wave function of He^+ has the same form as that of H but with a greater amplitude at the origin, as in Fig. 8.5e. Evidently ψ is going to be something like the wave function sketched in Fig. 8.5d when R is comparable with a_0. There is an enhanced likelihood of finding the electron in the region between the protons, which corresponds to the sharing of the electron by the protons. Thus there is on the average an excess of negative charge between the protons, and this attracts the protons together. We have still to establish whether this attraction is strong enough to overcome the mutual repulsion of the protons.

The combination of ψ_a and ψ_b in Fig. 8.5 is symmetric, since exchanging a and b does not affect ψ (see Sec. 7.3). However, it is also conceivable that we could have an *antisymmetric* combination of ψ_a and ψ_b, as in Fig. 8.6. Here there is a node between a and b where $\psi = 0$, which implies a reduced likelihood of finding the electron between the protons. Now there is on the average a deficiency of negative charge between the protons and in consequence a repulsive force. With only repulsive forces acting, bonding cannot occur.

An interesting question concerns the behavior of the antisymmetric H_2^+ wave function ψ_A as $R \rightarrow 0$. Obviously ψ_A does not become the 1s wave function of He^+ when $R = 0$. However, ψ_A *does* approach the 2p wave function of He^+ (Fig. 8.6e), which has a node at the origin. But the 2p state of He^+ is an excited state whereas the 1s state is the ground state. Hence H_2^+ in the antisymmetric state ought to have more energy than when it is in the symmetric state, which agrees with our inference from the shapes of the wave functions ψ_A and ψ_S that in the former case there is a repulsive force and in the latter, an attractive one.

System Energy

A line of reasoning similar to the preceding one lets us estimate how the total energy of the H_2^+ system varies with R. We first consider the symmetric state. When R is large, the electron energy E_S must be the -13.6-eV energy of the hydrogen atom, while the electric potential energy U_p of the protons,

$$U_p = \frac{e^2}{4\pi\epsilon_0 R} \tag{8.1}$$

falls to 0 as $R \rightarrow \infty$. (U_p is a positive quantity, corresponding to a repulsive force.) When $R \rightarrow 0$, $U_p \rightarrow \infty$ as $1/R$. At $R = 0$, the electron energy must equal that of the He^+ ion, which is Z^2, or 4 times, that of the H atom. (See Exercise 35 of Chap. 4; the same result is obtained from the quantum theory of one-electron atoms.) Hence $E_S = -54.4$ eV when $R = 0$.

Both E_S and U_p are sketched in Fig. 8.7 as functions of R. The shape of the curve for E_S can only be approximated without a detailed calculation, but we do have its value for both $R = 0$ and $R = \infty$ and, of course, U_p obeys Eq. (8.1).

The total energy E_S^{total} of the system is the sum of the electron energy E_S and the potential energy U_p of the protons. Evidently E_S^{total} has a minimum, which corresponds to a stable molecular state. This result is confirmed by the experimental data on H_2^+ which indicate a bond energy of 2.65 eV and an equilibrium separation R of 0.106 nm. By "bond energy" is meant the energy needed to break H_2^+ into $H + H^+$. The *total*

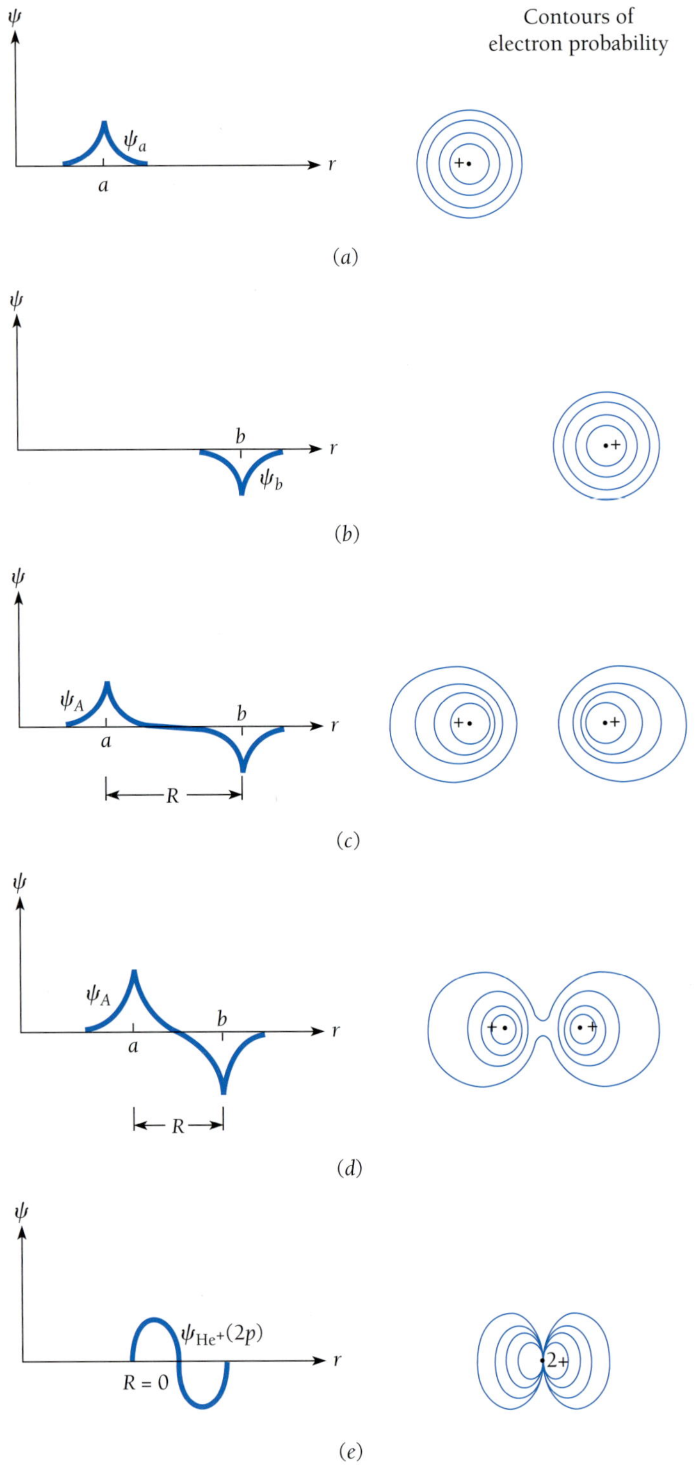

Figure 8.6 (*a*)–(*d*) The combination of two hydrogen-atom 1*s* wave functions to form the antisymmetric H_2^+ wave function ψ_A. A stable H_2^+ molecular ion is not formed because now the electron has a smaller probability of being between the protons than outside them. (*e*) If the protons could join together, the resulting wave function would be the same as the 2*p* wave function of a He^+ ion. In the 2*p* state a He^+ ion has more energy than in the 2*s* state.

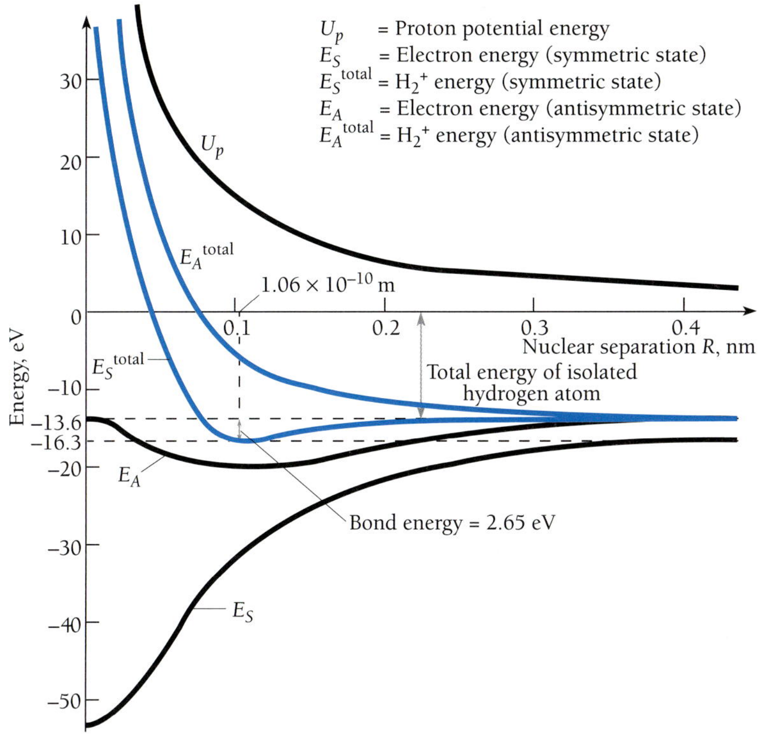

Figure 8.7 Electron, proton repulsion, and total energies in H_2^+ as a function of nuclear separation R for the symmetric and antisymmetric states. The antisymmetric state has no minimum in its total energy.

energy of H_2^+ is the -13.6 eV of the hydrogen atom plus the -2.65-eV bond energy, or -16.3 eV in all.

In the case of the antisymmetric state, the analysis proceeds in the same way except that the electron energy E_A when $R = 0$ is that of the $2p$ state of He^+. This energy is proportional to Z^2/n^2. With $Z = 2$ and $n = 2$, E_A is just equal to the -13.6 eV of the ground-state hydrogen atom. Since $E_A \rightarrow 13.6$ eV also as $R \rightarrow \infty$, we might think that the electron energy is constant, but actually there is a small dip at intermediate distances. However, the dip is not nearly enough to yield a minimum in the total energy curve for the antisymmetric state, as shown in Fig. 8.7, and so in this state no bond is formed.

8.4 THE HYDROGEN MOLECULE

The spins of the electrons must be antiparallel

The H_2 molecule has two electrons instead of the single electron of H_2^+. According to the exclusion principle, both electrons can share the same **orbital** (that is, be described by the same wave function ψ_{nlm_l}) provided their spins are antiparallel.

With two electrons to contribute to the bond, H_2 ought to be more stable than H_2^+—at first glance, twice as stable, with a bond energy of 5.3 eV compared with

2.65 eV for H_2^+. However, the H_2 orbitals are not quite the same as those of H_2^+ because of the electric repulsion between the two electrons in H_2, a factor absent in the case of H_2^+. This repulsion weakens the bond in H_2, so that the actual energy is 4.5 eV instead of 5.3 eV. For the same reason, the bond length in H_2 is 0.074 nm, which is somewhat larger than the use of unmodified H_2^+ wave functions would indicate. The general conclusion in the case of H_2^+ that the symmetric wave function ψ_S leads to a bound state and the antisymmetric wave function ψ_A to an unbound one remains valid for H_2.

In Sec. 7.3 the exclusion principle was formulated in terms of the symmetry and antisymmetry of wave functions, and it was concluded that systems of electrons are always described by antisymmetric wave functions (that is, by wave functions that reverse sign upon the exchange of any pair of electrons). However, the bound state in H_2 corresponds to both electrons being described by a symmetrical wave function ψ_S, which seems to contradict the above conclusion.

A closer look shows that there is really no contradiction. The *complete* wave function $\Psi(1, 2)$ of a system of two electrons is the product of a spatial wave function $\psi(1, 2)$ which describes the coordinates of the electrons and a spin function $s(1, 2)$ which describes the orientations of their spins. The exclusion principle requires that the complete wave function

$$\Psi(1, 2) = \psi(1, 2)\, s(1, 2)$$

be antisymmetric to an exchange of both coordinates and spins, not $\psi(1, 2)$ by itself. An antisymmetric complete wave function Ψ_A can result from the combination of a symmetric coordinate wave function ψ_S and an antisymmetric spin function s_A or from the combination of an antisymmetric coordinate wave function ψ_A and a symmetric spin function s_S. That is, only

$$\Psi(1, 2) = \psi_S s_A \quad \text{and} \quad \Psi(1, 2) = \psi_A s_S$$

are acceptable.

If the spins of the two electrons are parallel, their spin function is symmetric since it does not change sign when the electrons are exchanged. Hence the coordinate wave function ψ for two electrons whose spins are parallel must be antisymmetric:

Spins parallel

$$\Psi(1, 2) = \psi_A s_S$$

On the other hand, if the spins of the two electrons are antiparallel, their spin function is antisymmetric since it reverses sign when the electrons are exchanged. Hence the coordinate wave function ψ for two electrons whose spins are antiparallel must be symmetric:

Spins antiparallel

$$\Psi(1, 2) = \psi_S s_A$$

Schrödinger's equation for the H_2 molecule has no exact solution. In fact, only for H_2^+ is an exact solution possible, and all other molecular systems must be treated approximately. The results of a detailed analysis of the H_2 molecule are shown in Fig. 8.8 for the case when the electrons have their spins parallel and the case when their spins are antiparallel. The difference between the two curves is due to the exclusion principle, which leads to a dominating repulsion when the spins are parallel.

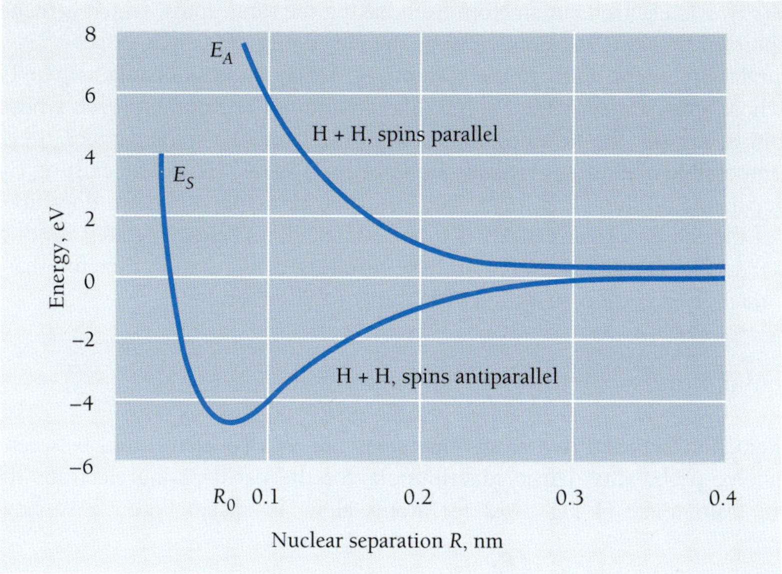

Figure 8.8 The variation of the energy of the system H + H with their distances apart when the electron spins are parallel and antiparallel.

8.5 COMPLEX MOLECULES

Their geometry depends on the wave functions of the outer electrons of their atoms

Covalent bonding in molecules other than H_2, diatomic as well as polyatomic, is usually a more complicated story. It would be yet more complicated but for the fact that any alteration in the electronic structure of an atom due to the proximity of another atom is confined to its outermost, or **valence**, electron shell. There are two reasons for this:

1 The inner electrons are much more tightly bound and hence less responsive to external influences, partly because they are closer to their parent nucleus and partly because they are shielded from the nuclear charge by fewer intervening electrons.
2 The repulsive interatomic forces in a molecule become predominant while the inner shells of its atoms are still relatively far apart.

The idea that only the valence electrons are involved in chemical bonding is supported by x-ray spectra that arise from transitions to inner-shell electron states. These spectra are virtually independent of how the atoms are combined in molecules or solids.

We have seen that two H atoms can combine to form an H_2 molecule; and, indeed, hydrogen molecules in nature always consist of two H atoms. The exclusion principle is what prevents molecules such as He_2 and H_3 from existing, while permitting such other molecules as H_2O to be stable.

Every He atom in its ground state has a 1*s* electron of each spin. If it is to join with another He atom by exchanging electrons, each atom will have two electrons with the same spin for part of the time. That is, one atom will have both electron spins up (↑↑) and the other will have both spins down (↓↓). The exclusion principle, of course, prohibits two 1*s* electrons in an atom from having the same spins, which is manifested in a repulsion between He atoms. Hence the He_2 molecule cannot exist.

A similar argument holds in the case of H_3. An H_2 molecule contains two 1*s* electrons whose spins are antiparallel (↑↓). Should another H atom approach whose electron spin is, say, up, the resulting molecule would have two spins parallel (↑↑↓), and this is impossible if all three electrons are to be in 1*s* states. Hence the existing H_2 molecule repels the additional H atom. The exclusion-principle argument does not apply if one of the three electrons in H_3 is in an excited state. All such states are of higher energy than the 1*s* state, however, and the resulting configuration therefore has more energy than H_2 + H and so will decay rapidly to H_2 + H.

Molecular Bonds

The interaction between two atoms that gives rise to a covalent bond between them may involve probability-density distributions for the participating electrons that are different from those of Fig. 6.12 for atoms alone in space. Figure 8.9 shows the

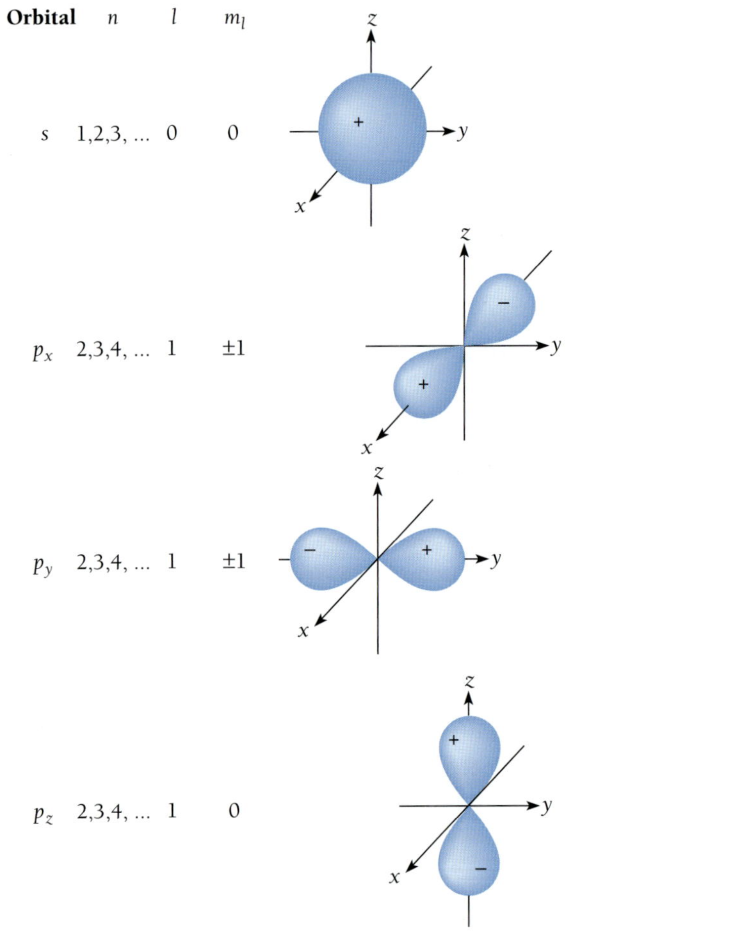

Figure 8.9 Boundary surface diagrams for *s* and *p* atomic orbitals. Each orbital can "contain" two electrons. There is a high probability of finding an electron described by one of these orbitals in the shaded regions. The sign of the wave function in each lobe is indicated.

configurations of the s and p atomic orbitals important in bond formation. What are drawn are boundary surfaces of constant $|\psi|^2 = |R\Theta\Phi|^2$ that outline the regions within which the probability of finding the electron has some definite value, say 90 or 95 percent. The diagrams thus show $|\Theta\Phi|^2$ in each case; Fig. 6.11 gives the corresponding radial probabilities. The sign of the wave function ψ is indicated in each lobe of the orbitals.

In Fig. 8.9 the s and p_z orbitals are the same as the hydrogen-atom wave functions for s and p ($m_l = 0$) states. The p_x and p_y orbitals are linear combinations of the p ($m_l = +1$) and p ($m_l = -1$) orbitals, where

$$\psi_{p_x} = \frac{1}{\sqrt{2}}(\psi_{+1} + \psi_{-1}) \qquad \psi_{p_y} = \frac{1}{\sqrt{2}}(\psi_{+1} - \psi_{-1}) \tag{8.2}$$

The $1/\sqrt{2}$ factors are needed to normalize the wave functions. Because the energies of the $m_l = +1$ and $m_l = -1$ orbitals are the same, the superpositions of the wave functions in Eq. (8.2) are also solutions of Schrödinger's equation (see Sec. 5.4).

When two atoms come together, their orbitals overlap. If the result is an increased $|\psi|^2$ between them, the combined orbitals constitute a bonding molecular orbital. In Sec. 8.4 we saw how the 1s orbitals of two hydrogen atoms could join to form the bonding orbital ψ_S. Molecular bonds are classified by Greek letters according to their angular momenta L about the bond axis, which is taken to be the z axis: σ (the Greek equivalent of s) corresponds to $L = 0$, π (the Greek equivalent of p) corresponds to $L = \hbar$, and so on in alphabetic order.

Figure 8.10 shows the formation of σ and π bonding molecular orbitals from s and p atomic orbitals. Evidently ψ_S for H_2 is an $ss\sigma$ bond. Since the lobes of p_z orbitals are on the bond axis, they form σ molecular orbitals; the p_x and p_y orbitals usually form π molecular orbitals.

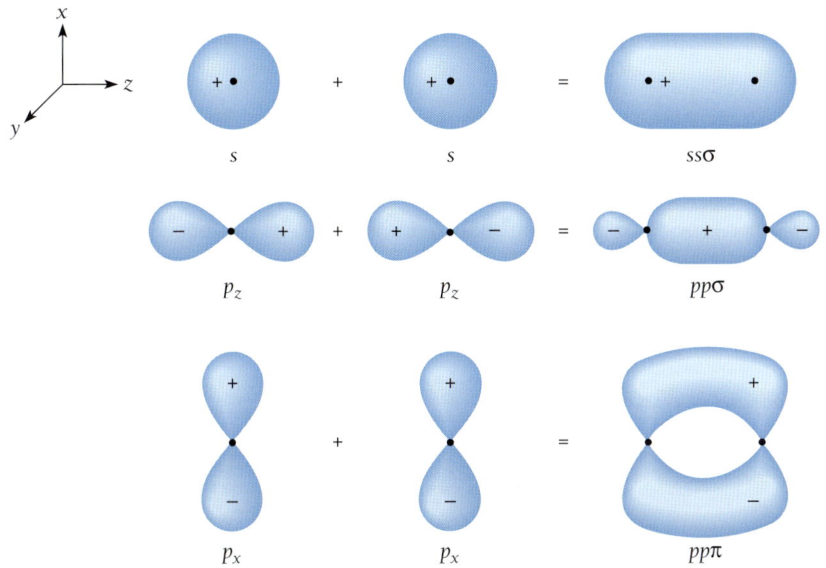

Figure 8.10 The formation of $ss\sigma$, $pp\sigma$, and $pp\pi$ bonding molecular orbitals. Two p_y atomic orbitals can combine to form a $pp\sigma$ molecular orbital in the same way as shown for two p_x atomic orbitals but with a different orientation.

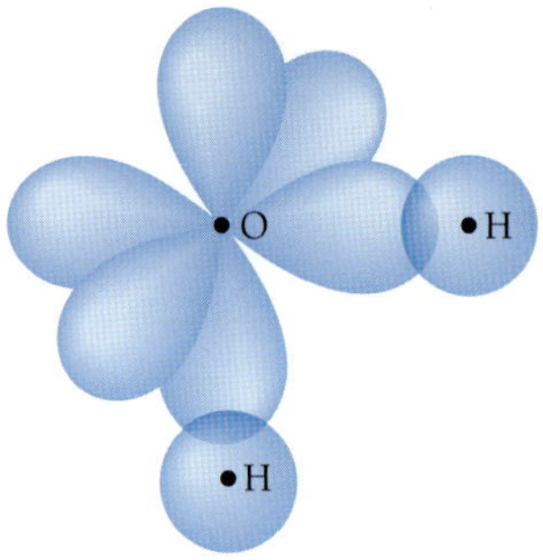

Figure 8.11 Formation of an H_2O molecule. Overlaps represent $sp\sigma$ covalent bonds. The angle between the bonds is 104.5°.

The atomic orbitals that combine to form a molecular orbital may be different in the two atoms. An example is the water molecule H_2O. Although one $2p$ orbital in O is fully occupied by two electrons, the other two $2p$ orbitals are only singly occupied and so can join with the $1s$ orbitals of two H atoms to form $sp\sigma$ bonding orbitals (Fig. 8.11). The mutual repulsion between the H nuclei (which are protons) widens the angles between the bond axes from 90° to the observed 104.5°.

Hybrid Orbitals

The straightforward way in which the shape of the H_2O molecule is explained fails in the case of methane, CH_4. A carbon atom has two electrons in its $2s$ orbital and one electron in each of two $2p$ orbitals. Thus we would expect the hydride of carbon to be CH_2, with two $sp\sigma$ bonding orbitals and a bond angle of a little over 90°. The $2s$ electrons should not participate in the bonding at all. Yet CH_4 exists and is perfectly symmetrical in structure with tetrahedral molecules whose C—H bonds are exactly equivalent to one another.

The problem of CH_4 (and those of many other molecules) was solved by Linus Pauling in 1928. He proposed that linear combinations of *both* the $2s$ and $2p$ atomic orbitals of C contribute to *each* molecular orbital in CH_4. The $2s$ and $2p$ wave functions are both solutions of the same Schrödinger's equation if the corresponding energies are the same, which is not true in the isolated C atom. However, in an actual CH_4 molecule the electric field experienced by the outer C electrons is affected by the nearby H nuclei, and the energy difference between $2s$ and $2p$ states then can disappear. **Hybrid orbitals** that consist of mixtures of s and p orbitals occur when the bonding energies they produce are greater than those which pure orbitals would produce. In CH_4 the four hybrid orbitals are mixtures of one $2s$ and three $2p$ orbitals, and accordingly are called sp^3 hybrids (Fig. 8.12). The wave functions of these hybrid orbitals are

$$\psi_1 = \frac{1}{2}(\psi_s + \psi_{p_x} + \psi_{p_y} + \psi_{p_z}) \qquad \psi_3 = \frac{1}{2}(\psi_s + \psi_{p_x} - \psi_{p_y} - \psi_{p_z})$$

$$\psi_2 = \frac{1}{2}(\psi_s - \psi_{p_x} - \psi_{p_y} + \psi_{p_z}) \qquad \psi_4 = \frac{1}{2}(\psi_s - \psi_{p_x} + \psi_{p_y} - \psi_{p_z})$$

Figure 8.13 shows the resulting structure of the CH_4 molecule.

Two other types of a hybrid orbital in addition to sp^3 can occur in carbon atoms. In sp^2 hybridization, one outer electron is in a pure p orbital and the other three are

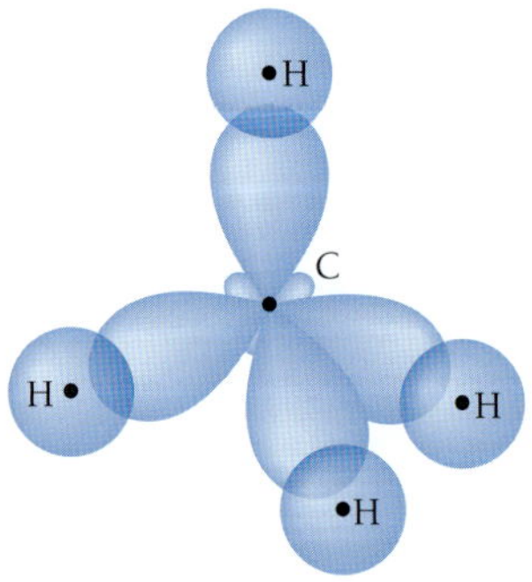

Figure 8.13 The bonds in the CH_4 (methane) molecule involve sp^3 hybrid orbitals.

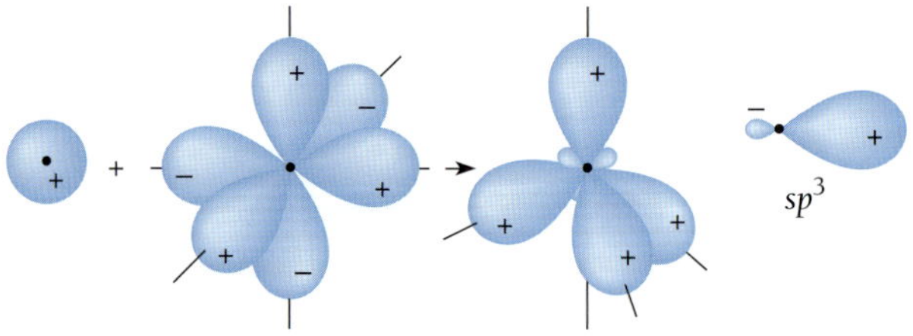

Figure 8.12 In sp^3 hybridization, an s orbital and three p orbitals in the same atom combine to form four sp^3 hybrid orbitals.

About Scientists

Linus Pauling (1901–1994), a native of Oregon, received his Ph.D. from the California Institute of Technology and remained there for his entire scientific career except for a period in the middle 1920s when he was in Germany to study the new quantum mechanics. A pioneer in the application of quantum theory to chemistry, he provided many of the key insights that permitted the details of chemical bonding to be understood. His *The Nature of the Chemical Bond* has been one of the most influential books in the history of science. Pauling also did important work in molecular biology, in particular in the study of protein structure: with the help of x-ray diffraction, he discovered the helical and pleated sheet forms that protein molecules can have. It was Pauling who realized that sickle cell anemia is a "molecular disease" due to hemoglobin with one wrong amino acid resulting from a genetic fault. He received the Nobel Prize in chemistry in 1954.

In 1923 Pauling met Ava Helen Miller in a chemistry class, and she married him despite his admission that "If I had to choose between you and science, I'm not sure that I would choose you." She introduced him to the world outside the laboratory, and he became more and more politically active in his later years. Pauling fought to stop the atmospheric testing of nuclear weapons with its attendant radioactive fallout, a crusade that did not endear him to Caltech or to the FBI, whose file on him grew to 2500 pages. Elsewhere his ideas were better received in the forms of a nuclear test ban treaty and the Nobel Peace Prize. Pauling championed large daily doses of vitamin C as an aid to good health, an idea rejected at first by the medical establishment but eventually shown to have much in its favor. He died at ninety-three of cancer, certain that vitamin C had prolonged his life.

in hybrid orbitals that are $\frac{1}{3}s$ and $\frac{2}{3}p$ in character. In *sp* hybridization, two outer electrons are in pure *p* orbitals and the other two are in hybrid orbitals that are $\frac{1}{2}s$ and $\frac{1}{2}p$ in character.

Ethylene, C_2H_4, is an example of sp^2 hybridization in which the two C atoms are joined by two bonds, one a σ bond and one a π bond (Fig. 8.14). The conventional structural formula of ethylene shows these two bonds:

Ethylene

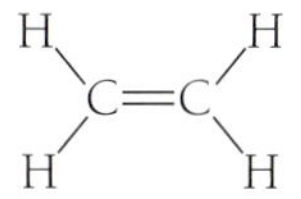

The electrons in the π bond are "exposed" outside the molecule, so ethylene and similar compounds are much more reactive chemically than compounds whose molecules have only σ bonds between their C atoms.

In benzene, C_6H_6, the six C atoms are arranged in a flat hexagonal ring, as in Fig. 8.15, with three sp^2 orbitals per C atom forming σ bonds with each other and with the H atoms. This leaves each C atom with one 2*p* orbital. The total of six 2*p* orbitals in the molecule combine into bonding π orbitals that are continuous above and below the plane of the ring. The six electrons involved belong to the molecule as a whole and not to any particular pair of atoms; these electrons are **delocalized.** An appropriate structural formula for benzene is therefore

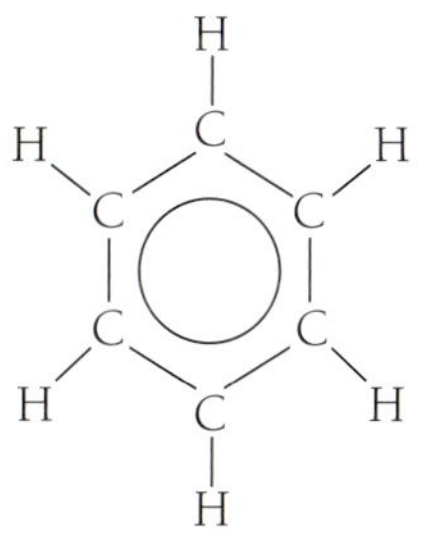

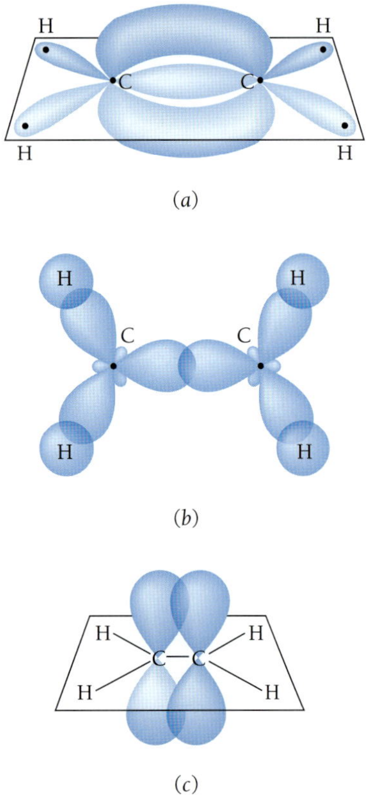

Figure 8.14 (*a*) The ethylene (C_2H_4) molecule. All the atoms lie in a plane perpendicular to the plane of the paper. (*b*) Top view, showing the sp^2 hybrid orbitals that form σ bonds between the C atoms and between each of the C atoms. (*c*) Side view, showing the pure p_x orbitals that form a π bond between the C atoms.

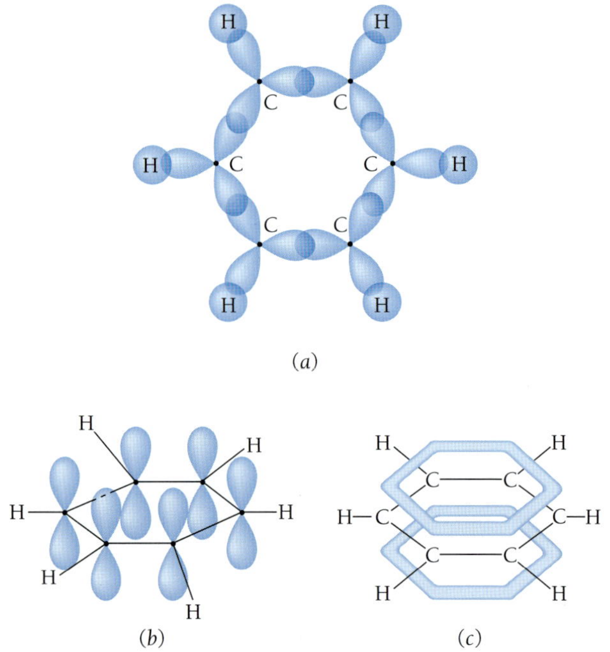

Figure 8.15 The benzene molecule. (*a*) The overlaps between the sp^2 hybrid orbitals in the C atoms with each other and with the *s* orbitals of the H atoms lead to σ bonds. (*b*) Each C atom has a pure p_x orbital occupied by one electron. (*c*) The bonding π molecular orbitals formd by the six p_x atomic orbitals constitute a continuous electron probability distribution around the molecule that contains six delocalized electrons.

About Scientists

Dorothy Crowfoot Hodgkin (1910– 1994) was fascinated at the age of ten by the growth of crystals in alum and copper sulfate solutions as their solvent water evaporated. This fascination with crystals never left her. She studied chemistry at Oxford University despite the difficulties women students of science had to face in those days, and as an undergraduate had mastered x-ray crystallography well enough to have a research paper published. In this technique a narrow beam of x-rays is directed at a crystal from various angles and the resulting interference patterns are analyzed to yield the arrangement of the atoms in the crystal. Dorothy Crowfoot (as she then was) went on to Cambridge University to work with J. D. Bernal, who had just begun to use x-rays to investigate biological molecules. Under the right conditions many such molecules form crystals from whose structures the structures of the molecules themselves can be inferred. In particular, the structures of protein molecules are important because they are closely related to their biological functions. She and Bernal were the first to map the arrangement of the atoms in a protein, the digestive enzyme pepsin.

After two intense years at Cambridge, Dorothy Crowfoot returned to Oxford where she married Thomas Hodgkin and had three children while continuing active research. Her most notable work was on penicillin (then the most complex molecule to be successfully analyzed), vitamin B_{12}, and insulin (it took thirty-five years of on-and-off effort to finish the job). She was a pioneer in using computers to interpret x-ray data, an arduous task for all but the simplest molecules. For all her achievements and their recognition in the scientific world, Hodgkin was for many years shabbily treated at Oxford: poor laboratory facilities, the lowest possible official status, half the pay of her male colleagues with continual worries about making ends meet until outside support (much of it from the Rockefeller Foundation of the United States) became available. She received the Nobel Prize in chemistry in 1964, the third woman to do so.

8.6 ROTATIONAL ENERGY LEVELS

Molecular rotational spectra are in the microwave region

Molecular energy states arise from the rotation of a molecule as a whole, from the vibrations of its atoms relative to one another, and from changes in its electronic configuration:

1 *Rotational states* are separated by quite small energy intervals (10^{-3} eV is typical). The spectra that arise from transitions between these states are in the microwave region with wavelengths of 0.1 mm to 1 cm. The absorption by water molecules of rotational energy from microwaves underlies the operation of microwave ovens.
2 *Vibrational states* are separated by somewhat larger energy intervals (0.1 eV is typical). Vibrational spectra are in the infrared region with wavelengths of 1 μm to 0.1 mm.
3 *Molecular electronic states* have the highest energies, with typical separations between the energy levels of outer electrons of several eV. The corresponding spectra are in the visible and ultraviolet regions.

A detailed picture of a particular molecule can often be obtained from its spectrum, including bond lengths, force constants, and bond angles. For simplicity, the treatment here will cover only diatomic molecules, but the main ideas apply to more complicated ones as well.

The lowest energy levels of a diatomic molecule arise from rotation about its center of mass. We may picture such a molecule as consisting of atoms of masses m_1 and m_2 a distance R apart, as in Fig. 8.16. The moment of inertia of this molecule about an axis passing through its center of mass and perpendicular to a line joining the atoms is

$$I = m_1 r_1^2 + m_2 r_2^2 \tag{8.3}$$

where r_1 and r_2 are the distances of atoms 1 and 2, respectively, from the center of mass. From the definition of center of mass,

$$m_1 r_1 = m_2 r_2 \tag{8.4}$$

Hence the moment of inertia may be written

Moment of inertia
$$I = \frac{m_1 m_2}{m_1 + m_2}(r_1 + r_2)^2 = m' R^2 \tag{8.5}$$

Here

Reduced mass
$$m' = \frac{m_1 m_2}{m_1 + m_2} \tag{8.6}$$

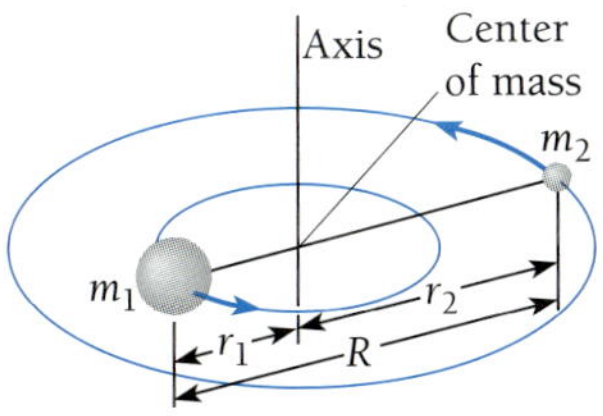

Figure 8.16 A diatomic molecule can rotate about its center of mass.

is the **reduced mass** of the molecule. Equation (8.5) states that the rotation of a diatomic molecule is equivalent to the rotation of a single particle of mass m' about an axis located a distance R away.

The angular momentum **L** of the molecule has the magnitude

$$L = I\omega \tag{8.7}$$

where ω is its angular velocity. Angular momentum is always quantized in nature, as we know. If we denote the **rotational quantum number** by J, we have here

Angular momentum

$$L = \sqrt{J(J+1)}\hbar \qquad J = 0, 1, 2, 3, \ldots \tag{8.8}$$

The energy of a rotating molecule is $\frac{1}{2}I\omega^2$, and so its energy levels are specified by

Rotational energy levels

$$E_J = \frac{1}{2}I\omega^2 = \frac{L^2}{2I}$$

$$= \frac{J(J+1)\hbar^2}{2I} \tag{8.9}$$

Rotations about the Bond Axis

We have been considering only rotation about an axis perpendicular to the bond axis of a diatomic molecules, as in Fig 8.16—end-over-end rotations. What about rotations about the axis of symmetry itself?

Such rotations can be neglected because the mass of an atom is located almost entirely in its nucleus, whose radius is only $\sim 10^{-4}$ of the radius of the atom itself. The main contribution to the moment of inertia of a diatomic molecule about its bond axis therefore comes from its electrons, which are concentrated in a region whose radius about the axis is roughly half the bond length R but whose total mass is only about $\frac{1}{4000}$ of the total molecular mass. Since the allowed rotational energy levels are proportional to $1/I$, rotation about the symmetry axis must involve energies $\sim 10^4$ times the E_J values for end-over-end rotations. Hence energies of at least several eV would be involved in any rotation about the symmetry axis of a diatomic molecule. Bond energies are also of this order of magnitude, so the molecule would be likely to dissociate in any environment in which such a rotation could be excited.

Example 8.1

The carbon monoxide (CO) molecule has a bond length R of 0.113 nm and the masses of the ^{12}C and ^{16}O atoms are respectively 1.99×10^{-26} kg and 2.66×10^{-26} kg. Find (*a*) the energy and (*b*) the angular velocity of the CO molecule when it is in its lowest rotational state.

Solution

(*a*) The reduced mass m' of the CO molecule is

$$m' = \frac{m_1 m_2}{m_1 + m_2} = \left[\frac{(1.99)(2.66)}{1.99 + 2.66}\right] \times 10^{-26}\ \text{kg}$$

$$= 1.14 \times 10^{-26}\ \text{kg}$$

and its moment of inertia I is

$$I = m'R^2 = (1.14 \times 10^{-26}\ \text{kg})(1.13 \times 10^{-10}\ \text{m})^2$$
$$= 1.46 \times 10^{-46}\ \text{kg} \cdot \text{m}^2$$

The lowest rotational energy level corresponds to $J = 1$, and for this level in CO

$$E_{J=1} = \frac{J(J+1)\hbar^2}{2I} = \frac{\hbar^2}{I} = \frac{(1.054 \times 10^{-34}\ \text{J} \cdot \text{s})^2}{1.46 \times 10^{-46}\ \text{kg} \cdot \text{m}^2}$$
$$= 7.61 \times 10^{-23}\ \text{J} = 4.76 \times 10^{-4}\ \text{eV}$$

This is not a lot of energy, and at room temperature, when $kT \approx 2.6 \times 10^{-2}$ eV, nearly all the molecules in a sample of CO are in excited rotational states.
(*b*) The angular velocity of the CO molecule when $J = 1$ is

$$\omega = \sqrt{\frac{2E}{I}} = \sqrt{\frac{(2)(7.61 \times 10^{-23}\ \text{J})}{1.46 \times 10^{-46}\ \text{kg} \cdot \text{m}^2}}$$
$$= 1.02 \times 10^{12}\ \text{rad/s}$$

Rotational Spectra

Rotational spectra arise from transitions between rotational energy states. Only molecules that have electric dipole moments can absorb or emit electromagnetic photons in such transitions. For this reason nonpolar diatomic molecules such as H_2 and symmetric polyatomic molecules such as CO_2 (O=C=O) and CH_4 (Fig. 8.13) do not exhibit rotational spectra. Transitions between rotational states in molecules like H_2, CO_2, and CH_4 can take place during collisions, however.

Even in molecules with permanent dipole moments, not all transitions between rotational states involve radiation. As in the case of atomic spectra, certain selection rules summarize the conditions for a radiative transition between rotational states to be possible. For a rigid diatomic molecule the selection rule for rotational transitions is

Selection rule
$$\Delta J = \pm 1 \tag{8.10}$$

In practice, rotational spectra are always obtained in absorption, so that each transition that is found involves a change from some initial state of quantum number J to the next higher state of quantum number $J + 1$. In the case of a rigid molecule, the frequency of the absorbed photon is

$$\nu_{J \to J+1} = \frac{\Delta E}{h} = \frac{E_{J+1} - E_J}{h}$$

Rotational spectra
$$= \frac{\hbar}{2\pi I}(J+1) \tag{8.11}$$

where I is the moment of inertia for end-over-end rotations. The spectrum of a rigid molecule therefore consists of equally spaced lines, as in Fig. 8.17. The frequency of each line can be measured, and the transition it corresponds to can often be found from the sequence of lines. From these data the moment of inertia of the molecule can be calculated. Alternatively, the frequencies of any two successive lines may be used to determine I if the lowest-frequency lines in a particular spectral sequence are not recorded.

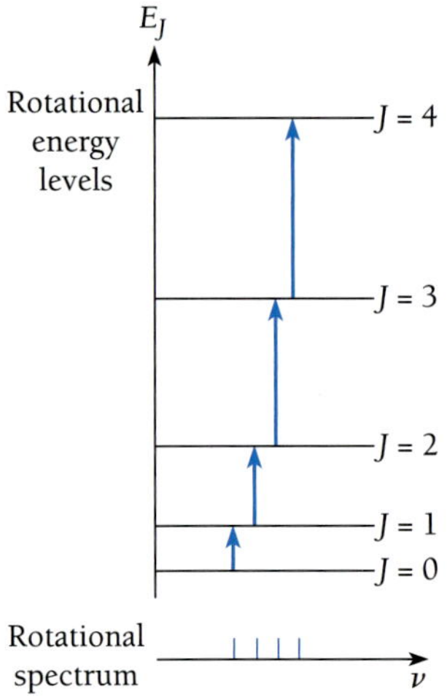

Figure 8.17 Energy levels and spectrum of molecular rotation.

Example 8.2

In CO the $J = 0 \rightarrow J = 1$ absorption line occurs at a frequency of 1.15×10^{11} Hz. What is the bond length of the CO molecule?

Solution

First we find the moment of inertia of this molecule from Eq. (8.11):

$$I_{CO} = \frac{\hbar}{2\pi\nu}(J+1) = \frac{1.054 \times 10^{-34}\ \text{J}\cdot\text{s}}{(2\pi)(1.15 \times 10^{11}\ \text{s}^{-1})} = 1.46 \times 10^{-46}\ \text{kg}\cdot\text{m}^2$$

In Example 8.1 we saw that the reduced mass of the CO molecule is $m' = 1.14 \times 10^{-26}$ kg. From Eq. (8.5), $I = m'R^2$, we obtain

$$R_{CO} = \sqrt{\frac{I}{m'}} = \sqrt{\frac{1.46 \times 10^{-46}\ \text{kg}\cdot\text{m}^2}{1.14 \times 10^{-26}\ \text{kg}}} = 1.13 \times 10^{-10}\ \text{m} = 0.113\ \text{nm}$$

This is the way in which the bond length for CO quoted earlier was determined.

8.7 VIBRATIONAL ENERGY LEVELS

A molecule may have many different modes of vibration

When sufficiently excited, a molecule can vibrate as well as rotate. Figure 8.18 shows how the potential energy of a diatomic molecule varies with the internuclear distance R. Near the minimum of this curve, which corresponds to the normal configuration of the molecule, the shape of the curve is very nearly a parabola. In this region, then,

Parabolic approximation $$U = U_0 + \frac{1}{2}k(R - R_0)^2 \qquad (8.12)$$

where R_0 is the equilibrium separation of the atoms.

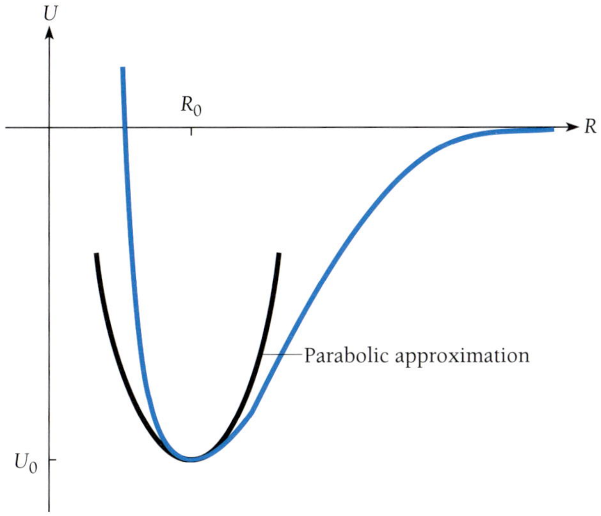

Figure 8.18 The potential energy of a diatomic molecule as a function of internuclear distance.

The interatomic force that gives rise to this potential energy is given by differentiating U:

$$F = -\frac{dU}{dR} = -k(R - R_0) \tag{8.13}$$

The force is just the restoring force that a stretched or compressed spring exerts—a Hooke's law force—and, as with a spring, a molecule suitably excited can undergo simple harmonic oscillations.

Classically, the frequency of a vibrating body of mass m connected to a spring of force constant k is

$$\nu_0 = \frac{1}{2\pi}\sqrt{\frac{k}{m}} \tag{8.14}$$

What we have in the case of a diatomic molecule is the somewhat different situation of two bodies of masses m_1 and m_2 joined by a spring, as in Fig. 8.19. In the absence

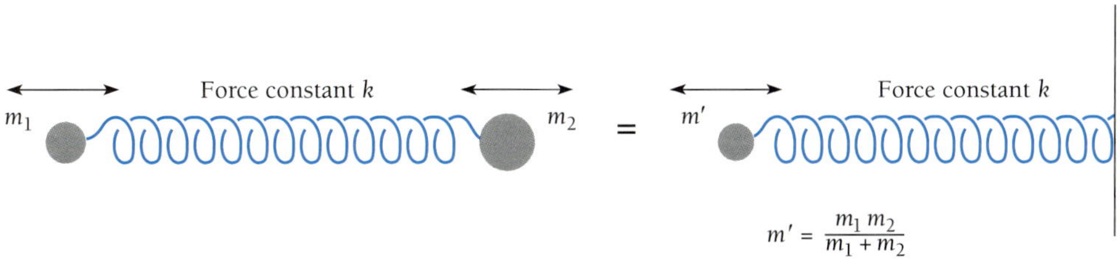

Figure 8.19 A two-body oscillator behaves like an ordinary harmonic oscillator with the same spring constant but with the reduced mass m′.

of external forces the linear momentum of the system remains constant, and the oscillations of the bodies therefore cannot effect the motion of their center of mass. For this reason m_1 and m_2 vibrate back and forth relative to their center of mass in opposite directions, and both reach the extremes of their respective motions at the same time. The frequency of oscillation of such a two-body oscillator is given by Eq. (8.14) with the reduced mass m' of Eq. (8.6) substituted for m:

Two-body oscillator

$$\nu_0 = \frac{1}{2\pi}\sqrt{\frac{k}{m'}} \tag{8.15}$$

When the harmonic-oscillator problem is solved quantum mechanically (see Sec. 5.11), the energy of the oscillator turns out to be restricted to the values

Harmonic oscillator

$$E_v = (v + \tfrac{1}{2})h\nu_0 \tag{8.16}$$

where v, the **vibrational quantum number**, may have the values

Vibrational quantum number

$$v = 0, 1, 2, 3, \ldots$$

The lowest vibrational state ($v = 0$) has the zero-point energy $\frac{1}{2}h\nu_0$, not the classical value of 0. This result is in accord with the uncertainty principle, because if the oscillating particle were stationary, the uncertainty in its position would be $\Delta x = 0$ and its momentum uncertainty would then have to be infinite—and a particle with $E = 0$ cannot have an infinitely uncertain momentum. In view of Eq. (8.15) the vibrational energy levels of a diatomic molecule are specified by

Vibrational energy levels

$$E_v = (v + \tfrac{1}{2})\hbar\sqrt{\frac{k}{m'}} \tag{8.17}$$

The higher vibrational states of a molecule do not obey Eq. (8.16) because the parabolic approximation to its potential-energy curve becomes less and less valid with increasing energy. As a result, the spacing between adjacent energy levels of high v is less than the spacing between adjacent levels of low v, which is shown in Fig. 8.20.

About Scientists

Gerhard Herzberg (1904–1999) was born in Hamburg, Germany, and received his doctorate from the Technical University of Darmstadt in 1928. The rise to power of the Nazis led Herzberg to leave Germany in 1935 for Canada, where he joined the University of Saskatchewan. From 1945 to 1948 he was at Yerkes Observatory in Wisconsin, and after that he directed the Division of Pure Physics of Canada's National Research Council in Ottawa until he retired in 1969. Herzberg was a pioneer in using spectra to determine molecular structures, and also did important work in analyzing the spectra of stars, interstellar gas, comets, and planetary atmospheres. His books under the general title *Molecular Spectra and Molecular Structure* are classics in the field. He received the Nobel Prize in chemistry in 1971.

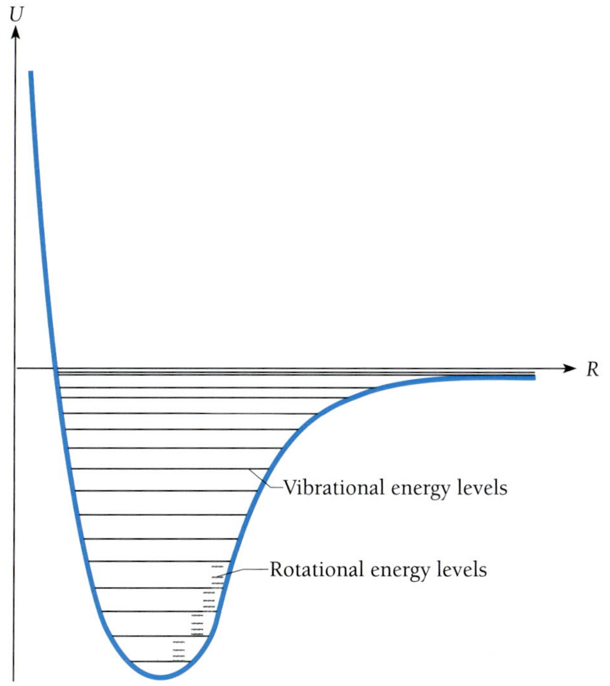

Figure 8.20 The potential energy of a diatomic molecule as a function of interatomic distance, showing vibrational and rotational energy levels.

This diagram also shows the fine structure in the vibrational levels caused by the simultaneous excitation of rotational levels.

Vibrational Spectra

The selection rule for transitions between vibrational states is

Selection rule
$$\Delta v = \pm 1 \tag{8.18}$$

in the harmonic-oscillator approximation. This rule is easy to understand. An oscillating dipole whose frequency is ν_0 can absorb or emit only electromagnetic radiation of the same frequency and all quanta of frequency ν_0 have the energy $h\nu_0$. The oscillating dipole accordingly can only absorb $\Delta E = h\nu_0$ at a time, in which case its energy increases from $(v + \frac{1}{2})h\nu_0$ to $(v + \frac{1}{2} + 1)h\nu_0$. It can also emit only $\Delta E = h\nu_0$ at a time, in which case its energy decreases from $(v + \frac{1}{2})h\nu_0$ to $(v + \frac{1}{2} - 1)h\nu_0$. Hence the selection rule $\Delta v = \pm 1$.

Example 8.3

When CO is dissolved in liquid carbon tetrachloride, infrared radiation of frequency 6.42×10^{13} Hz is absorbed. Carbon tetrachloride by itself is transparent at this frequency, so the absorption must be due to the CO. (*a*) What is the force constant of the bond in the CO molecule? (*b*) What is the spacing between its vibrational energy levels?

Solution

(*a*) As we know, the reduced mass of the CO molecule is $m' = 1.14 \times 10^{-26}$ kg. From Eq. (8.15), $\nu_0 = (1/2\pi)\sqrt{k/m'}$, the force constant is

$$k = 4\pi^2\nu_0^2 m' = (4\pi^2)(6.42 \times 10^{13}\ \text{Hz})^2(1.14 \times 10^{-26}\ \text{kg})$$
$$= 1.86 \times 10^3\ \text{N/m}$$

This is about 10 lb/in.

(*b*) The separation ΔE between the vibrational levels in CO is

$$\Delta E = E_{v+1} - E_v = h\nu_0 = (6.63 \times 10^{-34}\ \text{J}\cdot\text{s})(6.42 \times 10^{13}\ \text{Hz})$$
$$= 4.26 \times 10^{-20}\ \text{J} = 0.266\ \text{eV}$$

This is considerably more than the spacing between its rotational energy levels. Because $\Delta E > kT$ for vibrational states in a sample at room temperature, most of the molecules in such a sample exist in the $v = 0$ state with only their zero-point energies. This situation is very different from that characteristic of rotational states, where the much smaller energies mean that the majority of the molecules in a room-temperature sample are excited to higher states.

A complex molecule may have many different modes of vibration. Some of these modes involve the entire molecule (Figs. 8.21 and 8.22), but others ("local modes") involve only groups of atoms whose vibrations occur more or less independently of the rest of the molecule. Thus the —OH group has a characteristic vibrational frequency of 1.1×10^{14} Hz and the —NH_2 gssroup has a frequency of 1.0×10^{14} Hz.

The characteristic vibrational frequency of a carbon-carbon group depends upon the number of bonds between the C atoms: the >C—C< group vibrates at about 3.3×10^{13} Hz, the >C=C< group vibrates at about 5.0×10^{13} Hz, and the —C≡C— group vibrates at about 6.7×10^{13} Hz. (As we would expect, the more carbon-carbon bonds, the larger the force constant k and the higher the frequency.) In each case the frequency does not depend strongly on the particular molecule or the location in the molecule of the group, which makes vibrational spectra a valuable tool in determining molecular structures.

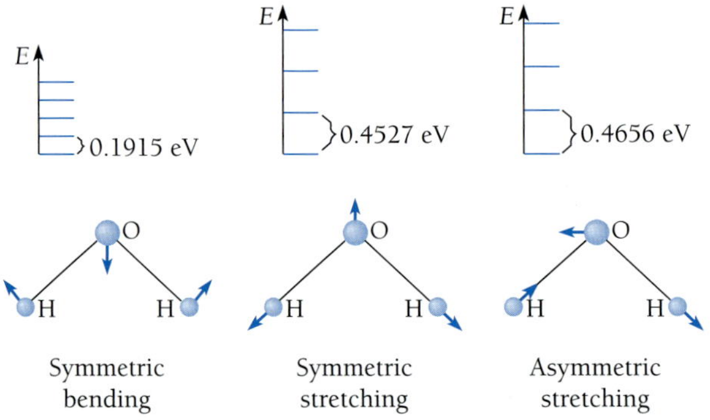

Figure 8.21 The normal modes of vibration of the H_2O molecule and the energy levels of each mode. More energy is needed to stretch the molecule than to bend it, which is generally true.

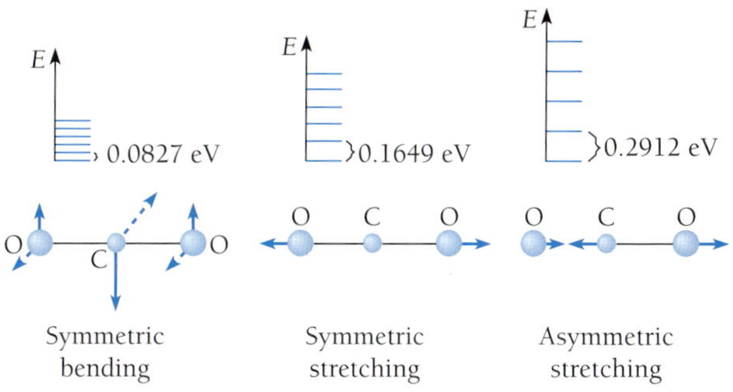

Figure 8.22 The normal modes of vibration of the CO_2 molecule and the energy levels of each mode. The symmetric bending mode can occur in two perpendicular planes. In this molecule the O atoms are negatively charged and the C atom is positively charged. The symmetric streching mode cannot be initiated by the absorption of a photon because the overall charge distribution in the molecule does not change in this mode. In the other modes of vibration, however, the charge distribution does change and the molecule can absorb photons of appropriate wavelength (4.26 μm and 15.00 μm for the asymmetric stretching and symmetric bending modes, respectively). The absorption of infrared radiation from the earth by atmospheric CO_2 molecules is partly responsible for the greenhouse effect (see Fig. 9.8), and the increase in the CO_2 content of the atmosphere due to the burning of fossil fuels seems to be the chief cause of the global warming trend now under way. Other molecules in the atmosphere, such as H_2O and CH_4 (methane), also contribute to the greenhouse effect, but N_2 and O_2 do not because, since their overall charge distributions do not change when they vibrate, they do not absorb infrared radiation.

An example is thioacetic acid, whose structure might conceivably be either $CH_3CO—SH$ or $CH_3CS—OH$. The infrared absorption spectrum of thioacetic acid contains lines at frequencies equal to the vibrational frequencies of the $>C{=}O$ and and —SH groups, but no lines corresponding to the $>C{=}S$ or —OH groups. The first alternative is evidently the correct one.

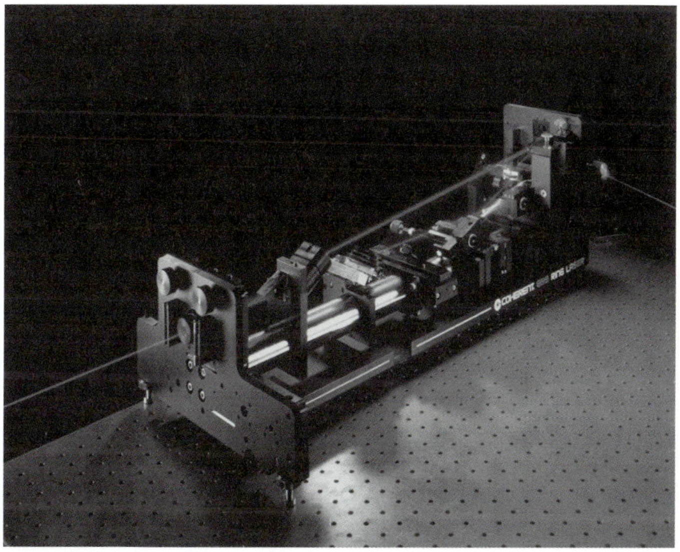

This tunable dye laser emits light with wavelengths from 370 to 900 nm, which includes the entire visible spectrum. The bandwidth can be as narrow as 500 kHz.

Vibration-Rotation Spectra

Pure vibrational spectra are observed only in liquids where interactions between adjacent molecules inhibit rotation. Because the excitation energies involved in molecular rotation are much smaller than those involved in vibration, the freely moving molecules in a gas or vapor nearly always are rotating, regardless of their vibrational state. The spectra of such molecules do not show isolated lines corresponding to each vibrational transition, but instead a large number of closely spaced lines due to transitions between the various rotational states of one vibrational level and the rotational states of the other. In spectra obtained using a spectrometer with inadequate resolution, the lines appear as a broad streak called a vibration-rotation band.

Tunable Dye Lasers

The existence of bands of extremely closely spaced lines in molecular spectra underlies the operation of the **tunable dye laser.** Such a laser uses an organic dye whose molecules are "pumped" to excited states by light from another laser. The dye then fluoresces in a broad emission band. From this band, light of the desired wavelength λ can be selected for laser amplification with the help of a pair of facing mirrors, one of them partly transparent. The separation of the mirrors is set to an integral multiple of $\lambda/2$. As in the case of the lasers discussed in Sec. 4.9, the trapped laser light forms an optical standing wave that emerges through the partly transparent mirror. A dye laser of this kind can be tuned to a precision of better than one part in a million by adjusting the spacing of the mirrors.

8.8 ELECTRONIC SPECTRA OF MOLECULES

How fluorescence and phosphorescence occur

The energies of rotation and vibration in a molecule are due to the motion of its atomic nuclei, which contain virtually all the molecule's mass. The molecule's electrons also can be excited to higher energy levels than those corresponding to its ground state. However, the spacing of these levels is much greater than the spacing of rotational or vibrational levels.

Electronic transitions involve radiation in the visible or ultraviolet parts of the spectrum. Each transition appears as a series of closely spaced lines, called a band, due to the presence of different rotational and vibrational states in each electronic state (Fig. 8.23). All molecules exhibit electronic spectra, since a dipole moment change always accompanies a change in the electronic configuration of a molecule. Therefore homonuclear molecules, such as H_2 and N_2, which have neither rotational nor

Figure 8.23 A portion of the band spectrum of PN.

vibrational spectra because they lack permanent dipole moments, nevertheless have electronic spectra whose rotational and vibrational fine structures enable moments of inertia and bond force constants to be found.

Electronic excitation in a polyatomic molecule often leads to a change in the molecule's shape, which can be determined from the rotational fine structure in its band spectrum. The origin of such changes lies in the different characters of the wave functions of electrons in different states, which lead to correspondingly different bond geometries. For example, the molecule beryllium hydride, BeH_2, is linear (H—Be—H) in one state and bent (H—Be with H attached below Be) in another.

Fluorescence

A molecule in an excited electronic state can lose energy and return to its ground state in various ways. The molecule may, of course, simply emit a photon of the same frequency as that of the photon it absorbed, thereby returning to the ground state in a single step. Another possibility is **fluorescence.** Here the molecule gives up some of its vibrational energy in collisions with other molecules, so that the downward radiative transition originates from a lower vibrational level in the upper electronic state (Fig. 8.24). Fluorescent radiation is therefore of lower frequency than that of the absorbed radiation.

Fluorescence excited by ultraviolet light has many applications, for instance to help identify minerals and biochemical compounds. Fabrissc "brighteners" that are some-

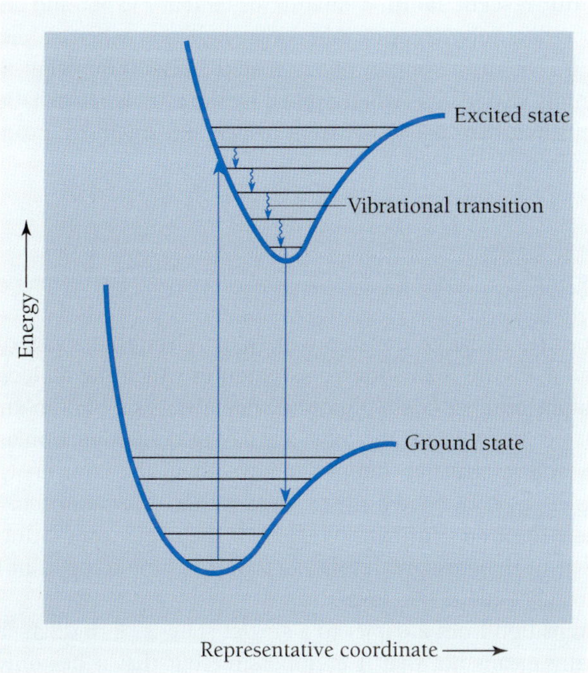

Figure 8.24 The origin of fluorescence. The emitted radiation is lower in frequency than the absorbed radiation.

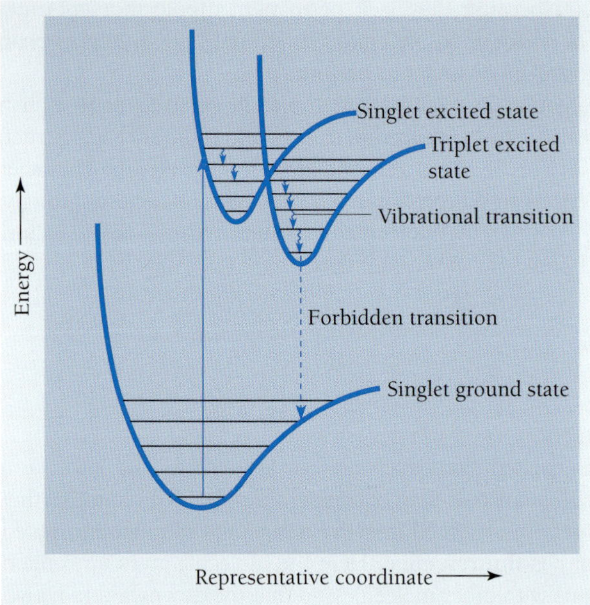

Figure 8.25 The origin of phosphorescence. The final transition is delayed because it violates the selection rules for electronic transitions.

times added to detergents absorb ultraviolet radiation in daylight and then fluoresce blue light. In a **fluorescent lamp,** a mixture of mercury vapor and an inert gas such as argon inside a glass tube gives off ultraviolet radiation when an electric current is passed through it. The inside of the tube is coated with a fluorescent material called a phosphor that emits visible light when excited by the ultraviolet radiation. The process is much more efficient than using a current to heat a filament to incandescence, as in ordinary light bulbs.

Phosphorescence

In molecular spectra, radiative transitions between electronic states of different total spin are prohibited. Figure 8.25 shows a situation in which the molecule in its singlet (total spin quantum number $S = 0$) ground state absorbs a photon and is raised to a singlet excited state. In collisions the molecule can undergo radiationless transitions to a lower vibrational level that may happen to have about the same energy as one of the levels in the triplet ($S = 1$) excited state. There is then a certain probability for a shift to the triplet state to occur. Further collisions in the triplet state bring the molecule's energy below that of the crossover point, so that it is now trapped in the triplet state and ultimately reaches the $v = 0$ level.

A radiative transition from a triplet to a singlet state is "forbidden" by the selection rules, which really means not that it is impossible but that it has only a small likelihood of occurring. Such transitions accordingly have long half-lives, and the resulting **phosphorescent radiation** may be emitted minutes or even hours after the initial absorption.

EXERCISES*

We are wiser than we know. —Ralph Waldo Emerson

8.4 The Hydrogen Molecule

1. The energy needed to detach the electron from a hydrogen atom is 13.6 eV, but the energy needed to detach an electron from a hydrogen molecule is 15.7 eV. Why do you think the latter energy is greater?

2. The protons in the H_2^+ molecular ion are 0.106 nm apart, and the binding energy of H_2^+ is 2.65 eV. What negative charge must be placed halfway between two protons this distance apart to give the same binding energy?

3. At what temperature would the average kinetic energy of the molecules in a hydrogen sample be equal to their binding energy?

8.6 Rotational Energy Levels

4. Microwave communication systems operate over long distances in the atmosphere. The same is true for radar, which locates objects such as ships and aircraft by means of microwave pulses they reflect. Molecular rotational spectra are in the microwave region. Can you think of the reason why atmospheric gases do not absorb microwaves to any great extent?

5. When a molecule rotates, inertia causes its bonds to stretch. (This is why the earth bulges at the equator.) What effects does this stretching have on the rotational spectrum of the molecule?

6. Find the frequencies of the $J = 1 \rightarrow J = 2$ and $J = 2 \rightarrow J = 3$ rotational absorption lines in NO, whose molecules have the moment of inertia 1.65×10^{-46} kg · m^2.

7. The $J = 0 \rightarrow J = 1$ rotational absorption line occurs at 1.153×10^{11} Hz in $^{12}C^{16}O$ and at 1.102×10^{11} Hz in $^{?}C^{16}O$. Find the mass number of the unknown carbon isotope.

8. Calculate the energies of the four lowest non-zero rotational energy states of the H_2 and D_2 molecules, where D represents the deuterium atom ^2_1H.

9. The rotational spectrum of HCl contains the following wavelengths:

$$12.03 \times 10^{-5} \text{ m}$$
$$9.60 \times 10^{-5} \text{ m}$$
$$8.04 \times 10^{-5} \text{ m}$$
$$6.89 \times 10^{-5} \text{ m}$$
$$6.04 \times 10^{-5} \text{ m}$$

   If the isotopes involved are ^1H and ^{35}Cl, find the distance between the hydrogen and chlorine nuclei in an HCl molecule.

10. The lines of the rotational spectrum of HBr are 5.10×10^{11} Hz apart in frequency. Find the internuclear distance in HBr. (*Note:* Since the Br atom is about 80 times more massive than the proton, the reduced mass of an HBr molecule can be taken as just the ^1H mass.)

11. A $^{200}Hg^{35}Cl$ molecule emits a 4.4-cm photon when it undergoes a rotational transition from $J = 1$ to $J = 0$. Find the interatomic distance in this molecule.

12. The lowest frequency in the rotational absorption spectrum of $^1H^{19}F$ is 1.25×10^{12} Hz. Find the bond length in this molecule.

13. In Sec. 4.6 it was shown that, for large quantum numbers, the frequency of the radiation from a hydrogen atom that drops from an initial state of quantum number n to a final state of quantum number $n - 1$ is equal to the classical frequency of revolution of an electron in the nth Bohr orbit. This is an example of Bohr's correspondence principle. Show that a similar correspondence holds for a diatomic molecule rotating about its center of mass.

14. Calculate the classical frequency of rotation of a rigid body whose energy is given by Eq. (8.9) for states of $J = J$ and $J = J + 1$, and show that the frequency of the spectral line associated with a transition between these states is intermediate between the rotational frequencies of the states.

8.7 Vibrational Energy Levels

15. The hydrogen isotope deuterium has an atomic mass approximately twice that of ordinary hydrogen. Does H_2 or HD have the greater zero-point energy? How does this affect the binding energies of the two molecules?

16. Can a molecule have zero vibrational energy? Zero rotational energy?

17. The force constant of the $^1H^{19}F$ molecule is approximately 966 N/m. (*a*) Find the frequency of vibration of the molecule. (*b*) The bond length in $^1H^{19}F$ is approximately 0.92 nm. Plot the potential energy of this molecule versus internuclear distance in the vicinity of 0.92 nm and show the vibrational energy levels as in Fig. 8.20.

18. Assume that the H_2 molecule behaves exactly like a harmonic oscillator with a force constant of 573 N/m. (*a*) Find the energy (in eV) of its ground and first excited vibrational states. (*b*) Find the vibrational quantum number that approximately corresponds to its 4.5-eV dissociation energy.

19. The lowest vibrational states of the $^{23}Na^{35}Cl$ molecule are 0.063 eV apart. Find the approximate force constant of this molecule.

20. Find the amplitude of the ground-state vibrations of the CO molecule. What percentage of the bond length is this? Assume the molecule vibrates like a harmonic oscillator.

21. The bond between the hydrogen and chlorine atoms in a $^1H^{35}Cl$ molecule has a force constant of 516 N/m. Is it likely that an HCl molecule will be vibrating in its first excited vibrational state at room temperature?

*Atomic masses are given in the Appendix.

22. The observed molar specific heat of hydrogen gas at constant volume is plotted in Fig. 8.26 versus absolute temperature. (The temperature scale is logarithmic.) Since each degree of freedom (that is, each mode of energy possession) in a gas molecule contributes ~1 kcal/kmol · K to the specific heat of the gas, this curve is interpreted as indicating that only translational motion, with three degrees of freedom, is possible for hydrogen molecules at very low temperatures. At higher temperatures the specific heat rises to ~5 kcal/kmol · K, indicating that two more degrees of freedom are available, and at still higher temperatures the specific heat is ~7 kcal/kmol · K, indicating two further degrees of freedom. The additional pairs of degrees of freedom represent respectively rotation, which can take place about two independent axes perpendicular to the axis of symmetry of the H_2 molecule, and vibration, in which the two degrees of freedom correspond to the kinetic and potential modes of energy possession by the molecule. (*a*) Verify this interpretation of Fig. 8.26 by calculating the temperatures at which kT is equal to the minimum rotational energy and to the minimum vibrational energy an H_2 molecule can have. Assume that the force constant of the bond in H_2 is 573 N/m and that the H atoms are 7.42×10^{-11} m apart. (At these temperatures, approximately half the molecules are rotating or vibrating, respectively, though in each case some are in higher states than $J = 1$ or $v = 1$.) (*b*) To justify considering only two degrees of rotational freedom in the H_2 molecule, calculate the temperature at which kT is equal to the minimum nonzero rotational energy an H_2 molecule can have for rotation about its axis of symmetry. (*c*) How many vibrations does an H_2 molecule with $J = 1$ and $v = 1$ make per rotation?

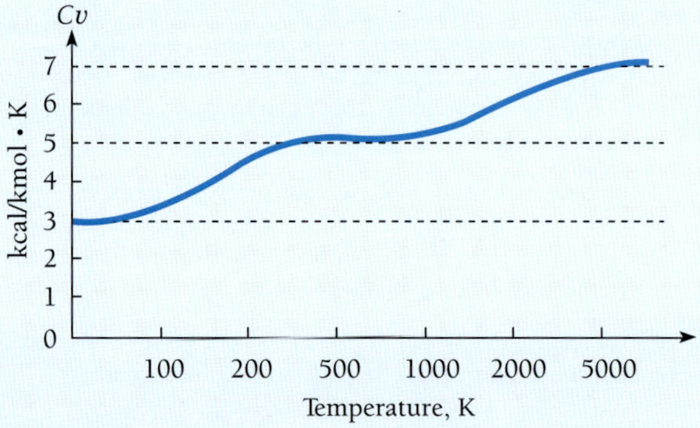

Figure 8.26 Molar specific heat of hydrogen at constant volume.

CHAPTER 9

Statistical Mechanics

The Crab Nebula is the result of a supernova explosion that was observed in A.D. 1054. The explosion left behind a star believed to consist entirely of neutrons. Statistical mechanics is needed to understand the properties of neutron stars.

The branch of physics called **statistical mechanics** considers how the overall behavior of a system of many particles is related to the properties of the particles themselves. As its name implies, statistical mechanics is not concerned with the actual motions or interactions of individual particles, but instead with what is most likely to happen. While statistical mechanics cannot help us find the life history of one of the particles in a system, it is able to tell us, for instance, the probability that the particle has a certain amount of energy at a certain moment.

Because so many phenomena in the physical world involve systems of great numbers of particles, the value of a statistical approach is clear. Owing to the generality of its arguments, statistical mechanics can be applied equally well to classical systems (notably molecules in a gas) and to quantum-mechanical systems (notably photons in a cavity and free electrons in a metal), and it is one of the most powerful tools of the theoretical physicist.

9.1 STATISTICAL DISTRIBUTIONS

Three different kinds

What statistical mechanics does is determine the most probable way in which a certain total amount of energy E is distributed among the N members of a system of particles in thermal equilibrium at the absolute temperature T. Thus we can establish how many particles are likely to have the energy ϵ_1, how many to have the energy ϵ_2, and so on.

The particles are assumed to interact with one another and with the walls of their container to an extent sufficient to establish thermal equilibrium but not so much that their motions are strongly correlated. More than one particle state may correspond to a certain energy ϵ. If the particles are not subject to the exclusion principle, more than one particle may be in a certain state.

A basic premise of statistical mechanics is that the greater the number W of different ways in which the particles can be arranged among the available states to yield a particular distribution of energies, the more probable is the distribution. It is assumed that each state of a certain energy is equally likely to be occupied. This assumption is plausible but its ultimate justification (as in the case of Schrödinger's equation) is that the conclusions arrived at with its help agree with results found through experimentation.

The program of statistical mechanics begins by finding a general formula for W for the kind of particles being considered. The most probable distribution, which corresponds to the system's being in thermal equilibrium, is the one for which W is a maximum, subject to the condition that the system consists of a fixed number N of particles (except when they are photons or their acoustic equivalent called **phonons**) whose total energy is some fixed amount E. The result in each case is an expression for $n(\epsilon)$, the number of particles with the energy ϵ, that has the form

Number of particles of energy ϵ

$$n(\epsilon) = g(\epsilon)f(\epsilon) \tag{9.1}$$

where $g(\epsilon)$ = number of states of energy ϵ
= statistical weight corresponding to energy ϵ
$f(\epsilon)$ = distribution function
= average number of particles in each state of energy ϵ
= probability of occupancy of each state of energy ϵ

When a continuous rather than a discrete distribution of energies is involved, $g(\epsilon)$ is replaced by $g(\epsilon)\,d\epsilon$, the number of states with energies between ϵ and $\epsilon + d\epsilon$.

We shall consider systems of three different kinds of particles:

1 Identical particles that are sufficiently far apart to be distinguishable, for instance, the molecules of a gas. In quantum terms, the wave functions of the particles overlap to a negligible extent. The **Maxwell-Boltzmann distribution function** holds for such particles.

2 Identical particles of 0 or integral spin that cannot be distinguished one from another because their wave functions overlap. Such particles, called **bosons** in Chap. 7, do not obey the exclusion principle, and the **Bose-Einstein distribution function** holds for them. Photons are in this category, and we shall use Bose-Einstein statistics to account for the spectrum of radiation from a blackbody.

3 Identical particles with odd half-integral spin ($\frac{1}{2}, \frac{3}{2}, \frac{5}{2}, \ldots$) that also cannot be distinguished one from another. Such particles, called **fermions,** obey the exclusion principle, and the **Fermi-Dirac distribution function** holds for them. Electrons are in this category, and we shall use Fermi-Dirac statistics to study the behavior of the free electrons in a metal that are responsible for its ability to conduct electric current.

9.2 MAXWELL-BOLTZMANN STATISTICS

Classical particles such as gas molecules obey them

The Maxwell-Boltzmann distribution function states that the average number of particles $f_{MB}(\epsilon)$ in a state of energy ϵ in a system of particles at the absolute temperature T is

About Scientists

Ludwig Boltzmann (1844–1906) was born in Vienna and attended the university there. He then taught and carried out both experimental and theoretical research at a number of institutions in Austria and Germany, moving from one to another every few years. Boltzmann was interested in poetry, music, and travel as well as in physics; he visited the United States three times, something unusual in those days.

Of Boltzmann's many contributions to physics, the most important were to the kinetic theory of gases, which he developed independently of Maxwell, and to statistical mechanics, whose foundations he established. The constant k in the formula $\frac{3}{2}kT$ for the average energy of a gas molecule is named after him in honor of his work on the distribution of molecular energies in a gas. In 1884 Boltzmann derived from thermodynamic considerations the Stefan-Boltzmann law $R = \sigma T^4$ for the radiation rate of a blackbody. Josef Stefan, who had been one of Boltzmann's teachers, had discovered this law experimentally 5 years earlier. One of Boltzmann's major achievements was the interpretation of the second law of thermodynamics in terms of order and disorder. A monument to Boltzmann in Vienna is inscribed with his formula $S = k \log W$, which relates the entropy S of a system to its probability W.

Boltzmann was a champion of the atomic theory of matter, still controversial in the late nineteenth century because there was then only indirect evidence for the existence of atoms and molecules. Battles with nonbelieving scientists deeply upset Boltzmann, and in his later years asthma, headaches, and increasingly poor eyesight further depressed his spirits. He committed suicide in 1906, not long after Albert Einstein published a paper on brownian motion that was to convince the remaining doubters of the atomic theory of its correctness. Boltzmann had not been alone in his despair over doubters of the reality of atoms. Planck was driven to an extreme of pessimism: "A new scientific truth does not triumph by convincing its opponents and making them see the light, but rather because its opponents eventually die and a new generation grows up that is familiar with it."

Maxwell-Boltzmann distribution function

$$f_{MB}(\epsilon) = Ae^{-\epsilon/kT} \tag{9.2}$$

The value of A depends on the number of particles in the system and plays a role here analogous to that of the normalization constant of a wave function. As usual, k is Boltzmann's constant, whose value is

Boltzmann's constant

$$k = 1.381 \times 10^{-23}\ \text{J/K} = 8.617 \times 10^{-5}\ \text{eV/K}$$

Combining Eqs. (9.1) and (9.2) gives us the number $n(\epsilon)$ of identical, distinguishable particles in an assembly at the temperature T that have the energy ϵ:

Maxwell-Boltzmann

$$n(\epsilon) = Ag(\epsilon)e^{-\epsilon/kT} \tag{9.3}$$

Example 9.1

A cubic meter of atomic hydrogen at 0°C and at atmospheric pressure contains about 2.7×10^{25} atoms. Find the number of these atoms in their first excited states ($n = 2$) at 0°C and at 10,000°C.

Solution

(*a*) The constant A in Eq. (9.3) is the same for atoms in both states, so the ratio between the numbers of atoms in the $n = 1$ and $n = 2$ states is

$$\frac{n(\epsilon_2)}{n(\epsilon_1)} = \frac{g(\epsilon_2)}{g(\epsilon_1)}e^{-(\epsilon_2-\epsilon_1)/kT}$$

From Eq. (7.14) we know that the number of possible states that correspond to the quantum number n is $2n^2$. Thus the number of states of energy ϵ_1 is $g(\epsilon_1) = 2$; a 1*s* electron has $l = 0$ and $m_l = 0$ but m_s can be $-\frac{1}{2}$ or $+\frac{1}{2}$. The number of states of energy ϵ_2 is $g(\epsilon_2) = 8$; a 2*s* ($l = 0$) electron can have $m_s = \pm\frac{1}{2}$ and a 2*p* ($l = 1$) electron can have $m_l = 0, \pm 1$, in each case with $m_s = \pm\frac{1}{2}$. Since the ground-state energy is $\epsilon_1 = -13.6$ eV, $\epsilon_2 = \epsilon_1/n^2 = -3.4$ eV and $\epsilon_1 - \epsilon_2 = 10.2$ eV. Here $T = 0°\text{C} = 273$ K, so

$$\frac{\epsilon_2 - \epsilon_1}{kT} = \frac{10.2\ \text{eV}}{(8.617 \times 10^{-5}\ \text{eV/K})(273\ \text{K})} = 434$$

The result is

$$\frac{n(\epsilon_2)}{n(\epsilon_1)} = \left(\frac{8}{2}\right)e^{-434} = 1.3 \times 10^{-188}$$

Thus about 1 atom in every 10^{188} is in its first excited state at 0°C. With only 2.7×10^{25} atoms in our sample, we can be confident that all are in their ground states. (If all the known matter in the universe were in the form of hydrogen atoms, there would be about 10^{78} of them, and if they were at 0°C the same conclusion would still hold.)

(*b*) When $T = 10{,}000°\text{C} = 10{,}273$ K,

$$\frac{\epsilon_2 - \epsilon_1}{kT} = 11.5$$

and

$$\frac{n(\epsilon_2)}{n(\epsilon_1)} = \left(\frac{8}{2}\right)e^{-11.5} = 4.0 \times 10^{-5}$$

Now the number of excited atoms is about 10^{21}, a substantial number even though only a small fraction of the total.

Example 9.2

Obtain a formula for the populations of the rotational states of a rigid diatomic molecule.

Solution

For such a molecule Eq. (8.9) gives the energy states in terms of the rotational quantum number J as

$$\epsilon_J = J(J+1)\frac{\hbar^2}{2I}$$

More than one rotational state may correspond to a particular J because the component L_z in any specified direction of the angular momentum L may have any value in multiples of $\hbar$ from $J\hbar$ through 0 to $-J\hbar$, for a total of $2J + 1$ possible values. Each of these $2J + 1$ possible orientations of L constitutes a separate quantum state, and so

$$g(\epsilon) = 2J + 1$$

If the number of molecules in the $J = 0$ state is n_0, the normalization constant A in Eq. (9.3) is just n_0, and the number of molecules in the $J = J$ state is

$$n_J = Ag(\epsilon)e^{-\epsilon/kT} = n_0(2J+1)e^{-J(J+1)\hbar^2/2IkT}$$

In carbon monoxide, to give an example, this formula shows that the $J = 7$ state is the most highly populated at 20°C. The intensities of the rotational lines in a molecular spectrum are proportional to the relative populations of the various rotational energy levels.

9.3 MOLECULAR ENERGIES IN AN IDEAL GAS

They vary about an average of $\frac{3}{2}kT$

We now apply Maxwell-Boltzmann statistics to find the distribution of energies among the molecules of an ideal gas. Energy quantization is inconspicuous in the translational motion of gas molecules, and the total number of molecules N in a sample is usually very large. It is therefore reasonable to consider a continuous distribution of molecular energies instead of the discrete set $\epsilon_1, \epsilon_2, \epsilon_3, \ldots$ If $n(\epsilon)\, d\epsilon$ is the number of molecules whose energies lie between ϵ and $\epsilon + d\epsilon$, Eq. (9.1) becomes

Number of molecules with energies between ϵ and $\epsilon + d\epsilon$

$$n(\epsilon)\, d\epsilon = [g(\epsilon)\, d\epsilon][f(\epsilon)] = Ag(\epsilon)e^{-\epsilon/kT}\, d\epsilon \tag{9.4}$$

The first task is to find $g(\epsilon)\, d\epsilon$, the number of states that have energies between ϵ and $\epsilon + d\epsilon$. This is easiest to do in an indirect way. A molecule of energy ϵ has a momentum $\mathbf{p}$ whose magnitude p is specified by

$$p = \sqrt{2m\epsilon} = \sqrt{p_x^2 + p_y^2 + p_z^2}$$

Each set of momentum components p_x, p_y, p_z specifies a different state of motion. Let us imagine a **momentum space** whose coordinate axes are p_x, p_y, p_z, as in Fig. 9.1.

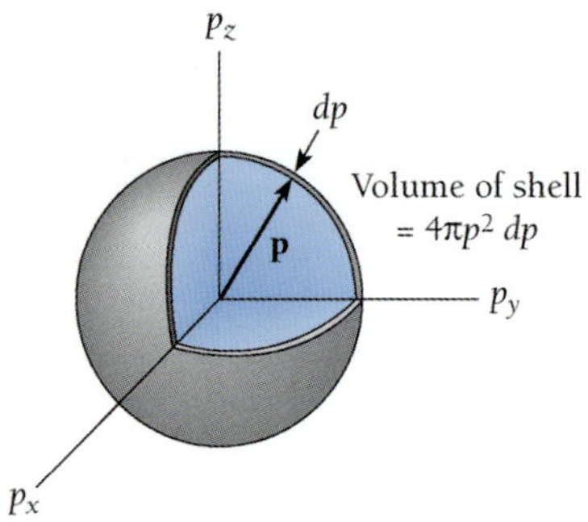

Figure 9.1 The coordinates in momentum space are p_x, p_y, p_z. The number of momentum states available to a particle with a momentum whose magnitude is between p and $p + dp$ is proportional to the volume of a spherical shell in momentum space of radius p and thickness dp.

The number of states $g(p)\, dp$ with momenta whose magnitudes are between p and $p + dp$ is proportional to the volume of a spherical shell in momentum space p in radius and dp thick, which is $4\pi p^2\, dp$. Hence

Number of momentum states

$$g(p)\, dp = Bp^2\, dp \tag{9.5}$$

where B is some constant. [The function $g(p)$ here is not the same as the function $g(\epsilon)$ in Eq. (9.4).]

Since each momentum magnitude p corresponds to a single energy ϵ, the number of energy states $g(\epsilon)\, d\epsilon$ between ϵ and $\epsilon + d\epsilon$ is the same as the number of momentum states $g(p)\, dp$ between p and $p + dp$, and so

$$g(\epsilon)\, d\epsilon = Bp^2\, dp \tag{9.6}$$

Because

$$p^2 = 2m\epsilon \quad \text{and} \quad dp = \frac{m\, d\epsilon}{\sqrt{2m\epsilon}}$$

Eq. (9.6) becomes

Number of energy states

$$g(\epsilon)\, d\epsilon = \sqrt{2}\, m^{3/2}\, B\sqrt{\epsilon}\, d\epsilon \tag{9.7}$$

The number of molecules with energies between ϵ and $d\epsilon$ is therefore

$$n(\epsilon)\, d\epsilon = C\sqrt{\epsilon}\, e^{-\epsilon/kT}\, d\epsilon \tag{9.8}$$

where $C(=\sqrt{2}\, m^{3/2}\, AB)$ is a constant to be evaluated.

To find C we make use of the normalization condition that the total number of molecules is N, so that

Normalization

$$N = \int_0^\infty n(\epsilon)\, d\epsilon = C\int_0^\infty \sqrt{\epsilon}\, e^{-\epsilon/kT}\, d\epsilon \tag{9.9}$$

From a table of definite integrals we find that

$$\int_0^\infty \sqrt{x}\, e^{-ax}\, dx = \frac{1}{2a}\sqrt{\frac{\pi}{a}}$$

Here $a = 1/kT$, and the result is

$$N = \frac{C}{2}\sqrt{\pi}\,(kT)^{3/2}$$

$$C = \frac{2\pi N}{(\pi kT)^{3/2}} \tag{9.10}$$

and, finally,

Molecular energy distribution

$$n(\epsilon)\, d\epsilon = \frac{2\pi N}{(\pi kT)^{3/2}}\sqrt{\epsilon}\, e^{-\epsilon/kT}\, d\epsilon \tag{9.11}$$

This formula gives the number of molecules with energies between ϵ and $\epsilon + d\epsilon$ in a sample of an ideal gas that contains N molecules and whose absolute temperature is T.

Equation (9.11) is plotted in Fig. 9.2 in terms of kT. The curve is not symmetrical about the most probable energy because the lower limit to ϵ is $\epsilon = 0$ while there is, in principle, no upper limit (although the likelihood of energies many times greater than kT is small).

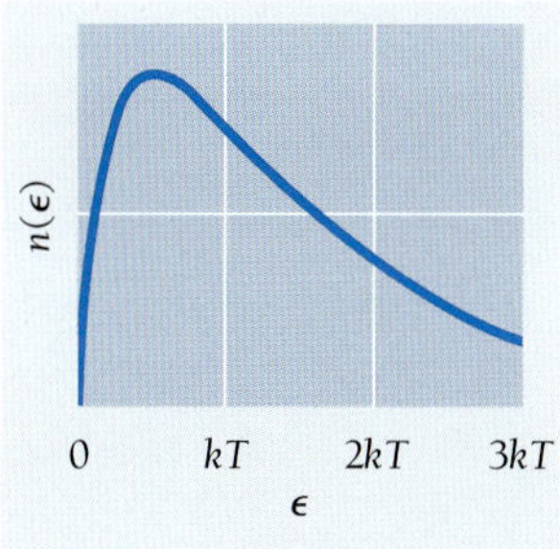

Figure 9.2 Maxwell-Boltzmann energy distribution for the molecules of an ideal gas. The average molecular energy is $\bar{\epsilon} = \frac{3}{2}kT$.

Average Molecular Energy

To find the average energy per molecule we begin by calculating the total internal energy of the system. To do this we multiply $n(\epsilon)d\epsilon$ by the energy ϵ and then integrate over all energies from 0 to ∞:

$$E = \int_0^\infty \epsilon\, n(\epsilon)\, d\epsilon = \frac{2\pi N}{(\pi kT)^{3/2}}\int_0^\infty \epsilon^{3/2}\, e^{-\epsilon/kT}\, d\epsilon$$

Making use of the definite integral

$$\int_0^\infty x^{3/2}\, e^{-ax}\, dx = \frac{3}{4a^2}\sqrt{\frac{\pi}{a}}$$

we have

Total energy of N gas molecules

$$E = \left[\frac{2\pi N}{(\pi kT)^{3/2}}\right]\left[\frac{3}{4}(kT)^2\sqrt{\pi kT}\right] = \frac{3}{2}NkT \tag{9.12}$$

The average energy of an ideal-gas molecule is E/N, or

Average molecular energy

$$\bar{\epsilon} = \frac{3}{2}kT \tag{9.13}$$

which is independent of the molecule's mass: a light molecule has a greater average speed at a given temperature than a heavy one. The value of $\bar{\epsilon}$ at room temperature is about 0.04 eV, $\frac{1}{25}$ eV.

Equipartition of Energy

A gas molecule has three **degrees of freedom** that correspond to motions in three independent (that is, perpendicular) directions. Since the average kinetic energy of the molecule is $\frac{3}{2}kT$ we can associate $\frac{1}{2}kT$ with the average energy of each degree of freedom: $\frac{1}{2}m\overline{v_x^2} = \frac{1}{2}m\overline{v_y^2} = \frac{1}{2}m\overline{v_z^2} = \frac{1}{2}kT$. This association turns out to be quite general and is called the **equipartition theorem:**

> The average energy per degree of freedom of any classical object that is a member of a system of such objects in thermal equilibrium at the temperature T is $\frac{1}{2}kT$.

Degrees of freedom are not limited to linear velocity components—each variable that appears squared in the formula for the energy of a particular object represents a degree of freedom. Thus each component ω_i of angular velocity (provided it involves a moment of inertia I_i), is a degree of freedom, so that $\frac{1}{2}I_i\overline{\omega_i^2} = \frac{1}{2}kT$. A rigid diatomic molecule of the kind described in Sec. 8.6 therefore has five degrees of freedom, one each for motions in the x, y, and z directions and two for rotations about axes perpendicular to its symmetry axis.

A degree of freedom is similarly associated with each component Δs_i of the displacement of an object that gives rise to a potential energy proportional to $(\Delta s_i)^2$. For example, a one-dimensional harmonic oscillator has two degrees of freedom, one that corresponds to its kinetic energy $\frac{1}{2}mv_x^2$ and the other to its potential energy $\frac{1}{2}K(\Delta x)^2$, where K is the force constant. Each oscillator in a system of them in thermal equilibrium accordingly has a total average energy of $2(\frac{1}{2}kT) = kT$ provided that quantization can be disregarded. To a first approximation, the constituent particles (atoms, ions, or molecules) of a solid behave thermally like a system of classical harmonic oscillators, as we shall see shortly.

The equipartition theorem also applies to nonmechanical systems, for instance to thermal fluctuations ("noise") in electrical circuits.

Distribution of Molecular Speeds

The distribution of molecular speeds in an ideal gas can be found from Eq. (9.11) by making the substitutions

$$\epsilon = \tfrac{1}{2}mv^2 \qquad d\epsilon = mv\,dv$$

The result for the number of molecules with speeds between v and $v + dv$ is

Molecular-speed distribution

$$n(v)\,dv = 4\pi N\left(\frac{m}{2\pi kT}\right)^{3/2} v^2 e^{-mv^2/2kT}\,dv \tag{9.14}$$

This formula, which was first obtained by Maxwell in 1859, is plotted in Fig. 9.3.

The speed of a molecule with the average energy of $\frac{3}{2}kT$ is

RMS speed

$$v_{\mathrm{rms}} = \sqrt{\overline{v^2}} = \sqrt{\frac{3kT}{m}} \tag{9.15}$$

since $\overline{\frac{1}{2}mv^2} = \frac{3}{2}kT$. This speed is denoted v_{rms} because it is the square root of the average of the squared molecular speeds—the root-mean-square speed—and is not the same as the simple arithmetical average speed $\overline{v}$. The relationship between $\overline{v}$ and v_{rms} depends on the distribution law that governs the molecular speeds in a particular system. For a Maxwell-Boltzmann distribution the rms speed is about 9 percent greater than the arithmetical average speed.

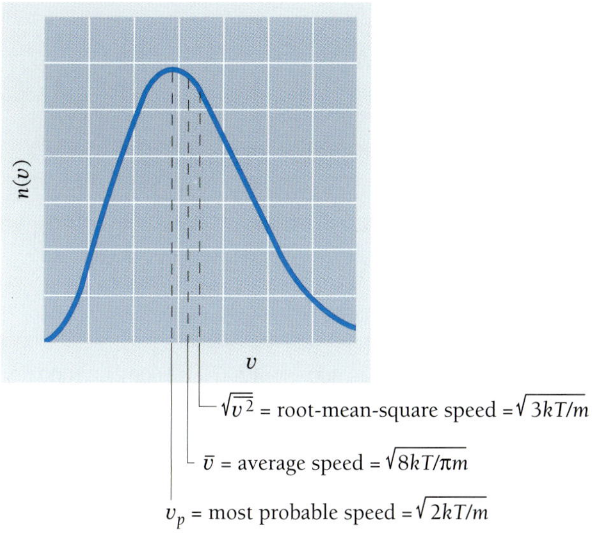

Figure 9.3 Maxwell-Boltzmann speed distribution.

Example 9.3

Verify that the rms speed of an ideal-gas molecule is about 9 percent greater than its average speed.

Solution

Equation (9.14) gives the number of molecules with speeds between v and $v + dv$ in a sample of N molecules. To find their average speed $\bar{v}$, we multiply $n(v)\,dv$ by v, integrate over all values of v from 0 to ∞, and then divide by N. (See the discussion of expectation values in Sec. 5.5.) This procedure gives

$$\bar{v} = \frac{1}{N}\int^{\infty} v\, n(v)\, dv = 4\pi\left(\frac{m}{2\pi kT}\right)^{3/2}\int^{\infty} v^3 e^{-mv^2/2kT}\, dv$$

If we let $a = m/2kT$, we see that the integral is the standard one

$$\int_0^{\infty} x^3 e^{-ax^2}\, dx = \frac{1}{2a^2}$$

and so

$$\bar{v} = \left[4\pi\left(\frac{m}{2\pi kT}\right)^{3/2}\right]\left[\frac{1}{2}\left(\frac{2kT}{m}\right)^2\right] = \sqrt{\frac{8kT}{\pi m}}$$

Comparing $\bar{v}$ with v_{rms} from Eq. (9.15) shows that

$$v_{\text{rms}} = \sqrt{\frac{3kT}{m}} = \sqrt{\frac{3\pi}{8}}\,\bar{v} \approx 1.09v$$

Because the speed distribution of Eq. (9.14) is not symmetrical, the most probable speed v_p is smaller than either $\bar{v}$ or v_{rms}. To find v_p, we set equal to zero the derivative of $n(v)$ with respect to v and solve the resulting equation for v. The result is

Most probable speed

$$v_p = \sqrt{\frac{2kT}{m}} \tag{9.16}$$

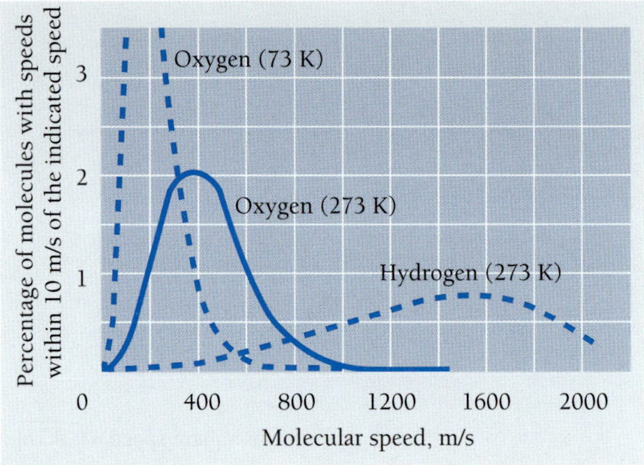

Figure 9.4 The distributions of molecular speeds in oxygen at 73 K, in oxygen at 273 K, and in hydrogen at 273 K.

Molecular speeds in a gas vary considerably on either side of v_p. Figure 9.4 shows the distribution of speeds in oxygen at 73 K (−200°C), in oxygen at 273 K (0°C), and in hydrogen at 273 K. The most probable speed increases with temperature and decreases with molecular mass. Accordingly molecular speeds in oxygen at 73 K are on the whole less than at 273 K, and at 273 K molecular speeds in hydrogen are on the whole greater than in oxygen at the same temperature. The average molecular *energy* is the same in both oxygen and hydrogen at 273 K, of course.

Example 9.4

Find the rms speed of oxygen molecules at 0°C.

Solution

Oxygen molecules have two oxygen atoms each. Since the atomic mass of oxygen is 16.0 u, the molecular mass of O_2 is 32.0 u which is equivalent to

$$m = (32.0 \text{ u})(1.66 \times 10^{-27} \text{ kg/u}) = 5.31 \times 10^{-26} \text{ kg}$$

At an absolute temperature of 273 K, the rms speed of an O_2 molecule is

$$v_{\text{rms}} = \sqrt{\frac{3kT}{m}} = \sqrt{\frac{3(1.38 \times 10^{-23} \text{ J/K})(273 \text{ K})}{5.31 \times 10^{-26} \text{ kg}}} = 461 \text{ m/s}$$

This is a little over 1000 mi/h.

9.4 QUANTUM STATISTICS

Bosons and fermions have different distribution functions

As mentioned in Sec. 9.1, the Maxwell-Boltzmann distribution function holds for systems of identical particles that can be distinguished one from another, which means particles whose wave functions do not overlap very much. Molecules in a gas fit this description and obey Maxwell-Boltzmann statistics. If the wave functions do overlap appreciably, the situation changes because the particles cannot now be distinguished, although they can still be counted. The quantum-mechanical consequences of indistinguishability were discussed in Sec. 7.3, where we saw that systems of particles with overlapping wave functions fall into two categories:

1 Particles with 0 or integral spins, which are **bosons.** Bosons do not obey the exclusion principle, and the wave function of a system of bosons is not affected by the exchange of any pair of them. A wave function of this kind is called **symmetric.** Any number of bosons can exist in the same quantum state of the system.
2 Particles with odd half-integral spins ($\frac{1}{2}, \frac{3}{2}, \frac{5}{2}, \ldots$), which are **fermions.** Fermions obey the exclusion principle, and the wave function of a system of fermions changes sign upon the exchange of any pair of them. A wave function of this kind is called **antisymmetric.** Only one fermion can exist in a particular quantum state of the system.

We shall now see what difference all this makes in the probability $f(\epsilon)$ that a particular state of energy ϵ will be occupied.

Let us consider a system of two particles, 1 and 2, one of which is in state a and the other in state b. When the particles are distinguishable there are two possibilities for occupancy of the states, as described by the wave functions

$$\psi_{\mathrm{I}} = \psi_a(1)\psi_b(2) \tag{9.17}$$

$$\psi_{\mathrm{II}} = \psi_a(2)\psi_b(1) \tag{9.18}$$

When the particles are not distinguishable, we cannot tell which of them is in which state, and the wave function must be a combination of ψ_{I} and ψ_{II} to reflect their equal likelihoods. As we found in Sec. 7.3, if the particles are bosons, the system is described by the symmetric wave function

Bosons

$$\psi_B = \frac{1}{\sqrt{2}}[\psi_a(1)\psi_b(2) + \psi_a(2)\psi_b(1)] \tag{9.19}$$

and if they are fermions, the system is described by the antisymmetric wave function

Fermions

$$\psi_F = \frac{1}{\sqrt{2}}[\psi_a(1)\psi_b(2) - \psi_a(2)\psi_b(1)] \tag{9.20}$$

The $1/\sqrt{2}$ factors are needed to normalize the wave functions.

Now we ask what the likelihood in each case that both particles be in the same state, say a. For distinguishable particles, both ψ_{I} and ψ_{II} become

$$\psi_M = \psi_a(1)\psi_a(2) \tag{9.21}$$

to give a probability density of

Distinguishable particles

$$\psi_M^*\psi_M = \psi_a^*(1)\psi_a^*(2)\psi_a(1)\psi_a(2) \tag{9.22}$$

For bosons the wave function becomes

$$\psi_B = \frac{1}{\sqrt{2}}[\psi_a(1)\psi_a(2) + \psi_a(1)\psi_a(2)] = \frac{2}{\sqrt{2}}\psi_a(1)\psi_a(2) = \sqrt{2}\psi_a(1)\psi_a(2) \quad (9.23)$$

to give a probability density of

Bosons

$$\psi_B^*\psi_B = 2\psi_a^*(1)\psi_a^*(2)\psi_a(1)\psi_a(2) = 2\psi_M^*\psi_M \quad (9.24)$$

Thus the probability that both bosons are in the same state is twice what it is for distinguishable particles!

For fermions the wave function becomes

Fermions

$$\psi_F = \frac{1}{\sqrt{2}}[\psi_a(1)\psi_a(2) - \psi_a(1)\psi_a(2)] = 0 \quad (9.25)$$

It is impossible for both particles to be in the same state, which is a statement of the exclusion principle.

These results can be generalized to apply to systems of many particles:

1 In a system of bosons, the presence of a particle in a certain quantum state *increases* the probability that other particles are to be found in the same state;
2 In a system of fermions, the presence of a particle in a certain state *prevents* any other particles from being in that state.

Bose-Einstein and Fermi-Dirac Distribution Functions

The probability $f(\epsilon)$ that a boson occupies a state of energy ϵ turns out to be

Bose-Einstein distribution function

$$f_{BE}(\epsilon) = \frac{1}{e^{\alpha}e^{\epsilon/kT} - 1} \quad (9.26)$$

and the probability for a fermion turns out to be

Fermi-Dirac distribution function

$$f_{FD}(\epsilon) = \frac{1}{e^{\alpha}e^{\epsilon/kT} + 1} \quad (9.27)$$

Names of the Functions

The Indian physicist S. N. Bose in 1924 derived Planck's radiation formula on the basis of the quantum theory of light with indistinguishable photons whose number is not conserved. His paper was rejected by a leading British journal. He then sent it to Einstein, who translated the paper into German and submitted it to a German journal where it was published. Because Einstein extended Bose's treatment to material particles whose number is conserved, both names are attached to Eq. 9.26. Two years later Enrico Fermi and Paul Dirac independently realized that Pauli's exclusion principle would lead to different statistics for electrons, and so Eq. 9.27 is named after them.

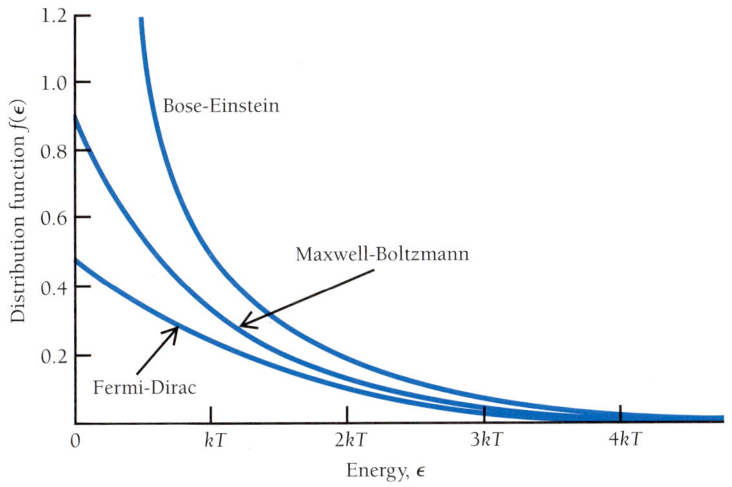

Figure 9.5 A comparison of the three distribution functions for the same value of α. The Bose-Einstein function is always higher than the Maxwell-Boltzmann one, which is a pure exponential, and the Fermi-Dirac function is always lower. The functions give the probability of occupancy of a state of energy ϵ at the absolute temperature T.

The quantity α depends on the properties of the particular system and may be a function of T. Its value is determined by the normalization condition that the sum over all energy states of $n(\epsilon) = g(\epsilon)f(\epsilon)$ be equal to the total number of particles in the system. If the number of particles is not fixed, as in the case of a photon gas, then from the way α is defined in deriving Eqs. (9.26) and (9.27), $\alpha = 0$, $e^{\alpha} = 1$.

The -1 term in the denominator of Eq. (9.26) expresses the increased likelihood of multiple occupancy of an energy state by bosons compared with the likelihood for distinguishable particles such as molecules. The $+1$ term in the denominator of Eq. (9.27) is a consequence of the uncertainty principle: No matter what the values of α, ϵ, and T, $f(\epsilon)$ can never exceed 1. In both cases, when $\epsilon \gg kT$ the functions $f(\epsilon)$ approach that of Maxwell-Boltzmann statistics, Eq. (9.2). Figure 9.5 is a comparison of the three distribution functions. Clearly $f_{BE}(\boldsymbol{\epsilon})$ for bosons is always greater at a given ratio of $\boldsymbol{\epsilon}/kT$ than it is for molecules, and $f_{FD}(\boldsymbol{\epsilon})$ for fermions is always smaller.

From Eq. (9.27) we see that $f_{FD}(\epsilon) = \frac{1}{2}$ for an energy of

Fermi energy

$$\epsilon_F = -\alpha kT \tag{9.28}$$

This energy, called the **Fermi energy**, is a very important quantity in a system of fermions, such as the electron gas in a metal. In terms of $\boldsymbol{\epsilon}_F$ the Fermi-Dirac distribution function becomes

Fermi-Dirac

$$f_{FD}(\epsilon) = \frac{1}{e^{(\epsilon-\epsilon_F)/kT} + 1} \tag{9.29}$$

To appreciate the significance of the Fermi energy, let us consider a system of fermions at $T = 0$ and investigate the occupancy of states whose energies are less than $\boldsymbol{\epsilon}_F$ and greater than $\boldsymbol{\epsilon}_F$. What we find is this:

$$T = 0,\ \epsilon < \epsilon_F: \qquad f_{FD}(\epsilon) = \frac{1}{e^{(\epsilon-\epsilon_F)/kT} + 1} = \frac{1}{e^{-\infty} + 1} = \frac{1}{0+1} = 1$$

$$T = 0,\ \epsilon > \epsilon_F: \qquad f_{FD}(\epsilon) = \frac{1}{e^{(\epsilon-\epsilon_F)/kT} + 1} = \frac{1}{e^{\infty} + 1} = 0$$

Bose-Einstein Condensate

Under ordinary conditions, the wave packets that correspond to individual atoms in a gas of atoms are sufficiently small in size relative to their average spacing for the atoms to move independently and be distinguishable. If the temperature of the gas is reduced, the wave packets grow larger as the atoms lose momentum, in accord with the uncertainty principle. When the gas becomes very cold, the dimensions of the wave packets exceed the average atomic spacing so that the wave packets overlap. If the atoms are bosons, the eventual result is that all the atoms fall into the lowest possible energy state and their separate wave packets merge into a single wave packet. The atoms in such a **Bose-Einstein condensate** are barely moving, are indistinguishable, and form one entity—a superatom.

Although such condensates were first visualized by Einstein in 1924, not until 1995 was one actually created. The problem was to achieve a cold enough gas without it becoming a liquid or solid first. This was accomplished by Eric Cornell, Carl Wieman, and their coworkers in Colorado using a gas of rubidium atoms. The atoms were first cooled and trapped by six intersecting beams of laser light. The frequency of the light was adjusted so that the atoms moving against one of the beams would "see" light whose frequency was doppler-shifted to that of one of rubidium's absorption lines. Thus the atoms would only absorb photons coming toward them, which would slow the atoms and thereby cool the assembly as well as pushing the atoms together and away from the warm walls of the chamber. To get the assembly still colder, the lasers were turned off and a magnetic field held the slower atoms together while allowing the faster ones to escape. (Such evaporative cooling is familiar in everyday life when the faster molecules of a liquid, for instance perspiration, leave its surface and so reduce the average energy of the remaining molecules.) Finally, when the temperature was down to under 10^{-7} K—a tenth of a millionth of a degree above absolute zero—about 2000 rubidium atoms came together in a Bose-Einstein condensate 10 μm long that lasted for 10 s.

Soon after this achievement other groups succeeded in creating Bose-Einstein condensates in lithium and sodium. One condensate in sodium contained about 5 million atoms, was shaped like a pencil 8 μm across and 150 μm long, and lasted for 20 s. Still larger condensates were later produced, including one that consisted of 10^8 hydrogen atoms. It proved possible to extract from condensates beams of atoms whose behavior confirmed that they were coherent, with all the atomic wave functions in phase just like the light waves in the coherent beam from a laser. Bose-Einstein condensates are extremely interesting from a number of points of view both fundamental and applied—for example, for possible use in ultrasensitive measurements of various kinds.

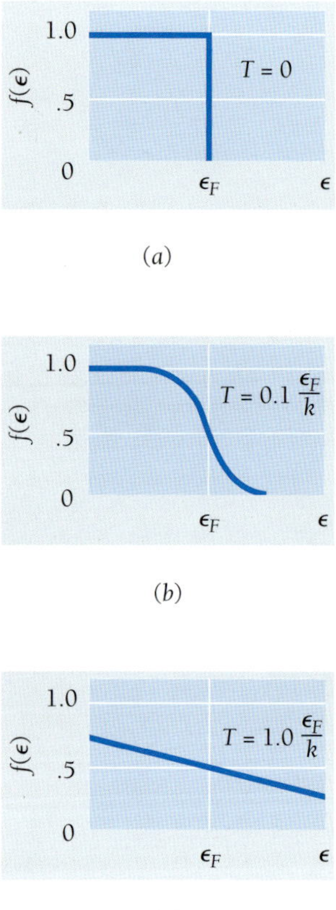

Figure 9.6 Distribution function for fermions at three different temperatures. (*a*) At $T = 0$, all the energy states up to the Fermi energy ϵ_F are occupied. (*b*) At a low temperature, some fermions will leave states just below ϵ_F and move into states just above ϵ_F. (*c*) At a higher temperature, fermions from any state below ϵ_F may move into states above ϵ_F.

Thus at absolute zero all energy states up to ϵ_F are occupied, and none above ϵ_F (Fig. 9.6*a*). If a system contains N fermions, we can calculate its Fermi energy ϵ_F by filling up its energy states with the N particles in order of increasing energy starting from $\epsilon = 0$. The highest state to be occupied will then have the energy $\epsilon = \epsilon_F$. This calculation will be made for the electrons in a metal in Sec. 9.9.

As the temperature is increased above $T = 0$ but with kT still smaller than ϵ_F, fermions will leave states just below ϵ_F to move into states just above it, as in Fig. 9.6*b*. At higher temperatures, fermions from even the lowest state will begin to be excited to higher ones, so $f_{FD}(0)$ will drop below 1. In these circumstances $f_{FD}(\epsilon)$ will assume a shape like that in Fig. 9.6*c*, which corresponds to the lowest curve in Fig. 9.5.

The properties of the three distribution functions are summarized in Table 9.1. It is worth recalling that to find the *actual number* $n(\epsilon)$ of particles with an energy ϵ, the functions $f(\epsilon)$ must be multiplied by the number of states $g(\epsilon)$ with this energy:

$$n(\epsilon) = g(\epsilon)f(\epsilon) \tag{9.1}$$

Table 9.1 The Three Statistical Distribution Functions

	Maxwell-Boltzmann	Bose-Einstein	Fermi-Dirac
Applies to systems of	Identical, distinguishable particles	Identical, indistinguishable particles that do not obey exclusion principle	Identical, indistinguishable particles that obey exclusion principle
Category of particles	Classical	Bosons	Fermions
Properties of particles	Any spin, particles far enough apart so wave functions do not overlap	Spin 0, 1, 2, . . . ; wave functions are symmetric to interchange of particle labels	Spin $\frac{1}{2}, \frac{3}{2}, \frac{5}{2}, \ldots$; wave functions are antisymmetric to interchange of particle labels
Examples	Molecules of a gas	Photons in a cavity; phonons in a solid; liquid helium at low temperatures	Free electrons in a metal; electrons in a star whose atoms have collapsed (white dwarf stars)
Distribution function (number of particles in each state of energy ϵ at the temperature T)	$f_{MB}(\epsilon) = Ae^{-\epsilon/kT}$	$f_{BE}(\epsilon) = \dfrac{1}{e^{\alpha}e^{\epsilon/kT} - 1}$	$f_{FD}(\epsilon) = \dfrac{1}{e^{(\epsilon-\epsilon_F)/kT} + 1}$
Properties of distribution	No limit to number of particles per state	No limit to number of particles per state; more particles per state than f_{MB} at low energies; approaches f_{MB} at high energies	Never more than 1 particle per state; fewer particles per state than f_{MB} at low energies; approaches f_{MB} at high energies

About Scientists

Paul A. M. Dirac (1902–1984) was born in Bristol, England, and studied electrical engineering there. He then switched his interest to mathematics and finally to physics, obtaining his Ph.D. from Cambridge in 1926. After reading Heisenberg's first paper on quantum mechanics in 1925, Dirac soon devised a more general theory and the next year formulated Pauli's exclusion principle in quantum-mechanical terms. He investigated the statistical behavior of particles that obey the Pauli principle, such as electrons, which Fermi had done independently a little earlier, and the result is called Fermi-Dirac statistics in honor of both. In 1928 Dirac joined special relativity to quantum theory to give a theory of the electron that not only permitted its spin and magnetic moment to be calculated but also predicted the existence of positively charged electrons, or positrons, which were discovered by Carl Anderson in the United States in 1932.

In an attempt to explain why charge is quantized, Dirac in 1931 found it necessary to postulate the existence of **magnetic monopoles,** isolated N or S magnetic poles. More recent theories show that magnetic monopoles should have been created in profusion just after the Big Bang that marked the beginning of the universe; the predicted monopole mass is $\sim 10^{16}$ GeV/c^2 ($\sim 10^{-8}$ g!). As Dirac said in 1981, "From the theoretical point of view one would think that monopoles should exist, because of the prettiness of the mathematics. Many attempts to find them have been made, but all have been unsuccessful. One should conclude that pretty mathematics by itself is not an adequate reason for nature to have made use of a theory."

In 1932 Dirac became Lucasian Professor of Mathematics at Cambridge, the post Newton had held two and a half centuries earlier, and in 1933 shared the Nobel Prize in physics with Schrödinger. He remained active in physics for the rest of his life, after 1969 in the warmer climate of Florida, but as is often the case in science he will be remembered for the brilliant achievements of his youth.

9.5 RAYLEIGH-JEANS FORMULA

The classical approach to blackbody radiation

Blackbody radiation was discussed briefly in Sec. 2.2, where we learned about the failure of classical physics to account for the shape of the blackbody spectrum—the "ultraviolet catastrophe"—and how Planck's introduction of energy quantization led to the correct formula for this spectrum. Because the origin of blackbody radiation is such a fundamental question, it deserves a closer look.

Figure 2.6 shows the blackbody spectrum for two temperatures. To explain this spectrum, the classical calculation by Rayleigh and Jeans begins by considering a blackbody as a radiation-filled cavity at the temperature T (Fig. 2.5). Because the cavity walls are assumed to be perfect reflectors, the radiation must consist of standing em waves, as in Fig. 2.7. In order for a node to occur at each wall, the path length from wall to wall, in any direction, must be an integral number j of half-wavelengths. If the cavity is a cube L long on each edge, this condition means that for standing waves in the x, y, and z directions respectively, the possible wavelengths are such that

$$\begin{aligned} j_x &= \frac{2L}{\lambda} = 1, 2, 3, \ldots = \text{number of half-wavelengths in } x \text{ direction} \\ j_y &= \frac{2L}{\lambda} = 1, 2, 3, \ldots = \text{number of half-wavelengths in } y \text{ direction} \\ j_z &= \frac{2L}{\lambda} = 1, 2, 3, \ldots = \text{number of half-wavelengths in } z \text{ direction} \end{aligned} \tag{9.30}$$

For a standing wave in any arbitrary direction, it must be true that

Standing waves in a cubic cavity

$$j_x^2 + j_y^2 + j_z^2 = \left(\frac{2L}{\lambda}\right)^2 \qquad \begin{aligned} j_x &= 0, 1, 2, \ldots \\ j_y &= 0, 1, 2, \ldots \\ j_z &= 0, 1, 2, \ldots \end{aligned} \tag{9.31}$$

in order that the wave terminate in a node at its ends. (Of course, if $j_x = j_y = j_z = 0$, there is no wave, though it is possible for any one or two of the j's to equal 0.)

To count the number of standing waves $g(\lambda)\,d\lambda$ within the cavity whose wavelengths lie between λ and $\lambda + d\lambda$, what we have to do is count the number of permissible sets of j_x, j_y, j_z values that yield wavelengths in this interval. Let us imagine a j-space whose coordinate axes are j_x, j_y, *and* j_z; Fig. 9.7 shows part of the j_x-j_y plane of such a space. Each point in the j-space corresponds to a permissible set of j_x, j_y, j_z values and thus to a standing wave. If $\mathbf{j}$ is a vector from the origin to a particular point j_x, j_y, j_z, its magnitude is

$$j = \sqrt{j_x^2 + j_y^2 + j_z^2} \tag{9.32}$$

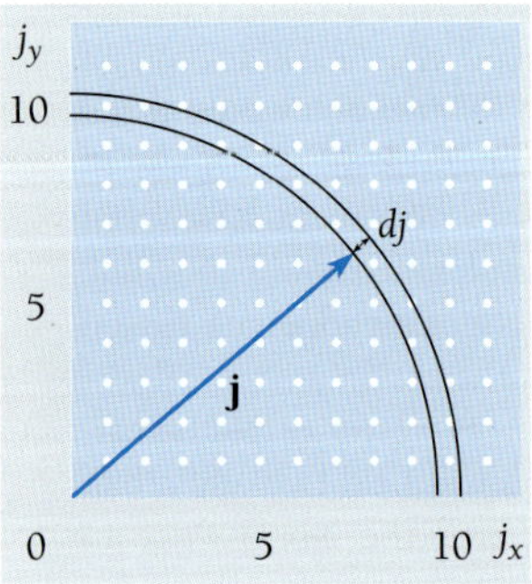

Figure 9.7 Each point in j space corresponds to a possible standing wave.

The total number of wavelengths between λ and $\lambda + d\lambda$ is the same as the number of points in j space whose distances from the origin lie between j and $j + dj$. The volume of a spherical shell of radius j and thickness dj is $4\pi j^2\,dj$, but we are only interested in the octant of this shell that includes non-negative values of j_x, j_y, and j_z. Also, for each standing wave counted in this way, there are two perpendicular directions of polarization. Hence the number of independent standing waves in the cavity is

Number of standing waves

$$g(j)\,dj = (2)(\tfrac{1}{8})(4\pi j^2\,dj) = \pi j^2\,dj \tag{9.33}$$

What we really want is the number of standing waves in the cavity as a function of their frequency ν instead of as a function of j. From Eqs. (9.31) and (9.32) we have

$$j = \frac{2L}{\lambda} = \frac{2L\nu}{c} \qquad dj = \frac{2L}{c}\,d\nu$$

and so

Number of standing waves

$$g(\nu)\,d\nu = \pi\left(\frac{2L\nu}{c}\right)^2 \frac{2L}{c}\,d\nu = \frac{8\pi L^3}{c^3}\nu^2\,d\nu \tag{9.34}$$

The cavity volume is L^3, which means that the number of independent standing waves per unit volume is

Density of standing waves in a cavity

$$G(\nu)\,d\nu = \frac{1}{L^3}\,g(\nu)\,d\nu = \frac{8\pi\nu^2\,d\nu}{c^3} \tag{9.35}$$

Equation (9.35) is independent of the shape of the cavity, even though we used a cubical cavity to facilitate the derivation. The higher the frequency, the shorter the wavelength and the greater the number of standing waves that are possible, as must be the case.

The next step is to find the average energy per standing wave. Here is where classical and quantum physics diverge. According to the classical theorem of equipartition of energy, as already mentioned, the average energy per degree of freedom of an entity that is part of a system of such entities in thermal equilibrium at the temperature T is $\frac{1}{2}kT$. Each standing wave in a radiation-filled cavity corresponds to two degrees of freedom, for a total $\overline{\epsilon}$ of kT, because each wave originates in an oscillator in the cavity wall. Such an oscillator has two degrees of freedom, one that represents its kinetic energy and one that represents its potential energy. The energy $u(\nu)\,d\nu$ per unit volume in the cavity in the frequency interval from ν to $\nu + d\nu$ is therefore, according to classical physics,

Rayleigh-Jeans formula

$$\begin{aligned} u(\nu)\,d\nu &= \overline{\epsilon}G(\nu)\,d\nu = kT\,G(\nu)\,d\nu \\ &= \frac{8\pi\nu^2 kT\,d\nu}{c^3} \end{aligned} \tag{9.36}$$

The Rayleigh-Jeans formula, which has the spectral energy density of blackbody radiation increasing as ν^2 without limit, is obviously wrong. Not only does it predict a spectrum different from the observed one (see Fig. 2.8), but integrating Eq. (9.36) from $\nu = 0$ to $\nu = \infty$ gives the total energy density as infinite at all temperatures. The discrepancy between theory and observation was at once recognized as fundamental. This is the failure of classical physics that led Max Planck in 1900 to discover that only if light emission is a quantum phenomenon can the correct formula for $u(\nu)\,d\nu$ be obtained.

About Scientists

Lord Rayleigh (1842–1919) was born John William Strutt to a wealthy English family and inherited his title on the death of his father. After being educated at home, he went on to be an outstanding student at Cambridge University and then spent some time in the United States. On his return Rayleigh set up a laboratory in his home. There he carried out both experimental and theoretical research except for a five-year period when he directed the Cavendish Laboratory at Cambridge following Maxwell's death in 1879.

For much of his life Rayleigh's work concerned the behavior of waves of all kinds, and he made many contributions to acoustics and optics. One of the types of wave an earthquake produces is named after him. In 1871 Rayleigh explained the blue color of the sky in terms of the preferential scattering of short-wavelength sunlight in the atmosphere. The formula for the resolving power of an optical instrument is another of his achievements.

At the Cavendish Laboratory, Rayleigh completed the standardization of the volt, the ampere, and the ohm, a task Maxwell had begun. Back at home, he found that nitrogen prepared from air is very slightly denser than nitrogen prepared from nitrogen-containing compounds. Together with the chemist William Ramsay, Rayleigh showed that the reason for the discrepancy was a hitherto unknown gas that makes up about 1 percent of the atmosphere. They called the gas argon, from the Greek word for "inert," because argon did not react with other substances. Ramsay went on to discover the other inert gases neon ("new"), krypton ("hidden"), and xenon ("stranger"). He was also able to isolate the lightest inert gas, helium, which had thirty years earlier been identified in the sun by its spectral lines; *helios* means "sun" in Greek. Rayleigh and Ramsay won Nobel Prizes in 1904 for their work on argon.

What was possibly Rayleigh's greatest contribution to science came after the discovery of argon and took the form of an equation that did not agree with experimentation. The problem was accounting for the spectrum of blackbody radiation, that is, the relative intensities of the different wavelengths present in such radiation. Rayleigh calculated the shape of this spectrum; because the astronomer James Jeans pointed out a small error Rayleigh had made, the result is called the Rayleigh-Jeans formula. The formula follows directly from the laws of physics known at the end of the nineteenth century—and it is hopelessly incorrect, as Rayleigh and Jeans were aware. (For instance, the formula predicts that a blackbody should radiate energy at an infinite rate.) The search for a correct blackbody formula led to the founding of the quantum theory of radiation by Max Planck and Albert Einstein, a theory that was to completely revolutionize physics.

Despite the successes of quantum theory and of Einstein's theory of relativity that followed soon afterward, Rayleigh, after a lifetime devoted to classical physics, never really accepted them. He died in 1919.

9.6 PLANCK RADIATION LAW

How a photon gas behaves

Planck found that he had to assume that the oscillators in the cavity walls were limited to energies of $\epsilon_n = nh\nu$, where $n = 0, 1, 2, \ldots$. He then used the Maxwell-Boltzmann distribution law to find that the number of oscillators with the energy ϵ_n is proportional to $e^{-\epsilon_n/kT}$ at the temperature T. In this case the average energy per oscillator (and so per standing wave in the cavity) is

$$\overline{\epsilon} = \frac{h\nu}{e^{h\nu/kT} - 1} \tag{9.37}$$

instead of the energy-equipartition average of kT which Rayleigh and Jeans had used. The result was

Planck radiation formula

$$u(\nu)\, d\nu = \overline{\epsilon} G(\nu)\, d\nu = \frac{8\pi h}{c^3} \frac{\nu^3\, d\nu}{e^{h\nu/kT} - 1} \tag{9.38}$$

which agrees with the experimental findings.

Although Planck got the right formula, his derivation is, from today's perspective, seriously flawed. We now know that the harmonic oscillators in the cavity walls have the energies $\epsilon_n = (n + \frac{1}{2})h\nu$, not $nh\nu$. Including the zero-point energy of $\frac{1}{2}h\nu$ does not lead to the average energy of Eq. (9.37) when Maxwell-Boltzmann statistics are used. The proper procedure is to consider the em waves in a cavity as a photon gas subject to Bose-Einstein statistics, since the spin of a photon is 1. The average number of photons $f(\nu)$ in each state of energy $\epsilon = h\nu$ is therefore given by the Bose-Einstein distribution function of Eq. (9.26).

The value of α in Eq. (9.26) depends on the number of particles in the system being considered. But the number of photons in a cavity need not be conserved: unlike gas molecules or electrons, photons are created and destroyed all the time. Although the total radiant energy in a cavity at a given temperature remains constant, the number of photons that incorporate this energy can change. As mentioned in Sec. 9.4, the nonconservation of photons means that $\alpha = 0$. Hence the Bose-Einstein distribution function for photons is

Photon distribution function

$$f(\nu) = \frac{1}{e^{h\nu/kT} - 1} \tag{9.39}$$

Equation (9.35) for the number of standing waves of frequency ν per unit volume in a cavity is valid for the number of quantum states of frequency ν since photons also have two directions of polarization, which corresponds to two orientations of their spins relative to their directions of motion. The energy density of photons in a cavity is accordingly

$$u(\nu)\,d\nu = h\nu G(\nu) f(\nu)\,d\nu = \frac{8\pi h}{c^3}\frac{\nu^3\,d\nu}{e^{h\nu/kT} - 1}$$

which is Eq. (9.38).

Example 9.5

How many photons are present in 1.00 cm^3 of radiation in thermal equilibrium at 1000 K? What is their average energy?

Solution

(*a*) The total number of photons per unit volume is given by

$$\frac{N}{V} = \int_0^\infty n(\nu)\,d\nu$$

where $n(\nu)\,d\nu$ is the number of photons per unit volume with frequencies between ν and $\nu + d\nu$. Since such photons have energies of $h\nu$,

$$n(\nu)\,d\nu = \frac{u(\nu)\,d\nu}{h\nu}$$

with $u(\nu)\,d\nu$ being the energy density given by Planck's formula, Eq. (9.38). Hence the total number of photons in the volume V is

$$N = V\int_0^\infty \frac{u(\nu)\,d\nu}{h\nu} = \frac{8\pi V}{c^3}\int_0^\infty \frac{\nu^2\,d\nu}{e^{h\nu/kT} - 1}$$

If we let $h\nu/kT = x$, then $\nu = kTx/h$ and $d\nu = (kT/h)\,dx$, so that

$$N = 8\pi V\left(\frac{kT}{hc}\right)^3 \int_0^\infty \frac{x^2\,dx}{e^x - 1}$$

The definite integral is a standard one equal to 2.404. Inserting the numerical values of the other quantities, with $V = 1.00\text{ cm}^3 = 1.00 \times 10^{-6}\text{ m}^3$, we find that

$$N = 2.03 \times 10^{10}\text{ photons}$$

(*b*) The average energy $\bar{\epsilon}$ of the photons is equal to the total energy per unit volume divided by the number of photons per unit volume:

$$\bar{\epsilon} = \frac{\int_0^\infty u(\nu)\,d\nu}{\int_0^\infty n(\nu)\,d\nu} = \frac{aT^4}{N/V}$$

Since $a = 4\sigma/c$ (see the discussion of the Stefan-Boltzmann law later in this section) and $N = (2.405)\,[8\pi V(kT/hc)^3]$,

$$\bar{\epsilon} = \frac{\sigma c^2 h^3 T}{(2.405)(2\pi k^3)} = 3.73 \times 10^{-20}\text{ J} = 0.233\text{ eV}$$

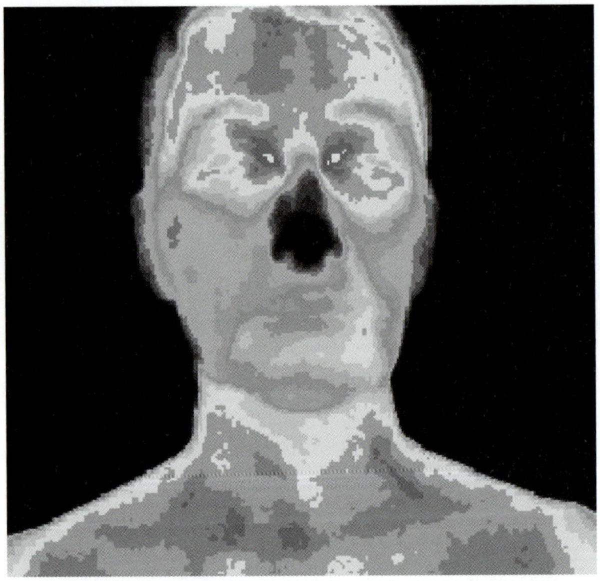

A thermograph measures the amount of infrared radiation each small portion of a person's skin emits and presents this information in pictorial form by different shades of gray or different colors in a thermogram. The skin over a tumor is warmer than elsewhere (perhaps because of increased blood flow or a higher metabolic rate), and thus a thermogram is a valuable diagnostic aid for detecting such maladies as breast and thyroid cancer. A small difference in skin temperature leads to a significant difference in radiation rate.

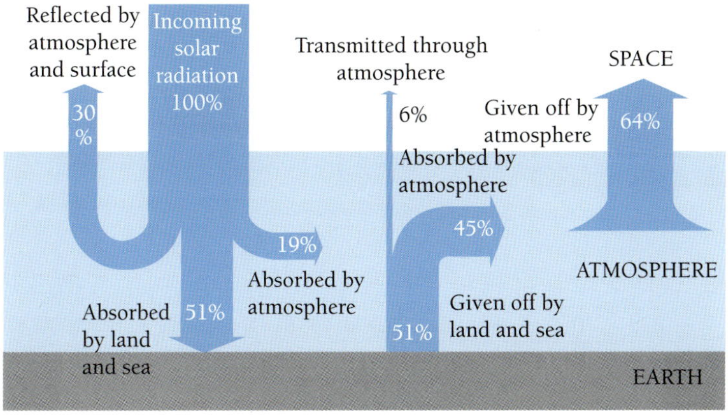

Figure 9.8 The greenhouse effect is important in heating the earth's atmosphere. Much of the short-wavelength visible light from the sun that reaches the earth's surface is reradiated as long-wavelength infrared light that is readily absorbed by CO_2 and H_2O in the atmosphere. This means that the atmosphere heated mainly from below by the earth rather than from above by the sun. The total energy that the earth and its atmosphere radiate into space on the average equals the total energy that they receive from the sun.

It is worth noting again that every body of condensed matter radiates according to Eq. (9.38), regardless of its temperature. An object need not be so hot that it glows conspicuously in the visible region in order to be radiating. The radiation from an object at room temperature, for instance, is chiefly in the infrared part of the spectrum to which the eye is not sensitive. Thus the interior of a greenhouse is warmer than the outside air because sunlight can enter through its windows but the infrared radiation given off by the interior cannot escape through them (Fig. 9.8).

Wien's Displacement Law

An interesting feature of the blackbody spectrum at a given temperature is the wavelength λ_{max} for which the energy density is the greatest. To find λ_{max} we first express Eq. (9.38) in terms of wavelength and solve $du(\lambda)/d\lambda = 0$ for $\lambda = \lambda_{max}$. We obtain in this way

$$\frac{hc}{kT\lambda_{max}} = 4.965$$

which is more conveniently expressed as

Wien's displacement law

$$\lambda_{max}T = \frac{hc}{4.965k} = 2.898 \times 10^{-3}\ \text{m} \cdot \text{K} \tag{9.40}$$

Equation (9.40) is known as **Wien's displacement law.** It quantitatively expresses the empirical fact that the peak in the blackbody spectrum shifts to progressively shorter wavelengths (higher frequencies) as the temperature is increased, as in Fig. 2.6.

Example 9.6

Radiation from the Big Bang has been doppler-shifted to longer wavelengths by the expansion of the universe and today has a spectrum corresponding to that of a blackbody at 2.7 K. Find the wavelength at which the energy density of this radiation is a maximum. In what region of the spectrum is this radiation?

Solution

From Eq. (9.40) we have

$$\lambda_{\max} = \frac{2.898 \times 10^{-3}\ \text{m} \cdot \text{K}}{T} = \frac{2.898 \times 10^{-3}\ \text{m} \cdot \text{K}}{2.7\ \text{K}} = 1.1 \times 10^{-3}\ \text{m} = 1.1\ \text{mm}$$

This wavelength is in the microwave region (see Fig. 2.2). The radiation was first detected in a microwave survey of the sky in 1964.

Stefan-Boltzmann Law

Another result we can obtain from Eq. (9.38) is the total energy density u of the radiation in a cavity. This is the integral of the energy density over all frequencies,

$$u = \int_0^\infty u(\nu)\, d\nu = \frac{8\pi^5 k^4}{15c^3h^3} T^4 = aT^4$$

where a is a universal constant. The total energy density is proportional to the fourth power of the absolute temperature of the cavity walls. We therefore expect that the energy R radiated by an object per second per unit area is also proportional to T^4, a conclusion embodied in the **Stefan-Boltzmann law:**

Stefan-Boltzmann law

$$R = e\sigma T^4 \tag{9.41}$$

The value of **Stefan's constant** σ is

Stefan's constant

$$\sigma = \frac{ac}{4} = 5.670 \times 10^{-8}\ \text{W/m}^2 \cdot \text{K}^4$$

The emissivity e depends on the nature of the radiating surface and ranges from 0, for a perfect reflector which does not radiate at all, to 1, for a blackbody. Some typical values of e are 0.07 for polished steel, 0.6 for oxidized copper and brass, and 0.97 for matte black paint.

Example 9.7

Sunlight arrives at the earth at the rate of about 1.4 kW/m^2 when the sun is directly overhead. The average radius of the earth's orbit is 1.5×10^{11} m and the radius of the sun is 7.0×10^8 m. From these figures find the surface temperature of the sun on the assumption that it radiates like a blackbody, which is approximately true.

Solution

We begin by finding the total power P radiated by the sun. The area of a sphere whose radius r_e is that of the earth's orbit is $4\pi r_e^2$. Since solar radiation falls on this sphere at a rate of $P/A = 1.4\ \text{kW/m}^2$,

$$P = \left(\frac{P}{A}\right)(4\pi r_e^2) = (1.4 \times 10^3\ \text{W/m}^2)(4\pi)(1.5 \times 10^{11}\ \text{m})^2 = 3.96 \times 10^{26}\ \text{W}$$

Next we find the radiation rate R of the sun. If r_s is the sun's radius, its surface area is $4\pi r_s^2$ and

$$R = \frac{\text{power output}}{\text{surface area}} = \frac{P}{4\pi r_s^2} = \frac{3.96 \times 10^{26}\text{W}}{(4\pi)(7.0 \times 10^8\ \text{m})^2} = 6.43 \times 10^7\ \text{W/m}^2$$

The emissivity of a blackbody is $e = 1$, so from Eq. (9.41) we have

$$T = \left(\frac{R}{e\sigma}\right)^{1/4} = \left(\frac{6.43 \times 10^7\ \text{W/m}^2}{(1)(5.67 \times 10^{-8}\ \text{W/m}^2 \cdot \text{K}^4)}\right)^{1/4} = 5.8 \times 10^3\,\text{K}$$

9.7 EINSTEIN'S APPROACH

Introducing stimulated emission

The stimulated emission of radiation was mentioned in Sec. 4.9 as the key concept behind the laser. In a 1917 paper Einstein introduced stimulated emission and used it to arrive at the form of Planck's radiation law in an elegantly simple manner. By the early 1920s this idea together with what had become known about the physics of the atom would have enabled the laser to have been invented then, but somehow nobody connected the dots until over thirty years later.

Let us consider two energy states in a particular atom, a lower one i and an upper one j (Fig. 9.9). If the atom is initially in state i, it can be raised to state j by absorbing a photon of frequency

$$\nu = \frac{E_j - E_i}{h} \tag{9.42}$$

Now we imagine an assembly of N_i atoms in state i and N_j atoms in state j, all in thermal equilibrium at the temperature T with light of frequency ν and energy density $u(\nu)$. The probability that an atom in state i absorbs a photon is proportional to the

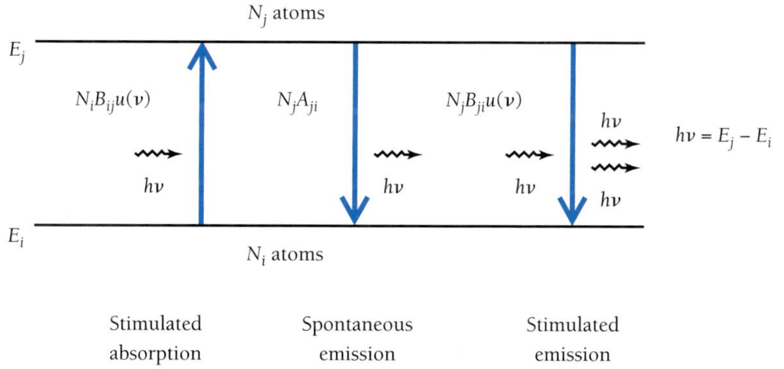

Figure 9.9 Three kinds of transition between states of energies E_i and E_j in an atom. In spontaneous emission, the photon leaves the atom in a random direction. In stimulated emission, the photons that leave are in phase with each other and with the incident photon, and all the photons move in the same direction. The number of atoms that undergo each transition per second is indicated, where the quantity $u(\nu)$ is the density of photons of frequency ν and A_{ji}, B_{ij}, and B_{ji} are constants that depend on the properties of the atomic states.

energy density $u(\nu)$ and also to the properties of states i and j, which we can include in some constant B_{ij}. Hence the number $N_{i\to j}$ of atoms per second that absorb photons is given by

Number of atoms that absorb photons

$$N_{i\to j} = N_i B_{ij} u(\nu) \tag{9.43}$$

An atom in the upper state j has a certain probability A_{ji} to spontaneously drop to state i by emitting a photon of frequency ν. We also suppose that light of frequency ν can somehow interact with an atom in state j to induce its transition to the lower state i. An energy density of $u(\nu)$ therefore means a probability for stimulated emission of $B_{ji}u(\nu)$, where B_{ji}, like B_{ij} and A_{ji}, depends on the properties of states i and j. Since N_j is the number of atoms in state j, the number of atoms per second that fall to the lower state i is

Number of atoms that emit photons

$$N_{j\to i} = N_j[A_{ji} + B_{ji}u(\nu)] \tag{9.44}$$

As discussed in Sec. 4.9, stimulated emission has a classical analog in the behavior of a harmonic oscillator. Of course, classical physics often does not apply on an atomic scale, but we have not assumed that stimulated emission *does* occur, only that it *may* occur. If we are wrong, we will ultimately find merely that $B_{ji} = 0$.

Since the system here is in equilibrium, the number of atoms per second that go from state i to j must equal the number that go from j to i. Therefore

$$N_{i\to j} = N_{j\to i}$$

$$N_i\, B_{ij}u(\nu) = N_j\,[A_{ji} + B_{ji}u(\nu)]$$

Dividing both sides of the latter equation by N_jB_{ji} and solving for $u(\nu)$ gives

$$\left(\frac{N_i}{N_j}\right)\left(\frac{B_{ij}}{B_{ji}}\right)u(\nu) = \frac{A_{ji}}{B_{ji}} + u(\nu)$$

$$u(\nu) = \frac{A_{ji}/B_{ji}}{\left(\frac{N_i}{N_j}\right)\left(\frac{B_{ij}}{B_{ji}}\right) - 1} \tag{9.45}$$

Finally we draw on Eq. (9.2) for the numbers of atoms of energies E_i and E_j in a system of these atoms at the temperature T, which we can write as

$$N_i = Ce^{-E_i/kT}$$

$$N_j = Ce^{-E_j/kT}$$

Hence

$$\frac{N_i}{N_j} = e^{-(E_i-E_j)/kT} = e^{(E_j-E_i)/kT} = e^{h\nu/kT} \tag{9.46}$$

and so

$$u(\nu) = \frac{A_{ji}/B_{ji}}{\left(\frac{B_{ij}}{B_{ji}}\right)e^{h\nu/kT} - 1} \tag{9.47}$$

This formula gives the energy density of photons of frequency ν in equilibrium at the temperature T with atoms whose possible energies are E_i and E_j.

Equation (9.47) is consistent with the Planck radiation law of Eq. (9.38) if

$$B_{ij} = B_{ji} \tag{9.48}$$

and

$$\frac{A_{ji}}{B_{ji}} = \frac{8\pi h\nu^3}{c^3} \tag{9.49}$$

We can draw these conclusions:

1 Stimulated emission does occur and its probability for a transition between two states is equal to the probability for absorption.
2 The ratio between the probabilities for spontaneous and stimulated emission varies with ν^3, so the relative likelihood of spontaneous emission increases rapidly with the energy difference between the two states.
3 All we need to know is one of the probabilities A_{ji}, B_{ij}, B_{ji} to find the others.

Conclusion 3 suggests that the process of spontaneous emission is intimately related to the processes of absorption and stimulated emission. Absorption and stimulated emission can be understood classically by considering the interaction between an atom and an electromagnetic wave, but spontaneous emission can occur in the absence of any such wave, yet apparently by a comparable interaction. This paradox is removed by the theory of quantum electrodynamics. As briefly described in Sec. 6.9, this theory shows that "vacuum fluctuations" in **E** and **B** occur even when **E** = **B** = 0 classically, and these fluctuations, analogs of the zero-point vibrations of a harmonic oscillator, stimulate what is apparently spontaneous emission.

9.8 SPECIFIC HEATS OF SOLIDS

Classical physics fails again

Blackbody radiation is not the only familiar phenomenon whose explanation requires quantum statistical mechanics. Another is the way in which the internal energy of a solid varies with temperature.

Let us consider the molar specific heat of a solid at constant volume, c_V. This is the energy that must be added to 1 kmol of the solid, whose volume is held fixed, to raise its temperature by 1 K. The specific heat at constant pressure c_p is 3 to 5 percent higher than c_V in solids because it includes the work associated with a volume change as well as the change in internal energy.

The internal energy of a solid resides in the vibrations of its constituent particles, which may be atoms, ions, or molecules; we shall refer to them as atoms here for convenience. These vibrations may be resolved into components along three perpendicular axes, so that we may represent each atom by three harmonic oscillators. As we know, according to classical physics a harmonic oscillator in a system of them in thermal equilibrium at the temperature T has an average energy of kT. On this basis each atom in a solid should have $3kT$ of energy. A kilomole of a solid contains Avogadro's number N_0 of atoms, and its total internal energy E at the temperature T accordingly ought to be

Classical internal energy of solid

$$E = 3N_0kT = 3RT \tag{9.50}$$

where $R = N_0k = 8.31 \times 10^3$ J/Kmol · K = 1.99 kcal/kmol · K

is the universal gas constant. (We recall that in an ideal-gas sample of n kilomoles, $pV = nRT$.)

The specific heat at constant volume is given in terms of E by

Specific heat at constant volume

$$c_V = \left(\frac{\partial E}{\partial T}\right)_V$$

and so here

Dulong-Petit law

$$c_V = 3R = 5.97 \text{ kcal/kmol} \cdot \text{K} \tag{9.51}$$

Over a century ago Dulong and Petit found that, indeed, $c_V \approx 3R$ for most solids at room temperature and above, and Eq. (9.51) is known as the **Dulong-Petit law** in their honor.

However, the Dulong-Petit law fails for such light elements as boron, beryllium, and carbon (as diamond), for which c_V = 3.34, 3.85, and 1.46 kcal/kmol · K respectively at 20°C. Even worse, the specific heats of *all* solids drop sharply at low temperatures and approach 0 as T approaches 0 K. Figure 9.10 shows how c_V varies with T for several elements. Clearly something is wrong with the analysis leading up to Eq. (9.51), and it must be something fundamental because the curves of Fig. 9.10 share the same general character.

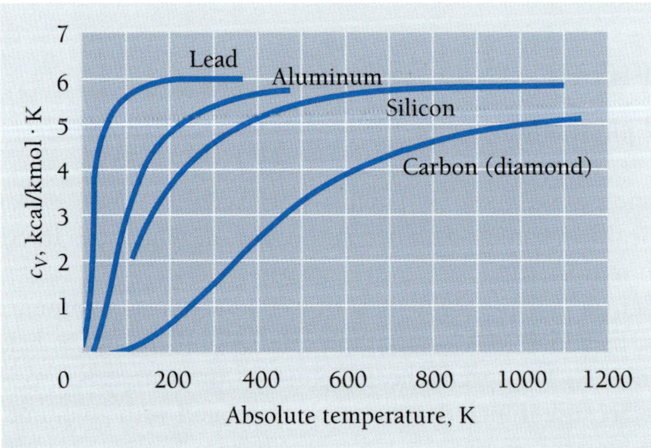

Figure 9.10 The variation with temperature of the molar specific heat at constant volume C_V for several elements.

Einstein's Formula

In 1907 Einstein discerned that the basic flaw in the derivation of Eq. (9.51) lies in the figure of kT for the average energy per oscillator in a solid. This flaw is the same as that responsible for the incorrect Rayleigh-Jeans formula for blackbody radiation. According to Einstein, the probability $f(\nu)$ that an oscillator have the frequency ν is given by Eq. (9.39), $f(\nu) = 1/(e^{h\nu/kT} - 1)$. Hence the average energy for an oscillator whose frequency of vibration is ν is

Average energy per oscillator

$$\bar{\epsilon} = h\nu f(\nu) = \frac{h\nu}{e^{h\nu/kT} - 1} \tag{9.52}$$

and not $\bar{\epsilon} = kT$. The total internal energy of a kilomole of a solid therefore becomes

Internal energy of solid

$$E = 3N_0\bar{\epsilon} = \frac{3N_0 h\nu}{e^{h\nu/kT} - 1} \tag{9.53}$$

and its molar specific heat is

Einstein specific heat formula

$$c_V = \left(\frac{\partial E}{\partial T}\right)_V = 3R\left(\frac{h\nu}{kT}\right)^2 \frac{e^{h\nu/kT}}{(e^{h\nu/kT} - 1)^2} \tag{9.54}$$

We can see at once that this approach is on the right track. At high temperatures, $h\nu \ll kT$, and

$$e^{h\nu/kT} \approx 1 + \frac{h\nu}{kT}$$

since

$$e^x = 1 + x + \frac{x^2}{2!} + \frac{x^3}{3!} + \cdots$$

Hence Eq. (9.52) becomes $\bar{\epsilon} \approx h\nu/(h\nu/kT) = kT$, which leads to $c_V \approx 3R$, the Dulong-Petit value, as it should. At high temperatures the spacing $h\nu$ between possible energies is small relative to kT, so ϵ is effectively continuous and classical physics holds.

As the temperature decreases, the value of c_V given by Eq. (9.54) decreases. The reason for the change from classical behavior is that now the spacing between possible energies is becoming large relative to kT, which inhibits the possession of energies above the zero-point energy. The natural frequency ν for a particular solid can be determined by comparing Eq. (9.54) with an empirical curve of its c_V versus T. The result in the case of aluminum is $\nu = 6.4 \times 10^{12}$ Hz, which agrees with estimates made in other ways, for instance on the basis of elastic moduli.

Why is it that the zero-point energy of a harmonic oscillator does not enter this analysis? As we recall, the permitted energies of a harmonic oscillator are $(n + \frac{1}{2})h\nu$, $n = 0, 1, 2, \ldots$. The ground state of each oscillator in a solid is therefore $\epsilon_0 = \frac{1}{2}h\nu$, the zero-point value, and not $\epsilon_0 = 0$. But the zero-point energy merely adds a constant, temperature-independent term of $\epsilon_0 = (3N_0)(\frac{1}{2}h\nu)$ to the molar energy of a solid, and this term vanishes when the partial derivative $(\partial E/\partial T)_V$ is taken to find c_V.

The Debye Theory

Although Einstein's formula predicts that $c_V \to 0$ as $T \to 0$, as observed, the precise manner of this approach does not agree too well with the data. The inadequacy of Eq. (9.54) at low temperatures led Peter Debye to look at the problem in a different way in 1912. In Einstein's model, each atom is regarded as vibrating independently of its neighbors at a fixed frequency ν. Debye went to the opposite extreme and considered a solid as a continuous elastic body. Instead of residing in the vibrations of individual atoms, the internal energy of a solid according to the new model resides in elastic standing waves.

The elastic waves in a solid are of two kinds, longitudinal and transverse, and range in frequency from 0 to a mximum ν_m. (The interatomic spacing in a solid sets a lower limit to the possible wavelengths and hence an upper limit to the frequencies.) Debye assumed that the total number of different standing waves in a kilomole of a solid is equal to its $3N_0$ degrees of freedom. These waves, like em waves, have energies quantized in units of $h\nu$. A quantum of acoustic energy in a solid is called a **phonon**, and it travels with the speed of sound since sound waves are elastic in nature. The concept of phonons is quite general and has applications other than in connection with specific heats.

Debye finally asserted that a phonon gas has the same statistical behavior as a photon gas or a system of harmonic oscillators in thermal equilibrium, so that the average energy $\overline{\epsilon}$ per standing wave is the same as in Eq. (9.52). The resulting formula for c_V reproduces the observed curves of c_V versus T quite well at all temperatures.

Peter Debye, who was Dutch, did original work in many aspects of both physics and chemistry, at first in Germany and later at Cornell University. Although Heisenberg, a colleague for a time, thought him lazy ("I could frequently see him walking around in his garden and watering the roses even during duty hours of the Institute"), he published nearly 250 papers and received the Nobel Prize in chemistry in 1936.

9.9 FREE ELECTRONS IN A METAL

No more than one electron per quantum state

The classical, Einstein, and Debye theories of specific heats of solids apply with equal degrees of success to both metals and nonmetals, which is strange because they ignore the presence of free electrons in metals.

As discussed in Chap. 10, in a typical metal each atom contributes one electron to the common "electron gas," so in 1 kilomole of the metal there are N_0 free electrons. If these electrons behave like the molecules of an ideal gas, each would have $\frac{3}{2}kT$ of kinetic energy on the average. The metal would then have

$$E_e = \frac{3}{2}N_0kT = \frac{3}{2}RT$$

of internal energy per kilomole due to the electrons. The molar specific heat due to the electrons should therefore be

$$c_{Ve} = \left(\frac{\partial E_e}{\partial T}\right)_V = \frac{3}{2}R$$

and the total specific heat of the metal should be

$$c_V = 3R + \frac{3}{2}R = \frac{9}{2}R$$

at high temperatures where a classical analysis is valid. Actually, of course, the Dulong-Petit value of $3R$ holds at high temperatures, from which we conclude that the free electrons do not in fact contribute to the specific heat. Why not?

If we reflect on the characters of the entities involved in the specific heat of a metal, the answer begins to emerge. Both the harmonic oscillators of Einstein's model and the phonons of Debye's model are bosons and obey Bose-Einstein statistics, which place no upper limit on the occupancy of a particular quantum state. Electrons, however, are fermions and obey Fermi-Dirac statistics, which means that no more than one electron can occupy each quantum state. Although both systems of bosons and systems of fermions approach Maxwell-Boltzmann statistics with average energies $\overline{\epsilon} = \frac{1}{2}kT$ per degree of freedom at "high" temperatures, how high is high enough for classical behavior is not necessarily the same for the two kinds of systems in a metal.

According to Eq. (9.29), the distribution function that gives the average occupancy of a state of energy ϵ in a system of fermions is

Average occupancy per state

$$f_{FD}(\epsilon) = \frac{1}{e^{(\epsilon - \epsilon_F)/kT} + 1} \tag{9.29}$$

What we also need is an expression for $g(\epsilon)\,d\epsilon$, the number of quantum states available to electrons with energies between ϵ and $\epsilon + d\epsilon$.

We can use exactly the same reasoning to find $g(\epsilon)\,d\epsilon$ that we used to find the number of standing waves in a cavity with the wavelength λ in Sec. 9.5. The correspondence is exact because there are two possible spin states, $m_s = +\frac{1}{2}$ and $m_s = -\frac{1}{2}$ ("up" and "down"), for electrons, just as there are two independent directions of polarization for otherwise identical standing waves.

We found earlier that the number of standing waves in a cubical cavity L on a side is

$$g(j)\,dj = \pi j^2\,dj \tag{9.33}$$

where $j = 2L/\lambda$. In the case of an electron, λ is its de Broglie wavelength of $\lambda = h/p$. Electrons in a metal have nonrelativistic velocities, so $p = \sqrt{2m\epsilon}$ and

$$j = \frac{2L}{\lambda} = \frac{2Lp}{h} = \frac{2L\sqrt{2m\epsilon}}{h} \qquad dj = \frac{L}{h}\sqrt{\frac{2m}{\epsilon}}\,d\epsilon$$

Using these expressions for j and dj in Eq. (9.33) gives

$$g(\epsilon)\,d\epsilon = \frac{8\sqrt{2}\pi L^3 m^{3/2}}{h^3}\sqrt{\epsilon}\,d\epsilon$$

As in the case of standing waves in a cavity the exact shape of the metal sample does not matter, so we can substitute its volume V for L^3 to give

Number of electron states

$$g(\epsilon)\,d\epsilon = \frac{8\sqrt{2}\pi V m^{3/2}}{h^3}\sqrt{\epsilon}\,d\epsilon \tag{9.55}$$

Fermi Energy

The final step is to calculate the value of ϵ_F, the Fermi energy. As mentioned in Sec. 9.4, we can do this by filling up the energy states in the metal sample at $T = 0$ with the N free electrons it contains in order of increasing energy starting from $\epsilon = 0$. The highest state to be filled will then have the energy $\epsilon = \epsilon_F$ by definition. The number of electrons that can have the same energy ϵ is equal to the number of states that have this energy, since each state is limited to one electron. Hence

$$N = \int_0^{\epsilon_F} g(\epsilon)\, d\epsilon = \frac{8\sqrt{2}\pi V m^{3/2}}{h^3}\int_0^{\epsilon_F}\sqrt{\epsilon}\, d\epsilon = \frac{16\sqrt{2}\pi V m^{3/2}}{3h^3}\epsilon_F^{3/2}$$

and so

Fermi energy

$$\epsilon_F = \frac{h^2}{2m}\left(\frac{3N}{8\pi V}\right)^{2/3} \tag{9.56}$$

The quantity N/V is the density of free electrons.

Example 9.8

Find the Fermi energy in copper on the assumption that each copper atom contributes one free electron to the electron gas. (This is a reasonable assumption since, from Table 7.4, a copper atom has a single 4*s* electron outside closed inner shells.) The density of copper is 8.94×10^3 kg/m^3 and its atomic mass is 63.5 u.

Solution

The electron density N/V in copper is equal to the number of copper atoms per unit volume. Since 1 u = 1.66×10^{-27} kg,

$$\frac{N}{V} = \frac{\text{atoms}}{\text{m}^3} = \frac{\text{mass/m}^3}{\text{mass/atom}} = \frac{8.94 \times 10^3\ \text{kg/m}^3}{(63.5\ \text{u}) \times (1.66 \times 10^{-27}\ \text{kg/u})}$$

$$= 8.48 \times 10^{28}\ \text{atoms/m}^3 = 8.48 \times 10^{28}\ \text{electrons/m}^3$$

The corresponding Fermi energy is, from (9.56),

$$\epsilon_F = \frac{(6.63 \times 10^{-34}\ \text{J}\cdot\text{s})^2}{(2)(9.11 \times 10^{-31}\ \text{kg/electron})}\left[\frac{(3)(8.48 \times 10^{28}\ \text{electrons/m}^3)}{8\pi}\right]^{2/3}$$

$$= 1.13 \times 10^{-18}\ \text{J} = 7.04\ \text{eV}$$

At absolute zero, $T = 0$ K, there would be electrons with energies of up to 7.04 eV in copper (corresponding to speeds of up to 1.6×10^6 m/s!). By contrast, *all* the molecules in an ideal gas at 0 K would have zero energy. The electron gas in a metal is said to be **degenerate.**

9.10 ELECTRON-ENERGY DISTRIBUTION

Why the electrons in a metal do not contribute to its specific heat except at very high and very low temperatures

With the help of Eqs. (9.29) and (9.55) we have for the number of electrons in an electron gas that have energies between ϵ and $\epsilon + d\epsilon$

$$n(\epsilon)\, d\epsilon = g(\epsilon)f(\epsilon)\, d\epsilon = \frac{(8\sqrt{2}\pi V m^{3/2}/h^3)\sqrt{\epsilon}\, d\epsilon}{e^{(\epsilon-\epsilon_F)/kT} + 1} \tag{9.57}$$

If we express the numerator of Eq. (9.57) in terms of the Fermi energy ϵ_F we get

Electron energy distribution

$$n(\epsilon)\, d\epsilon = \frac{(3N/2)\, \epsilon_F^{-3/2}\sqrt{\epsilon}\, d\epsilon}{e^{(\epsilon-\epsilon_F)/kT} + 1} \tag{9.58}$$

This formula is plotted in Fig. 9.11 for $T = 0$, 300, and 1200 K.

It is interesting to determine the average electron energy at 0 K. To do this, we first find the total energy E_0 at 0 K, which is

$$E_0 = \int_0^{\epsilon_F} \epsilon n(\epsilon)\, d\epsilon$$

Since at $T = 0$ K all the electrons have energies less than or equal to the Fermi energy ϵ_F, we may let

$$e^{(\epsilon-\epsilon_F)/kT} = e^{-\infty} = 0$$

and

$$E_0 = \frac{3N}{2}\epsilon_F^{-3/2}\int_0^{\epsilon_F} \epsilon^{3/2}\, d\epsilon = \frac{3}{5}N\epsilon_F$$

The average electron energy $\bar{\epsilon}_0$ is this total energy divided by the number N of electrons present, which gives

Average electron energy at $T = 0$

$$\bar{\epsilon}_0 = \frac{3}{5}\epsilon_F \tag{9.59}$$

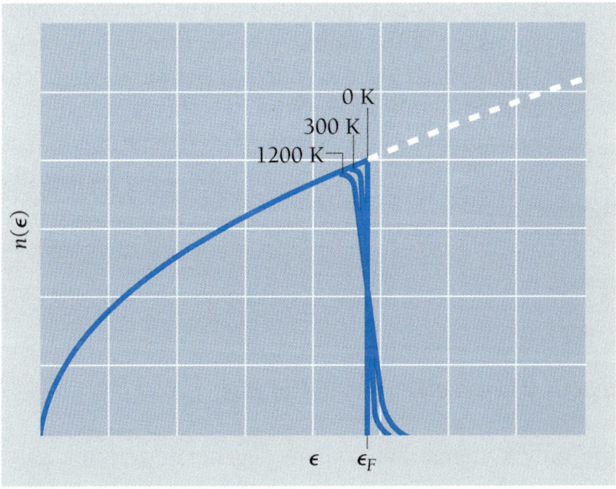

Figure 9.11 Distribution of electron energies in a metal at various temperatures.

Table 9.2 Some Fermi Energies

Metal		Fermi Energy, eV
Lithium	Li	4.72
Sodium	Na	3.12
Aluminum	Al	11.8
Potassium	K	2.14
Cesium	Cs	1.53
Copper	Cu	7.04
Zinc	Zn	11.0
Silver	Ag	5.51
Gold	Au	5.54

Since Fermi energies for metals are usually several electronvolts (Table 9.2), the average electron energy in them at 0 K will also be of this order of magnitude. The temperature of an ideal gas whose molecules have an average kinetic energy of 1 eV is 11,600 K. If free electrons behaved classically, a sample of copper would have to be at a temperature of about 50,000 K for its electrons to have the same average energy they actually have at 0 K!

The failure of the free electrons in a metal to contribute appreciably to its specific heat follows directly from their energy distribution. When a metal is heated, only those electrons near the very top of the energy distribution—those within about kT of the Fermi energy—are excited to higher energy states. The less energetic electrons cannot absorb more energy because the states above them are already filled. It is unlikely that an electron with, say, an energy ϵ that is 0.5 eV below ϵ_F can leapfrog the filled states above it to the nearest vacant state when kT at room temperature is 0.025 eV and even at 500 K is only 0.043 eV.

A detailed calculation shows that the specific heat of the electron gas in a metal is given by

Electron specific heat

$$c_{Ve} = \frac{\pi^2}{2}\left(\frac{kT}{\epsilon_F}\right)R \tag{9.60}$$

At room temperature, kT/ϵ_F ranges from 0.016 for cesium to 0.0021 for aluminum for the metals listed in Table 9.2, so the coefficient of R is very much smaller than the classical figure of $\frac{3}{2}$. The dominance of the atomic specific heat c_V in a metal over the electronic specific heat is pronounced over a wide temperature range. However, at very low temperatures c_{Ve} becomes significant because c_V is then approximately proportional to T^3 whereas c_{Ve} is proportional to T. At very high temperatures c_V has leveled out at about $3R$ while c_{Ve} has continued to increase, and the contribution of c_{Ve} to the total specific heat is then detectable.

9.11 DYING STARS

What happens when a star runs out of fuel

Metals are not the only systems that contain degenerate fermion gases—many dead and dying stars fall into this category also.

White Dwarfs

Perhaps 10 percent of the stars in our galaxy are believed to be **white dwarfs.** These are stars in the final stages of their evolution with original masses that were less than about 8 solar masses. After the nuclear reactions that provided it with energy run out of fuel, such a star becomes unstable, swells to become a red giant, and eventually throws off its outer layer. The remaining core then cools and contracts gravitationally until its atoms collapse into nuclei and electrons packed closely together. A typical white dwarf has a mass of two-thirds that of the sun but is only about the size of the earth; a handful of its matter would weigh over a ton on the earth.

As a prospective white dwarf contracts, its volume V decreases and as a result the Fermi energy ϵ_F of its electrons increases; see Eq. (9.56). When ϵ_F exceeds kT, the electrons form a degenerate gas. A reasonable estimate for the Fermi energy in a typical white dwarf is 0.5 MeV. The nuclei present are much more massive than the electrons, and because ϵ_F is inversely proportional to m, they continue to behave classically.

With the star's nuclear reactions at an end, the nuclei cool down and come together under the influence of gravitation. The electrons, however, cannot cool down since most of the low-energy states available to them are already filled; the situation corresponds to Fig. 9.6*b*. The electron gas becomes hotter and hotter as the star shrinks. Even though the total electron mass is only a small fraction of the star's mass, in time it exerts enough pressure to stop the gravitational contraction. Thus the size of a white dwarf is determined by a balance between the inward gravitational pull of its atomic nuclei and the pressure of its degenerate electron gas.

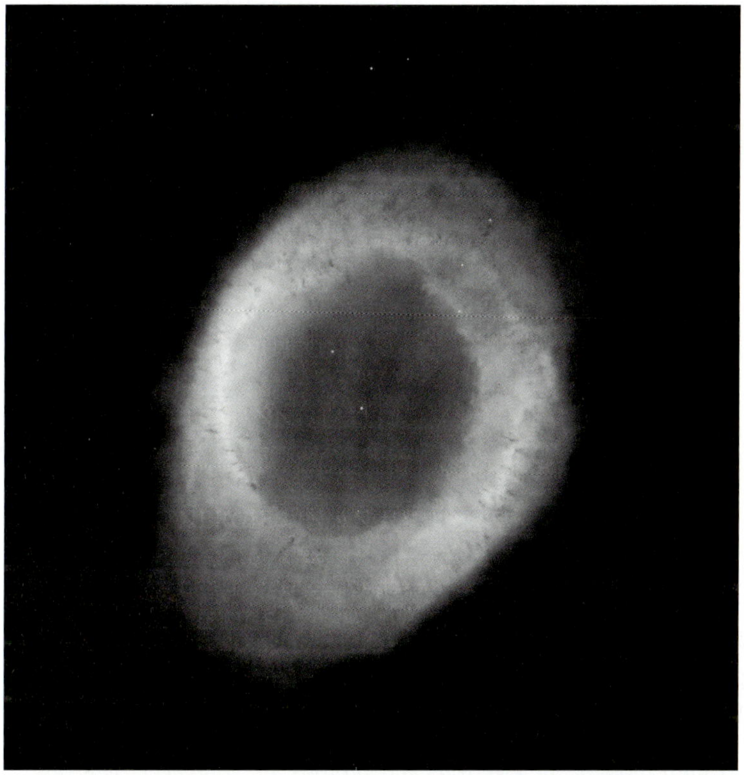

The Ring nebula in the constellation Lyra is a shell of gas moving outward from the star at its center, which is in the process of becoming a white dwarf.

The Chandrasekhar Limit

The maximum white dwarf mass of $1.4M_{sun}$ is called the **Chandrasekhar limit** after its discoverer, Subrahmanyan Chandrasekhar, who calculated it in 1930 at the age of nineteen on the ship bringing him from his native India to take up a fellowship at Cambridge. Two observations underlie the existence of the limit:

1 Both the internal energy of a dwarf and its gravitational potential energy vary in the same way ($1/R$) with its radius.
2 Its internal energy is proportional to the mass M of the dwarf but its gravitational potential energy is proportional to M^2.

Because of (2), the inward gravitational pressure dominates for a sufficiently massive dwarf, which causes a contraction that cannot be stopped by the pressure of its electron gas as R decreases because of (1).

What becomes of dying stars with $M > 1.4M_{sun}$? The answer then seemed to be total collapse into what today is called a black hole. (We know now that a neutron star can be somewhat more massive than a white dwarf and still be stable.) The noted Cambridge astrophysicist Arthur Eddington, one of Chandrasekhar's heroes, publicly derided the idea of total collapse as absurd, a humiliation that was one of the reasons Chandrasekhar later moved to the University of Chicago where he had a distinguished career. His work on white dwarfs led to a Nobel Prize in 1983.

In a white dwarf, only electrons with the highest energies can radiate, since only such electrons have empty lower states to fall into. As the states lower than ϵ_F become filled, the star becomes dimmer and dimmer and in a few billion years ceases to radiate at all. It is now a **black dwarf,** a dead lump of matter, since the energies of its electrons are forever locked up below the Fermi level.

The greater the mass of a shrinking star, the greater the electron pressure needed to keep it in equilibrium. If the mass is more than about $1.4M_{sun}$, gravity is so overwhelming that the electron gas can never counteract it. Such a star cannot become a stable white dwarf.

Neutron Stars

A star too heavy—more than about 8 solar masses—to follow the evolutionary path that leads to a white dwarf has a different fate. The large mass of such a star causes it to collapse abruptly when out of fuel, and then to explode violently. The explosion flings into space most of the star's mass. An event of this kind, called a **supernova,** is billions of times brighter than the original star ever was.

What is left after a supernova explosion may be a remnant whose mass is greater than $1.4M_{sun}$. As this star contracts gravitationally, its electrons become more and more energetic. When the Fermi energy reaches about 1.1 MeV, the average electron energy is 0.8 MeV, which is the minimum energy needed for an electron to react with a proton to produce a neutron. (The neutron mass exceeds the combined mass of an electron and a proton by the mass equivalent of 0.8 MeV.) This point is reached when the star's density is perhaps 20 times that of a white dwarf. From then on neutrons are produced until most of the electrons and protons are gone. The neutrons, which are fermions, end up as a degenerate gas, and their pressure supports the star against further gravitational shrinkage.

Discovery of Neutron Stars

In a paper published in 1934, only two years after the discovery of the neutron, the astronomers Walter Baade and Fritz Zwicky proposed that, at the end of its active life, an exceptionally heavy star undergoes a cataclysmic explosion that appears in the sky as a brilliant supernova. "We advance the view that a supernova represents the transition of an ordinary star into a neutron star, consisting mainly of neutrons. Such a star may possess a very small radius and an extremely high density [and would] represent the most stable configuration of matter as such."

Although several physicists developed the theory of neutron stars further in the next few years, it was not until pulsars were detected in 1967 that their existence was confirmed. In that year unusual radio signals with an extremely regular period of exactly 1.33730113 s were picked up that came from a source in the direction of the constellation Vulpecula. They were found by Jocelyn Bell (now Jocelyn Bell Burnell), then a graduate student at Cambridge University; her thesis advisor received the Nobel Prize in physics for the discovery. At first only radio emissions from pulsars were observed, but later flashes of visible light were detected from some pulsars that were synchronized with the radio signals.

The power output of a pulsar is about 10^{26} W, which is comparable with the total power output of the sun. So strong a source of energy cannot possibly be switched on and off in a fraction of a second, which is the period of some pulsars, nor can it be the size of the sun. Even if the sun were to suddenly stop radiating, it would take an interval of 2.3 s before light stopped reaching us, because all parts of the sun that we see are not the same distance away. Nor could a sun-sized pulsar spin around in less than a second per turn. The conclusion is that a pulsar must have the mass of a star, in order to be able to emit so much energy, but it must be very much smaller than a star, in order that its signals fluctuate so rapidly. From these and other considerations it seems clear that pulsars are neutron stars in rapid rotation.

Neutron stars are thought to be 10 to 15 km in radius with masses between 1.4 and $\sim 3M_{\text{sun}}$ (Fig. 9.12). If the earth were this dense, it would fit into a large apartment house. Stars called **pulsars** are believed to be neutron stars that are rotating rapidly. Most stars have magnetic fields, and as a star contracts into a neutron star, its surface field increases enormously. The magnetic field is produced by motions of the electrons that remain in its interior, and since they cannot lose energy (the gas they form is degenerate, with all the lowest states filled), the field should persist for a time long compared with the age of the universe.

The magnetic field of a pulsar traps tails of ionized gas that radiate light, radio waves, and x-rays. If the magnetic axis is not aligned with the rotational axis, a distant observer, such as an astronomer on the earth, will receive bursts of radiation as the pulsar spins. Thus a pulsar is like a lighthouse whose flashes are due to a rotating beam of light.

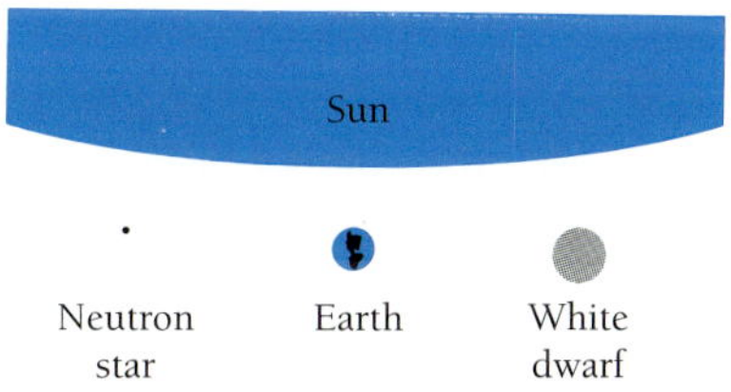

Figure 9.12 A comparison of a white dwarf and a neutron star with the sun and the earth. Both white dwarfs and neutron stars are thought to have masses similar to that of the sun.

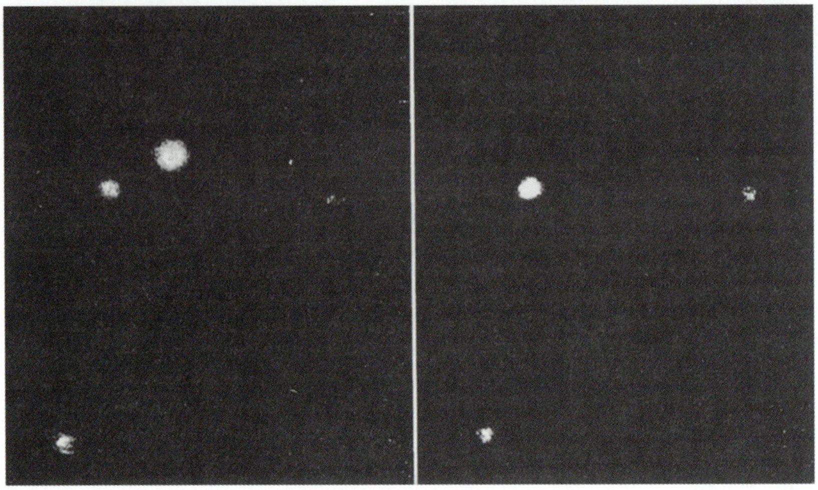

The pulsar at the center of the Crab nebula flashes 30 times per second and is throught to be a rotating neutron star. These photographs were taken at maximum and minimum emission. The nebula itself is shown in the photograph at the start of this chapter; it is now about 10 light-years across and is still expanding.

Over 1000 pulsars have been discovered, all with periods between 0.0016 and 4 s. The best known pulsar, which is at the center of the Crab nebula, has a period of 0.033 s that is increasing at a rate of 10^{-5} s per year as the pulsar loses angular momentum.

Black Holes

An old star whose mass is less than $1.4M_{sun}$ becomes a white dwarf and one whose mass is between 1.4 and $\sim 3M_{sun}$ becomes a neutron star. What about still heavier old stars? Neither a degenerate electron gas nor a degenerate neutron gas can resist gravitational collapse when $M > \sim 3M_{sun}$. Does such a star end up as a point in space? This does not seem likely. One argument comes from the uncertainty principle, $\Delta x\ \Delta p \geq \hbar/2$. This principle prevents a hydrogen atom from collapsing beyond a certain size under the inward pull of the proton's electric field. The same principle ought to prevent a massive old star from collapsing beyond a certain size under an inward gravitational pull. Or perhaps the quarks of which neutrons and protons are composed (Chap. 13) have special properties that stabilize such a star when it reaches a certain density.

Whatever its final nature, as an old star of $M > 3M_{sun}$ contracts it passes the Schwarzschild radius of Eq. (2.30) and from then on is a black hole (Sec. 2.9). We can receive no further information from the star because its gravitational field is too intense to permit anything, even photons, to escape past its event horizon.

Not only heavy stars end up as black holes. As time goes on, both white dwarfs and neutron stars attract more and more cosmic dust and gas. When they have gathered up enough additional mass, they too will become black holes. If the universe lasts long enough, then everything in it may be in the form of black holes.

EXERCISES

By the pricking of my thumbs / Something wicked this way comes. —William Shakespeare, *Macbeth*

9.2 Maxwell-Boltzmann Statistics

1. At what temperature would one in a thousand of the atoms in a gas of atomic hydrogen be in the $n = 2$ energy level?

2. The temperature in part of the sun's atmosphere is 5000 K. Find the relative numbers of hydrogen atoms in this region that are in the $n = 1, 2, 3,$ and 4 energy levels. Be sure to take into account the multiplicity of each level.

3. The $3^2P_{1/2}$ first excited state in sodium is 2.093 eV above the $3^2S_{1/2}$ ground state. Find the ratio between the numbers of atoms in each state in sodium vapor at 1200 K. (See Example 7.6.)

4. The frequency of vibration of the H_2 molecule is 1.32×10^{14} Hz. (*a*) Find the relative populations of the $v = 0, 1, 2, 3,$ and 4 vibrational states at 5000 K. (*b*) Can the populations of the $v = 2$ and $v = 3$ states ever be equal? If so, at what temperature does this occur?

5. The moment of inertia of the H_2 molecule is 4.64×10^{-48} $\text{kg} \cdot \text{m}^2$. (*a*) Find the relative populations of the $J = 0, 1, 2, 3,$ and 4 rotational states at 300 K. (*b*) Can the populations of the $J = 2$ and $J = 3$ states ever be equal? If so, at what temperature does this occur?

6. In a certain four-level laser (Sec. 4.9), the final state of the laser transition is 0.03 eV above the ground state. What fraction of the atoms are in this state at 300 K in the absence of external excitation? What is the minimum fraction of the atoms that must be excited in order for laser amplification to occur at this temperature? Why? How is the situation changed at 100 K? Would you expect cooling a three-level laser to have the same effect?

9.3 Molecular Energies in an Ideal Gas

7. Find $\bar{v}$ and v_{rms} for an assembly of two molecules, one with a speed of 1.00 m/s and the other with a speed of 3.00 m/s.

8. Show that the average kinetic energy per molecule at room temperature (20°C) is much less than the energy needed to raise a hydrogen atom from its ground state to its first excited state.

9. At what temperature will the average molecular kinetic energy in gaseous hydrogen equal the binding energy of a hydrogen atom?

10. Show that the de Broglie wavelength of an oxygen molecule in thermal equilibrium in the atmosphere at 20°C is smaller than its diameter of about 4×10^{-10} m.

11. Find the width due to the Doppler effect of the 656.3-nm spectral line emitted by a gas of atomic hydrogen at 500 K.

12. Verify that the most probable speed of an ideal-gas molecule is $\sqrt{2kT/m}$.

13. Verify that the average value of $1/v$ for an ideal-gas molecule is $\sqrt{2m/\pi kT}$. [*Note:* $\int_0^\infty ve^{-av^2}\,dv = 1/(2a)$]

14. A flux of 10^{12} neutrons/m^2 emerges each second from a port in a nuclear reactor. If these neutrons have a Maxwell-Boltzmann energy distribution corresponding to $T = 300$ K, calculate the density of neutrons in the beam.

9.4 Quantum Statistics

15. At the same temperature, will a gas of classical molecules, a gas of bosons, or a gas of fermions exert the greatest pressure? The least pressure? Why?

16. What is the significance of the Fermi energy in a fermion system at 0 K? At $T > 0$ K?

9.5 Rayleigh-Jeans Formula

17. How many independent standing waves with wavelengths between 9.5 and 10.5 mm can occur in a cubical cavity 1 m on a side? How many with wavelengths between 99.5 and 100.5 mm? (*Hint:* First show that $g(\lambda)\,d\lambda = 8\pi L^3\,d\lambda/\lambda^4$.)

9.6 Planck Radiation Law

18. If a red star and a white star radiate energy at the same rate, can they be the same size? If not, which must be the larger?

19. A thermograph measures the rate at which each small portion of a person's skin emits infrared radiation. To verify that a small difference in skin temperature means a significant difference in radiation rate, find the percentage difference between the total radiation from skin at 34° and at 35°C.

20. Sunspots appear dark, although their temperatures are typically 5000 K, because the rest of the sun's surface is even hotter, about 5800 K. Compare the radiation rates of surfaces of the same emissivity whose temperatures are respectively 5000 and 5800 K.

21. At what rate would solar energy arrive at the earth if the solar surface had a temperature 10 percent lower than it is?

22. The sun's mass is 2.0×10^{30} kg, its radius is 7.0×10^8 m, and its surface temperature is 5.8×10^3 K. How many years are needed for the sun to lose 1.0 percent of its mass by radiation?

23. An object is at a temperature of 400°C. At what temperature would it radiate energy twice as fast?

24. A copper sphere 5 cm in diameter whose emissivity is 0.3 is heated in a furnace to 400°C. At what rate does it radiate?

25. At what rate does radiation escape from a hole 10 cm^2 in area in the wall of a furnace whose interior is at 700°C?

26. An object at 500°C is just hot enough to glow perceptibly; at 750°C it appears cherry-red in color. If a certain blackbody radiates 1.00 kW when its temperature is 500°C, at what rate will it radiate when its temperature is 750°C?

27. Find the surface area of a blackbody that radiates 1.00 kW when its temperature is 500°C. If the blackbody is a sphere, what is its radius?

28. The microprocessors used in computers produce heat at rates as high as 30 W per square centimeter of surface area. At what temperature would a blackbody be if it had such a radiance? (Microprocessors are cooled to keep from being damaged by the heat they give off.)

29. Considering the sun as a blackbody at 6000 K, estimate the proportion of its total radiation that consists of yellow light between 570 and 590 nm.

30. Find the peak wavelength in the spectrum of the radiation from a blackbody at a temperature of 500°C. In what part of the em spectrum is this wavelength?

31. The brightest part of the spectrum of the star Sirius is located at a wavelength of about 290 nm. What is the surface temperature of Sirius?

32. The peak wavelength in the spectrum of the radiation from a cavity is 3.00 μm. Find the total energy density in the cavity.

33. A gas cloud in our galaxy emits radiation at a rate of 1.0×10^{27} W. The radiation has its maximum intensity at a wavelength of 10 μm. If the cloud is spherical and radiates like a blackbody, find its surface temperature and its diameter.

34. (*a*) Find the energy density in the universe of the 2.7-K radiation mentioned in Example 9.6. (*b*) Find the approximate number of photons per cubic meter in this radiation by assuming that all the photons have the wavelength of 1.1 mm at which the energy density is a maximum.

35. Find the specific heat at constant volume of 1.00 cm^3 of radiation in thermal equilibrium at 1000 K.

9.10 Electron-Energy Distribution

36. What is the connection between the fact that the free electrons in a metal obey Fermi statistics and the fact that the photoelectric effect is virtually temperature-independent?

37. Show that the median energy in a free-electron gas at $T = 0$ is equal to $\epsilon_F/2^{2/3} = 0.630\epsilon_F$.

38. The Fermi energy in copper is 7.04 eV. Compare the approximate average energy of the free electrons in copper at room temperature ($kT = 0.025$ eV) with their average energy if they followed Maxwell-Boltzmann statistics.

39. The Fermi energy in silver is 5.51 eV. (*a*) What is the average energy of the free electrons in silver at 0 K? (*b*) What temperature is necessary for the average molecular energy in an ideal gas to have this value? (*c*) What is the speed of an electron with this energy?

40. The Fermi energy in copper is 7.04 eV. (*a*) Approximately what percentage of the free electrons in copper are in excited states at room temperature? (*b*) At the melting point of copper, 1083°C?

41. Use Eq. (9.29) to show that, in a system of fermions at $T = 0$, all states of $\epsilon < \epsilon_F$ are occupied and all states of $\epsilon > \epsilon_F$ are unoccupied.

42. An electron gas at the temperature T has a Fermi energy of ϵ_F. (*a*) At what energy ϵ is there a 5.00 percent probability that a state of that energy is occupied? (*b*) At what energy is there a 95.00 percent probability that a state of that energy is occupied? Express the answers in terms of ϵ_F and kT.

43. Show that, if the average occupancy of a state of energy $\epsilon_F + \Delta\epsilon$ is f_1 at any temperature, then the average occupancy of a state of energy $\epsilon_F - \Delta\epsilon$ is $f_2 = 1 - f_1$. (This is the reason for the symmetry of the curves in Fig. 9.11 about ϵ_F.)

44. The density of aluminum is 2.70 g/cm^3 and its atomic mass is 26.97 u. The electronic structure of aluminum is given in Table 7.4 (the energy difference between 3*s* and 3*p* electrons is very small), and the effective mass of an electron in aluminum is 0.97 m_e Calculate the Fermi energy in aluminum. (Effective mass is discussed at the end of Sec. 10.8.)

45. The density of zinc is 7.13 g/cm^3 and its atomic mass is 65.4 u. The electronic structure of zinc is given in Table 7.4, and the effective mass of an electron in zinc is 0.85 m_e. Calculate the Fermi energy in zinc.

46. Find the number of electrons each lead atom contributes to the electron gas in solid lead by comparing the density of free electrons obtained from Eq. (9.56) with the number of lead atoms per unit volume. The density of lead is 1.1×10^4 kg/m^3 and the Fermi energy in lead is 9.4 eV.

47. Find the number of electron states per electronvolt at $\epsilon = \epsilon_F/2$ in a 1.00-g sample of copper at 0 K. Are we justified in considering the electron energy distribution as continuous in a metal?

48. The specific heat of copper at 20°C is 0.0920 kcal/kg · °C. (*a*) Express this in joules per kilomole per kelvin (J/kmol · K). (*b*) What proportion of the specific heat can be attributed to the electron gas, assuming one free electron per copper atom?

49. The Bose-Einstein and Fermi-Dirac distribution functions both reduce to the Maxwell-Boltzmann function when $e^{\alpha}e^{\epsilon/kT} \gg 1$. For energies in the neighborhood of kT, this approximation holds if $e^{\alpha} \gg 1$. Helium atoms have spin 0 and so obey Bose-Einstein statistics. Verify that $f(\epsilon) \approx 1/e^{\alpha}e^{\epsilon/kT} \approx Ae^{-\epsilon/kT}$ is valid for He at STP (20°C and atmospheric pressure, when the volume of 1 kmol of any gas is ≈ 22.4 m^3) by showing that $A \ll 1$ under these circumstances. To do this, use Eq. (9.55) for $g(\epsilon)\,d\epsilon$ with a coeffficient of 4 instead of 8 since a He atom does not have the two spin states of an electron, and employing the approximation, find A from the normalization condition $\int_0^\infty n(\epsilon)\,d\epsilon = N$, where N is the total number of atoms in the sample. (A kilomole of He contains Avogadro's number N_0 of atoms, the atomic mass of He is 4.00 u, and $\int_0^\infty \sqrt{x}e^{-ax}\,dx = \sqrt{\pi/a}/2a$.)

50. Helium is a liquid of density 145 kg/m^3 at atmospheric pressure and temperatures under 4.2 K. Use the method of Exercise 49 to show that $A > 1$ for liquid helium, so that it cannot be satisfactorily described by Maxwell-Boltzmann statistics.

51. The Fermi-Dirac distribution function for the free electrons in a metal cannot be approximated by the Maxwell-Boltzmann function at STP (see Exercise 49) for energies in the neighborhood of kT. Verify this by using the method of Exercise 49 to show that $A > 1$ in copper if $f(\epsilon) \approx Ae^{-\epsilon/kT}$. As calculated in Sec. 9.9 $N/V = 8.48 \times 10^{28}$ electrons/m^3 for copper. Note that Eq. (9.55) must be used unchanged here.

9.11 Dying Stars

52. The sun has a mass of 2.0×10^{30} kg and a radius of 7.0×10^8 m. Assume it consists of completely ionized hydrogen at a temperature of 10^7 K. (*a*) Find the Fermi energies of the proton gas and of the electron gas in the sun. (*b*) Compare these energies with kT to see whether each gas is degenerate ($kT \ll \epsilon_F$, so that few particles have energies over ϵ_F) or nondegenerate ($kT \gg \epsilon_F$, so that few particles have energies below ϵ_F and the gas behaves classically).

53. Consider a white dwarf star whose mass is half that of the sun and whose radius is 0.01 that of the sun. Assume it consists of completely ionized carbon atoms (mass 12 u), so that there are six electrons per nucleus, and its interior temperature is 10^7 K. (*a*) Find the Fermi energies of the carbon nucleus gas and of the electron gas. (*b*) Compare these energies with kT to see whether each gas is degenerate or nondegenerate, as in Exercise 52.

54. The gravitational potential energy of a uniform-density sphere of mass M and radius R is $E_g = -\frac{3}{5}GM^2/R$. Consider a white dwarf star that contains N electrons whose Fermi energy is ϵ_F. Since $kT \ll \epsilon_F$, the average electron energy is, from Eq.(9.59), about $\frac{3}{5}\epsilon_F$ and the total electron energy is $E_e = \frac{3}{5}N\epsilon_F$. The energies of the nuclei can be neglected compared with E_e. Hence the total energy of the star is $E = E_g + E_e$. (*a*) Find the equilibrium radius of the star by letting $dE/dR = 0$ and solving for R. (*b*) Evaluate R for a star whose mass is half that of the sun and consists of completely ionized carbon atoms, as in Exercise 53.

CHAPTER 10

The Solid State

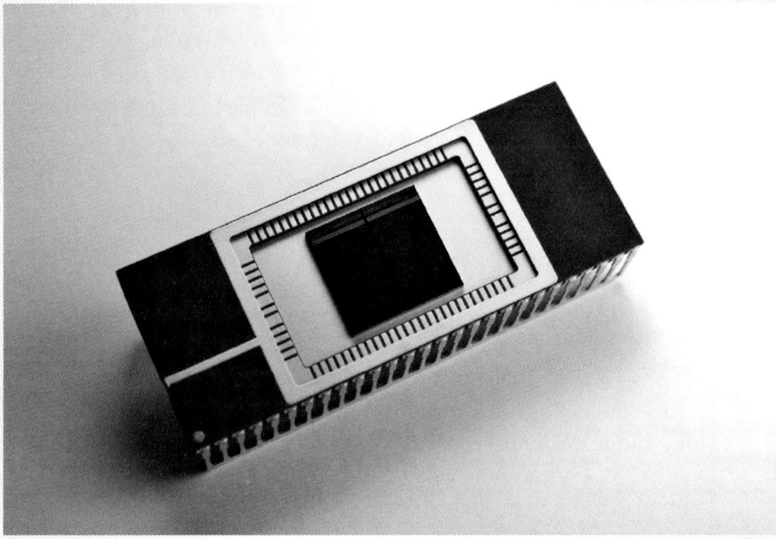

In a 3D NAND flash memory chip of Samsung Electronics, memory cells are vertically stacked to increase capacity. Understanding solids through quantum mechanics makes it possible to utilize properties of various materials and is at the heart of technological innovation.

A solid consists of atoms, ions, or molecules packed closely together, and the forces that hold them in place give rise to the distinctive properties of the various kinds of solids. The covalent bonds that can link a fixed number of atoms to form a certain molecule can also link an unlimited number of them to form a solid. In addition, ionic, van der Waals, and metallic bonds provide the cohesive forces in solids whose structural elements are respectively ions, molecules, and metal atoms. All these bonds involve electric forces, with the chief differences among them being in the ways in which the outer electrons of the structural elements are distributed. Although very little of the matter in the universe is in the solid state, solids constitute much of the physical world around us, and a large part of modern technology is based on the special characteristics of various solid materials.

10.1 CRYSTALLINE AND AMORPHOUS SOLIDS

Long-range and short-range order

Most solids are **crystalline,** with the atoms, ions, or molecules of which they are composed falling into regular, repeated three-dimensional patterns. The presence of **long-range order** is thus the defining property of a crystal, although relatively few samples of crystalline solids consist of single crystals. Most are polycrystalline and are composed of a great many small crystals (sometimes called crystallites).

Other solids lack the definite arrangements of their member particles so conspicuous in crystals. They may be regarded as supercooled liquids whose stiffness is due to an exceptionally high viscosity. Glass, pitch, and many plastics are examples of such **amorphous** ("without form") solids.

Amorphous solids do exhibit **short-range order** in their structures, however. The distinction between the two kinds of order is nicely exhibited in boron trioxide (B_2O_3), which can occur in both crystalline and amorphous forms. In each case every boron atom is surrounded by three oxygen atoms, which represents a short-range order. In a B_2O_3 crystal a long-range order is also present, as shown in a two-dimensional representation in Fig. 10.1. Amorphous B_2O_3, a vitreous or "glassy" substance, lacks this additional regularity. Crystallization from the vitreous state is so sluggish that it ordinarily does not occur, but it is not unknown. Glass may devitrify when heated until it has not quite begun to soften, and extremely old glass specimens are sometimes found to have crystallized.

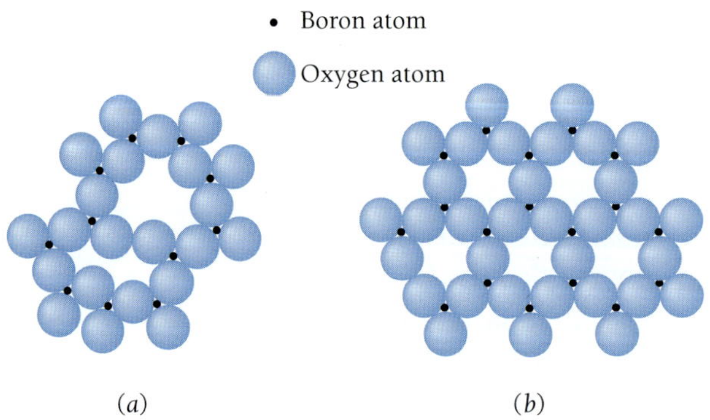

Figure 10.1 Two-dimensional representation of B_2O_3. (*a*) Amorphous B_2O_3 exhibits only short-range order. (*b*) Crystalline B_2O_3 exhibits long-range order as well.

The analogy between an amorphous solid and a liquid helps in understanding both states of matter. The density of a given liquid is usually close to that of the corresponding solid, for instance, which suggests that the degree of packing is similar. This inference is supported by the compressibilities of these states. Furthermore, x-ray diffraction indicates that many liquids have definite short-range structures at any instant, quite similar to those of amorphous solids except that the groupings of liquid molecules are continually shifting. A conspicuous example of short-range order in a liquid occurs in water just above the melting point, where the result is a lower density than at higher temperatures because H_2O molecules are less tightly packed when linked in crystals than when free to move.

The bonds in an amorphous solid vary in strength because of the lack of long-range order. When an amorphous solid is heated, the weakest bonds break at lower temperatures than the others, and the solid softens gradually. In a crystalline solid the bonds break simultaneously, and melting has a sudden onset. Metallic "glasses" have been made from mixtures of metals whose atoms differ greatly in size, which prevents them from forming the ordered structures of crystals when cooled from a molten state. One such metallic glass has half the density of steel but twice its strength, and is hard but can be deformed without breaking. Its gradual softening when heated make the material exceptionally easy to shape.

Crystal Defects

In a perfect crystal each atom has a definite equilibrium location in a regular array. Actual crystals are never perfect. Defects such as missing atoms, atoms out of place, irregularities in the spacing of rows of atoms and the presence of impurities have a considerable bearing on the physical properties of a crystal. Thus the behavior of a solid under stress is largely determined by the nature and concentration of defects in its structure, as is the electrical behavior of a semiconductor.

The simplest category of crystal imperfection is the **point defect.** Figure 10.2 shows the basic kinds of point defect. Both vacancies and interstitials, which require about 1 to 2 eV to be created, occur in all crystals as a result of thermal excitation, and their number accordingly increases rapidly with temperature. Of much importance is the production of such defects by particle radiation. In a nuclear reactor, for instance, energetic

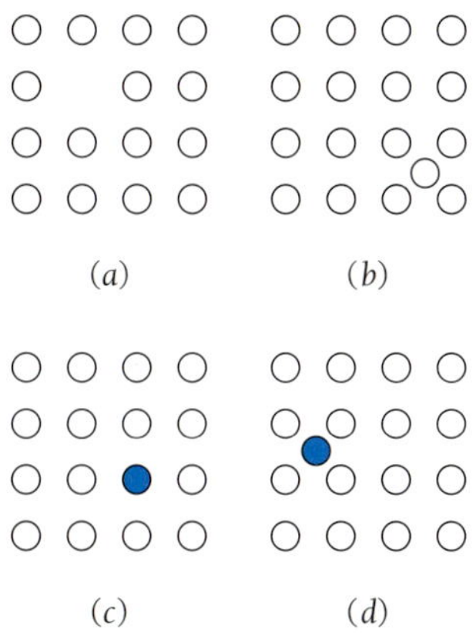

Figure 10.2 Point defects in a crystal. (*a*) Vacancy. (*b*) Interstitial. (*c*) Substitutional impurity. (*d*) Interstitial impurity.

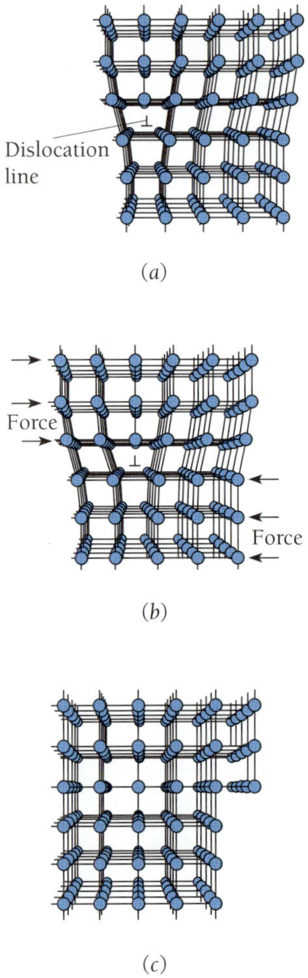

Figure 10.3 A crystal under stress becomes permanently deformed when dislocations in its structure shift their positions. (*a*) Initial configuration of a crystal with an edge dislocation. (*b*) The dislocation moves to the right as the atoms in the layer under it successively shift their bonds with those of the upper layer one line at a time. (*c*) The crystal has taken on a permanent deformation. The forces needed for this step-by-step process are much smaller than those needed to slide one entire layer of atoms past another layer.

neutrons readily knock atoms out of their normal locations. The result is a change in the properties of the bombarded material; most metals, for instance, become more brittle.

The effects of impurity atoms on the electrical properties of semiconductors, which underlie the operation of such devices as transistors, are discussed later in this chapter.

A **dislocation** is a type of crystal defect in which a line of atoms is not in its proper position. Dislocations are of two basic kinds. Figure 10.3 shows an **edge dislocation,** which we can visualize as the result of removing part of a layer (here vertical) of atoms. Edge dislocations enable a solid to be permanently deformed without breaking, a property called **ductility.** Metals are the most ductile solids. In the figure the bonds between atoms are represented by lines. The other kind of dislocation is the **screw dislocation.** We can visualize the formation of a screw dislocation by imagining that a cut is made partway into a perfect crystal and one side of the cut is then displaced relative to the other, as in Fig. 10.4. The atomic layers spiral around the dislocation, which accounts for its name. Actual dislocations in crystals are usually combinations of the edge and screw varieties.

Dislocations multiply when a solid is deformed. When the dislocations become so numerous and tangled together that they impede one another's motion, the material is then less easy to deform. This effect is called work hardening. Strongly heating (annealing) a work-hardened solid tends to return its disordered lattice to regularity and it becomes more ductile as a result. Steel bars and sheets formed by cold rolling are much harder than those formed by hot rolling.

10.2 IONIC CRYSTALS

The attraction of opposites can produce a stable union

Ionic bonds come into being when atoms that have low ionization energies, and hence lose electrons readily, interact with other atoms that tend to acquire excess electrons. The former atoms give up electrons to the latter, and they thereupon become positive and negative ions respectively (Fig. 8.2). In an ionic crystal these ions assemble themselves in an equilibrium configuration in which the attractive forces between positive and negative ions balance the repulsive forces between the ions.

As in the case of molecules, crystals of all types are prevented from collapsing under the influence of the cohesive forces present by the action of the exclusion principle, which requires the occupancy of higher energy states when electron shells of different atoms overlap and mesh together.

In general, in an ionic crystal each ion is surrounded by as many ions of the opposite sign as can fit closely, which leads to maximum stability. The relative sizes of the ions involved therefore govern the type of structure that occurs. Two common types of structure found in ionic crystals are shown in Figs. 10.5 and 10.6.

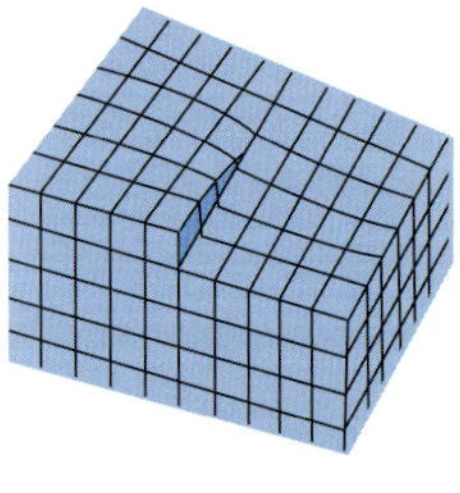

Figure 10.4 A screw dislocation.

Ionic bonds between the atoms of two elements can form when one element has a low ionization energy, so that its atoms tend to become positive ions, and the other

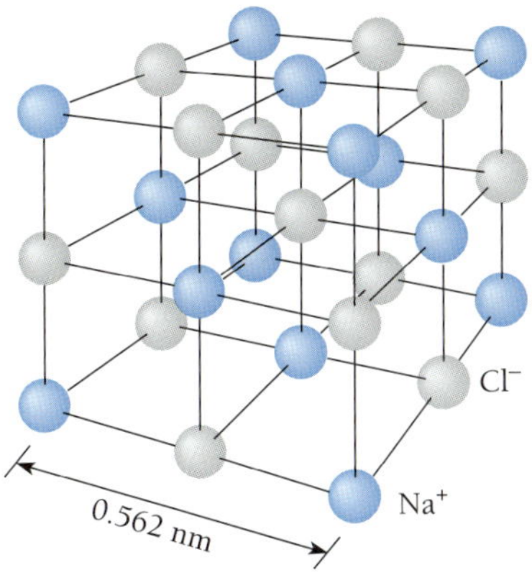

Figure 10.5 The face-centered cubic structure of NaCl. The coordination number (the number of nearest neighbors about each ion) is 6.

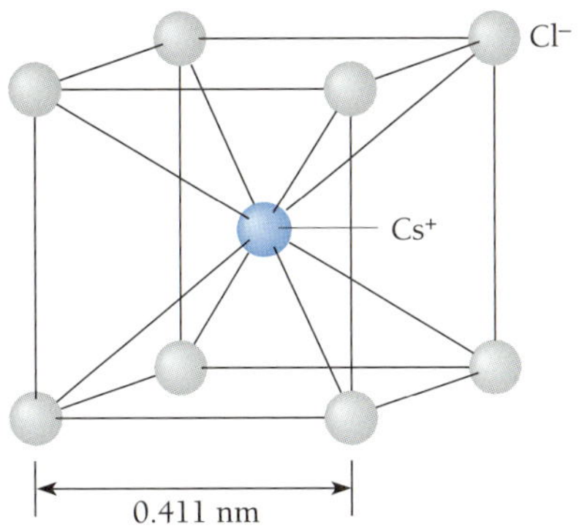

Figure 10.6 The body-centered cubic structure of CsCl. The coordination number is 8.

element has a high **electron affinity.** Electron affinity is the energy released when an electron is added to an atom of a given element; the greater the electron affinity, the more such atoms tend to become negative ions. Sodium, with an ionization energy of 5.14 eV, tends to form Na^+ ions, and chlorine, with an electron affinity of 3.61 eV, tends to form Cl^- ions. The condition for a stable crystal of NaCl is simply that the total energy of a system of Na^+ and Cl^- ions be less than the total energy of a system of Na and Cl atoms.

The **cohesive energy** of an ionic crystal is the energy per ion needed to break the crystal up into individual atoms. Part of the cohesive energy is the electric potential energy U_{coulomb} of the ions. Let us consider a Na^+ ion in NaCl. From Fig. 10.5 its nearest neighbors are six Cl^- ions, each one the distance r away. The potential energy

Common salt. Scanning electron micrograph (SEM) of BDH Analar sodium chloride recrystallised from distilled water. The crystal is built up from a cubic lattice of sodium and chloride ions. In the absence of impurities the exact crystal form (cubic) is formed. The micrographs show that in practice this basic cube is most often disrupted by dislocations giving rise to crystals with a wide variety of shapes - all retaining the cubic symmetry.

of the Na^+ ion due to these six Cl^- ions is therefore

$$U_1 = -\frac{6e^2}{4\pi\epsilon_0 r}$$

The next nearest neighbors are 12 Na^+ ions, each one the distance $\sqrt{2}\,r$ away since the diagonal of a square r long on a side is $\sqrt{2}\,r$. The potential energy of the Na^+ ion due to the 12 Na^+ ions is

$$U_2 = +\frac{12e^2}{4\pi\epsilon_0 \sqrt{2}\,r}$$

When the summation is continued over all the + and − ions in a crystal of infinite size, the result is

$$U_{\text{coulomb}} = -\frac{e^2}{4\pi\epsilon_0 r}\left(6 - \frac{12}{\sqrt{2}} + \cdots\right) = -1.748\,\frac{e^2}{4\pi\epsilon_0 r}$$

or, in general,

Coulomb energy

$$U_{\text{coulomb}} = -\alpha\,\frac{e^2}{4\pi\epsilon_0 r} \tag{10.1}$$

This result holds for the potential energy of a Cl^- ion as well, of course.

The quantity α is called the **Madelung constant** of the crystal, and it has the same value for all crystals of the same structure. Similar calculations for other crystal varieties yield different Madelung constants. Crystals whose structures are like that of cesium chloride (Fig. 10.6), for instance, have $\alpha = 1.763$. Simple crystal structures have Madelung constants that lie between 1.6 and 1.8.

The potential energy contribution of the repulsive forces due to the action of the exclusion principle has the approximate form

Repulsive energy

$$U_{\text{repulsive}} = \frac{B}{r^n} \tag{10.2}$$

The sign of $U_{\text{repulsive}}$ is positive, which corresponds to a repulsion. The dependence on r^{-n} implies a short-range force that increases as the interionic distance r decreases. The total potential energy of each ion due to its interactions with all the other ions is therefore

$$U_{\text{total}} = U_{\text{coulomb}} + U_{\text{repulsive}} = -\frac{\alpha e^2}{4\pi\epsilon_0 r} + \frac{B}{r^n} \tag{10.3}$$

How can we find the value of B? At the equilibrium separation r_0 of the ions, U is a minimum by definition, and so $dU/dr = 0$ when $r = r_0$. Hence

$$\left(\frac{dU}{dr}\right)_{r=r_0} = \frac{\alpha e^2}{4\pi\epsilon_0 r_0^2} - \frac{nB}{r_0^{n+1}} = 0$$

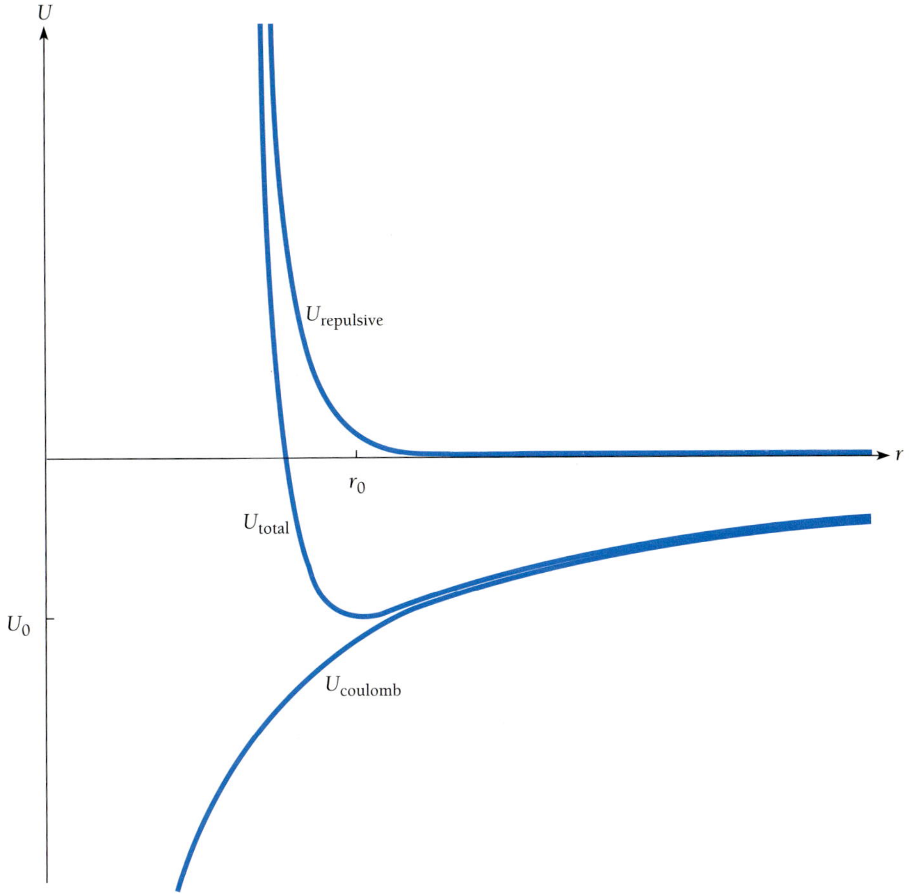

Figure 10.7 How the ionic potential energies in an ionic crystal vary with ionic separation r. The minimum value U_0 of U_{total} occurs at an equilibrium separation of r_0.

$$B = \frac{\alpha e^2}{4\pi\epsilon_0 n} r_0^{n-1} \tag{10.4}$$

The total potential energy at the equilibrium separation is therefore given by

Total potential energy

$$U_0 = -\frac{\alpha e^2}{4\pi\epsilon_0 r_0}\left(1 - \frac{1}{n}\right) \tag{10.5}$$

We must add this amount of energy per ion pair to separate an ionic crystal into individual ions (Fig. 10.7). For the cohesive energy, which corresponds to separating the crystal into atoms, we must take into account the energy involved in shifting an electron from a Na atom to a Cl atom to give a Na^+-Cl^- ion pair.

The exponent n can be found from the observed compressibilities of ionic crystals. The average result is $n \approx 9$, which means that the repulsive force varies sharply with r. The ions are "hard" rather than "soft" and strongly resist being packed too tightly. At the equilibrium ion spacing, the mutual repulsion due to the exclusion principle (as distinct from the electric repulsion between like ions) decreases the potential energy by about 11 percent. A really precise knowledge of n is not essential; if $n = 10$ instead of $n = 9$, U_0 would change by only 1 percent.

Example 10.1

In an NaCl crystal, the equilibrium distance r_0 between ions is 0.281 nm. Find the cohesive energy in NaCl.

Solution

Since $\alpha = 1.748$ and $n \approx 9$, the potential energy per ion pair is

$$U_0 = -\frac{\alpha e^2}{4\pi\epsilon_0 r_0}\left(1 - \frac{1}{n}\right) = -\frac{(9 \times 10^9\ \text{N}\cdot\text{m}^2/\text{C}^2)(1.748)(1.60 \times 10^{-19}\ \text{C})^2}{2.81 \times 10^{-10}\ \text{m}}\left(1 - \frac{1}{9}\right)$$

$$= -1.27 \times 10^{-18}\ \text{J} = -7.96\ \text{eV}$$

Half this figure, -3.98 eV, represents the contribution per ion to the cohesive energy of the crystal.

Now we need the electron transfer energy, which is the sum of the +5.14-eV ionization energy of Na and the −3.61-eV electron affinity of Cl, or +1.53 eV. Each atom therefore contributes +0.77 eV to the cohesive energy from this source. The total cohesive energy per atom is thus

$$E_{\text{cohesive}} = (-3.98 + 0.77)\ \text{eV} = -3.21\ \text{eV}$$

which is not far from the experimental value of −3.28 eV.

Most ionic solids are hard, owing to the strength of the bonds between their constituent ions, and have high melting points. They are usually brittle as well, since the slipping of atoms past one another that accounts for the ductility of metals is prevented by the ordering of positive and negative ions imposed by the nature of the bonds. Polar liquids such as water are able to dissolve many ionic crystals, but covalent liquids such as gasoline generally cannot. Because the outer electrons of their ions are tightly bound, ionic crystals are good electrical insulators and are transparent to visible light. However, such crystals strongly absorb infrared radiation at the frequencies at which the ions vibrate about their equilibrium positions.

10.3 COVALENT CRYSTALS

Shared electrons lead to the strongest bonds

The cohesive forces in covalent crystals arise from the sharing of electrons by adjacent atoms. Each atom that participates in a covalent bond contributes an electron to the bond. Figure 10.8 shows the tetrahedral structure of a diamond crystal, each of whose carbon atoms is linked by covalent bonds to four other carbon atoms.

Another crystalline form of carbon is graphite. Graphite consists of layers of carbon atoms in a hexagonal network in which each atom is joined to three others by covalent bonds 120° apart, as in Fig. 10.9. One electron per atom participates in

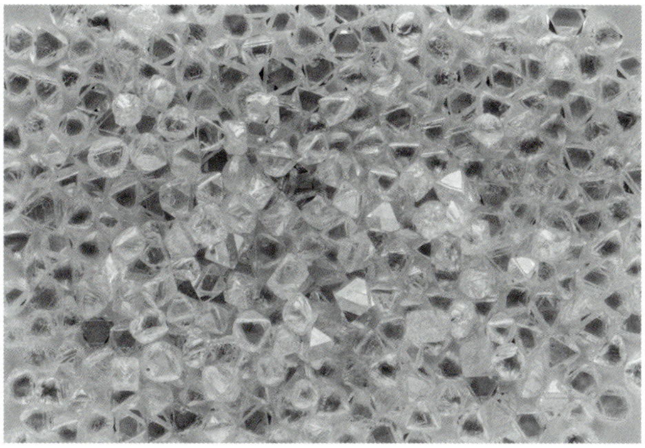

Uncut diamonds. The strength of the covalent bonds between adjacent carbon atoms gives diamonds their hardness.

each bond. This leaves one outer electron in each carbon atom free to circulate through the network, thereby accounting for graphite's near-metallic luster and electrical conductivity. Although each layer is quite strong, weak van der Waals forces (Sec. 10.4) bond the layers together. As a result, the layers can slide past each other readily and are easily flaked apart, which is why graphite is so useful as a lubricant and in pencils.

Under ordinary conditions graphite is more stable than diamond, so crystallizing carbon normally produces only graphite. Because graphite is less dense than diamond (2.25 g/cm^3 versus 3.51 g/cm^3), high pressures favor the formation of diamond. Natural diamonds originated deep in the earth where pressures are enormous. To synthesize diamonds, graphite is dissolved in molten cobalt or nickel and the mixture is compressed at about 1600 K to about 60,000 bar. The resulting diamonds are less than 1 mm across and are widely used industrially for cutting and grinding tools.

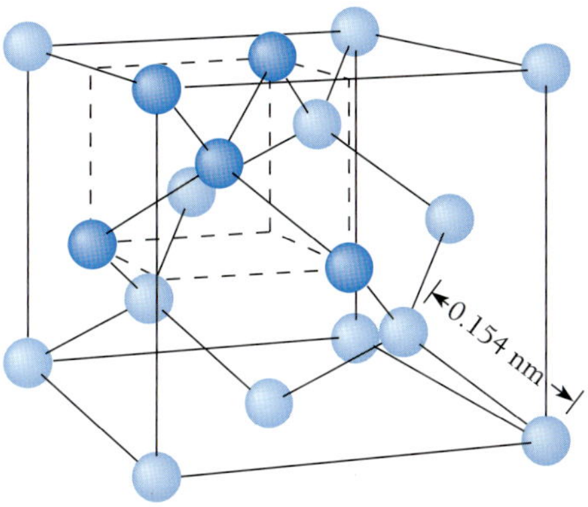

Figure 10.8 The tetrahedral structure of diamond. The coordination number is 4.

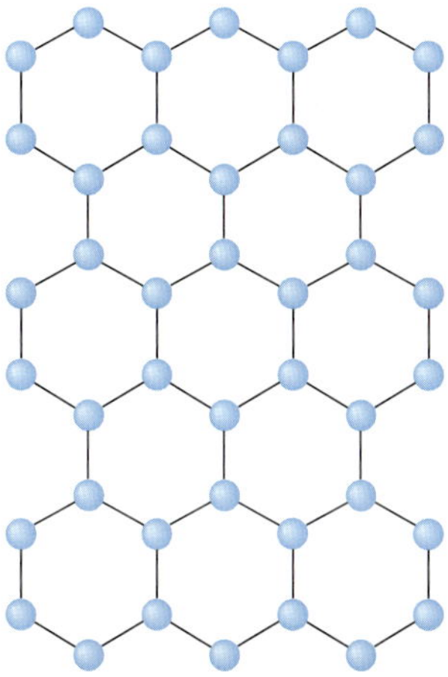

Figure 10.9 Graphite consists of layers of carbon atoms in hexagonal arrays, with each atom bonded to three others. The layers are held together by weak van der Waals forces.

Purely covalent crystals are relatively few in number. In addition to diamond, some examples are silicon, germanium, and silicon carbide, all of which have the same tetrahedral structure as diamond; in SiC each atom is surrounded by four atoms of the other kind. Cohesive energies are usually greater in covalent crystals than in ionic ones. As a result covalent crystals are hard (diamond is the hardest substance known, and SiC is the industrial abrasive carborundum), have high melting points, and are insoluble in all ordinary liquids. The optical and electrical properties of covalent solids are discussed later.

Buckyballs and Nanotubes

An unexpected form of carbon was accidentally discovered in 1985 at Rice University in Texas. The most common version consists of 60 carbon atoms arranged in a cage structure of 12 pentagons and 20 hexagons whose geometry is like that of a soccer ball (Fig. 10.10). This extraordinary molecule was called "buckminsterfullerene" in honor of the American architect R. Buckminster Fuller, whose geodesic domes it resembles; the name is usually shortened to **buckyball.**

Buckyballs, which are stable and chemically unreactive, can be made in the laboratory from graphite and are present in small quantities in ordinary soot and in a carbon-rich rock found in Russia. The original C_{60} buckyball is not the only form of fullerene known: C_{28}, C_{32}, C_{50}, C_{70}, and still larger ones have been made. Fullerene molecules are held together to form solids by van der Waals bonds like those that hold together the layers of C atoms in graphite. Since their discovery, the fullerenes and their offshoots have shown some remarkable properties. For instance, the combination of C_{60} with potassium to form K_3C_{60} yields a superconductor at low temperatures.

Carbon nanotubes, cousins of buckyballs, consist of tiny cylinders of carbon atoms arranged in hexagons, like rolled-up chicken wire. Depending on whether their rows of hexagons are straight or wind around in a helix, such nanotubes act either as electrical conductors or as semiconductors and their use is being explored in such electronic applications as transistors and flat-panel displays. If carbon nanotubes can be made long enough, they will form exceedingly strong fibers, ten times stronger than steel while six times lighter, that are flexible as well. Fibers like this would be ideal in composite materials to reinforce epoxy resins. Nanotubes also have promise for storing the hydrogen needed for the fuel cells of future electric cars, which would make heavy steel containers unnecessary.

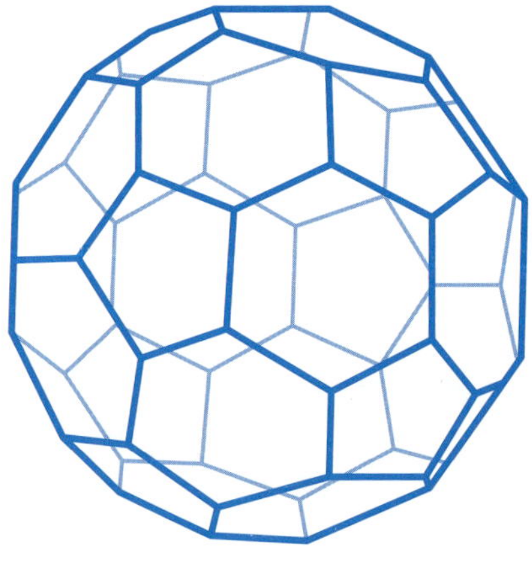

Figure 10.10 In a buckyball, carbon atoms form a closed cagelike structure in which each atom is bonded to three others. Shown here is the C_{60} buckyball that contains 60 carbon atoms. The lines represent carbon-carbon bonds; their pattern of hexagons and pentagons closely resembles the pattern made by the seams of a soccer ball. Other buckyballs have different numbers of carbon atoms.

10.4 VAN DER WAALS BOND

Weak but everywhere

All atoms and molecules—even inert-gas atoms such as those of helium and argon—exhibit weak, short-range attractions for one another due to **van der Waals forces.** These forces were proposed over a century ago by the Dutch physicist Johannes van der Waals to explain departures of real gases from the ideal-gas law. The explanation of the actual mechanism of the forces, of course, is more recent.

Van der Waals forces are responsible for the condensation of gases into liquids and the freezing of liquids into solids in the absence of ionic, covalent, or metallic bonding mechanisms. Such familiar aspects of the behavior of matter in bulk as friction, surface tension, viscosity, adhesion, cohesion, and so on, also arise from these forces. As we shall find, the van der Waals attraction between two molecules the distance r apart is proportional to r^{-7}, so that it is significant only for molecules very close together.

We begin by noting that many molecules, called **polar molecules,** have permanent electric dipole moments. An example is the H_2O molecule, in which the concentration of electrons around the oxygen atom makes that end of the molecule more negative than the end where the hydrogen atoms are. Such molecules tend to clump together with ends of opposite sign adjacent, as in Fig. 10.11.

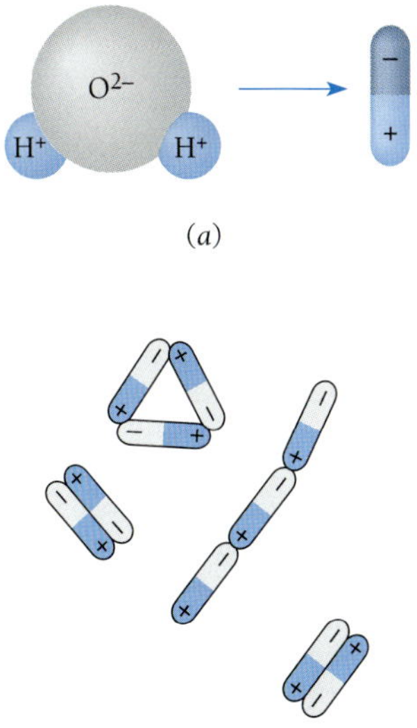

Figure 10.11 (*a*) The water molecule is polar because the end where the H atoms are attached behaves as if positively charged and the opposite end behaves as if negatively charged. (*b*) Polar molecules attract each other.

A polar molecule can also attract molecules which lack a permanent dipole moment. The process is illustrated in Fig. 10.12. The electric field of the polar molecule causes a separation of charge in the other molecule, with the induced moment the same in direction as that of the polar molecule. The result is an attractive force. The effect is the same as that involved in the attraction of an unmagnetized piece of iron by a magnet.

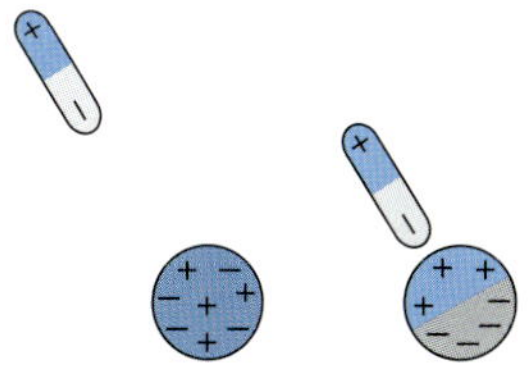

Figure 10.12 Polar molecules attract polarizable molecules.

Let us see what the characteristics of the attractive force between a polar and a nonpolar molecule depend on. The electric field $\mathbf{E}$ a distance r from a dipole of moment $\mathbf{p}$ is given by

Dipole electric field

$$\mathbf{E} = \frac{1}{4\pi\epsilon_0}\left[\frac{\mathbf{p}}{r^3} - \frac{3(\mathbf{p}\cdot\mathbf{r})}{r^5}\mathbf{r}\right] \tag{10.6}$$

We recall from vector analysis that $\mathbf{p}\cdot\mathbf{r} = pr\cos\theta$, where θ is the angle between $\mathbf{p}$ and $\mathbf{r}$. The field $\mathbf{E}$ induces in the other, normally nonpolar molecule an electric dipole moment $\mathbf{p}'$ proportional to $\mathbf{E}$ in magnitude and ideally in the same direction. Hence

Induced dipole moment

$$\mathbf{p}' = \alpha\mathbf{E} \tag{10.7}$$

where α is a constant called the **polarizability** of the molecule. The energy of the induced dipole in the electric field $\mathbf{E}$ is

$$U = -\mathbf{p}'\cdot\mathbf{E} = -\alpha\mathbf{E}\cdot\mathbf{E}$$

$$= -\frac{\alpha}{(4\pi\epsilon_0)^2}\left(\frac{p^2}{r^6} - \frac{3p^2}{r^6}\cos^2\theta - \frac{3p^2}{r^6}\cos^2\theta + \frac{9p^2}{r^6}\cos^2\theta\right)$$

Interaction energy

$$= -\frac{\alpha}{(4\pi\epsilon_0)^2}(1 + 3\cos^2\theta)\frac{p^2}{r^6} \tag{10.8}$$

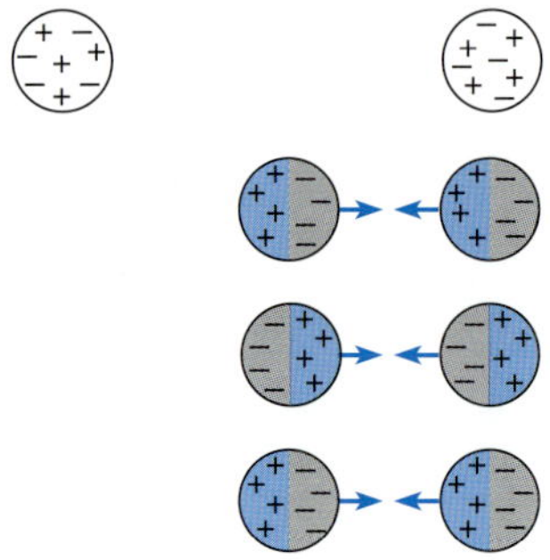

Figure 10.13 On the average, nonpolar molecules have symmetrical charge distributions, but at any moment the distributions are asymmetric. The fluctuations in the charge distributions of nearby molecules are coordinated as shown. This situation leads to an attractive force between them whose magnitude varies as $1/r^7$, where r is their distance apart.

The potential energy of the two molecules that arises from their interaction is negative, signifying that the force between them is attractive, and is proportional to r^{-6}. The force itself is equal to $-dU/dr$ and so is proportional to r^{-7}, which means that it drops rapidly with increasing separation. Doubling the distance between two molecules reduces the attractive force between them to only 0.8 percent of its original value.

More remarkably, two nonpolar molecules can attract each other by the above mechanism. The electron distribution in a nonpolar molecule is symmetric *on the average*. However, the electrons themselves are in constant motion and at *any given instant* one part or another of the molecule has an excess of them. Instead of the fixed charge asymmetry of a polar molecule, a nonpolar molecule has a constantly shifting asymmetry. When two nonpolar molecules are close enough, their fluctuating charge distributions tend to shift together with adjacent ends always having opposite signs (Fig. 10.13), which leads to an attractive force.

Van der Waals forces occur not only between all molecules but also between all atoms, including those of the rare gases which do not otherwise interact. Without such forces these gases would not condense into liquids or solids. The values of p^2 (or $\overline{p^2}$, the average of p^2, which applies for molecules with no permanent dipole moment) and the polarizability α are comparable for most molecules. This is part of the reason why the densities and heats of vaporization of liquids, properties that depend on the strength of intermolecular forces, have a rather narrow range.

Van der Waals forces are much weaker than those found in ionic and covalent bonds, and as a result molecular crystals generally have low melting and boiling points and little mechanical strength. Cohesive energies are low, only 0.08 eV/atom in solid argon (melting point −189°C), 0.01 eV/molecule in solid hydrogen (mp −259°C), and 0.1 eV/molecule in solid methane, CH_4 (mp −183°C).

Hydrogen Bonds

An especially strong type of van der Waals bond called a **hydrogen bond** occurs between certain molecules containing hydrogen atoms. The electron distribution in such a molecule is severely distorted by the affinity of a heavier atom for electrons. Each hydrogen atom in effect donates most of its negative charge to the other atom, to leave behind a poorly shielded proton. The result is a molecule with a localized positive charge which can link up with the concentration of negative charge elsewhere in another molecule of the same kind. The key factor here is the small effective size of the poorly shielded proton, since electric forces vary as $1/r^2$. Hydrogen bonds are typically about a tenth as strong as covalent bonds.

Water molecules are exceptionally prone to form hydrogen bonds because the electrons around the O atom in H_2O are not symmetrically distributed but are more likely to be found in certain regions of high probability density. These regions project outward as though toward the vertices of a tetrahedron, as shown in Fig. 10.14. Hydrogen atoms are at two of these vertices, which accordingly exhibit localized positive charges, while the other two vertices exhibit somewhat more diffused negative charges.

Each H_2O molecule can therefore form hydrogen bonds with *four* other H_2O molecules. In two of these bonds the central molecule provides the bridging protons, and in the other two the attached molecules provide them. In the liquid state, the hydrogen bonds between adjacent H_2O molecules are continually being broken and re-formed owing to thermal agitation, but even so at any instant the molecules are combined in definite clusters. In the solid state, these clusters are large and stable and

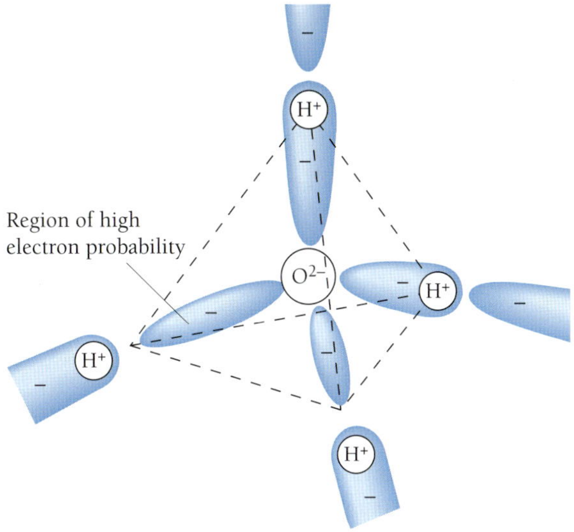

Figure 10.14 In an H_2O molecule, the four pairs of valence electrons around the oxygen atom (six electrons contributed by the O atom and one each by the H atoms) preferentially occupy four regions that form a tetrahedral pattern. Each H_2O molecule can form hydrogen bonds with four other H_2O molecules.

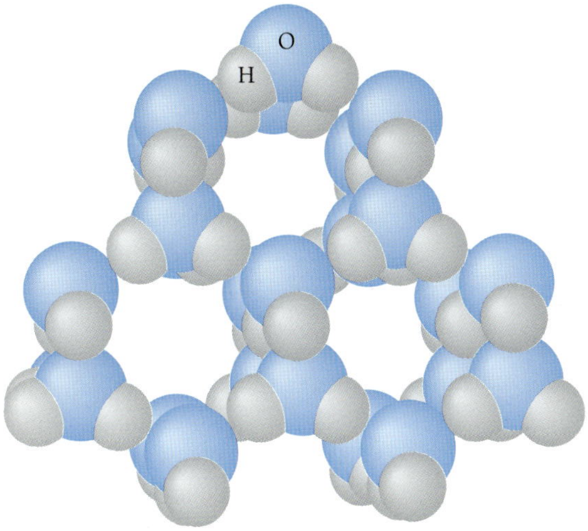

Figure 10.15 The structure of an ice crystal, showing the open hexagonal arrangement of the H_2O molecules. There is less order in liquid water, which allows the molecules to be closer together on the average than they are in ice. Thus the density of ice is less than that of water, and ice floats.

constitute ice crystals (Fig. 10.15). With only four nearest neighbors around each molecule, instead of as many as twelve in other solids, ice crystals have extremely open structures, which is why ice has a relatively low density.

Hydrogen bonds occur widely in biological materials. The peptide bonds that join amino acids to form proteins are hydrogen bonds, for example, as are the bonds that hold together the two strands of the double helix of DNA. The bonds in DNA are strong enough for it to be a reliable store of genetic information but weak enough to permit its strands to be unzipped temporarily for the information to be transcribed ultimately into proteins and also permanently for DNA replication.

The water molecules in a snowflake are held together by hydrogen bonds.

10.5 METALLIC BOND

A gas of free electrons is responsible for the characteristic properties of a metal

The valence (outer) electrons of metal atoms are only weakly bound, as Fig. 7.10 shows. When such atoms interact to become a solid, their valence electrons form a "gas" of electrons that move with relative freedom through the resulting assembly of metal ions. The electron gas acts to hold the ions together and also provides the high electric and thermal conductivities, opacity, surface luster, and other characteristic properties of metals. Because the free electrons do not belong to particular atom-atom bonds, different metals can be alloyed together in more-or-less arbitrary proportions if their atoms are similar in size. In contrast, the components of ionic solids and of covalent solids such as SiC combine only in specific proportions.

As in any other solid, metal atoms cohere because their total energy is lower when they are bound together than when they are separate atoms. This energy reduction occurs in a metal crystal because each valence electron is on the average closer to one ion or another than it would be if it belonged to an isolated atom. Hence the electron's potential energy is less in the crystal than in the atom.

Another factor is involved here: although the potential energy of the free electrons is reduced in a metal crystal, their kinetic energy is increased. The valence energy levels of the metal atoms are all slightly altered by their interactions to give as many different energy levels as the total number of atoms present. The levels are so closely spaced as to form an essentially continuous **energy band.** As discussed in Chap. 9, the free electrons in this band have a Fermi-Dirac energy distribution in which, at 0 K, their kinetic energies range from 0 to a maximum of ϵ_F, the Fermi energy. The Fermi energy in copper, for example, is 7.04 eV, and the average KE of the free electrons in metallic copper at 0 K is 4.22 eV.

Metallic Hydrogen

Hydrogen is in group 1 of the periodic table, all the other elements of which are metals. Hydrogen is the exception, which is not surprising when it is in the gaseous state, but it does not behave as a metal (for instance by being a good electrical conductor) even when it has been cooled to the liquid or solid states. The reason is that both liquid and solid hydrogen at atmospheric pressure consist of hydrogen molecules, H_2, and these molecules hold their electrons so tightly that none can break loose and move about freely as in the case of the atomic electrons of metals.

However, extremely high pressures—several million times atmospheric pressure—turn hydrogen into a conducting liquid. What the pressure does is force the H_2 molecules so close together that their electron wave functions overlap, which allows electrons to migrate from one molecule to the next. Pressures inside the giant planet Jupiter, which consists largely of hydrogen, are sufficient for Jupiter apparently to have a hydrogen core that is in the form of a liquid metal. Electric currents in Jupiter's core produce its magnetic field; this field is about 20 times stronger than the earth's field, which is due to currents in its molten iron core.

Conceivably someday solid metallic hydrogen could be created, perhaps combined with other substances to help stabilize it, that would survive ordinary temperatures and pressures. The possible properties of such metallic hydrogen include superconductivity and light weight combined with mechanical strength. The energy that would be released by allowing solid hydrogen to turn into a gas could be used to propel spacecraft—it might give five times as much thrust per kilogram as current rocket fuels. Because solid hydrogen would be much denser than ordinary hydrogen, in the form of its isotopes deuterium and tritium it would make an extremely efficient fuel for fusion reactors. All in all, wonderful prospects, but how realistic they are remains to be seen.

Table 10.1 Types of Crystalline Solids. The cohesive energy is the work needed to remove an atom (or molecule) from the crystal and so indicates the strength of the bonds holding it in place.

Type	Ionic	Covalent	Molecular	Metallic
Lattice	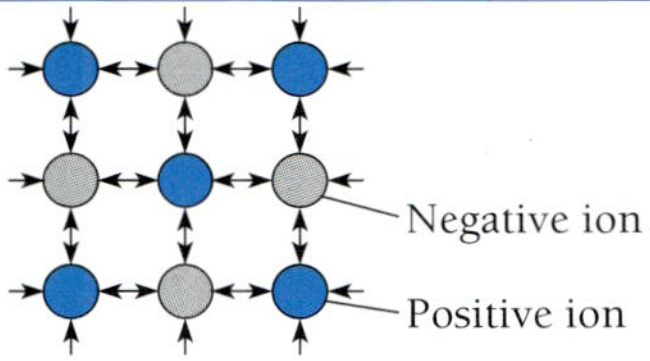	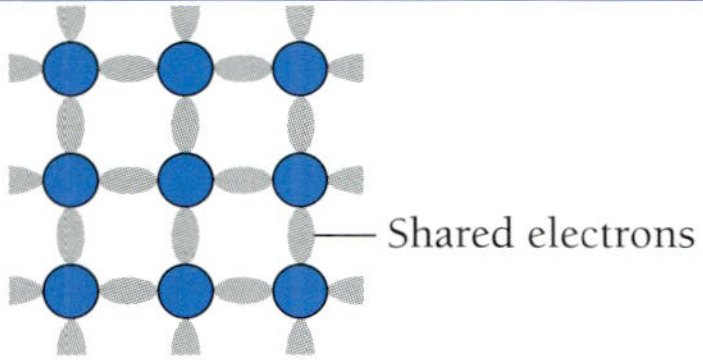	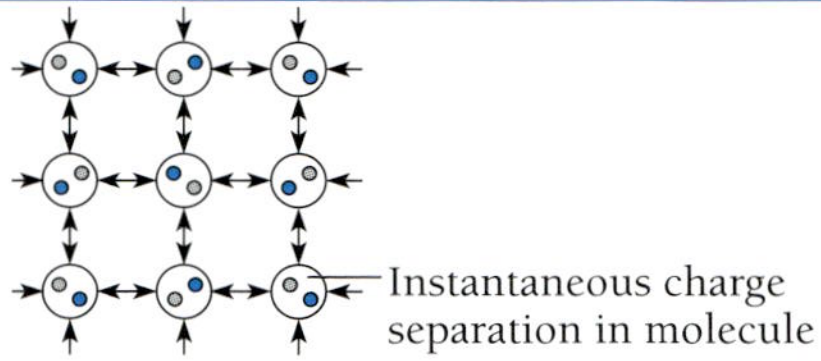	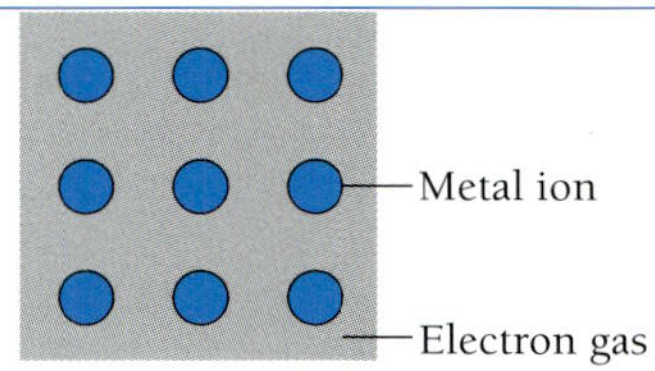
Bond	Electric attraction	Shared electrons	Van der Waals forces	Electron gas
Properties	Hard; high melting points; may be soluble in polar liquids such as water; electrical insulators (but conductors in solution)	Very hard; high melting points; insoluble in nearly all liquids; semiconductors (except diamond, which is an insulator)	Soft; low melting and boiling points; soluble in covalent liquids; electrical insulators	Ductile; metallic luster; high electrical and thermal conductivity
Example	Sodium chloride, NaCl $E_{cohesive}$ = 3.28 eV/atom	Diamond, C $E_{cohesive}$ = 7.4 eV/atom	Methane, CH_4 $E_{cohesive}$ = 0.1 eV/molecule	Sodium, Na $E_{cohesive}$ = 1.1 eV/atom

Metallic bonding occurs when the reduction in electron potential energy outbalances the increase in electron KE that accompanies it. The more valence electrons per atom, the higher the average KE of the free electrons, but without a commensurate drop in their potential energy. For this reason nearly all the metallic elements are found in the first three groups of the periodic table.

Ohm's Law

When the potential difference across the ends of a metal conductor is V, the resulting current I is, within wide limits, directly proportional to V. This empirical observation, called **Ohm's law,** is usually expressed as

Ohm's law
$$I = \frac{V}{R} \tag{10.9}$$

Here R, the **resistance** of the conductor, depends on its dimensions, composition, and temperature, but is independent of V. Ohm's law follows from the free-electron model of a metal.

We begin by assuming that the free electrons in a metal, like the molecules in a gas, move in random directions and undergo frequent collisions. The collisions here, however, are not billiard-ball collisions with other electrons but represent the scattering of electron waves by irregularities in the crystal structure, defects such as impurity atoms and also atoms temporarily out of place as they vibrate. As we will see later, the atoms of a perfect crystal lattice do not scatter free electron waves except under certain specific circumstances.

If λ is the mean free path between the collisions of a free electron, the average time τ between collisions is

Collision time
$$\tau = \frac{\lambda}{v_F} \tag{10.10}$$

The quantity v_F is the electron velocity that corresponds to the Fermi energy ϵ_F, since only electrons at or near the top of their energy distribution can be accelerated (see Sec. 9.10). This average time is virtually independent of an applied electric field $\mathbf{E}$ because v_F is extremely high compared with the velocity change such a field produces. In copper, for instance, $\epsilon_F = 7.04$ eV and so

$$v_F = \sqrt{\frac{2\epsilon_F}{m}} = \sqrt{\frac{(2)(7.04\text{ eV})(1.60 \times 10^{-19}\text{ J/eV})}{9.11 \times 10^{-31}\text{ kg}}} = 1.57 \times 10^6\text{ m/s}$$

The superimposed **drift velocity** v_d due to an applied electric field, however, is usually less than 1 mm/s.

Example 10.2

Find the drift velocity v_d of the free electrons in a copper wire whose cross-sectional area is $A = 1.0\text{ mm}^2$ when the wire carries a current of 1.0 A. Assume that each copper atom contributes one electron to the electron gas.

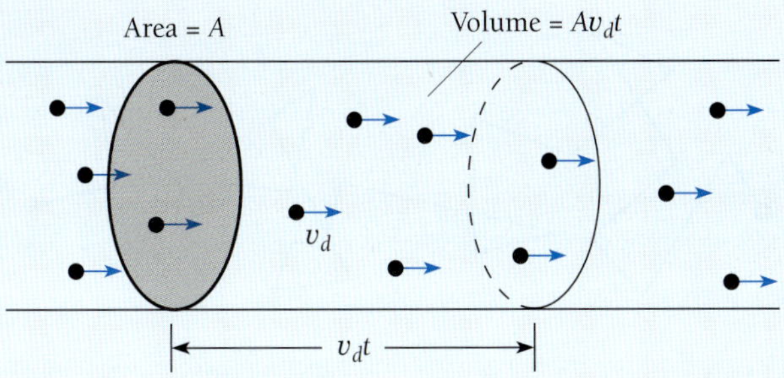

Figure 10.16 The number of free electrons in a wire that drift past a cross-section of the wire in the time t is $nV = nA\boldsymbol{v}_dt$, where n is the number of free electrons/m^3 in the wire.

Solution

The wire contains n free electrons per unit volume. Each electron has the charge e and in the time t it travels the distance $\boldsymbol{v}_dt$ along the wire, as in Fig. 10.16. The number of free electrons in the volume $A\boldsymbol{v}_dt$ is $nA\boldsymbol{v}_dt$, and all of them pass through any cross section of the wire in the time t. Thus the charge that passes through this cross section in t is $Q = nAe\boldsymbol{v}_dt$, and the corresponding current is

$$I = \frac{Q}{t} = nAe\boldsymbol{v}_d$$

The drift velocity of the electrons is therefore

$$\boldsymbol{v}_d = \frac{I}{nAe}$$

From Example 9.8 we know that, in copper, $n = N/V = 8.5 \times 10^{28}$ electrons/m^3, and here $I = 1.0$ A and $A = 1.0\ \text{mm}^2 = 1.0 \times 10^{-6}\ \text{m}^2$. Hence

$$v_d = \frac{1.0\ \text{A}}{(8.5 \times 10^{28}\ \text{m}^{-3})(1.0 \times 10^{-6}\ \text{m}^2)(1.6 \times 10^{-19}\ \text{C})} = 7.4 \times 10^{-4}\ \text{m/s}$$

But if the free electrons have so small a drift velocity, why does an electric appliance go on as soon as its switch is closed and not minutes or hours later? The answer is that applying a potential difference across a circuit very rapidly creates an electric field in the circuit, and as a result all the free electrons begin their drift almost simultaneously.

A potential difference V across the ends of a conductor of length L produces an electric field of magnitude $E = V/L$ in the conductor. This field exerts a force of eE on a free electron in the conductor, whose acceleration is

$$a = \frac{F}{m} = \frac{eE}{m} \tag{10.11}$$

When the electron undergoes a collision, it rebounds in an arbitrary direction and, on the average, no longer has a component of velocity parallel to **E**. Imposing the field **E** on the free electron gas in a metal superimposes a general drift on the faster but random motions of the electron (Fig. 10.17). We can therefore ignore the electron's motion at the Fermi velocity $\boldsymbol{v}_F$ in calculating the drift velocity $\boldsymbol{v}_d$.

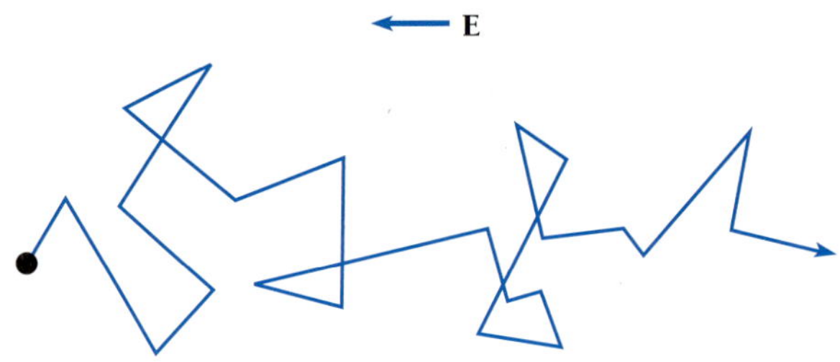

Figure 10.17 An electric field produces a general drift superimposed on the random motion of a free electron. The electron's path between collisions is actually slightly curved because of the acceleration due to the field.

After each collision, the electron is accelerated for some time interval Δt before the next collision, and at the end of the interval has traveled $\frac{1}{2}a\ \Delta t^2$. When the electron has made many collisions, its average displacement will be $\overline{X} = \frac{1}{2}a\ \overline{\Delta t^2}$, where $\overline{\Delta t^2}$ is the average of the squared time intervals. Because of the way Δt varies, $\overline{\Delta t^2} = 2\tau^2$. Hence $\overline{X} = a\tau^2$ and the drift velocity is $\overline{X}/\tau = a\tau$, so that

Drift velocity

$$v_d = a\tau = \left(\frac{eE}{m}\right)\left(\frac{\lambda}{v_F}\right) = \frac{eE\lambda}{mv_F} \tag{10.12}$$

In Example 10.2 we found that the current I in a conductor of cross-sectional area A in which the free electron density is n is given by

$$I = nAev_d \tag{10.13}$$

Using the value of v_d from Eq. (10.12) gives

$$I = \frac{nAe^2E\lambda}{mv_F}$$

Since the electric field in the conductor is $E = V/L$,

$$I = \left(\frac{ne^2\lambda}{mv_F}\right)\left(\frac{A}{L}\right)V \tag{10.14}$$

This formula becomes Ohm's law if we set

Resistance of metal conductor

$$R = \left(\frac{mv_F}{ne^2\lambda}\right)\frac{L}{A} \tag{10.15}$$

The quantity in parentheses is known as the **resistivity** ρ of the metal and is a constant for a given sample at a given temperature:

Resistivity

$$\rho = \frac{mv_F}{ne^2\lambda} \tag{10.16}$$

Example 10.3

The resistivity of copper at 20°C is $\rho = 1.72 \times 10^{-8}\ \Omega \cdot \text{m}$. Estimate the mean free path λ between collisions of the free electrons in copper at 20°C.

Solution

In Example 9.8 we found that the free electron density in copper is $n = 8.48 \times 10^{28}\ \text{m}^{-3}$, and earlier in this section we saw that the Fermi velocity there is $v_F = 1.57 \times 10^6$ m/s. Solving Eq. (10.16) for λ gives

$$\lambda = \frac{mv_F}{ne^2\rho} = \frac{(9.11 \times 10^{-31}\ \text{kg})(1.57 \times 10^6\ \text{m/s})}{(8.48 \times 10^{28}\ \text{m}^{-3})(1.60 \times 10^{-19}\ \text{C})^2(1.72 \times 10^{-8}\ \Omega \cdot \text{m})}$$

$$= 3.83 \times 10^{-8}\ \text{m} = 38.3\ \text{nm}$$

The ions in solid copper are 0.26 nm apart, so a free electron travels past nearly 150 of them, on the average, before being scattered.

The scattering of free electron waves in a metal that leads to its electric resistance is caused both by structural defects and by ions out of place as they vibrate. Imperfections of the former kind do not depend on temperature but on the purity of the metal and on its history. The resistivities of cold-worked metals (such as "hard drawn" wires) are lowered by annealing because the number of defects is thereby decreased. On the other hand, lattice vibrations increase in amplitude with increasing temperature, and their contribution to resistivity accordingly goes up with temperature. Thus the resistivity of a metal is the sum $\rho = \rho_i + \rho_t$, where ρ_i depends on the concentration of defects and ρ_t depends on temperature.

Figure 10.18 shows how the resistivities of two sodium samples vary with temperature. The top curve corresponds to the sample with the higher concentration of defects, which accounts for its upward displacement. In very pure and almost defect-free samples, ρ_i is small, and at low temperatures, ρ_t is also small. When both these conditions hold in copper, for example, the mean free path may be 10^5 times the value found in Example 10.3.

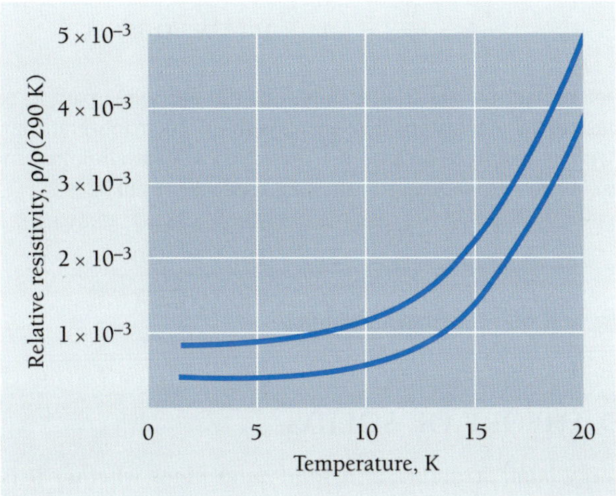

Figure 10.18 Resistivities of two sodium samples at low temperatures relative to their resistivities at 290 K. The upper curve corresponds to the sample with the higher concentration of impurities.

Weidemann-Franz Law

The free-electron model of metallic conduction was proposed by Paul Drude in 1900, only three years after the discovery of the electron by J. J. Thomson, and was later elaborated by Hendrik Lorentz. Fermi-Dirac statistics were unknown then, and Drude and Lorentz assumed that the free electrons were in thermal equilibrium with a Maxwell-Boltzmann velocity distribution. This meant that the v_F in Eq. (10.16) was replaced by the rms electron velocity v_{rms}. In addition, Drude and Lorentz assumed that the free electrons collide with the metal ions, not with the much farther apart lattice defects. The net result was resistivity values on the order of 10 times greater than the measured ones.

The theory was nevertheless considered to be on the right track, both because it gave the correct form of Ohm's law and also because it accounted for the **Weidemann-Franz law.** This empirical law states that the ratio K/σ (where $\sigma = 1/\rho$) between thermal and electric conductivities is the same for all metals and is a function only of temperature. If there is a temperature difference ΔT between the sides of a slab of material Δx thick whose cross-sectional area is A, the rate $\Delta Q/\Delta t$ at which heat passes through the slab is given by

$$\frac{\Delta Q}{\Delta t} = -KA\frac{\Delta T}{\Delta x}$$

where K is the thermal conductivity. According to the kinetic theory of a classical gas applied to the electron gas in the Drude-Lorentz model,

$$K = \frac{knv_{\text{rms}}\lambda}{2}$$

From Eq. (10.16) with v_F replaced by v_{rms},

$$\sigma = \frac{1}{\rho} = \frac{ne^2\lambda}{mv_{\text{rms}}}$$

Hence the ratio between the thermal and electric resistivities of a metal is

$$\frac{K}{\sigma} = \left(\frac{knv_{\text{rms}}\lambda}{2}\right)\left(\frac{mv_{\text{rms}}}{ne^2\lambda}\right) = \frac{kmv_{\text{rms}}^2}{2e^2}$$

According to Eq. (9.15), $v_{\text{rms}}^2 = 3kT/m$, which gives

$$\frac{K}{\sigma T} = \frac{3k^2}{2e^2} = 1.11 \times 10^{-8}\ \text{W}\cdot\Omega/\text{K}^2$$

This ratio does not contain the electron density n or the mean free path λ, so $K/\sigma T$ ought to have the same constant value for all metals, which is the Weidemann-Franz law. To be sure, the above value of $K/\sigma T$ is incorrect because it is based on a Maxwell-Boltzmann distribution of electron velocities. When Fermi-Dirac statistics are used, the result is

$$\frac{K}{\sigma T} = \frac{\pi^2k^2}{3e^2} = 2.45 \times 10^{-8}\ \text{W}\cdot\Omega/\text{K}^2$$

which agrees quite well with experimental findings.

10.6 BAND THEORY OF SOLIDS

The energy band structure of a solid determines whether it is a conductor, an insulator, or a semiconductor

No property of solids varies as widely as their ability to conduct electric current. Copper, a good conductor, has a resistivity of $\rho = 1.7 \times 10^{-8}\ \Omega \cdot \text{m}$ at room temperature,

About Scientists

Felix Bloch (1905–1983) was born in Zurich, Switzerland, and did his undergraduate work in engineering there. He went to Leipzig in Germany for his Ph.D. in physics, remaining there until the rise of Hitler. In 1934 Bloch joined the faculty of Stanford University where he stayed until his retirement except for the war years, which he spent at Los Alamos helping develop the atomic bomb, and for 1954 to 1955, when he was the first director of CERN, the European center for nuclear and elementary-particle research in Geneva.

In 1928 in his doctoral thesis Bloch showed how allowed and forbidden bands arise by solving Schrödinger's equation for an electron moving in the periodic potential of a crystal. This important step in the development of the theory of solids supplemented earlier work by Walter Heitler and Fritz London, who showed how energy levels broaden into bands when atoms are brought together to form a solid. Later Bloch studied the magnetic behavior of atomic nuclei in solids and liquids, which led to the extremely sensitive nuclear magnetic resonance method of analysis. Bloch received the Nobel Prize in physics in 1952 together with Edward Purcell of Harvard, who had also done important work in nuclear magnetism.

whereas for quartz, a good insulator, $\rho = 7.5 \times 10^{17}\ \Omega \cdot \text{m}$, more than 25 powers of ten greater. The existence of electron energy bands in solids makes it possible to understand this remarkable span.

There are two ways to consider how energy bands arise. The simplest is to look at what happens to the energy levels of isolated atoms as they are brought closer and closer together to form a solid. We will begin in this way and then examine the significance of energy bands. Later we will consider the origin of energy bands in terms of the restrictions the periodicity of a crystal lattice imposes on the motion of electrons.

The atoms in every solid, not just in metals, are so near one another that their valence electron wave functions overlap. In Sec. 8.3 we saw the result when two H atoms are brought together. The original 1*s* wave functions can combine to form symmetric or antisymmetric joint wave functions, as in Figs. 8.5 and 8.6, whose energies are different. The splitting of the 1*s* energy level in an isolated H atom into two levels, marked E_A^{total} and E_S^{total}, is shown as a function of internuclear distance in Fig. 8.7.

The greater the number of interacting atoms, the greater the number of levels produced by the mixing of their respective valence wave functions (Fig. 10.19). In a solid, because the splitting is into as many levels as there are atoms present (nearly 10^{23} in a cubic centimeter of copper, for instance), the levels are so close together that they form an energy band that consists of a virtually continuous spread of permitted energies. The energy bands of a solid, the gaps between them, and the extent to which they are filled by electrons not only govern the electrical behavior of the solid but also have important bearing on others of its properties.

Conductors

Figure 10.20 shows the energy levels and bands in sodium. The 3*s* level is the first occupied level to broaden into a band. The lower 2*p* level does not begin to spread out until a much smaller internuclear distance because the 2*p* wave functions are closer to the nucleus than are the 3*s* wave functions. The average energy in the 3*s* band drops at first, which signifies attractive forces between the atoms. The actual internuclear distance in solid sodium corresponds to the minimum average 3*s* electron energy.

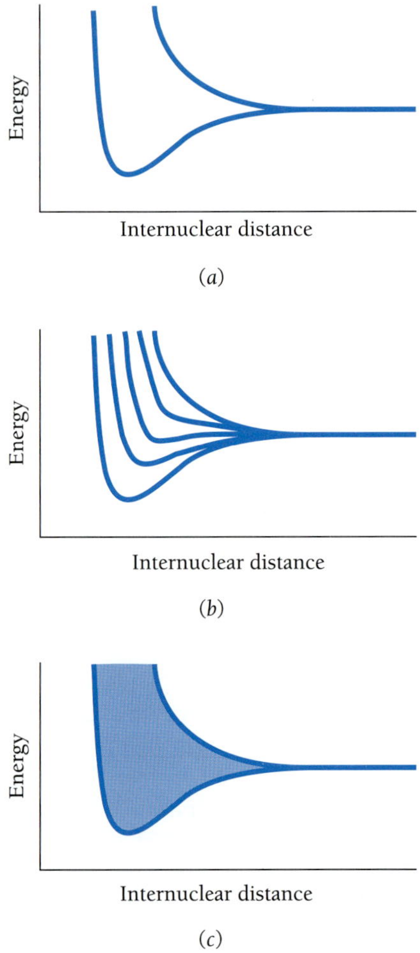

Figure 10.19 The 3*s* level is the highest occupied level in a ground-state sodium atom. (*a*) When two sodium atoms come close together, their 3*s* levels, initially equal, become two separate levels because of the overlap of the corresponding electron wave functions. (*b*) The number of new levels equals the number of interacting atoms, here 5. (*c*) When the number of interacting atoms is very large, as in solid sodium, the result is an energy band of very closely spaced levels.

An electron in a solid can only have energies that fall within its energy bands. The various outer energy bands in a solid may overlap, as in Fig. 10.21*a*, in which case its valence electrons have available a continuous distribution of permitted energies. In other solids the bands may not overlap, as in Fig. 10.21*b*, and the intervals between them represent energies their electrons cannot have. Such intervals are called **forbidden bands** or **band gaps.**

Figure 9.11 shows the distribution of electron energies in a band at various temperatures. At 0 K all levels in the band are filled by electrons up to the Fermi energy ϵ_F, and those above ϵ_F are empty. At temperatures above 0 K, electrons with energies below ϵ_F can move into higher states, in which case ϵ_F represents a level with a 50 percent likelihood of being occupied.

A sodium atom has a single 3*s* valence electron. Each *s* ($l = 0$) atomic level can hold $2(2l + 1) = 2$ electrons, so each *s* band formed by N atoms can hold $2N$ electrons.

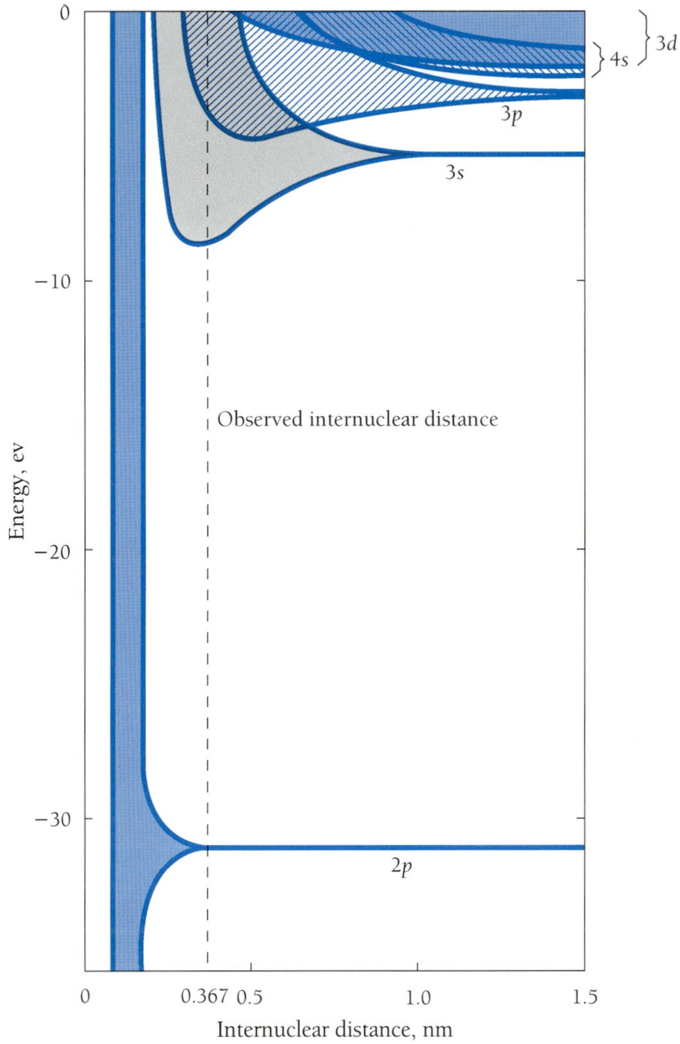

Figure 10.20 The energy levels of sodium atoms become bands as their internuclear distance decreases. The observed internuclear distance in solid sodium is 0.367 nm.

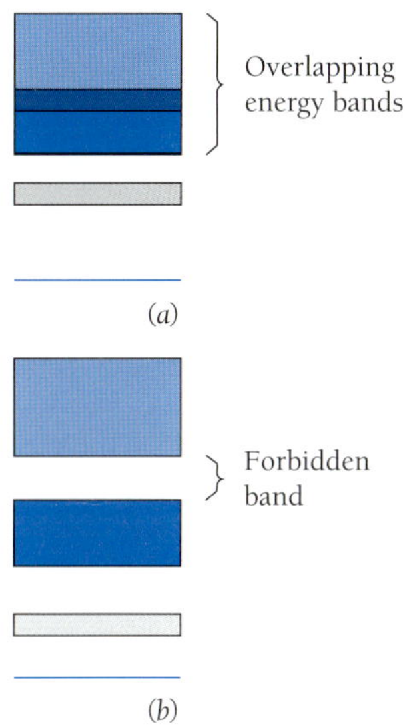

Figure 10.21 (*a*) The energy bands in some solids may overlap to give a continuous band. (*b*) A forbidden band separate nonover-lapping energy bands in other solids.

Thus the 3*s* band in solid sodium is only half filled by electrons (Fig. 10.22) and the Fermi energy ϵ_F lies in the middle of the band.

When a potential difference is applied across a piece of solid sodium, 3*s* electrons can pick up additional energy while remaining in their original band. The additional energy is in the form of KE, and the drift of the electrons constitutes an electric current. Sodium is therefore a good conductor, as are other solids with partly filled energy bands.

Magnesium atoms have filled 3*s* shells. If the 3*s* level simply spreads into a 3*s* band in solid magnesium, as in Fig. 10.21*b*, there would be a forbidden band above it and the 3*s* electrons could not easily pick up enough energy to jump the forbidden band to the empty band above it. Nevertheless magnesium is a metal. What actually happens is that the 3*p* and 3*s* bands overlap as magnesium atoms become close together to give the structure shown in Fig. 10.21*a*. A *p* ($l = 1$) atomic level can hold $2(2l + 1) = 2(2 + 1) = 6$ electrons, so a *p* band formed by *N* atoms can hold 6*N* electrons. Together

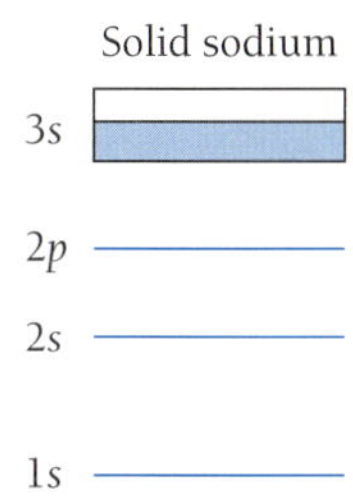

Figure 10.22 The 3*s* energy band in solid sodium is half filled with electrons. The Fermi energy ϵ_F is in the middle of the band.

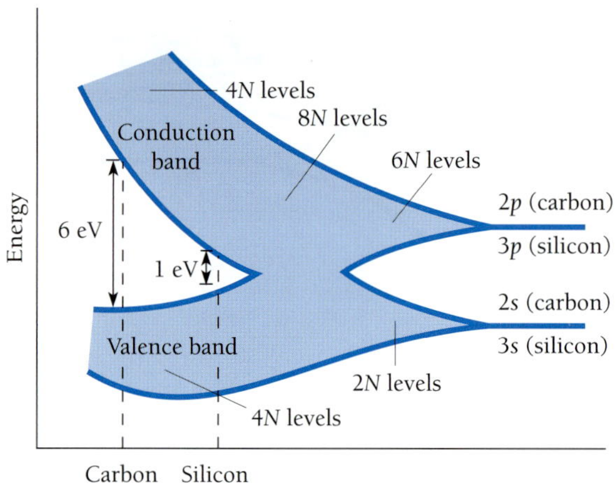

Figure 10.23 Origin of the energy bands of carbon and silicon. The 2*s* and 2*p* levels of carbon atoms and the 3*s* and 3*p* levels of silicon atoms spread into bands that first overlap with decreasing atomic separation and then split into two diverging bands. The lower band is occupied by valence electrons and the upper conduction band is empty. The energy gap between the bands depends on the internuclear separation and is greater for carbon than for silicon.

with the 2*N* electrons the 3*s* band can hold, the 3*s* + 3*p* band in magnesium can hold 8*N* electrons in all. With only 2*N* electrons in the band, it is only one-quarter filled and so magnesium is a conductor.

Insulators

In a carbon atom the 2*p* shell contains only two electrons. Because a *p* shell can hold six electrons, we might think that carbon is a conductor, just as sodium is. What actually happens is that, although the 2*s* and 2*p* bands that form when carbon atoms come together overlap at first (as the 3*s* and 3*p* bands in sodium do), at smaller separations the combined band splits into two bands (Fig. 10.23), each able to contain 4*N* electrons. Because a carbon atom has two 2*s* and two 2*p* electrons, in diamond there are 4*N* valence electrons that completely fill the lower (or **valence**) band, as in Fig. 10.24. The empty **conduction band** above the valence band is separated from it by a forbidden band 6 eV wide. Here the Fermi energy ϵ_F is at the top of the valence band. At least 6 eV of additional energy must be provided to an electron in diamond

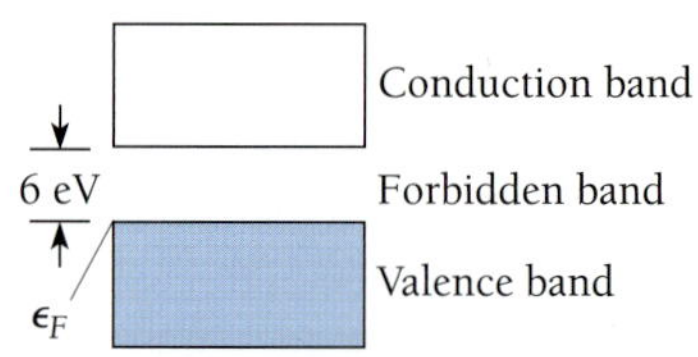

Figure 10.24 Energy bands in diamond. The Fermi energy is at the top of the filled lower band. Because an electron in the valence band needs at least 6 eV to reach the empty conduction band, diamond is an insulator.

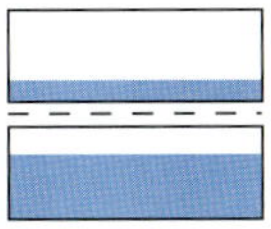

Figure 10.25 The valence and conduction bands in a semiconductor are separated by a smaller gap than in the case of an insulator. Here a small number of electrons near the top of the valence band can acquire enough thermal energy to jump the gap and enter the conduction band. The Fermi energy is therefore in the middle of the gap.

if it is to climb to the conduction band where it can move about freely. With $kT =$ 0.025 eV at room temperature, valence electrons in diamond do not have enough thermal energy to jump the 6 eV gap.

Nor can an energy increment of 6 eV be given to a valence electron in diamond by an electric field, because such an electron undergoes frequent collisions with crystal imperfections during which it loses most of the energy it gains from the field. An electric field of over 10^8 V/m is needed for an electron to gain 6 eV in a typical mean free path of 5×10^{-8} m. This is billions of times stronger than the field needed for a current to flow in a metal. Diamond is therefore a very poor conductor and is classed as an insulator.

Semiconductors

Silicon has a crystal structure like that of diamond and, as in diamond, a gap separates the top of its filled valence band from an empty conduction band above it (see Fig. 10.23). The forbidden band in silicon, however, is only about 1 eV wide. At low temperatures silicon is little better than diamond as a conductor, but at room temperature a small number of its valence electrons have enough thermal energy to jump the forbidden band and enter the conduction band (Fig. 10.25). These electrons, though few, are still enough to allow a small amount of current to flow when an electric field is applied. Thus silicon has a resistivity intermediate between those of conductors and those of insulators, and it and other solids with similar band structures are classed as **semiconductors**.

Impurity Semiconductors

Small amounts of impurity can drastically change the conductivity of a semiconductor. Suppose we incorporate a few arsenic atoms in a silicon crystal. Arsenic atoms have five electrons in their outer shells, silicon atoms have four. (These shells have the configurations $4s^2 4p^3$ and $3s^2 3p^2$ respectively.) When an arsenic atom replaces a silicon atom in a silicon crystal, four of its electrons participate in covalent bonds with its nearest neighbors. The fifth electron needs very little energy—only about 0.05 eV in silicon, about 0.01 eV in germanium—to be detached and move about freely in the crystal.

As shown in Fig. 10.26, arsenic as an impurity in silicon provides energy levels just below the conduction band. Such levels are called **donor levels**, and the substance is called an ***n*-type** semiconductor because electric current in it is carried by negative charges (Fig. 10.27). The presence of donor levels below the conduction band raises the Fermi energy above the middle of the forbidden band between the valence and conduction bands.

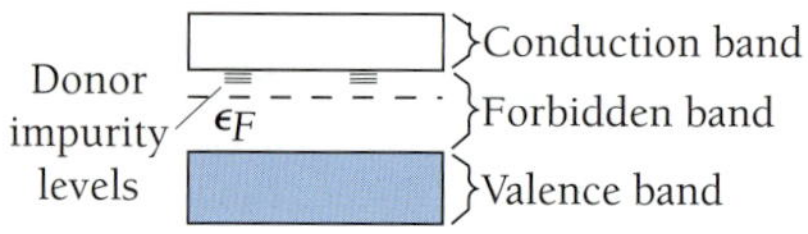

Figure 10.26 A trace of arsenic in a silicon crystal provides donor levels in the normally forbidden band, producing an n-type semiconductor.

Optical Properties of Solids

The optical properties of solids are closely related to their energy-band structures. Photons of visible light have energies from about 1 to 3 eV. A free electron in a metal can readily absorb such an amount of energy without leaving its valence band, and metals are accordingly opaque. The characteristic luster of a metal is due to the reradiation of light absorbed by its free electrons. If the metal surface is smooth, the reradiated light appears as a reflection of the original incident light.

For a valence electron in an insulator to absorb a photon, on the other hand, the photon energy must be over 3 eV if the electron is to jump across the forbidden band to the conduction band. Insulators therefore cannot absorb photons of visible light and are transparent. Of course, most samples of insulating materials do not appear transparent, but this is due to the scattering of light by irregularities in their structures. Insulators are opaque to ultraviolet light, whose higher frequencies mean high enough photon energies to allow electrons to cross the forbidden band.

Because the forbidden bands in semiconductors are about the same in width as the photon energies of visible light, they are usually opaque to visible light. However, they are transparent to infrared light whose lower frequencies mean photon energies too low to be absorbed. For this reason infrared lenses can be made from the semiconductor germanium, whose appearance in visible light is that of an opaque solid.

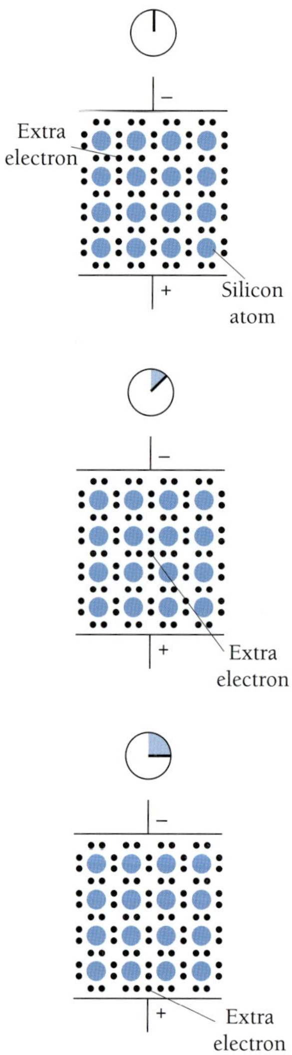

Figure 10.27 Current in an n-type semiconductor is carried by surplus electrons that do not fit into the electron structure of a pure crystal.

If we instead incorporate gallium atoms in a silicon crystal, a different effect occurs. Gallium atoms have only three electrons in their outer shells, whose configuration is $4s^2 4p$, and their presence leaves vacancies called **holes** in the electron structure of the crystal. An electron needs relatively little energy to enter a hole, but as it does so, it leaves a new hole in its former location. When an electric field is applied across a silicon crystal containing a trace of gallium, electrons move toward the anode by successively filling holes (Fig. 10.28). The flow of current here is conveniently described with reference to the holes, whose behavior is like that of positive charges since they move toward the negative electrode. A substance of this kind is called a ***p*-type** semiconductor.

In the energy-band diagram of Fig. 10.29 we see that gallium as an impurity in silicon provides energy levels, called **acceptor levels,** just above the valence band. Any electrons that occupy these levels leave behind them vacancies in the valence band that permit electric current to flow. The Fermi energy in a p-type semiconductor lies below the middle of the forbidden band.

Adding an impurity to a semiconductor is called **doping.** Phosphorus, antimony, and bismuth as well as arsenic have atoms with five valence electrons and so can be used as donor impurities in doping silicon and germanium to yield an n-type semiconductor. Similarly, indium and thallium as well as gallium have atoms with three valence electrons and so can be used as acceptor impurities. A minute amount of impurity can produce a dramatic change in the conductivity of a semiconductor. As an example, 1 part

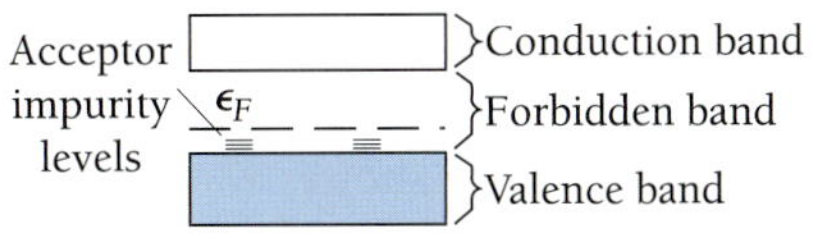

Figure 10.29 A trace of gallium in a silicon crystal provides acceptor levels in the normally forbidden band, producing a *p*-type semiconductor.

of a donor impurity per 10^9 parts of germanium increases its conductivity by a factor of nearly 10^3. Silicon and germanium are not the only semiconducting materials with practical applications: another important class of semiconductors consists of compounds of trivalent and pentavalent elements, such as GaAs, GaP, InSb, and InP.

10.7 SEMICONDUCTOR DEVICES

The properties of the* p-n *junction are responsible for the microelectronics industry

The operation of most semiconductor devices is based upon the nature of junctions between *p*- and *n*-type materials. Such junctions can be made in several ways. A method especially adapted to the production of integrated circuits involves diffusing impurities in vapor form into a semiconductor wafer in regions defined by masks. A series of diffusion steps using donor and acceptor impurities is part of the procedure for manufacturing circuits that can contain millions of resistors, capacitors, diodes, and transistors on a chip a few millimeters across. The limiting factor in this method is the wavelength of the light that is shined through masks to expose and thereby harden the

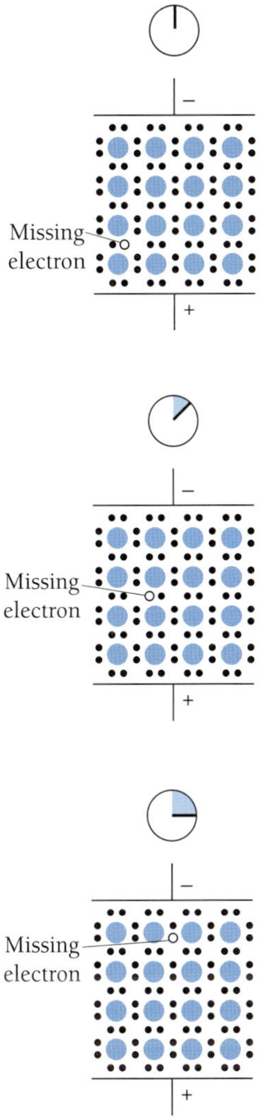

Figure 10.28 Current in a *p*-type semiconductor is carried by the motion of "holes," which are sites of missing electrons. Holes move toward the negative electrode as a succession of electrons move into them.

A number of semiconductor chips for various functions are mounted on the electronic circuit board.

photoresist compound on the wafer surface. (The unexposed photoresist is then washed away to leave areas open to the next diffusion step.) The shortest wavelength that can be used with conventional optical systems is 193 nm (which is in the ultraviolet) because no suitable material transparent to shorter wavelengths is known that can be made into lenses. Features as small as 130 nm might possibly be created with 193-nm light, but the demands of the electronics industry for ever-more components per chip will soon have to be met with some other technology. X-rays, electron and ion beams are being studied for this purpose, with an immediate goal of chips with 200 million circuit elements each 100 nm across.

Junction Diode

A characteristic property of a *p-n* junction is that electric current can pass through it much more readily in one direction than in the other. In the diode shown in Fig. 10.30, the left-hand end is a *p*-type region in which conduction involves the motion of holes, and the right-hand end is an *n*-type region in which conduction occurs by means of the motion of electrons. Three situations can occur:

1 No bias This is illustrated in Fig. 10.30*a*. Electron-hole pairs are created spontaneously by thermal excitation in the valence band of the *p*-region. Some of the electrons have enough energy to jump the gap to the conduction band and then migrate to the *n* region. There they lose energy in collisions. At the same time, some electrons in the *n* region are sufficiently energetic to climb the energy hill and enter the *p* region, where they recombine with holes there. At thermal equilibrium the two processes occur at the same low rate, so there is no net current. The Fermi energy is the same in both *p* and *n* regions; if it were not, electrons would flow to the region with vacant states of lower energy until ϵ_F is the same.

2 Reverse bias As in Fig. 10.30*b*, an external voltage V is applied across the diode with the *p* end negative and the *n* end positive. The energy difference across the junction is greater by Ve than in part *a*, which impedes the recombination current i_r: the holes in the *p* region migrate to the left and are filled at the negative terminal, while the electrons in the *n* region migrate to the right and leave the diode at the positive terminal. New electron-hole pairs are still being created as before by thermal excitation, but because they are relatively few in number the resulting net current $i_t - i_r$ is very small even when the applied voltage V is high. (We note that the conventional current I, which flows from + to −, is opposite in direction to the electron current i.)

3 Forward bias As in Fig. 10.30*c*, the external voltage is applied with the *p* end of the diode positive and the *n* end negative. The energy difference across the junction is now *less* by Ve than in part *a*, which increases the recombination current i_r since the electrons have a smaller energy hill to climb. Under these circumstances new holes are created continuously by the removal of electrons at the positive terminal while new electrons are added at the negative terminal. The holes migrate to the right and the electrons to the left under the influence of the applied potential. The holes and electrons meet in the vicinity of the *p-n* junction and recombine there.

Thus current can flow readily in one direction through a *p-n* junction but hardly at all in the other direction, which makes such a junction an ideal rectifier in an electric circuit. The greater the applied voltage, the greater the current in the forward direction. Figure 10.31 shows how I varies with V for a *p-n* junction rectifier.

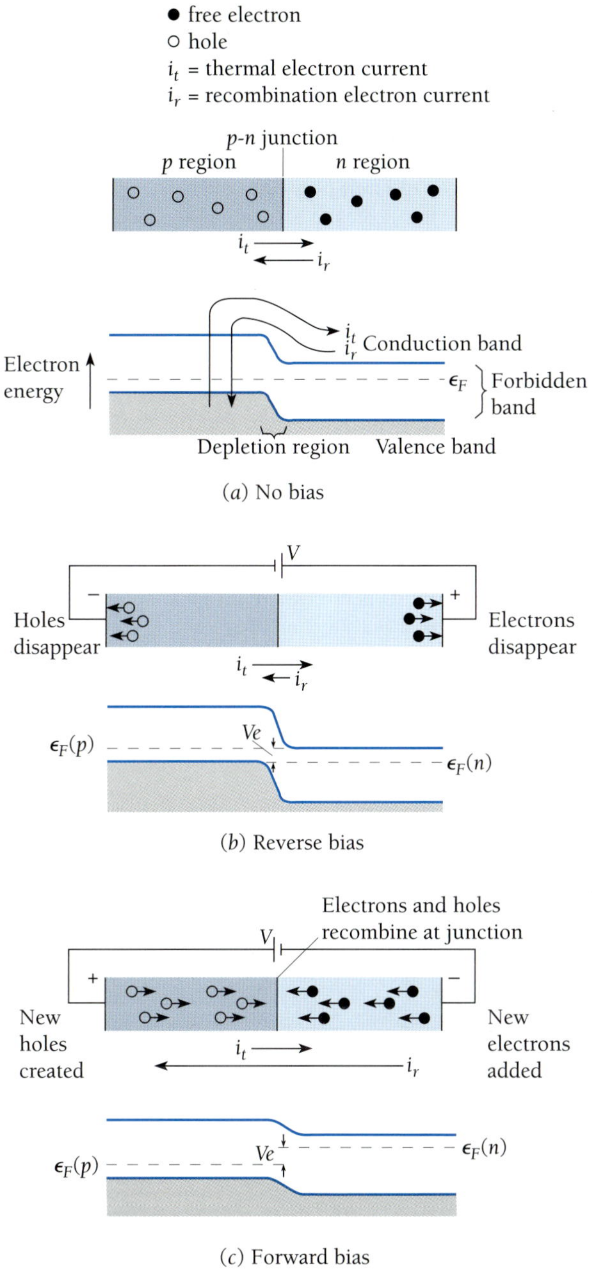

Figure 10.30 Operation of a semiconductor diode.

(a) When there is no applied voltage, the thermal electron current to the right equals the recombination electron current to the left and there is no net current. Both these currents are small.

(b) When an external voltage is applied so that the *p* end of the diode is negative, the recombination electron current is less than the thermal electron current. The result is a very small net electron current to the right.

(c) When an external voltage is applied so that the *p* end of the diode is positive, the recombination current can be much larger than the thermal electron current to give a large net electron current to the left. The conventional current is in the opposite direction to the electron current.

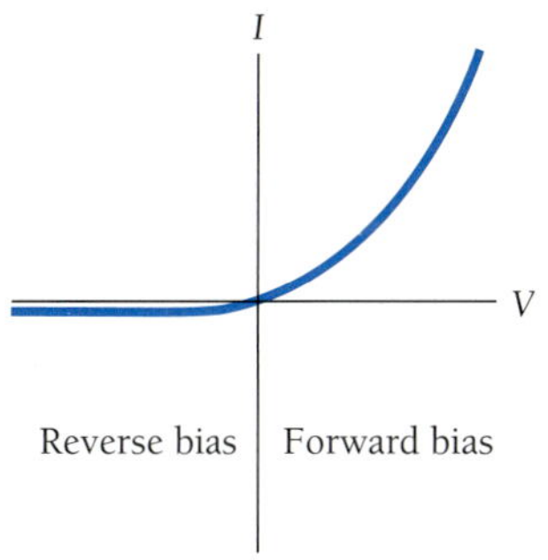

Figure 10.31 Voltage-current characteristic of a *p-n* semiconductor diode.

Photodiodes

Energy is needed to create an electron-hole pair, and this energy is released when an electron and a hole recombine. In silicon and germanium the recombination energy is absorbed by the crystal as heat, but in certain other semiconductors, for instance gallium arsenide, a photon is emitted when recombination occurs. This is the basis of the **light-emitting diode** (LED). Forward bias is used in an LED, so the electrons and holes both move toward the *p-n* junction, as in Fig. 10.30*c*, where they recombine to create photons.

A fairly small current is used in an LED and the photons are produced by spontaneous emission. When the current is high, spontaneous emission may not keep up with the rate of arrival of electrons and holes in the depletion region, and the result is a substantial population inversion there. This is the condition for laser action to occur, with spontaneously emitted photons causing avalanches of additional photons by stimulated emission. In a **semiconductor laser** opposite ends of the *p-n* junction are made parallel and partly reflecting. The coherent light produced by the stimulated emission is intensified as it moves back and forth in the thin depletion region, and emerges through the ends (Fig. 10.32).

The process that occurs in an LED is reversed in a silicon **solar cell.** Here photons arriving at or near the depletion region of a *p-n* junction after passing through a thin ($<1\ \mu$m) outer layer of silicon produce electron-hole pairs if sufficiently energetic. The electrons are raised to the conduction band, leaving holes in the valence band. The potential difference across the depletion region provides an electric field that pulls the electrons to the *n* region and the holes to the *p* region. The newly freed electrons can then flow from the *n* region through an external

The Hubble Space Telescope being launched from the Space Shuttle Discovery. One of the two arrays of solar cells that power the telescope has been deployed.

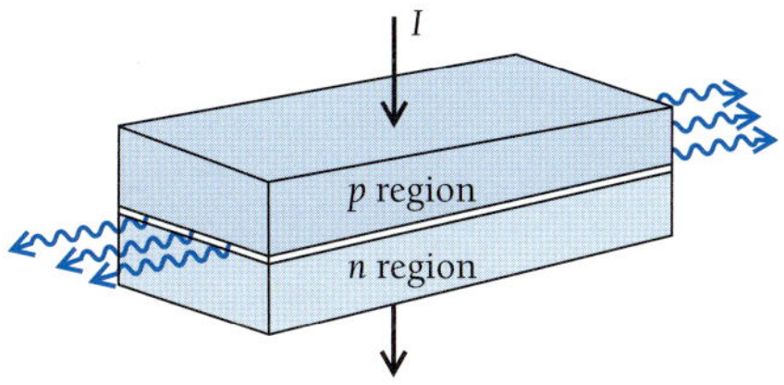

Figure 10.32 A semiconductor laser. Each dimension is less than a millimeter and its light output, as in all lasers, is coherent. The junction between the p and n regions from which the light emerges is only a few micrometers thick.

circuit to the p region where they recombine with the newly created holes. In this way the energy of incident photons can be converted to electric energy. Diodes of this kind are widely used to detect photons in such devices as light meters in cameras as well as to produce electric energy from solar radiation.

The only charge carriers shown in Fig. 10.30 were the electrons. Actually, of course, what was said also applies to the holes, which act as positive charges and behave in exactly the opposite way to add their current to the conventional current.

When a p material joins an n material, a **depletion region** occurs between them instead of a sharp interface, as shown in the lower part of Fig. 10.30*a*. In this region electrons from the donor levels of the n material fill the holes of the acceptor levels of the p material, so that few charge carriers of either kind are present there. The width of the depletion region depends on exactly how the diode is produced, and is typically about 10^{-6} m.

Tunnel Diode

The p and n parts of a diode can be heavily doped to give the energy band structure of Fig. 10.33*a*. The depletion region is very narrow, $\sim 10^{-8}$ m, and the bottom of the n conduction band overlaps the top of the p valence band. The large concentration of impurities causes the donor levels to merge into the bottom of the n conduction band, which moves the Fermi energy there upward into the band. Similarly the acceptor levels merge into the top of the p valence band, which lowers the Fermi energy below the top of the band.

Because the depletion region is so narrow, only a few electron wavelengths across, electrons can "tunnel" through the forbidden band there by the mechanism described in Sec. 5.9. For this reason such a diode is called a **tunnel diode.** When no external voltage is applied to the diode, electrons tunnel in both directions across the gap in equal numbers and the Fermi energy is constant across the diode.

Figure 10.33*b* shows what happens when a small forward voltage is applied to the diode. Now the filled lower part of the n conduction band is opposite the empty upper part of the p valence band, and the tunneling is from n to p only. This gives an electron current to the left, which corresponds to a conventional current to the right.

When the external voltage is increased further, the two bands no longer overlap, as in Fig. 10.33*c*. The tunnel current therefore ceases. From now on the diode behaves exactly like the ordinary junction diode of Fig. 10.30. Figure 10.34 shows the voltage-current characteristic curve of a tunnel diode.

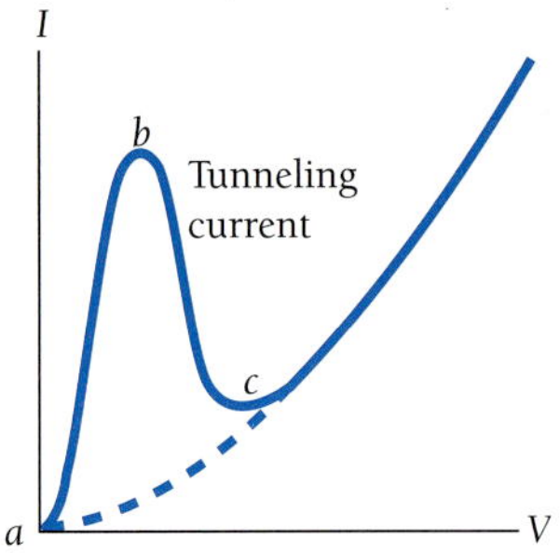

Figure 10.34 Voltage-current characteristic of a tunnel diode. The points *a*, *b*, and *c* correspond to parts *a*, *b*, and *c* of Fig. 10.33. The dashed line indicates the behavior of an ordinary junction diode, as in Fig. 10.30.

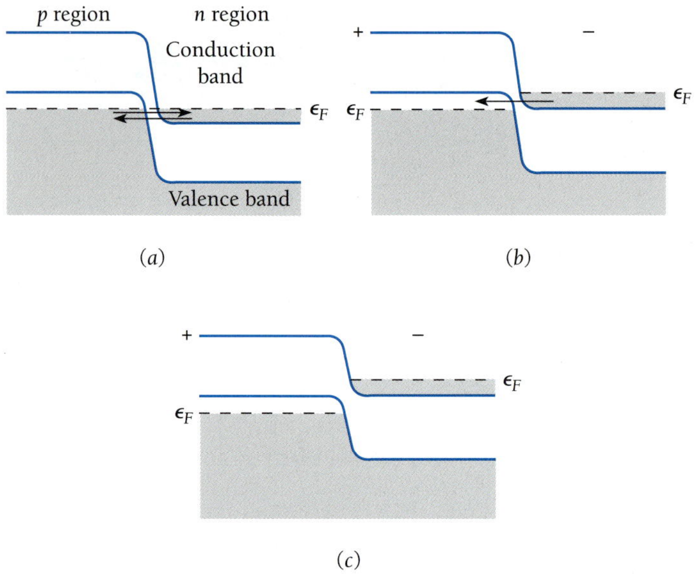

Figure 10.33 Operation of a tunnel diode. (*a*) No bias. Electrons tunnel both ways between the *p* and *n* regions. (*b*) Small forward bias. Electrons tunnel from the *n* to the *p* region only. (*c*) Larger forward bias. Now the valence band of the *p* region does not overlap the conduction band of the *n* region and so no tunneling can occur. At higher voltages the diode behaves like the ordinary diode of Fig. 10.30.

The importance of the tunnel diode lies in the rapidity with which a voltage change between *a* and *b* or between *b* and *c* in Fig. 10.34 can alter the current. In ordinary diodes and transistors, the response time depends on the diffusion speed of the charge carriers, which is low. Hence such devices operate slowly. Tunnel diodes, on the other hand, respond quickly to appropriate voltage changes and can be used in high-frequency oscillators and as fast switches in computers.

Zener Diode

Although the reverse current in many semiconductor diodes remains virtually constant even at high voltages, as in Fig. 10.31, in certain diodes the reverse current increases abruptly when a particular voltage is reached, as in Fig. 10.35. Such diodes are called **Zener diodes** and are widely used in voltage-regulation circuits.

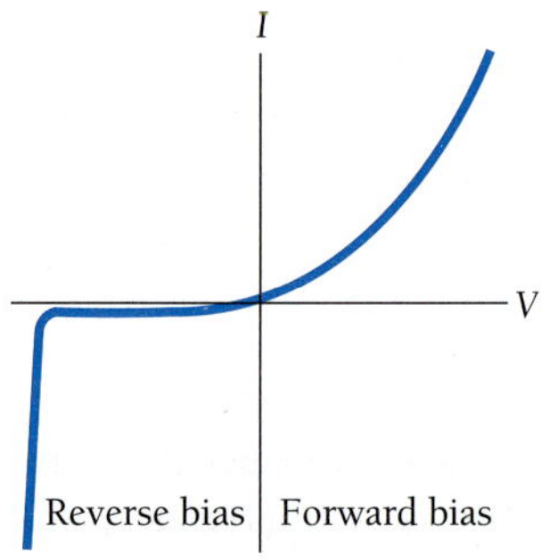

Figure 10.35 Voltage-current characteristic of a Zener diode.

Two mechanisms contribute to the sharp rise in current. One, called **avalanche multiplication,** occurs when an electron near the junction is sufficiently accelerated by the electric field to ionize atoms it collides with, thereby creating fresh electron-hole pairs. The new electrons in their turn continue the process to produce a flood of charge carriers in the diode.

The other mechanism, called **Zener breakdown,** involves the tunneling of valence-band electrons on the *p* side of the junction to the conduction band on the *n* side even though these electrons do not have enough energy to first enter the conduction band on the *p* side. (Such tunneling is in the opposite direction to that occurring in a tunnel diode.) Zener breakdown can occur in heavily doped diodes at voltages of 6 V or less. In lightly doped diodes the necessary voltage is higher, and avalanche multiplication is then the chief process involved.

Junction Transistor

A **transistor** is a semiconductor device that can amplify a weak signal into a strong one when appropriately connected. Figure 10.36 shows an *n-p-n* junction transistor, which consists of a thin *p*-type region called the **base** that is sandwiched between two *n*-type regions called the **emitter** and the **collector.** (A *p-n-p* transistor behaves in a similar manner, except that the current then is carried by holes rather than by electrons.) The energy-band structure of an *n-p-n* transistor is given in Fig. 10.37.

The transistor is given a forward bias across the emitter-base junction and a reverse bias across the base-collector junction. The emitter is more heavily doped than the base, so nearly all the current across the emitter-base junction consists of electrons moving from left to right. Because the base is very thin (1 μm or so) and the concentration of holes there is low, most of the electrons entering the base diffuse through it to the base-collector junction where the high positive potential attracts them into the collector. Changes in the input-circuit current are thus mirrored by changes in the output-circuit current, which is only a few percent smaller.

The ability of the transistor of Fig. 10.36 to produce amplification comes about because the reverse bias across the base-collector junction permits a much higher voltage in the output circuit than that in the input circuit. Since electric power = (current)(voltage), the power of the output signal can greatly exceed the power of the input signal.

Field-Effect Transistor

Although its advent revolutionized electronics, the low input impedance of the junction transistor is a handicap in certain applications. In addition, it is difficult

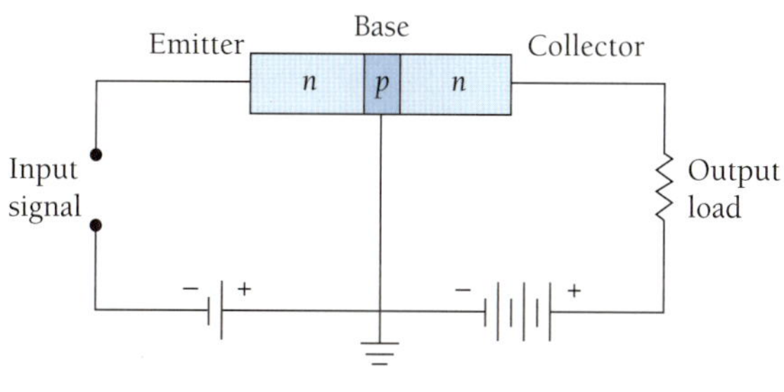

Figure 10.36 A simple junction-transistor amplifier.

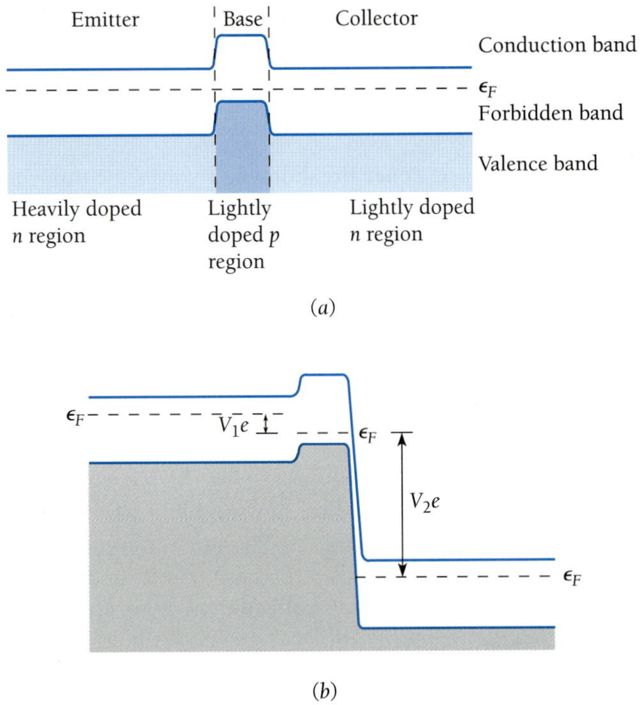

Figure 10.37 (*a*) Isolated *n-p-n* transistor. (*b*) Transistor connected as in Fig. 10.36. The forward bias V_1 between emitter and base is small; the reverse bias V_2 between base and collector is large. Because the base is very thin, electrons can pass through it from emitter to collector without recombining with holes there. Once the electrons are in the collector, they undergo collisions in which they lose energy, and afterward cannot return to the base because the potential hill V_2e is too high.

to incorporate large numbers of them in an integrated circuit and they consume relatively large amounts of power. The **field-effect transistor** (FET) lacks these disadvantages and is widely used today although slower in operation than junction transistors.

As in Fig. 10.38, an **n-channel** FET consists of a block of *n*-type material with contacts at each end together with a strip of *p*-type material on one side that is called the **gate.** When connected as shown, electrons move from the **source** terminal to the **drain** terminal through the *n*-type channel. The *p-n* junction is given a reverse bias, and as a result both the *n* and *p* materials near the junction are depleted of charge carriers (see Fig. 10.30*b*). The higher the reverse potential on the gate, the larger the

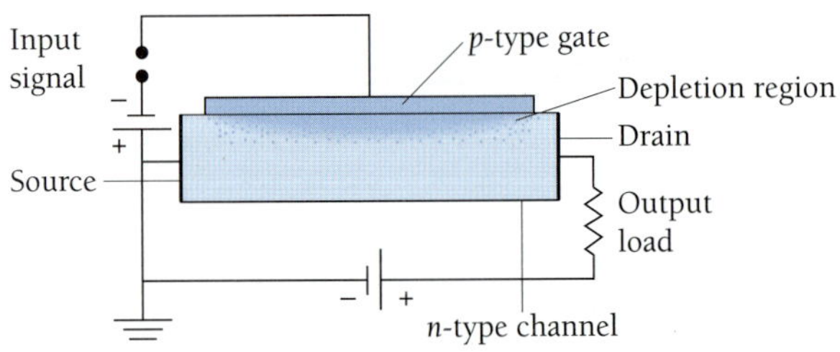

Figure 10.38 A field-effect transistor.

depleted region in the channel and the fewer the electrons available to carry the current. Thus the gate voltage controls the channel current. Very little current passes through the gate circuit owing to the reverse bias, and the result is an extremely high input impedance.

Even higher input impedances (up to 10^{15} Ω) together with greater ease of manufacture are characteristic of the metal-oxide-semiconductor FET (MOSFET), a FET in which the semiconductor gate is replaced by a metal film separated from the channel by an insulating layer of silicon dioxide. The metal film is thus capacitively coupled to the channel, and its potential controls the drain current through the number of induced charges in the channel. A MOSFET occupies only a small percentage of the area needed for a junction transistor.

10.8 ENERGY BANDS: ALTERNATIVE ANALYSIS

How the periodicity of a crystal lattice leads to allowed and forbidden bands

A very different approach can be taken to the origin of energy bands from that described in Sec. 10.6. There we saw that bringing together isolated atoms to form a solid has the effect of broadening their energy levels into bands of allowed electron energies. Alternatively we can start with the idea that an electron in a crystal moves in a region of periodically varying potential (Fig. 10.39) rather than one of constant potential. As a result diffraction effects occur that limit the electron to certain ranges of momenta that correspond to allowed energy bands. In this way of thinking, the interactions among the atoms influence the behavior of their valence electrons indirectly through the crystal lattice these interactions bring about, rather than directly through the atomic interactions themselves. An intuitive approach will be used here to bring out more clearly the physics of the situation, instead of a formal treatment based on Schrödinger's equation.

The de Broglie wavelength of a free electron of momentum p is

Free electron
$$\lambda = \frac{h}{p} \qquad (10.17)$$

Unbound low-energy electrons can travel freely through a crystal since their wavelengths are long relative to the lattice spacing a. More energetic electrons, such as those with the Fermi energy in a metal, have wavelengths comparable with a, and such electrons are diffracted in precisely the same way as x-rays (Sec. 2.6) or electrons in a beam

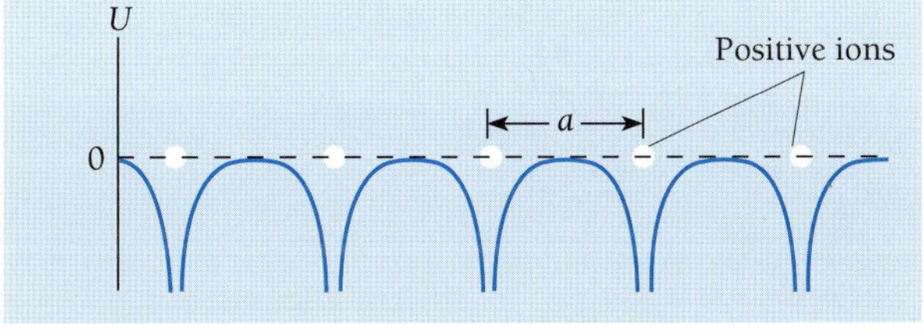

Figure 10.39 The potential energy of an electron in a periodic array of positive ions.

(Sec. 3.5) directed at the crystal from the outside. [When λ is near a, $2a$, $3a$, . . . in length, Eq. (10.17) no longer holds, as discussed later.] An electron of wavelength λ undergoes Bragg reflection from one of the atomic planes in a crystal when it approaches the plane at an angle θ, where from Eq. (2.13)

$$n\lambda = 2a \sin\theta \qquad n = 1, 2, 3, \ldots \tag{10.18}$$

It is customary to treat the situation of electron waves in a crystal by replacing λ by the wave number k introduced in Sec. 3.3, where

Wave number

$$k = \frac{2\pi}{\lambda} = \frac{p}{\hbar} \tag{10.19}$$

The wave number is equal to the number of radians per meter in the wave train it describes, and is proportional to the momentum p of the electron. Since the wave train moves in the same direction as the particle, we can describe the wave train by means of a vector $\mathbf{k}$. Bragg's formula in terms of k is

Bragg reflection

$$k = \frac{n\pi}{a \sin\theta} \qquad n = 1, 2, 3, \ldots \tag{10.20}$$

Figure 10.40 shows Bragg reflection in a two-dimensional square lattice. Evidently we can express the Bragg condition by saying that reflection from the vertical rows of ions occurs when the component of $\mathbf{k}$ in the x direction, k_x, is equal to $n\,\pi/a$. Similarly, reflection from the horizontal rows occurs when $k_y = n\,\pi/a$.

Let us consider first electrons whose wave numbers are sufficiently small for them to avoid reflection. If k is less than π/a, the electron can move freely through the lattice in any direction. When $k = \pi/a$, they are prevented from moving in the x or y directions by reflection. The more k exceeds π/a, the more limited the possible directions of motion, until when $k = \pi/a \sin 45° = \sqrt{2}\pi/a$ the electrons are reflected, even when they move diagonally through the lattice.

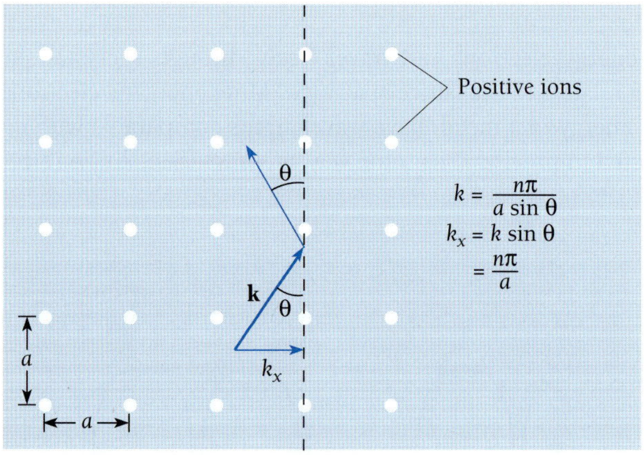

Figure 10.40 Bragg reflection from the vertical rows of ions occurs when $k_x = n\pi/a$.

Brillouin Zones

The region in k-space (here an imaginary plane whose rectangular coordinates are k_x and k_y) that low-k electrons can occupy without being diffracted is called the **first Brillouin zone,** shown in Fig. 10.41. The second Brillouin zone is also shown; it contains electrons with $k > \pi/a$ that do not fit into the first zone yet which have sufficiently small wave numbers to avoid diffraction by the diagonal sets of atomic planes in Fig. 10.40. The second zone contains electrons with k values from π/a to $2\pi/a$ for electrons moving in the $\pm x$ and $\pm y$ directions, with the possible range of k values narrowing as the diagonal directions are approached. Further Brillouin zones can be constructed in the same manner. The extension of this analysis to actual three-dimensional structures leads to Brillouin zones such as those shown in Fig. 10.42.

The significance of the Brillouin zones becomes apparent when we look at the energies of the electrons in each zone.

The energy of a free electron is related to its momentum p by

Energy and momentum

$$E = \frac{p^2}{2m} \tag{10.21}$$

and hence to its wave number k by

Energy and wave number

$$E = \frac{\hbar^2 k^2}{2m} \tag{10.22}$$

In the case of an electron in a crystal for which $k \ll \pi/a$, there is practically no interaction with the lattice, and Eq. (10.22) is valid. Since the energy of such an electron

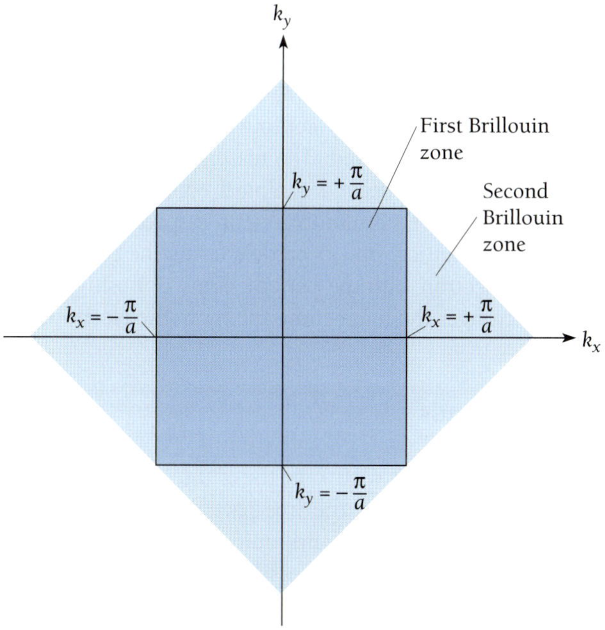

Figure 10.41 The first and second Brillouin zones of a two-dimensional square lattice.

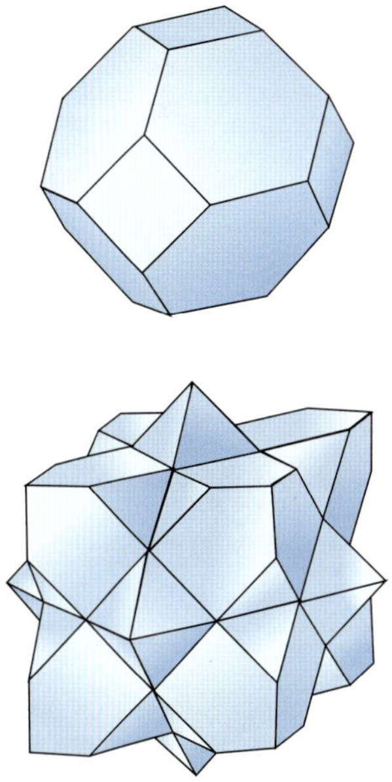

Figure 10.42 First and second Brillouin zones in a face-centered crystal.

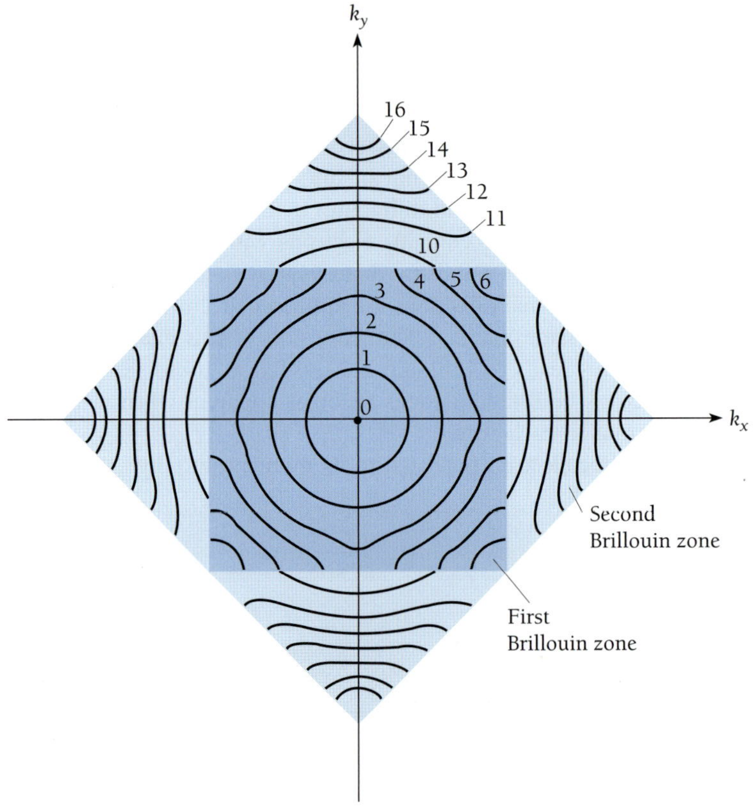

Figure 10.43 Energy contours in electronvolts in the first and second Brillouin zones of a hypothetical square lattice.

depends on k^2, the contour lines of constant energy in a two-dimensional k space are simply circles of constant k, as in Fig. 10.43, for such k values.

With increasing k the constant-energy contour lines become progressively closer together and also more and more distorted. The reason for the first effect is merely that E varies with k^2. The reason for the second is almost equally straightforward. The closer an electron is to the boundary of a Brillouin zone in k-space, the closer it is to being reflected by the actual crystal lattice. But in particle terms the reflection occurs by virtue of the interaction of the electron with the periodic array of positive ions that occupy the lattice points, and the stronger the interaction, the more the electron's energy is affected.

Origin of Forbidden Bands

Figure 10.44 shows how E varies with k in the x direction. As k approaches π/a, E increases more slowly than $\hbar^2k^2/2m$, the free-particle figure. At $k = \pi/a$, E has two values, the lower belonging to the first Brillouin zone and the higher to the second zone. There is a definite gap between the possible energies in the first and second Brillouin zones which corresponds to a forbidden band. The same pattern continues as successively higher Brillouin zones are reached.

The energy discontinuity at the boundary of a Brillouin zone follows from the fact that the limiting values of k correspond to standing waves rather than traveling

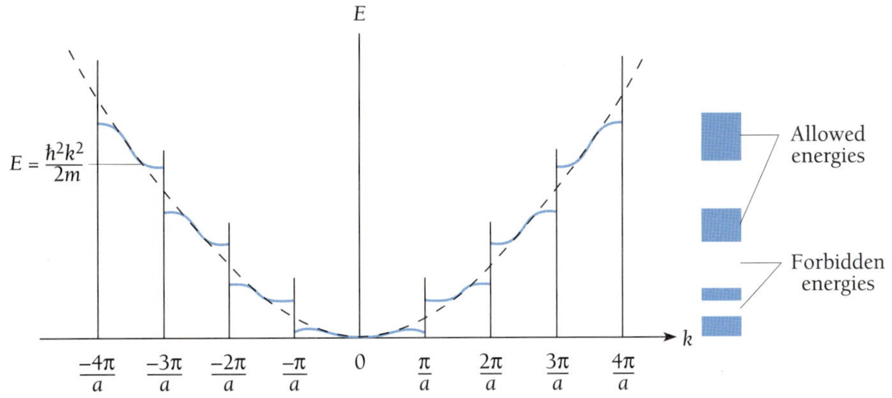

Figure 10.44 Electron energy E versus wave number k in the k_x direction. The dashed line shows how E varies with k for a free electron, as given by Eq. (10.22).

waves. For clarity we consider electrons moving in the x direction; extending the argument to any other direction is straightforward. When $k = \pm\pi/a$, as we have seen, the waves are Bragg-reflected back and forth, and so the only solutions of Schrödinger's equation consist of standing waves whose wavelength is equal to the periodicity of the lattice. There are two possibilities for these standing waves for $n = 1$, namely,

$$\psi_1 = A \sin \frac{\pi x}{a} \tag{10.23}$$

$$\psi_2 = A \cos \frac{\pi x}{a} \tag{10.24}$$

The probability densities $|\psi_1|^2$ and $|\psi_2|^2$ are plotted in Fig. 10.45. Evidently $|\psi_1|^2$ has its minima at the lattice points occupied by the positive ions, while $|\psi_2|^2$ has its

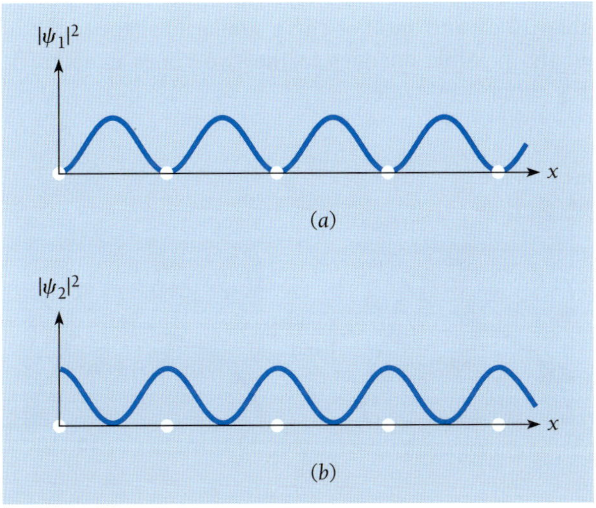

Figure 10.45 Distributions of the probability densities $|\psi_1|^2$ and $|\psi_2|^2$.

Table 10.2 Effective Mass Ratios m^*/m at the Fermi Surface in Some Metals

Metal		m^*/m
Lithium	Li	1.2
Beryllium	Be	1.6
Sodium	Na	1.2
Aluminum	Al	0.97
Cobalt	Co	14
Nickel	Ni	28
Copper	Cu	1.01
Zinc	Zn	0.85
Silver	Ag	0.99
Platinum	Pt	13

Effective Mass

Because an electron in a crystal interacts with the crystal lattice, its response to an external electric field is not the same as that of a free electron. Remarkably enough, the most important results of the free-electron theory of metals discussed in Secs. 9.9 and 9.10 can be incorporated in the more realistic band theory merely by replacing the electron mass m by an average **effective mass** m^*. For example, Eq. (9.56) for the Fermi energy is equally valid in the band theory when m^* is used in place of m. Table 10.2 is a list of effective mass ratios m^*/m for several metals.

maxima at the lattice points. Since the charge density corresponding to an electron wave function ψ is $-e|\psi|^2$, the charge density in the case of ψ_1 is concentrated *between* the positive ions, while in the case of ψ_2, it is concentrated *at* the positive ions. The potential energy of an electron in a lattice of positive ions is greatest midway between each pair of ions and least at the ions themselves, so the electron energies E_1 and E_2 associated with the standing waves ψ_1 and ψ_2 are different. No other solutions are possible when $k = \pm\pi/a$ and accordingly no electron can have an energy between E_1 and E_2.

Figure 10.46 shows the distribution of electron energies that corresponds to the Brillouin zones pictured in Fig. 10.43. At low energies (in this hypothetical situation for $E < \sim 2$ eV) the curve is almost exactly the same as that of Fig. 9.11 based on the free-electron theory. This is not surprising since at low energies k is small and the electrons in a periodic lattice then *do* behave like free electrons.

With increasing energy, however, the number of available energy states goes beyond that of the free-electron theory owing to the distortion of the energy contours by the lattice. Hence there are more different k values for each energy. Then, when $k = \pm\pi/a$, the energy contours reach the boundaries of the first zone and energies higher than about 4 eV (in this particular model) are forbidden for electrons in the k_x and k_y directions although permitted in other directions. As the energy goes farther and farther beyond 4 eV, the available energy states become restricted more and more to the corners of the zone, and $n(E)$ falls. Finally, at approximately $6\frac{1}{2}$ eV, there are no more states and $n(E) = 0$. The lowest possible energy in the second zone is somewhat less than 10 eV and another curve similar in shape to the first begins. Here the gap between the possible energies in the two zones is about 3 eV, and so the forbidden band is about 3 eV wide.

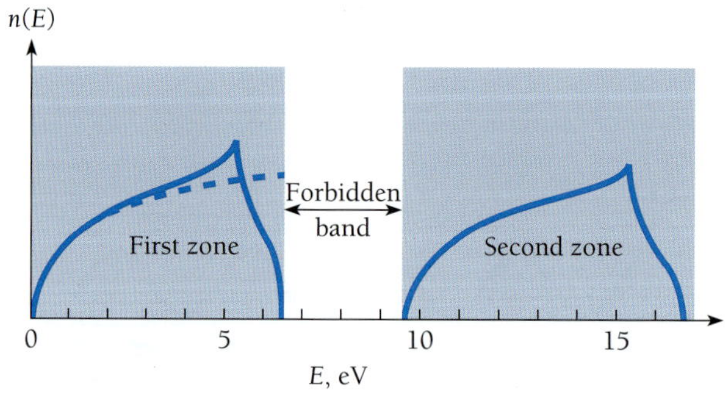

Figure 10.46 The distributions of electron energies in the Brillouin zones of Fig. 10.43. The dashed line is the distribution predicted by the free-electron theory.

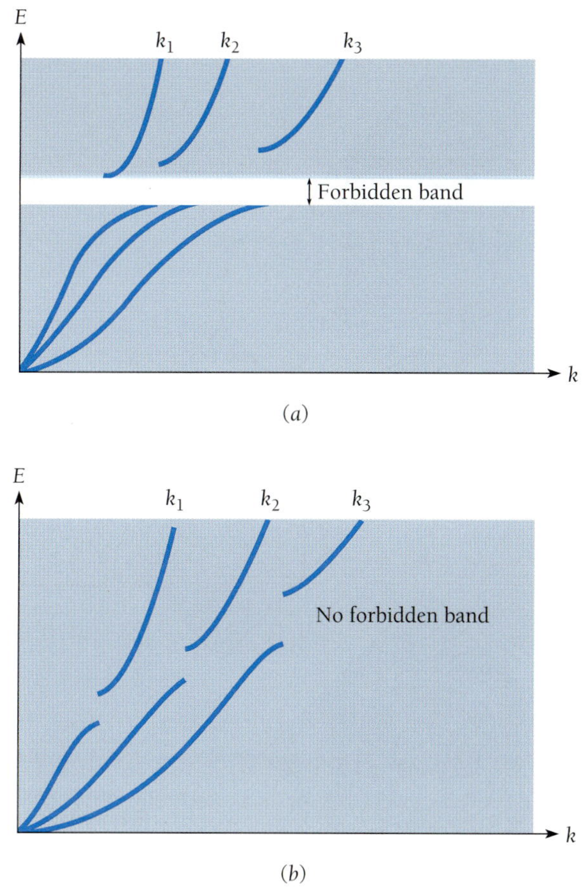

Figure 10.47 E versus k curves for three directions in two crystals. In (*a*) there is a forbidden band, in (*b*) the allowed energy bands overlap and there is no forbidden band.

Although there must be an energy gap between successive Brillouin zones in any given direction, the various gaps may overlap permitted energies in other directions so that there is no forbidden band in the crystal as a whole. Figure 10.47 contains graphs of E versus k for three directions (*a*) in a crystal that has a forbidden band and (*b*) in a crystal whose allowed bands overlap sufficiently to avoid having a forbidden band.

As we know, the electrical behavior of a solid depends on the degree of occupancy of its energy bands as well as on its band structure. Figure 10.48*a* shows the first and second Brillouin zones of a hypothetical two-dimensional insulator. The first zone is filled with electrons, and the energy gap between this zone and the second is much wider than kT. This corresponds to the situation shown in Fig. 10.24 where the insulator is diamond. In Fig. 10.48*b* the zones are the same, but the first zone is only half filled. This corresponds to the situation shown in Fig. 10.22, and the material is analogous to a metal such as sodium whose atoms have one valence electron each. In Fig. 10.48*c* the energies in the second zone overlap those in the first zone, so the valence electrons partly occupy both zones. This corresponds to the situation shown in Fig. 10.47*b*, and the material is analogous to a metal such as magnesium which has two valence electrons per atom.

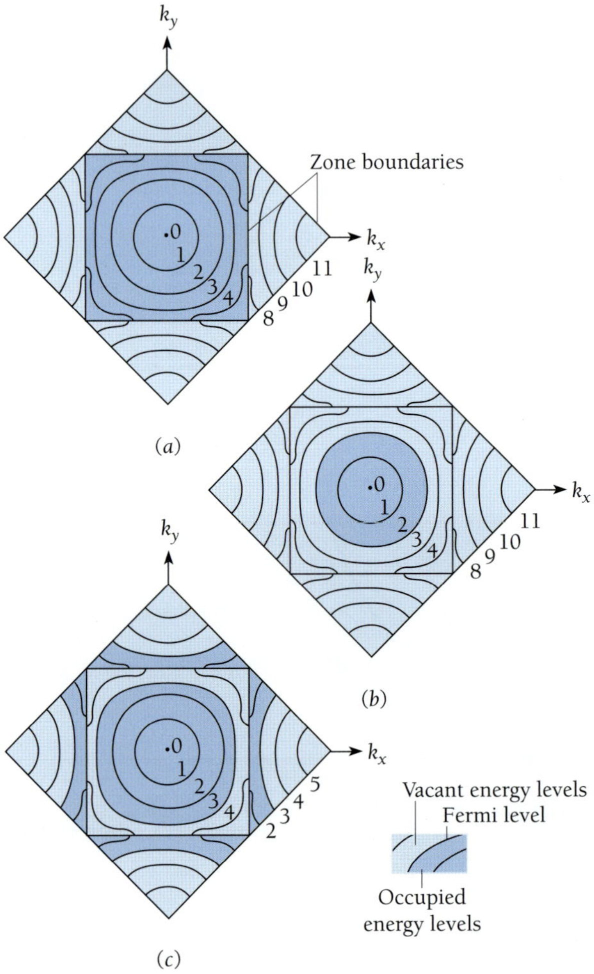

Figure 10.48 Electron energy contours and Fermi levels in three types of solid: (*a*) insulator; (*b*) monovalent metal; (*c*) divalent metal. Energies are in electronvolts.

10.9 SUPERCONDUCTIVITY

No resistance at all, but only at very low temperatures (so far)

Electrical conductors, even the very best, resist to some extent the flow of charge through them at ordinary temperatures. At very low temperatures, however, most metals, many alloys, and some chemical compounds all allow current to pass freely through them. This phenomenon is called **superconductivity.**

Superconductivity was discovered in 1911 by the Dutch physicist Heike Kamerlingh Onnes. He found that, down to 4.15 K, the resistance of a mercury sample decreased with temperature as other metals do (see Fig. 10.18). At $T_c = 4.15$ K, though, the resistance fell sharply to as close to zero as his instruments could measure (Fig. 10.49). The **critical temperature** T_c for other superconducting elements varies from less than 0.1 K to nearly 10 K. As we shall see later, it is significant that elements which are ordinarily good conductors, such as copper and silver, do not become

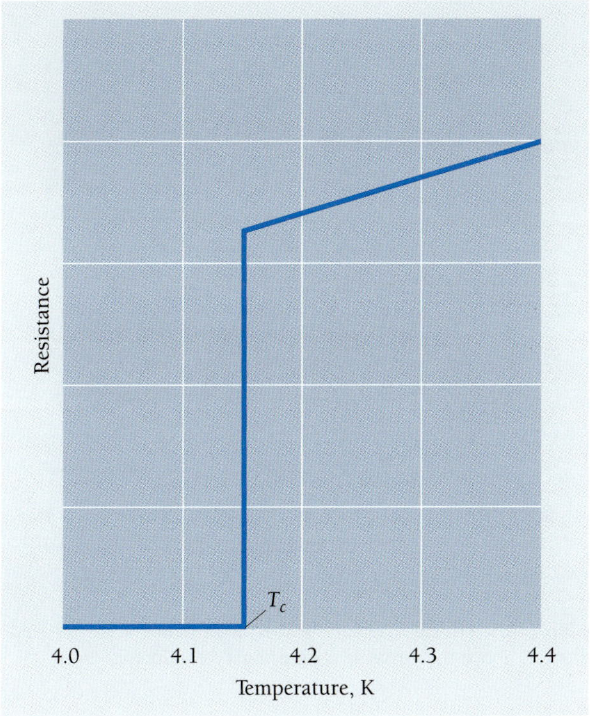

Figure 10.49 Resistance of a mercury sample at low temperature. Below the critical temperature of $T_c = 4.15$ K mercury is a superconductor with zero electrical resistance.

superconducting when cooled. The highest critical temperatures, as much as 134 K, are found in certain ceramic materials.

Does a superconductor actually have zero resistance or just very little? To find out, currents have been set up in superconducting wire loops and the resulting magnetic fields monitored, sometimes for years. No decrease in such currents has ever been found: superconductors do have no resistance at all.

Magnetic Effects

The presence of a magnetic field causes the critical temperature of **type I superconductors** to decrease in the manner shown in Fig. 10.50. If the magnetic field exceeds a certain critical value B_c, which depends on the material and its temperature, its superconductivity disappears altogether. Such materials are superconductors only for values of T and B below their respective curves and are normal conductors for values of T and B above these curves. The critical field B_c would be a maximum at 0 K.

Table 10.3 gives critical temperatures and critical magnetic fields $B_c(0)$ extrapolated to 0 K for several type I superconductors. The critical fields are all quite low, less than 0.1 T, so type I superconductors cannot be used for the coils of strong electromagnets.

Superconductors are perfectly diamagnetic—no magnetic field can exist inside them under any circumstance. If we put a sample of a superconductor in a magnetic field weaker than the critical field and then reduce the temperature below T_c, the field is expelled from the interior of the sample (Fig. 10.51). What happens is that currents appear on the

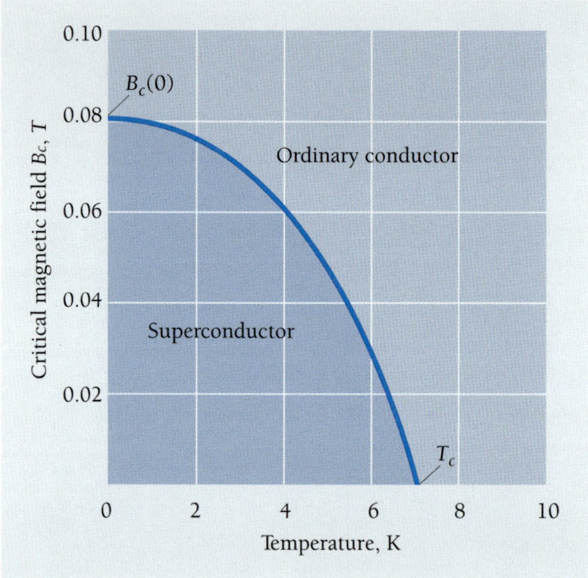

Figure 10.50 Variation of the critical magnetic field B_c with temperature for lead. Below the curve, lead is a superconductor; above the curve, it is an ordinary conductor.

surface of the sample whose magnetic fields exactly cancel the original field inside it. This **Meissner effect** would not occur in an ordinary conductor whose resistance we can imagine reduced to zero; it is characteristic only of superconductivity, which is evidently a unique state of matter in respects other than ability to conduct electric current.

Type I superconductors exist only in two states, normal and superconducting. **Type II superconductors**, which were discovered several decades later and are usually alloys, have an intermediate state as well. Such materials have two critical magnetic fields, B_{c1} and B_{c2} (Fig. 10.52). For an applied magnetic field less than B_{c1}, a type II superconductor behaves just like its type I counterpart when $B < B_c$: it is superconducting with no magnetic field in its interior. When $B > B_{c2}$, a type II superconductor exhibits normal behavior, again like a type I superconductor. However, in applied fields between B_{c1} and B_{c2}, a type II superconductor is in a mixed state in which it

Table 10.3 Critical Temperatures and Critical Magnetic Fields (at $T = 0$) of Some Type I Superconductors

Superconductor	T_c, K	$B_c(0)$, T
Al	1.18	0.0105
Hg	4.15	0.0411
In	3.41	0.0281
Pb	7.19	0.0803
Sn	3.72	0.0305
Zn	0.85	0.0054

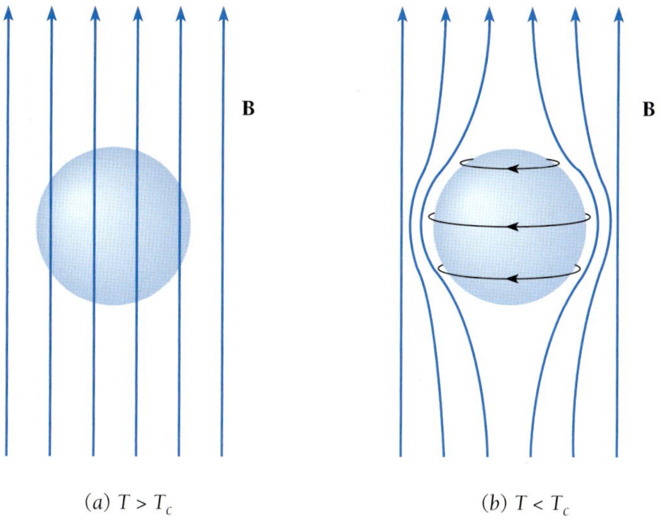

Figure 10.51 The Meissner effect. (a) An applied magnetic field can exist inside a superconductor at temperatures above its critical temperature T_c. (b) When the superconductor is then cooled below T_c, surface currents appear whose effect is to expel the magnetic field from the interior of the superconductor.

contains some magnetic flux but is superconducting. The stronger the external field, the more flux penetrates the material, up to the higher critical field B_{c2}.

A type II superconductor behaves as though it consists of filaments of normal and of superconducting matter mixed together. A magnetic field can exist in the normal filaments, while the superconducting filaments are diamagnetic and resistanceless like

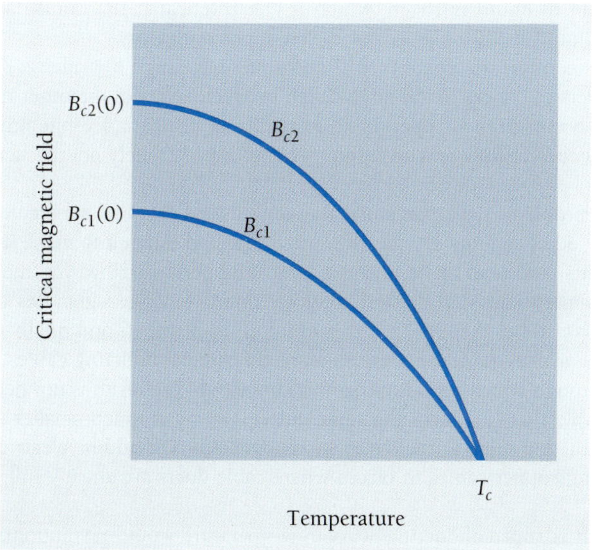

Figure 10.52 Variation of the critical magnetic fields B_{c1} and B_{c2} with temperature for a type II superconductor. For magnetic fields between B_{c1} and B_{c2} the material is in a mixed state in which it is superconducting but a magnetic field can exist in its interior.

Table 10.4 Critical Temperatures and Upper Critical Magnetic Fields (at $T = 0$) of Some Type II Superconductors

Superconductor	T_c, K	$B_{c2}(0)$, T
Nb_3Sn	18.0	24.5
Nb_3Ge	23.2	38
Nb_3Al	18.7	32.4
$Nb_3(AlGe)$	20.7	44
V_3Ge	14.8	2.08
V_3Si	16.9	2.35
PbMoS	14.4	6.0

type I superconductors. Because B_{c2} can be quite high (Table 10.4), type II superconductors are used to make high-field (up to 20 T) magnets for particle accelerators, fusion reactors, magnetic resonance imagery, and experimental **maglev** (magnetic levitation) trains in which magnetic forces provide both propulsion and frictionless support.

High-Temperature Superconductors

Despite much effort, until 1986 no superconductor was known whose critical temperature was higher than 27 K. In that year Alex Muller and Georg Bednorz, working in Switzerland, studied a class of ceramic materials that had never before been suspected of superconducting behavior. They discovered an oxide of lanthanum, barium, and copper for which T_c was 30 K, and soon afterward others extended their approach to produce superconductors with critical temperatures of as high as 134 K (-139°C) for an oxide of mercury, barium, calcium, and copper. (This material has an even higher critical temperature when under pressure.) Although still extremely cold by everyday standards, such temperatures are above the 77-K boiling point of liquid nitrogen, which is cheap (cheaper than milk) and readily available, unlike the liquid helium needed for earlier superconductors.

The new superconductors are all type II and some have high B_{c2} values. The ceramic crystals consist of layers of copper oxide sandwiched between layers of the other metal oxides. The superconduction occurs in the copper oxide, normally an insulator. Despite much study, the exact mechanism of current flow remains unknown, but it is definitely not the same as in ordinary superconductors.

A number of problems have prevented the wide use of the new superconductors thus far. For instance, like other ceramic crystals they are brittle and difficult to make into wires, cannot carry high currents, and tend to be unstable over long periods. However, methods have been devised to overcome or sidestep these difficulties; one is to encase granules of superconducting material in silver tubes that are then drawn into thin filaments and finally bundled into cables or ribbons. For electric power transmission, the superconducting cables are placed in an insulated pipe through which liquid nitrogen is circulated. The result is not necessarily cheaper than a copper cable that can carry the same current but it is much smaller and lighter. This makes superconducting pipes attractive in such applications as adding electric distribution capacity by replacing copper cables in places where cable ducts are already full, a common situation in cities.

A material that is superconducting at room temperature would revolutionize technology. In addition, by reducing the waste of electrical energy (about 10 percent of the electrical energy generated in the United States is lost as heat in transmission lines), the rate at which the world's resources are being depleted would be reduced. Since 1986 such a material no long seems inconceivable.

Magnetic levitation. A small permanent magnet is floating freely above a high-temperature superconductor cooled with liquid nitrogen. The magnetic field of the magnet induces electric currents in the superconductor which lead to a zero resultant field inside the superconductor. The magnetic field of these current outside the superconductor repels the magnet.

10.10 BOUND ELECTRON PAIRS

The key to superconductivity

The origin of superconductivity remained a mystery until the Bardeen-Cooper-Schrieffer (BCS) theory of 1957. An earlier hint of the direction such a theory should take was the discovery that the critical temperatures T_c of the isotopes of a superconducting element decrease with increasing atomic mass. For instance, in mercury T_c is 4.161 K in ^{199}Hg but only 4.126 K in ^{204}Hg. This **isotope effect** suggests that the current-carrying electrons in a superconductor do not move independently of the ion lattice (as we might think when we recall that the resistance of ordinary conductors arises from the scattering of these electrons by lattice defects and vibrations) but instead are somehow interacting with the lattice.

The nature of the interaction became clear when Leon Cooper showed how two electrons in a superconductor could form a bound state despite their coulomb repulsion. What happens is that the lattice is slightly deformed as an electron moves through it, with the positive ions in the electron's path being displaced toward it. The deformation produces a region of increased positive charge. Another electron moving through this polarized region will be attracted by the greater concentration of positive charge there. If the attraction is stronger than the repulsion between the electrons, the electrons are effectively coupled together into a **Cooper pair** with the deformed lattice as the intermediary.

The electron-lattice-electron interaction does not keep the electrons a fixed distance apart. In fact, the theory shows that they must be moving in opposite directions, and their correlations may persist over lengths as great as 10^{-6} m. The binding energy of

About Scientists

John Bardeen (1908–1991) was born in Madison, Wisconsin, and studied electrical engineering at the University of Wisconsin and solid-state physics at Princeton University. After working at several universities and, during World War II, at the Naval Ordnance Laboratory, he went to Bell Telephone Laboratories in 1945 where he joined a semiconductor research group led by William Shockley. In 1948 the group produced the first transistor, for which Shockley, Bardeen, and their collaborator Walter Brattain received a Nobel Prize in 1956. Bardeen later said, "I knew the transistor was important, but I never foresaw the revolution in electronics it would bring."

In 1951 Bardeen left Bell Labs for the University of Illinois where, together with Leon Cooper and J. Robert Schrieffer, he developed the theory of superconductivity. Compared with his earlier work on the transistor, "Superconductivity was more difficult to solve, and it required some radically new concepts." According to the theory, the motions of two electrons can become correlated through their interactions with a crystal lattice, which enables the pair to move with complete freedom through the crystal. Bardeen received his second Nobel Prize in 1972 for this theory along with Cooper and Schrieffer; he was the first person to receive two such prizes in the same field.

a Cooper pair, called the **energy gap** E_g, is of the order of 10^{-3} eV, which is why superconductivity is a low-temperature phenomenon. The energy gap can be measured by directing microwave radiation of frequency ν at a superconductor. When $h\nu \geq E_g$, strong absorption occurs as the Cooper pairs break apart.

The BCS theory relates the energy gap of a superconductor at 0 K to its critical temperature T_c by the formula

Energy gap at 0 K

$$E_g(0) = 3.53kT_c \tag{10.25}$$

Equation (10.25) agrees fairly well with the observed values of E_g and T_c. At temperatures above 0 K, some Cooper pairs break up. The resulting individual electrons interact with the remaining Cooper pairs and reduce the energy gap (Fig. 10.53). Finally,

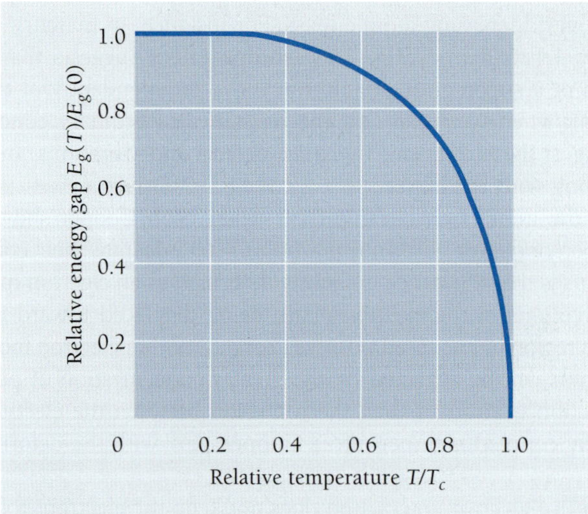

Figure 10.53 Variation of the superconducting energy gap with temperature. Here $E_g(T)$ is the energy gap at the temperature T and $E_g(0)$ is the gap at $T = 0$; T_c is the critical temperature of the material.

at the critical temperature T_c, the energy gap disappears, there are no more Cooper pairs, and the material is no longer superconducting.

The electrons in a Cooper pair have opposite spins, so the pair has a total spin of zero. As a result, the electron pairs in a superconductor are bosons (unlike individual electrons, which have spins of $\frac{1}{2}$ and are fermions), and any number of them can exist in the same quantum state at the same time. When there is no current in the superconductor, the linear momenta of the electrons in a Cooper pair are equal and opposite for a total of zero. All the pairs are then in the same ground state and make up a giant system the size of the superconductor. A single wave function represents this system, whose total energy is less than that of a system of the same number of electrons with a Fermi energy distribution.

A current in a superconductor involves the entire system of electron pairs acting as a unit. Every pair now has non-zero momentum. To alter such a current means that the correlated states of motion of *all* the electron pairs, not just the states of motion of some individual electrons as in an ordinary conductor, must be changed. Because such a change requires a relatively large amount of energy, the current persists indefinitely

Flux Quantization

Figure 10.54 shows a superconducting ring of area A that carries a current. The amount of magnetic flux $\Phi = BA$ passes through the ring as a result. According to Faraday's law of electromagnetic induction, any change in the flux will change the current in the ring so as to oppose the change in flux. Because the ring has no resistance, the change in flux will be perfectly canceled out. The flux Φ therefore is permanently trapped.

Because the phase of the wave function of the Cooper pairs in the ring must be continuous around the ring, it turns out that Φ is quantized. The only values that Φ can have are

Flux quantization
$$\Phi = n\left(\frac{h}{2e}\right) = n\Phi_0 \qquad n = 1, 2, 3, \ldots \tag{10.26}$$

The quantum of magnetic flux is

Flux quantum
$$\Phi_0 = \frac{h}{2e} = 2.068 \times 10^{-15}\ \mathrm{T \cdot m^2}$$

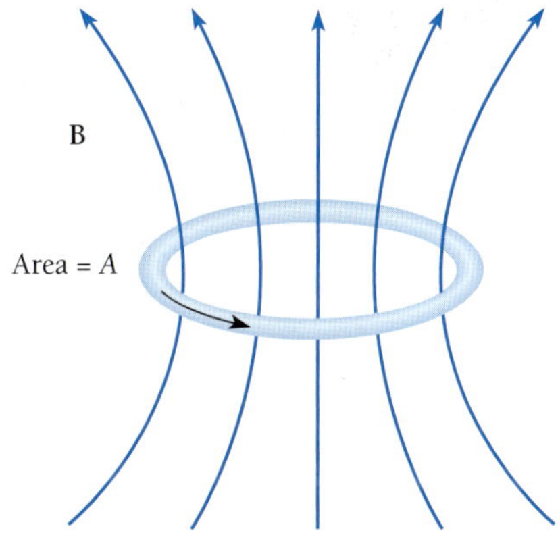

Figure 10.54 The magnetic flux $\Phi = BA$ that passes through a superconducting ring can only have the values $\Phi = n\Phi_0$ where Φ_0 is the flux quantum and $n = 1, 2, 3, \ldots$

if undisturbed, and the electron scattering that leads to resistance in an ordinary conductor does not occur.

A material with large-amplitude lattice vibrations may be only a fair conductor at ordinary temperatures because electron scattering takes place frequently. However, the same ease of lattice deformation means more strongly bound Cooper pairs at low temperatures, and hence the material is more likely to be a superconductor then. Good conductors, such as copper and silver, have small lattice vibrations at ordinary temperatures, which means their lattices are unable to mediate the formation of Cooper pairs at low temperatures and so they do not become superconducting. Such metals as mercury, tin, and lead have large lattice vibrations at ordinary temperatures and so are poorer conductors than copper and silver, but they are superconductors at low temperatures.

Josephson Junctions

As we learned in Chap. 5, the wave nature of a moving particle allows it to tunnel through a barrier that, in classical physics, it could not penetrate. Thus a small but detectable current of electrons can tunnel through a thin insulating layer between two metals. In 1962 Brian Josephson, then a graduate student at Cambridge University, predicted that Cooper pairs could tunnel through what is now called a **Josephson junction,** a thin insulating layer between two superconductors. The wave functions of the Cooper pairs on each side of the junction penetrate the insulating layer with exponentially decreasing amplitudes, just as the wave functions of individual electrons would. If the layer is thin enough, less than 2 nm in practice, the wave functions overlap sufficiently to become coupled together, and the Cooper pairs they describe can then pass through the junction. Josephson shared the 1975 Nobel Prize in physics for his work.

In the **dc Josephson effect,** the current through a Josephson junction that has no voltage across it is given by

dc Josephson effect
$$I_J = I_{\max} \sin \phi \tag{10.27}$$

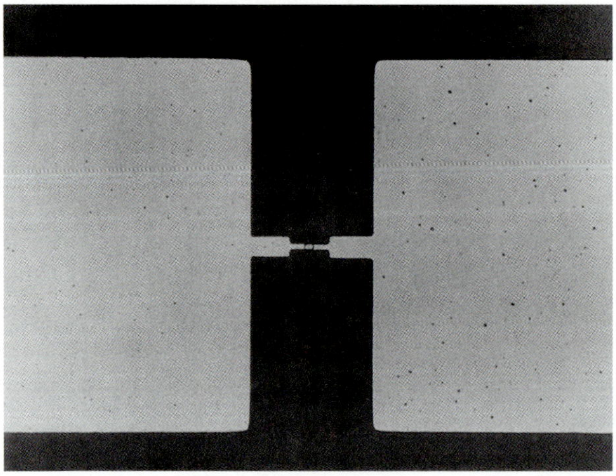

The small rectangle at the center of this photograph is a Josephson junction 1.25 μm wide.

Here ϕ is the phase difference between the wave functions of the Cooper pairs on either side of the junction. The value I_{max} of the maximum junction current depends on the thickness of the insulating layer and is quite small, between 1 μA and 1 mA in a Nb-NbO-Nb junction, for example.

When a voltage V is applied across a Josephson junction, the phase difference ϕ increases with time at the rate

ac Josephson effect
$$\nu = \frac{d\phi}{dt} = \frac{2Ve}{h} \tag{10.28}$$

As a result, I_J varies sinusoidally with time, which constitutes the **ac Josephson effect.** The value of $2e/h$ is 483.5979 THz/volt. Because ν is proportional to V and can be measured accurately, for instance by finding the frequency of the em radiation emitted by the junction, the ac Josephson effect enables very precise voltage determinations to be made. In fact, the effect is the basis for the present definition of the volt: one volt is the potential difference across a Josephson junction that produces oscillations at a frequency of 483.5979 THz.

Josephson junctions are used in extremely sensitive magnetometers called **SQUIDs**—superconducting *q*uantum *i*nterference *d*evices. SQUIDs vary in detail, but all make use of the fact that the maximum current in a superconducting ring that contains a Josephson junction varies periodically as the magnetic flux through the ring changes. The periodicity is interpreted as an interference effect involving the wave functions of the Cooper pairs. Magnetic field changes as small as 10^{-21} T can be detected by SQUIDs, which among other applications permits sensing the weak magnetic fields produced by biological currents such as those in the brain.

EXERCISES

I pass with relief from the tossing sea of Cause and Theory to the firm ground of Result and Fact. —Winston Churchill

10.2 Ionic Crystals

1. The ion spacings and melting points of the sodium halides are as follows:

	NaF	NaCl	NaBr	NaI
Ion spacing, nm	0.23	0.28	0.29	0.32
Melting point, °C	988	801	740	660

Explain the regular variation in these quantities with halogen atomic number.

2. Show that the first five terms in the series for the Madelung constant of NaCl are

$$\alpha = 6 - \frac{12}{\sqrt{2}} + \frac{8}{\sqrt{3}} - \frac{6}{2} + \frac{24}{\sqrt{5}} - \cdots$$

3. (*a*) The ionization energy of potassium is 4.34 eV and the electron affinity of chlorine is 3.61 eV. The Madelung constant for the KCl structure is 1.748 and the distance between ions of opposite sign is 0.314 nm. On the basis of these data only, compute the cohesive energy of KCl. (*b*) The observed cohesive energy of KCl is 6.42 eV per ion pair. On the assumption that the difference between this figure and that obtained in *a* is due to the exclusion-principle repulsion, find the exponent n in the formula Br^{-n} for the potential energy arising from this source.

4. Repeat Exercise 3 for LiCl, in which the Madelung constant is 1.748, the ion spacing is 0.257 nm, and the observed cohesive energy is 6.8 eV per ion pair. The ionization energy of Li is 5.4 eV.

10.4 Van der Waals Bond

5. The **Joule-Thomson effect** refers to the drop in temperature a gas undergoes when it passes slowly from a full container to an empty one through a porous plug. Since the expansion is into a rigid container, no mechanical work is done. Explain the Joule-Thomson effect in terms of the van der Waals attraction between molecules.

6. Van der Waals forces can hold inert gas atoms together to form solids at low temperatures, but they cannot hold such atoms together to form molecules in the gaseous state. Why not?

7. What is the effect on the cohesive energy of ionic and covalent crystals of (*a*) van der Waals forces and (*b*) zero-point oscillations of the ions and atoms about their equilibrium positions?

10.5 Metallic Bond

8. Lithium atoms, like hydrogen atoms, have only a single electron in their outer shells, yet lithium atoms do not join together to form Li_2 molecules the way hydrogen atoms form H_2 molecules. Instead, lithium is a metal with each atom part of a crystal lattice. Why?

9. Does the "gas" of freely moving electrons in a metal include all the electrons present? If not, which electrons are members of the "gas"?

10. Gold has an atomic mass of 197 u, a density of 19.3×10^3 kg/m^3, a Fermi energy of 5.54 eV, and a resistivity of $2.04 \times 10^{-8}\ \Omega \cdot$ m. Estimate the mean free path in atom spacings between collisions of the free electrons in gold under the assumption that each gold atom contributes one electron to the electron gas.

11. Silver has an atomic mass of 108 u, a density of 10.5×10^3 kg/m^3, and a Fermi energy of 5.51 eV. On the assumptions that each silver atom contributes one electron to the electron gas and that the mean free path of the electrons is 200 atom spacings, estimate the resistivity of silver. (The actual resistivity of silver at 20°C is $1.6 \times 10^{-8}\ \Omega \cdot$ m.)

10.6 Band Theory of Solids

12. What is the basic physical principle responsible for the presence of energy bands rather than specific energy levels in a solid?

13. How are the band structures of insulators and semiconductors similar? How are they different?

14. What are the two combinations of band structure and occupancy by electrons that can cause a solid to be a metal?

15. (*a*) Why are some solids transparent to visible light and others opaque? (*b*) The forbidden band is 1.1 eV in silicon and 6 eV in diamond. To what wavelengths of light are these substances transparent?

16. The forbidden band is 0.7 eV in germanium and 1.1 eV in silicon. How does the conductivity of germanium compare with that of silicon at (*a*) very low temperatures and (*b*) room temperature?

17. (*a*) When germanium is doped with aluminum, is the result an *n*-type or a *p*-type semiconductor? (*b*) Why?

10.8 Energy Bands: Alternative Analysis

18. Compare the de Broglie wavelength of an electron in copper with the 7.04-eV Fermi energy with the 0.256-nm spacing of the copper atoms.

19. Draw the third Brillouin zone of the two-dimensional square lattice whose first two Brillouin zones are shown in Fig. 10.41.

20. Find the ratio between the kinetic energies of an electron in a two-dimensional square lattice which has $k_x = k_y = \pi/a$ and an electron which has $k_x = \pi/a$, $k_y = 0$.

21. Phosphorus is present in a germanium sample. Assume that one of its five valence electrons revolves in a Bohr orbit around each P^+ ion in the germanium lattice. (*a*) If the effective mass of the electron is 0.17 m_e and the dielectric constant of germanium is 16, find the radius of the first Bohr orbit of the electron. (*b*) The energy gap between the valence and conduction bands in germanium is 0.65 eV. How does the ionization energy of the above electron compare with this energy and with kT at room temperature?

22. Repeat Exercise 21 for a silicon sample that contains arsenic. The effective mass of an electron in silicon is about 0.31 m_e, the dielectric constant of silicon is 12, and the energy gap in silicon is 1.1 eV.

23. The effective mass m^* of a current carrier in a semiconductor can be directly determined by means of a **cyclotron resonance** experiment in which the carriers (whether electrons or holes) move in helical orbits about the direction of an externally applied magnetic field **B**. An alternating electric field is applied perpendicular to **B**, and resonant absorption of energy from this field occurs when its frequency ν is equal to the frequency of revolution ν_c of the carrier. (*a*) Derive an equation for ν_c in terms of m^*, e, and B. (*b*) In a certain experiment, $B = 0.1$ T and maximum absorption is found to occur at $\nu = 1.4 \times 10^{10}$ Hz. Find m^*. (*c*) Find the maximum orbital radius of a charge carrier in this experiment whose speed is 3×10^4 m/s.

10.10 Bound Electron Pairs

24. The actual energy gap at 0 K in lead is 2.73×10^{-3} eV. (*a*) What is the prediction of the BCS theory for this energy gap? (*b*) Radiation of what minimum frequency could break apart Cooper pairs in lead at 0 K? In what part of the em spectrum is such radiation?

25. A voltage of 5.0 μV is applied across a Josephson junction. What is the frequency of the radiation emitted by the junction?

26. A SQUID magnetometer that uses a superconducting ring 2.0 mm in diameter indicates a change in the magnetic flux through it of 5 flux quanta. What is the corresponding magnetic field change?

Nuclear Structure

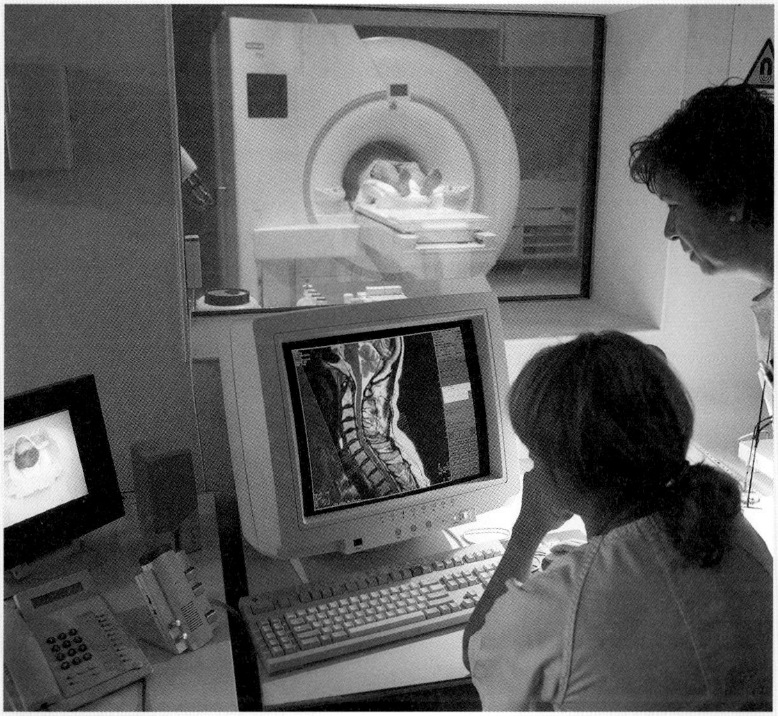

Nuclear magnetic resonance is the basis of a high-resolution method of imaging body tissues. The screen shows a computer-constructed cross section of the head of the person lying inside the powerful magnet at the rear.

Thus far we have been able to regard the nucleus of an atom merely as a tiny, positively charged object whose only roles are to provide the atom with most of its mass and to hold its electrons in thrall. The chief properties (except mass) of atoms, molecules, solids, and liquids can all be traced to the behavior of atomic electrons, not to the behavior of nuclei. Nevertheless, the nucleus turns out to be of paramount importance in the grand scheme of things. To begin with, the very existence of the various elements is due to the ability of nuclei to possess multiple electric charges. Furthermore, the energy involved in almost all natural processes can be traced to nuclear reactions and transformations. And the liberation of nuclear energy in reactors and weapons has affected all our lives in one way or another.

11.1 NUCLEAR COMPOSITION

Atomic nuclei of the same element have the same numbers of protons but can have different numbers of neutrons

The electron structure of the atom was understood before even the composition of its nucleus was known. The reason is that the forces that hold the nucleus together are vastly stronger than the electric forces that hold the electrons to the nucleus, and it is correspondingly harder to break apart a nucleus to find out what is inside. Changes in the electron structure of an atom, such as those that occur when a photon is emitted or absorbed or when a chemical bond is formed or broken, involve energies of only a few electronvolts. Changes in nuclear structure, on the other hand, involve energies in the MeV range, a million times greater.

An ordinary hydrogen atom has as its nucleus a single proton, whose charge is $+e$ and whose mass is 1836 times that of the electron. All other elements have nuclei that contain neutrons as well as protons. As its name suggests, the neutron is uncharged; its mass is slightly greater than that of the proton. Neutrons and protons are jointly called **nucleons.**

The **atomic number** of an element is the number of protons in each of its atomic nuclei, which is the same as the number of electrons in a neutral atom of the element. Thus the atomic number of hydrogen is 1, of helium 2, of lithium 3, and of uranium 92. All nuclei of a given element do not necessarily have equal numbers of neutrons. For instance, although over 99.9 percent of hydrogen nuclei are just single protons, a few also contain a neutron, and a very few two neutrons, along with the proton (Fig. 11.1). The varieties of an element that differ in the numbers of neutrons their nuclei contain are called its **isotopes.**

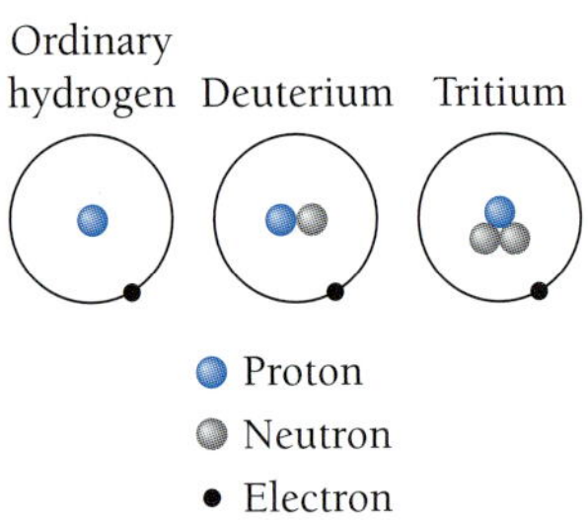

Figure 11.1 The isotopes of hydrogen.

The hydrogen isotope **deuterium** is stable, but **tritium** is radioactive and eventually changes into an isotope of helium. The flux of cosmic rays from space continually replenishes the earth's tritium by nuclear reactions in the atmosphere. Only about 2 kg of tritium of natural origin is present at any time on the earth, nearly all of it in the oceans. **Heavy water** is water in which deuterium atoms instead of ordinary hydrogen atoms are combined with oxygen atoms.

The conventional symbols for nuclear species, or **nuclides,** follow the pattern ${}^{A}_{Z}X$,

where X = chemical symbol of the element
Z = atomic number of the element
= number of protons in the nucleus
A = mass number of the nuclide
= number of nucleons in the nucleus

James Chadwick (1891–1974) was educated at the University of Manchester in England and remained there to work on gamma-ray emission under Rutherford. In Germany to investigate beta decay when World War I broke out, Chadwick was interned as an enemy alien. After the war he joined Rutherford at Cambridge, where he used alpha-particle scattering to show that the atomic number of an element equals its nuclear charge. Rutherford and Chadwick suggested an uncharged particle as a nuclear constituent but could not find a way to detect it experimentally.

Then, in 1930, the German physicists W. Bothe and H. Becker found that an uncharged radiation able to penetrate lead is emitted by beryllium bombarded with alpha particles from polonium (Fig. 11.2). Irene Curie and her husband Frederic Joliot, working in France in 1932, discovered that this mysterious radiation could knock protons with energies up to 5.7 MeV out of a paraffin slab. They assumed the radiation consisted of gamma rays (photons more energetic than x-rays) and, on the basis that the protons were knocked out of the hydrogen-rich paraffin in Compton collisions, calculated that the gamma-ray photon energy had to be at least 55 MeV. But this was far too much energy to be produced by the alpha particles interacting with beryllium nuclei.

Chadwick proposed instead that neutral particles with about the same mass as the proton are responsible, in which case their energy need be only 5.7 MeV since a particle colliding head on with another particle of the same mass can transfer all of its KE to the latter. Other experiments confirmed his hypothesis, and he received the Nobel Prize in 1935 for his part in the discovery of the neutron. (Chadwick did not immediately regard the neutron as an elementary particle but instead as "a small dipole, or perhaps better as a proton embedded in an electron." The idea that the neutron is actually an elementary particle was first put forward by the Russian physicist Dmitri Iwanenko.) During World War II Chadwick headed the British group that participated in developing the atomic bomb.

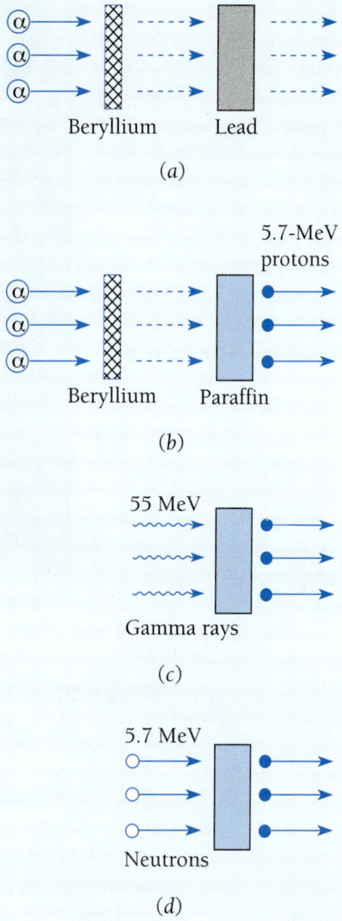

Figure 11.2 (*a*) Alpha particles incident on a beryllium foil cause the emission of a very penetrating radiation. (*b*) Protons of up to 5.7 MeV are ejected when the radiation strikes a paraffin slab. (*c*) If the radiation consists of gamma rays, their energies must be at least 55 MeV. (*d*) If the radiation consists of neutral particles of approximately proton mass, their energies need not exceed 5.7 MeV.

Hence ordinary hydrogen is ${}^1_1H$, deuterium is ${}^2_1H$, and the two isotopes of chlorine ($Z = 17$), whose nuclei contain 18 and 20 neutrons respectively, are ${}^{35}_{17}Cl$ and ${}^{37}_{17}Cl$. Because every element has a characteristic atomic number, Z is often omitted from the symbol for a nuclide: ${}^{35}Cl$ (read as "chlorine 35") instead of ${}^{35}_{17}Cl$.

Atomic Masses

Atomic masses refer to the masses of neutral atoms, not of bare nuclei. Thus an atomic mass always includes the masses of its Z electrons. Atomic masses are expressed in

Mass spectrometer being used to study the composition of semiconductor crystals.

mass units (u), which are so defined that the mass of a $^{12}_{6}C$ atom, the most abundant isotope of carbon, is exactly 12 u. The value of a mass unit is

Atomic mass unit $$1\ u = 1.66054 \times 10^{-27}\ kg$$

The energy equivalent of a mass unit is 931.49 MeV. Table 11.1 gives the masses of the proton, neutron, electron, and $^{1}_{1}H$ atom in various units, including the MeV/c^2. The advantage of using this unit is that the energy equivalent of a mass of, say, 10 MeV/c^2 is simply $E = mc^2 = 10$ MeV.

Table 11.2 gives the compositions of the isotopes of hydrogen and chlorine. Chlorine in nature consists of about three-quarters of the ^{35}Cl isotope and one-quarter of the ^{37}Cl isotope, which yields the average atomic mass of 35.46 u that chemists use (see Table 7.2). The chemical properties of an element are determined by the number and arrangement of the electrons in its atoms. Since the isotopes of an element have almost identical electron structures in their atoms, it is not surprising that the two isotopes of chlorine, for instance, have the same yellow color, the same suffocating odor, the same efficiency as poisons and bleaching agents, and the same ability to combine with metals. Because boiling and freezing points depend somewhat on atomic mass, they are slightly different for the two isotopes, as are their densities. Other physical

Table 11.1 Some Masses in Various Units

Particle	Mass (kg)	Mass (u)	Mass (MeV/c^2)
Proton	1.6726×10^{-27}	1.007276	938.28
Neutron	1.6750×10^{-27}	1.008665	939.57
Electron	9.1095×10^{-31}	5.486×10^{-4}	0.511
$^{1}_{1}H$ atom	1.6736×10^{-27}	1.007825	938.79

Table 11.2 The Isotopes of Hydrogen and Chlorine Found in Nature

	Properties of Element		Properties of Isotope				
Element	Atomic Number	Average Atomic Mass, u	Protons in Nucleus	Neutrons in Nucleus	Mass Number	Atomic Mass, u	Relative Abundance, Percent
Hydrogen	1	1.008	1	0	1	1.008	99.985
			1	1	2	2.014	0.015
			1	2	3	3.016	Very small
Chlorine	17	35.46	17	18	35	34.97	75.53
			17	20	37	36.97	24.47

properties of isotopes may very more dramatically with mass number: tritium is radioactive, for instance, whereas ordinary hydrogen and deuterium are not.

Nuclear Electrons

Nuclide masses are always very close to being integral multiples of the mass of the hydrogen atom, as we can see in Table 11.2. Before the discovery of the neutron, it was tempting to regard all nuclei as consisting of protons together with enough electrons to neutralize the positive charge of some of them. This hypothesis is buttressed by the fact that certain radioactive nuclei spontaneously emit electrons, a phenomenon called beta decay. However, there are some strong arguments against the idea of nuclear electrons.

1 Nuclear size. In Example 3.7 we saw that an electron confined to a box of nuclear dimensions must have an energy of more than 20 MeV, whereas electrons emitted during beta decay have energies of only 2 or 3 MeV, an order of magnitude smaller. A similar calculation for protons gives a minimum energy of around 0.2 MeV, which is entirely plausible.
2 Nuclear spin. Protons and electrons are fermions with spins (that is, spin quantum numbers) of $\frac{1}{2}$. Thus nuclei with an even number of protons plus electrons should have 0 or integral spin, those with an odd number of protons plus electrons should have half-integral spins. This prediction is not obeyed. For instance, if a deuterium nucleus, ${}^2_1H$, consisted of two protons and an electron, its nuclear spin should be $\frac{1}{2}$ or $\frac{3}{2}$, but in fact is observed to be 1.
3 Magnetic moment. The proton has a magnetic moment only about 0.15 percent that of the electron. If electrons are part of a nucleus, its magnetic moment ought to be of the order of magnitude of that of the electron. However, observed nuclear magnetic moments are comparable with that of the proton, not with that of the electron.
4 Electron-nuclear interaction. The forces that hold the constituents of a nucleus together lead to typical binding energies of around 8 MeV per particle. If some electrons can bind this strongly to protons in the nucleus of an atom, how can the other electrons in the atom remain outside the nucleus? Furthermore, when fast electrons are scattered by nuclei, they behave as though acted upon solely by electric forces, whereas the scattering of fast protons shows that a different force also acts on them.

Despite these difficulties, the hypothesis of nuclear electrons was not universally abandoned untill the discovery of the neutron in 1932. When he wrote a book on nuclear physics published the year before, George Gamow felt so uneasy about the accepted proton-electron model of the nucleus that he marked each section dealing with nuclear electrons with a skull and crossbones. When the publisher objected, Gamow replied that "It has never been my intention to scare the poor readers more than the text itself will undoubtedly do," and replaced the skull and crossbones with a less dramatic symbol.

11.2 SOME NUCLEAR PROPERTIES

Small in size, a nucleus may have angular momentum and a magnetic moment

The Rutherford scattering experiment provided the first estimates of nuclear sizes. In that experiment, as we saw in Chap. 4, an incident alpha particle is deflected by a target nucleus in a manner consistent with Coulomb's law provided the distance between them exceeds about 10^{-14} m. For smaller separations Coulomb's law is not obeyed because the nucleus no longer appears as a point charge to the alpha particle.

Since Rutherford's time a variety of experiments have been performed to determine nuclear dimensions, with particle scattering still a favored technique. Fast electrons and neutrons are ideal for this purpose, since an electron interacts with a nucleus only through electric forces while a neutron interacts only through specifically nuclear forces. Thus electron scattering provides information on the distribution of charge in a nucleus and neutron scattering provides information on the distribution of nuclear matter. In both cases the de Broglie wavelength of the particle must be smaller than the radius of the nucleus under study. What is found is that the volume of a nucleus is directly proportional to the number of nucleons it contains, which is its mass number A. This suggests that the density of nucleons is very nearly the same in the interiors of all nuclei.

If a nuclear radius is R, the corresponding volume is $\frac{4}{3}\pi R^3$ and so R^3 is proportional to A. This relationship is usually expressed in inverse form as

Nuclear radii
$$R = R_0 A^{1/3} \tag{11.1}$$

The value of R_0 is

$$R_0 \approx 1.2 \times 10^{-15} \text{ m} \approx 1.2 \text{ fm}$$

It is necessary to be indefinite in expressing R_0 because, as Fig. 11.3 shows, nuclei do not have sharp boundaries. Despite this, the values of R from Eq. (11.1) are representative of effective nuclear sizes. The value of R_0 is slightly smaller when it is deduced from electron scattering, which implies that nuclear matter and nuclear charge are not identically distributed through a nucleus.

Nuclei are so small that the unit of length appropriate in describing them is the **femtometer** (fm), equal to 10^{-15} m. The femtometer is sometimes called the **fermi** in

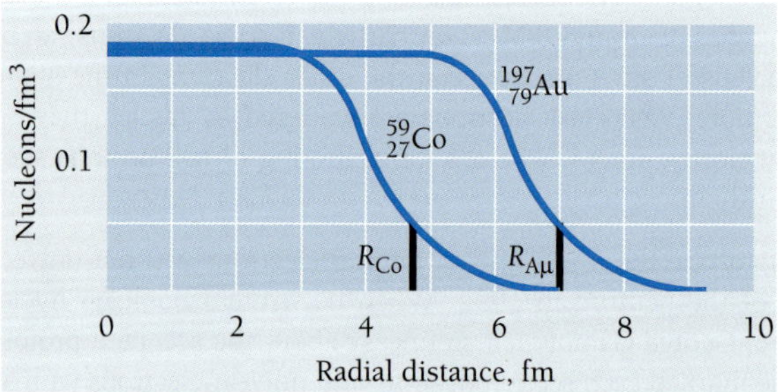

Figure 11.3 The density of nucleons in $^{59}_{27}$Co (cobalt) and $^{197}_{79}$Au (gold) nuclei plotted versus radial distance from the center. The values of the nuclear radius given by $R = 1.2A^{1/3}$ fm are indicated.

honor of Enrico Fermi, a pioneer in nuclear physics. From Eq. (11.1) we find that the radius of the $^{12}_{6}$C nucleus is

$$R \approx (1.2)(12)^{1/3} \text{ fm} \approx 2.7 \text{ fm}$$

Similarly, the radius of the $^{107}_{47}$Ag nucleus is 5.7 fm and that of the $^{238}_{92}$U nucleus is 7.4 fm.

Example 11.1

Find the density of the $^{12}_{6}$C nucleus.

Solution

The atomic mass of $^{12}_{6}$C is 12 u. Neglecting the masses and binding energies of the six electrons, we have for the nuclear density

$$\rho = \frac{m}{\frac{4}{3}\pi R^3} = \frac{(12 \text{ u})(1.66 \times 10^{-27} \text{ Kg/u})}{(\frac{4}{3}\pi)(2.7 \times 10^{-15} \text{ m})^3} = 2.4 \times 10^{17} \text{ kg/m}^3$$

This figure—equivalent to 4 billion tons per cubic inch!—is essentially the same for all nuclei. We learned in Sec. 9.11 of the existence of neutron stars, which consist of atoms that have been so compressed that their protons and electrons have interacted to become neutrons. Neutrons in such an assembly, as in a stable nucleus, do not undergo radioactive decay as do free neutrons. The densities of neutron stars are comparable with that of nuclear matter: a neutron star packs the mass of 1.4 to 3 suns into a sphere only about 10 km in radius.

Example 11.2

Find the repulsive electric force on a proton whose center is 2.4 fm from the center of another proton. Assume the protons are uniformly charged spheres of positive charge. (Protons actually have internal structures, as we shall learn in Chapter 13.)

Solution

Everywhere outside a uniformly charged sphere the sphere is electrically equivalent to a point charge located at the center of the sphere. Hence

$$F = \frac{1}{4\pi\epsilon_0}\frac{e^2}{r^2} = \frac{(8.99 \times 10^9 \text{ N} \cdot \text{m}^2/\text{C}^2)(1.60 \times 10^{-19} \text{ C})^2}{(2.4 \times 10^{-15} \text{ m})^2} = 40 \text{ N}$$

This is equivalent to 9 lb, a familiar enough amount of force—but it acts on a particle whose mass is less than 2×10^{-27} kg! Evidently the attractive forces that bind protons into nuclei despite such repulsions must be very strong indeed.

Spin and Magnetic Moment

Protons and neutrons, like electrons, are fermions with spin quantum numbers of $s = \frac{1}{2}$. This means they have spin angular momenta **S** of magnitude

$$S = \sqrt{s(s+1)}\hbar = \sqrt{\frac{1}{2}\left(\frac{1}{2}+1\right)}\hbar = \frac{\sqrt{3}}{2}\hbar \tag{11.2}$$

and spin magnetic quantum numbers of $m_s = \pm\frac{1}{2}$ (see Fig. 7.2).

As in the case of electrons, magnetic moments are associated with the spins of protons and neutrons. In nuclear physics, magnetic moments are expressed in **nuclear magnetons** (μ_N), where

Nuclear magneton
$$\mu_N = \frac{e\hbar}{2m_p} = 5.051 \times 10^{-27}\ \text{J/T} = 3.152 \times 10^{-8}\ \text{eV/T} \tag{11.3}$$

Here m_p is the proton mass. The nuclear magneton is smaller than the Bohr magneton of Eq. (6.42) by the ratio of the proton mass to the electron mass, which is 1836. The spin magnetic moments of the proton and neutron have components in any direction of

Proton
$$\mu_{pz} = \pm 2.793\ \mu_N$$

Neutron
$$\mu_{nz} = \mp 1.913\ \mu_N$$

There are two possibilities for the signs of μ_{pz} and μ_{nz} depending on whether m_s is $-\frac{1}{2}$ or $+\frac{1}{2}$. The $\pm$ sign is used for μ_{pz} because $\boldsymbol{\mu}_{pz}$ is in the same direction as the spin **S**, whereas $\mp$ is used for μ_{nz} because $\boldsymbol{\mu}_{nz}$ is opposite to **S** (Fig. 11.4).

At first glance it seems odd that the neutron, with no net charge, has a spin magnetic moment. But if we assume that the neutron contains equal amounts of positive and negative charge, a spin magnetic moment could arise even with no net charge. As we shall find in Chap. 13, such a picture has experimental support.

The hydrogen nucleus $^1_1$H consists of a single proton, and its total angular momentum is given by Eq. (11.2). A nucleon in a more complex nucleus may have orbital angular momentum due to motion inside the nucleus as well as spin angular momentum. The total angular momentum of such a nucleus is the vector sum of the spin and orbital angular momenta of its nucleons, as in the analogous case of the electrons of an atom. This subject will be considered further in Sec. 11.6.

When a nucleus whose magnetic moment has the z component μ_z is in a constant magnetic field **B**, the magnetic potential energy of the nucleus is

Magnetic energy
$$U_m = -\mu_z B \tag{11.4}$$

This energy is negative when $\boldsymbol{\mu}_z$ is in the same direction as **B** and positive when $\boldsymbol{\mu}_z$ is opposite to **B**. In a magnetic field, each angular momentum state of the nucleus is therefore split into components, just as in the Zeeman effect in atomic electron states. Figure 11.5 shows the splitting when the angular momentum of the nucleus is due to the spin of a single proton. The energy difference between the sublevels is

$$\Delta E = 2\mu_{pz}B \tag{11.5}$$

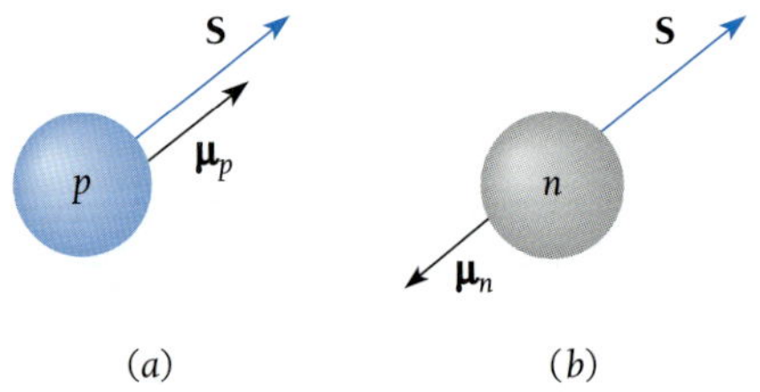

Figure 11.4 (*a*) The spin magnetic moment $\boldsymbol{\mu}_p$ of the proton is in the same direction as its spin angular momentum **S**. (*b*) In the case of the neutron, $\boldsymbol{\mu}_n$ is opposite to **S**.

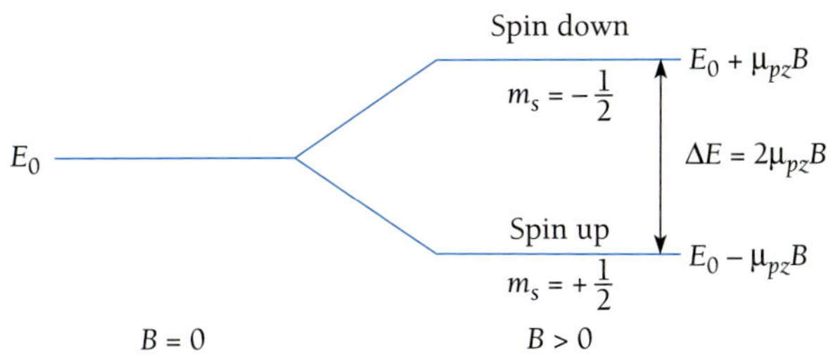

Figure 11.5 The energy levels of a proton in a magnetic field are split into spin-up ($\mathbf{S}_z$ parallel to $\mathbf{B}$) and spin-down ($\mathbf{S}_z$ antiparallel to $\mathbf{B}$) sublevels.

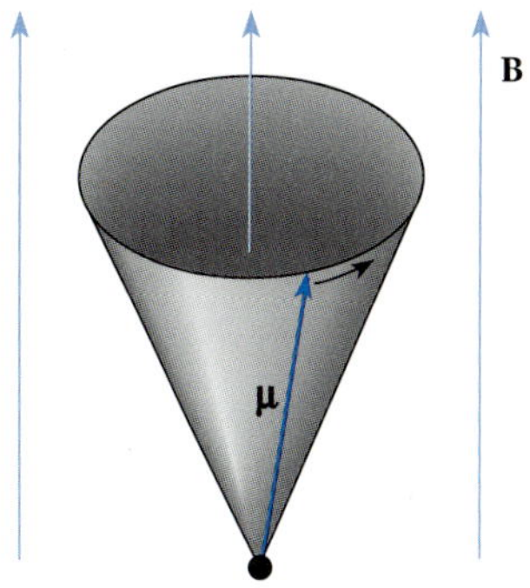

Figure 11.6 A nuclear magnetic moment $\boldsymbol{\mu}$ precesses around an external magnetic field $\mathbf{B}$ with a frequency called the Larmor frequency that is proportional to B.

A photon with this energy will be emitted when a proton in the upper state flips its spin to fall to the lower state. A proton in the lower state can be raised to the upper one by absorbing a photon of this energy. The photon frequency ν_L that corresponds to ΔE is

Larmor frequency for protons

$$\nu_L = \frac{\Delta E}{h} = \frac{2\mu_{pz}B}{h} \tag{11.6}$$

This is equal to the frequency with which a magnetic dipole precesses around a magnetic field (Fig. 11.6). It is named for Joseph Larmor, who derived ν_L from classical physics for an orbiting electron in a magnetic field; his result can be generalized to any magnetic dipole.

Example 11.3

(*a*) Find the energy difference between the spin-up and spin-down states of a proton in a magnetic field of $B = 1.000$ T (which is quite strong). (*b*) What is the Larmor frequency of a proton in this field?

Solution

(*a*) The energy difference is

$$\Delta E = 2\mu_{pz}B = (2)(2.793)(3.153 \times 10^{-8} \text{ eV/T})(1.000 \text{ T}) = 1.761 \times 10^{-7} \text{ eV}$$

If an electron rather than a proton were involved, ΔE would be considerably greater.
(*b*) The Larmor frequency of the proton in this field is

$$\nu_L = \frac{\Delta E}{h} = \frac{1.761 \times 10^{-7} \text{ eV}}{4.136 \times 10^{-15} \text{ eV} \cdot \text{s}} = 4.258 \times 10^{7} \text{ Hz} = 42.58 \text{ MHz}$$

From Fig. 2.2 we see that em radiation of this frequency is in the lower end of the microwave part of the spectrum.

Nuclear Magnetic Resonance

Suppose we put a sample of some substance that contains nuclei with spins of $\frac{1}{2}$ in a magnetic field $\mathbf{B}$. The spins of most of these nuclei will become aligned parallel to $\mathbf{B}$

Applications of NMR

NMR turns out to be far more useful than just a way to find nuclear magnetic moments. The electrons around a nucleus partly shield it from an external magnetic field to an extent that depends on the chemical environment of the nucleus. The **relaxation time** needed for the nuclei to drop to the lower state after having been excited also depends on this environment. These properties of NMR enable chemists to use NMR spectroscopy to help unravel details of chemical structures and reactions. For instance, the hydrogen nuclei in the CH_3, CH_2 and OH groups have slightly different resonant frequencies in the same magnetic field. All of these frequencies appear in the NMR spectrum of ethanol with a 3:2:1 ratio of intensities. Ethanol molecules are known to contain two C atoms, six H atoms, and one O atom, so they must consist of the three above groups linked together. The formula CH_3CH_2OH thus better represents methanol than C_2H_6O, which merely lists the atoms in its molecules. The intensity ratio 3:2:1 corroborates this picture since the CH_3 group has three H atoms, CH_2 has two, and OH has one. The NMR spectra of other spin-$\frac{1}{2}$ nuclei, such as ^{13}C and ^{32}P, are also of great help to chemists.

In medicine, NMR is the basis of an imaging method with higher resolution than x-ray tomography. In addition, NMR imaging is safer because rf radiation, unlike x radiation, has too little quantum energy to disrupt chemical bonds and so cannot harm living tissue. What is done is to use a nonuniform magnetic field, which means that the resonance frequency for a particular nucleus depends on the position of the nucleus in the field. Because our bodies are largely water, H_2O, proton NMR is usually employed. By changing the direction of the field gradient, an image that shows the proton density in a thin (3–4 mm) slice of the body can then be constructed by a computer. Relaxation times can also be mapped, which is useful because they are different in diseased tissue. In medicine, NMR imaging is called just magnetic resonance imaging, or MRI, to avoid frightening patients with the word "nuclear."

(spin-up) because this is the lowest energy state; see Fig. 11.5. If we now supply em radiation at the Larmor frequency ν_L to the sample, the nuclei will receive the right amount of energy to flip their spins to the higher state (spin-down). This phenomenon is called **nuclear magnetic resonance** (NMR) and it gives a way to determine nuclear magnetic moments experimentally. In one method, radio frequency (rf) radiation is supplied at a fixed frequency by a coil around the sample, and B is varied until the energy absorbed is a maximum. The resonance frequency is then the Larmor frequency for that value of B, from which μ can be calculated. Another method is to apply a broad-spectrum rf pulse and then measure the frequency (which will be ν_L) of the radiation the sample gives off as its excited nuclei return to the lower energy state.

11.3 STABLE NUCLEI

Why some combinations of neutrons and protons are more stable than others

Not all combinations of neutrons and protons form stable nuclei. In general, light nuclei ($A < 20$) contain approximately equal numbers of neutrons and protons, while in heavier nuclei the proportion of neutrons becomes progressively greater. This is evident from Fig. 11.7, which is a plot of N versus Z for stable nuclides.

The tendency for N to equal Z follows from the existence of nuclear energy levels. Nucleons, which have spins of $\frac{1}{2}$, obey the exclusion principle. As a result, each nuclear energy level can contain two neutrons of opposite spins and two protons of opposite

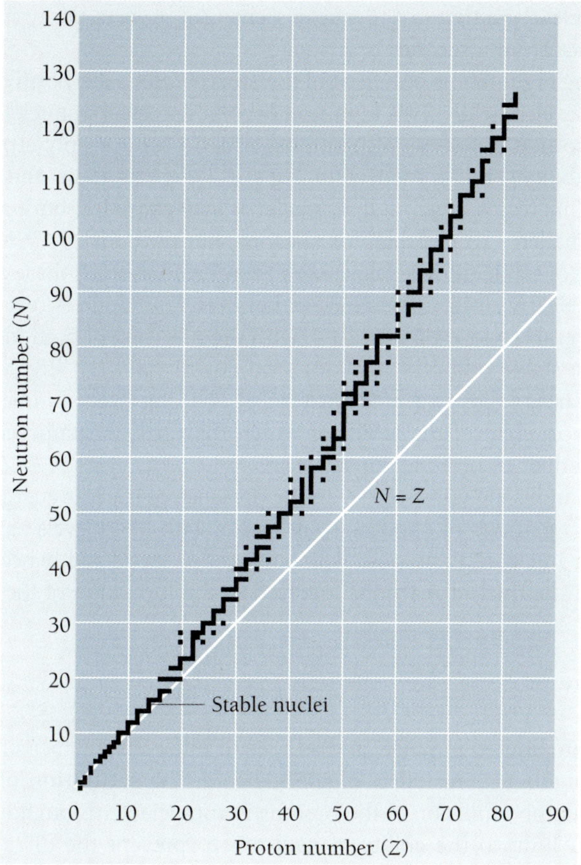

Figure 11.7 Neutron-proton diagram for stable nuclides. There are no stable nuclides with $Z = 43$ or 61, with $N = 19, 35, 39, 45, 61, 89, 115, 126$, or with $A = Z + N = 5$ or 8. All nuclides with $Z > 83$, $N > 126$, and $A > 209$ are unstable.

spins. Energy levels in nuclei are filled in sequence, just as energy levels in atoms are, to achieve configurations of minimum energy and therefore maximum stability. Thus the boron isotope $^{12}_{5}B$ has more energy than the carbon isotope $^{12}_{6}C$ because one of its neutrons is in a higher energy level, and $^{12}_{5}B$ is accordingly unstable (Fig. 11.8). If

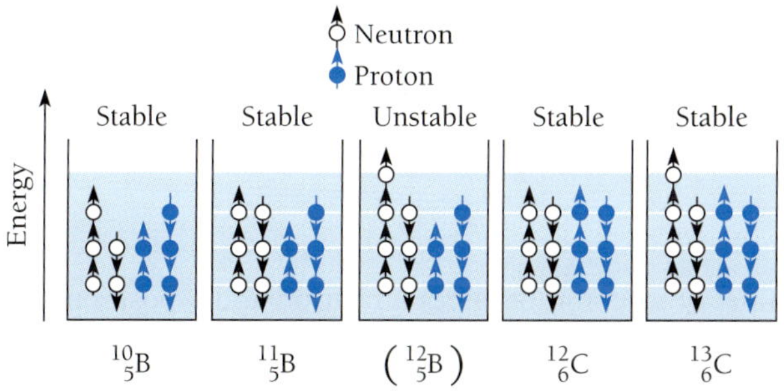

Figure 11.8 Simplified energy-level diagrams of some boron and carbon isotopes. The exclusion principle limits the occupancy of each level to two neutrons of opposite spin and two protons of opposite spin. Stable nuclei have configurations of minimum energy.

created in a nuclear reaction, a $^{12}_{5}B$ nucleus changes by beta decay into a stable $^{12}_{6}C$ nucleus in a fraction of a second.

The preceding argument is only part of the story. Protons are positively charged and repel one another electrically. This repulsion becomes so great in nuclei with more than 10 protons or so that an excess of neutrons, which produce only attractive forces, is required for stability. Thus the curve of Fig. 11.7 departs more and more from the $N = Z$ line as Z increases. Even in light nuclei N may exceed Z, but (except in $^{1}_{1}H$ and $^{3}_{2}He$) is never smaller; $^{11}_{5}B$ is stable, for instance, but not $^{11}_{6}C$.

Sixty percent of stable nuclides have both even Z and even N; these are called "even-even" nuclides. Nearly all the others have either even Z and odd N (even-odd nuclides) or odd Z and even N (odd-even nuclides), with the numbers of both kinds being about equal. Only five stable odd-odd nuclides are known: $^{2}_{1}H$, $^{6}_{3}Li$, $^{10}_{5}Be$, $^{14}_{7}N$, and $^{180}_{73}Ta$. Nuclear abundances follow a similar pattern of favoring even numbers for Z and N. Only about one in eight of the atoms of which the earth is composed has a nucleus with an odd number of protons, for instance.

These observations are consistent with the presence of nuclear energy levels that can each contain two particles of opposite spin. Nuclei with filled levels have less tendency to pick up other nucleons than those with partly filled levels and hence were less likely to participate in the nuclear reactions involved in the formation of the elements.

Nuclear Decay

Nuclear forces are limited in range, and as a result nucleons interact strongly only with their nearest neighbors. This effect is referred to as the **saturation** of nuclear forces. Because the coulomb repulsion of the protons is appreciable throughout the entire nucleus, there is a limit to the ability of neutrons to prevent the disruption of a large nucleus. This limit is represented by the bismuth isotope $^{209}_{83}Bi$, which is the heaviest stable nuclide. All nuclei with $Z > 83$ and $A > 209$ spontaneously transform themselves into lighter ones through the emission of one or more alpha particles, which are $^{4}_{2}He$ nuclei:

Alpha decay

$$^{A}_{Z}X \rightarrow {}^{A-4}_{Z-2}Y + {}^{4}_{2}He$$

$$\text{Parent nucleus} \rightarrow \text{Daughter nucleus} + \text{Alpha particle}$$

Since an alpha particle consists of two protons and two neutrons, an alpha decay reduces the Z and the N of the original nucleus by two each. If the resulting daughter nucleus has either too small or too large a neutron/proton ratio for stability, it may beta-decay to a more appropriate configuration. In negative beta decay, a neutron is transformed into a proton and an electron is emitted:

Beta decay

$$n^0 \rightarrow p^+ + e^-$$

In positive beta decay, a proton becomes a neutron and a positron is emitted:

Positron emission

$$p^+ \rightarrow n^0 + e^+$$

Thus negative beta decay decreases the proportion of neutrons and positive beta decay increases it. A process that competes with positron emission is the capture by a

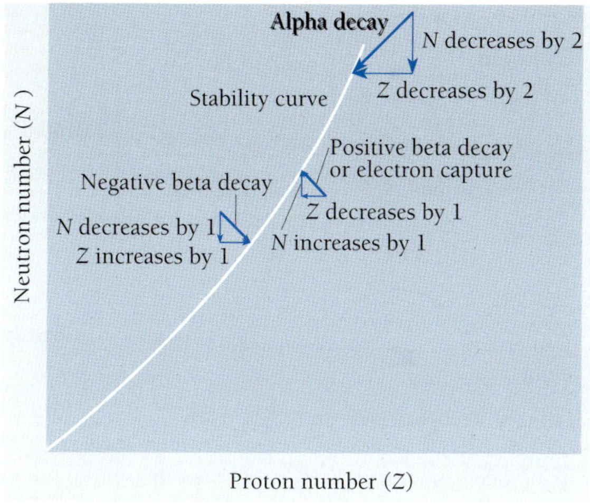

Figure 11.9 Alpha and beta decays permit an unstable nucleus to reach a stable configuration.

nucleus of an electron from its innermost shell. The electron is absorbed by a nuclear proton which is thereby transformed into a neutron:

Electron capture $$p^+ + e^- \rightarrow n^0$$

Figure 11.9 shows how alpha and beta decays enable stability to be achieved. Radioactivity is considered in more detail in Chap. 12, where we will find that another particle, the neutrino, is also involved in beta decay and electron capture.

11.4 BINDING ENERGY

The missing energy that keeps a nucleus together

The hydrogen isotope deuterium, ${}^2_1\text{H}$, has a neutron as well as a proton in its nucleus. Thus we would expect the mass of the deuterium atom to be equal to that of an ordinary ${}^1_1\text{H}$ atom plus the mass of a neutron:

Mass of ${}^1_1\text{H}$ atom	1.007825 u
+ mass of neutron	+1.008665 u
Expected mass of ${}^2_1\text{H}$ atom	2.016490 u

However, the measured mass of the ${}^2_1\text{H}$ atom is only 2.014102 u, which is 0.002388 u *less* than the combined masses of a ${}^1_1\text{H}$ atom and a neutron (Fig. 11.10).

What comes to mind is that the "missing" mass might correspond to energy given off when a ${}^2_1\text{H}$ nucleus is formed from a free proton and neutron. The energy equivalent of the missing mass is

$$\Delta E = (0.002388 \text{ u})(931.49 \text{ MeV/u}) = 2.224 \text{ MeV}$$

To test this interpretation of the missing mass, we can perform experiments to see how much energy is needed to break apart a deuterium nucleus into a separate neutron and

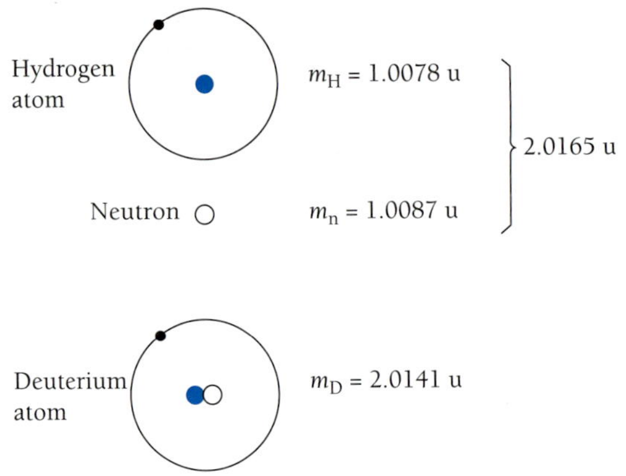

Figure 11.10 The mass of a deuterium atom ($^2_1$H) is less than the sum of the masses of a hydrogen atom ($^1_1$H) and a neutron. The energy equivalent of the missing mass is called the binding energy of the nucleus.

proton. The required energy indeed turns out to be 2.224 MeV (Fig. 11.11). When less energy than 2.224 MeV is given to a $^2_1$H nucleus, the nucleus stays together. When the added energy is more than 2.224 MeV, the extra energy goes into kinetic energy of the neutron and proton as they fly apart.

Deuterium atoms are not the only ones that have less mass than the combined masses of the particles they are composed of—*all* atoms are like that. The energy equivalent of the missing mass of a nucleus is called the **binding energy** of the nucleus. The greater its binding energy, the more the energy that must be supplied to break up the nucleus.

The binding energy E_b in MeV of the nucleus ^A_ZX, which has $N = A - Z$ neutrons, is given by

$$E_b = [Zm(^1_1\text{H}) + Nm(n) - m(^A_ZX)](931.49 \text{ MeV/u}) \qquad (11.7)$$

where $m(^1_1\text{H})$ is the atomic mass of $^1_1$H, $m(n)$ is the neutron mass, and $m(^A_ZX)$ is the atomic mass of ^A_ZX, all in mass units. As mentioned before, atomic masses, not nuclear masses, are used in such calculations; the electron masses subtract out.

Nuclear binding energies are strikingly high. The range for stable nuclei is from 2.224 MeV for $^2_1$H (deuterium) to 1640 MeV for $^{209}_{83}$Bi (an isotope of the metal bismuth). To appreciate how high binding energies are, we can compare them with more familiar energies in terms of kilojoules of energy per kilogram of mass. In these units, a typical binding energy is 8×10^{11} kJ/kg—800 billion kJ/kg. By contrast, to boil water

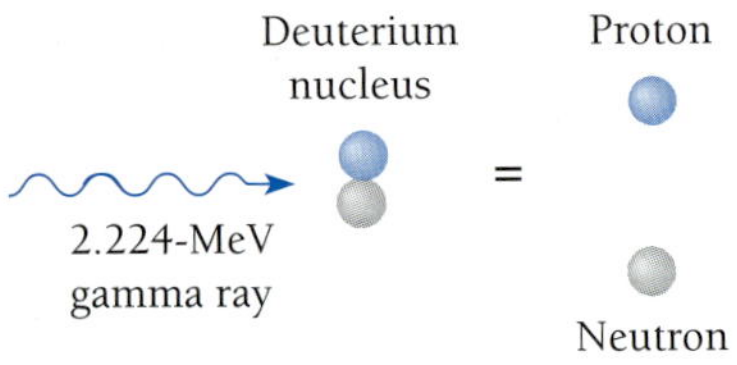

Figure 11.11 The binding energy of the deuterium nucleus is 2.224 MeV. A gamma ray whose energy is 2.224 MeV or more can split a deuterium nucleus into a proton and neutron. A gamma ray whose energy is less than 2.224 MeV cannot do this.

involves a heat of vaporization of a mere 2260 kJ/kg, and even the heat given off by burning gasoline is only 4.7×10^4 kJ/kg, 17 million times smaller.

Example 11.4

The binding energy of the neon isotope $^{20}_{10}\text{Ne}$ is 160.647 MeV. Find its atomic mass.

Solution

Here $Z = 10$ and $N = 10$. From Eq. (11.7),

$$m(^A_Z\text{X}) = [Zm(^1_1\text{H}) + Nm(n)] - \frac{E_b}{931.49 \text{ MeV/u}}$$

$$m(^{20}_{10}\text{Ne}) = [10(1.007825 \text{ u}) + 10(1.008665)] - \frac{160.647 \text{ MeV}}{931.49 \text{ MeV/u}} = 19.992 \text{ u}$$

Binding Energy per Nucleon

The **binding energy per nucleon** for a given nucleus is an average found by dividing its total binding energy by the number of nucleons it contains. Thus the binding energy per nucleon for ^2_1H is $(2.2 \text{ MeV})/2 = 1.1$ MeV/nucleon, and for $^{209}_{83}\text{Bi}$ it is $(1640 \text{ MeV})/209 = 7.8$ MeV/nucleon.

Figure 11.12 shows the binding energy per nucleon plotted against the number of nucleons in various atomic nuclei. The greater the binding energy per nucleon, the

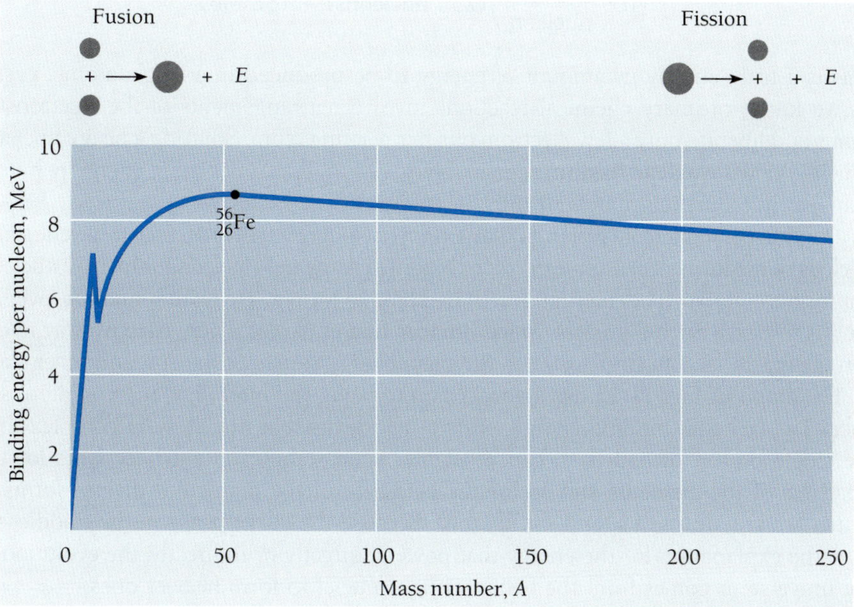

Figure 11.12 Binding energy per nucleon as a function of mass number. The peak at $A = 4$ corresponds to the exceptionally stable ^4_2He nucleus, which is the alpha particle. The binding energy per nucleon is a maximum for nuclei of mass number $A = 56$. Such nuclei are the most stable. When two light nuclei join to form a heavier one, a process called fusion, the greater binding energy of the product nucleus causes energy to be given off. When a heavy nucleus is split into two lighter ones, a process called fission, the greater binding energy of the product nuclei also causes energy to be given off.

The Strong Interaction

The short-range attractive forces between nucleons arise from the **strong interaction.** (There is another fundamental interaction affecting nucleons called the **weak interaction** that will be discussed in Chaps. 12 and 13.) The strong interaction is what holds nucleons together to form nuclei, and it is powerful enough to overcome the electric repulsion of the positively charged protons in nuclei provided neutrons are also present to help. If the strong interaction were a little stronger—perhaps only 1 percent would be enough—two protons could stick together without any neutrons needed. In this case, when the universe came into being in the Big Bang (Sec. 13.8), all its protons would have joined into diprotons almost as soon as they appeared. Then there would be no individual protons to undergo the fusion reactions that power the stars and have created the chemical elements. The universe would be a very different place from what it is today, and we would not exist.

more stable the nucleus is. The graph has its maximum of 8.8 MeV/nucleon when the total number of nucleons is 56. The nucleus that has 56 protons and neutrons is ${}^{56}_{26}\text{Fe}$, an iron isotope. This is the most stable nucleus of them all, since the most energy is needed to pull a nucleon away from it.

Two remarkable conclusions can be drawn from the curve of Fig. 11.12. The first is that if we can somehow split a heavy nucleus into two medium-sized ones, each of the new nuclei will have *more* binding energy per nucleon than the original nucleus did. The extra energy will be given off, and it can be a lot. For instance, if the uranium nucleus ${}^{235}_{92}\text{U}$ is broken into two smaller nuclei, the binding energy difference per nucleon is about 0.8 MeV. The total energy given off is therefore

$$\left(0.8\,\frac{\text{MeV}}{\text{nucleon}}\right)(235 \text{ nucleons}) = 188 \text{ MeV}$$

This is a truly enormous amount of energy to be produced in a single atomic event. As we know, ordinary chemical reactions involve rearrangements of the electrons in atoms and liberate only a few electronvolts per reacting atom. Splitting a heavy nucleus, which is called **nuclear fission,** thus involves 100 million times more energy per atom than, say, the burning of coal or oil.

The other notable conclusion is that joining two light nuclei together to give a single nucleus of medium size also means more binding energy per nucleon in the new nucleus. For instance, if two ${}^{2}_{1}\text{H}$ deuterium nuclei combine to form a ${}^{4}_{2}\text{He}$ helium nucleus, over 23 MeV is released. Such a process, called **nuclear fusion,** is also a very effective way to obtain energy. In fact, nuclear fusion is the main energy source of the sun and other stars.

The graph of Fig. 11.12 has a good claim to being the most significant in all of science. The fact that binding energy exists at all means that nuclei more complex than the single proton of hydrogen can be stable. Such stability in turn accounts for the existence of the elements and so for the existence of the many and diverse forms of matter we see around us (and for us, too). Because the curve peaks in the middle, we have the explanation for the energy that powers, directly or indirectly, the evolution of the universe: it comes from the fusion of light nuclei to form heavier ones.

Example 11.5

(*a*) Find the energy needed to remove a neutron from the nucleus of the calcium isotope ${}^{42}_{20}\text{Ca}$. (*b*) Find the energy needed to remove a proton from this nucleus. (*c*) Why are these energies different?

Solution

(*a*) Removing a neutron from $^{42}_{20}$Ca leaves $^{41}_{20}$Ca. From the table of atomic masses in the Appendix the mass of $^{41}_{20}$Ca plus the mass of a free neutron is

$$40.962278 \text{ u} + 1.008665 \text{ u} = 41.970943 \text{ u}$$

The difference between this mass and the mass of $^{42}_{20}$Ca is 0.012321 u, so the binding energy of the missing neutron is

$$(0.012321 \text{ u})(931.49 \text{ MeV/u}) = 11.48 \text{ MeV}$$

(*b*) Removing a proton from $^{42}_{20}$Ca leaves the potassium isotope $^{41}_{19}$K. A similar calculation gives a binding energy of 10.27 MeV for the missing proton.
(*c*) The neutron was acted upon only by attractive nuclear forces whereas the proton was also acted upon by repulsive electric forces that decrease its binding energy.

11.5 LIQUID-DROP MODEL

A simple explanation for the binding-energy curve

The short-range force that binds nucleons so securely into nuclei is by far the strongest type of force known. Unfortunately the nuclear force is not as well understood as the electromagnetic force, and the theory of nuclear structure is less complete than the theory of atomic structure. However, even without a full understanding of the nuclear force, much progress has been made in devising nuclear models able to account for prominent aspects of nuclear properties and behavior. We shall examine some of the concepts embodied in these models in this section and the next.

While the attractive forces that nucleons exert upon one another are very strong, their range is short. Up to a separation of about 3 fm, the nuclear attraction between two protons is about 100 times stronger than the electric repulsion between them. The nuclear interactions between protons and protons, between protons and neutrons, and between neutrons and neutrons appear to be identical.

As a first approximation, we can think of each nucleon in a nucleus as interacting solely with its nearest neighbors. This situation is the same as that of atoms in a solid, which ideally vibrate about fixed positions in a crystal lattice, or that of molecules in a liquid, which ideally are free to move about while maintaining a fixed intermolecular distance. The analogy with a solid cannot be pursued because a calculation shows that the vibrations of the nucleons about their average positions would be too great for the nucleus to be stable. The analogy with a liquid, on the other hand, turns out to be extremely useful in understanding certain aspects of nuclear behavior. This analogy was proposed by George Gamow in 1929 and developed in detail by C. F. von Weizsäcker in 1935.

Let us see how the picture of a nucleus as a drop of liquid accounts for the observed variation of binding energy per nucleon with mass number. We start by assuming that the energy associated with each nucleon-nucleon bond has some value U. This energy is actually negative since attractive forces are involved, but is usually written as positive because binding energy is considered a positive quantity for convenience.

Because each bond energy U is shared by two nucleons, each has a binding energy of $\frac{1}{2}U$. When an assembly of spheres of the same size is packed together into the smallest volume, as we suppose is the case of nucleons within a nucleus, each interior sphere

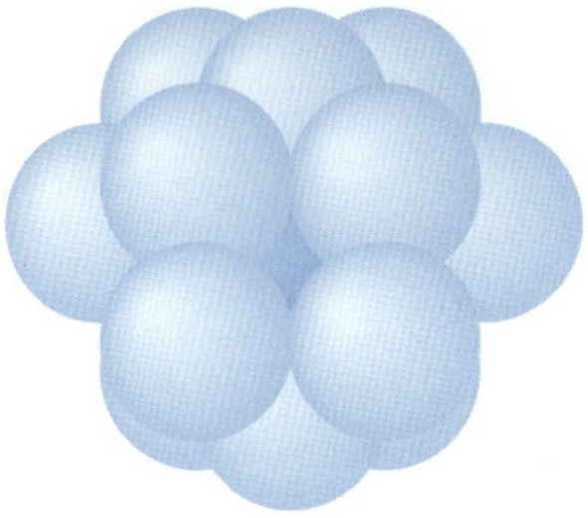

Figure 11.13 In a tightly packed assembly of identical spheres, each interior sphere is in contact with 12 others.

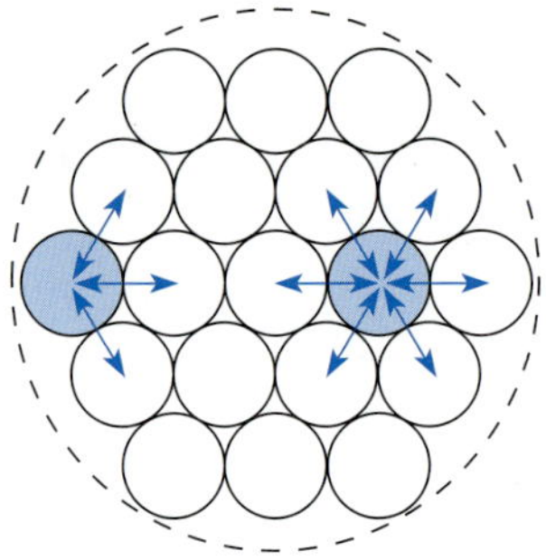

Figure 11.14 A nucleon at the surface of a nucleus interacts with fewer other nucleons than one in the interior of the nucleus and hence its binding energy is less. The larger the nucleus, the smaller the proportion of nucleons at the surface.

has 12 other spheres in contact with it (Fig. 11.13). Hence each interior nucleon in a nucleus has a binding energy of $(12)(\frac{1}{2}U)$ or 6 U. If all A nucleons in a nucleus were in its interior, the total binding energy of the nucleus would be

$$E_v = 6\ AU \tag{11.8}$$

Equation (11.8) is often written simply as

Volume energy

$$E_v = a_1 A \tag{11.9}$$

The energy E_v is called the **volume energy** of a nucleus and is directly proportional to A.

Actually, of course, some nucleons are on the surface of every nucleus and therefore have fewer than 12 neighbors (Fig. 11.14). The number of such nucleons depends on the surface area of the nucleus in question. A nucleus of radius R has an area of $4\pi R^2 = 4\pi R_0^2 A^{2/3}$. Hence the number of nucleons with fewer than the maximum number of bonds is proportional to $A^{2/3}$, reducing the total binding energy by

Surface energy

$$E_s = -a_2 A^{2/3} \tag{11.10}$$

The negative energy E_s is called the **surface energy** of a nucleus. It is most significant for the lighter nuclei since a greater fraction of their nucleons are on the surface. Because natural systems always tend to evolve toward configurations of minimum potential energy, nuclei tend toward configurations of maximum binding energy. Hence a nucleus should exhibit the same surface-tension effects as a liquid drop, and in the absence of other effects it should be spherical, since a sphere has the least surface area for a given volume.

The electric repulsion between each pair of protons in a nucleus also contributes toward decreasing its binding energy. The **coulomb energy** E_c of a nucleus is the work that must be done to bring together Z protons from infinity into a spherical aggregate the size of the nucleus. The potential energy of a pair of protons r apart is equal to

$$V = -\frac{e^2}{4\pi\epsilon_0 r}$$

Since there are $Z(Z-1)/2$ pairs of protons,

$$E_c = \frac{Z(Z-1)}{2} V = -\frac{Z(Z-1)e^2}{8\pi\epsilon_0}\left(\frac{1}{r}\right)_{\text{av}} \tag{11.11}$$

where $(1/r)_{\text{av}}$ is the value of $1/r$ averaged over all proton pairs. If the protons are uniformly distributed throughout a nucleus of radius R, $(1/r)_{\text{av}}$ is proportional to $1/R$ and hence to $1/A^{1/3}$, so that

Coulomb energy

$$E_c = -a_3 \frac{Z(Z-1)}{A^{1/3}} \tag{11.12}$$

The coulomb energy is negative because it arises from an effect that opposes nuclear stability.

This is as far as the liquid-drop model itself can go. Let us now see how the result compares with reality.

The total binding energy E_b of a nucleus ought to be the sum of its volume, surface, and coulomb energies:

$$E_b = E_v + E_s + E_c = a_1 A - a_2 A^{2/3} - a_3 \frac{Z(Z-1)}{A^{1/3}} \tag{11.13}$$

The binding energy per nucleon is therefore

$$\frac{E_b}{A} = a_1 - \frac{a_2}{A^{1/3}} - a_3 \frac{Z(Z-1)}{A^{4/3}} \tag{11.14}$$

Each of the terms of Eq. (11.14) is plotted in Fig. 11.15 versus A, together with their sum E_b/A. The coefficients were chosen to make the E_b/A curve resemble as closely as possible the empirical binding energy per nucleon curve of Fig. 11.12. The fact that the theoretical curve can be made to agree so well with the empirical one means that the analogy between a nucleus and a liquid drop has at least some validity.

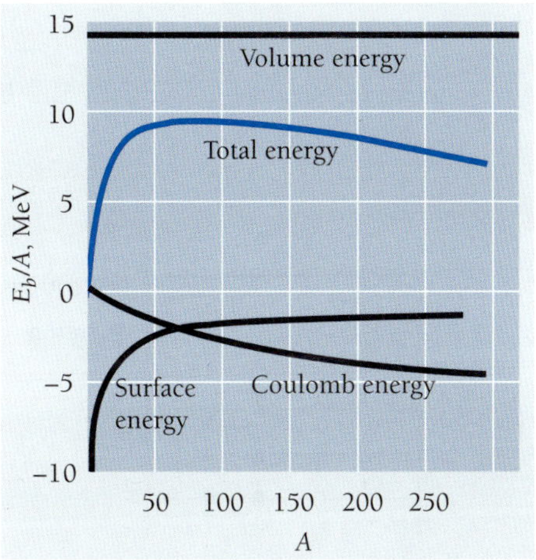

Figure 11.15 The binding energy per nucleon is the sum of the volume, surface, and coulomb energies.

Corrections to the Formula

The binding-energy formula of Eq. (11.13) can be improved by taking into account two effects that do not fit into the simple liquid-drop model but which make sense in terms of a model that provides for nuclear energy levels. (We will see in the next section how these apparently very different approaches can be reconciled.) One of these effects occurs when the neutrons in a nucleus outnumber the protons, which means that higher energy levels have to be occupied than would be the case if N and Z were equal.

Let us suppose that the uppermost neutron and proton energy levels, which the exclusion principle limits to two particles each, have the same spacing ϵ, as in Fig. 11.16. In order to produce a neutron excess of, say, $N - Z = 8$ without changing A, $\frac{1}{2}(N - Z) = 4$ neutrons would have to replace protons in an original nucleus in which $N = Z$. The new neutrons would occupy levels higher in energy by $2\epsilon = 4\epsilon/2$ than those of the protons they replace. In the general case of $\frac{1}{2}(N - Z)$ new neutrons, each must be raised in energy by $\frac{1}{2}(N - Z)\epsilon/2$. The total work needed is

$$\Delta E = (\text{number of new neutrons})\left(\frac{\text{energy increase}}{\text{new neutron}}\right)$$

$$= \left[\frac{1}{2}(N - Z)\right]\left[\frac{1}{2}(N - Z)\frac{\epsilon}{2}\right] = \frac{\epsilon}{8}(N - Z)^2$$

Because $N = A - Z$, $(N - Z)^2 = (A - 2Z)^2$, and

$$\Delta E = \frac{\epsilon}{8}(A - 2Z)^2 \tag{11.15}$$

As it happens, the greater the number of nucleons in a nucleus, the smaller is the energy level spacing ϵ, with ϵ proportional to $1/A$. This means that the **asymmetry energy** E_a due to the difference between N and Z can be expressed as

Asymmetry energy

$$E_a = -\Delta E = -a_4 \frac{(A - 2Z)^2}{A} \tag{11.16}$$

The asymmetry energy is negative because it reduces the binding energy of the nucleus.

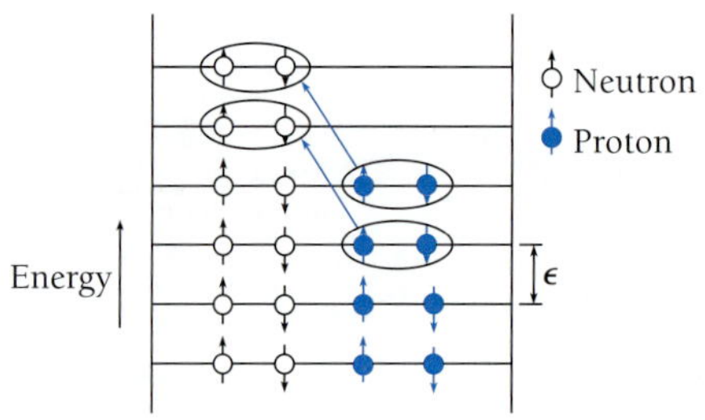

Figure 11.16 In order to replace 4 protons in a nucleus with $N = Z$ by 4 neutrons, the work $(4)(4\epsilon/2)$ must be done. The resulting nucleus has 8 more neutrons than protons.

The last correction term arises from the tendency of proton pairs and neutron pairs to occur (Sec. 11.3). Even-even nuclei are the most stable and hence have higher binding energies than would otherwise be expected. Thus such nuclei as ${}^{4}_{2}He$, ${}^{12}_{6}C$, and ${}^{16}_{8}O$ appear as peaks on the empirical curve of binding energy per nucleon. At the other extreme, odd-odd nuclei have both unpaired protons and neutrons and have relatively low binding energies. The **pairing energy** E_p is positive for even-even nuclei, 0 for odd-even and even-odd nuclei, and negative for odd-odd nuclei, and seems to vary with A as $A^{-3/4}$. Hence

Pairing energy

$$E_p = (\pm, 0)\frac{a_5}{A^{3/4}} \tag{11.17}$$

The final expression for the binding energy of a nucleus of atomic number Z and mass number A, which was first obtained by C. F. von Weizsäcker in 1935, is

Semiempirical binding-energy formula

$$E_b = a_1A - a_2A^{2/3} - a_3\frac{Z(Z-1)}{A^{1/3}} - a_4\frac{(A-2Z)^2}{A}(\pm, 0)\frac{a_5}{A^{3/4}} \tag{11.18}$$

A set of coefficients that gives a good fit with the data is as follows:

$$a_1 = 14.1 \text{ MeV} \qquad a_2 = 13.0 \text{ MeV} \qquad a_3 = 0.595 \text{ MeV}$$

$$a_4 = 19.0 \text{ MeV} \qquad a_5 = 33.5 \text{ MeV}$$

Other sets of coefficients have also been proposed. Equation (11.18) agrees better with observed binding energies than does Eq. (11.13), which suggests that the liquid-drop model, though a good approximation, is not the last word on the subject.

Example 11.6

The atomic mass of the zinc isotope ${}^{64}_{30}Zn$ is 63.929 u. Compare its binding energy with the prediction of Eq. (11.18).

Solution

The binding energy of ${}^{64}_{30}Zn$ is, from Eq. (11.7),

$$E_b = [(30)(1.007825 \text{ u}) + (34)(1.008665 \text{ u}) - 63.929 \text{ u}](931.49 \text{ MeV/u}) = 559.1 \text{ MeV}$$

The semiempirical binding energy formula, using the coefficients in the text, gives

$$E_b = (14.1 \text{ MeV})(64) - (13.0 \text{ MeV})(64)^{2/3} - \frac{(0.595 \text{ MeV})(30)(29)}{(64)^{1/3}} - \frac{(19.0 \text{ MeV})(16)}{64} + \frac{33.5 \text{ MeV}}{(64)^{3/4}} = 561.7 \text{ MeV}$$

The plus sign is used for the last term because ${}^{64}_{30}Zn$ is an even-even nucleus. The difference between the observed and calculated binding energies is less than 0.5 percent.

Example 11.7

Isobars are nuclides that have the same mass number A. Derive a formula for the atomic number of the most stable isobar of a given A and use it to find the most stable isobar of $A = 25$.

Solution

To find the value of Z for which the binding energy E_b is a maximum, which corresponds to maximum stability, we must solve $dE_b/dZ = 0$ for Z. From Eq. (11.18) we have

$$\frac{dE_b}{dZ} = -\frac{a_3}{A^{1/3}}(2Z - 1) + \frac{4a_4}{A}(A - 2Z) = 0$$

$$Z = \frac{a_3A^{-1/3} + 4a_4}{2a_3A^{-1/3} + 8a_4A^{-1}} = \frac{0.595A^{-1/3} + 76}{1.19A^{-1/3} + 152A^{-1}}$$

For $A = 25$ this formula gives $Z = 11.7$, from which we conclude that $Z = 12$ should be the atomic number of the most stable isobar of $A = 25$. This nuclide is $^{25}_{12}\text{Mg}$, which is in fact the only stable $A = 25$ isobar. The other isobars, $^{25}_{11}\text{Na}$ and $^{25}_{13}\text{Al}$, are both radioactive.

11.6 SHELL MODEL

Magic numbers in the nucleus

The basic assumption of the liquid-drop model is that each nucleon in a nucleus interacts only with its nearest neighbors, like a molecule in a liquid. At the other extreme, the hypothesis that each nucleon interacts chiefly with a general force field produced by all the other nucleons also has a lot of support. The latter situation is like that of electrons in an atom, where only certain quantum states are permitted and no more than two electrons, which are fermions, can occupy each state. Nucleons are also fermions, and several nuclear properties vary periodically with Z and N in a manner reminiscent of the periodic variation of atomic properties with Z.

The electrons in an atom may be thought of as occupying positions in "shells" designated by the various principal quantum numbers. The degree of occupancy of the outermost shell is what determines certain important aspects of an atom's behavior. For instance, atoms with 2, 10, 18, 36, 54, and 86 electrons have all their electron shells completely filled. Such electron structures have high binding energies and are exceptionally stable, which accounts for the chemical inertness of the rare gases.

The same kind of effect is observed with respect to nuclei. Nuclei that have 2, 8, 20, 28, 50, 82, and 126 neutrons or protons are more abundant than other nuclei of similar mass numbers, suggesting that their structures are more stable. Since complex nuclei arose from reactions among lighter ones, the evolution of heavier and heavier nuclei became retarded when each relatively inert nucleus was formed, which accounts for their abundance.

Other evidence also points to the significance in nuclear structure of the numbers 2, 8, 20, 28, 50, 82, and 126, which have become known as **magic numbers.** An example is the observed pattern of nuclear electric quadrupole moments, which are measures of how much nuclear charge distributions depart from sphericity. A spherical nucleus has no quadrupole moment, while one shaped like a football has a positive

About Scientists

Maria Goeppert-Mayer (1906–1972) was the daughter of the pediatrician of Max Born's children, and she studied at Göttingen under Born. As Born recalled, "She went through all my courses with great industry and conscientiousness, yet remained at the same time a gay and witty member of Göttingen society, fond of parties, of laughter, dancing, and jokes. . . . After she got her doctor's degree with a very good thesis on a problem of quantum mechanics, she married a young American, Joseph Mayer, who worked with me on problems of crystal theory. Both had brilliant careers in the U.S.A., always remaining together." At the University of Chicago in 1948 Goeppert-Mayer reopened the question of periodicities in nuclear stability, which had remained a mystery since their discovery in the early 1930s, and devised a shell model that agreed with the data. J. H. D. Jensen in Germany published a similar theory independently at the same time, and both received the Nobel Prize in 1963 for their work.

moment and one shaped like a pumpkin has a negative moment. Nuclei of magic N and Z are found to have zero quadrupole moments and hence are spherical, while other nuclei are distorted in shape.

The **shell model** of the nucleus is an attempt to account for the existence of magic numbers and certain other nuclear properties in terms of nucleon behavior in a common force field.

Because the precise form of the potential-energy function for a nucleus is not known, unlike the case of an atom, a suitable function $U(r)$ has to be assumed. A reasonable guess on the basis of the nuclear density curves of Fig. 11.3 is a square well with rounded corners. Schrödinger's equation for a particle in a potential well of this kind is then solved, and it is found that stationary states of the system occur that are characterized by quantum numbers n, l, and m_l whose significance is the same as in the analogous case of stationary states of atomic electrons. Neutrons and protons occupy separate sets of states in a nucleus because the latter interact electrically as well as through the specifically nuclear charge. However, the energy levels that come from such a calculation do not agree with the observed sequence of magic numbers. Using other potential-energy functions, for instance that of the harmonic oscillator, gives no better results. Something essential is missing from the picture.

How Magic Numbers Arise

The problem was finally solved independently by Maria Goeppert-Mayer and J. H. D. Jensen in 1949. They realized that it is necessary to incorporate a spin-orbit interaction whose magnitude is such that the consequent splitting of energy levels into sublevels is many times larger than the analogous splitting of atomic energy levels. The exact form of the potential-energy function then turns out not to be critical, provided that it more or less resembles a square well.

The shell theory assumes that *LS* coupling holds only for the very lightest nuclei, in which the l values are necessarily small in their normal configurations. In this scheme, as we saw in Chap. 7, the intrinsic spin angular momenta $\mathbf{S}_i$ of the particles concerned (the neutrons form one group and the protons another) are coupled together into a total spin momentum **S**. The orbital angular momenta $\mathbf{L}_i$ are separately coupled together into a total orbital momentum **L**. Then **S** and **L** are coupled to form a total angular momentum **J** of magnitude $\sqrt{J(J+1)}\hbar$.

After a transition region in which an intermediate coupling scheme holds, the heavier nuclei exhibit ***jj* coupling.** In this case the $\mathbf{S}_i$ and $\mathbf{L}_i$ of each particle are first coupled to form a $\mathbf{J}_i$ for that particle of magnitude $\sqrt{j(j+1)}\hbar$. The various $\mathbf{J}_i$ then couple together to form the total angular momentum $\mathbf{J}$. The *jj* coupling scheme holds for the great majority of nuclei.

When an appropriate strength is assumed for the spin-orbit interaction, the energy levels of either class of nucleon fall into the sequence shown in Fig. 11.17. The levels

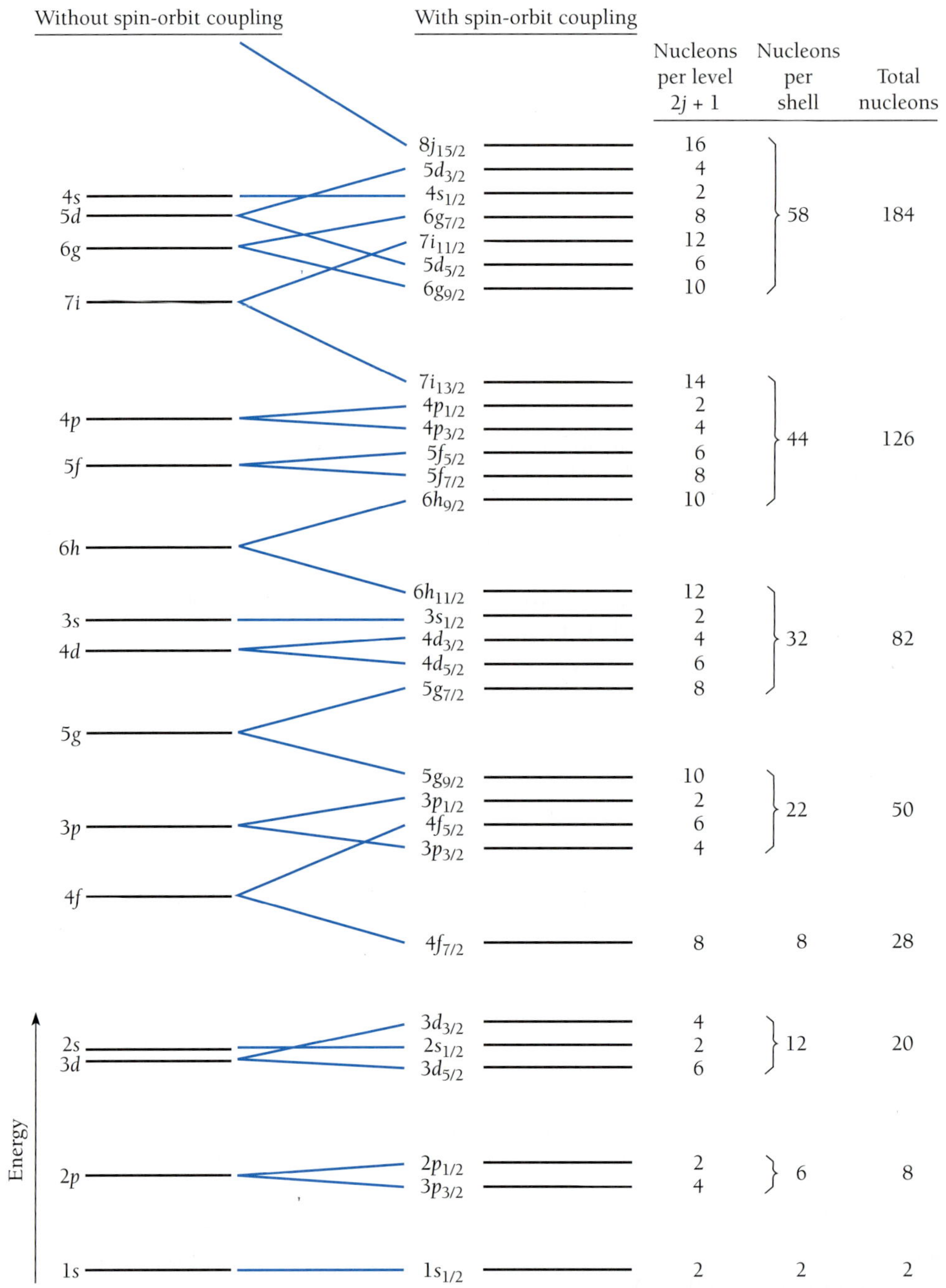

Figure 11.17 Sequence of neutron and proton energy levels according to the shell model (not to scale). The numbers in the right-hand column correspond to the magic numbers expected on the basis of this sequence.

are designated by a prefix equal to the total quantum number n, a letter that indicates l for each particle in that level according to the usual pattern (s, p, d, f, g, . . . corresponding, respectively, to $l = 0, 1, 2, 3, 4, \ldots$), and a subscript equal to j. The spin-orbit interaction splits each state of given j into $2j + 1$ substates, since there are $2j + 1$ allowed orientations of $\mathbf{J}_i$. Large energy gaps appear in the spacing of the levels at intervals that are consistent with the notion of separate shells. The number of available nuclear states in each nuclear shell is, in ascending order of energy, 2, 6, 12, 8, 22, 32, and 44. Hence shells are filled when there are 2, 8, 20, 28, 50, 82, and 126 neutrons or protons in a nucleus.

The shell model accounts for several nuclear phenomena in addition to magic numbers. To begin with, the very existence of energy sublevels that can each be occupied by two particles of opposite spin explains the tendency of nuclear abundances to favor even Z and even N as discussed in Sec. 11.3.

The shell model can also predict nuclear angular momenta. In even-even nuclei, all the protons and neutrons should pair off to cancel out one another's spin and orbital angular momenta. Thus even-even nuclei ought to have zero nuclear angular momenta, as observed. In even-odd and odd-even nuclei, the half-integral spin of the single "extra" nucleon should be combined with the integral angular momentum of the rest of the nucleus for a half-integral total angular momentum. Odd-odd nuclei each have an extra neutron and an extra proton whose half-integral spins should yield integral total angular momenta. Both these predictions are experimentally confirmed.

Reconciling the Models

If the nucleons in a nucleus are so close together and interact so strongly that the nucleus can be considered as analogous to a liquid drop, how can these same nucleons be regarded as moving independently of each other in a common force field as required by the shell model? It would seem that the points of view are mutually exclusive, since a nucleon moving about in a liquid-drop nucleus must surely undergo frequent collisions with other nucleons.

A closer look shows that there is no contradiction. In the ground state of a nucleus, the neutrons and protons fill the energy levels available to them in order of increasing energy in such a way as to obey the exclusion principle (see Fig. 11.8). In a collision, energy is transferred from one nucleon to another, leaving the former in a state of reduced energy and the latter in one of increased energy. But all the available levels of lower energy are already filled, so such an energy transfer can take place only if the exclusion principle is violated. Of course, it is possible for two indistinguishable nucleons of the same kind to merely exchange their respective energies, but such a collision is hardly significant since the system remains in exactly the same state it was in initially. In essence, then, the exclusion principle prevents nucleon-nucleon collisions even in a tightly packed nucleus and thereby justifies the independent-particle approach to nuclear structure.

Both the liquid-drop and shell models of the nucleus are, in their very different ways, able to account for much that is known of nuclear behavior. The **collective model** of Aage Bohr (Niels Bohr's son) and Ben Mottelson combines features of both models in a consistent scheme that has proved quite successful. The collective model takes into account such factors as the nonspherical shape of all but even-even nuclei and the centrifugal distortion experienced by a rotating nucleus. The detailed theory is able to account for the spacing of excited nuclear levels inferred from the gamma-ray spectra of nuclei and in other ways.

Island of Stability

As mentioned in Sec. 11.3, the short range of the strong interaction means that the largest stable nucleus is that of the bismuth isotope $^{209}_{83}Bi$. All nuclei with $Z > 83$ and $A > 209$ undergo radioactive decays until they reach a stable configuration. We can think of the stable nuclei in Fig. 11.7 as representing a peninsula of stability in a sea of instability.

In general, the farther from the peninsula of stability a nucleus is, the faster it decays. For nuclei heavier than $^{209}_{83}Bi$, lifetimes become shorter and shorter with increasing size until they are only milliseconds for $Z = 107$, 108, and 109. (Such superheavy nuclei are created in the laboratory by bombarding targets of heavy atoms with beams of lighter ones.) Since a nucleus with magic numbers of protons or neutrons is exceptionally stable, the question arises whether there might be an island of relative stability among the superheavy nuclei.

In the case of neutrons, Fig. 11.17 shows that the next magic number after $N = 126$ is $N = 184$. For protons the situation is complicated by their electric potential energy, which becomes significant relative to the purely nuclear potential energy (which is independent of charge) when Z is large. The electric potential has a greater effect on proton levels of low l because it is stronger near the nuclear center where the probability densities of such levels are concentrated (see Fig. 6.8). In consequence, the order of proton levels changes from that shown in Fig. 11.17 to make $Z = 114$ a proton magic number instead of $Z = 126$.

A nucleus with $Z = 114$ and $N = 184$ would therefore be doubly magic. This nucleus and nuclei near it in Z and N ought to form an island of stability in the sea of instability that is (so to speak) northeast of the tip of the peninsula of stability in Fig. 11.7.

In 1998 Russian physicists directed a beam of the calcium isotope $^{48}_{20}Ca$ at a target of the plutonium isotope $^{244}_{94}Pu$ to create a nucleus of $Z = 114$ and $N = 175$. Magic in proton number and not far from the middle of the island of stability, this nucleus has a half-life (the time needed for half a sample to decay; see Sec. 12.2) of 30.4 s. As expected, this half-life is much longer than those of nuclei near but outside the island of stability.

When the idea of an island of stability first came up in 1966, it was thought that perhaps the nucleus of $Z = 114$, $N = 184$ might have a half-life in the billions of years. Later calculations gave more modest estimates that range from less than a hundred years to millions of years. When this doubly magic nucleus is eventually produced, we will know. In the meantime, physicists at the Lawrence Berkeley National Laboratory in California have managed to sail past the island of stability to create nuclei of $Z = 116$.

11.7 MESON THEORY OF NUCLEAR FORCES

Particle exchange can produce either attraction or repulsion

In Chap. 8 we saw how a molecule is held together by the exchange of electrons between adjacent atoms. Is it possible that a similar mechanism operates inside a nucleus, with its component nucleons being held together by the exchange of particles of some kind among them?

The first approach to this question was made in 1932 by Heisenberg, who suggested that electrons and positrons shift back and forth between nucleons. A neutron, for instance, might emit an electron and become a proton, while a proton absorbing the electron would become a neutron. However, calculations based on beta-decay data showed that the forces resulting from electron and positron exchange by nucleons would be too small by the huge factor of 10^{14} to be significant in nuclear structure.

About Scientists

Hideki Yukawa (1907–1981) grew up in Kyoto, Japan, and attended the university there. After receiving his doctorate at Osaka, he returned to Kyoto where he spent the rest of his career. In the early 1930s Yukawa tackled the problem of what keeps an atomic nucleus together despite the repulsive forces its protons exert on one another. The interaction must be extremely strong but limited in range, and Yukawa found it could be explained on the basis of the exchange between nucleons of particles whose mass is in the neighborhood of 200 electron masses: "Could the neutrons and protons be playing catch?" In 1936, the year after Yukawa published his proposal, a particle of such intermediate mass was found in cosmic rays by C. D. Anderson, who had earlier discovered the positron, and others. But, this particle, today called the muon, did not interact strongly with nuclei, as it should have. The mystery was not cleared up until 1947 when British physicist C. F. Powell discovered the pion, which has the properties Yukawa predicted but decays rapidly into the longer-lived (and hence easier-to-detect) muon. (The pion and muon were originally called the π and μ mesons by Powell because, according to legend, these were the only Greek letters on his typewriter.) Yukawa received the Nobel Prize in 1949, the first Japanese to do so.

The Japanese physicist Hideki Yukawa was more successful with his 1935 proposal that particles intermediate in mass between electrons and nucleons are responsible for nuclear forces. Today these particles are called **pions.** Pions may be charged (π^+, π^-) or neutral (π^0), and are members of a class of elementary particles collectively called **mesons.** The word pion is a contraction of the original name π meson.

According to Yukawa's theory, every nucleon continually emits and reabsorbs pions. If another nucleon is nearby, an emitted pion may shift across to it instead of returning to its parent nucleon. The associated transfer of momentum is equivalent to the action of a force. Nuclear forces are repulsive at very short range as well as being attractive at greater nucleon-nucleon distances; otherwise the nucleons in a nucleus would mesh together. One of the strengths of the meson theory of such forces is that it can account for both these properties. Although there is no simple way to explain how this comes about, a rough analogy may make it less mysterious.

Let us imagine two boys exchanging basketballs (Fig. 11.18). If they throw the balls at each other, the boys move backward, and when they catch the balls thrown at them,

Figure 11.18 Attractive and repulsive forces can both arise from particle exchange.

their backward momentum increases. Thus this method of exchanging basketballs has the same effect as a repulsive force between the boys. If the boys snatch the basketballs from each other's hands, however, the result will be equivalent to an attractive force acting between them.

A fundamental problem presents itself at this point. If nucleons constantly emit and absorb pions, why are neutrons and protons never found with other than their usual masses? The answer is based upon the uncertainty principle. The laws of physics refer to measurable quantities only, and the uncertainty principle limits the accuracy with which certain combinations of measurements can be made. The emission of a pion by a nucleon which does not change in mass—a clear violation of the law of conservation of energy—can take place provided that the nucleon reabsorbs it or absorbs another pion emitted by a neighboring nucleon so soon afterward that *even in principle* it is impossible to determine whether or not any mass change has actually been involved.

From the uncertainty principle in the form

$$\Delta E\, \Delta t \geq \frac{\hbar}{2} \tag{3.26}$$

an event in which an amount of energy ΔE is not conserved is not prohibited so long as the duration of the event does not exceed $\hbar/2\Delta E$. This condition lets us estimate the pion mass.

Let us assume that a pion travels between nucleons at a speed of $v \sim c$ (actually $v < c$, of course); that the emission of a pion of mass m_π represents a temporary energy discrepancy of $\Delta E \sim m_\pi c^2$ (this neglects the pion's kinetic energy); and that $\Delta E\, \Delta t \sim \hbar$. Nuclear forces have a maximum range r of about 1.7 fm, and the time Δt needed for the pion to travel this far (Fig. 11.19) is

$$\Delta t = \frac{r}{v} \sim \frac{r}{c}$$

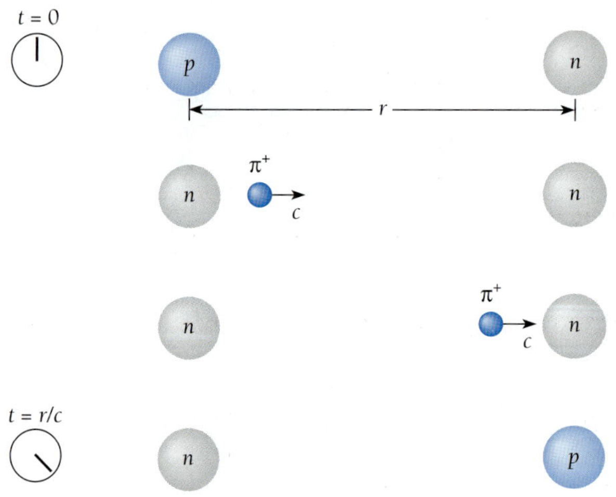

Figure 11.19 The uncertainty principle permits the creation, transfer, and disappearance of a pion to occur without violating conservation of energy provided that the sequence takes place fast enough. Here a positive pion emitted by a proton is absorbed by a neutron; as a result, the proton becomes a neutron and the neutron becomes a proton.

We therefore have

$$\Delta E\,\Delta t \sim \hbar$$

$$(m_\pi c^2)\left(\frac{r}{c}\right) \sim \hbar$$

$$m_\pi \sim \frac{\hbar}{rc} \tag{11.19}$$

which gives a value for m_π of

$$m_\pi \sim \frac{1.05 \times 10^{-34}\,\text{J}\cdot\text{s}}{(1.7 \times 10^{-15}\,\text{m})(3 \times 10^{8}\,\text{m/s})} \sim 2 \times 10^{-28}\,\text{kg}$$

This rough figure is about 220 times the rest mass m_e of the electron.

Discovery of the Pion

A dozen years after Yukawa's proposal, particles with the properties he had predicted were actually discovered. The rest mass of charged pions is 273 m_e and that of neutral pions is 264 m_e, not far from the above estimate.

Two factors contributed to the belated discovery of the free pion. First, enough energy must be supplied to a nucleon so that its emission of a pion conserves energy. Thus at least $m_\pi c^2$ of energy, about 140 MeV, is required. To furnish a stationary nucleon with this much energy in a collision, the incident particle must have considerably more kinetic energy than $m_\pi c^2$ in order that momentum as well as energy be conserved. Particles with kinetic energies of several hundred MeV are therefore required to produce free pions, and such particles are found in nature only in the diffuse stream of cosmic radiation that bombards the earth. Hence the discovery of the pion had to await the

Virtual Photons

Some years before Yukawa's work, particle exchange had been suggested as the mechanism of electromagnetic forces. In this case the particles are photons which, being massless, are not limited in range by Eq.(11.19). However, the greater the distance between two charges, the smaller must be the energies of the photons that pass between them (and hence the less the momenta of the photons and the weaker the resulting force) in order that the uncertainty principle not be violated. For this reason electric forces decrease with distance. Because the photons exchanged in the interactions of electric charges cannot be detected, they are called **virtual photons.** As in the case of pions, they can become actual photons if enough energy is somehow supplied to liberate them from the energy-conservation constraint.

The idea of photons as carriers of electromagnetic forces is attractive on many counts, an obvious one being that it explains why such forces are transmitted with the speed of light and not, say, instantaneously. As subsequently developed, the full theory is called quantum electrodynamics (see Sec. 6.9). Its conclusions have turned out to be in extraordinarily precise agreement with the data on such phenomena as the photoelectric and Compton effects, pair production and annihilation, bremsstrahlung, and photon emission by excited atoms. Unfortunately the details of the theory are too mathematically complex to consider here.

development of sufficiently sensitive and precise methods of investigating cosmic-ray interactions. Later high-energy accelerators were placed in operation which gave the necessary particle energies, and the profusion of pions that were created with their help could be studied readily.

The second reason for the lag between the prediction and experimental discovery of the pion is its instability; the mean lifetime of the charged pion is only 2.6×10^{-8} s and that of the neutral pion is 8.4×10^{-17} s. The lifetime of the π^0 is so short, in fact, that its existence was not established until 1950. The modes of decay of the π^+, π^-, and π^0 are described in Chap. 13. Heavier mesons than the pion have also been discovered, some over a thousand times the electron mass. The contribution of these mesons to nuclear forces is, by Eq. (11.19), limited to shorter distances than those characteristic of pions.

EXERCISES

I hear, and I forget. I see, and I remember. I do, and I understand. —Anon.

11.1 Nuclear Composition

1. State the number of neutrons and protons in each of the following: ${}^{6}_{3}\text{Li}$; ${}^{22}_{10}\text{Ne}$; ${}^{94}_{40}\text{Zr}$; ${}^{180}_{72}\text{Hf}$.

2. Ordinary boron is a mixture of the ${}^{10}_{5}\text{B}$ and ${}^{11}_{5}\text{B}$ isotopes and has a composite atomic mass of 10.82 u. What percentage of each isotope is present in ordinary boron?

11.2 Some Nuclear Properties

3. Electrons of what energy have wavelengths comparable with the radius of a ${}^{197}_{79}\text{Au}$ nucleus? (*Note*: A relativistic calculation is needed.)

4. The greater the atomic number of an atom, the larger its nucleus and the closer its inner electrons are to the nucleus. Compare the radius of the ${}^{238}_{92}\text{U}$ nucleus with the radius of its innermost Bohr orbit.

5. It is believed possible on the basis of the shell model that the nuclide of $Z = 110$ and $A = 294$ may be exceptionally long-lived. Estimate its nuclear radius.

6. Show that the nuclear density of ${}^{1}_{1}\text{H}$ is over 10^{14} times greater than its atomic density. (Assume the atom to have the radius of the first Bohr orbit.)

7. Compare the magnetic potential energies (in eV) of an electron and of a proton in a magnetic field of 0.10 T.

8. One type of magnetometer is based on proton precession. What is the Larmor frequency of a proton in the earth's magnetic field where its magnitude is 3.00×10^{-5} T? In what part of the em spectrum is radiation of this frequency?

9. A system of a million distinguishable protons is in thermal equilibrium at 20°C in a 1.00-T magnetic field. More of the protons are in the lower-energy spin-up state than in the higher-energy spin-down state. (*a*) On the average, how many more? (*b*) Repeat the calculation for a temperature of 20 K. (*c*) What do these results suggest about how strongly such a system will absorb em radiation at the Larmor frequency? (*d*) Could such a system in principle be used as the basis of a laser? If not, why not?

11.3 Stable Nuclei

10. The Appendix at the back of the book lists all known stable nuclides. Are there any for which $Z > N$? Why are such nuclides so rare (or absent)?

11. What limits the size of a stable nucleus?

12. What happens to the atomic number and mass number of a nucleus when it (*a*) emits an alpha particle, (*b*) emits an electron, (*c*) emits a position, (*d*) captures an electron?

13. Which nucleus would you expect to be more stable, ${}^{7}_{3}\text{Li}$ or ${}^{8}_{3}\text{Li}$; ${}^{13}_{6}\text{C}$ or ${}^{15}_{6}\text{C}$?

14. Both ${}^{14}_{8}\text{O}$ and ${}^{19}_{8}\text{O}$ undergo beta decay. Which would you expect to emit a positron and which an electron? Why?

11.4 Binding Energy

15. Find the binding energy per nucleon in ${}^{20}_{10}\text{Ne}$ and in ${}^{56}_{26}\text{Fe}$.

16. Find the binding energy per nucleon in ${}^{79}_{35}\text{Br}$ and in ${}^{197}_{79}\text{Au}$.

17. Find the energies needed to remove a neutron from ${}^{4}_{2}\text{He}$, then to remove a proton, and finally to separate the remaining neutron and proton. Compare the total with the binding energy of ${}^{4}_{2}\text{He}$.

18. The binding energy of ${}^{24}_{12}\text{Mg}$ is 198.25 MeV. Find its atomic mass.

19. Show that the potential energy of two protons 1.7 fm (the maximum range of nuclear forces) apart is of the correct order of magnitude to account for the difference in binding energy between ${}^{3}_{1}\text{H}$ and ${}^{3}_{2}\text{He}$. How does this result bear upon the question of the dependence of nuclear forces on electric charge?

20. The neutron decays in free space into a proton and an electron. What must be the minimum binding energy contributed by a neutron to a nucleus in order that the neutron not decay inside the nucleus? How does this figure compare with the observed binding energies per nucleon in stable nuclei?

11.5 Liquid-Drop Model

21. Use the semiempirical binding-energy formula to calculate the binding energy of $^{40}_{20}\text{Ca}$. What is the percentage discrepancy between this figure and the actual binding energy?

22. Two nuclei with the same mass number for which $Z_1 = N_2$ and $Z_2 = N_1$, so that their atomic numbers differ by 1, are called **mirror isobars;** for example, $^{15}_{7}\text{N}$ and $^{15}_{8}\text{O}$. The constant a_3 in the coulomb energy term of Eq. (11.18) can be evaluated from the mass difference between two mirror isobars, one of which is odd-even and the other even-odd (so that their pairing energies are zero). (*a*) Derive a formula for a_3 in terms of the mass difference between two such nuclei, their mass number A, the smaller atomic number Z of the pair, and the masses of the hydrogen atom and the neutron. (*Hint:* First show that $(A - 2Z)^2 = 1$ for both nuclei.) (*b*) Evaluate a_3 for the case of the mirror isobars $^{15}_{7}\text{N}$ and $^{15}_{8}\text{O}$.

23. The coulomb energy of Z protons uniformly distributed throughout a spherical nucleus of radius R is given by

$$E_C = \frac{3}{5}\frac{Z(Z-1)e^2}{4\pi\epsilon_0 R}$$

(*a*) On the assumption that the mass difference ΔM between a pair of mirror isobars is entirely due to the difference Δm between the $^{1}_{1}\text{H}$ and neutron masses and to the difference between their coulomb energies, derive a formula for R in terms of ΔM, Δm, and Z, where Z is the atomic number of the nucleus with the smaller number of protons. (*b*) Use this formula to find the radii of the mirror isobars $^{15}_{7}\text{N}$ and $^{15}_{8}\text{O}$.

24. Use the formula for E_c of Exercise 23 to calculate a_3 in Eq. (11.12). If this figure is not the same as the value of 0.60 MeV quoted in the text, can you think of any reasons for the difference?

25. (*a*) Find the energy needed to remove a neutron from ^{81}Kr, from ^{82}Kr, and from ^{83}Kr. (*b*) Why is the figure for ^{82}Kr so different from the others?

26. Which isobar of $A = 75$ does the liquid-drop model suggest is the most stable?

27. Use the liquid-drop model to establish which of the mirror isobars $^{127}_{52}\text{Te}$ and $^{127}_{53}\text{I}$ decays into the other. What kind of decay occurs?

11.6 Shell Model

28. According to the **Fermi gas model** of the nucleus, its protons and neutrons exist in a box of nuclear dimensions and fill the lowest available quantum states to the extent permitted by the exclusion principle. Since both protons and neutrons have spins of $\frac{1}{2}$ they are fermions and obey Fermi-Dirac statistics. (*a*) Find an equation for the Fermi energy in a nucleus under the assumption that $A = 2Z$. Note that the protons and neutrons must be considered separately. (*b*) What is the Fermi energy in such a nucleus for $R_0 = 1.2$ fm? (*c*) In heavier nuclei, $A > 2Z$. What effect will this have on the Fermi energies for each type of particle?

29. A simplified model of the deuteron consists of a neutron and a proton in a square potential well 2 fm in radius and 35 MeV deep. Is this model consistent with the uncertainty principle?

11.7 Meson Theory of Nuclear Forces

30. Van der Waals forces are limited to very short ranges and do not have an inverse-square dependence on distance, yet nobody suggests that the exchange of a special mesonlike particle is responsible for such forces. Why not?

CHAPTER 12

Nuclear Transformations

Interior of the Tokamak Fusion Test Reactor at the Princeton Plasma Physics Laboratory. In December 1993 this reactor produced 6.2 MW of fusion power for 4 s from a deuterium-tritium plasma confined by strong magnetic fields.

Despite the strength of the forces that hold nucleons together to form an atomic nucleus, many nuclides are unstable and spontaneously change into other nuclides by radioactive decay. And all nuclei can be transformed by reactions with nucleons or other nuclei that collide with them. In fact, all complex nuclei came into being in the first place through successive nuclear reactions, some in the first few minutes after the Big Bang and the rest in stellar interiors. The principal aspects of radioactivity and nuclear reactions are considered in this chapter.

12.1 RADIOACTIVE DECAY

Five kinds

No single phenomenon has played so significant a role in the development of nuclear physics as radioactivity, which was discovered in 1896 by Antoine Becquerel. Three features of radioactivity are extraordinary from the perspective of classical physics:

1 When a nucleus undergoes alpha or beta decay, its atomic number Z changes and it becomes the nucleus of a different element. Thus the elements are not immutable, although the mechanism of their transformation would hardly be recognized by an alchemist.
2 The energy liberated during radioactive decay comes from *within* individual nuclei without external excitation, unlike the case of atomic radiation. How can this happen? Not until Einstein proposed the equivalence of mass and energy could this puzzle be understood.
3 Radioactive decay is a statistical process that obeys the laws of chance. No cause-effect relationship is involved in the decay of a particular nucleus, only a certain probability per unit time. Classical physics cannot account for such behavior, although it fits naturally into the framework of quantum physics.

The radioactivity of an element arises from the radioactivity of one or more of its isotopes. Most elements in nature have no radioactive isotopes, although such isotopes can be prepared artificially and are useful in biological and medical research as "tracers." (The procedure is to incorporate a radionuclide in a chemical compound and follow what happens to the compound in a living organism by monitoring the radiation from the nuclide.) Other elements, such as potassium, have some stable isotopes and some radioactive ones; a few, such as uranium, have only radioactive isotopes.

The early experimenters, among them Rutherford and his coworkers, distinguished three components in the radiations from radionuclides (Figs. 12.1 and 12.2). These components

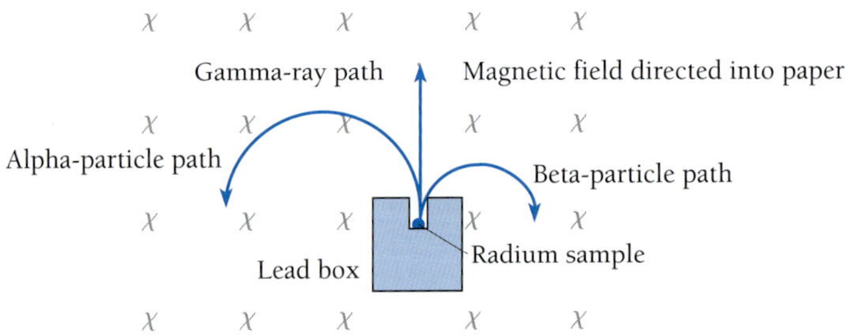

Figure 12.1 The radiations from a radium sample may be analyzed with the help of a magnetic field. Alpha particles are deflected to the left, hence they are positively charged; beta particles are deflected to the right, hence they are negatively charged; and gamma rays are not affected, hence they are unchanged.

About Scientists

Antoine-Henri Becquerel (1852–1908) was born and educated in Paris. His grandfather, father, and son were also physicists, all of them in turn professors at the Paris Museum of Natural History. Like his grandfather and father, Becquerel specialized in fluorescence and phosphorescence, phenomena in which a substance absorbs light at one frequency and reemits it at another, lower frequency.

In 1895 Roentgen had detected x-rays by the fluorescence they cause in an appropriate material. When he learned of this early in 1896, Becquerel wondered whether the reverse process might not occur, with intense light stimulating a fluorescent material to give off x-rays. He placed a fluorescent uranium salt on a photographic plate covered with black paper, exposed the arrangement to the sun, and indeed found the plate fogged when he had developed it. Becquerel then tried to repeat the experiment, but clouds obscured the sun for several days. He developed the plates anyway, expecting them to be clear, but to his surprise they were just as fogged as before. In a short time he had identified the source of the penetrating radiation as the uranium in the fluorescent salt. He was also able to show that the radiation ionized gases and that part of it consisted of fast charged particles.

Although Becquerel's discovery was accidental, he realized its importance at once and explored various aspects of the radioactivity of uranium for the rest of his life. He received the Nobel Prize in physics in 1903.

were called alpha, beta, and gamma, which were eventually identified as ${}^4_2\text{He}$ nuclei, electrons, and high-energy photons respectively. Later, positron emission and electron capture were added to the list of decay modes. Figure 12.3 shows the five ways in which an unstable nucleus can decay, together with the reason for the instability. (The neutrinos given off when nuclei emit or absorb electrons are discussed in Sec. 12.5.) Examples of the nuclear transformations that accompany the various decays are given in Table 12.1.

Table 12.1 Radioactive Decay†

Decay	Transformation	Example
Alpha decay	${}^A_ZX \rightarrow {}^{A-4}_{Z-2}Y + {}^4_2\text{He}$	${}^{238}_{92}\text{U} \rightarrow {}^{234}_{90}\text{Th} + {}^4_2\text{He}$
Beta decay	${}^A_ZX \rightarrow {}_{Z+1}^{A}Y + e^-$	${}^{14}_6\text{C} \rightarrow {}^{14}_7\text{N} + e^-$
Positron emission	${}^A_ZX \rightarrow {}_{Z-1}^{A}Y + e^+$	${}^{64}_{29}\text{Cu} \rightarrow {}^{64}_{28}\text{Ni} + e^+$
Electron capture	${}^A_ZX + e^- \rightarrow {}_{Z-1}^{A}Y$	${}^{64}_{29}\text{Cu} + e^- \rightarrow {}^{64}_{28}\text{Ni}$
Gamma decay	${}^A_ZX^* \rightarrow {}^A_ZX + \gamma$	${}^{87}_{38}\text{Sr}^* \rightarrow {}^{87}_{38}\text{Sr} + \gamma$

†The * denotes an excited nuclear state and γ denotes a gamma-ray photon.

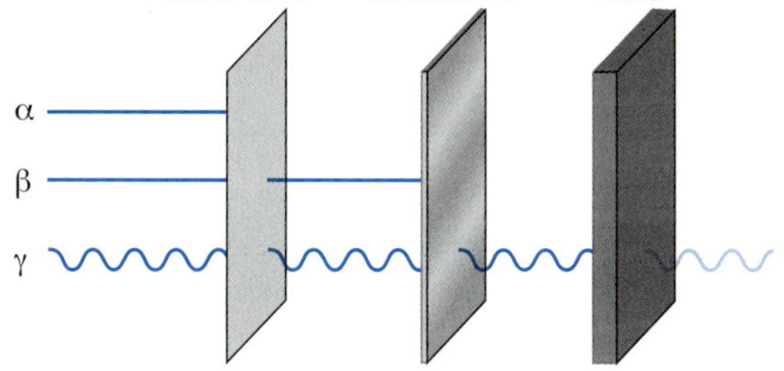

Figure 12.2 Alpha particles from radioactive materials are stopped by a piece of cardboard. Beta particles penetrate the cardboard but are stopped by a sheet of aluminum. Even a thick slab of lead may not stop all the gamma rays.

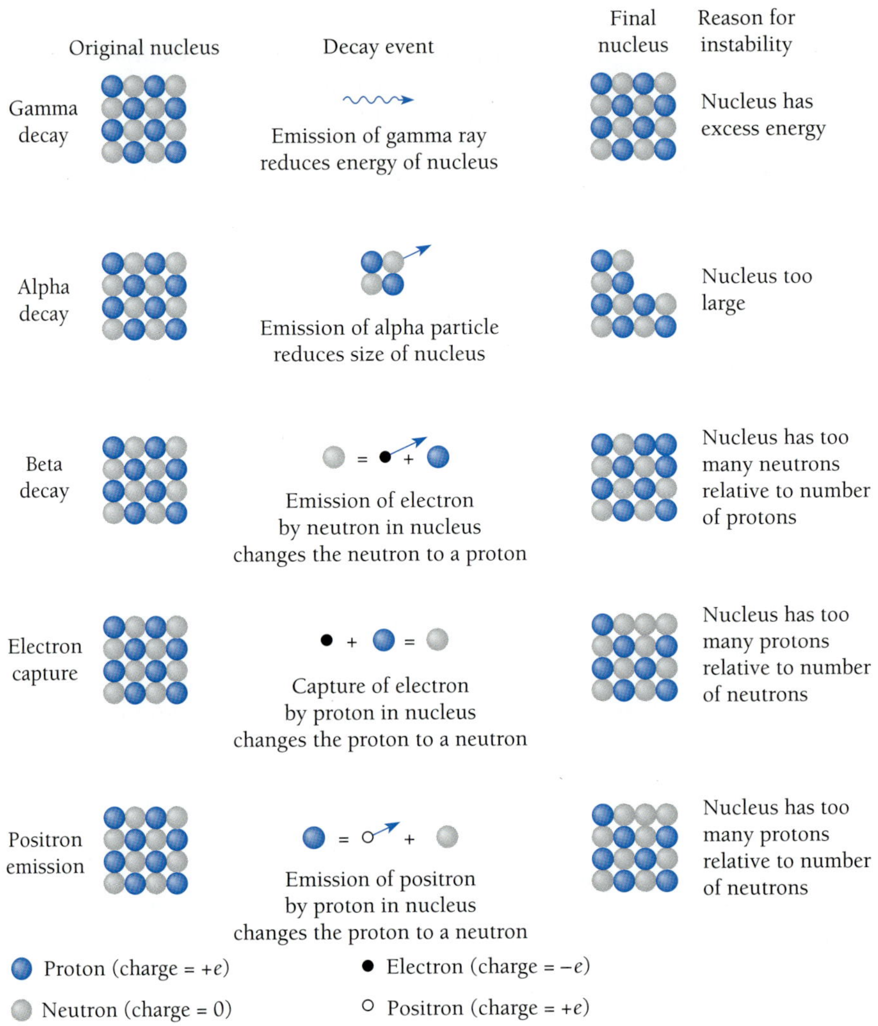

Figure 12.3 Five kinds of radioactive decay.

Example 12.1

The helium isotope ^6_2He is unstable. What kind of decay would you expect it to undergo?

Solution

The most stable helium nucleus is ^4_2He, all of whose neutrons and protons are in the lowest possible energy levels (see Sec. 11.3). Since ^6_2He has four neutrons whereas ^4_2He has only two, the instability of ^6_2He must be due to an excess of neutrons. This suggests that ^6_2He undergoes negative beta decay to become the lithium isotope ^6_3Li whose neutron/proton ratio is more consistent with stability:

$$^6_2\text{He} \rightarrow {}^6_3\text{Li} + e^-$$

This is, in fact, the manner in which ^6_2He decays.

Radioactivity and the Earth

Most of the energy responsible for the geological history of the earth can be traced to the decay of the radioactive uranium, thorium, and potassium isotopes it contains. The earth is believed to have come into being perhaps 4.5 billion years ago as a cold aggregate of smaller bodies that consisted largely of metallic iron and silicate minerals that had been circling the sun. Heat of radioactive origin accumulated in the interior of the infant earth and in time led to partial melting. The influence of gravity then caused the iron to migrate inward to form the molten core of today's planet; the geomagnetic field comes from electric currents in this core. The lighter silicates rose to form the rocky mantle around the core that makes up about 80 percent of the earth's volume. Most of the earth's radioactivity is now concentrated in the upper mantle and the crust (the relatively thin outer shell), where the heat it produces escapes and cannot collect to remelt the earth. The steady stream of heat is more than enough to power the motions of the giant plates into which the earth's surface is divided and the mountain building, earthquakes, and volcanoes associated with these motions.

Activity

The **activity** of a sample of any radioactive nuclide is the rate at which the nuclei of its constituent atoms decay. If N is the number of nuclei present in the sample at a certain time, its activity R is given by

Activity

$$R = -\frac{dN}{dt} \tag{12.1}$$

The minus sign is used to make R a positive quantity since dN/dt is, of course, intrinsically negative. The SI unit of activity is named after Becquerel:

$$1 \text{ becquerel} = 1 \text{ Bq} = 1 \text{ decay/s}$$

The activities encountered in practice are usually so high that the megabecquerel ($1 \text{ MBq} = 10^6 \text{ Bq}$) and gigabecquerel ($1 \text{ GBq} = 10^9 \text{ Bq}$) are more often appropriate.

The traditional unit of activity is the **curie** (Ci), which was originally defined as the activity of 1 g of radium, $^{226}_{88}\text{Ra}$. Because the precise value of the curie changed as methods of measurement improved, it is now defined arbitrarily as

$$1 \text{ curie} = 1 \text{ Ci} = 3.70 \times 10^{10} \text{ decays/s} = 37 \text{ GBq}$$

The activity of 1 g of radium is a few percent smaller. Ordinary potassium has an activity of about 0.7 microcurie ($1\ \mu\text{Ci} = 10^{-6}$ Ci) per kilogram because it contains a small proportion of the radioisotope $^{40}_{19}\text{K}$.

Radiation Hazards

The various radiations from radionuclides ionize matter through which they pass. X-ray ionize matter, too. All ionizing radiation is harmful to living tissue, although if the damage is slight, the tissue can often repair itself with no permanent effect. Radiation hazards are easy to underestimate because there is usually a delay, sometimes of many years, between an exposure and some of its possible consequences. These consequences include cancer, leukemia, and changes in the DNA of reproductive cells that lead to children with physical deformities and mental handicaps.

Radiation dosage is measured in **sieverts** (Sv), where 1 Sv is the amount of any radiation that has the same biological effect as those produced when 1 kg of body tissue absorbs 1 joule of x-rays or gamma rays. Although radiobiologists disagree about the exact relationship between radiation exposure and the likelihood of developing cancer, there is no question that such a link exists. The International Commission on Radiation Protection estimates an average risk factor of $0.05\ Sv^{-1}$. This means that the chances of dying from cancer as a result of radiation are 1 in 20 for a dose of 1 Sv, 1 in 20,000 for a dose of 1 mSv (1 mSv = 0.001 Sv), and so on.

Figure 12.4 shows the chief sources of radiation dosage on a worldwide basis. The most important single source is the radioactive gas radon, a decay product of radium whose own origin traces back to the decay of uranium. Uranium is found in many common rocks, notably granite. Hence radon, colorless and odorless, is present nearly everywhere, though usually in amounts too small to endanger health. Problems arise when houses are built in uranium-rich regions, since it is impossible to prevent radon from entering such houses from the ground under them. Surveys show that millions of American homes have radon concentrations high enough to pose a nonnegligible cancer risk. As a cause of lung cancer, radon is second only to cigarette smoking. The most effective method of reducing radon levels in an existing house in a hazardous region seems to be to extract air with fans from underneath the ground floor and disperse it into the atmosphere before it can enter the house.

Other natural sources of radiation dosage include cosmic rays from space and radionuclides present in rocks, soil, and building materials. Food, water, and the human body itself contain small amounts of radionuclides of such elements as potassium and carbon.

Many useful processes involve ionizing radiation. Some employ such radiation directly, as in the x-rays and gamma rays used in medicine and industry. In other cases the radiation is an unwanted but inescapable byproduct, notably in the operation of nuclear reactors and in the disposal of their wastes. In many countries the dose limit for workers (about 9 million wordwide) whose jobs involve ionizing radiation is 20 mSv per year. For the general public, which has no choice in the matter, the dose limit for nonbackground radiation is 1 mSv per year.

An appropriate balance between risk and benefit is not always easy to find where radiation is concerned. This seems particularly true for medical x-ray exposures, many of which are made for no strong reason and do more harm than good. The once "routine" x-raying of symptomless young women to search for breast cancer is now generally believed to have increased, not decreased, the overall death rate due to cancer. Particularly dangerous is the x-raying of pregnant women, until not long ago another "routine" procedure, which dramatically increases the chance of cancer in their children. Of course, x-rays have many valuable applications in medicine. The

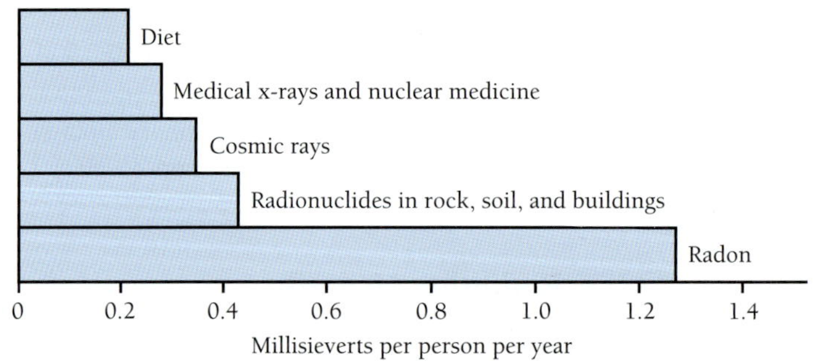

Figure 12.4 The chief sources of radiation dosage averaged around the world. The total is 2.7 mSv, but actual dosages vary widely. For instance, radon concentrations are not the same everywhere, some people receive much more medical radiation than others, cosmic rays are more intense at high altitudes (frequent fliers may get double the sea-level dose, residents of high-altitude cities up to five times as much), and so on. Nuclear power stations are responsible for less than 0.1 percent of the total, though accidents can raise the amount in affected areas to dangerous levels.

point is that every exposure should have a definite justification that outweights the risk involved. An ordinary chest x-ray using modern equipment involves a radiation dose of about 0.017 mSv, much less than in the past. However, a CT chest scan (Sec. 2.5) involves the considerable dose of 8 mSv. CT scans of children pose especially serious risks and need equally serious justification.

12.2 HALF-LIFE

Less and less, but always some left

Measurements of the activities of radioactive samples show that, in every case, they fall off exponentially with time. Figure 12.5 is a graph of R versus t for a typical radionuclide. We note that in every 5.00-h period, regardless of when the period starts, the activity drops to half of what it was at the start of the period. Accordingly the **half-life** $T_{1/2}$ of the nuclide is 5.00 h.

Every radionuclide has a characteristic half-life. Some half-lives are only a millionth of a second, others are billions of years. One of the major problems faced by nuclear power plants is the safe disposal of radioactive wastes since some of the nuclides present have long half-lives.

The behavior illustrated in Fig. 12.5 means that the time variation of activity follows the formula

Activity law
$$R = R_0 e^{-\lambda t} \tag{12.2}$$

where λ, called the **decay constant**, has a different value for each radionuclide. The connection between decay constant λ and half-life $T_{1/2}$ is easy to find. After a half-life has elapsed, that is, when $t = T_{1/2}$, the activity R drops to $\frac{1}{2}R_0$ by definition. Hence

$$\tfrac{1}{2}R_0 = R_0 e^{-\lambda T_{1/2}}$$

$$e^{\lambda T_{1/2}} = 2$$

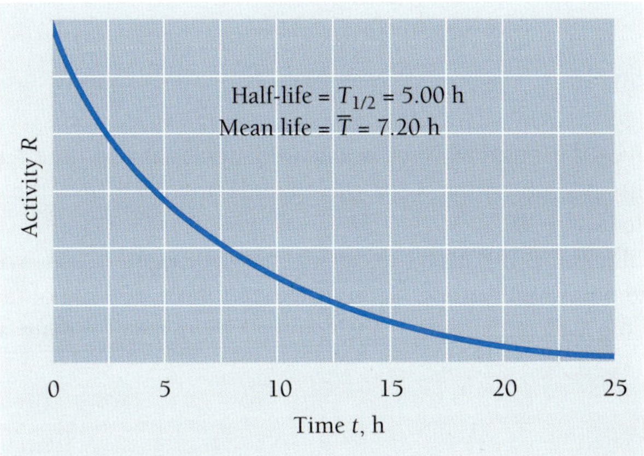

Figure 12.5 The activity of a radionuclide decreases exponentially with time. The half-life is the time needed for an initial activity to drop by half. The mean life of a radionuclide is 1.44 times its half-life [Eq. (12.7)].

Taking natural logarithms of both sides of this equation,

$$\lambda T_{1/2} = \ln 2$$

Half-life

$$T_{1/2} = \frac{\ln 2}{\lambda} = \frac{0.693}{\lambda} \tag{12.3}$$

The decay constant of the radionuclide whose half-life is 5.00 h is therefore

$$\lambda = \frac{0.693}{T_{1/2}} = \frac{0.693}{(5.00\text{ h})(3600\text{ s/h})} = 3.85 \times 10^{-5}\text{ s}^{-1}$$

The larger the decay constant, the greater the chance a given nucleus will decay in a certain period of time.

The activity law of Eq. (12.2) follows if we assume a constant probability λ per unit time for the decay of each nucleus of a given nuclide. With λ as the probability per unit time, $\lambda\, dt$ is the probability that any nucleus will undergo decay in a time interval dt. If a sample contains N undecayed nuclei, the number dN that decay in a time dt is the product of the number of nuclei N and the probability $\lambda\, dt$ that each will decay in dt. That is,

$$dN = -N\lambda\, dt \tag{12.4}$$

where the minus sign is needed because N decreases with increasing t.

Equation (12.4) can be rewritten

$$\frac{dN}{N} = -\lambda\, dt$$

and each side can now be integrated:

$$\int_{N_0}^{N} \frac{dN}{N} = -\lambda \int_0^t dt$$

$$\ln N - \ln N_0 = -\lambda t$$

Radioactive decay

$$N = N_0 e^{-\lambda t} \tag{12.5}$$

This formula gives the number N of undecayed nuclei at the time t in terms of the decay probability per unit time λ of the nuclide involved and the number N_0 of undecayed nuclei at $t = 0$.

Figure 12.6 illustrates the alpha decay of the gas radon, $^{222}_{86}Rn$, whose half-life is 3.82 days, to the polonium isotope $^{218}_{84}Po$. If we start with 1.00 mg of radon in a closed container, 0.50 mg will remain after 3.82 days, 0.25 mg will remain after 7.64 days, and so on.

Example 12.2

How long does it take for 60.0 percent of a sample of radon to decay?

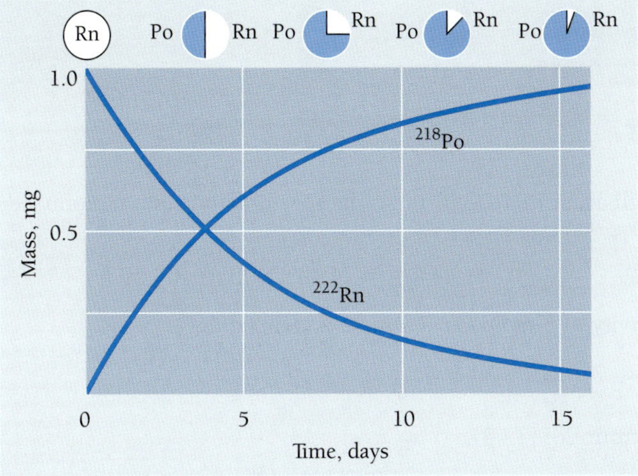

Figure 12.6 The alpha decay of ^{222}Rn to ^{218}Po has a half-life of 3.8 d. The sample of radon whose decay is graphed here had an initial mass of 1.0 mg.

Solution

From Eq. (12.5)

$$\frac{N}{N_0} = e^{-\lambda t} \qquad -\lambda t = \ln\frac{N}{N_0} \qquad \lambda t = \ln\frac{N_0}{N}$$

$$t = \frac{1}{\lambda}\ln\frac{N_0}{N}$$

Here $\lambda = 0.693/T_{1/2} = 0.693/3.82$ d and $N = (1 - 0.600)\, N_0 = 0.400N_0$, so that

$$t = \frac{3.82 \text{ d}}{0.693}\ln\frac{1}{0.400} = 5.05 \text{ d}$$

The fact that radioactive decay follows the exponential law of Eq. (12.2) implies that this phenomenon is statistical in nature. Every nucleus in a sample of a radionuclide has a certain probability of decaying, but there is no way to know in advance *which* nuclei will actually decay in a particular time span. If the sample is large enough—that is, if many nuclei are present—the actual fraction of it that decays in a certain time span will be very close to the probability for any individual nucleus to decay.

To say that a certain radioisotope has a half-life of 5 h, then, signifies that every nucleus of this isotope has a 50 percent change of decaying in every 5-h period. This does *not* mean a 100 percent probability of decaying in 10 h. A nucleus does not have a memory, and its decay probability per unit time is constant until it actually does decay. A half-life of 5 h implies a 75 percent probability of decay in 10 h, which increases to 87.5 percent in 15 h, to 93.75 percent in 20 h, and so on, because in every 5-h interval the probability is 50 percent.

It is worth keeping in mind that the half-life of a radionuclide is not the same as its **mean lifetime** $\overline{T}$. The mean lifetime of a nuclide is the reciprocal of its decay probability per unit time:

$$\overline{T} = \frac{1}{\lambda} \tag{12.6}$$

Hence

Mean lifetime
$$\overline{T} = \frac{1}{\lambda} = \frac{T_{1/2}}{0.693} = 1.44T_{1/2} \tag{12.7}$$

$\overline{T}$ is nearly half again more than $T_{1/2}$. The mean lifetime of a radionuclide whose half-life is 5.00 h is

$$\overline{T} = 1.44T_{1/2} = (1.44)(5.00\text{ h}) = 7.20\text{ h}$$

Since the activity of a radioactive sample is defined as

$$R = -\frac{dN}{dt}$$

we see that, from Eq. (12.5),

$$R = \lambda N_0 e^{-\lambda t}$$

This agrees with the activity law of Eq. (12.2) if $R_0 = \lambda N_0$, or, in general, if

Activity
$$R = \lambda N \tag{12.8}$$

Example 12.3

Find the activity of 1.00 mg of radon, ^{222}Rn, whose atomic mass is 222 u.

Solution

The decay constant of radon is

$$\lambda = \frac{0.693}{T_{1/2}} = \frac{0.693}{(3.8\text{ d})(86{,}400\text{ s/d})} = 2.11 \times 10^{-6}\text{ s}^{-1}$$

The number N of atoms in 1.00 mg of ^{222}Rn is

$$N = \frac{1.00 \times 10^{-6}\text{ kg}}{(222\text{ u})(1.66 \times 10^{-27}\text{ kg/u})} = 2.71 \times 10^{18}\text{ atoms}$$

Hence

$$\begin{aligned} R = \lambda N &= (2.11 \times 10^{-6}\text{ s}^{-1})(2.71 \times 10^{18}\text{ nuclei}) \\ &= 5.72 \times 10^{12}\text{ decays/s} = 5.72\text{ TBq} = 155\text{ Ci} \end{aligned}$$

Example 12.4

What will the activity of the above radon sample be exactly one week later?

Solution

The activity of the sample decays according to Eq. (12.2). Since R_0 = 155 Ci here and

$$\lambda t = (2.11 \times 10^{-6}\text{ s}^{-1})(7.00\text{ d})(86{,}400\text{ s/d}) = 1.28$$

we find that

$$R = R_0 e^{-\lambda t} = (155\text{ Ci})e^{-1.28} = 43\text{ Ci}$$

Radiometric Dating

Radioactivity makes it possible to establish the ages of many geological and biological specimens. Because the decay of any particular radionuclide is independent of its environment, the ratio between the amounts of that nuclide and its stable daughter in a specimen depends on the latter's age. The greater the proportion of the daughter nuclide, the older the specimen. Let us see how this procedure is used to date objects of biological origin using **radiocarbon**, the beta-active carbon isotope $^{14}_{6}C$.

Cosmic rays are high-energy atomic nuclei, chiefly protons, that circulate through the Milky Way galaxy of which the sun is a member. About 10^{18} of them reach the earth each second. When they enter the atmosphere, they collide with the nuclei of atoms in their paths to produce showers of secondary particles. Among these secondaries are neutrons that can react with nitrogen nuclei in the atmosphere to form radiocarbon with the emission of a proton:

Formation of radiocarbon

$$^{14}_{7}N + ^{1}_{0}n \rightarrow ^{14}_{6}C + ^{1}_{1}H$$

The proton picks up an electron and becomes a hydrogen atom. Radiocarbon has too many neutrons for stability and beta decays into $^{14}_{7}N$ with a half-life of about 5760 years. Although the radiocarbon decays steadily, the cosmic-ray bombardment constantly replenishes the supply. A total of perhaps 90 tons of radiocarbon is distributed around the world at the present time.

Shortly after their formation, radiocarbon atoms combine with oxygen molecules to form carbon dioxide molecules. Green plants take in carbon dioxide and water which they convert into carbohydrates in the process of photosynthesis, so that every plant contains some radiocarbon. Animals eat plants and thereby become radioactive themselves. Because the mixing of radiocarbon is efficient, living plants and animals all have the same ratio of radiocarbon to ordinary carbon (^{12}C).

When plants and animals die, however, they no longer take in radiocarbon atoms, but the radiocarbon they contain keeps decaying away to ^{14}N. After 5760 years, then, they have only one-half as much radiocarbon left—relative to their total carbon content—as they had as living matter, after 11,520 years only one-fourth as much, and so on. By determining the proportion of radiocarbon to ordinary carbon it is therefore possible to evaluate the ages of ancient objects and remains of organic origin. This elegant method permits the dating of mummies, wooden implements, cloth, leather, charcoal from campfires, and similar artifacts from ancient civilizations as much as 50,000 years old, about nine half-lives of ^{14}C.

Example 12.5

A piece of wood from the ruins of an ancient dwelling was found to have a ^{14}C activity of 13 disintegrations per minute per gram of its carbon content. The ^{14}C activity of living wood is 16 disintegrations per minute per gram. How long ago did the tree die from which the wood sample came?

Solution

If the activity of a certain mass of carbon from a plant or animal that was recently alive is R_0 and the activity of the same mass of carbon from the sample to be dated is R, then from Eq. (12.2)

$$R = R_0 e^{-\lambda t}$$

To solve for the age t we proceed as follows:

$$e^{\lambda t} = \frac{R_0}{R} \qquad \lambda t = \ln \frac{R_0}{R} \qquad t = \frac{1}{\lambda} \ln \frac{R_0}{R}$$

From Eq. (12.3) the decay constant λ of radiocarbon is $\lambda = 0.693/T_{1/2} = 0.693/5760$ y. Here $R_0/R = 16/13$ and so

$$t = \frac{1}{\lambda} \ln \frac{R_0}{R} = \frac{5760 \text{ y}}{0.693} \ln \frac{16}{13} = 1.7 \times 10^3 \text{ y}$$

Radiocarbon dating is limited to about 50,000 years whereas the earth's history goes back 4.5 or so billion years. Geologists accordingly use radionuclides of much longer half-lives to date rocks (Table 12.2). In each case it is assumed that all the stable daughter nuclides found in a particular rock sample came from the decay of the parent nuclide. Although the thorium and uranium isotopes in the table do not decay in a single step as do ^{40}K and ^{87}Rb, the half-lives of the intermediate products are so short compared with those of the parents that only the latter need be considered.

If the number of atoms of a parent nuclide in a sample is N and the number of atoms of both parent and daughter is N_0, then from Eq. (12.5)

Geological dating

$$t = \frac{1}{\lambda} \ln \frac{N_0}{N}$$

The precise significance of the time t depends on the nature of the rock involved. It may refer to the time at which the minerals of the rock crystallized, for instance, or it may refer to the most recent time at which the rock cooled below a certain temperature.

The most ancient rocks whose ages have been determined are found in Greenland and are believed to be 3.8 billion years old. Lunar rocks and meteorites as well as terrestrial rocks have been dated by the methods of Table 12.2. Some lunar samples apparently solidified 4.6 billion years ago, which is very soon after the solar system came into being. Because the youngest rocks found on the moon are 3 billion years old, the inference is that although the lunar surface was once molten and there were widespread volcanic eruptions for some time afterwards, all such activity must have ceased 3 billion years ago. To be sure, the lunar surface has been disturbed in a variety of small-scale ways since it cooled, but apparently meteoroid bombardment was responsible for most of them.

Table 12.2 Geological Dating Methods

Method	Parent Radionuclide	Stable Daughter Nuclide	Half-Life, Billion Years
Potassium-argon	^{40}K	^{40}Ar	1.3
Rubidium-strontium	^{87}Rb	^{87}Sr	47
Thorium-lead	^{232}Th	^{208}Pb	13.9
Uranium-lead	^{235}U	^{207}Pb	0.7
Uranium-lead	^{238}U	^{206}Pb	4.5

Astronaut Charles M. Duke, Jr., collecting rocks from the surface of the moon during the Apollo 16 expedition in 1972. The rocks were dated radiometrically. The youngest was found to be 3 billion years old, so igneous activity such as volcanic eruptions must have stopped at that time.

12.3 RADIOACTIVE SERIES

Four decay sequences that each end in a stable daughter

Most of the radionuclides found in nature are members of four **radioactive series,** with each series consisting of a succession of daughter products all ultimately derived from a single parent nuclide.

The reason that there are exactly four series follows from the fact that alpha decay reduces the mass number of a nucleus by 4. Thus the nuclides whose mass numbers are all given by $A = 4n$, where n is an integer, can decay into one another in descending order of mass number. The other three series have mass numbers specified by $A = 4n + 1$, $4n + 2$, and $4n + 3$. The members of these series, too, can decay into one another.

Table 12.3 lists the four radioactive series. The half-life of neptunium is so short compared with the age of the solar system that members of this series are not found on the earth today. They have, however, been produced in the laboratory by bombarding other heavy nuclei with neutrons, as described later. The sequence of alpha and beta decays that lead from parent to stable end product is shown in Fig. 12.7 for the uranium series. The decay chain branches at ^{214}Bi, which may decay either by alpha or beta emission. The alpha decay is followed by a beta decay and the beta decay is followed by an alpha decay, so both branches lead to ^{210}Pb.

Table 12.3 Four Radioactive Series

Mass Numbers	Series	Parent	Half-Life, Years	Stable End Product
$4n$	Thorium	$^{232}_{90}Th$	1.39×10^{10}	$^{208}_{82}Pb$
$4n + 1$	Neptunium	$^{237}_{93}Np$	2.25×10^{6}	$^{209}_{83}Bi$
$4n + 2$	Uranium	$^{238}_{92}U$	4.47×10^{9}	$^{206}_{82}Pb$
$4n + 3$	Actinium	$^{235}_{92}U$	7.07×10^{8}	$^{207}_{82}Pb$

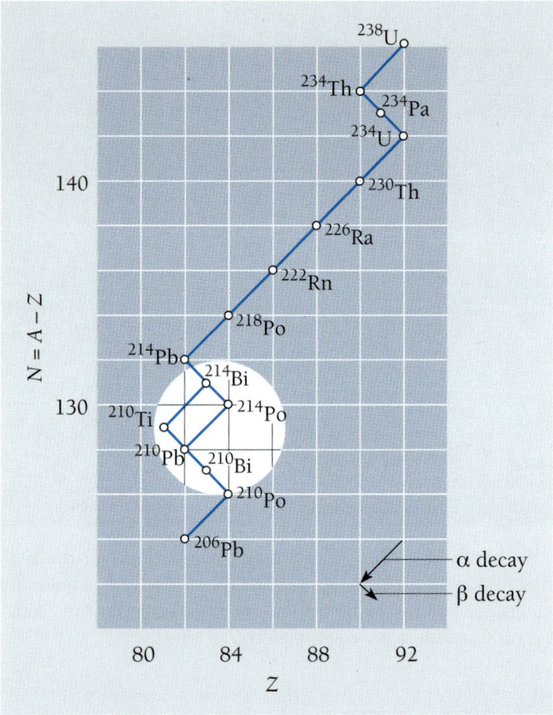

Figure 12.7 The uranium decay series ($A = 4n + 2$). The decay of $^{214}_{83}Bi$ may proceed either by alpha emission and then beta emission or in the reverse order.

About Scientists

Marie Sklodowska Curie (1867–1934) was born in Poland, at that time under Russia's oppressive domination. Following high school, she worked as a governess until she was twenty-four so that she could study science in Paris, where she had barely enough money to survive. In 1894 Marie married Pierre Curie, eight years older and already a noted physicist. In 1897, just after the birth of her daughter Irene (who was to win a Nobel Prize in physics herself in 1935), Marie began to investigate the newly discovered phenomenon of radioactivity—her word—for her doctoral thesis.

The year before, Becquerel had found that uranium emitted a mysterious radiation. Marie, after a search of all the known elements, learned that thorium did so as well. She then examined various minerals for radioactivity. Her studies showed that the uranium ore pitchblende was far more radioactive than its uranium content would suggest. Marie and Pierre together went on to identify first polonium, named for her native Poland, and then radium as the sources of the additional activity. With the primitive facilities that were all they could afford (they had to use their own money), they had succeeded by 1902 in purifying a tenth of a gram of radium from several tons of ore, a task that involved immense physical as well as intellectual labor.

Together with Becquerel, the Curies shared the 1903 Nobel Prize in physics. Pierre ended his acceptance speech with these words: "One may also imagine that in criminal hands radium might become very dangerous, and here one may ask if humanity has anything to gain by learning the secrets of nature, if it is ready to profit from them, or if this knowledge is not harmful. . . . I am among those who think . . . that humanity will obtain more good than evil from the new discoveries."

In 1906 Pierre was struck and killed by a horse-drawn carriage in a Paris street. Marie continued work on radioactivity, still in an inadequate laboratory, and won the Nobel Prize in chemistry in 1911. Not until her scientific career was near an end did she have proper research facilities. Even before Pierre's death, both Curies had suffered from ill health because of their exposure to radiation, and much of Marie's later life was marred by radiation-induced ailments, including the leukemia from which she died.

Several alpha-radioactive nuclides whose atomic numbers are less than 82 are found in nature, though they are not very abundant.

The intermediate members of each decay series have much shorter half-lives than their parent nuclide. As a result, if we start with a sample of N_A nuclei of a parent nuclide A, after a period of time an equilibrium situation will come about in which each successive daughter B, C, . . . decays at the same rate as it is formed. Thus the activities R_A, R_B, R_C, . . . are all equal at equilibrium, and since $R = \lambda N$ we have

Radioactive equilibrium

$$N_A\lambda_A = N_B\lambda_B = N_C\lambda_C = \cdots \qquad (12.9)$$

Each number of atoms N_A, N_B, N_C, . . . decreases exponentially with the decay constant λ_A of the parent nuclide, but Eq. (12.9) remains valid at any time. Equation (12.9) can be used to establish the decay constant (or half-life) of any member of the series if the decay constant of another member and their relative proportions in a sample are known.

Example 12.6

The atomic ratio between the uranium isotopes ^{238}U and ^{234}U in a mineral sample is found to be 1.8×10^4. The half-life of ^{234}U is $T_{1/2}(234) = 2.5 \times 10^5$ y. Find the half-life of ^{238}U.

Solution

Since $T_{1/2} = 0.693/\lambda$, from Eq. (12.9) we have

$$T_{1/2}(238) = \frac{N(238)}{N(234)} T_{1/2}(234)$$

$$= (1.8 \times 10^4)(2.5 \times 10^5 \text{ y}) = 4.5 \times 10^9 \text{ y}$$

This method is convenient for finding the half-lives of very long-lived and very short-lived radionuclides that are in equilibrium with other radionuclides whose half-lives are easier to measure.

12.4 ALPHA DECAY

Impossible in classical physics, it nevertheless occurs

Because the attractive forces between nucleons are of short range, the total binding energy in a nucleus is approximately proportional to its mass number A, the number of nucleons it contains. The repulsive electric forces between protons, however, are of unlimited range, and the total disruptive energy in a nucleus is approximately proportional to Z^2 [Eq. (11.12)]. Nuclei which contain 210 or more nucleons are so large that the short-range nuclear forces that hold them together are barely able to counterbalance the mutual repulsion of their protons. Alpha decay occurs in such nuclei as a means of increasing their stability by reducing their size.

Why are alpha particles emitted rather than, say, individual protons or ^3_2He nuclei? The answer follows from the high binding energy of the alpha particle. To escape from a nucleus, a particle must have kinetic energy, and only the alpha-particle mass is sufficiently smaller than that of its constituent nucleons for such energy to be available.

To illustrate this point, we can compute, from the known masses of each particle and the parent and daughter nuclei, the energy Q released when various particles are emitted by a heavy nucleus. This is given by

Disintegration energy

$$Q = (m_i - m_f - m_x)c^2 \tag{12.10}$$

where m_i = mass of initial nucleus
m_f = mass of final nucleus
m_x = particle mass

We find that the emission of an alpha particle in some cases is energetically possible, but other decay modes would need energy supplied from outside the nucleus. Thus alpha decay in $^{232}_{92}U$ is accompanied by the release of 5.4 MeV, while 6.1 MeV would be needed for a proton to be emitted and 9.6 MeV for a $^{3}_{2}He$ nucleus to be emitted. The observed disintegration energies in alpha decay agree with the predicted values based upon the nuclear masses involved.

The kinetic energy KE_α of the emitted alpha particle is never quite equal to the disintegration energy Q because, since momentum must be conserved, the nucleus recoils with a small amount of kinetic energy when the alpha particle emerges. It is easy to show (see Exercise 23) from momentum and energy conservation that KE_α is related to Q and the mass number A of the original nucleus by

Alpha-particle energy

$$KE_\alpha \approx \frac{A-4}{A} Q \tag{12.11}$$

The mass numbers of nearly all alpha emitters exceed 210, and so most of the disintegration energy appears as the kinetic energy of the alpha particle.

Example 12.7

The polonium isotope $^{210}_{84}Po$ is unstable and emits a 5.30-MeV alpha particle. The atomic mass of $^{210}_{84}Po$ is 209.9829 u and that of $^{4}_{2}He$ is 4.0026 u. Identify the daughter nuclide and find its atomic mass.

Solution

(*a*) The daughter nuclide has an atomic number of $Z = 84 - 2 = 82$ and a mass number of $A = 210 - 4 = 206$. Since $Z = 82$ is the atomic number of lead, the symbol of the daughter nuclide is $^{206}_{82}Pb$.
(*b*) The disintegration energy that follows from an alpha-particle energy of 5.30 MeV is

$$Q = \frac{A}{A-4} KE_\alpha = \left(\frac{210}{210-4}\right)(5.30 \text{ MeV}) = 5.40 \text{ MeV}$$

The mass equivalent of this Q value is

$$m_Q = \frac{5.40 \text{ MeV}}{931 \text{ MeV/u}} = 0.0058 \text{ u}$$

Hence

$$m_f = m_i - m_\alpha - m_Q = 209.9829 \text{ u} - 4.0026 \text{ u} - 0.0058 \text{ u} = 205.9745 \text{ u}$$

Tunnel Theory of Alpha Decay

While a heavy nucleus can, in principle, spontaneously reduce its bulk by alpha decay, there remains the problem of *how* an alpha particle can actually escape the nucleus. Figure 12.8 is a plot of the potential energy U of an alpha particle as a function of its distance r from the center of a certain heavy nucleus. The height of the potential barrier is about 25 MeV, which is equal to the work that must be done against the repulsive electric force to bring an alpha particle from infinity to a position adjacent to the nucleus but just outside the range of its attractive forces. We may therefore regard an alpha particle in such a nucleus as being inside a box whose walls require an energy of 25 MeV to be surmounted. However, decay alpha particles have energies that range from 4 to 9 MeV, depending on the particular nuclide involved—16 to 21 MeV short of the energy needed for escape.

Although alpha decay is inexplicable classically, quantum mechanics provides a straightforward explanation. In fact, the theory of alpha decay, developed independently in 1928 by Gamow and by Gurney and Condon, was greeted as an especially striking confirmation of quantum mechanics.

In the Appendix to this chapter we shall find that even a simplified treatment of the problem of the escape of an alpha particle from a nucleus gives results in agreement with experiment. Gurney and Condon made these observations in their paper: "It has hitherto been necessary to postulate some special arbitrary 'instability' of the nucleus; but in the following note it is pointed out that disintegration is a natural consquence of the laws of quantum mechanics without any special hypothesis. . . . Much has been written about the explosive violence with which the α-particle is hurled from its place in the nucleus. But from the process pictured above, one would rather say that the particle slips away almost unnoticed."

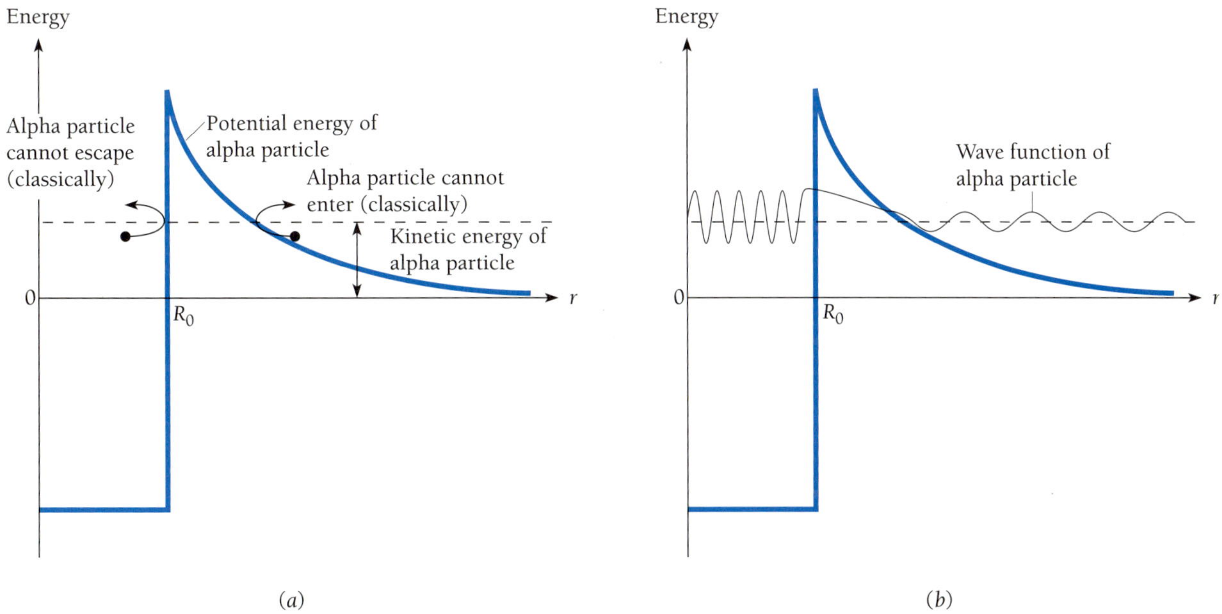

Figure 12.8 (*a*) In classical physics, an alpha particle whose kinetic energy is less than the height of the potential barrier around a nucleus cannot enter or leave the nucleus, whose radius is R_0. (*b*) In quantum physics, such an alpha particle can tunnel through the potential barrier with a probability that decreases with the height and thickness of the barrier.

About Scientists

George Gamow (1904–1968), born and educated in Russia, did his first important work at Göttingen in 1928 when he developed the theory of alpha decay, the first application of quantum mechanics to nuclear physics. (Edward U. Condon and Ronald W. Gurney, working together, arrived at the same theory independently of Gamow at about the same time). In 1929 he proposed the liquid-drop model of the nucleus. After periods in Copenhagen, Cambridge, and Leningrad, Gamow went to the United States in 1934 where he was first at George Washington University and later at the University of Colorado. In 1936 Gamow collaborated with Edward Teller on an extension of Fermi's theory of beta decay. Much of his later research was concerned with astrophysics, notably on the evolution of stars, where he showed that as a star uses up its supply of hydrogen in thermonuclear reactions, it becomes hotter, not cooler. Gamow also did important work on the origin of the universe (he and his students predicted the 2.7-K remnant radiation from the Big Bang) and on the formation of the elements. His books for the general public introduced many people to the concepts of modern physics.

The basic notions of this theory are:

1 An alpha particle may exist as an entity within a heavy nucleus.
2 Such a particle is in constant motion and is held in the nucleus by a potential barrier.
3 There is a small—but definite—likelihood that the particle may tunnel through the barrier (despite its height) each time a collision with it occurs.

According to the last assumption, the decay probability per unit time λ can be expressed as

Decay constant

$$\lambda = \nu T \tag{12.12}$$

Here ν is the number of times per second an alpha particle within a nucleus strikes the potential barrier around it and T is the probability that the particle will be transmitted through the barrier.

If we suppose that at any moment only one alpha particle exists as such in a nucleus and that it moves back and forth along a nuclear diameter,

Collision frequency

$$\nu = \frac{v}{2R_0} \tag{12.13}$$

where v is the alpha-particle velocity when it eventually leaves the nucleus and R_0 is the nuclear radius. Typical values of v and R_0 might be 2×10^7 m/s and 10^{-14} m respectively, so that

$$\nu \approx 10^{21}\ \text{s}^{-1}$$

The alpha particle knocks at its confining wall 10^{21} times per second and yet may have to wait an average of as much as 10^{10} y to escape from some nuclei!

As developed in the Appendix to this chapter, the tunnel theory for the decay constant λ gives the formula

Alpha decay constant

$$\log_{10} \lambda = \log_{10}\left(\frac{v}{2R_0}\right) + 1.29Z^{1/2}R_0^{1/2} - 1.72ZE^{-1/2} \tag{12.14}$$

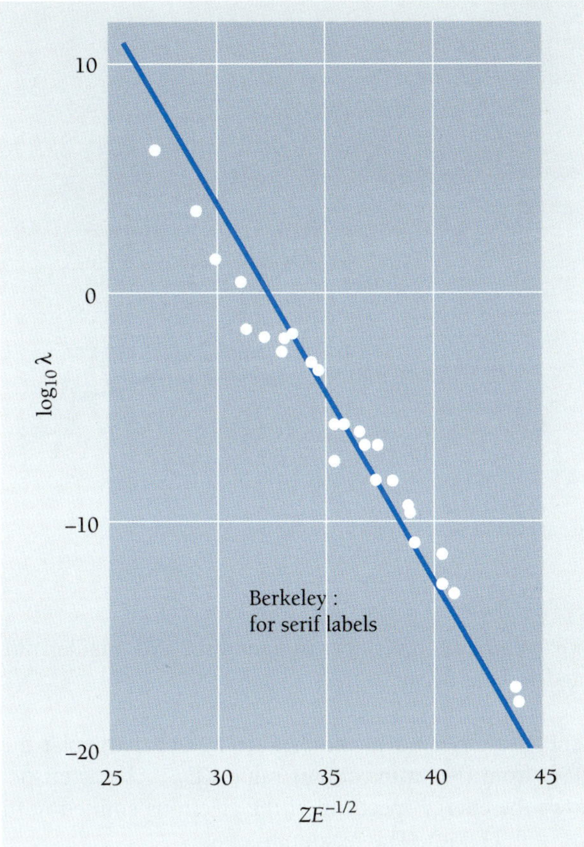

Figure 12.9 Experimental verification of the theory of alpha decay.

Here v is the alpha-particle velocity in m/s and E its energy in MeV, R_0 is the nuclear radius in fermis, and Z is the atomic number of the daughter nucleus. Figure 12.9 is a plot of $\log_{10} \lambda$ versus $ZE^{-1/2}$ for a number of alpha-radioactive nuclides. The straight line fitted to the experimental data has the -1.72 slope predicted throughout the entire range of decay constants. We can use the position of the line to determine R_0, the nuclear radius. The result is just about what is obtained from nuclear scattering experiments. This approach thus constitutes an independent means of determining nuclear sizes.

Equation (12.14) predicts that the decay constant λ, and hence the half-life, should vary strongly with the alpha-particle energy E. This is indeed the case. The slowest decay is that of ${}^{232}_{90}Th$, whose half-life is 1.3×10^{10} y, and the fastest decay is that of ${}^{212}_{84}Po$, whose half-life is 3.0×10^{-7} s. Whereas its half-life is 10^{24} greater, the alpha-particle energy of ${}^{232}_{90}Th$ (4.05 MeV) is only about half that of ${}^{212}_{84}Po$ (8.95 MeV).

12.5 BETA DECAY

Why the neutrino should exist and how it was discovered

Like alpha decay, beta decay is a means whereby a nucleus can alter its composition to become more stable. Also like alpha decay, beta decay has its puzzling aspects: the

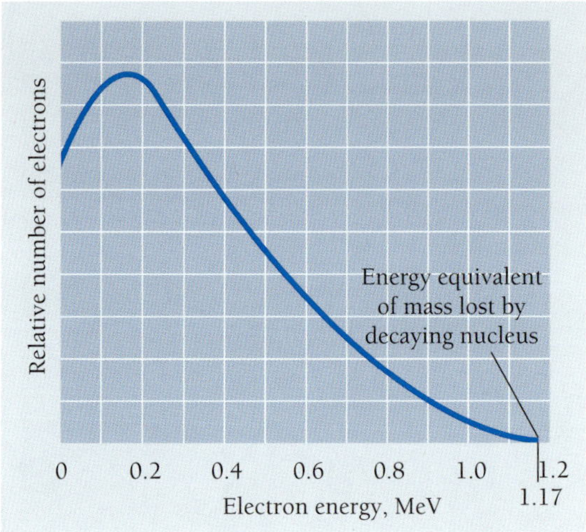

Figure 12.10 Energy spectrum of electrons from the beta decay of ${}^{210}_{83}Bi$.

conservation principles of energy, linear momentum, and angular momentum are all apparently violated in beta decay.

1 The electron energies observed in the beta decay of a particular nuclide are found to vary *continuously* from 0 to a maximum value KE_{max} characteristic of the nuclide. Figure 12.10 shows the energy spectrum of the electrons emitted in the beta decay of ${}^{210}_{83}Bi$; here KE_{max} = 1.17 MeV. The maximum energy

$$E_{max} = mc^2 + KE_{max}$$

carried off by the decay electron is equal to the energy equivalent of the mass difference between the parent and daughter nuclei. Only seldom, however, is an emitted electron found with an energy of KE_{max}.

2 When the directions of the emitted electrons and of the recoiling nuclei are observed, they are almost never exactly opposite as required for linear momentum to be conserved.

3 The spins of the neutron, proton, and electron are all $\frac{1}{2}$. If beta decay involves just a neutron becoming a proton and an electron, spin (and hence angular momentum) is not conserved.

In 1930 Pauli proposed a "desperate remedy": if an uncharged particle of small or zero rest mass and spin $\frac{1}{2}$ is emitted in beta decay together with the electron, the above discrepancies would not occur. This particle, later called the **neutrino** ("little neutral one") by Fermi, would carry off an energy equal to the difference between KE_{max} and the actual KE of the electron (the recoiling nucleus carries away negligible KE). The neutrino's linear momentum also exactly balances those of the electron and the recoiling daughter nucleus.

Subsequently it was found that *two* kinds of neutrinos are involved in beta decay, the neutrino itself (symbol ν) and the **antineutrino** (symbol $\overline{\nu}$). The distinction between them is discussed in Chap. 13. In ordinary beta decay it is an antineutrino that is emitted:

Beta decay $$n \rightarrow p + e^- + \overline{\nu} \qquad (12.15)$$

The neutrino hypothesis has turned out to be completely successful. The neutrino mass was not expected to be more than a small fraction of the electron mass because KE_{max} is observed to be equal (within experimental error) to the value calculated from the parent-daughter mass difference. The neutrino mass is now believed to be the mass equivalent of at most a few electronvolts. The interaction of neutrinos with matter is extremely feeble. Lacking charge and mass, and not electromagnetic in nature as is the photon, the neutrino can pass unimpeded through vast amounts of matter. A neutrino would have to pass through over 100 *light-years* of solid iron on the average before interacting! The only interaction with matter a neutrino can experience is through a process called inverse beta decay, which we shall consider shortly. Neutrinos are believed to outnumber protons in the universe by about a billion to one.

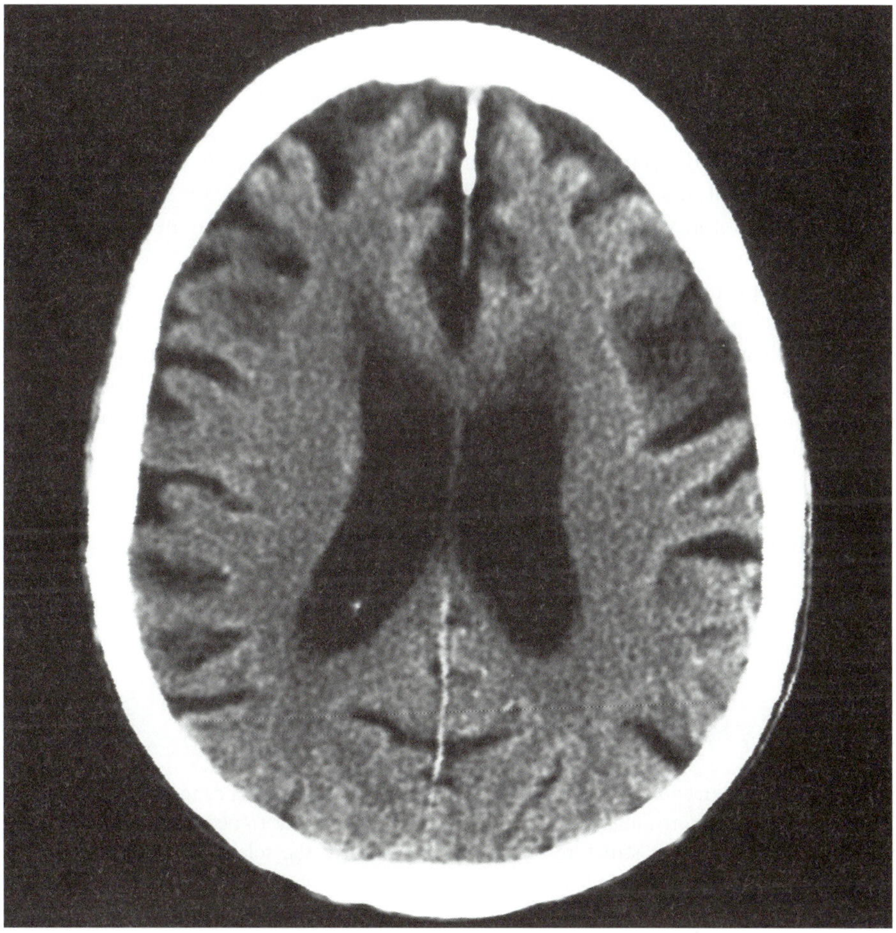

A positron emission tomography (PET) scan of the brain of a patient with Alzheimer's disease. The lighter the area, the higher the rate of metabolic activity. In PET, a suitable positron-emitting radionuclide (here the oxygen isotope ^{15}O) is injected and allowed to circulate in a patient's body. When a positron encounters an electron, which it does almost at once after being emitted, both are annihilated. From the directions of the resulting pair of gamma rays the location of the annihilation, and hence of the emitting nucleus, can be found. In this way, a map that is accurate to several millimeters of the concetration of the radionuclide can be built up. In a normal brain, metabolic activity produces a similar PET pattern in each hemisphere; here, the irregular appearance of the scan indicates degeneration of brain tissue.

Positions were discovered in 1932 and two years later were found to be spontaneously emitted by certain nuclei. The properties of the positron are identical with those of the electron except that it carries a charge of $+e$ instead of $-e$. Positron emission corresponds to the conversion of a nuclear proton into a neutron, a positron, and a neutrino:

Positron emission $$p \rightarrow n + e^+ + \nu \tag{12.16}$$

Whereas a neutron outside a nucleus undergoes negative beta decay into a proton (half-life = 10 min 16 s) because its mass is greater than that of the proton, the lighter proton cannot be transformed into a neutron except within a nucleus. Positron emission leads to a daughter nucleus of lower atomic number Z while leaving the mass number A unchanged.

Closely connected with positron emission is electron capture. In electron capture a nucleus absorbs one of its inner atomic electrons, with the result that a nuclear proton becomes a neutron and a neutrino is emitted:

Electron capture $$p + e^- \rightarrow n + \nu \tag{12.17}$$

Usually the absorbed electron comes from the K shell, and an x-ray photon is emitted when one of the atom's outer electrons falls into the resulting vacant state. The wavelength of the photon will be one of those characteristic of the daughter element, not of the original one, and the process can be recognized on this basis.

Electron capture is competitive with positron emission since both processes lead to the same nuclear transformation. Electron capture occurs more often than positron emission in heavy nuclides because the electrons in such nuclides are relatively close to the nucleus, which promotes their interaction with it. Since nearly all the unstable nuclei found in nature are of high Z, positron emission was not discovered until several decades after electron emission had been established.

Inverse Beta Decay

By comparing Eqs. (12.16) and (12.17) we see that electron capture by a nuclear proton is equivalent to a proton's emission of a positron. Similarly the absorption of an

The Weak Interaction

The nuclear interaction that holds nucleons together to form nuclei cannot account for beta decay. Another short-range fundamental interaction turns out to be responsible: the **weak interaction.** Insofar as the structure of matter is concerned, the role of the weak interaction seems to be confined to causing beta decays in nuclei whose neutron/proton ratios are not appropriate for stability. This interaction also affects elementary particles that are not part of a nucleus and can lead to their transformation into other particles. The name "weak interaction" arose because the other short-range force affecting nucleons is extremely strong, as the high binding energies of nuclei attest. The gravitational interaction is weaker than the weak interaction at distances where the latter is a factor.

Thus four fundamental interactions are apparently sufficient to govern the structure and behavior of the entire physical universe, from atoms to galaxies of stars. In order of increasing strength they are gravitational, weak nuclear, electromagnetic, and strong nuclear. These interactions and how they are related to one another and to the origin and evolution of the universe will be discussed in Chap. 13.

antineutrino is equivalent to the emission of a neutrino, and vice versa. The latter reactions are called **inverse beta decays:**

Inverse beta decay

$$p + \bar{\nu} \rightarrow n + e^+ \tag{12.18a}$$

$$n + \nu \rightarrow p + e^- \tag{12.18b}$$

Inverse beta decays have extremely low probabilities, which is why neutrinos and antineutrinos are able to pass through such vast amounts of matter, but these probabilities are not zero. Starting in 1953, a series of experiments was carried out by F. Reines, C. L. Cowan, and others to detect the considerable flux of neutrinos (actually antineutrinos) from the beta decays that occur in a nuclear reactor. A tank of water containing a cadmium compound in solution supplied the protons which were to interact with the incident neutrinos. Surrounding the tank were gamma-ray detectors. Immediately after a proton absorbed a neutrino to yield a positron and a neutron, as in Eq. (12.18*a*), the positron encountered an electron and both were annihilated. The gamma-ray detectors responded to the resulting pair of 0.51-MeV photons. Meanwhile the newly formed neutron migrated through the solution until, after a few microseconds, it was captured by a cadmium nucleus. The new, heavier cadmium nucleus then released about 8 MeV of excitation energy divided among three or four photons, which were picked up by the detectors several microseconds after those from the positron-electron annihilation. In principle, then, the arrival of this sequence of photons at the detector is a sure sign that the reaction of Eq. (12.18*a*) has occurred. To avoid any uncertainty, the experiment was performed with the reactor alternately on and off, and the expected variation in the frequency of neutrino-capture events was observed. In this way the neutrino hypothesis was confirmed.

12.6 GAMMA DECAY

Like an excited atom, an excited nucleus can emit a photon

A nucleus can exist in states whose energies are higher than that of its ground state, just as an atom can. An excited nucleus is denoted by an asterisk after its usual symbol, for instance $^{87}_{38}\text{Sr}^*$. Excited nuclei return to their ground states by emitting photons whose energies correspond to the energy differences between the various initial and final states in the transitions involved. The photons emitted by nuclei range in energy up to several MeV, and are traditionally called **gamma rays.**

A simple example of the relationship between energy levels and decay schemes is shown in Fig. 12.11, which pictures the beta decay of $^{27}_{12}\text{Mg}$ to $^{27}_{13}\text{Al}$. The half-life of the decay is 9.5 min, and it may take place to either of the two excited states of $^{27}_{13}\text{Al}$. The resulting $^{27}_{13}\text{Al}^*$ nucleus then undergoes one or two gamma decays to reach the ground state.

As an alternative to gamma decay, an excited nucleus in some cases may return to its ground state by giving up its excitation energy to one of the atomic electrons around it. While we can think of this process, which is known as **internal conversion,** as a kind of photoelectric effect in which a nuclear photon is absorbed by an atomic electron, it is in better accord with experiment to regard internal conversion as representing a direct transfer of excitation energy from a nucleus to an electron. The emitted electron has a kinetic energy equal to the lost nuclear excitation energy minus the binding energy of the electron in the atom.

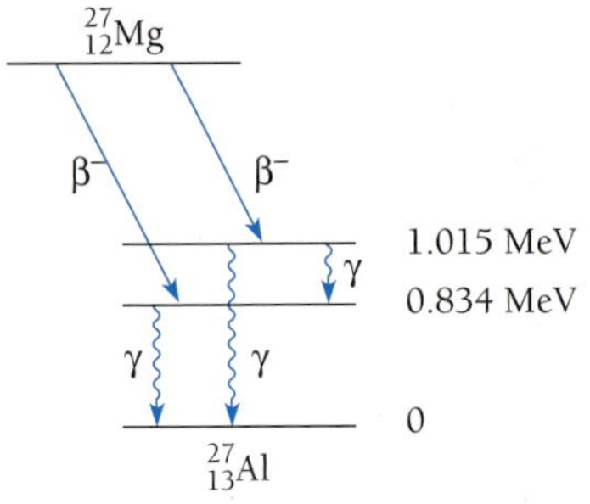

Figure 12.11 Successive beta and gamma emissions in the decay of $^{27}_{12}$Mg to $^{27}_{13}$Al via $^{27}_{13}$Al*.

Most excited nuclei have very short half-lives against gamma decay, but a few remain excited for as long as several hours. The analogy with metastable atomic states is a close one. A long-lived excited nucleus is called an **isomer** of the same nucleus in its ground state. The excited nucleus $^{87}_{38}$Sr* has a half-life of 2.8 h and is accordingly an isomer of $^{87}_{38}$Sr.

12.7 CROSS SECTION

A measure of the likelihood of a particular interaction

Most of what is known about atomic nuclei has come from experiments in which energetic bombarding particles collide with stationary target nuclei. A very convenient way to express the probability that a bombarding particle will interact in a certain way with a target particle employs the idea of **cross section** that was introduced in the Appendix to Chap. 4 in connection with the Rutherford scattering experiment.

What we do is imagine each target particle as presenting a certain area, called its cross section, to the incident particles, as in Fig. 12.12. Any incident particle that is directed at this area interacts with the target particle. Hence the greater the cross section, the greater the likelihood of an interaction. The interaction cross section of a target particle varies with the nature of the process involved and with the energy of the incident particle; it may be greater or less than the geometrical cross section of the particle.

Suppose we have a slab of some material whose area is A and whose thickness is dx (Fig. 12.13). If the material contains n atoms per unit volume, a total of $nA\,dx$ nuclei is in the slab, since its volume is $A\,dx$. Each nucleus has a cross section of σ for some particular interaction, so that the aggregate cross section of all the nuclei in the slab is $nA\sigma\,dx$. If there are N incident particles in a bombarding beam, the number dN that interact with nuclei in the slab is therefore specified by

$$\frac{\text{Interacting particles}}{\text{Incident particles}} = \frac{\text{aggregate cross section}}{\text{target area}}$$

$$\frac{dN}{N} = \frac{nA\sigma\,dx}{A}$$

Cross section

$$= n\sigma\,dx \tag{12.19}$$

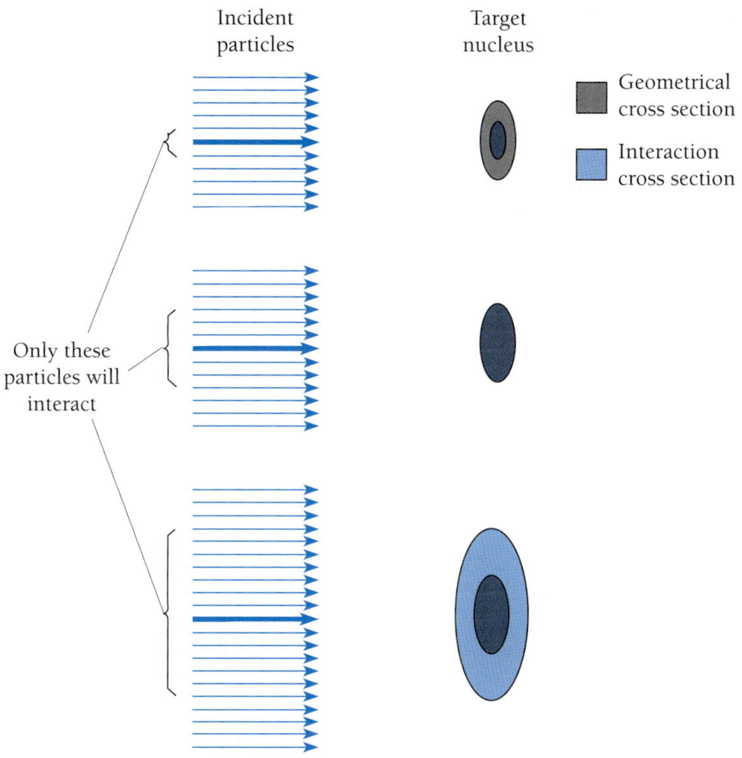

Figure 12.12 A geometrical interpretation of the concept of cross section. The interaction cross section may be smaller than, equal to, or larger than the geometrical cross section. The cross section of a nucleus for a particular interaction is a mathematical way to express the probability that the interaction will occur when a certain particle is incident on the nucleus; the diagram here is nothing more than a helpful visualization.

Now we consider the same beam of particles incident on a slab of finite thickness x. If each particle can interact only once, dN particles may be thought of as being removed from the beam in passing through the first dx of the slab. Hence we need a minus sign in Eq. (12.19), which becomes

$$-\frac{dN}{N} = n\sigma\, dx$$

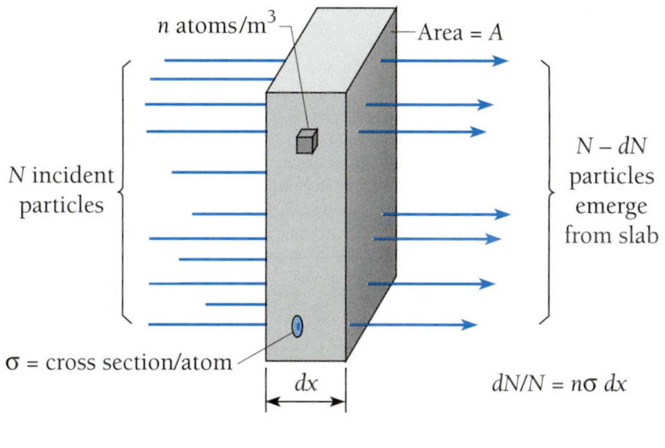

Figure 12.13 The relationship between cross section and beam intensity.

Denoting the initial number of incident particles by N_0, we have

$$\int_{N_0}^{N} \frac{dN}{N} = -n\sigma \int_0^x dx$$

$$\ln N - \ln N_0 = -n\sigma x \tag{12.20}$$

Surviving particles

$$N = N_0 e^{-n\sigma x}$$

The number of surviving particles N decreases exponentially with increasing slab thickness x.

The customary unit for nuclear cross sections is the **barn,** where

$$1 \text{ barn} = 1 \text{ b} = 10^{-28} \text{ m}^2 = 100 \text{ fm}^2$$

Although not an SI unit, the barn is handy because it is of the same order of magnitude as the geometrical cross section of a nucleus. The name comes from a more familiar target cross-sectional area, the side of a barn.

The cross sections for most nuclear reactions depend on the energy of the incident particle. Figure 12.14 shows how the neutron-capture cross section of $^{113}_{48}Cd$ varies with neutron energy. This reaction, in which the absorption of a neutron is followed by the emission of a gamma ray, is usually expressed in shorthand form as

$$^{113}\text{Cd}(n, \gamma)^{114}\text{Cd}$$

The narrow peak at 0.176 eV is a resonance effect associated with an excited state in the ^{114}Cd nucleus. Although the ^{113}Cd isotope constitutes only 12 percent of natural cadmium, its capture cross sections for slow neutrons are so great that cadmium is widely used in control rods for nuclear reactors.

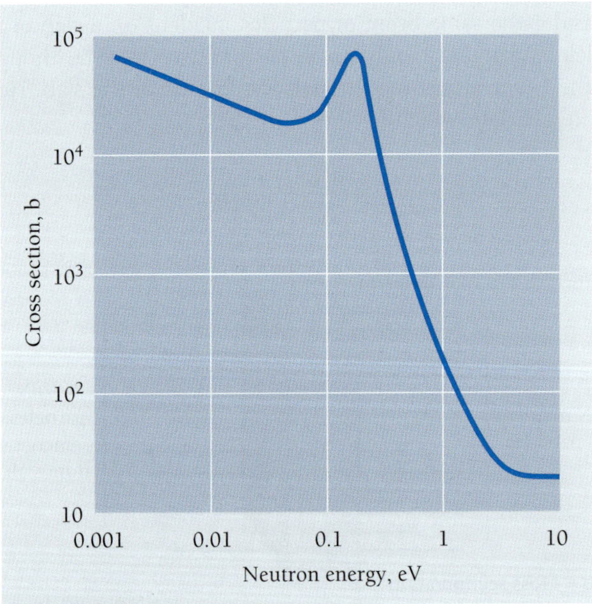

Figure 12.14 The cross section for the reaction $^{113}Cd(n, \gamma)^{114}Cd$ varies strongly with neutron energy. In this reaction a neutron is absorbed and a gamma ray is emitted.

Example 12.8

A neutron passing through a body of matter and not absorbed in a nuclear reaction undergoes frequent elastic collisions in which some of its kinetic energy is given up to nuclei in its path. Very soon the neutron reaches thermal equilibrium, which means that it is equally likely to gain or lose energy in further collisions. At room temperature such a **thermal neutron** has an average energy of $\frac{3}{2}kT = 0.04$ eV and a most probable energy of $kT = 0.025$ eV; the latter figure is usually quoted as the energy of such neutrons.

The cross section of ^{113}Cd for capturing thermal neutrons is 2×10^4 b, the mean atomic mass of natural cadmium is 112 u, and its density is 8.64 g/cm^3 = 8.64×10^3 kg/m^3. (*a*) What fraction of an incident beam of thermal neutrons is absorbed by a cadmium sheet 0.1 mm thick? (*b*) What thickness of cadmium is needed to absorb 99 percent of an incident beam of thermal neutrons?

Solution

(*a*) Since ^{113}Cd constitutes 12 percent of natural cadmium, the number of ^{113}Cd atoms per cubic meter is

$$n = (0.12)\left[\frac{8.64 \times 10^3 \text{ kg/m}^3}{(112 \text{ u/atom})(1.66 \times 10^{-27} \text{ kg/u})}\right]$$

$$= 5.58 \times 10^{27} \text{ atoms/m}^3$$

The capture cross section is $\sigma = 2 \times 10^4 \text{ b} = 2 \times 10^{-24} \text{ m}^2$, so

$$n\sigma = (5.58 \times 10^{27} \text{ m}^{-3})(2 \times 10^{-24} \text{ m}^2) = 1.12 \times 10^4 \text{ m}^{-1}$$

From Eq. (12.20), $N = N_0 e^{-n\sigma x}$, so the fraction of incident neutrons that is absorbed is

$$\frac{N_0 - N}{N_0} = \frac{N_0 - N_0 e^{-n\sigma x}}{N_0} = 1 - e^{-n\sigma x}$$

Since $x = 0.1 \text{ mm} = 10^{-4}$ m here,

$$\frac{N_0 - N}{N_0} = 1 - e^{(-1.12\times 10^4 \text{ m}^{-1})(10^{-4} \text{ m})} = 0.67$$

Two-thirds of the incident neutrons are absorbed.

(*b*) Since we are given that 1 percent of the incident neutrons pass through the cadmium sheet, $N = 0.01N_0$ and

$$\frac{N}{N_0} = 0.01 = e^{-n\sigma x}$$

$$\ln 0.01 = -n\sigma x$$

$$x = \frac{-\ln 0.01}{n\sigma} = \frac{-\ln 0.01}{1.12 \times 10^4 \text{ m}^{-1}} = 4.1 \times 10^{-4} \text{ m} = 0.41 \text{ mm}$$

Cadmium is evidently a very efficient absorber of thermal neutrons.

The mean free path λ of a particle in a material is the average distance it can travel in the material before interacting there. Since $e^{-n\sigma x}\,dx$ is the probability that a particle interact in the interval dx at the distance x, we have, by the same reasoning as that used in Sec. 5.4,

Mean free path
$$\lambda = \frac{\int_0^\infty xe^{-n\sigma x}\,dx}{\int_0^\infty e^{-n\sigma x}\,dx} = \frac{1}{n\sigma} \tag{12.21}$$

Example 12.9

Find the mean free path of thermal neutrons in ^{113}Cd.

Solution

Since $n\sigma = 1.12 \times 10^4\ \text{m}^{-1}$ here, the mean free path is

$$\lambda = \frac{1}{n\sigma} = \frac{1}{1.12 \times 10^4 \text{m}^{-1}} = 8.93 \times 10^{-5}\text{m} = 0.0893\ \text{mm}$$

Slow Neutron Cross Sections

Although neutrons interact with nuclei only through short-range nuclear forces, reaction cross sections for slow neutrons can be much greater than the geometrical cross sections of the nuclei involved.The geometrical cross section of ^{113}Cd is 1.06 b, for example but its cross section for the capture of thermal neutrons is 20,000 b.

When we recall the wave nature of a moving neutron, though, such discrepancies become less bizarre. The slower a neutron, the greater its de Broglie wavelength λ and the larger the region of space through which we must regard it as being spread out. A fast neutron with a wavelength smaller than the radius R of a target nucleus behaves more or less like a particle when it interacts with the nucleus. The cross section is then approximately geometrical, in the neighborhood of πR^2. Less energetic neutrons behave more like wave packets and interact over larger areas. Although cross sections in the latter case of $\pi\lambda^2$ (which is over 10^7 b for a thermal neutron) are rare, cross sections for nuclear reactions with slow neutrons greatly exceed πR^2, as we have seen.

Reaction Rate

When we know the cross section for a nuclear reaction caused by a beam of incident particles, we can find the rate $\Delta N/\Delta t$ at which the reaction occurs in a given sample of the target material. Let us consider a sample in the form of a slab of area A and thickness x that contains n atoms/m^3, with the particle beam incident normal to one face of the slab. From Eq. (12.20)

$$\frac{\Delta N}{\Delta t} = \frac{N_0 - N}{\Delta t} = \frac{N_0}{\Delta t}(1 - e^{-n\sigma x})$$

If the slab is thin enough so that none of the nuclear cross sections overlaps any others, $n\sigma x \ll 1$. Since $e^{-y} = 1 - y$ for $y \ll 1$, in this case

$$\frac{\Delta N}{\Delta t} = \left(\frac{N_0}{\Delta t}\right) n\sigma x$$

The flux Φ of the beam is the number of incident particles per unit area per unit time, so $\Phi A = N_0/\Delta t$ is their number per unit time. Because Ax is the volume of the sample, the total number of atoms it contains is $n' = nAx$. The reaction rate is therefore just

Reaction rate
$$\frac{\Delta N}{\Delta t} = (\Phi A)(n\sigma x) = \Phi n'\sigma \tag{12.22}$$

Example 12.10

Natural gold consists entirely of the isotope $^{197}_{79}Au$ whose cross section for thermal neutron capture is 99 b. When $^{197}_{79}Au$ absorbs a neutron, the product is $^{198}_{79}Au$ which is beta-radioactive with a half-life of 2.69 d. How long should a 10.0-mg gold foil be exposed to a flux of 2.00×10^{16} neutrons/m$^2 \cdot$ s in order for the sample to have an activity of 200 μCi? Assume that the irradiation period is much shorter than the half-life of $^{198}_{79}Au$ so the decays that occur during the irradiation can be neglected.

Solution

The decay constant of $^{198}_{79}Au$ is

$$\lambda = \frac{0.693}{(2.69 \text{ d})(86{,}400 \text{ s/d})} = 2.98 \times 10^{-6}\text{s}^{-1}$$

The required activity of $R = \Delta N\lambda = 200\mu\text{Ci} = 2.00 \times 10^{-4}$ Ci means that the number of $^{198}_{79}Au$ atoms must be

$$\Delta N = \frac{R}{\lambda} = \frac{(2.00 \times 10^{-4}\text{ Ci})(3.70 \times 10^{10}\text{ s}^{-1}/\text{Ci})}{2.98 \times 10^{-6}\text{s}^{-1}} = 2.48 \times 10^{12} \text{ atoms}$$

The number of atoms in 10.0 mg = 1.00×10^{-5} kg of $^{197}_{79}Au$ is

$$n' = \frac{1.00 \times 10^{-5}\text{ kg}}{(197 \text{ u/atom})(1.66 \times 10^{-27}\text{ kg/u})} = 3.06 \times 10^{19} \text{ atoms}$$

From Eq. (12.22) we find that

$$\Delta t = \frac{\Delta N}{\Phi n'\sigma} = \frac{2.48 \times 10^{12}\text{ atoms}}{(2.00 \times 10^{16}\text{ neutrons/m}^2 \cdot \text{s})(3.06 \times 10^{19}\text{ atoms})(99 \times 10^{-28}\text{ m}^2)}$$
$$= 409 \text{ s} = 6 \text{ min } 49 \text{ s}$$

As we assumed, $\Delta t \ll T_{1/2}$.

12.8 NUCLEAR REACTIONS

In many cases, a compound nucleus is formed first

When two nuclei come close together, a **nuclear reaction** can occur that results in new nuclei being formed. Nuclei are positively charged and the repulsion between them keeps them beyond the range where they can interact unless they are moving very fast to begin with. In the sun and other stars, whose internal temperatures range up to millions of kelvins, many nuclei present have high enough speeds for reactions to be frequent. Indeed, the reactions provide the energy that maintains these temperatures.

In the laboratory, it is easy to produce nuclear reactions on a small scale, either with alpha particles from radionuclides or with protons or heavier nuclei accelerated in various ways. But only one type of nuclear reaction has as yet proved to be a practical source of energy on the earth, namely the fission of certain nuclei when struck by neutrons.

Many nuclear reactions actually involve two separate stages. In the first, an incident particle strikes a target nucleus and the two combine to form a new nucleus, called a **compound nucleus,** whose atomic and mass numbers are respectively the sum of the atomic numbers of the original particles and the sum of their mass numbers. This idea was proposed by Bohr in 1936.

A compound nucleus has no memory of how it was formed, since its nucleons are mixed together regardless of origin and the energy brought into it by the incident particle is shared among all of them. A given compound nucleus may therefore be formed in a variety of ways. To illustrate this, Fig. 12.15 shows six reactions whose product is the compound nucleus ${}^{14}_{7}\text{N}^*$. (The asterisk signifies an excited state. Compound nuclei are always excited by amounts equal to at least the binding energies of the incident particles in them.) Compound nuclei have lifetimes on the order of 10^{-16} s or so. Although too short to permit actually observing such nuclei directly, such lifetimes are long relative to the 10^{-21} s or so a nuclear particle with an energy of several MeV would need to pass through a nucleus.

A given compound nucleus may decay in one or more ways, depending on its excitation energy. Thus ${}^{14}_{7}\text{N}^*$ with an excitation energy of, say, 12 MeV can decay in any of the four ways shown in Fig. 12.15. ${}^{14}_{7}\text{N}^*$ can also simply emit one or more gamma rays whose energies total 12 MeV. However, it *cannot* decay by the emission of a triton (${}^{3}_{1}\text{H}$) or a helium-3 (${}^{3}_{2}\text{He}$) particle since it does not have enough energy to liberate them. Usually a particular decay mode is favored by a compound nucleus in a specific excited state.

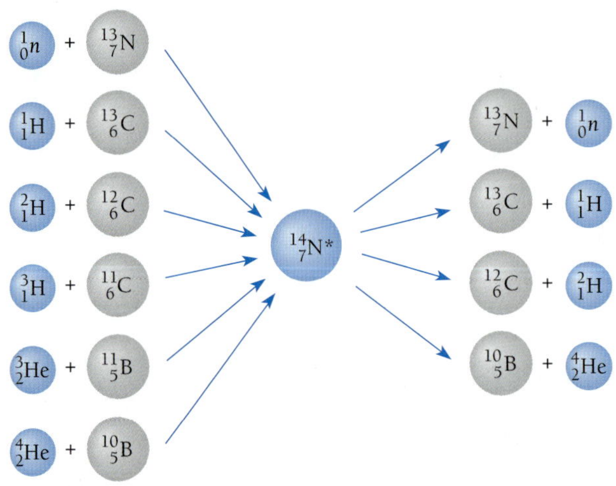

Figure 12.15 Six nuclear reactions whose product is the compound nucleus ${}^{14}_{7}\text{N}^*$ and four ways in which ${}^{14}_{7}\text{N}^*$ can decay if its excitation energy is 12 MeV. Other decay modes are possible if the excitation energy is greater, fewer are possible if this energy is less. In addition, ${}^{14}_{7}\text{N}^*$ can simply lose its excitation energy by emitting one or more gamma rays.

The formation and decay of a compound nucleus has an interesting interpretation on the basis of the liquid-drop nuclear model described in Sec. 11.5. In terms of this model, an excited nucleus is analogous to a drop of hot liquid, with the binding energy of the emitted particles corresponding to the heat of vaporization of the liquid molecules. Such a drop of liquid will eventually evaporate one or more molecules, thereby cooling down. The evaporation occurs when random fluctuations in the energy distribution within the drop cause a particular molecule to have enough energy to escape. Similarly, a compound nucleus persists in its excited state until a particular nucleon or group of nucleons happens to gain enough of the excitation energy to leave the nucleus. The time interval between the formation and decay of a compound nucleus fits in nicely with this picture.

Resonance

Information about the excited states of nuclei can be gained from nuclear reactions as well as from radioactive decay. The presence of an excited state may be detected by a peak in the cross section versus energy curve of a particular reaction, as in the neutron-capture reaction of Fig. 12.14. Such a peak is called a **resonance** by analogy with ordinary acoustic or ac circuit resonances. A compound nucleus is more likely to be formed when the excitation energy provided exactly matches one of its energy levels than if the excitation energy has some other value.

The reaction in Fig. 12.14 has a resonance at 0.176 eV whose width (at half-maximum) is $\Gamma = 0.115$ eV. This resonance corresponds to an excited state in ^{114}Cd that decays by the emission of a gamma ray. The mean lifetime τ of an excited state is related to its level width Γ by the formula

Mean lifetime of excited state

$$\tau = \frac{\hbar}{\Gamma} \tag{12.23}$$

This result is in accord with the uncertainty principle in the form $\Delta E \, \Delta t \geq \hbar/2$ if we associate Γ with the uncertainty ΔE in the excitation energy of the state and τ with the uncertainty Δt in the time the state will decay. In the case of the above reaction, the level width of 0.115 eV implies a mean lifetime for the compound nucleus of

$$\tau = \frac{1.054 \times 10^{-34}\ \text{J} \cdot \text{s}}{(0.115\ \text{eV})(1.60 \times 10^{-19}\ \text{J/eV})} = 5.73 \times 1^{-15}\ \text{s}$$

Center-of-Mass Coordinate System

Most nuclear reactions in the laboratory occur when a moving nucleon or nucleus strikes a stationary one. Analyzing such a reaction is simplified when we use a coordinate system that moves with the center of mass of the colliding particles.

To an observer located at the center of mass, the particles have equal and opposite momenta (Fig. 12.16). Hence if a particle of mass m_A and speed v approaches a stationary particle of mass m_B as viewed by an observer in the laboratory, the speed V of the center of mass is defined by the condition

$$m_A(v - V) = m_B V$$

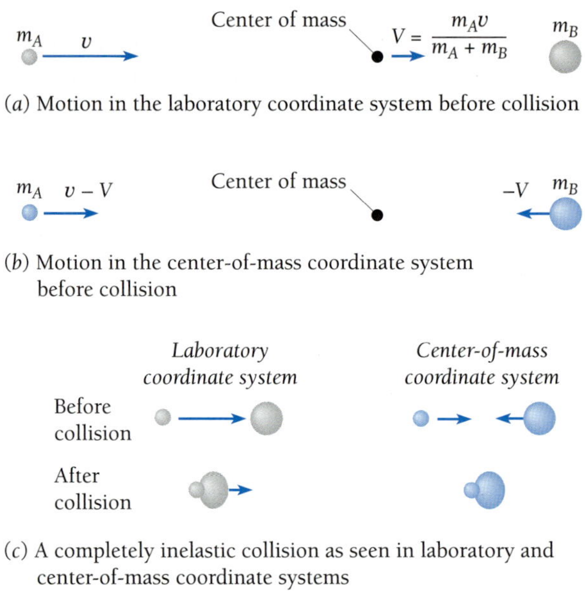

Figure 12.16 Laboratory and center-of-mass coordinate systems.

so that

Speed of center of mass

$$V = \left(\frac{m_A}{m_A + m_B}\right) \boldsymbol{v} \tag{12.24}$$

In most nuclear reactions, $\boldsymbol{v} \ll c$ and a nonrelativistic treatment is sufficient.

In the laboratory system, the total kinetic energy is that of the incident particle only:

Kinetic energy in lab system

$$\text{KE}_{\text{lab}} = \tfrac{1}{2} m_A \boldsymbol{v}^2 \tag{12.25}$$

In the center-of-mass system, both particles are moving and contribute to the total kinetic energy:

$$\begin{aligned} \text{KE}_{\text{cm}} &= \tfrac{1}{2} m_A(\boldsymbol{v} - V)^2 + \tfrac{1}{2} m_B V^2 \\ &= \tfrac{1}{2} m_A \boldsymbol{v}^2 - \tfrac{1}{2}(m_A + m_B)V^2 \\ &= \text{KE}_{\text{lab}} - \tfrac{1}{2}(m_A + m_B)V^2 \end{aligned}$$

Kinetic energy in CM system

$$\text{KE}_{\text{cm}} = \left(\frac{m_B}{m_A + m_B}\right) \text{KE}_{\text{lab}} \tag{12.26}$$

The total kinetic energy of the particles relative to the center of mass is their total kinetic energy in the laboratory system minus the kinetic energy $\frac{1}{2}(m_A + m_B)V^2$ of the moving center of mass. Thus we can regard KE_{cm} as the kinetic energy of the relative motion of the particles. When the particles collide, the maximum amount of kinetic energy that can be converted to excitation energy of the resulting compound nucleus while still conserving momentum is KE_{cm}, which is always less than KE_{lab}.

The **Q value** of the nuclear reaction

$$A + B \rightarrow C + D$$

is defined as the difference between the rest energies of A and B and the rest energies of C and D:

Q value of nuclear reaction

$$Q = (m_A + m_B - m_C - m_D)c^2 \tag{12.27}$$

If Q is a positive quantity, energy is given off by the reaction. If Q is a negative quantity, enough kinetic energy KE_{cm} in the center-of-mass system must be provided by the reacting particles so that $KE_{cm} + Q \geq 0$.

Example 12.11

Find the minimum kinetic energy in the laboratory system needed by an alpha particle to cause the reaction $^{14}N(\alpha, p)^{17}O$. The masses of ^{14}N, ^4He, ^1H, and ^{17}O are respectively 14.00307 u, 4.00260 u, 1.00783 u, and 16.99913 u.

Solution

Since the mases are given in atomic mass units, it is easiest to proceed by finding the mass difference between reactants and products in the same units and then multiplying by 931.5 MeV/u. Thus we have

$$Q = (14.00307\text{ u} + 4.00260\text{ u} - 1.00783\text{ u} - 16.99913\text{ u})\,(931.5\text{ MeV/u}) = -1.20\text{ MeV}$$

The minimum kinetic energy KE_{cm} in the center-of-mass system must therefore be 1.20 MeV in order for the reaction to occur. From Eq. (12.26) with the alpha particle as A,

$$KE_{lab} = \left(\frac{m_A + m_B}{m_B}\right) KE_{cm} = \left(\frac{4.00260 + 14.00307}{14.00307}\right)(1.20\text{ MeV}) = 1.54\text{ MeV}$$

The cross section for this reaction is another matter. Because both alpha particles and ^{14}N nuclei are positively charged and repel electrically, the greater KE_{cm} is above the threshold of 1.20 MeV, then the greater the cross section and the more likely the reaction will occur.

12.9 NUCLEAR FISSION

Divide and conquer

As we saw in Sec. 11.4, a lot of binding energy will be released if we can break a large nucleus into smaller ones. But nuclei are ordinarily not at all easy to split. What we need is a way to disrupt a heavy nucleus without using more energy than we get back from the process.

The answer came in 1938 with the realization by Lise Meitner that a nucleus of the uranium isotope $^{235}_{92}U$ undergoes fission when struck by a neutron. It is not the impact of the neutron that has this effect. Instead, the $^{235}_{92}U$ nucleus absorbs the neutron to become $^{236}_{92}U$, and the new nucleus is so unstable that almost at once it explodes into two fragments (Fig. 12.17). Later several other heavy nuclides were found to be fissionable by neutrons in similar processes.

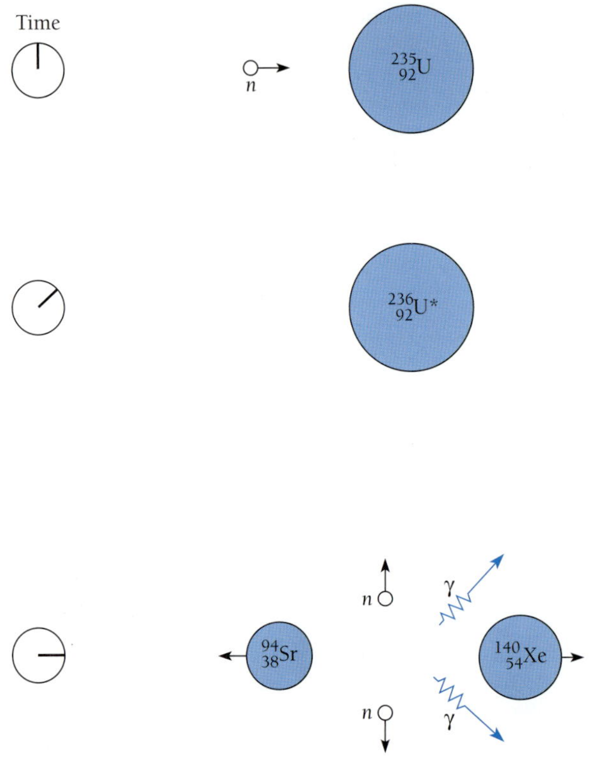

Figure 12.17 In nuclear fission, an absorbed neutron causes a heavy nucleus to split into two parts. Several neutrons and gamma rays are emitted in the process. The smaller nuclei shown here are typical of those produced in the fission of $^{235}_{92}U$ and are both radioactive.

Nuclear fission can be understood on the basis of the liquid-drop model of the nucleus (Sec. 11.5). When a liquid drop is suitably excited, it may oscillate in a variety of ways. A simple one is shown in Fig 12.18: the drop in turn becomes a prolate spheroid, a sphere, an oblate spheroid, a sphere, a prolate spheroid again, and so on. The restoring force of its surface tension always returns the drop to spherical shape, but the inertia of the moving liquid molecules causes the drop to overshoot sphericity and go to the opposite extreme of distortion.

Nuclei exhibit surface tension, and so can vibrate like a liquid drop when in an excited state. They also are subject to disruptive forces due to the mutual repulsion of their protons. When a nucleus is distorted from a spherical shape, the short-range restoring force of surface tension must cope with the long-range repulsive force as well as with the inertia of the nuclear matter. If the degree of distortion is small, the surface tension can do this, and the nucleus vibrates back and forth until it eventually loses its excitation energy by gamma decay. If the degree of distortion is too great, however, the surface tension is unable to bring back together the now widely separated groups of protons, and the nucleus splits into two parts. This picture of fission is illustrated in Fig. 12.19.

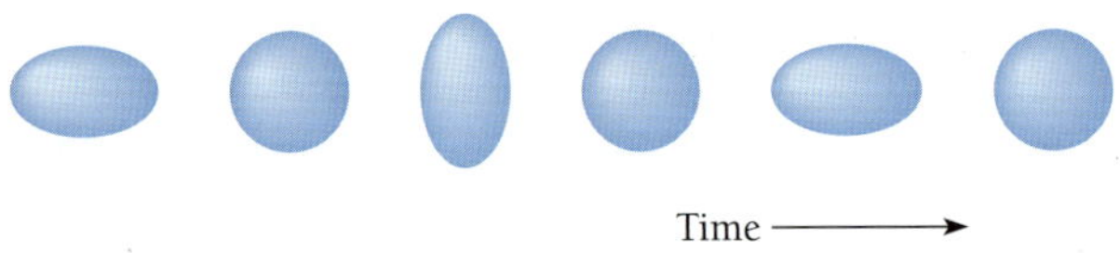

Figure 12.18 The oscillations of a liquid drop.

About Scientists

Lise Meitner (1878–1968), the daughter of a Viennese lawyer, became interested in science when she read about the Curies and radium. She earned her Ph.D. in physics in 1905 at the University of Vienna, only the second woman to obtain a doctorate there. She then went to Berlin where she began research on radioactivity with the chemist Otto Hahn. Their supervisor refused to have a woman in his laboratory, so they started their work in a carpentry shop. Ten years later she was a professor, a department head, and, with Hahn, the discoverer of a new element, protactinium.

In the 1930s the Italian physicist Enrico Fermi found that bombarding heavy elements with neutrons led to the production of other elements. What happened in the case of uranium was especially puzzling, and Meitner and Hahn tried to find out by repeating the experiment. At the time the German persecution of Jews had begun, but Meitner, who was Jewish, was protected by her Austrian citizenship. In 1938 Germany annexed Austria, and Meitner fled to Sweden but kept in touch with Hahn and their younger colleague Fritz Strassmann. Hahn and Strassmann finally concluded that neutrons interact with uranium to produce radium, but Meitner's calculations showed that this was impossible and she urged them to persist in their work. They did, and found to their surprise that the lighter element barium had in fact been created. Meitner surmised that the neutrons had caused the uranium nuclei to split apart and, with her nephew Otto Frisch, developed the theoretical picture of what they called fission.

In January 1939 Hahn and Strassmann published the discovery of fission in a German journal; because Meitner was Jewish, they thought it safer for themselves to ignore her contribution. Meitner and Frisch later published their own paper on fission in an English journal, but it was too late: Hahn disgracefully claimed full credit, and not once in the years that followed acknowledged her role. Hahn alone received the Nobel Prize in physics for discovering fission. Unfortunately Meitner did not live to see a measure of justice: the element of atomic number 109 is called meitnerium in her honor, while the tentative name of hahnium for element 105 was changed in 1997 to dubnium, after the Russian nuclear research center in Dubna.

Niels Bohr carried the news of the discovery of fission to the United States later in 1939, just before the start of World War II, where its military possibilities were immediately recognized. Expecting that German physicists would come to the same conclusion and would start work on an atomic bomb, such a program began in earnest in the United States. By the time it was successful, in 1945, Germany had been defeated, and two atomic bombs exploded over Hiroshima and Nagasaki then ended the war with Japan. It was later learned that the German atomic-bomb effort had amounted to very little. Not long afterward the Soviet Union, Great Britain, and France also developed nuclear weapons, and later China, Israel, South Africa, India, and Pakistan did so as well.

The new nuclei that result from fission are called **fission fragments.** Usually fission fragments are of unequal size (Fig. 12.20). Because heavy nuclei have a greater neutron/proton ratio than lighter ones, the fragments contain an excess of neutrons. To reduce this excess, two or three neutrons are emitted by the fragments as soon as they are formed, and subsequent beta decays bring their neutron/proton ratios to stable values. A typical fission reaction is

$$ {}^{235}_{92}\mathrm{U} + {}^{1}_{0}n \rightarrow {}^{236}_{92}\mathrm{U}^* \rightarrow {}^{140}_{54}\mathrm{Xe} + {}^{94}_{38}\mathrm{Sr} + {}^{1}_{0}n + {}^{1}_{0}n $$

which was illustrated in Fig. 12.17.

A heavy nucleus undergoes fission when it has enough excitation energy (5 MeV or so) to oscillate violently. A few nuclei, notably ^{235}U, are able to split in two merely by

Figure 12.19 Nuclear fission according to the liquid-drop model.

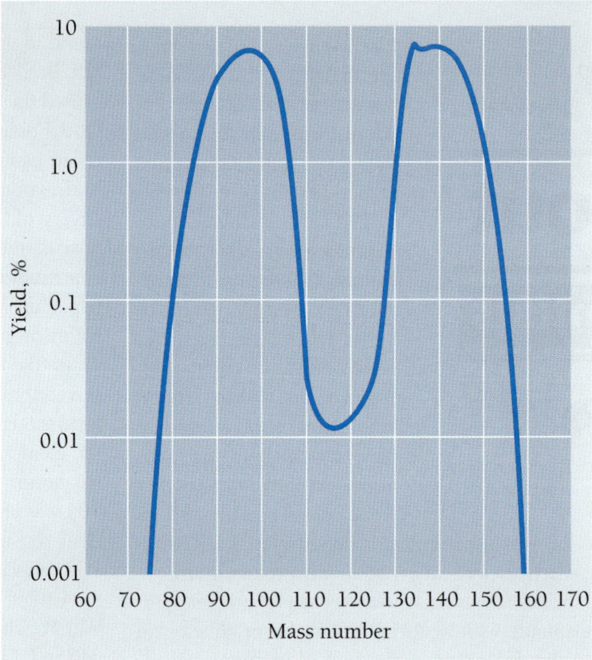

Figure 12.20 The distribution of mass numbers in the fragments from the fission of ${}^{235}_{92}U$.

absorbing an additional neutron. Other nuclei, notably ${}^{238}U$ (which makes up 99.3 percent of natural uranium, with ${}^{235}U$ as the remainder) need more excitation energy for fission than the binding energy released when another neutron is absorbed. Such nuclei undergo fission only by reaction with fast neutrons whose kinetic energies exceed about 1 MeV.

Fission can occur after a nucleus is excited by means other than neutron capture, for instance by gamma-ray or proton bombardment. Some nuclides are so unstable as to be capable of spontaneous fission, but they are more likely to undergo alpha decay before this takes place.

A striking aspect of nuclear fission is the magnitude of the energy given off. As we saw earlier, this is in the neighborhood of 200 MeV, a remarkable figure for a single atomic event; chemical reactions liberate only a few electronvolts per event. Most of the energy released in fission goes into the kinetic energy of the fission fragments. In the case of the fission of ${}^{235}U$, about 83 percent of the energy appears as kinetic energy of the fragments, about 2.5 percent as kinetic energy of the neutrons, and about 3.5 percent in the form of instantly emitted gamma rays. The remaining 11 percent is given off in the subsequent beta and gamma decays of the fission fragments.

Shortly after nuclear fission was discovered it was realized that, because fission leads to other neutrons being given off, a self-substaining sequence of fissions should be possible (Fig.12.21). The condition for such a **chain reaction** to occur in an assembly of fissionable material is simple: at least one neutron produced during each fission must, on the average, cause another fission. If too few neutrons cause fissions, the reaction will slow down and stop; if precisely one neutron per fission causes another fission, energy will be released at a constant rate. (which is the case in a **nuclear reactor**); and if the frequency of fissions increases, the energy release will be so rapid that an explosion will occur (which is the case in an **atomic bomb**).

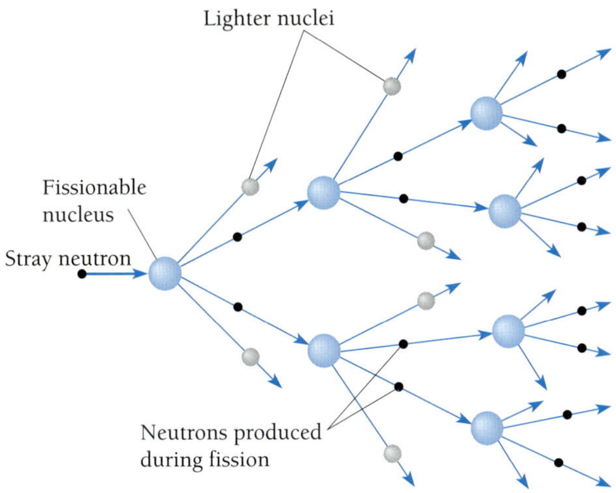

Figure 12.21 Sketch of a chain reaction. The reaction is self-sustaining if at least one neutron from each fission event on the average induces another fission event. If more than one neutron per fission on the average induces another fission, the reaction is explosive.

These situations are respectively called **subcritical, critical,** and **supercritical.** If two neutrons from each fission in an atomic bomb induce further fissions in 10^{-8} s, a chain reaction starting with a single fission will give off 2×10^{13} J of energy in less than 10^{-6} s.

12.10 NUCLEAR REACTORS

$E_0 = mc^2 + \$\$\$$

A nuclear reactor is a very efficient source of energy: the fission of 1 g of ^{235}U per day evolves energy at a rate of about 1 MW, whereas 2.6 tons of coal per day must be burned in a conventional power plant to produce 1 MW. The energy given off in a reactor becomes heat, which is removed by a liquid or gas coolant. The hot coolant is then used to boil water, and the resulting steam is fed to a turbine that can power an electric generator, a ship, or a submarine.

Each fission in ^{235}U releases an average of 2.5 neutrons, so no more than 1.5 neutrons per fission can be lost for a self-substaining chain reaction to occur. However, natural uranium contains only 0.7 percent of the fissionable isotope ^{235}U. The more abundant ^{238}U readily captures fast neutrons but usually does not undergo fission as a result. As it happens, ^{238}U has only a small cross section for the capture of *slow* neutrons, whereas the cross section of ^{235}U for slow neutron-induced fission is a whopping 582 barns. Slowing down the fast neutrons that are liberated in fission thus helps prevent their unproductive absorption by ^{238}U and at the same time promotes further fissions in ^{235}U.

To slow down fission neutrons, the uranium in a reactor is mixed with a **moderator,** a substance whose nuclei absorb energy from fast neutrons in collisions without much tendency to capture the neutrons. While the exact amount of energy lost by a moving body that collides elastically with another depends on the details of the interaction, in general the energy transfer is a maximum when the participants are of equal mass

Enrico Fermi (1901–1954) was born in Rome and obtained his doctorate at Pisa. After periods at Göttingen and Leiden working with leading figures in the new quantum mechanics, Fermi returned to Italy. At the University of Rome in 1926 he investigated the statistical mechanics of particles that obey Pauli's exclusion principle, such as electrons; the result is called Fermi-Dirac statistics because Dirac independently arrived at the same conclusions shortly afterward. In 1933 Fermi introduced the concept of the weak interaction and used it together with Pauli's newly postulated neutrino (as Fermi called it) to develop a theory of beta decay able to account for the shape of the electron energy spectrum and the decay half-life.

Later in the 1930s Fermi and a group of collaborators carried out a series of experiments in which radionuclides were produced artificially by bombarding various elements with neutrons; they found slow neutrons especially effective. Some of their results seemed to suggest the formation of transuranic elements. In fact, as Meitner and Hahn were to find later, what they were observing was nuclear fission. In 1938 Fermi received the Nobel Prize for this work, but instead of returning to Mussolini's Fascist Italy, he went to the United States. As part of the atomic-bomb program, Fermi directed the design and construction of the first nuclear reactor at the University of Chicago, which began operating in December 1942, four years after the discovery of fission. After the war Fermi shifted to a different field, high-energy particle physics, where he made important contributions. He died of cancer in 1954, one of the very few physicists of the modern era to combine virtuosity in both theory and experiment. The element of atomic number 100, discovered the year after his death, is called fermium in his honor.

(Fig. 12.22). The greater the difference between the masses, the greater the number of collisions needed to slow a neutron down, and the longer the period in which it is in danger of being captured by a ^{238}U nucleus. The majority of today's commercial reactors use light water both as moderator and as coolant. Each molecule of water contains two hydrogen atoms whose proton nuclei have masses almost identical with that of the neutron, so light water is an efficient moderator.

Unfortunately protons tend to capture neutrons to form deuterons in the reaction $^1\text{H}(n, \gamma)^2\text{H}$. Light-water reactors therefore cannot use natural uranium for fuel but need **enriched uranium** whose ^{235}U content has been increased to about 3 percent. Enriched uranium can be produced in several ways. Originally all enriched uranium was produced by gaseous diffusion, with uranium hexafluoride (UF_6) gas being passed through about 2000 successive permeable barriers. Molecules of $^{235}UF_6$ are slightly more likely to diffuse through each barrier than $^{238}UF_6$ because of their smaller mass. A more recent method uses high-speed gas centrifuges for the separation. Still other processes are possible.

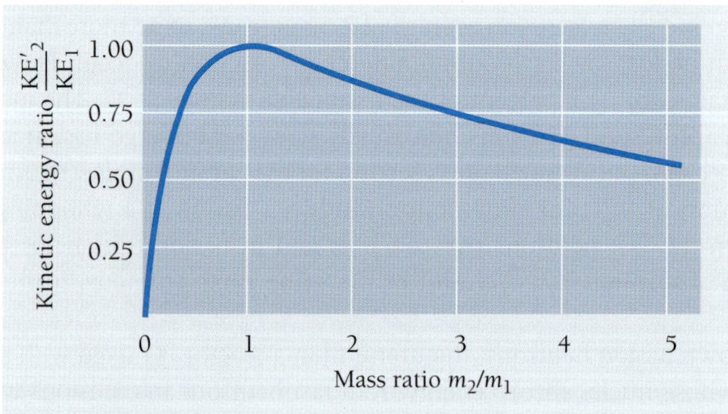

Figure 12.22 Energy transfer in an elastic head-on collision between a moving object of mass m_1 and a stationary object of mass m_2 (see Exercise 59).

Fuel rods being loaded into the core of a 1,129-MW reactor at the William McGuire Nuclear Power Plant in Cornelius, North Carolina.

The fuel for a water-moderated reactor consists of uranium oxide (UO_2) pellets sealed in long, thin tubes. Control rods of cadmium or boron, which are good absorbers of slow neutrons, can be slid in and out of the reactor core to adjust the rate of the chain reaction. In the most common type of reactor, the water that circulates around the fuel in the core is kept at a high pressure, about 155 atmospheres, to prevent boiling. The water, which acts as both moderator and coolant, is passed through a heat exchanger to produce steam that drives a turbine (Fig. 12.23). Such a reactor might contain 90 tons of UO_2 and operate at 3400 MW to yield 1100 MW of electric power. The reactor fuel must be replaced every few years as its ^{235}U content is used up.

Breeder Reactors

Some nonfissionable nuclides can be transmuted into fissionable ones by absorbing neutrons. A notable example is ^{238}U, which becomes ^{239}U when it captures a fast neutron. This uranium isotope beta-decays with a half-life of 24 min into $^{239}_{93}Np$, an isotope of the element neptunium, which is also beta-active. The decay of ^{239}Np has a half-life of 2.3 days and yields $^{239}_{94}Pu$, an isotope of plutonium whose half-life against

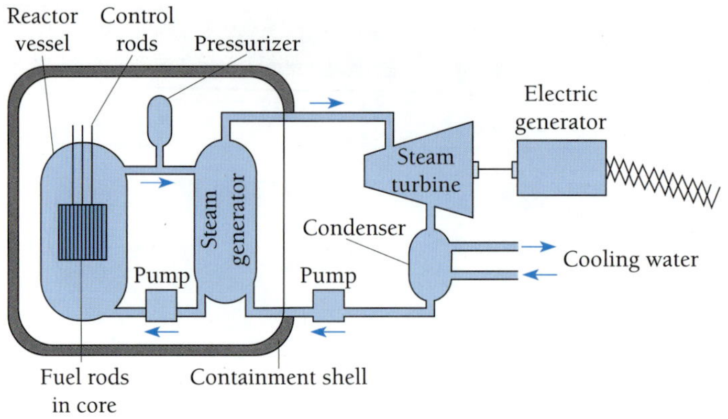

Figure 12.23 Basic design of the most common type of nuclear power plant. Water under pressure is both the moderator and coolant, and transfers heat from the chain reaction in the fuel rods of the core to a steam generator. The resulting steam then passes out of the containment shell, which serves as a barrier to protect the outside world from accidents to the reactor, and is directed to a turbine that drives an electric generator. In a typical plant, the reactor vessel is 13.5 m high and 4.4 m in diameter and weighs 385 tons. It contains about 90 tons of uranium oxide in the form of 50,952 fuel rods each 3.85 m long and 9.5 mm in diameter. Four steam generators are used, instead of the single one shown here, as well as a number of turbine generators.

alpha decay is 24,000 y. The entire sequence is shown in Fig. 12.24. Both neptunium and plutonium are **transuranic elements,** none of which are found on the earth because their half-lives are too short for them to have survived even if they had been present when the earth came into being 4.5 billion years ago.

The plutonium isotope ^{239}Pu is fissionable and can be used as a reactor fuel and for weapons. Plutonium is chemically different from uranium, and its separation from the remaining ^{238}U after neutron irradiation is more easily accomplished than the separation of ^{235}U from the much more abundant ^{238}U in natural uranium.

A **breeder reactor** is one especially designed to produce more plutonium than the ^{235}U it consumes. Because the otherwise useless ^{238}U is 140 times more abundant than the fissionable ^{235}U, the widespread use of breeder reactors would mean that known

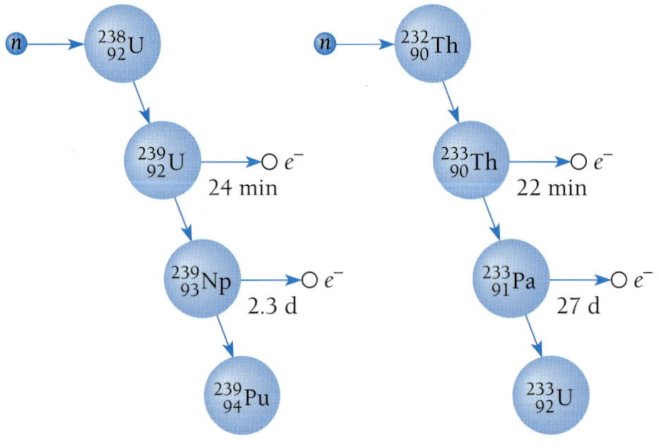

Figure 12.24 ^{238}U and ^{232}Th are "fertile" nuclides. Each becomes a fissionable nuclide after absorbing a neutron and undergoing two beta decays. These transformations are the basis of the breeder reactor, which produces more fuel in the form of ^{239}Pu or ^{233}U than it uses up in the form of ^{235}U.

reserves of uranium could fuel reactors for many centuries to come. Because plutonium can also be used in nuclear weapons (unlike the slightly enriched uranium that fuels ordinary reactors), the widespread use of breeder reactors would also complicate the control of nuclear weapons. Several breeder reactors are operating today, all of them outside the United States. They have proved to be extremely expensive and have had severe operating problems.

Actually, plutonium is already an important nuclear fuel. By the end of the usual three-year fuel cycle in a reactor, after which the fuel rods are replaced, so much plutonium has been produced from the ^{238}U present that more fissions occur in ^{239}Pu than in ^{235}U.

A Nuclear World?

In 1951 the first electricity from a nuclear plant was generated in Idaho. Today over 400 reactors in 26 countries produce about 200,000 MW of electric power—the equivalent of nearly 10 million barrels of oil per day. France, Belgium, and Taiwan obtain more than half of their electricity from reactors, with several other countries close behind (Fig. 12.25). In the

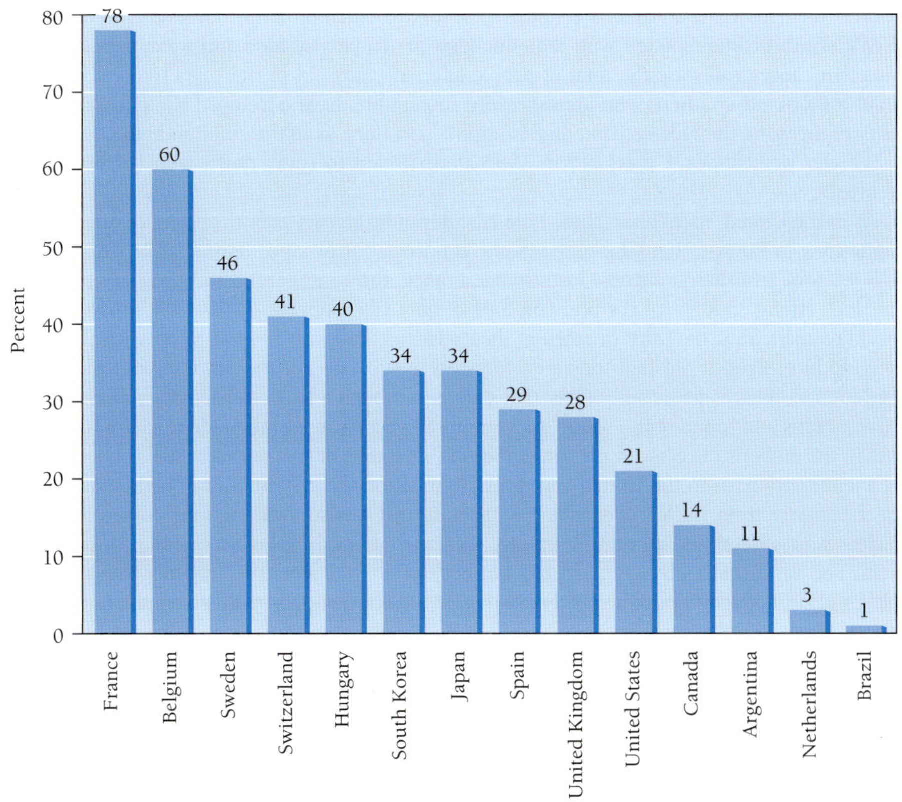

Figure 12.25 Percentage of electric energy in various countries that comes from nuclear power stations. Figures are for 1997.

United States, nuclear energy is responsible for 21 percent of generated electricity, slightly more than the world average; there are 103 reactors in 31 states. Yet for all the success of nuclear technology, no new nuclear power stations have been planned in this country since 1979. Why not?

In March 1979, failures in its cooling system disabled one of the reactors at Three Mile Island in Pennsylvania and a certain amount of radioactive material escaped. Although a nuclear reactor cannot explode in the way an atomic bomb does, breakdowns can occur that put a large population at risk. Although a true catastrophe was narrowly avoided, the Three Mile Island incident made it clear that the hazards associated with nuclear energy are real.

After 1979 it was inevitable that greater safety would have to be built into new reactors, adding to their already high cost. In addition, demand for electricity in the United States was not increasing as fast as expected, partly because of efforts toward greater efficiency and partly because of a decline in some of industries (such as steel, cars, and chemicals) that are heavy users of electricity. As a result of these factors, new reactors made less economic sense than before, which together with widespread public unease led to a halt in the expansion of nuclear energy in the United States.

Elsewhere the situation was different. Nuclear reactors still seemed the best way to meet the energy needs of many countries without abundant fossil fuel resources. Then in April 1986, a severe accident destroyed a 1000-MW reactor at Chernobyl in what is now Ukraine, then part of the Soviet Union. This was the worst environmental disaster of technological origin in history and contributed to the collapse of the Soviet Union. Over 50 tons of radioactive material escaped and was carried around the world by winds. The radiation released was nearly 200 times the total given off by the Hiroshima and Nagasaki atomic bombs in 1945. Radiation levels in much of Europe rose well above normal for a time and a quarter of a million people were permanently evacuated from the vicinity of Chernobyl. A number of reactor, rescue, and cleanup workers died soon afterward as a result of exposure to radiation, and thousands more became ill. Widespread contamination with radionuclides, particularly of food and water supplies, suggests that cancer will raise the total of people affected manyfold in the years to come. Already about a thousand children, who are especially susceptible, have developed thyroid cancer as a result of ingesting the radioactive iodine isotope ^{131}I; a third of all the children living near Chernobyl who were under 4 years old in 1986 are expected to come down with thyroid cancer eventually.

As in the United States after Three Mile Island, public anxiety over the safety of nuclear programs grew in Europe after Chernobyl. Some countries, for instance Italy, abandoned plans for new reactors. In other countries, for instance France, the logic behind their nuclear programs remained strong enough for them to continue despite Chernobyl.

Quite apart from the safty of reactors themselves is the issue of what to do with the waste they produce. Even if old fuel rods are processed to separate out the uranium and plutonium they contain, what is left is still highly radioactive. Although a lot of the activity will be gone in a few months and much of the rest in a few hundred years, some of the radionuclides have half-lives in the millions of years. At present perhaps 20,000 tons of spent nuclear fuel is being stored on a temporary basis in the United States (not to mention the vast amount of highly radioactive waste left over from nuclear weapons manufacture that is also awaiting safe storage). Burying nuclear wastes deep underground currently seems to be the best long-term way to dispose of them. The right location is easy to specify but not easy to find: stable geologically with no earthquakes likely, no nearby population centers, a type of rock that does not disintegrate in the presence of heat and radiation but is easy to drill into, and not near groundwater that might become contaminated.

From today's perspective, nuclear energy has important advantages not fully appreciated in the past: it does not produce the air pollution that fossil-fuel burning does, nor the huge quantities of carbon dioxide that are the main contributor to global warming via the greenhouse effect. Together with the rising cost of fossil fuels and increasing demand for electricity, these factors seem likely to lead to the construction of new nuclear reactors in the United States after a delay of over two decades.

12.11 NUCLEAR FUSION IN STARS

How the sun and stars get their energy

Here on the earth, 150 million km from the sun, a surface 1 m^2 in area exposed to the vertical rays of the sun receives energy at a rate of about 1.4 kW. Adding up all the energy radiated by the sun per second gives the enormous total of 4×10^{26} W. And the sun has been emitting energy at this rate for billions of years. Where does it all come from?

The basic energy-producing process in the sun is the fusion of hydrogen nuclei into helium nuclei. This can take place in two different reaction sequences, the most common of which, the **proton-proton cycle,** is shown in Fig. 12.26. The total evolved energy is 24.7 MeV per ${}^4_2\text{He}$ nucleus formed.

Since 24.7 MeV is 4×10^{-12} J, the sun's power output of 4×10^{26} W means the sequence of reactions in Fig. 12.26 must occur 10^{38} times per second. The sun consists of 70 percent hydrogen, 28 percent helium, and 2 percent of other elements, so plenty of hydrogen remains for billions of years of further energy production at its current rate. Eventually the hydrogen in the sun's core will be exhausted, and then, as the other reactions described below take over, the sun will swell to become a red giant star and later subside into a white dwarf.

Self-sustaining fusion reactions can occur only under conditions of extreme temperature and density. The high temperature ensures that some nuclei—those in the high-velocity tail of the Maxwell-Boltzmann distribution—have the energy needed to come close enough together to interact, which they do by tunnelling through the electric potential barrier between them. (At the 10^7 K temperature typical of the sun's interior, the average proton kinetic energy is only about 1 keV, whereas the barrier is about 1 MeV, a thousand times higher.) The high density ensures that such collisions are frequent. A further condition for the proton-proton and other multistep cycles is a large reacting mass, such as that of the sun, since much time may elapse between the initial fusion of a particular proton and its eventual incorporation in an alpha particle.

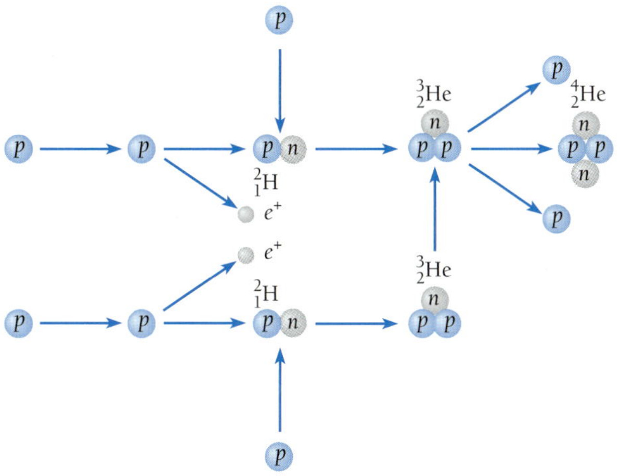

Figure 12.26 The proton-proton cycle. This is the chief nuclear reaction sequence that takes place in stars like the sun and cooler stars. Energy is given off at each step. The net result is the combination of four hydrogen nuclei to form a helium nucleus and two positrons. The neutrinos also produced are not shown.

Hans A. Bethe (1906–2005) was born in Strasbourg, then part of Germany but today part of France. He studied physics in Frankfurt and Munich and taught at various German universities until 1933, when Hitler came to power. After two years in England he came to the United States where he was professor of physics at Cornell University from 1937 to 1975. He remained active in research and in public affairs even after he formally retired.

Notable among Bethe's many and varied contributions to physics was his 1938 account of the sequences of nuclear reactions that power the sun and stars, for which he received the Nobel Prize in 1967. During World War II he directed the theoretical physics division of the laboratory at Los Alamos, New Mexico, where the atomic bomb was developed. A strong believer in nuclear energy—"it is more necessary now than ever before because of global warming"—Bethe was an effective advocate of nuclear disarmament.

The proton-proton cycle dominates in the sun and other stars with less than about 1.5 times the sun's mass. In more massive stars, whose interiors are hotter, the carbon cycle is the main energy source. This cycle proceeds as in Fig. 12.27. The net result again is the formation of an alpha particle and two positrons from four protons, with the evolution of 24.7 MeV. The initial $^{12}_{6}\text{C}$ acts as a kind of catalyst for the process, since it reappears at its end. The dependence of the two cycles on temperature is shown in Fig. 12.28.

Formation of Heavier Elements

Fusion reactions that produce helium are not the only ones that occur in the sun and other stars. When all the hydrogen in a star's core has become helium, gravitational

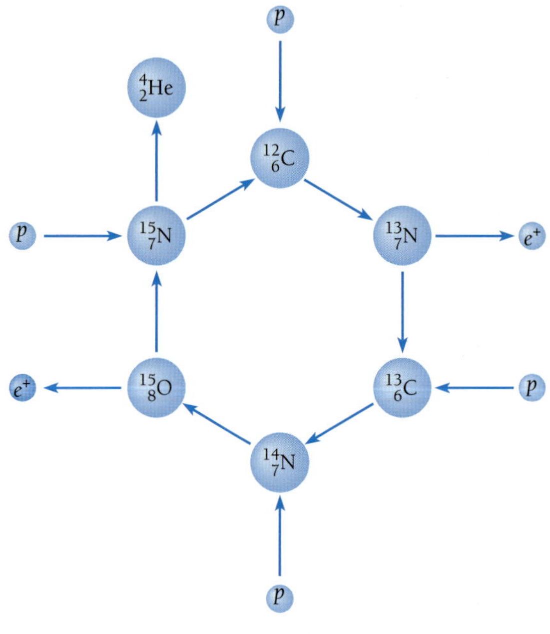

Figure 12.27 The carbon cycle also involves the combination of four hydrogen nuclei to form a helium nucleus with the evolution of energy. The $^{12}_{6}\text{C}$ nucleus is unchanged by the series of reactions. This cycle occurs in stars hotter than the sun.

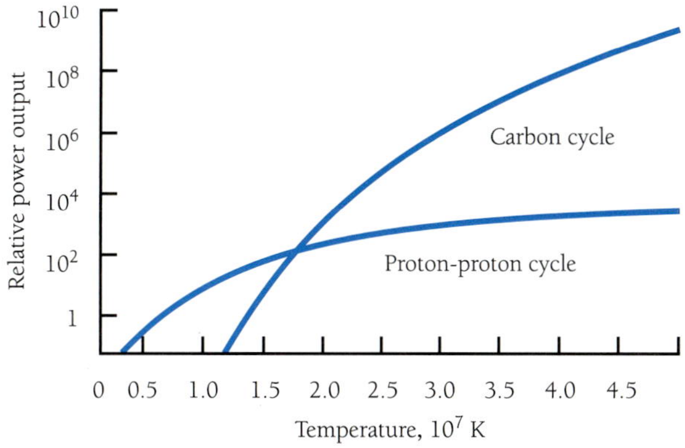

Figure 12.28 How the rates of energy generation for the carbon and proton-proton fusion cycles vary with the temperature of a star's interior. The rates are equal at about 1.8×10^7 K. Note that the power output scale is not linear.

contraction compresses the core and raises its temperature to the 10^8 K needed for helium fusion to begin. This involves the combination of three alpha particles to form a carbon nucleus with the evolution of 7.5 MeV:

$$^4_2\text{He} + ^4_2\text{He} \rightarrow ^8_4\text{Be} + \gamma$$

$$^4_2\text{He} + ^8_4\text{Be} \rightarrow ^{12}_6\text{C} + \gamma$$

Because the beryllium isotope $^8_4$Be is unstable and breaks apart into two alpha particles with a half-life of only 6.7×10^{-17} s, the second reaction must take place immediately after the first. The sequence is called the triple-alpha reaction.

The smallest stars do not get hot enough (over 10^7 K) to go beyond hydrogen fusion, and helium fusion is as far as a star with the sun's mass gets. But in heavier stars, core temperatures can go even higher, and fusion reactions that involve carbon then become possible. Some examples are

$$^4_2\text{He} + ^{12}_6\text{C} \rightarrow ^{16}_8\text{O}$$

$$^{12}_6\text{C} + ^{12}_6\text{C} \rightarrow ^{24}_{12}\text{Mg}$$

$$^{12}_6\text{C} + ^{12}_6\text{C} \rightarrow ^{20}_{10}\text{Ne} + ^4_2\text{He}$$

The heavier the star, the higher the eventual temperature of its core, and the larger the nuclei that can be formed. (The high temperatures, of course, are needed to overcome the greater electric repulsion of reacting nuclei with many protons.) In stars more than

The Triple-Alpha Reaction

Because no sufficiently stable nuclides with $A = 5$ or $A = 8$ exist, there is no simple way in which protons, neutrons, and alpha particles can add together in succession to form the nuclei of carbon and elements of still higher atomic number. Eventually it became clear that three alpha particles could react to produce a $^{12}_6$C nucleus in stars whose interiors are sufficiently hot. However, the cross section (Sec. 12.7) for the process seemed much too small for the reaction to be significant. Then, in 1953, the British astronomer Fred Hoyle realized that a resonance associated with the triple-alpha process would greatly enhance its likelihood. Hoyle's calculation indicated that the resonance would correspond to an excited state in $^{12}_6$C of 7.7 MeV. Experiments soon showed that this excited state indeed occurred and increased the cross section by a factor of 10^7, thereby removing the biggest obstacle to understanding the origin of the elements.

about 10 times as massive as the sun, the iron isotope ${}^{56}_{26}Fe$ is reached. This is the nucleus with the greatest binding energy per nucleon (Fig. 11.12). Any reaction between a ${}^{56}_{26}Fe$ nucleus and another nucleus will therefore lead to the breakup of the iron nucleus, not to the formation of a still heavier one.

Then how do nuclides beyond ${}^{56}_{26}Fe$ originate? The answer is through the successive capture of neutrons, with beta decays when needed for appropriate neutron/proton ratios. The neutrons are liberated in such sequences as

$$
\begin{aligned}
{}^{1}_{1}H + {}^{12}_{6}C &\rightarrow {}^{13}_{7}N + \gamma \\
{}^{13}_{7}N &\rightarrow {}^{13}_{6}C + e^{+} + \nu \\
{}^{4}_{2}He + {}^{13}_{6}C &\rightarrow {}^{16}_{8}O + {}^{1}_{0}n
\end{aligned}
$$

Neutron-capture reactions in a stellar interior can build up nuclides as far as ${}^{209}_{83}Bi$, the largest stable nucleus, but no further. The density of neutrons there is not sufficient for them to be captured in rapid enough succession by nuclei of $A > 209$ before such nuclei decay. However, when a very massive star has reached the end of its fuel supply, its core collapses and a violent explosion follows that appears in the sky as a supernova. During the collapse neutrons are produced in abundance, some by the disintegration of neutron-rich nuclei into alpha particles and neutrons in collisions and some by the reaction $e^{-} + p \rightarrow n + \nu$. The huge neutron flux lasts only a matter of seconds, but this is sufficient to produce nuclei with mass numbers up to perhaps 260.

A supernova explosion, which occurs once or twice per century in a galaxy of stars like our own Milky Way, flings into space a large part of the star's mass, which becomes dispersed in interstellar matter. New stars (and their planets, such as our own) that come into being from this matter thus contain the entire spectrum of nuclides, not just the hydrogen and helium of the early universe. We are all made of stardust.

12.12 FUSION REACTORS

The energy source of the future?

Enormous as the energy produced by fission is, the fusion of light nuclei to form heavier ones can give out even more per kilogram of starting materials. It seems possible that nuclear fusion could become the ultimate source of energy on the earth: safe, relatively nonpolluting, and with the oceans themselves supplying limitless fuel.

On the earth, where any reacting mass must be very limited in size, an efficient fusion process cannot involve more than a single step. Two reactions that may eventually power fusion reactors involve the combination of two deuterons to form a triton and a proton,

$$ {}^{2}_{1}H + {}^{2}_{1}H \rightarrow {}^{3}_{1}H + {}^{1}_{1}H + 4.0 \text{ MeV} \tag{12.28} $$

or their combination to form a ${}^{3}_{2}He$ nucleus and a neutron,

$$ {}^{2}_{1}H + {}^{2}_{1}H \rightarrow {}^{3}_{2}He + {}^{1}_{0}n + 3.3 \text{ MeV} \tag{12.29} $$

Both D-D reactions have about equal probabilities. A major advantage of these reactions is that deuterium is present in seawater and is cheap to extract. Although its concentration in seawater is only 33 g/m^3, this adds up to a total of about 10^{15} tons of deuterium in the world's oceans. The deuterium in a gallon of seawater can yield as much energy through fusion as 600 gallons of gasoline can through combustion.

The first fusion reactors are more likely to employ a deuterium-tritium mixture because the D-T reaction

$$^3_1\text{H} + {}^2_1\text{H} \rightarrow {}^4_2\text{He} + {}^1_0n + 17.6 \text{ MeV} \tag{12.30}$$

has a higher yield than the others and occurs at lower temperatures. Seawater contains too little tritium to be extracted economically, but it can be produced by the neutron bombardment of the two isotopes of natural lithium:

$$^6_3\text{Li} + {}^1_0n \rightarrow {}^3_1\text{H} + {}^4_2\text{He} \tag{12.31}$$

$$^7_3\text{Li} + {}^1_0n \rightarrow {}^3_1\text{H} + {}^4_2\text{He} + {}^1_0n \tag{12.32}$$

In fact, plans for future fusion reactors include lithium blankets that will make the tritium they need by absorbing neutrons liberated in the fusion reactions.

At the required temperatures, a fusion reactor's fuel will be in the form of a **plasma,** which is a fully ionized gas. **Breakeven** occurs when the energy produced equals the energy input to the reacting plasma. **Ignition,** a more difficult (and perhaps unnecessary) target, occurs when enough energy is produced for the reaction to be self-sustaining.

A successful fusion reactor has three basic conditions to meet:

1 The plasma temperature must be high so that an adequate number of the ions have the speeds needed to come close enough together to react despite their mutual repulsion. Taking into account that many ions have speeds well above the average and that tunneling through the potential barrier reduces the ion energy needed, the minimum temperature for igniting a D-T plasma is about 100 million K, which corresponds to an "ion temperature" of $kT \sim 10$ keV.
2 The plasma density n (in ions/m^3) must be high to ensure that collisions between nuclei are frequent.
3 The plasma of reacting nuclei must remain together for a sufficiently long time τ. How long depends on the product $n\tau$, the confinement quality parameter. In the case of a D-T plasma with $kT \sim 10$ keV, $n\tau$ must be greater than roughly 10^{20} s/m^3 for breakeven, more than that for ignition (Fig. 12.29).

Apart from stellar interiors, the combination of temperature, density, and confinement time needed for fusion thus far has occurred only in the explosion of fission ("atomic") bombs. Incorporating the ingredients for fusion reactions in such a bomb leads to an even more destructive weapon, the "hydrogen" bomb.

Confinement Methods

The approach to the controlled release of fusion energy that has thus far shown the most promise uses a strong magnetic field to confine the reactive plasma. In the Russian-designed **tokamak** scheme, the magnetic field is a modified torus (doughnut) in form (Fig. 12.30). (In Russian, tokamak stands for "toroidal magnetic chamber.") Because the field lines of a purely toroidal field are curved, an ion moving in

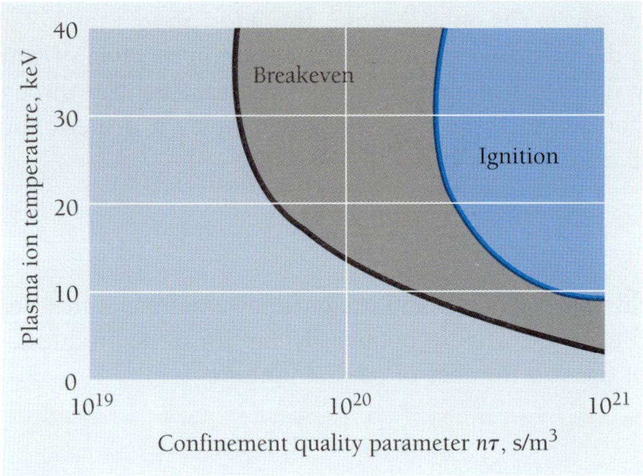

Figure 12.29 Conditions for breakeven (energy output equals energy input) and for ignition (a self-sustaining reaction) in a fusion reactor. Existing reactors have come close to breakeven; the projected International Thermonuclear Experimental Reactor is intended to go beyond it.

The KSTAR(Korea Superconducting Tokamak Advanced Research) is an experimental tokamak fusion reactor in Daejeon, Korea. It is the first reactor which maintained the plasma ion temperature of 100 million degrees for 1.5 seconds. It also sustained high-performance mode plasmas for 90 seconds. (Source: https://www.nfri.re.kr/kor/post/photo?clsf=photo02)

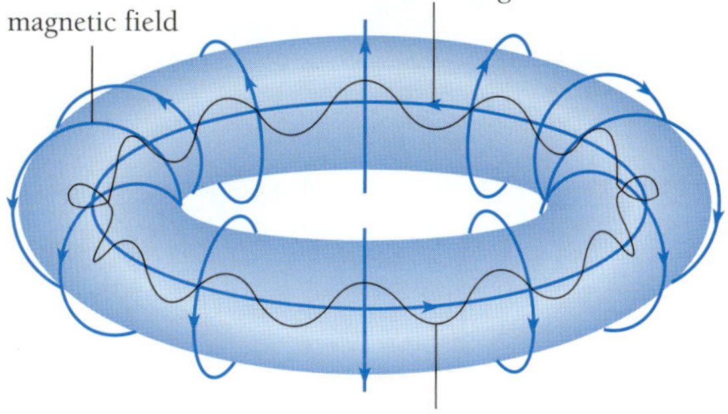

Figure 12.30 In a tokamak, combined toroidal and poloidal magnetic fields confine a plasma.

a helical path around its field lines will drift across the field and escape. To prevent this, a tokamak uses a poloidal field whose field lines are circles around the toroid axis. The poloidal field is produced by a current setup in the plasma itself by the changing field of an electromagnet in the center of the toroid. This current also heats the plasma; once the plasma is sufficiently hot, the current needs little help to continue.

ITER

The planned International Thermonuclear Experimental Reactor (ITER) represents what is hoped to be the final step before practical fusion energy becomes a reality. ITER is currently sponsored by Japan and several European countries; the United States pulled out of the project because of concerns about its original design and cost, and Russia withdrew (except for providing some staff) because it cannot afford to participate. The redesigned ITER is expected to generate 400 MW from deuterium-tritium reactions, to weigh 32,000 tons, to cost \$3 billion, and to take 10 to 15 years to build. Superconducting magnets (a large part of the cost) will keep the reacting ions in a doughnut-shaped region whose volume is that of a large house. About 80 percent of the energy released will be carried off by the neutrons that are produced, and these neutrons will be absorbed by lithium pellets in tubes that surround the reaction chamber. Circulating water will carry away the resulting heat; this is the heat that could be used in a commercial reactor to power turbines connected to electric generators.

Even if ITER works as planned, though, not every pessimistic observer of the fusion program is likely to become a convert to the cause. Fusion reactors will certainly be enormously complex and expensive and not wholly safe: lithium is an extremely reactive metal that burns or explodes on contact with water. Also, when lithium absorbs neutrons in the reactions of Eqs. (12.31) and (12.32), radioactive tritium is produced. Hence an accident could be catastrophic. Of course, the optimists could turn out to be correct, and fusion will become the preferred energy source of the future. But even if this happens, many decades lie ahead in which energy problems will remain. Fission reactors employ an established technology and ways exist to make them very safe, but memories of Three Mile Island and Chernobyl, plus continuing questions about the disposal of radioactive wastes, continue to affect their public image. Meanwhile fossil fuels are being used up and burning them produces enough CO_2 to affect weather and climate. Such "green" energy sources as solar cells and wind turbines are unlikely to provide more than a small (though welcome) fraction of energy needs. An energy strategy for the world that is both sensible and widely acceptable is not obvious.

The world's most powerful laser, located at the Lawrence National Laboratory in California, is used in inertial confinement experiments. Its output of 60 kJ per nanosecond (10^{-9} s) pulse is divided into 10 beams that are directed at tiny dueterium-tritium pellets in an effort to induce fusion reactions in them.

The most powerful tokamaks today have attained plasma temperatures of 30 keV and confinement quality $n\tau$ values of 2×10^{19} s/m^3, but not breakeven. Breakeven will probably have to wait for the planned International Thermonuclear Experimental Reactor (ITER).

An entirely different procedure, called **inertial confinement,** uses energetic beams to both heat and compress tiny deuterium-tritium pellets by blasting them from all sides. The result is, in effect, a miniature hydrogen-bomb explosion, and a succession of them could provide a steady stream of energy. If ten 0.1-mg pellets are ignited every second, the average thermal output would be about 1 GW and could yield 300 MW or so of electric power, enough for a city of 175,000 people.

Laser beams have received the most attention for inertial confinement, but electron and proton beams have promise as well. The beam energy is absorbed in the outer layer of the fuel pellet, which blows off outward. Conservation of momentum leads to an inward shock wave that must squeeze the rest of the pellet to about 10^4 times its original density to heat the fuel sufficiently to start fusion reactions. The required beam energy is well beyond the capacity of today's lasers, though perhaps not of future ones. Particle beams are closer to reaching the needed energy but are much harder to focus on the tiny fuel pellets. Research continues, but magnetic confinement seems closer to the goal of a working fusion reactor.

Appendix to Chapter 12

Theory of Alpha Decay

In the discussion of the tunnel effect in Sec. 5.10 a beam of particles of kinetic energy E was considered which was incident on a rectangular potential barrier whose height U was greater than E. An approximate value of the transmission probability—the ratio between the number of particles that pass through the barrier and the number that arrive—was found to be

Approximate transmission probability

$$T = e^{-2k_2L} \tag{5.60}$$

where L is the width of the barrier and

Wave number inside barrier

$$k_2 = \frac{\sqrt{2m(U - E)}}{\hbar} \tag{5.61}$$

Equation (5.60) was derived for a rectangular potential barrier, whereas an alpha particle inside a nucleus is faced with a barrier of varying height, as in Figs. 12.8 and 12.31. It is now our task to adapt Eq. (5.60) to the case of a nuclear alpha particle.

The first step is to rewrite Eq. (5.60) in the form

$$\ln T = -2k_2L \tag{12.33}$$

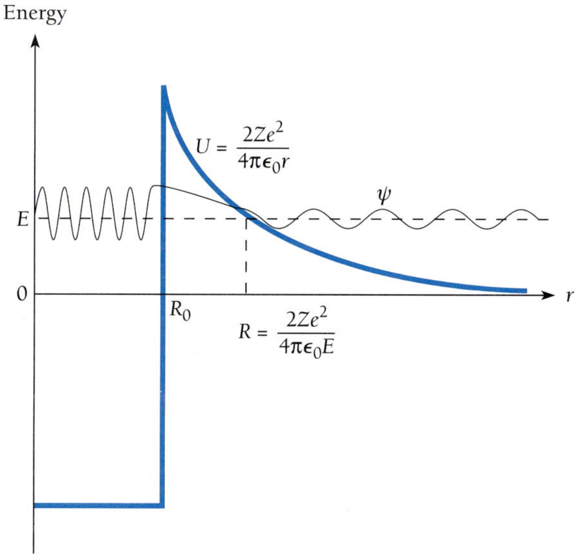

Figure 12.31 Alpha decay from the point of view of the quantum mechanics. The kinetic energy of alpha particle is E.

and then express it as the integral

$$\ln T = -2\int_0^L k_2(r)\, dr = -2\int_{R_0}^{R} k_2(r)\, dr \tag{12.34}$$

where R_0 is the radius of the nucleus and R is the distance from its center at which $U = E$. The kinetic energy E is greater than the potential energy U for $r > R$, so if it can get past R, the alpha particle will have permanently escaped from the nucleus.

The electric potential energy of an alpha particle at the distance r from the center of a nucleus of charge Ze is given by

$$U(r) = \frac{2Ze^2}{4\pi\epsilon_0 r}$$

Here Ze is the nuclear charge *minus* the alpha-particle charge of $2e$; thus Z is the atomic number of the daughter nucleus.

We therefore have

$$k_2 = \frac{\sqrt{2m(U-E)}}{\hbar} = \left(\frac{2m}{\hbar^2}\right)^{1/2}\left(\frac{2Ze^2}{4\pi\epsilon_0 r} - E\right)^{1/2}$$

Since $U = E$ when $r = R$,

$$E = \frac{2Ze^2}{4\pi\epsilon_0 R} \tag{12.35}$$

and we can write k_2 in the form

$$k_2 = \left(\frac{2mE}{\hbar^2}\right)^{1/2}\left(\frac{R}{r} - 1\right)^{1/2} \tag{12.36}$$

Hence

$$\begin{aligned}
\ln T &= -2\int_{R_0}^{R} k_2(r)\, dr \\
&= -2\left(\frac{2mE}{\hbar^2}\right)^{1/2}\int_{R_0}^{R}\left(\frac{R}{r} - 1\right)^{1/2} dr \\
&= -2\left(\frac{2mE}{\hbar^2}\right)^{1/2} R\left[\cos^{-1}\left(\frac{R_0}{R}\right)^{1/2} - \left(\frac{R_0}{R}\right)^{1/2}\left(1 - \frac{R_0}{R}\right)^{1/2}\right]
\end{aligned} \tag{12.37}$$

Because the potential barrier is relatively wide, $R \gg R_0$, and

$$\cos^{-1}\left(\frac{R_0}{R}\right)^{1/2} \approx \frac{\pi}{2} - \left(\frac{R_0}{R}\right)^{1/2}$$

$$\left(1 - \frac{R_0}{R}\right)^{1/2} \approx 1$$

with the result that

$$\ln T = -2\left(\frac{2mE}{\hbar^2}\right)^{1/2} R\left[\frac{\pi}{2} - 2\left(\frac{R_0}{R}\right)^{1/2}\right]$$

From Eq. (12.35),

$$R = \frac{2Ze^2}{4\pi\epsilon_0 E}$$

and so

$$\ln T = \frac{4e}{\hbar}\left(\frac{m}{\pi\epsilon_0}\right)^{1/2} Z^{1/2}R_0^{1/2} - \frac{e^2}{\hbar\epsilon_0}\left(\frac{m}{2}\right)^{1/2} ZE^{-1/2} \tag{12.38}$$

The result of evaluating the various constants in Eq. (12.38) is

$$\ln T = 2.97Z^{1/2}R_0^{1/2} - 3.95ZE^{-1/2} \tag{12.39}$$

where E (the alpha-particle kinetic energy) is expressed in MeV, R_0 (the nuclear radius) is expressed in fermis (1 fm $= 10^{-15}$ m), and Z is the atomic number of the nucleus minus the alpha particle. Since

$$\log_{10} A = (\log_{10} e)(\ln A) = 0.4343 \ln A$$

we have

$$\log_{10} T = 1.29Z^{1/2}R_0^{1/2} - 1.72ZE^{-1/2} \tag{12.40}$$

From Eqs. (12.12) and (12.13) the decay constant λ is given by

$$\lambda = \nu T = \frac{v}{2R_0}T$$

where v is the alpha-particle velocity. Taking the logarithm of both sides and substituting for the transmission probability T gives

Alpha decay constant

$$\log_{10}\lambda = \log_{10}\left(\frac{v}{2R_0}\right) + 1.29Z^{1/2}R_0^{1/2} - 1.72ZE^{-1/2} \tag{12.14}$$

This is the formula quoted at the end of Sec. 12.4 and plotted in Fig. 12.9.

EXERCISES

What we have to learn to do, we learn by doing. —Aristotle

12.2 Half-Life

1. Tritium ($^3_1$H) has a half-life of 12.5 y against beta decay. What fraction of a sample of tritium will remain undecayed after 25 y?

2. The most probable energy of a thermal neutron is 0.025 eV at room temperature. In what distance will half of a beam of 0.025-eV neutrons have decayed? The half-life of the neutron is 10.3 min.

3. Find the probability that a particular nucleus of ^{38}Cl will undergo beta decay in any 1.00-s period. The half-life of ^{38}Cl is 37.2 min.

4. The activity of a certain radionuclide decreases to 15 percent of its original value in 10 d. Find its half-life.

5. The half-life of ^{24}Na is 15.0 h. How long does it take for 80 percent of a sample of this nuclide to decay?

6. The radionuclide ^{24}Na beta-decays with a half-life of 15.0 h. A solution that contains 0.0500 μCi of ^{24}Na is injected into a person's bloodstream. After 4.50 h the activity of a sample of the person's blood is found to be 8.00 pCi/cm^3. How many liters of blood does the person's body contain?

7. One g of ^{226}Ra has an activity of nearly 1 Ci. Determine the half-life of ^{226}Ra.

8. The mass of a millicurie of ^{214}Pb is 3.0×10^{-14} kg. Find the decay constant of ^{214}Pb.

9. The half-life of $^{238}_{92}$U against alpha decay is 4.5×10^9 y. Find the activity of 1.0 g of ^{238}U.

10. Use the data in the Appendix to this book to verify the statement at the end of Sec. 12.1 that the activity of ordinary potassium is about 0.7 μCi per kilogram due to its ^{40}K content.

11. The half-life of the alpha-emitter ^{210}Po is 138 d. What mass of ^{210}Po is needed for a 10-mCi source?

12. The energy of the alpha particles emitted by ^{210}Po ($T_{1/2}$ = 138 d) is 5.30 MeV. (*a*) What mass of ^{210}Po is needed to power a thermoelectric cell of 1.00-W output if the efficiency of energy conversion is 8.00 percent? (*b*) What would the power output be after 1.00 y?

13. The activity R of a sample of an unknown radionuclide is measured at hourly intervals. The results, in MBq, are 80.5, 36.2, 16.3, 7.3, and 3.3. Find the half-life of the radionuclide in the following way. First, show that, in general, $\ln(R/R_0) = -\lambda t$. Next, plot $\ln(R/R_0)$ versus t and find λ from the resulting curve. Finally calculate $T_{1/2}$ from λ.

14. The activity of a sample of an unknown radionuclide is measured at daily intervals. The results, in MBq, are 32.1, 27.2, 23.0, 19.5, and 16.5. Find the half-life of the radionuclide.

15. A rock sample contains 1.00 mg of ^{206}Pb and 4.00 mg of ^{238}U, whose half-life is 4.47×10^9 y. How long ago was the rock formed?

16. In Example 12.5 it is noted that the present radiocarbon activity of living things is 16 disintegrations per minute per gram of their carbon content. From this figure find the ratio of ^{14}C to ^{12}C atoms in the CO_2 of the atmosphere.

17. The relative radiocarbon activity in a piece of charcoal from the remains of an ancient campfire is 0.18 that of a contemporary specimen. How long ago did the fire occur?

18. Natural thorium consists entirely of the alpha-radioactive isotope ^{232}Th which has a half-life of 1.4×10^{10} y. If a rock sample known to have solidified 3.5 billion years ago contains 0.100 percent of ^{232}Th today, what was the percentage of this nuclide it contained when the rock solidified?

19. As discussed in this chapter, the heaviest nuclides are probably created in supernova explosions and become distributed in the galactic matter from which later stars (and their planets) form. Under the assumption that equal amounts of the ^{235}U and ^{238}U now in the earth were created in this way in the same supernova, calculate how long ago this occurred from their respective observed relative abundances of 0.7 and 99.3 percent and respective half-lives of 7.0×10^8 y and 4.5×10^9 y.

12.3 Radioactive Series

20. In the uranium decay series that begins with ^{238}U, ^{214}Bi beta-decays into ^{214}Po with a half-life of 19.9 min. In turn ^{214}Po alpha-decays into ^{210}Pb with a half-life of 163 μs, and ^{210}Pb beta-decays with a half-life of 22.3 y. If these three nuclides are in radioactive equilibrium in a mineral sample that contains 1.00 g of ^{210}Pb, what are the masses of ^{214}Bi and ^{214}Po in the sample?

21. The radionuclide $^{238}_{92}$U decays into a lead isotope through the successive emissions of eight alpha particles and six electrons. What is the symbol of the lead isotope? What is the total energy released?

12.4 Alpha Decay

22. The radionuclide ^{232}U alpha-decays into ^{228}Th. (*a*) Find the energy released in the decay. (*b*) Is it possible for ^{232}U to decay into ^{231}U by emitting a neutron? (*c*) Is it possible for ^{232}U to decay into ^{231}Pa by emitting a proton? The atomic masses of ^{231}U and ^{231}Pa are respectively 231.036270 u and 231.035880 u.

23. Derive Eq. (12.11), $KE_\alpha = (A-4)Q/A$, for the kinetic energy of the alpha particle released in the decay of a nucleus of mass number A. Assume that the ratio M_α/M_d between the mass of an alpha particle and the mass of the daughter is $\approx 4/(A-4)$.

24. The energy liberated in the alpha decay of ^{226}Ra is 4.87 MeV. (*a*) Identify the daughter nuclide. (*b*) Find the energy of the alpha particle and the recoil energy of the daughter atom. (*c*) If the alpha particle has the energy in *b* within the nucleus, how many of its de Broglie wavelengths fit inside the nucleus? (*d*) How many times per second does the alpha particle strike the nuclear boundary?

12.5 Beta Decay

25. Positron emission resembles electron emission in all respects except that the shapes of their respective energy spectra are different: there are many low-energy electrons emitted, but few low-energy positrons. Thus the average electron energy in beta decay is about $0.3KE_{max}$, whereas the average positron energy is about $0.4KE_{max}$. Can you suggest a simple reason for this difference?

26. By how much must the atomic mass of a parent exceed the atomic mass of a daughter when (*a*) an electron is emitted, (*b*) a positron is emitted, and (*c*) an electron is captured?

27. The nuclide ^{7}Be is unstable and decays into ^{7}Li by electron capture. Why does it not decay by positron emission?

28. Show that it is energetically possible for ^{64}Cu to undergo beta decay by electron emission, positron emission, and electron capture and find the energy released in each case.

29. Carry out the calculations of Exercise 28 for ^{80}Br.

30. Calculate the maximum energy of the electrons emitted in the beta decay of ^{12}B.

31. Find the minimum antineutrino energy needed to produce the inverse beta-decay reaction $p + \overline{\nu} \rightarrow n + e^+$.

32. Find the neutrino energy required to initiate the reaction $\nu + {}^{37}\text{Cl} \rightarrow {}^{37}\text{Ar} + e^-$ by which solar neutrinos are detected in Davis's experiment.

12.6 Gamma Decay

33. Determine the ground and lowest excited states of the thirty-ninth proton in ^{89}Y with the help of Fig. 11.17. Use this information to explain the isomerism of ^{89}Y together with the fact, noted in Sec. 6.9, that radiative transitions between states with very different angular momenta are extremely improbable.

34. When an excited nucleus emits a gamma-ray photon, some of the excitation energy goes into the kinetic energy of the recoil of the nucleus. (*a*) Find the ratio between the recoil energy and the photon energy when the nucleus of an atom of mass 200 u emits a 2.0-MeV gamma ray. (*b*) The lifetime of an excited nuclear state is typically about 10^{-14} s. Compare the corresponding uncertainty in the energy of the excited state with the recoil energy. (See Exercise 53 of Chap. 2 to learn how the Mössbauer effect can minimize nuclear recoil.)

12.7 Cross Section

35. The cross sections for comparable neutron- and proton-induced nuclear reactions vary with energy in approximately the manner shown in Fig.12.32. Why does the neutron cross section decrease with increasing energy whereas the proton cross section increases?

36. A slab of absorber is exactly one mean free path thick for a beam of certain incident particles. What percentage of the particles will emerge from the slab?

37. The capture cross section of ^{59}Co for thermal neutrons is 37 b. (*a*) What percentage of a beam of thermal neutrons will penetrate a 1.0-mm sheet of ^{59}Co? The density of ^{59}Co is 8.9×10^3 kg/m^3. (*b*) What is the mean free path of thermal neutrons in ^{59}Co?

38. The cross section for the interaction of a neutrino with matter is $\sim 10^{-47}$ m^2. Find the mean free path of neutrinos in solid iron, whose density is 7.8×10^3 kg/m^3 and whose average atomic mass is 55.9 u. Express the answer in light-years, the distance light travels in free space in a year.

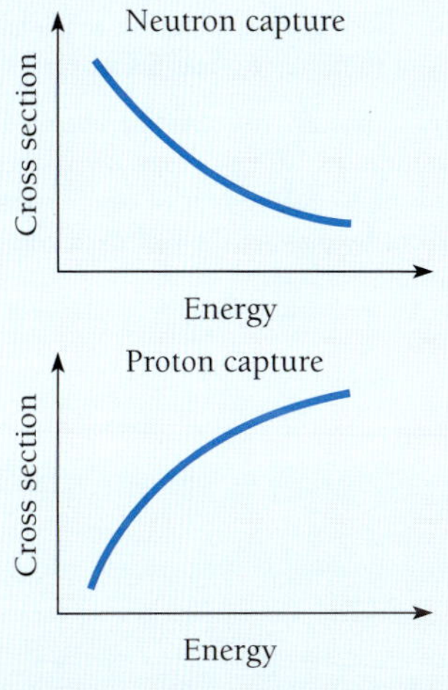

Figure 12.32 Cross sections for neutron and proton capture vary differently with particle energy.

39. The boron isotope ^{10}B captures neutrons in an (n, α)—neutron in, alpha particle out—reaction whose cross section for thermal neutrons is 4.0×10^3 b. The density of ^{10}B is 2.2×10^3 kg/m^3. What thickness of ^{10}B is needed to absorb 99 percent of an incident beam of thermal neutrons?

40. There are approximately 6×10^{28} atoms/m^3 in solid aluminum. A beam of 0.5-MeV neutrons is directed at an aluminum foil 0.1 mm thick. If the capture cross section for neutrons of this energy in aluminum is 2×10^{-31} m^2, find the fraction of incident neutrons that are captured.

41. Natural cobalt consists entirely of the isotope ^{59}Co whose cross section for thermal neutron capture is 37 b. When ^{59}Co absorbs a neutron, it becomes ^{60}Co, which is gamma-radioactive with a half-life of 5.27 y. If a 10.0-g cobalt sample is exposed to a thermal-neutron flux of 5.00×10^{17} neutrons/m$^2 \cdot$ s for 10.0 h, what is the activity of the sample afterward?

42. Natural sodium consists entirely of the isotope ^{23}Na whose cross section for thermal neutron capture is 0.53 b. When ^{23}Na absorbs a neutron, it becomes ^{24}Na, which is beta-radioactive with a half-life of 15.0 h. A sample of a material that contains sodium is placed in a thermal neutron beam whose flux is 2.0×10^{18} neutrons/m$^2 \cdot$ s for 1.00 h. The activity of the sample is then 5.0 μCi. How much sodium was present in the sample? (This is an example of **neutron activation analysis**, a very sensitive technique.)

12.8 Nuclear Reactions

43. Complete these nuclear reactions:

$$^{6}_{3}\text{Li} + ? \rightarrow {}^{7}_{4}\text{Be} + {}^{1}_{0}n$$

$$^{35}_{17}\text{Cl} + ? \rightarrow {}^{32}_{16}\text{S} + {}^{4}_{2}\text{He}$$

$$^{9}_{4}\text{Be} + {}^{4}_{2}\text{He} \rightarrow 3\,{}^{4}_{2}\text{He} + ?$$

$$^{79}_{35}\text{Br} + {}^{2}_{1}\text{H} \rightarrow ? + 2\,{}^{1}_{0}n$$

44. Find the minimum energy in the laboratory system that a neutron must have in order to initiate the reaction

$$^{1}_{0}n + {}^{16}_{8}\text{O} + 2.20\text{ MeV} \rightarrow {}^{13}_{6}\text{C} + {}^{4}_{2}\text{He}$$

45. Find the minimum energy in the laboratory system that a proton must have in order to initiate the reaction

$$p + d + 2.22\text{ MeV} \rightarrow p + p + n$$

46. Find the minimum kinetic energy in the laboratory system a proton must have to initiate the reaction ^{15}N (p, n)^{15}O.

47. A 5-MeV alpha particle strikes a stationary $^{16}_{8}$O target. Find the speed of the center of mass of the system and the kinetic energy of the particles relative to the center of mass.

48. A thermal neutron induces the reaction of Exercise 39. Find the kinetic energy of the alpha particle.

49. An alpha particle collides elastically with a stationary nucleus and continues on at an angle of 60° with respect to its original direction of motion. The nucleus recoils at an angle of 30° with respect to this direction. What is the mass number of the nucleus?

50. Neutrons were discovered with the help of the reaction ${}^{9}_{4}\text{Be}(\alpha, n){}^{12}_{6}\text{C}$ that occurs when alpha particles of 5.30 MeV energy (in the lab system) from the decay of the polonium isotope ${}^{210}\text{Po}$ are incident on ${}^{9}\text{Be}$ nuclei (see Fig. 11.2). What is the energy available for the reaction in the center-of-mass system?

51. (*a*) A particle of mass m_A and kinetic energy KE_A strikes a stationary nucleus of mass m_B to produce a compound nucleus of mass m_C. Express the excitation energy of the compound nucleus in terms of m_A, m_C, KE_A, and the Q value of the reaction. (*Note*: $|Q| \ll mc^2$.) (*b*) An excited state in ${}^{16}\text{O}$ occurs at an energy of 16.2 MeV. Find the kinetic energy needed by a proton to produce a ${}^{16}\text{O}$ nucleus in this state by reaction with a stationary ${}^{15}\text{N}$ nucleus.

52. (*a*) Find the minimum kinetic energy in the laboratory system a proton must have to react with ${}^{65}_{29}\text{Cu}$ to produce ${}^{65}_{30}\text{Zn}$ and a neutron. (*b*) Find the minimum kinetic energy a proton must have to come in contact with a ${}^{65}_{29}\text{Cu}$ nucleus. (*c*) If the energy in *b* is greater than the energy in *a*, is there any way in which a proton with the energy in *a* can react with ${}^{65}_{29}\text{Cu}$?

12.9 Nuclear Fission

53. When fission occurs, several neutrons are released and the fission fragments are beta-radioactive. Why?

54. ${}^{235}\text{U}$ loses about 0.1 percent of its mass when it undergoes fission. (*a*) How much energy is released when 1 kg of ${}^{235}\text{U}$ undergoes fission? (*b*) One ton of TNT releases about 4 GJ when it is detonated. How many tons of TNT are equivalent in destructive power to a bomb that contains 1 kg of ${}^{235}\text{U}$?

55. Assume that immediately after the fission event shown in Fig. 12.17 the fission fragment nuclei are spherical and in contact. What is the potential energy of this system?

56. Use the semiempirical binding-energy formula of Eq. (11.18) to calculate the energy that would be released if a ${}^{238}\text{U}$ nucleus were to split into two identical fragments.

12.10 Nuclear Reactors

57. What is the limitation on the fuel that can be used in a reactor whose moderator is ordinary water? Why is the situation different if the moderator is heavy water?

58. (*a*) How much mass is lost per day by a nuclear reactor operated at a 1.0-GW power level? (*b*) If each fission releases 200 MeV, how many fissions occur per second to yield this power level?

59. A particle of mass m_1 and kinetic energy KE_1 collides head on with a stationary particle of mass m_2. The two particles then move apart with the target particle having the kinetic energy KE_2'. (*a*) Use conservation of momentum and conservation of kinetic energy in a non-relativistic calculation to show that $\text{KE}_2'/\text{KE}_1 = 4(m_2/m_1)/(1 + m_2/m_1)^2$, which is what is plotted in Fig. 12.22. (*b*) What percentage of its initial KE does a neutron lose when it collides head on with a proton? With a deuteron? With a ${}^{12}\text{C}$ nucleus? With a ${}^{238}\text{U}$ nucleus? (Ordinary water, heavy water, and carbon in the form of graphite have all been used as moderators in nuclear reactors.)

12.11 Nuclear Fusion in Stars

60. In their old age, heavy stars obtain part of their energy by the reaction

$${}^{4}_{2}\text{He} + {}^{12}_{6}\text{C} \rightarrow {}^{16}_{8}\text{O}$$

How much energy does each such event give off?

61. The initial reaction in the carbon cycle from which stars hotter than the sun obtain their energy is

$${}^{1}_{1}\text{H} + {}^{12}_{6}\text{C} \rightarrow {}^{13}_{7}\text{N} + \gamma$$

Find the minimum energy the proton must have to come in contact with the ${}^{12}_{6}\text{C}$ nucleus.

62. Find the energy released in each step of the carbon cycle shown in Fig. 12.27 and add them up to find the total. Neglect the kinetic energies of the reacting particles, which are small compared with the Q values of the reactions. (*Hint*: Watch the electrons!)

12.12 Fusion Reactors

63. The electric repulsion between deuterons is a maximum when they are ~5 fm apart. (*a*) Find the temperature at which the deuterons in a plasma have average energies sufficient to surmount this potential barrier. (*b*) Fusion reactions between deuterons can take place at temperatures considerably below this figure. Can you think of two reasons why?

64. Show that the fusion energy that could be liberated in ${}^{2}_{1}\text{H} + {}^{2}_{1}\text{H}$ from the deuterium in 1.0 kg of seawater is about 600 times greater than the 47 MJ/kg heat of combustion of gasoline. About 0.015 percent by mass of the hydrogen content of seawater is deuterium.

CHAPTER 13

Elementary Particles

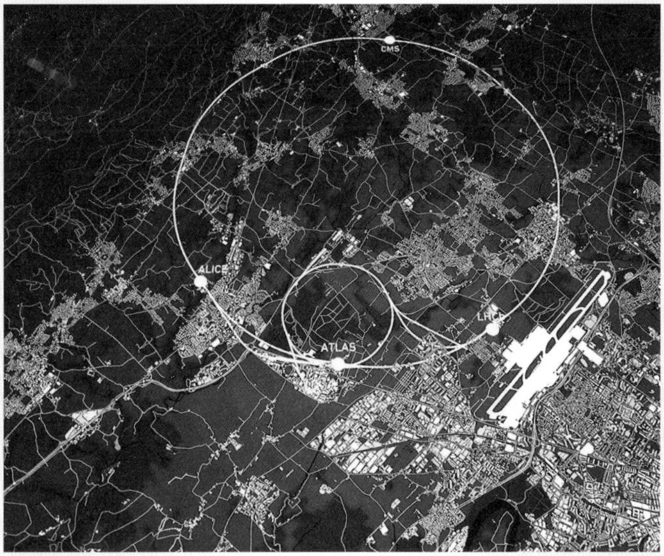

GENEVA, SWITZERLAND, 18 March 2017: Atlas map in CERN Interior

13.1 INTERACTIONS AND PARTICLES
Which affects which

13.2 LEPTONS
Three pairs of truly elementary particles

13.3 HADRONS
Particles subject to the strong interaction

13.4 ELEMENTARY PARTICLE QUANTUM NUMBERS
Finding order in apparent chaos

13.5 QUARKS
The ultimate constituents of hadrons

13.6 FIELD BOSONS
Carriers of the interactions

13.7 THE STANDARD MODEL AND BEYOND
Putting it all together

13.8 HISTORY OF THE UNIVERSE
It began with a bang

13.9 THE FUTURE
"In my beginning is my end." (T. S. Eliot, Four *Quartets)*

Ordinary matter is composed of protons, neutrons, and electrons, and at first glance these particles seem enough to account for the structure of the universe around us. Not all nuclides are stable, however, and neutrinos are needed for beta decay to take place—indeed, without neutrinos the reaction sequences that power the stars and that lead to the creation of elements heavier than hydrogen could not occur. Furthermore, as discussed in Sec. 11.7, the electromagnetic interaction between charged particles requires photons as its carrier, and the specifically nuclear interaction between nucleons requires pions for the same purpose. Even so, only a few particles seem to be needed, all of them with clearly defined roles to play.

But things are not nearly so straightforward. Hundreds of other "elementary" particles have been discovered, all of which decay rapidly after being created in high-energy collisions between other particles. It has become clear that some of these particles (called leptons) are more elementary than the others, and that the others (called hadrons) are composites of a far smaller number of rather unusual particles called quarks that have not been detected in isolation (and probably will never be).

13.1 INTERACTIONS AND PARTICLES

Which affects which

The four interactions we already know about—strong, electromagnetic, weak, and gravitational—are apparently enough to account for all the physical processes and structures in the universe on all scales of size from atoms and nuclei to galaxies of stars. The basic characteristics of these interactions are given in Table 13.1.

The list of fundamental interactions has changed over the years. Long ago, the strong and weak interactions were unknown and it was not even clear that the gravity that pulls things down to the earth, which we might call terrestrial gravity, is the same as the gravity that holds the planets to their orbits around the sun. One of Newton's great accomplishments was to show that both terrestrial and astronomical gravity are the same. Another notable unification was made by Maxwell when he demonstrated that

Table 13.1 The Four Fundamental Interactions. The graviton has not been experimentally detected as yet.

Interaction	Particles Affected	Range	Relative Strength	Particles Exchanged	Role in Universe
Strong	Quarks	$\sim 10^{-15}$ m	1	Gluons	Holds quarks together to form nucleons
	Hadrons			Mesons	Holds nucleons together to form atomic nuclei
Electromagnetic	Charged particles	∞	$\sim 10^{-2}$	Photons	Determines structures of atoms, molecules, solids, and liquids; is important factor in astronomical universe
Weak	Quarks and leptons	$\sim 10^{-18}$ m	$\sim 10^{-5}$	Intermediate bosons	Mediates transformations of quarks and leptons; helps determine compositions of atomic nuclei
Gravitational	All	∞	$\sim 10^{-39}$	Gravitons	Assembles matter into planets, stars, and galaxies

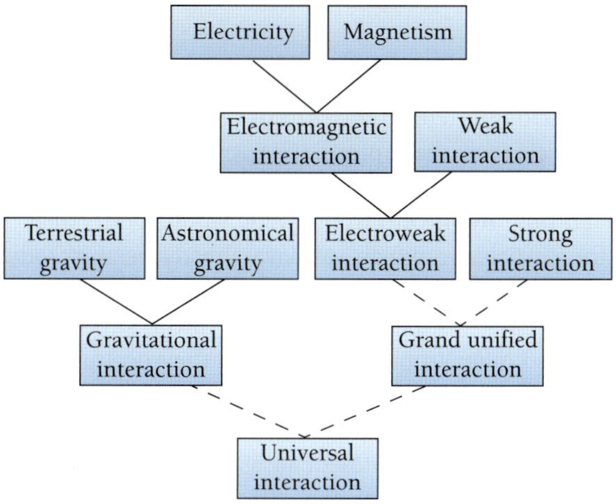

Figure 13.1 One of the goals of physics is a single theoretical picture that unites all the ways in which particles of matter interact with each other. Much progress has been made, but the task is not finished.

electric and magnetic forces can both be traced to a single interaction between charged particles.

As we shall see, the electromagnetic and weak interactions turn out to be different manifestations of a single **electroweak** interaction. This in turn seems to have links to the strong interaction, though the details of the relationship are still not entirely clear. The final step in understanding how nature operates would be a single picture that includes gravitation, and there are strong hints that such a Theory of Everything is not beyond reach (Fig. 13.1).

The relative strengths of the various interactions span 39 powers of 10 and the distances through which they are effective are also very different. While the strong force between nearby nucleons completely overwhelms the gravitational force between them, when they are a millimeter apart the reverse is true. The structures of nuclei are determined by the properties of the strong interaction, while the structures of atoms are determined by those of the electromagnetic interaction. Matter in bulk is electrically neutral, and the strong and weak interactions are severely limited in range. Hence the gravitational interaction, utterly insignificant on a small scale, becomes the dominant one on a large scale. The role of the weak force in the structure of matter is apparently that of a minor perturbation that sees to it that nuclei with inappropriate neutron/proton ratios undergo corrective beta decays.

The universe would be very different if the strengths of the various interactions had other values. For instance, as mentioned in Sec. 11.4, if the strong interaction were more than a trifle stronger, the universe would be filled with diprotons and the fusion reactions that give energy to the stars and create the chemical elements could not take place. If the strong interaction were weaker, protons could not combine with neutrons, also eliminating the exothermic fusion path to helium and heavier elements. The gravitational interaction is in a similar state of balance. If it were more powerful, stellar interiors would be hotter, their fusion reactions would occur more often, and stars would burn out sooner—perhaps too soon for life to have developed on their planets. Significantly weaker gravity, on the other hand, would not have clumped matter into stars to begin with. One of the tasks of a Theory of Everything is to establish why the fundamental interactions and the particles they affect have the properties they do.

Leptons and Hadrons

Elementary particles fall into two classes, **leptons** and **hadrons,** depending on whether they respond to the strong interaction (hadrons) or do not (leptons).

The simplest particles are the leptons (Greek: "light," "swift"), which seem to be truly elementary with no hint of internal structures or even of extension in space. Leptons are affected only by the electromagnetic (if charged), weak, and gravitational interactions. Of the particles to which we have already been introduced, the electron and the neutrino are leptons; there are four other types.

Hadrons (Greek: "heavy," "strong") are subject to the strong interaction as well as to the others. They also differ from leptons in that they occupy space, rather than being infinitesimal in size: hadrons seem to be a little over 1 fm (10^{-15} m) across. Hadrons are composed of either two or three **quarks,** which, like leptons, are structureless and as close to being point particles as present measurements can establish. Hadrons that consist of three quarks, such as the proton and neutron, are called **baryons; mesons,** such as the pion, consist of two quarks. Like nothing else in nature, quarks have charges of $\pm\frac{1}{3}e$ or $\pm\frac{2}{3}e$, and their combination in hadrons is always such that the hadron charges are either 0 or $\pm e$. Quarks have never been observed outside of hadrons, but, as we shall see, there is convincing evidence that they do exist. The strong force that acts between hadrons is the external manifestation of the more basic interactions among the quarks they contain and is mediated by the exchange of mesons, as described in Sec.11.7.

13.2 LEPTONS

Three pairs of truly elementary particles

Table 13.2 lists the six known leptons and their antiparticles. Because the neutrinos involved in beta decays, which were discussed in Chap. 12, are associated with electrons, their proper symbol is ν_e.

The electron was the first elementary particle for which a satisfactory theory was developed. This theory was proposed in 1928 by Paul A. M. Dirac, who obtained a relativistically correct wave equation for a charged particle in an electromagnetic field. When the observed mass and charge of the electron are inserted in the solutions of this equation, the intrinsic angular momentum of the electron is found to be $\frac{1}{2}\hbar$ (that

Table 13.2 **Leptons. All are unaffected by the strong interaction and are fermions. The neutrinos are uncharged; their masses are unknown but unlikely to exceed a few eV/c^2.**

Lepton	Symbol	Antiparticle	Mass, MeV/c^2	Mean Life, s	Spin
Electron	e^-	e^+	0.511	Stable	$\frac{1}{2}$
e-neutrino	ν_e	$\bar{\nu}_e$	Very small	Stable	$\frac{1}{2}$
Muon	μ^-	μ^+	106	2.2×10^{-6}	$\frac{1}{2}$
μ-neutrino	ν_μ	$\bar{\nu}_\mu$	Very small	Stable	$\frac{1}{2}$
Tau	τ^-	τ^+	1777	2.9×10^{-23}	$\frac{1}{2}$
τ-neutrino	ν_τ	$\bar{\nu}_\tau$	Very small	Stable	$\frac{1}{2}$

Antimatter

There seems to be no reason why atoms could not be composed of antiprotons, antineutrons, and positrons. Such **antimatter** ought to behave exactly like ordinary matter. If galaxies of antimatter stars existed, their spectra would not differ from the spectra of galaxies of matter stars. Thus we have no way to distinguish between the two kinds of galaxies—except when antimatter from one comes in contact with matter from the other. Mutual annihilation would then occur with the release of an immense amount of energy. (A postage stamp of antimatter annihilating a postage stamp of matter would give enough energy to send the space shuttle into orbit.) But the gamma rays of characteristic energies that such an event would create have never been observed, nor have antiparticles ever been identified in the cosmic rays that reach the earth from space. It seems the universe consists entirely of ordinary matter.

is, spin $\frac{1}{2}$) and its magnetic moment is found to be $e\hbar/2m$, one Bohr magneton. These predictions agree with experiment, and the agreement is strong evidence for the correctness of the Dirac theory.

An unexpected result of Dirac's theory was its requirement that an electron can have negative as well as positive energies. That is, when the relativistic formula for total energy

$$E = \sqrt{m^2c^4 + p^2c^2}$$

is applied to electrons, both the negative and positive roots are acceptable solutions. But if negative energy states going all the way to $E = -\infty$ are possible, what keeps all the electrons in the universe from ending up with negative energies? The existence of stable atoms is by itself evidence that electrons are not subject to such a fate.

Dirac rescued his theory by suggesting that all negative energy states are normally filled. The Pauli exclusion principle then prevents any other electrons from dropping into the negative states. But if an electron in the sea of filled negative states is given enough energy, say by absorbing a photon of energy $h\nu > 2mc^2$, it can jump out of this sea and become an electron with a positive energy (Fig. 13.2). This process leaves behind a hole in the negative-energy electron sea which, just like a hole in a semiconductor energy band, behaves as if it is a particle of positive charge—a positron. The result is the materialization of the photon into an electron-positron pair, $\gamma \rightarrow e^- + e^+$, as described in Sec. 2.8.

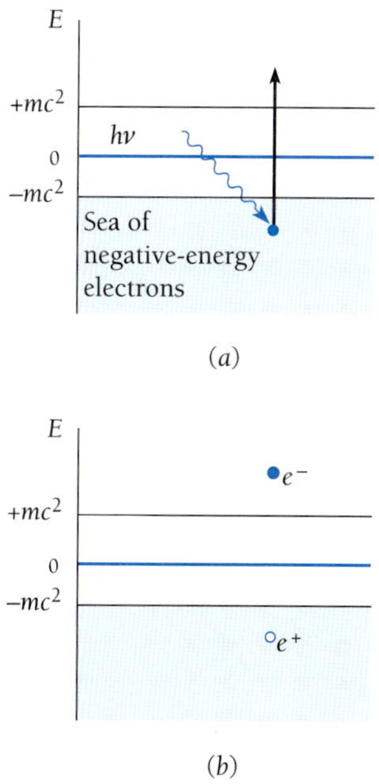

Figure 13.2 Electron-positron pair production. (*a*) A photon of energy $h\nu > 2mc^2$ ($>$1.02 MeV) is absorbed by a negative-energy electron, which gives the electron a positive energy. (*b*) The resulting hole in the negative-energy electron sea behaves like an electron of positive charge.

When Dirac developed his theory, the positron was unknown, and it was speculated that the proton might be the positive counterpart of the electron despite their difference in mass. Finally, in 1932, Carl Anderson unambiguously detected a positron in the stream of secondary particles that result from collisions between cosmic rays and atomic nuclei in the atmosphere.

The positron is the **antiparticle** of the electron. All other elementary particles also have antiparticles; a few, such as the neutral pion, are their own antiparticles. The antiparticle of a particle has the same mass, spin, and lifetime if unstable, but its charge (if any) has the opposite sign. The alignment or antialignment between its spin and magnetic moment is also opposite to that of the particle.

Neutrinos and Antineutrinos

The distinction between the neutrino ν and the antineutrino $\bar{\nu}$ is a particularly interesting one. The spin of the neutrino is opposite in direction to the direction of its

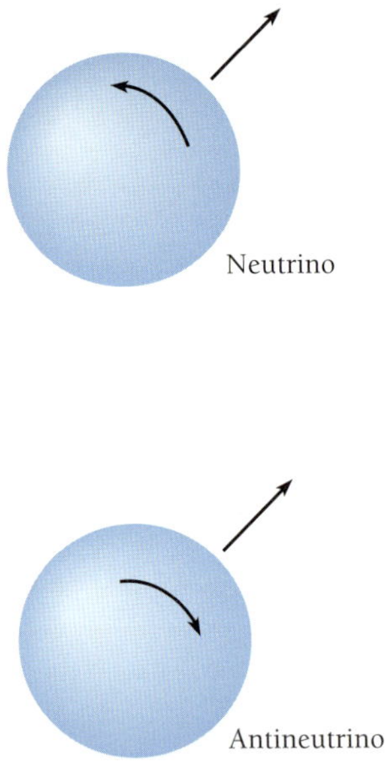

Figure 13.3 Neutrinos and antineutrinos have opposite directions of spin.

motion; viewed from behind, as in Fig. 13.3, the neutrino spins counterclockwise. The spin of the antineutrino, on the other hand, is in the same direction as its direction of motion; viewed from behind, it spins clockwise. Thus the neutrino moves through space in the manner of a left-handed screw, while the antineutrino does so in the manner of a right-handed screw.

Prior to 1956 it had been universally assumed that neutrinos could be either left-handed or right-handed. This implied that, since no difference was possible between them except one of spin direction, the neutrino and antineutrino are identical. The assumption had roots going all the way back to Leibniz, Newton's contemporary and an independent inventor of calculus. The argument is as follows. If we observe an object or a physical process of some kind both directly and in a mirror, we cannot ideally distinguish which object or process is being viewed directly and which by reflection. By definition, distinctions in physical reality must be capable of discernment or they are meaningless. But the only difference between something seen directly and the same thing seen in a mirror is the interchange of right-handedness and left-handedness, and so *all* objects and processes must occur with equal probability with right and left interchanged.

This plausible doctrine is indeed experimentally valid for the strong and electromagnetic interactions. However, until 1956 its applicability to neutrinos, which are subject only to the weak interaction, had never been actually tested. In that year Tsung Dao Lee and Chen Ning Yang suggested that several serious theoretical discrepancies would be removed if neutrinos and antineutrinos have different handedness, even though it meant that neither particle could therefore be reflected in a mirror. Experiments performed soon after their proposal showed unequivocally that neutrinos and antineutrinos are distinguishable, having left-handed and right-handed spins respectively.

Other Leptons

The **muon,** μ, and its associated neutrino ν_μ were first discovered in the decays of charged pions:

Charged pion decay $$\pi^+ \rightarrow \mu^+ + \nu_\mu \qquad \pi^- \rightarrow \mu^- + \bar{\nu}_\mu \tag{13.1}$$

The pion was discussed in Sec. 11.7 in connection with the strong force between nucleons, which it mediates. The pion's mass is intermediate between those of the electron and the proton, and it is unstable with a mean life of 2.6×10^{-8} s for $\pi^{\pm}$. The neutral pion has a mean life of 8.7×10^{-17} s and decays into two gamma rays:

Neutral pion decay $$\pi^0 \rightarrow \gamma + \gamma \tag{13.2}$$

The neutrinos involved in pion decays are not the same as those involved in beta decay. The existence of another class of neutrino was established in 1962. A metal target was bombarded with high-energy protons, and pions were created in profusion. Inverse reactions traceable to the neutrinos from the decay of these pions produced muons only, and no electrons. Hence these neutrinos must be different in some way from those associated with beta decay.

Positive and negative muons have the same rest mass of 106 MeV/c^2 (207 m_e) and the same spin of $\frac{1}{2}$. Both decay with a relatively long mean life of 2.2×10^{-6} s into electrons and neutrino-antineutrino pairs:

Muon decay $$\mu^+ \rightarrow e^+ + \nu_e + \bar{\nu}_\mu \qquad \mu^- \rightarrow e^- + \nu_\mu + \bar{\nu}_e \tag{13.3}$$

As with electrons, the positive-charge state of the muon represents the antiparticle. There is no neutral muon.

Because the decay of the muon is relatively slow and because, like all leptons, it is not subject to the strong interaction, muons readily penetrate considerable amounts of matter. The great majority of cosmic-ray secondary particles at sea level are muons. The muon lifetime is long enough for a negative muon sometimes to temporarily replace an atomic-electron to form a muonic atom (see Example 4.7).

The final pair of leptons is the **tau,** τ, which was discovered in 1975, and its associated neutrino ν_τ whose existence was not confirmed experimentally until 2000. The mass of the tau is 1777 MeV/c^2, almost double that of the proton, and its mean life is very short, only 2.9×10^{-23} s. All taus are charged and decay into electrons, muons, or pions along with appropriate neutrinos.

The Solar Neutrino Mystery

An immense number of neutrinos are produced in the sun and other stars in the course of the nuclear reactions that occur within them, and these neutrinos are apparently able to travel freely throughout the universe. Several percent of the energy released in such reactions is carried away by the neutrinos.

In the case of the sun, its observed luminosity implies a neutrino production rate of around 2×10^{38} per second, which means that 60 billion or so neutrinos should pass through each square centimeter of the earth's surface per second. To detect the most energetic of these neutrons, Raymond Davis installed a detector in an abandoned gold mine 1.5 km underground in South Dakota to prevent interference from cosmic rays. The detector contained 600 tons of the dry-cleaning liquid perchlorethylene, C_2Cl_4, and the reaction

$$\nu_e + {}^{37}_{17}\text{Cl} \rightarrow {}^{37}_{18}\text{Ar} + e^-$$

was looked for. The argon isotope ${}^{37}_{18}\text{Ar}$ remains in the liquid as a dissolved gas and can be separated out and identified by its beta decay back to ${}^{37}_{17}\text{Cl}$.

During eighteen years of operation only about a quarter as many neutrino interactions were observed (less than one per day) as were expected on the basis of an otherwise plausible model of the solar interior. The discrepancy was well beyond uncertainties in the measurements and in the calculations. More recent work with methods that respond to lower-energy neutrinos showed a smaller discrepancy, but still a major one. Something serious was wrong either with the theory of how stars produce energy, which in all other respects agrees well with observations, or with theories of how neutrinos come into being, travel through space, and interact with matter, which have also proved successful in their other predictions.

One speculation was based on the existence of muon and tau neutrinos as well as electron neutrinos. If neutrinos have masses (very little is needed), then after its creation a neutrino of one type (or **flavor**) could oscillate between that flavor and another one or perhaps both others. Since the sun gives off only electron neutrinos, if some of them have a different flavor when they reach the earth, the number of electron neutrinos recorded here will be less than the number expected. We can think of each neutrino flavor not as a particle with a distinct identity but as a mixture of mass states whose waves travel with different velocities. The waves interfere, and as time goes on the likelihood of being observed fluctuates in amplitude among the various flavors.

This hypothesis was confirmed in 1998 in measurements made in Japan with the Super Kamiokanda detector, which monitored the Cerenkov radiation (see Sec. 1.2) given off by the debris of interactions between incoming neutrinos and nuclei present in a tank of 50,000 tons of water. The results indicated that muon neutrinos (produced in the decays of cosmic-ray pions and muons in the earth's atmosphere) indeed metamorphose to and from tau neutrinos. Further experiments will no doubt provide a definitive answer to whether electron neutrinos also undergo oscillations into another flavor or flavors. In the meantime the solar neutrino mystery no longer seems so mysterious and it appears that neutrinos do have mass, settling a question seventy years old.

13.3 HADRONS

Particles subject to the strong interaction

Unlike leptons, hadrons are subject to the strong interaction. Table 13.3 lists the hadrons with the longest lifetimes against decay into other particles. **Mesons** are bosons and consist of a quark and an antiquark; about 140 types are known. The lightest meson is the pion, with other meson masses ranging beyond the proton mass. **Baryons** are fermions and consist of three quarks; about 120 types are known. Of the hadrons listed, the π^0 and η^0 are their own antiparticles. The charged pions differ in charge, so they are antiparticles of each other, but have no other attributes that are different, so each is both a particle and an antiparticle.

The lightest baryon is the proton, which is also the only hadron stable in free space. Or apparently stable—current theories call for the proton to decay with a very long lifetime, perhaps longer than the experimentally determined lower limit of 10^{32} years. Hence the ultimate stability of the proton is still an open question. (For comparison, the age of the universe is a little over 10^{10} years.) The neutron, although stable inside a nucleus, beta-decays in free space into a proton, an electron, and an antineutrino with a mean life of 14 min 49 s.

Table 13.3 Some hadrons and their properties. The symbol *S* stands for strangeness number, discussed in Sec. 13.4. Antiparticles have strangeness numbers the negative of those shown.

Class	Particle	Symbol	Antiparticle	Mass, MeV/c^2	Mean Life, s	Spin	*S*
Mesons	Pion	π^+	π^-	140	2.6×10^{-8}	0	0
		π^0	Self	135	8.7×10^{-17}		
		π^-	π^+	140	2.6×10^{-8}		
	Kaon	K^+	K^-	494	1.2×10^{-8}	0	+1
		K_S^0	$\overline{K_S^0}$	498	8.9×10^{-11}		
		K_L^0	$\overline{K_L^0}$	498	5.2×10^{-8}		
	Eta	η^0	Self	549	5×10^{-19}	0	0
		η'	Self	958	2.2×10^{-21}		
Baryons	Nucleon: Proton	p	$\overline{p}$	938.3	Stable	$\frac{1}{2}$	0
	Nucleon: Neutron	n	$\overline{n}$	939.6	889		
	Lambda	Λ^0	$\overline{\Lambda^0}$	1116	2.6×10^{-10}	$\frac{1}{2}$	−1
	Sigma	Σ^+	$\overline{\Sigma^-}$	1189	8.0×10^{-11}	$\frac{1}{2}$	−1
		Σ^0	$\overline{\Sigma^0}$	1193	6×10^{-20}		
		Σ^-	$\overline{\Sigma^+}$	1197	1.5×10^{-10}		
	Xi	Ξ^0	$\overline{\Xi^0}$	1315	2.9×10^{-10}	$\frac{1}{2}$	−2
		Ξ^-	$\overline{\Xi^+}$	1321	1.6×10^{-10}		
	Omega	Ω^-	$\overline{\Omega^+}$	1672	8.2×10^{-11}	$\frac{3}{2}$	−3

One of the accelerator sections of the Large Hadron Collider(LHC) at CERN. In these sections protons and irons are accelerated by alternating electric fields. Magnetic fields are used to focus the particles and to keep them in circular paths during the millions of orbits during which they gain energy.

All baryons other than nucleons decay with mean lives of less than 10^{-9} s in a variety of ways, but the end result is always a proton or neutron. For example, here is one sequence which the Ω^- baryon can follow in its decay:

$$\begin{aligned} \Omega^- &\rightarrow \Xi^0 + \pi^- \\ &\qquad \searrow \Lambda^0 + \pi^0 \\ &\qquad\qquad \searrow p^+ + \pi^- \end{aligned}$$

The Ξ^0 and Λ^0 particles are successively lighter baryons than the Ω^-. The π^- and π^0 mesons themselves decay as described earlier, so the final result of the decay of the Ω^- is a proton, two electrons, four neutrinos, and two photons.

Resonance Particles

Most of the particles in Table 13.3 exist long enough to travel as distinct entities along paths of measurable length, and their modes of decay can be observed in various devices.

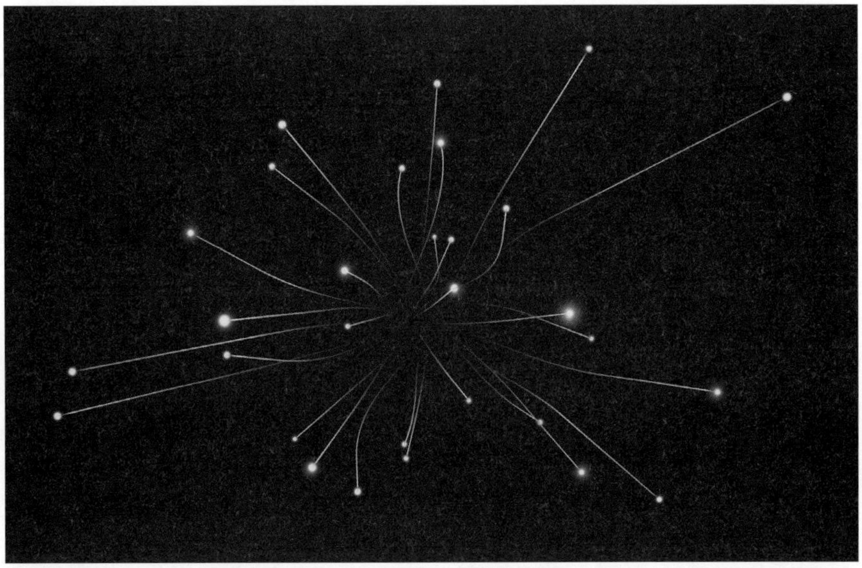

3D illustration of the results of the hadron collision. A variety of elementary particles are produced and their properties and decay schemes can be studied with giant detectors.

A large body of experimental evidence also points to the existence of many hadrons whose lifetimes may be only about 10^{-23} s. What can be meant by the idea of a particle that is in being for so brief an interval? Indeed, how can a time of 10^{-23} s be measured?

Ultra-short-lived particles cannot be detected by recording their creation and subsequent decay because the distance they cover in $\sim 10^{-23}$ s is only $\sim 3 \times 10^{-15}$ m even if they move at nearly the velocity of light—a length characteristic of hadron dimensions. Instead, such particles appear as resonant states in the interactions of longer-lived (and hence more readily observable) particles. Resonant states occur in atoms as energy levels; in Sec. 4.8 we reviewed the Franck-Hertz experiment, which demonstrated the existence of atomic energy levels by showing that inelastic electron scattering from atoms occurs only at certain energies.

An atom in a certain excited state is not the same as that atom in its ground state or in another excited state. However, such an excited atom is not spoken of as though it is a member of a special species only because the electromagnetic interaction that gives rise to the excited state is well understood. The situation is somewhat different for elementary particles because the weak and strong interactions that also govern them are more complicated and were not really understood until relatively recently.

Let us see what is involved in a resonance in the case of elementary particles. An experiment is performed, for instance the bombardment of protons by energetic π^+ mesons, and a certain reaction is studied, for instance

$$\pi^+ + p \rightarrow \pi^+ + p + \pi^+ + \pi^- + \pi^0$$

The effect of the interaction of the π^+ and the proton is the creation of three new pions. In each such reaction the new mesons have a certain total energy that consists of their rest energies plus their kinetic energies relative to their center of mass.

If we plot the number of events observed versus the total energy of the new mesons in each event, we obtain a graph like that of Fig. 13.4. Evidently there is a strong tendency for the total meson energy to be 783 MeV and a somewhat weaker tendency for it to be 549 MeV. We can say that the reaction exhibits resonances at 549 and 783 MeV or, equivalently, we can say that this reaction proceeds via the creation of an intermediate particle which can be either one whose mass is 549 MeV/c^2 or one whose mass is 783 MeV/c^2.

From Fig. 13.4 we can even estimate the mean lifetimes of these uncharged intermediate particles, which are known respectively as the η and ω mesons. In Chap. 12 we used the formula

Mean lifetime

$$\tau = \frac{\hbar}{\Gamma} \tag{12.23}$$

to relate the mean lifetime τ of an excited nuclear state to the width Γ at half-maximum of the corresponding resonance peak. Applying the same formula here gives a mean lifetime of 5×10^{-19} s for the η meson and one of 7×10^{-25} s for the ω meson.

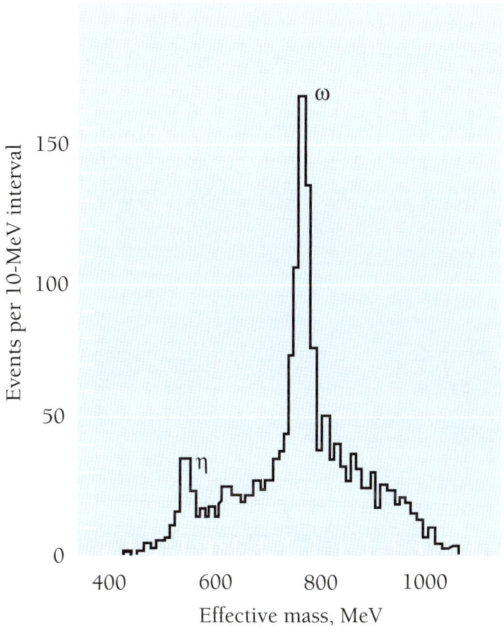

Figure 13.4 Resonant states in the reaction $\pi^+ + p \rightarrow \pi^+ + p + \pi^+ + \pi^- + \pi^0$ occur at effective masses of 549 and 783 MeV/c^2. By effective mass is meant the total energy, including mass energy, of the three new mesons relative to their center of mass.

13.4 ELEMENTARY PARTICLE QUANTUM NUMBERS

Finding order in apparent chaos

The interactions and decays of the hundreds of known elementary particles and resonances form what seems to be a bewildering array. Order can be brought into this situation by assigning certain quantum numbers to each entity and establishing which of these numbers are conserved and which can change in a given process. We are already familiar with two such quantum numbers, namely those that describe a particle's charge and spin. These quantum numbers are always conserved. In this section we shall look at some of the other quantum numbers that have proved useful in understanding the behavior of elementary particles.

Baryon and Lepton Numbers

One set of quantum numbers is used to characterize baryons and the three families of leptons. The **baryon number** $B = 1$ is assigned to all baryons, and $B = -1$ to all antibaryons; all other particles have $B = 0$. The **lepton number** $L_e = 1$ is assigned to the electron and the e-neutrino, and $L_e = -1$ to their antiparticles; all other particles have $L_e = 0$. In a similar way the lepton number $L_\mu = 1$ is assigned to the muon and the μ-neutrino, and the lepton number $L_\tau = 1$ to the tau lepton and its neutrino.

The significance of these numbers is that, in every process of whatever kind, the total values of B, L_e, L_μ, and L_τ separately remain constant: the number of baryons and of each kind of lepton, reckoning a particle as + and its antiparticles as −, never changes.

An example of particle-number conservation is the decay of the neutron, in which $B = 1$ and $L_e = 0$ before and after:

Neutron decay

$$\begin{array}{lcccc} & n^0 \rightarrow & p^+ + & e^- + & \bar{\nu}_e \\ L_e: & 0 & 0 & +1 & -1 \\ B: & +1 & +1 & 0 & 0 \end{array}$$

This is the only way in which the neutron can decay and still conserve both energy and baryon number B. The apparent stability of the proton is also a consequence of the need to conserve these quantities: there are no baryons of smaller mass, hence it cannot decay.

Example 13.1

Show that pion decay, muon decay, and pair production conserve the lepton numbers L_e and L_μ.

Solution

Pion decay

$$\begin{array}{lccc} & \pi^- \rightarrow & \mu^- + & \bar{\nu}_\mu \\ L_\mu: & 0 & +1 & -1 \end{array}$$

Muon decay

$$\begin{array}{lcccc} & \mu^- \rightarrow & e^- + & \nu_\mu + & \bar{\nu}_e \\ L_e: & 0 & +1 & 0 & -1 \\ L_\mu: & +1 & 0 & +1 & 0 \end{array}$$

Pair production

$$\begin{array}{lccc} & \gamma \rightarrow & e^- + & e^+ \\ L_e: & 0 & +1 & -1 \end{array}$$

Strangeness

Introducing baryon and lepton numbers still left some loose ends in the world of elementary particles. In particular, a number of particles were discovered that behaved so unexpectedly that they were called "strange particles." They were only created in pairs, for instance, and decayed only in certain ways but not in others that were allowed by existing conservation rules. To clarify the observations, M. Gell-Mann and, independently, K. Nishijina introduced the **strangeness number** S, whose assignments for the particles of Table 13.3 are shown there.

Strangeness number S is conserved in all processes mediated by the strong and electromagnetic interactions. The multiple creation of particles with $S \neq 0$ is the result of this conservation principle. An example is the result of this proton-proton collision:

$$\begin{array}{lcccccc} & p^+ & + p^+ \rightarrow & \Lambda^0 & + K^0 & + p^+ & + \pi^+ \\ S: & 0 & 0 & -1 & +1 & 0 & 0 \end{array}$$

On the other hand, S can change in an event mediated by the weak interaction. Decays that proceed via the weak interaction are relatively slow, a billion or more times slower than decays that proceed via the strong interaction (such as those of resonance particles). Even the weak interaction does not allow S to change by more than ± 1 in a decay. Thus the Ξ^- baryon does not decay directly into a neutron since

$$\begin{array}{lccc} & \Xi^- \rightarrow & n^0 & + \pi \\ S: & -2 & 0 & 0 \end{array}$$

but instead via the two steps

$$\begin{array}{lccc} & \Xi^- \rightarrow & \Lambda^0 & + \pi^- \\ S: & -2 & -1 & 0 \end{array} \qquad \begin{array}{ccc} \Lambda^0 \rightarrow & n^0 & + \pi^0 \\ -1 & 0 & 0 \end{array}$$

Symmetries and Conservation Principles

A remarkable theorem discovered early in this century by the German mathematician Emmy Noether states that

Every conservation principle corresponds to a symmetry in nature.

What is meant by a "symmetry"? In general, a symmetry of a particular kind exists when a certain operation leaves something unchanged. A candle is symmetric about a vertical axis because it can be rotated about that axis without changing in appearance or any other feature; it is also symmetric with respect to reflection in a mirror.

The simplest symmetry operation is translation in space, which means that the laws of physics do not depend on where we choose the origin of our coordinate system to be. Noether showed that the invariance of the description of nature to translations in space has as a consequence the conservation of linear momentum. Another simple symmetry operation is translation in time, which means that the laws of physics do not depend on when we choose $t = 0$ to be, and this invariance has as a consequence the conservation of energy. Invariance under rotations in space, which means that the laws of physics do not depend on the orientation of the coordinate system in which they are expressed, has as a consequence the conservation of angular momentum.

Conservation of electric charge is related to gauge transformations, which are shifts in the zeros of the scalar and vector electromagnetic potentials V and $\mathbf{A}$. (As elaborated in

About Scientists

Emmy Noether (1882–1935) was born in Germany and grew up among mathematicians, who included her father and brother. Her own mathematical work, mainly in algebra, was brilliant and original, and her papers and teaching had considerable influence. The atmosphere at the University of Göttingen, an outstanding center of mathematics where she went in 1919, was hostile to women, and she found it difficult to obtain a position there despite an appeal by the great mathematician David Hilbert: "I do not see why the sex of the candidate should be an argument against her appointment as Privatdocent; after all, we are not a bathhouse." The rise of Nazism in Germany led to her leaving in 1933 for the United States, where, after a period at the Institute for Advanced Study in Princeton, she became a professor at Bryn Mawr. Complications after what had seemed a successful operation ended her life at fifty-three while she was still full of ideas and energy.

electromagnetic theory, the electromagnetic field can be described in terms of the potentials V and $\mathbf{A}$ instead of in terms of $\mathbf{E}$ and $\mathbf{B}$, where the two descriptions are related by the vector calculus formulas $\mathbf{E} = -\nabla V$ and $\mathbf{B} = \nabla \times \mathbf{A}$.) Gauge transformations leave $\mathbf{E}$ and $\mathbf{B}$ unaffected since the latter are obtained by differentiating the potentials, and this invariance leads to charge conservation.

The interchange of identical particles in a system is a type of symmetry operation which leads to the preservation of the character of the wave function of a system. The wave function may be symmetric under such an interchange, in which case the particles do not obey the exclusion principle and the system follows Bose-Einstein statistics, or it may be antisymmetric, in which case the particles obey the exclusion principle and the system follows Fermi-Dirac statistics. **Conservation of statistics** (or, equivalently, of wave-function symmetry or antisymmetry) signifies that no process occurring within an isolated system can change the statistical behavior of that system. A system exhibiting Bose-Einstein statistical behavior cannot spontaneously alter itself to exhibit Fermi-Dirac statistical behavior, or vice versa. This conservation principle has applications in nuclear physics, where it is found that nuclei that contain an odd number of nucleons (odd mass number A) obey Fermi-Dirac statistics while those with even A obey Bose-Einstein statistics. Conservation of statistics is thus a further condition a nuclear reaction must observe.

More subtle and abstract than those mentioned above are the symmetries associated with the conservation of such quantities as baryon and lepton numbers and strangeness. These symmetries were important in the thinking that led to current theories of elementary particles, notably the quark model of hadrons.

The Eightfold Way

From Table 13.3 we can see that there are hadron families whose members have similar masses but different charges. These families are called **multiplets**, and it is natural to think of the members of a multiplet as representing different charge states of a single fundamental entity.

A classification system for hadrons, called the **eightfold way**, was proposed independently by Murray Gell-Mann and Yuval Ne'eman to encompass the many short-lived resonance particles as well as the relatively stable hadrons of Table 13.3. This scheme collects multiplets into supermultiplets whose members have the same spin but differ in charge and strangeness. The two supermultiplets shown in Figs. 13.5 and 13.6 consist respectively of spin $\frac{1}{2}$ baryons and spin 0 mesons, all stable against decay by the strong interaction. The supermultiplet of Fig. 13.7 consists of spin $\frac{3}{2}$ baryons which, except for the Ω^-, are resonance particles. The Ω^- was unknown when this supermultiplet was worked out, and its later discovery confirmed the validity of this classification method.

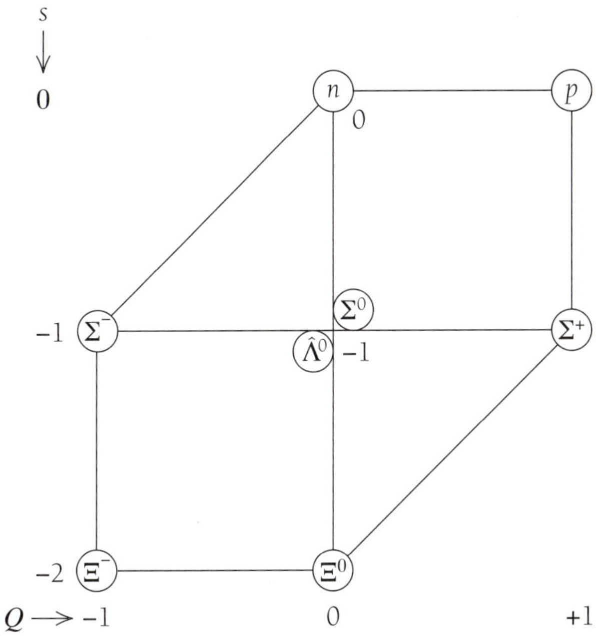

Figure 13.5 Supermultiplet of spin $\frac{1}{2}$ baryons on a plot of strangeness S versus charge Q (in units of e).

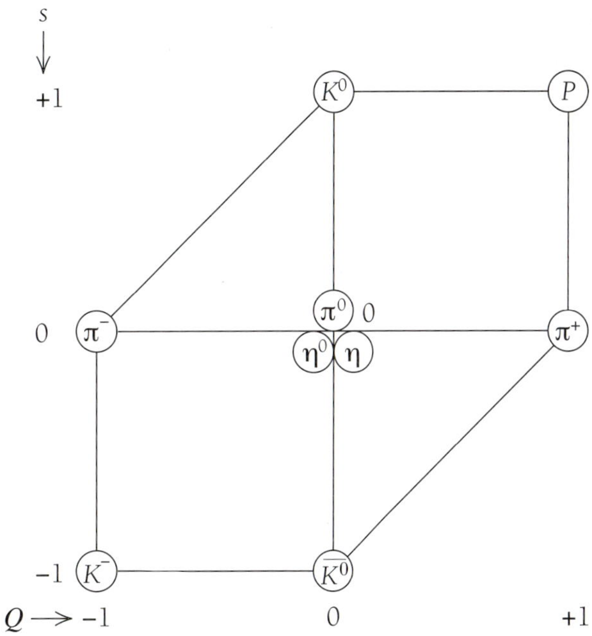

Figure 13.6 Supermultiplet of spin 0 mesons.

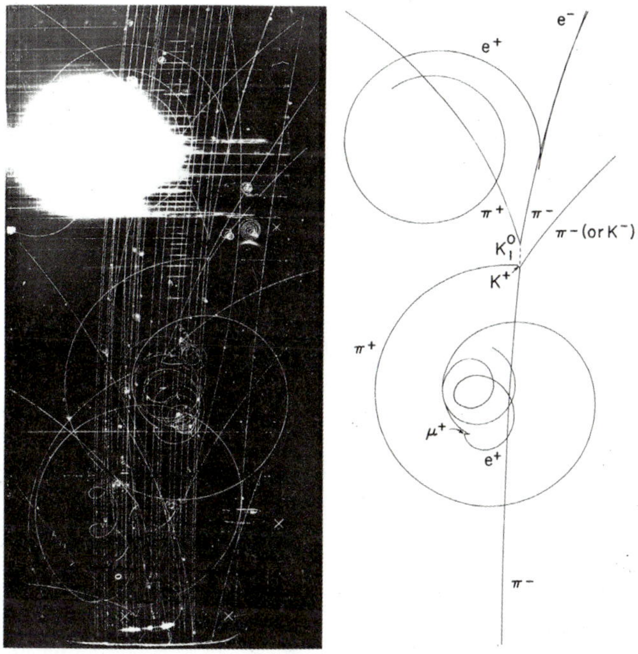

Liquid hydrogen bubble chamber photograph of a π— meson interacting with a proton (hydrogen nucleus) to produce a K+ meson, a neutral meson, shown by the dotted line, a negative meson, and a neutron (neutral or uncharged particles do not leave tracks). Image courtesy US Department of Energy.

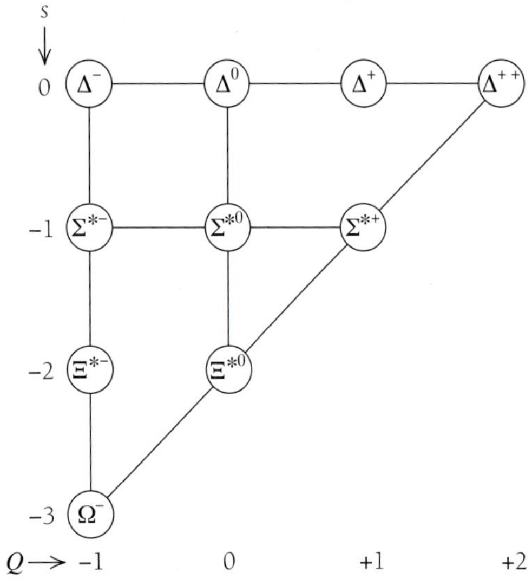

Figure 13.7 Baryon supermultiplet whose members have spin $\frac{3}{2}$ and (except Ω^-), are short-lived resonance particles. The Ξ^* and Σ^* particles here are heavier and have different spins from the ones in Table 13.3. The Ω^- particle was predicted from this scheme.

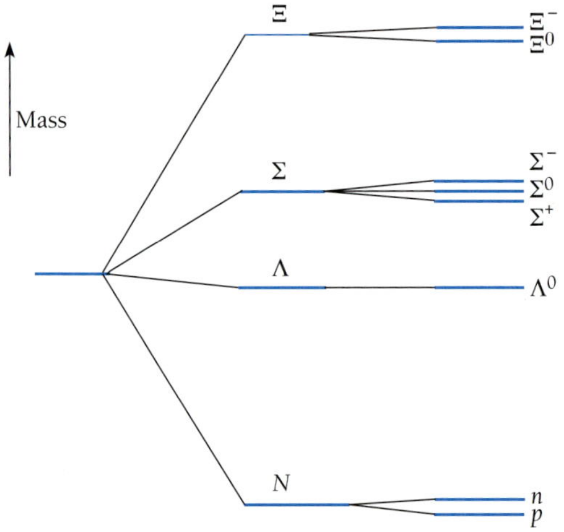

Figure 13.8 Origin of the baryon supermultiplet shown in Fig. 13.5.

The members of each supermultiplet would all be the same in the absence of any interactions, which are responsible for the differences that occur. Figure 13.8 shows how this idea applies to the baryon supermultiplet of Fig. 13.5. The strong interaction splits the basic baryon state into the four components Ξ, Σ, Λ, and N (for nucleon), and the electromagnetic interaction further splits the Ξ, Σ, and N components into multiplets. Because the strong interaction is more powerful than the electromagnetic one, the mass differences between multiplets are greater than those between members of a multiplet. Thus there is only 1.3-MeV difference between the p and n masses, but 176 MeV separates them from the Λ mass.

13.5 QUARKS

The ultimate constituents of hadrons

An effort to explain why only certain hadron families, such as those shown in Figs. 13.5, 13.6, and 13.7, occur but not others led Gell-Mann and, independently, George Zweig to propose in 1963 that all baryons consist of three still more fundamental

About Scientists

Murray Gell-Mann (1929–) was born in New York and entered Yale University at fifteen. After obtaining his Ph.D. from the Massachusetts Institute of Technology in 1951 he was at the Institute for Advanced Study in Princeton and at the University of Chicago before joining the faculty of the California Institute of Technology. In 1953 Gell-Mann introduced strangeness number and its conservation in certain interactions to help understand the properties of elementary particles. In 1961 he formulated a method of classifying elementary particles that enabled him to predict the Ω^- particle, which was later discovered. Two years later Gell-Mann came up with the idea of quarks, the ultimate entities from which particles subject to the strong interaction are composed. He received the Nobel Prize in Physics in 1969.

Table 13.4 Quarks. All have spin $\frac{1}{2}$ and baryon number $B = \frac{1}{3}$. Antiquarks have charges that are the negatives of those shown and baryon number $B = -\frac{1}{3}$. The strange antiquark has a strangeness number of $S = 1$.

Quark	Symbol	Mass, GeV/c^2	Charge, e	Strangeness
Up	u	0.3	$+\frac{2}{3}$	0
Down	d	0.3	$-\frac{1}{3}$	0
Strange	s	0.5	$-\frac{1}{3}$	−1
Charmed	c	1.5	$+\frac{2}{3}$	0
Top	t	174	$+\frac{2}{3}$	0
Bottom	b	4.3	$-\frac{1}{3}$	0

particles. Gell-Mann called these particles **quarks** from the phrase "three quarks for Muster Mark" that appears in James Joyce's novel *Finnegan's Wake*. The original three quarks were called **up** (symbol u), **down** (d), and **strange** (s); whereas u and d quarks have the strangeness number $S = 0$, the s quark has $S = -1$ (Table 13.4).

Because each baryon ($B = 1$) is made up of three quarks, the baryon number of a quark must be $B = \frac{1}{3}$. Antibaryons ($B = -1$) are made up of three antiquarks, so the baryon number of an antiquark must be $B = -\frac{1}{3}$. Mesons, for which $B = 0$, consist of a quark and an antiquark. Quarks all have spins of $\frac{1}{2}$, which accounts for the observed half-integral spins of baryons and the 0 or integral spins of mesons.

In order for hadrons to have charges of 0 or integral multiples of $\pm e$, the various quarks must have the fractional charges shown in Table 13.4. No other particles in nature have fractional charges, which made the quark hypothesis hard to accept at first, but soon the evidence for it proved overwhelming. The most direct experiments that point to the reality of quarks involved the scattering of high-energy (hence short-wavelength) electrons by protons, which revealed that there are indeed three pointlike concentrations of charge inside a proton. Quarks are thought to be elementary in the same sense as leptons, essentially point particles with no internal structures. Figure 13.9 shows the quark compositions of the hadrons of Fig. 13.5 and Table 13.5 details how the properties of several hadrons are derived from those of the quarks they contain. Figure 13.10 illustrates the quark models of nucleons and antinucleons.

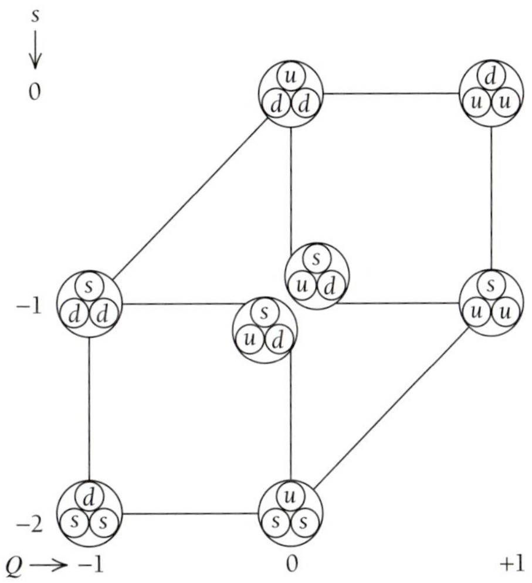

Figure 13.9 Quark compositions of the spin $\frac{1}{2}$ baryons shown in Fig. 13.5.

Table 13.5 Compositions of some hadrons according to the quark model

Hadron	Quark Content	Baryon Number	Charge, e	Spin	Strangeness
π^+	$u\bar{d}$	$\frac{1}{3} - \frac{1}{3} = 0$	$+\frac{2}{3} + \frac{1}{3} = +1$	$\uparrow\downarrow = 0$	$0 + 0 = 0$
K^+	$u\bar{s}$	$\frac{1}{3} - \frac{1}{3} = 0$	$+\frac{2}{3} + \frac{1}{3} = +1$	$\uparrow\downarrow = 0$	$0 + 1 = +1$
p^+	uud	$\frac{1}{3} + \frac{1}{3} + \frac{1}{3} = +1$	$+\frac{2}{3} + \frac{2}{3} - \frac{1}{3} = +1$	$\uparrow\uparrow\downarrow = \frac{1}{2}$	$0 + 0 + 0 = 0$
n^0	ddu	$\frac{1}{3} + \frac{1}{3} + \frac{1}{3} = +1$	$-\frac{1}{3} - \frac{1}{3} + \frac{2}{3} = 0$	$\downarrow\downarrow\uparrow = \frac{1}{2}$	$0 + 0 + 0 = 0$
Ω^-	sss	$\frac{1}{3} + \frac{1}{3} + \frac{1}{3} = +1$	$-\frac{1}{3} - \frac{1}{3} - \frac{1}{3} = -1$	$\uparrow\uparrow\uparrow = \frac{3}{2}$	$-1 - 1 - 1 = -3$

Color

A serious problem with the idea that baryons are composed of quarks was that the presence of two or three quarks of the same kind in a particular particle (for instance, two u quarks in a proton, three s quarks in an Ω^- baryon) violates the exclusion principle. Quarks ought to be subject to this principle since they are fermions with spins of $\frac{1}{2}$. To get around this problem, it was suggested that quarks and antiquarks have an additional property of some kind that can be manifested in a total of six different ways, rather as electric charge is a property that can be manifested in the two different ways that have come to be called positive and negative. In the case of quarks, this property became known as "color," and its three possibilities were called red, green, and blue. The antiquark colors are antired, antigreen, and antiblue.

According to the color hypothesis, all three quarks in a baryon have different colors, which satisfies the exclusion principle since all are then in different states even if two or three are otherwise identical. Such a combination can be thought of as white by analogy with the way red, green, and blue light combine to make white light (but there is no connection whatever except on this metaphorical level between quark colors and actual visual colors). Similarly, an antibaryon consists of an antired, an antigreen, and an antiblue quark. A meson consists of a quark of one color and an antiquark of the corresponding anticolor, which has the effect of canceling out the color. The result is that

Hadrons and antihadrons are colorless.

Quark color is thus a property that is significant within hadrons but is never directly observable in the outside world.

The notion of quark color is more than just a way around the exclusion principle. For one thing, it has turned out to be the key to explaining why the neutral pion has

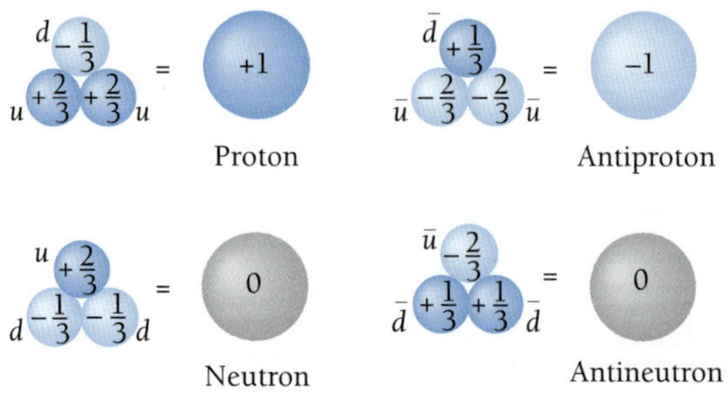

Figure 13.10 Quark models of the proton, antiproton, neutron, and antineutron. Electric charges are given in units of e.

its observed lifetime. On a deeper level, the strong interaction can be considered as being based on quark color, just as the electromagnetic interaction is based on electric charge.

Flavor

Not only do quarks come in three colors, but additional varieties (or "flavors") of quarks have had to be included in the scheme to supplement the original *u*, *d*, and *s* trio; see Table 13.4. The first of the new ones, the **charm** quark *c*, was proposed largely by analogy with the existence of lepton pairs: if quarks are elementary particles in the same sense as leptons, then there ought to be pairs of them, too. This may not appear to be very much of an argument, but so significant have symmetries of various kinds proved to be in physics that it is actually quite reasonable. Such a quark has a charge of $+\frac{2}{3}e$ and a charm quantum number of +1; other quarks have 0 charm. Charm apparently influences the likelihood of certain hadron decays, and both charmed baryons and mesons that contain c and $\bar{c}$ quarks have been found.

Amazingly, all the properties of ordinary matter can be understood on the basis of only two leptons, the electron and its associated neutrino, and two quarks, up and down, which constitute the first generation of Table 13.6.

The second generation of two leptons and two quarks—the muon and its neutrino, the charm and strange quarks—is responsible for most of the unstable particles and resonances created in high-energy collisions, all of which decay into members of the first generation. In the third generation the leptons are the tau meson, whose mass of 1.74 GeV is nearly twice that of the proton, and its neutrino. The quarks are called **top** and **bottom.** Both are extremely heavy, many times the proton mass, which is why hadrons that contain them can be produced only in the highest-energy events. The existence of the bottom quark was verified in 1977, that of the top quark not until 1995.

Are there further generations? Apparently not. Experiments sensitive to the number of generations of leptons and quarks unambiguously point to exactly three generations.

Quark Confinement

But for all the persuasiveness of the quark model of hadrons, and for all the searching that has gone on since 1963, no quark has ever been isolated. The present status of quarks may seem like that of neutrinos for twenty-five years after they were proposed: their reality is suggested by a wealth of indirect evidence, but something in their basic character impedes their detection. The parallel is not really accurate, however. The elusiveness of the neutrino was due merely to its feeble interaction with matter. On the other hand, a fundamental aspect of the color force seems to prevent quarks from existing independently outside hadrons. Indeed, the detection of a free quark would represent a failure of the theory, called **quantum chromodynamics,** that describes them and their behavior.

The explanation for quark confinement begins with the idea that, as though they were connected by a spring, the attractive force between two quarks goes up as the quarks move apart from their normal spacing. This means that more and more energy is needed to increase their separation. But with enough energy added, instead of a quark breaking free from the others in a hadron, the excess energy goes into producing a quark-antiquark pair. This results in a meson that does escape. To illustrate the

Table 13.6 Quarks and leptons and the interactions that affect them. Ordinary matter involves only the first generation. For each quark and lepton there is an antiquark and antilepton.

		Quarks		Leptons	
Generation	First	Up u	Down d	Electron e	e-neutrino ν_e
	Second	Charm c	Strange s	Muon μ	μ-neutrino ν_μ
	Third	Top t	Bottom b	Tau τ	τ-neutrino ν_τ
Charge	Electric	$+\frac{2}{3}$	$-\frac{1}{3}$	-1	0
	Color	Red / Green / Blue		Colorless	

Interaction: Color, Electro-magnetic, Weak

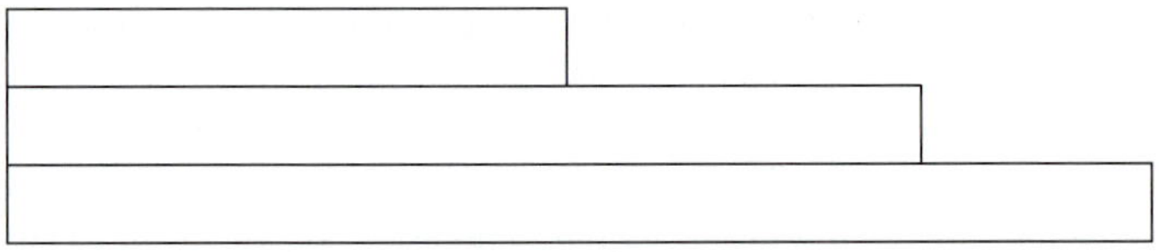

effect, Fig. 13.11 shows what happens when an energetic gamma-ray photon impinges on a neutron (composition *udd*) and causes a $u\bar{u}$ quark-antiquark pair to come into being. The quarks $udd + u\bar{u}$ then rearrange themselves into a proton (*duu*) and a negative pion ($\bar{u}d$), so that the net reaction is

$$\gamma + n^0 \rightarrow p^+ + \pi^-$$

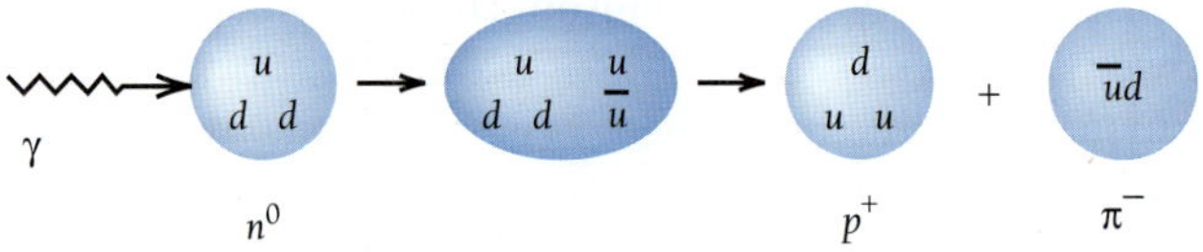

Figure 13.11 No matter how much energy is imparted to a hadron, an individual quark never emerges. Here energy is given to a neutron by a photon, and the result is a quark-antiquark pair created inside the neutron. The various quarks may then rearrange themselves into a proton and a negative pion.

Quark confinement is not the only example in physics of things that cannot be separated—the north and south poles of a magnet cannot be freed from each other either. If we pull apart a magnet so that it breaks we then have two magnets, each having a north and a south pole, instead of independent north and south poles.

13.6 FIELD BOSONS

Carriers of the interactions

As we saw in Sec. 11.7, the mutual forces between two particles can be regarded as being transmitted by the exchange of other particles between them. This concept applies to all the fundamental interactions. The particles exchanged, which are all bosons, are listed in Table 13.1. The **graviton** is the carrier of the gravitational field. The graviton should be massless and stable, have a spin of 2, and travel with the speed of light. Its zero mass can be inferred from the unlimited range of gravitational forces. If energy is to be conserved, the uncertainty principle requires that the range of the forces be inversely proportional to the mass of the particles being exchanged (see Eq. 11.19). Hence the gravitational interaction can have an infinite range only if the graviton mass is zero. The interaction of the graviton with matter should be quite feeble, making it extremely hard to detect. There is no definite experimental evidence either for or against the existence of the graviton.

The carriers of the weak interaction are called **intermediate vector bosons,** of which there are two kinds. Because the weak interaction has so short a range, the masses of such particles are large. One kind, called W, has a spin of 1 and a charge of $\pm e$ and is responsible for ordinary beta decays. Its mass is 85 times the proton mass. The other kind, called Z, also has a spin of 1 but is electrically neutral and heavier than the W ($97m_p$); its effects seem confined to certain high-energy events. Both decay in $\sim 10^{-25}$ s. Although the W particle is a natural concomitant of the weak interaction and was proposed many years ago, the idea of the Z particle originated more recently in a theory that unites the weak and electromagnetic interactions, and its discovery helped confirm the theory.

The connection between the weak and electromagnetic interactions was independently developed in the 1960s by Steven Weinberg and Abdus Salam. The key problem to be overcome in constructing the theory was that the carriers of the weak force have mass whereas the carriers of the electromagnetic force, namely photons, are massless. What Weinberg and Salam did was to show that, at a certain primitive level, both forces are aspects of a single interaction mediated by four massless bosons. Through a subtle process called spontaneous symmetry breaking, three of the bosons acquired mass and became the W and Z particles, with a consequent reduction in the range of what now appears as the weak part of the total interaction. One way to look at the situation is to regard the masses of the W and Z bosons as being attributes of the states they happen to occupy rather than as intrinsic attributes. The fourth electroweak boson, the photon, remained massless and the range of the electromagnetic part of the total interaction accordingly stayed infinite.

Since hadrons seem to be composed of quarks, the strong interaction between hadrons should ultimately be traceable to an interaction between quarks. The particles that quarks exchange to produce this interaction are called **gluons,** of which eight have been postulated. Gluons are massless and travel at the speed of light, and each one carries a color and an anticolor. The emission or absorption of a gluon by

About Scientists

Sheldon Lee Glashow (1932–) grew up in New York City and received his Ph.D. in 1958 from Harvard University, where he is now professor of Physics. Glashow was a student of Julian Schwinger, one of the pioneers of quantum electrodynamics, who had become interested in the weak interaction and its possible connection with the electromagnetic interaction. In 1961 Glashow took the first step in what was to prove the correct path to unifying these interactions, which was finally done in 1967 by Steven Weinberg and Abdus Salam working independently. All three received the Nobel Prize in 1979 for their contributions to the electroweak theory, which was given its final confirmation in 1983 when the predicted *W* and *Z* "carriers" of the weak interaction were experimentally observed at the CERN laboratory in Geneva. In 1970 Glashow and two collaborators proposed the existence of the charm quark; the discovery of particles that contain charm quarks and antiquarks followed a few years later. What is now called the Standard Model combining the strong and electroweak interactions that Glashow and Howard Georgi pioneered in 1974 accounts reasonably well for a number of otherwise unexplained observations.

The Compact Muon Solenoid (CMS) is a general-purpose detector at CERN's Large Hadron Collider. It is used in a range of physics programs from studying the Standard Model (including the Higgs boson) to searching for extra dimensions and particles that could make up dark matter.

a quark changes the quark's color. For instance, a blue quark that emits a blue-antired gluon becomes a red quark, and a red quark that absorbs this gluon becomes a blue quark. Because gluons have color changes, they should be able to interact with one another to form separate particles—"glueballs." The search for glueballs has thus far been fruitless, however.

13.7 THE STANDARD MODEL AND BEYOND

Putting it all together

The theory of how quarks interact with one another is known as quantum chromodynamics because it is modeled on quantum electrodynamics, the well-established theory of how charged particles interact, with quark color taking the place of electric charge. Quantum chromodynamics attempts to explain how quarks endow hadrons with their properties and has predicted a number of effects that have been observed in high-energy particle experiments.

The theory of the strong interaction has been added to that of the electroweak interaction to make a composite picture called the **Standard Model** that describes the structure of matter down to 10^{-18} m. It includes all the known constituents of matter—six leptons and six quarks—and the three strongest of the four forces that govern their behavior. As its name suggests, the Standard Model has been a considerable success, and its founders received over twenty Nobel Prizes over the years for their work.

But the Standard Model contains too many loose ends to be the last word. To begin with, important elements of the model have to be inserted arbitrarily. Instead of telling us the values of 18 basic quantities, such as the masses of the leptons and quarks, the model requires us to measure them ourselves; indeed, the essential fact that there are exactly three generations of leptons and quarks comes from experiment, not theory. The strong force that binds nucleons into nuclei and is mediated by meson exchange is the external manifestation of the color force between quarks in the nucleons that is mediated by gluon exchange, but nobody has been able to actually derive the details of the strong hadron force from the color quark force.

The Higgs Boson

In order for the Standard Model of leptons and quarks to be mathematically consistent, the Scottish physicist Peter Higgs showed that a field, now called the **Higgs field**, must exist everywhere in space. The Higgs field has an additional significance: by interacting with it, particles acquire their characteristic masses. The stronger the interaction, the greater the mass. We can think of the Higgs field as exerting a kind of viscous drag on particles that move through it; this drag appears as inertia, the defining property of mass.

As with other fields, a particle—here the **Higgs boson**—mediates the action of the Higgs field. The mass of the Higgs boson cannot be predicted from the Standard Model, but it is thought to be substantial, perhaps as much as 1 TeV/c^2, a thousand times the proton mass. Finding the Higgs boson would be a major step in validating the Standard Model, and knowing its mass and behavior would help to tie up loose ends in the model. Looking for the Higgs boson is one of the motivations for building particle accelerators more powerful than existing ones, which are inadequate for this search. Of course, nobody really knows what such accelerators will turn up— which is the best reason to build them. One such machine is the 4-billon-dollar Large Hadron Collider at CERN in Switzerland. Another, an upgraded accelerator called the Tevatron at the Fermi National Laboratory near Chicago, ought to be ready earlier, but it will be less powerful. (The top quark was discovered with the help of the original Tevatron.)

The next step is to weave together the electroweak and color interactions into a **grand unified theory** (GUT) that reveals the exact relationship between leptons and quarks. Among other things, a valid GUT should explain why the electron, a lepton, and the proton, a composite of quarks, have electric charges of the same magnitude. In order to do this, proposed GUTs require the existence of a lepton-quark interaction that would eventually cause protons to decay with a half-life of 10^{30} to 10^{33} years, which means that today's matter is inherently unstable. As mentioned earlier, experiments show that the proton half-life is at least 10^{32} years, so the question of ultimate proton stability has no answer as yet.

The search for a satisfactory GUT has led to a new symmetry principle called **supersymmetry.** If the universe is supersymmetric, it turns out that every particle must have a supersymmetric counterpart—a **sparticle**—whose spin differs from its own by $\frac{1}{2}$. Thus every fermion must be paired with a boson and every boson with a fermion. The boson superpartners of the fermion leptons and quarks are called sleptons and squarks, and the fermion superpartners of the field bosons γ, $W^{\pm}$, and gluons are called photinos, winos, and gluinos. The two salient aspects of supersymmetry (apart from the fun of naming the supposed new particles) are first, it integrates the separate theories in the Standard Model to form a much more satisfactory whole and second, no sparticle has ever been found despite much searching. Sparticles may well be too massive to be created in existing accelerators, and future accelerators may be able to produce them. And it is conceivable that the "missing" mass in the universe discussed in Sec. 13.9 consists of sparticles, though there has been no sign of them thus far.

A long-standing issue, one of the most basic in contemporary physics, is how gravitation connects with the other fundamental interactions. General relativity accounts for gravity in terms of the properties of spacetime and its conclusions have been verified whenever they have been tested. But general relativity is not a quantum-mechanical theory, unlike the components of the Standard Model and the proposed GUTs, so it cannot hold in its present form on very small scales of size.

According to its proponents, **string theory** can come to the rescue and be the basis of a final Theory of Everything. In this theory, leptons, quarks, and field bosons are not points in the four dimensions (x, y, z, t) of spacetime but vibrating loops of string in a space of ten dimensions. Each particle type represents a different mode of vibration of the string loops, which are supposed to be only about 10^{-35} m across and so appear as point particles to us. We are unaware of the additional six space dimensions because they are somehow "rolled up" by analogy with the way a two-dimensional surface (such as a sheet of paper) can be curled tightly to become a one-dimensional line. String theory, which is mathematically very difficult, incorporates the main features of GUTs, including in particular supersymmetry.

The notion that there may be additional hidden space dimensions goes all the way back to 1919, when the Polish mathematician Theodor Kaluza came close to successfully extending general relativity to include electromagnetism by postulating an extra dimension to provide a structure to every point in ordinary space. Kaluza's proposal was further developed by the Swedish physicist Oskar Klein, but some conclusions of the resulting theory, such as the ratio between the charge and mass of the electron, disagreed with measurements. With the ferment in physics in the 1920s that accompanied the advent of quantum mechanics, the Kaluza-Klein idea faded away until reborn and expanded into string theory starting over half a century later.

String theory has many attractive elements, notably that general relativity emerges from it in a natural way. An enormous amount of research into strings has been carried out, with results that encourage in many physicists a belief that it represents the road to a Theory of Everything. But thus far its predictions cannot be directly confronted with the results of experiment, so there is no way to know whether minute loops of string in ten dimensions actually exist with their vibrations making up the world we see around us.

13.8 HISTORY OF THE UNIVERSE

It began with a bang

The observed uniform expansion of the universe points to a Big Bang around 13 billion years ago that started from a singularity in spacetime, a point whose energy density and spacetime curvature were both infinite. In the absence of a quantum-mechanical theory of gravity, nothing can be said about the immediate aftermath of the Big Bang. After 10^{-43} s, however, the theory that ties together the strong, electromagnetic, and weak interactions, even though incomplete, permits a general picture to be sketched of what may well have happened.

As the initial compact, intensely hot fireball of matter and radiation from the Big Bang expanded, it cooled and underwent a series of transitions at specific temperatures. An analogy is with the cooling of steam, which becomes water and then ice as its temperature falls. Figure 13.12 shows the different phases of the universe on a graph of temperature (actually kT) versus time, both on logarithmic scales. The unit of kT here is the electronvolt, where 10^{-4} eV corresponds to ~1 K.

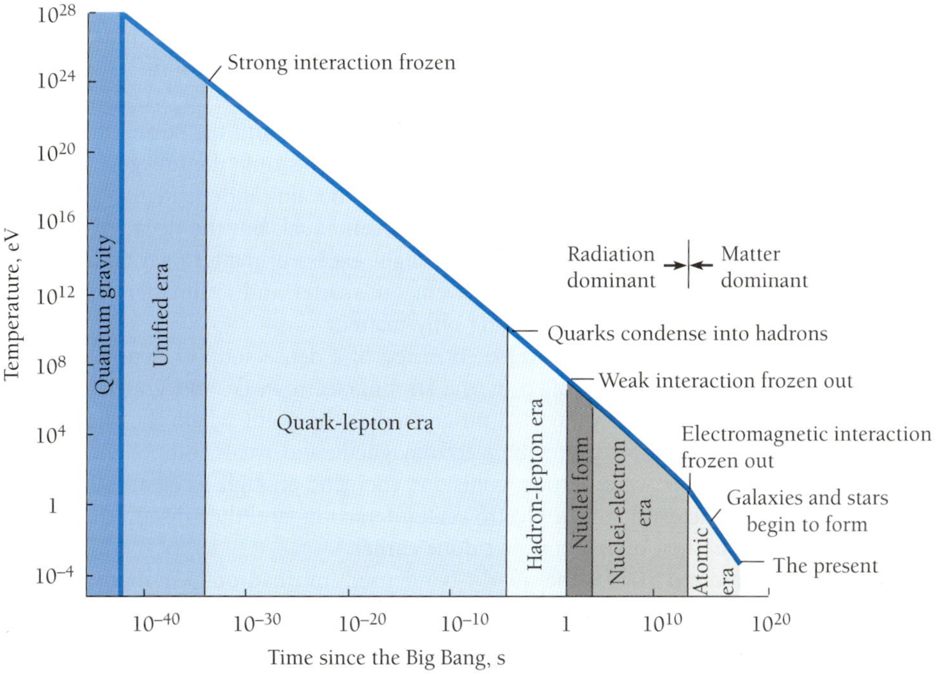

Figure 13.12 Thermal history of the universe on the basis of current theories. Nothing can be said about the state of the universe until 10^{-43} s after the Big Bang in the absence of a quantum-mechanical theory of gravity.

Planck Length and Time

Two fundamental constants are involved in general relativity: the gravitational constant G and the speed of light c. Similarly, Planck's constant h is the fundamental constant of quantum theory. We can combine G, c, and h to arrive at a "natural" unit of length, called the **Planck length** λ_P, given by

Planck length
$$\lambda_P = \sqrt{\frac{Gh}{c^3}} = 4.05 \times 10^{-35} \text{ m}$$

The Planck length is significant because, at shorter distances, quantum fluctuations allowed by the uncertainty principle disrupt the smooth geometry of space that is central to general relativity. On larger scales of length, quantum theory and general relativity each describe well different aspects of physical reality. For lengths less than about λ_P, however, both fail, leaving us ignorant about structures and events in this realm of size.

The time needed by something moving at the speed of light to travel λ_P is the **Planck time** t_P, given by

Planck time
$$t_P = \frac{\lambda_P}{c} = \sqrt{\frac{Gh}{c^5}} = 1.35 \times 10^{-43} \text{ s}$$

To deal with time intervals smaller than t_P we also require a theory that unifies quantum theory and general relativity. No such theory is yet adequate for such a purpose. What this lack means is that today we have no way at all to inquire into what the universe was like earlier than about 10^{-43} s after the Big Bang.

From 10^{-43} to 10^{-35} s the universe cooled from 10^{28} to 10^{23} eV. At energies like these the strong, electromagnetic, and weak interactions are merged into a single interaction mediated by extremely heavy field particles, the X bosons. Quarks and leptons are not distinguished from one another. At 10^{-35} s, however, particle energies became too low for free X bosons to be created any longer and the strong interaction became separated from the electroweak interaction. At this time the universe was only about a millimeter across. Quarks and leptons now became independent. Up to this time the amounts of matter and antimatter had been equal, but the decay of the field bosons was not symmetric and resulted in a slight excess of matter over antimatter—perhaps one part in 30 billion. As time went on, matter and antimatter annihilated each other to leave a universe containing only matter.

From 10^{-35} to 10^{-10} s the universe consisted of a dense soup of quarks and leptons whose behavior was controlled by the strong, electroweak, and gravitational interactions. At 10^{-10} s the cooling had progressed to the point where the electroweak interaction became separated into the electromagnetic and weak components we observe today. No longer were particle collisions energetic enough to create the free W and Z bosons characteristic of the electroweak interaction, and they disappeared as the X bosons of the unified interaction had done earlier.

Somewhere around 10^{-6} s the quarks condensed into hadrons. At about 1 s neutrino energies fell sufficiently for them to be unable to interact with the hadron-lepton soup—the "freezing out" of the weak interaction. The neutrinos and antineutrinos that existed remained in the universe but did not participate any further in its evolution. From then on protons could no longer be transformed into neutrons by inverse beta-decay events, but the free neutrons could beta-decay into protons. However, nuclear reactions were starting to occur that managed to incorporate many of the neutrons into

Radio waves thought to have originated in the primeval fireball that marked the start of the expansion of the universe were first detected by Arno Penzias and Robert Wilson with a sensitive receiver attached to this 15-m-long antenna at Holmdel, New Jersey. (Courtesy of NASA)

helium nuclei before their decay. Nuclear synthesis stopped at about $T = 5$ min when the ratio of helium mass to total mass should have been, according to theory, between 23 and 24 percent, which is indeed the ratio in most of the universe today. No stars or galaxies or gas clouds have been found with less than this proportion of helium. As a star ages, of course, its helium content increases as the result of nuclear reactions; in the sun's outer layers, which are accessible to measurement, the helium proportion is close to 28 percent. To be sure, some ^{2}H and ^{3}He were originally left over from incomplete synthesis of ^{4}He, and a little lithium also was produced, but ^{1}H and ^{4}He have been by far the main constituents of the universe after the first 5 min.

From 5 min to around 100,000 years after the Big Bang, the universe consisted of a plasma of hydrogen and helium nuclei and electrons in thermal equilibrium with radiation. Once the temperature fell below 13.6 eV, the ionization energy of hydrogen, hydrogen atoms could form and not be disrupted. Now matter and radiation were decoupled and the universe became transparent. The electromagnetic interaction was frozen out, as the strong and weak interactions had been before: photons had too little energy to materialize into particle-antiparticle pairs and, in a universe of neutral atoms, bremsstrahlung could not be produced by accelerated ions.

The radiation left behind then continued to spread out with the rest of the universe, undergoing doppler shifts to longer and longer wavelengths. An observer today would expect this remnant radiation to come equally strongly from all directions and to have a spectrum like that of a blackbody at 2.7 K—and such radiation has actually been found in microwave measurements made from the earth and from satellites. Thus we have three observations that strongly support Big-Bang cosmology:

1 The uniform expansion of the universe
2 The relative abundances of hydrogen and helium in the universe
3 The cosmic background radiation

Once matter and radiation were decoupled, gravity became the dominant influence on the evolution of the universe. Density fluctuations (whose existence is confirmed by irregularities—"ripples"—in the sea of 2.7-K radiation that were discovered in 1992) led to the formation of the galaxies and stars that adorn the night sky. Early supernovas spewed out the various elements heavier than helium that later became incorporated in other stars and in their satellite planets. Living things developed on at least one of these planets, and quite possibly on a great many others as well, which brings us to the present.

13.9 THE FUTURE

***"In my beginning is my end."* (*T. S. Eliot,* Four Quartets)**

Will the universe continue to expand forever? This depends on how much matter the universe contains and on how fast it is expanding. There are three possibilities:

1 If the average density ρ of the universe is smaller than a certain critical density ρ_c that is a function of the expansion rate, the universe is **open** and the expansion will never stop (Fig. 13.13). Eventually new galaxies and stars will cease to form and existing ones will end up as black dwarfs, neutron stars, and black holes—an icy death.
2 If ρ is greater than ρ_c, the universe is **closed** and sooner or later gravity will stop the expansion. The universe will then begin to contract. The progression of events will be the reverse of those that took place after the Big Bang, with an ultimate Big Crunch—a fiery death. And after that another Big Bang? If so, then the universe is cyclic, with no beginning and no end.
3 If $\rho = \rho_c$, the expansion will continue at an ever-decreasing rate but the universe will not contract. In this case the universe is said to be **flat** because of the geometry of space in such a universe (Fig. 13.14). If $\rho < \rho_c$, space is negatively curved; a two-dimensional analogy is a saddle. If $\rho > \rho_c$, space is positively curved; a two-dimensional analogy is the surface of a sphere. In all cases, however, spacetime is curved (Sec. 1.10).

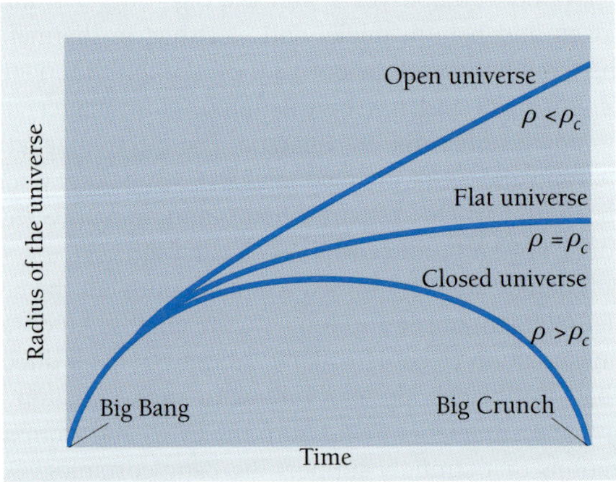

Figure 13.13 Three cosmological models that follow from the equations of general relativity. The quantity ρ is the average density of the universe and ρ_c, the critical density, is in the neighborhood of 9×10^{-27} kg/m^3, equivalent to about 5 hydrogen atoms per cubic meter.

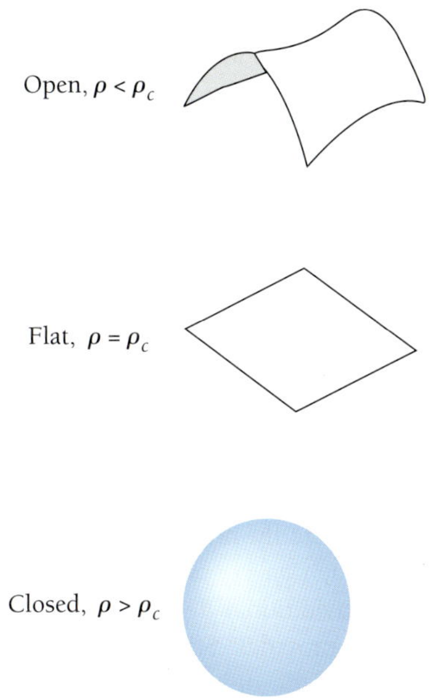

Figure 13.14 Two-dimensional analogies of the geometry of space in open, flat, and closed universes.

To find the value of the critical density ρ_c we begin the same way we would to find the escape velocity from the earth. The gravitational potential energy U of a spacecraft of mass m on the surface of the earth, whose mass is M and radius is R, is $U = -GmM/R$. (A negative potential energy corresponds to an attractive force.) To escape permanently from the earth, the spacecraft must have a minimum kinetic energy $\frac{1}{2}mv^2$ such that its total energy E is 0:

$$E = \text{KE} + U = \frac{1}{2}mv^2 - \frac{GmM}{R} = 0 \tag{13.9}$$

This gives $v = \sqrt{2GM/R} = 11.2$ km/s for the escape velocity.

Now we consider a spherical volume of the universe of radius R whose center is the earth. Only the mass inside this volume affects the motion of a galaxy on the surface of the sphere provided the distribution of matter in the universe is uniform, which it seems to be on a large enough scale. If the density of matter inside this volume is ρ, the volume contains a total mass of $M = \frac{4}{3}\pi R^3\rho$. According to Hubble's law (Sec.1.11), the outward velocity v of a galaxy R from the earth due to the expansion of the universe is proportional to R. Hence $v = HR$, where H is **Hubble's parameter.** Calling the galaxy's mass m, if it has just enough speed never to return, we have from Eq. (13.9)

$$\frac{1}{2}mv^2 = \frac{GmM}{R}$$

$$\frac{1}{2}m(HR)^2 = \frac{Gm}{R}\left(\frac{4}{3}\pi R^3\rho_c\right)$$

Critical density

$$\rho_c = \frac{3H^2}{8\pi G} \tag{13.10}$$

The critical density for a flat universe depends only on Hubble's parameter H, which is not accurately known. A reasonable value for H is 21 km/s per million light-years, which gives $\rho_c = 8.9 \times 10^{-27}$ kg/m^3. The mass of a hydrogen atom is 1.67×10^{-27} kg, so the critical density is equivalent to somewhere near 5.3 hydrogen atoms per cubic meter.

Dark Matter

The actual density of the luminous matter in the universe is just a few percent of ρ_c. Adding in the mass equivalent of the radiation in the universe increases the density only a little. But is luminous matter—the stars and galaxies we see in the sky—the only matter in the universe? Apparently not. Very strong evidence indicates that a large amount of **dark matter** is also present; so much, in fact, that at least 90 percent of all matter in the universe is nonluminous. For instance, the rotation speeds of the outer stars in spiral galaxies are unexpectedly high, which suggest that a spherical halo of invisible matter must surround each galaxy. Similarly, the motions of individual galaxies in clusters of them imply gravitational fields about 10 times more powerful than the visible matter of the galaxies provides. Still other observations support the idea of a preponderance of dark matter in the universe.

What can the dark matter be? The most obvious candidate is ordinary matter in various established forms, ranging from planetlike lumps too small to support the fusion reactions that would make them stars, through burnt-out dwarf stars, to black holes. The snag here is that, in the required numbers, such objects would certainly have been detected already. Another possibility rooted in what we already know is the sea of neutrinos (over 100 million per cubic meter) that pervades space. Neutrinos appear to have mass, but very little, nowhere near enough to account for all the dark matter. Indeed, if neutrinos were responsible for the dark matter, the universe could not have evolved to what it is today; galaxies, for example, would have to be much younger than they are. So neutrinos, too, may be part of the answer, but only part.

There is no shortage of other possibilities, all classed as cold dark matter. "Cold" means that the particles involved are relatively slow-moving, unlike, say, neutrinos, which constitute hot dark matter. Two main kinds of cold dark matter have been proposed, **WIMPs** and **axions.** WIMPs (weakly interacting massive particles) are hypothetical leftovers from the early moments of the universe. An example is the photino, one of the particles predicted by the supersymmetry approach to elementary particles. The photino is supposed to be stable and to have a mass of between 10 and 10^3 GeV/c^2, much more than the proton mass of 0.938 GeV/c^2. Axions are weakly interacting bosons associated with a field introduced to solve a major difficulty in the Standard Model. WIMPs and axions are being sought experimentally, thus far without success.

The dark matter needed to account for the motions of stars in galaxies and of galaxies in galactic clusters brings the total density of the universe up to about $0.1\rho_c$. There may be still more dark matter, however. In 1980 the American physicist Alan Guth proposed that, 10^{-35} s after the Big Bang, the universe underwent an extremely rapid expansion triggered by the separation of the single unified interaction into the strong and electroweak interactions. During the expansion the universe blew up from smaller than a proton to about a grapefruit in size in 10^{-30} s (Fig. 13.15). The **inflationary universe** automatically takes care of a number of previously troublesome problems in the Big Bang picture, and its basic concept is widely accepted. One of Guth's conclusions was that the density of matter in the universe must be exactly the critical density ρ_c. If the inflationary scenario is correct, then, the universe is not only perfectly

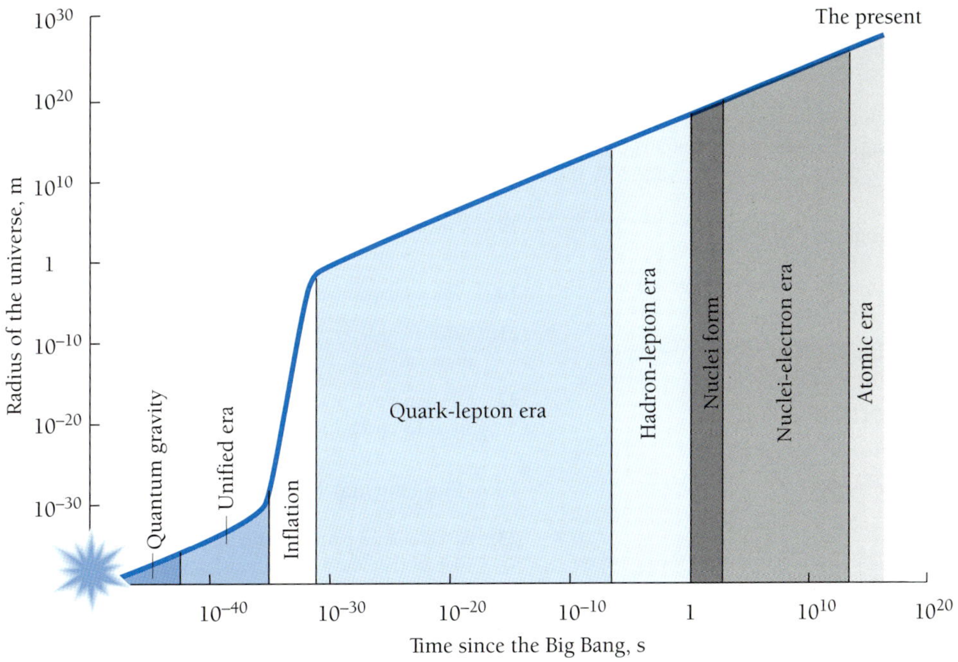

Figure 13.15 The inflationary universe.

flat but as much as perhaps 99 percent, not merely 90 percent, of the matter in it is dark matter. Finding the nature of the dark matter is clearly one of the most fundamental of all outstanding scientific problems.

EXERCISES

I have yet to see any problem, however complicated, which, when you looked at it in the right way, did not become still more complicated.—Poul Anderson

13.2 Leptons

1. The interaction of one photon with another can be understood by assuming that each photon can temporarily become a "virtual" electron-positron pair in free space, and the respective pairs can then interact electromagnetically. (*a*) How long does the uncertainty principle allow a virtual electron-positron pair to exist if $h\nu \ll 2mc^2$, where *m* is the electron mass? (*b*) If $h\nu > 2mc^2$, can you use the notion of virtual electron-positron pairs to explain the role of a nucleus in the production of an actual pair, apart from its function in ensuring the conservation of both energy and momentum?

2. The τ^+ lepton can decay in any of the following ways:

$$\tau^+ \rightarrow e^+ + \nu_e + \bar{\nu}_\tau$$

$$\tau^+ \rightarrow \mu^+ + \nu_\mu + \bar{\nu}_\tau$$

$$\tau^+ \rightarrow \pi^+ + \bar{\nu}_\tau$$

Why is only one neutrino emitted when the τ^+ decays into a pion?

13.3 Hadrons

3. Find the energy of the photon emitted in the decay $\Sigma^0 \rightarrow \Lambda^0 + \gamma$.

4. Find the energy of each of the gamma-ray photons produced in the decay of a neutral pion at rest. Why must their energies be the same?

5. Show that $4m_ec^2$, where m_e is the electron mass, is the minimum energy needed by a photon to produce an electron-positron pair when it collides with an electron in the process $\gamma + e^- \rightarrow e^- + e^+ + e^-$.

6. The π^0 meson has neither charge nor magnetic moment, which makes it hard to understand how it can decay into a pair of electromagnetic quanta. One way to account for this process is to assume that the π^0 first becomes a "virtual" nucleon-antinucleon pair, the members of which then interact electromagnetically to yield two photons whose energies total the mass

energy of the π^0. How long does the uncertainty principle allow the virtual nucleon-antinucleon pair to exist? Is this long enough for the process to be observed?

7. A neutral pion whose kinetic energy is equal to its rest energy decays in flight. Find the angle between the two gamma-ray photons that are produced if their energies are the same.

13.4 Elementary Particle Quantum Numbers

8. Why does a free neutron not decay into an electron and a positron? Into a proton-antiproton pair?

9. Which of the following reactions can occur? State the conservation principles violated by the others.

   (a) $\Lambda^0 \rightarrow \pi^+ + \pi^-$
   (b) $\pi^- + p \rightarrow n + \pi^0$
   (c) $\pi^+ + p \rightarrow \pi^+ + p + \pi^- + \pi^0$
   (d) $\gamma + n \rightarrow \pi^- + p$

10. Which of the following reactions can occur? State the conservation principles violated by the others.

    (a) $p + p \rightarrow n + p + \pi^+$
    (b) $p + p \rightarrow p + \Lambda^0 + \Sigma^+$
    (c) $e^+ + e^+ \rightarrow \mu^+ + \pi^-$
    (d) $p + p \rightarrow p + \pi^+ + K^0 + \Lambda^0$

11. According to the theory of the continuous creation of matter (which has turned out to be inconsistent with astronomical observations), the evolution of the universe can be traced to the spontaneous appearance of neutrons and antineutrons in free space. Which conservation law(s) would this process violate?

12. The products of a collision between a fast proton and a neutron are a neutron, a Σ^0 particle, and another particle. What is the other particle?

13. A μ^- muon collides with a proton, and a neutron plus another particle are created. What is the other particle?

14. A positive pion collides with a proton and two protons plus another particle are created. What is the other particle?

15. A negative kaon collides with a proton and a positive kaon and another particle are created. What is the other particle?

16. The hypercharge Y of a particle is defined as the sum of its strangeness and baryon numbers: $Y = S + B$. Verify from Table 13.3 that the hypercharge Y of each hadron group is equal to twice the average charge (in units of e) of the members of the group.

13.5 Quarks

17. Why must the quarks in a hadron have different colors? Would they have to have different colors if their spins were 0 or 1 rather than $\frac{1}{2}$?

18. The Λ particle consists of a u quark, a d quark, and an s quark. What is its charge?

19. A member of the Σ group of particles consists of two u quarks and an s quark. What is its charge?

20. Which quarks make up the negative pion? The Ξ^- hyperon?

21. What particle in Table 13.3 corresponds to the quark compositions uus?

22. One kind of D meson consists of a c and a $\overline{u}$ quark. What is its spin? Its charge? Its baryon number? Its strangeness? Its charm?

13.6 Fundamental Interactions

23. All resonance particles have very short lifetimes. Why does this suggest they must be hadrons?

24. The gravitational interaction is the weakest of all by far, yet it alone governs the motions of the planets around the sun and the motions of the stars of a galaxy around the galactic center. Why?

25. The initial reaction of the proton-proton cycle that provides most of the sun's energy is

$$^1_1\text{H} + {}^1_1\text{H} \rightarrow {}^2_1\text{H} + e^+ + \nu$$

This reacton occurs relatively infrequently in the sun for two reasons, one of which is the coulomb "barrier" the protons must overcome if they are to get close enough together to react. What do you think the other reason is?

26. The "carriers" of the weak interaction are the $W^\pm$, whose mass is 82 GeV/c^2, and the Z^0, whose mass is 93 GeV/c^2. Use the method of Sec. 11.7 to find an approximate figure for the range of the weak interaction.

13.9 The Future

27. Figure 1.8 shows the expanding-balloon analogy of the expanding universe. As the balloon expands, the angular separations of the spots (as measured from the center of the balloon) remain constant. (*a*) If s is the distance between any two spots, show that the recession speed ds/dt is proportional to s, which is the equivalent of Hubble's law in this situation. (*b*) Find an expression for Hubble's parameter H for the expanding balloon. Is H necessarily constant?

Atomic Masses

The masses of neutral atoms of all stable and some unstable nuclides are given here together with the relative abundances of nuclides found in nature and the half-lives of the listed radionuclides. Many other radionuclides are known.

Z	Element	Symbol	*A*	Atomic Mass, u	Relative Abundance, %	Half-Life
0	Neutron	*n*	1	1.008 665		10.6 min
1	Hydrogen	H	1	1.007 825	99.985	
			2	2.014 102	0.015	
			3	3.016 050		12.3 y
2	Helium	He	3	3.016 029	0.0001	
			4	4.002 603	99.9999	
			6	6.018 891		805 ms
3	Lithium	Li	6	6.015 123	7.5	
			7	7.016 004	92.5	
			8	8.022 487		844 ms
4	Beryllium	Be	7	7.016 930		53.3 d
			8	8.005 305		6.7×10^{-17} s
			9	9.012 182	100	
			10	10.013 535		1.6×10^{6} y
5	Boron	B	10	10.012 938	20	
			11	11.009 305	80	
			12	12.014 353		20.4 ms
6	Carbon	C	10	10.016 858		19.3 s
			11	11.011 433		20.3 min
			12	12.000 000	98.89	
			13	13.003 355	1.11	
			14	14.003 242		5760 y
			15	15.010 599		2.45 s
7	Nitrogen	N	12	12.018 613		11.0 ms
			13	13.005 739		9.97 min
			14	14.003 074	99.63	
			15	15.000 109	0.37	
			16	16.006 099		7.10 s
			17	17.008 449		4.17 s
8	Oxygen	O	14	14.008 597		70.5 s
			15	15.003 065		122 s
			16	15.994 915	99.758	
			17	16.999 131	0.038	
			18	17.999 159	0.204	
			19	19.003 576		26.8 s

Z	Element	Symbol	A	Atomic Mass, u	Relative Abundance, %	Half-Life
9	Fluorine	F	17	17.002 095		64.5 s
			18	18.000 937		109.8 min
			19	18.998 403	100	
			20	19.999 982		11.0 s
			21	20.999 949		4.33 s
10	Neon	Ne	18	18.005 710		1.67 s
			19	19.001 880		17.2 s
			20	19.992 439	90.51	
			21	20.993 845	0.57	
			22	21.991 384	9.22	
			23	22.994 466		37.5 s
			24	23.993 613		3.38 min
11	Sodium	Na	22	21.994 435		2.60 y
			23	22.989 770	100	
			24	23.990 963		15.0 h
12	Magnesium	Mg	23	22.994 127		11.3 s
			24	23.985 045	78.99	
			25	24.985 839	10.00	
			26	25.982 595	11.01	
13	Aluminum	Al	27	26.981 541	100	
14	Silicon	Si	28	27.976 928	92.23	
			29	28.976 496	4.67	
			30	29.973 772	3.10	
15	Phosphorus	P	30	29.978 310		2.50 min
			31	30.973 763	100	
16	Sulfur	S	32	31.972 072	95.02	
			33	32.971 459	0.75	
			34	33.967 868	4.21	
			35	34.969 032		87.2 d
			36	35.967 079	0.017	
17	Chlorine	Cl	35	34.968 853	75.77	
			36	35.968 307		3.01×10^5 y
			37	36.965 903	24.23	
18	Argon	Ar	36	35.967 546	0.337	
			37	36.966 776		34.8 d
			38	37.962 732	0.063	
			39	38.964 315		269 y
			40	39.962 383	99.60	
19	Potassium	K	39	38.963 708	93.26	
			40	39.963 999	0.01	1.28×10^9 y
			41	40.961 825	6.73	
20	Calcium	Ca	40	39.962 591	96.94	
			41	40.962 278		1.3×10^5 y
			42	41.958 622	0.647	
			43	42.958 770	0.135	
			44	43.955 485	2.09	

Z	Element	Symbol	A	Atomic Mass, u	Relative Abundance, %	Half-Life
			45	44.956 189		163 d
			46	45.953 689	0.0035	
			47	46.954 543		4.5 d
			48	47.952 532	0.187	
21	Scandium	Sc	45	44.955 914	100	
22	Titanium	Ti	46	45.952 633	8.25	
			47	46.951 765	7.45	
			48	47.947 947	73.7	
			49	48.947 871	5.4	
			50	49.944 786	5.2	
23	Vanadium	V	48	47.952 257		16 d
			50	49.947 161	0.25	$\sim 10^{17}$ y
			51	50.943 962	99.75	
24	Chromium	Cr	48	47.954 033		21.6 h
			50	49.946 046	4.35	
			52	51.940 510	83.79	
			53	52.940 651	9.50	
			54	53.938 882	2.36	
25	Manganese	Mn	54	53.940 360		312.5 d
			55	54.938 046	100	
26	Iron	Fe	54	53.939 612	5.8	
			56	55.934 939	91.8	
			57	56.935 396	2.1	
			58	57.933 278	0.3	
			59	58.934 878		44.6 d
27	Cobalt	Co	58	57.935 755		70.8 d
			59	58.933 198	100	
			60	59.933 820		5.3 y
28	Nickel	Ni	58	57.935 347	68.3	
			60	59.930 789	26.1	
			61	60.931 059	1.1	
			62	61.928 346	3.6	
			64	63.927 968	0.9	
29	Copper	Cu	63	62.929 599	69.2	
			64	63.929 766		12.7 h
			65	64.927 792	30.8	
30	Zinc	Zn	64	63.929 145	48.6	
			65	64.929 244		244 d
			66	65.926 035	27.9	
			67	66.927 129	4.1	
			68	67.924 846	18.8	
			70	69.925 325	0.6	
31	Gallium	Ga	69	68.925 581	60.1	
			71	70.924 701	39.9	
32	Germanium	Ge	70	69.924 250	20.5	
			72	71.922 080	27.4	

Z	Element	Symbol	A	Atomic Mass, u	Relative Abundance, %	Half-Life
			73	72.923 464	7.8	
			74	73.921 179	36.5	
			76	75.921 403	7.8	
33	Arsenic	As	74	73.923 930		17.8 d
			75	74.921 596	100	
34	Selenium	Se	74	73.922 477	0.9	
			76	75.919 207	9.0	
			77	76.919 908	7.6	
			78	77.917 304	23.5	
			80	79.916 520	49.8	
			82	81.916 709	9.2	
35	Bromine	Br	79	78.918 336	50.7	
			80	79.918 528		17.7 min
			81	80.916 290	49.3	
36	Krypton	Kr	78	77.920 397	0.35	
			80	79.916 375	2.25	
			81	80.916 578		2.1×10^5 y
			82	81.913 483	11.6	
			83	82.914 134	11.5	
			84	83.911 506	57.0	
			86	85.910 614	17.3	
37	Rubidium	Rb	85	84.911 800	72.2	
			87	86.909 184	27.8	4.9×10^{10} y
38	Strontium	Sr	84	83.913 428	0.6	
			86	85.909 273	9.8	
			87	86.908 890	7.0	
			88	87.905 625	82.6	
39	Yttrium	Y	89	88.905 856	100	
40	Zirconium	Zr	90	89.904 708	51.5	
			91	90.905 644	11.2	
			92	91.905 039	17.1	
			94	93.906 319	17.4	
			96	95.908 272	2.8	
41	Niobium	Nb	93	92.906 378	100	
42	Molybdenum	Mo	92	91.906 809	14.8	
			94	93.905 086	9.3	
			95	94.905 838	15.9	
			96	95.904 675	16.7	
			97	96.906 018	9.6	
			98	97.905 405	24.1	
			100	99.907 473	9.6	
43	Technetium	Tc	99	98.906 252		2.1×10^5 y
44	Ruthenium	Ru	96	95.907 596	5.5	
			98	97.905 287	1.9	
			99	98.905 937	12.7	
			100	99.904 217	12.6	

Z	Element	Symbol	A	Atomic Mass, u	Relative Abundance, %	Half-Life
			101	100.905 581	17.0	
			102	101.904 347	31.6	
			104	103.905 422	18.7	
45	Rhodium	Rh	103	102.905 503	100	
46	Palladium	Pd	102	101.905 609	1.0	
			104	103.904 026	11.0	
			105	104.905 075	22.2	
			106	105.903 475	27.3	
			108	107.903 894	26.7	
			110	109.905 169	11.8	
47	Silver	Ag	107	106.905 095	51.8	
			108	107.905 956		2.41 min
			109	108.904 754	48.2	
48	Cadmium	Cd	106	105.906 461	1.3	
			108	107.904 186	0.9	
			110	109.903 007	12.5	
			111	110.904 182	12.8	
			112	111.902 761	24.1	
			113	112.904 401	12.2	9×10^{15} y
			114	113.903 361	28.7	
			116	115.904 758	7.5	
49	Indium	In	113	112.904 056	4.3	
			115	114.903 875	95.7	5×10^{14} y
50	Tin	Sn	112	111.904 823	1.0	
			114	113.902 781	0.7	
			115	114.903 344	0.4	
			116	115.901 743	14.7	
			117	116.902 954	7.7	
			118	117.901 607	24.3	
			119	118.903 310	8.6	
			120	119.902 199	32.4	
			122	121.903 440	4.6	
			124	123.905 271	5.6	
51	Antimony	Sb	121	120.903 824	57.3	
			123	122.904 222	42.7	
52	Tellerium	Te	120	119.904 021	0.1	
			122	121.903 055	2.5	
			123	122.904 278	0.9	$\sim 1.2 \times 10^{13}$ y
			124	123.902 825	4.6	
			125	124.904 435	7.0	
			126	125.903 310	18.7	
			127	126.905 222		9.4 h
			128	127.904 464	31.7	
			130	129.906 229	34.5	
53	Iodine	I	127	126.904 477	100	
			131	130.906 119		8.0 d
54	Xenon	Xe	124	123.906 12	0.1	
			126	125.904 281	0.1	

Z	Element	Symbol	A	Atomic Mass, u	Relative Abundance, %	Half-Life
			128	127.903 531	1.9	
			129	128.904 780	26.4	
			130	129.903 509	4.1	
			131	130.905 076	21.2	
			132	131.904 148	26.9	
			134	133.905 395	10.4	
			136	135.907 219	8.9	
55	Cesium	Cs	133	132.905 433	100	
56	Barium	Ba	130	129.906 277	0.1	
			132	131.905 042	0.1	
			134	133.904 490	2.4	
			135	134.905 668	6.6	
			136	135.904 556	7.9	
			137	136.905 816	11.2	
			138	137.905 236	71.7	
57	Lanthanum	La	138	137.907 114	0.1	1×10^{11} y
			139	138.906 355	99.9	
58	Cerium	Ce	136	135.907 14	0.2	
			138	137.905 996	0.2	
			140	139.905 442	88.5	
			142	141.909 249	11.1	5×10^{16} y
59	Praseodymium	Pr	141	140.907 657	100	
60	Neodymium	Nd	142	141.907 731	27.2	
			143	142.909 823	12.2	
			144	143.910 096	23.8	2.1×10^{15} y
			145	144.912 582	8.3	$> 10^{17}$ y
			146	145.913 126	17.2	
			148	147.916 901	5.7	
			150	149.920 900	5.6	
61	Promethium	Pm	147	146.915 148		2.6 yr
62	Samarium	Sm	144	143.912 009	3.1	
			147	146.914 907	15.1	1.1×10^{11} y
			148	147.914 832	11.3	8×10^{15} y
			149	148.917 193	13.9	$> 10^{16}$ y
			150	149.917 285	7.4	
			152	151.919 741	26.7	
			154	153.922 218	22.6	
63	Europium	Eu	151	150.919 860	47.9	
			153	152.921 243	52.1	
64	Gadolinium	Gd	152	151.919 803	0.2	1.1×10^{14} y
			154	153.920 876	2.1	
			155	154.922 629	14.8	
			156	155.922 130	20.6	
			157	156.923 967	15.7	
			158	157.924 111	24.8	
			160	159.927 061	21.8	
65	Terbium	Tb	159	158.925 350	100	

Z	Element	Symbol	A	Atomic Mass, u	Relative Abundance, %	Half-Life
66	Dysprosium	Dy	156	155.924 287	0.1	$> 1 \times 10^{18}$ y
			158	157.924 412	0.1	
			160	159.925 203	2.3	
			161	160.926 939	19.0	
			162	161.926 805	25.5	
			163	162.928 737	24.9	
			164	163.929 183	28.1	
67	Holmium	Ho	165	164.930 332	100	
68	Erbium	Er	162	161.928 787	0.1	
			164	163.929 211	1.6	
			166	165.930 305	33.4	
			167	166.932 061	22.9	
			168	167.932 383	27.1	
			170	169.935 476	14.9	
69	Thulium	Tm	169	168.934 225	100	
70	Ytterbium	Yb	168	167.933 908	0.1	
			170	169.934 774	3.2	
			171	170.936 338	14.4	
			172	171.936 393	21.9	
			173	172.938 222	16.2	
			174	173.938 873	31.6	
			176	175.942 576	12.6	
71	Lutetium	Lu	175	174.940 785	97.4	
			176	175.942 694	2.6	2.9×10^{10} y
72	Hafnium	Hf	174	173.940 065	0.2	2.0×10^{15} y
			176	175.941 420	5.2	
			177	176.943 233	18.6	
			178	177.943 710	27.1	
			179	178.945 827	13.7	
			180	179.946 561	35.2	
73	Tantalum	Ta	180	179.947 489	0.01	$> 1.6 \times 10^{13}$ y
			181	180.948 014	99.99	
74	Tungsten	W	180	179.946 727	0.1	
			182	181.948 225	26.3	
			183	182.950 245	14.3	
			184	183.950 953	30.7	
			186	185.954 377	28.6	
75	Rhenium	Re	185	184.952 977	37.4	
			187	186.955 765	62.6	5×10^{10} y
76	Osmium	Os	184	183.952 514	0.02	
			186	185.953 852	1.6	2×10^{15} y
			187	186.955 762	1.6	
			188	187.955 850	13.3	
			189	188.958 156	16.1	
			190	189.958 455	26.4	
			192	191.961 487	41.0	
77	Iridium	Ir	191	190.960 603	37.3	
			193	192.962 942	62.7	

Z	Element	Symbol	A	Atomic Mass, u	Relative Abundance, %	Half-Life
78	Platinum	Pt	190	189.959 937	0.01	6.1×10^{11} y
			192	191.961 049	0.79	
			194	193.962 679	32.9	
			195	194.964 785	33.8	
			196	195.964 947	25.3	
			198	197.967 879	7.2	
79	Gold	Au	197	196.966 560	100	
80	Mercury	Hg	196	195.965 812	0.2	
			198	197.966 760	10.0	
			199	198.968 269	16.8	
			200	199.968 316	23.1	
			201	200.970 293	13.2	
			202	201.970 632	29.8	
			204	203.973 481	6.9	
81	Thallium	Tl	203	202.972 336	29.5	
			205	204.974 410	70.5	
82	Lead	Pb	204	203.973 037	1.4	1.4×10^{17} y
			206	205.974 455	24.1	
			207	206.975 885	22.1	
			208	207.976 641	52.4	
			210	209.984 178		22.3 y
			214	213.999 764		26.8 min
83	Bismuth	Bi	209	208.980 388	100	
			212	211.991 267		60.6 min
84	Polonium	Po	210	209.982 876		138 d
			214	213.995 191		0.16 ms
			216	216.001 790		0.15 s
			218	218.008 930		3.05 min
85	Astatine	At	218	218.008 607		1.3 s
86	Radon	Rn	220	220.011 401		56 s
			222	222.017 574		3.824 d
87	Francium	Fr	223	223.019 73		22 min
88	Radium	Ra	226	226.025 406		1.60×10^{3} y
89	Actinium	Ac	227	227.027 751		21.8 y
90	Thorium	Th	228	228.028 750		1.9 y
			230	230.033 131		7.7×10^{4} y
			232	232.038 054	100	1.4×10^{10} y
			233	233.041 580		22.2 min
91	Protactinium	Pa	233	233.040 244		27 d
92	Uranium	U	232	232.037 168		72 y
			233	233.039 629		1.6×10^{5} y
			234	234.040 947		2.4×10^{5} y
			235	235.043 925	0.72	7.04×10^{8} y
			238	238.050 786	99.28	4.47×10^{9} y

Z	Element	Symbol	A	Atomic Mass, u	Relative Abundance, %	Half-Life
93	Neptunium	Np	237	237.048 169		2.14×10^6 y
			239	239.052 932		2.4 d
94	Plutonium	Pu	239	239.052 158		2.4×10^4 y
			240	240.053 809		6.6×10^3 y
95	Americium	Am	243	243.061 374		7.7×10^3 y
96	Curium	Cm	247	247.070 349		1.6×10^7 y
97	Berkelium	Bk	247	247.070 300		1.4×10^3 y
98	Californium	Cf	251	251.079 581		900 y
99	Einsteinium	Es	252	252.082 82		472 d
100	Fermium	Fm	257	257.095 103		100.5 d
101	Mendelevium	Md	258	258.098 57		56 d
102	Nobelium	No	259	259.100 941		58 m
103	Lawrencium	Lr	260	260.105 36		3.0 m
104	Rutherfordium	Rf	261	261.108 69		1.1 m
105	Dubnium	Db	262	262.114 370		0.7 m
106	Seaborgium	Sg	263	263.118 218		0.9 s
107	Nielsbohrium	Ns	262	262.123 120		115 ms
108	Hassium	Hs	264	264.128 630		0.08 ms
109	Meitnerium	Mt	266	266.137 830		3.4 ms

Elements with atomic numbers 110, 111, 112, 114, and 116 have been created in nuclear reactions but not yet named.

Answers to Odd-Numbered Exercises

CHAPTER 1

1. More conspicuous.
3. 1.97 ms.
5. (*a*) $\theta' = \tan^{-1} \dfrac{\sin\theta\sqrt{1 - v^2/c^2}}{\cos\theta + v/c}$.

   (*b*) As $v \to c$, $\tan\theta' \to 0$ and $\theta' \to 0$. This means that the stars appear farther forward in the field of view of the porthole than they do when $v = 0$.
7. (*a*) 0.800*c*; 0.988*c*. (*b*) 0.900*c*; 0.988*c*.
9. 6 ft; 2.6 ft.
11. 3.32×10^{-8} s.
13. 14°.
15. No, because the observer in the spacecraft will find a longer time interval than an observer on the ground, not a shorter time interval.
17. (*a*) 3.93 s. (*b*) To *B*, *A*'s watch runs slow.
19. 2.6×10^8 m/s.
21. 5.0 y.
23. If $\mathbf{p} = m\mathbf{v}$, an event that conserves momentum in one inertial frame would not conserve momentum to observers in other inertial frames in relative motion, so momentum would not then be a useful quantity in physics.
27. 6.0×10^{-11}.
29. $(\sqrt{3}/2)c$.
31. 1.88×10^8 m/s; 1.64×10^8 m/s.
33. 0.9989*c*.
35. 0.294 MeV.
39. $\sim 10^{19}$ eV; $\sim 10^5$ y.
41. 0.383 MeV/*c*.
43. 885 keV/*c*.
45. 0.963*c*; 3.372 GeV/*c*.
47. 874 MeV/c^2; 0.37*c*.
51. 578 nm.
53. 1.34×10^4 m/s.

CHAPTER 2

1. Less conspicuous.
3. KE_{max} is proportional to ν minus the threshold frequency ν_0.
5. 1.77 eV.
7. 1.72×10^{30} photons/s.
9. (*a*) 4.2×10^{21} photons/m^2. (*b*) 4.0×10^{26} W; 1.2×10^{45} photons/s. (*c*) 1.4×10^{13} photons/m^3.
11. 180 nm.
13. 539 nm; 3.9 eV.
15. 0.48 μA.
17. 6.64×10^{-34} J · s; 3.0 eV.
19. In the reference frame of the electron at rest, the photon momentum must equal the final electron momentum *p*. The corresponding photon energy is *pc* but the

electron's final kinetic energy is $\sqrt{p^2c^2 + m^2c^4} - mc^2 \neq pc$, so the process can-occur while conserving both momentum and energy.

21. 2.4×10^{18} Hz; x-rays.
23. 2.9°.
25. 5.0×10^{18} Hz.
27. $\lambda_C = 5.8 \times 10^{-8}$ nm $\ll 0.1$ nm.
29. 1.5 pm.
31. 2.4×10^{19} Hz.
33. 64°.
37. 335 keV.
39. 0.821 pm.
43. (*b*) $2.3/\mu$.
45. 8.9 mm.
47. 11 cm.
49. 0.015 mm.
51. 1.06 pm.
53. (*a*) 1.9×10^{-3} eV. (*b*) 1.8×10^{-25} eV. (*c*) 3.5×10^{18} Hz; 7.6 kHz.
55. (*a*) $v_e = \sqrt{2GM/R}$. (*b*) $R = 2GM/c^2$.

CHAPTER 3

1. The momenta are the same; the particle's total energy exceeds the photon energy; the particle's kinetic energy is less than the photon energy.
3. 3.3×10^{-29} m.
5. 4.8 percent too high.
7. 0.0103 eV; a relativistic calculation is not needed.
9. 5.0 μV.
13. The electron has the longer wavelength. Both particles have the same phase and group velocities.
17. $v_p/2$.
19. $1.16c$; $0.863c$.
21. (*b*) $v_p = 1.00085c$; $v_g = 0.99915c$.
23. Increasing the electron energy increases the electron momentum and so decreases the de Broglie wavelength, which in turn reduces the scattering angle θ.
25. (*a*) 4.36×10^6 m/s outside; 5.30×10^6 m/s inside. (*b*) 0.167 nm outside; 0.137 nm inside.
27. $2.05n^2$ MeV; 2.05 MeV.
29. 45.3 fm.
31. Each atom in a solid is limited to a certain definite region of space—otherwise the assembly of atoms would not be a solid. The uncertainty in position of each atom is therefore finite, and its momentum and hence energy cannot be zero. The position of an ideal-gas molecule is not restricted, so the uncertainty in its position is effectively infinite and its momentum and hence energy can be zero.
33. 3.1 percent.
35. 1.44×10^{-13} m.
37. (*a*) 24 m; 752 waves. (*b*) 12.5 MHz.

CHAPTER 4

1. Most of an atom consists of empty space.
3. 1.14×10^{-13} m.
5. 1.46 μm.

7. A negative total energy signifies that the electron is bound to the nucleus; the kinetic energy of the electron is a positive quantity.
11. 2.6×10^{74}.
13. Δp calculated in this way is half the electron's linear momentum in orbit.
15. The Doppler effect shifts the frequencies of the emitted light to both higher and lower frequencies to produce wider lines than atoms at rest would give rise to.
17. 91.2 nm.
19. 92.1 nm; ultraviolet.
21. 12.1 V.
23. 91.13 nm.
25. $n = \sqrt{\lambda R/(\lambda R - 1)}$; $n_i = 3$.
27. (*a*) $E_i - E_f = h\nu(1 + h\nu/2Mc^2)$. (*b*) $\text{KE}/h\nu = 1.0 \times 10^{-9}$, so the effect is negligible for atomic radiation.
29. $f_n/\nu = (2n^2 + 4n + 2)/(2n^2 + n)$, which is greater than 1; $f_{n+1}/\nu = 2n^2/(2n^2 + 3n + 1)$, which is less than 1.
31. 0.653 nm; x-ray.
33. 0.238 nm.
35. (*a*) $E_n = -(m'Z^2e^4/8\epsilon_0^2h^2)(1/n^2)$.
(*b*)

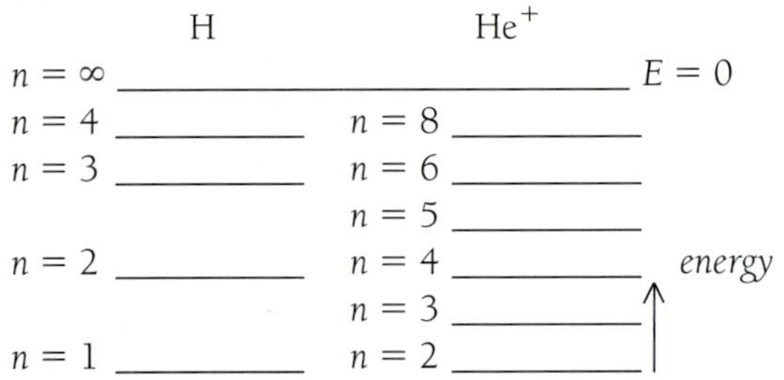

(*c*) 2.28×10^{-8} m.
37. 3.49×10^{18} ions.
39. Small θ implies a large impact parameter, in which case the full nuclear charge of the target atom is partially screened by its electrons.
41. 10°.
43. 0.84.
45. *Hint*: $f(\geq 60°, \leq 90°)/f(\geq 90°) = [f(\geq 60°) - f(\geq 90°)]/f(\geq 90°)$, where $f(\geq \theta)$ is proportional to $\cot^2 \theta/2$.
47. 0.87″.

CHAPTER 5

1. *b* is double-valued; *c* has a discontinuous derivative; *d* goes to infinity; *f* is discontinuous.
3. *a* and *b* are discontinuous and become infinite at $\pi/2, 3\pi/2, 5\pi/2, \ldots$; *c* becomes infinite as *x* goes to $\pm\infty$.
5. (*a*) $\sqrt{8/3\pi}$. (*b*) 0.462.
7. The wave function cannot be normalized, so it cannot represent a real particle. However, a linear superposition of such waves could give a wave group and be normalizable with $\psi \to 0$ at both ends of the group. Such a wave group would correspond to a real particle.
13. Near $x = 0$ the particle has more kinetic energy, hence more momentum, and ψ has a correspondingly shorter wavelength. The particle is less likely to be found

in this region because of its higher speed, hence ψ has a smaller amplitude there than near $x = L$.

17. $L^2/3 - L^2/2n^2\pi^2$.
19. $1/n$.
21. $(2/L)^{3/2}$
23. $(n_x^2 + n_y^2 + n_z^2)(\pi^2\hbar^2/2mL^2)$; $E_{3D} = 3E_{1D}$.
25. 0.95 eV.
27. The oscillator cannot have zero energy because this would mean it is at rest in a definite position, whereas according to the uncertainty principle a definite position corresponds to an infinite momentum (and hence energy) uncertainty.
31. $\langle x \rangle = 0$ and $\langle x^2 \rangle = E/k$ for both states.
33. (*a*) 2.07×10^{-15} eV; no. (*b*) 1.48×10^{28}.
37. (*a*) There is nothing in region II to reflect the particles, hence there is no wave moving to the left. (*b*) *Hint:* Make use of the boundary conditions that $\psi_I = \psi_{II}$ and $d\psi_I/dx = d\psi_{II}/dx$ at $x = 0$. (*c*) Transmitted current/incident current $= T = \frac{8}{9}$, hence the transmitted current is $\frac{8}{9}$ mA = 0.889 mA and the reflected current is $\frac{1}{9}$ mA = 0.111 mA.

CHAPTER 6

1. An atomic electron is free to move in three dimensions; hence, as in the case of a particle in a three-dimensional box, three quantum numbers are needed to describe its motion.
7. Bohr model: $L = mvr = \hbar$. Quantum theory: $L = 0$.
9. Only when $L = 0$, since L_z is otherwise always less than L.
11. 0, ±1, ±2, ±3, ±4.
13. 29 percent, 18 percent, 13 percent.
15. *Hint:* Solve $dP/dr = 0$ for r.
17. $9a_0$.
19. 1.85
21. (*a*) 68 percent. (*b*) 24 percent.
31. 1.34 T.

CHAPTER 7

1. (*a*) 1.39×10^{-4} eV (*b*) 8.93 mm.
3. 54.7°; 125.3°.
5. $^4_2$He atoms contain even numbers of spin $-\frac{1}{2}$ particles, which pair off to give zero or integral spins for the atoms. Such atoms do not obey the exclusion principle. $^3_2$He atoms contain odd numbers of spin-$\frac{1}{2}$ particles and so have net spins of $\frac{1}{2}$, $\frac{3}{2}$, or $\frac{5}{2}$, and they obey the exclusion principle.
7. An alkali metal atom has one electron outside closed inner shells; a halogen atom lacks one electron of having a closed outer shell; an inert gas atom has a closed outer shell.
9. 14.
11. 182.
13. The outermost of these electrons are, in the stated order, farther and farther from their respective nuclei with hence less and less tightly bound.
15. (*a*) $+2e$, relatively easy. (*b*) $+6e$, relatively hard.
17. Cl^- ions have closed shells, whereas a Cl atom lacks an electron of having a closed shell and the relatively poorly shielded nuclear charge tends to attract an

electron from another atom to fill the shell. Na^+ ions have closed shells, whereas a Na atom has a single outer electron that can be detached relatively easily in a chemical reaction with another atom.

19. The Li atom is larger because the effective nuclear charge acting on its outer electron is less than that acting on the outer electrons of the F atom. The Na atom is larger because it has an additional electron shell. The Cl atom is larger because it has an additional electron shell. The Na atom is larger than the Si atom for the same reason as given for the Li atom.
21. Only then is it possible for all the electrons to pair off with opposite spins to leave no net spin to produce an anomalous Zeeman effect.
23. 18.5 T.
25. 2, 3.
27. All its subshells are filled.
29. (*a*) There are no other allowed states. (*b*) This state has the lowest possible values of L and J, and is the only possible ground state.
31. $^2P_{1/2}$.
33. Since $\mathsf{L} < n$, a D ($\mathsf{L} = 2$) state is impossible for $n = 2$.
35. (*a*) $\frac{5}{2}, \frac{7}{2}$; (*b*) $\sqrt{35}\,\hbar/2, \sqrt{63}\,\hbar/2$; (*c*) 60°, 132°; (*d*) $^2F_{5/2}$, $^2F_{7/2}$.
37. $2\mathsf{J} + 1$; $\Delta E = g_J \mu_B B$
39. The transitions that give rise to x-ray spectra are the same in all elements since the transitions involve only inner, closed-shell electrons. Optical spectra, however, depend upon the possible states of the outermost electrons, which, together with the transitions permitted for them, are different for atoms of different atomic number.
41. 1.47 keV; 0.844 nm.
43. In a singlet state, the spins of the outer electrons are antiparallel. In a triplet state, they are parallel.

CHAPTER 8

1. The additional attractive force of the two protons exceeds the mutual repulsion of the electrons to increase the binding energy.
3. 3.5×10^4 K.
5. The increase in bond lengths in the molecule increases its moment of inertia and accordingly decreases the frequencies in its rotational spectrum. In addition, the higher the quantum number J, the faster the rotation and the greater the centrifugal distortion, so the spectral lines are no longer evenly spaced.
7. 13.
9. 0.129 nm.
11. 0.22 nm.
15. HD has the greater reduced mass, hence the smaller frequency of vibration and the smaller zero-point energy. HD therefore has the greater binding energy since its zero-point energy can contribute less energy to the splitting of the molecule.
17. (*a*) 1.24×10^{14} Hz.
19. 2.1×10^2 N/m.
21. Not very likely since $E_1 \gg kT$.

CHAPTER 9

1. 1.43×10^4 K.
3. 4.86×10^{-9}.
5. (*a*) 1 (by definition); 1.68:0.882:0.218:0.0277. (*b*) Yes; 1.55×10^3 K.
7. 2.00 m/s; 2.24 m/s.

9. 1.05×10^5 K.
11. 15.4 pm.
13. $(1/v)_{av} = (1/N) \int_0^\infty (1/v)n(v)\, dv.$
15. A fermion gas will exert the greatest pressure because the Fermi distribution has a larger proportion of high-energy particles than the other distributions; a boson gas will exert the least pressure because the Bose distribution has a larger proportion of low-energy particles than the others.
17. 2.5×10^6; 2.5×10^2.
19. 1.3 percent.
21. 0.92 kW/m^2.
23. 527°C.
25. 51 W.
27. 494 cm^2; 6.27 cm.
29. 2.5 percent.
31. 1.0×10^4 K.
33. 2.9×10^2 K; 8.9×10^{11} m.
35. 3.03×10^{-12} J/K.
39. (*a*) 3.31 eV. (*b*) 2.56×10^4 K. (*c*) 1.08×10^6 m/s.
45. 11 eV.
47. 1.43×10^{21} states/eV; yes.
49. At 20°C, $A = (Nh^3/V)(2\pi m_{He}kT)^{-3/2} = 3.56 \times 10^{-6}$, so $A \ll 1$.
51. At 20°C, $A = (Nh^3/2V)(2\pi m_e kT)^{-3/2} = 3.50 \times 10^{+3}$, so $A \gg 1$.
53. (*a*) 1.78 eV; 128 keV. (*b*) $kT = 862$ eV, so the gas of nuclei is nondegenerate but the electron gas is degenerate.

CHAPTER 10

1. The greater the atomic number Z of a halogen ion, the larger it is, hence the increase in interionic spacing with Z. The larger the ion spacing, the smaller the cohesive energy, hence the lower the melting point.
3. (*a*) 7.29 eV. (*b*) 9.26.
5. The heat lost by the expanding gas is equal to the work done against the attractive van der Waals forces between its molecules.
7. (*a*) Van der Waals forces increase the cohesive energy since they are attractive. (*b*) Zero-point oscillations decrease the cohesive energy since they represent a mode of energy possession present in a solid but not in individual atoms or ions.
9. Only the outer shell electrons in the atoms of a metal are members of its "gas" of free electrons.
11. $1.64 \times 10^{-8}\ \Omega \cdot$ m.
13. In both, a forbidden band separates a filled valence band from the conduction band above it. In semiconductors the band gap is smaller than in insulators, small enough so that some valence electrons have enough thermal energy to jump across the gap to the conduction band.
15. (*a*) Photons of visible light have energies of 1 − 3 eV, which can be absorbed by free electrons in a metal without leaving its valence band. Hence metals are opaque. The forbidden bands in insulators and semiconductors are too wide for valence electrons to jump across them by absorbing only 1 − 3 eV. Hence such solids are transparent. (*b*) Silicon, ≥1130 nm; diamond, ≥207 nm.
17. (*a*) *p*-type. (*b*) Aluminum atoms have 3 electrons in their outer shells, germanium atoms have 4. Replacing a germanium atom with an aluminum atom leaves a hole, so the result is a *p*-type semiconductor.

21. (*a*) 5.0 nm. (*b*) The ionization energy of the electron is 0.009 eV, which is much smaller than the energy gap and not very far from the 0.025-eV value of kT at 20°C.
23. (*a*) $\nu_c = eB/2\pi m^*$. (*b*) 0.2 m_e. (*c*) 3.4×10^{-7} m.
25. 2.4 GHz.

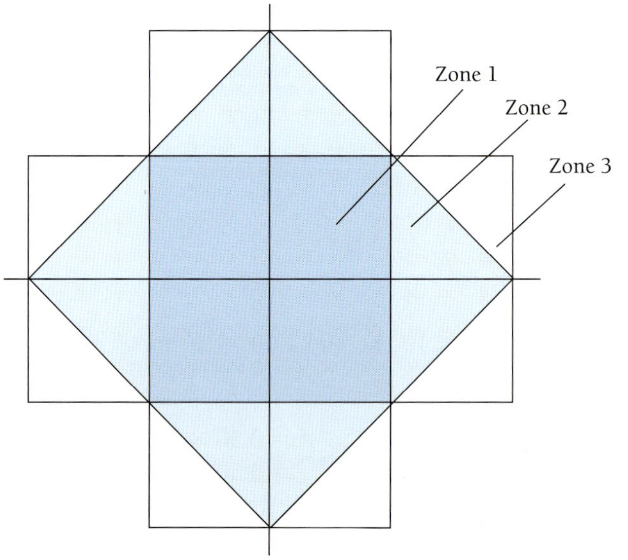

CHAPTER 11

1. $3n$, $3p$; $12n$, $10p$; $54n$, $40p$; $108n$, $72p$.
3. 177 MeV.
5. 7.9 fm.
7. Electron: 5.8×10^{-6} eV; proton: 8.8×10^{-9} eV.
9. (*a*) 3.5. (*b*) 51. (*c*) Because the populations are so close, induced emission will nearly equal induced absorption, so there will be very little net absorption of the radiation. The higher the temperature of the system, the less the absorption. (*d*) Because this is a two-level system, it could not be used as the basis for a laser.
11. The limited range of the strong nuclear interaction.
13. ${}^{7}_{3}\text{Li}$; ${}^{13}_{6}\text{C}$.
15. 8.03 MeV; 8.79 MeV.
17. 20.6 MeV; 5.5 MeV; 2.2 MeV; both calculations give 28.3 MeV.
19. $U = 0.85$ MeV and $\Delta E_b = 0.76$ MeV. Since the two figures are so close, nuclear forces must be very nearly independent of charge.
21. Calculated, 347.95 MeV; actual, 342.05 MeV, which is 1.7 percent less.
23. (*a*) $R = 3Ze^2/10\pi\epsilon_0(\Delta M + \Delta m)c^2$. (*b*) 3.42 fm.
25. (*a*) 7.88 MeV; 10.95 MeV; 7.46 MeV. (*b*) More energy is needed to remove a neutron from ^{82}Kr because of the tendency of neutrons to pair together.
27. ${}^{127}_{53}\text{I}$ is stable; ${}^{127}_{52}\text{Te}$ undergoes negative beta decay.
29. Yes. The nucleon kinetic energy that corresponds to the Δp implied by $\Delta x = 2$ fm is 1.3 MeV, which is consistent with a potential well 35 MeV deep.

CHAPTER 12

1. 1/4.
3. 3.10×10^{-4}.
5. 34.8 h.
7. 1.6×10^3 y.
9. 1.23×10^4 Bq.
11. 2.22×10^{-9} kg.
13. 52 min.
15. 1.64×10^9 y.
17. 1.4×10^4 y.
19. 5.9×10^9 y.
21. ${}^{206}_{82}\text{Pb}$; 48.64 MeV.
25. An electron leaving a nucleus is attracted by the positive nuclear charge, which reduces its energy. A positron leaving a nucleus, on the other hand, is repelled and is accordingly accelerated outward.
27. The energy available is less than $2m_ec^2$.
29. 2.01 MeV; 0.85 MeV; 1.87 MeV.
31. 1.80 MeV.
33. The thirty-ninth proton in ${}^{89}\text{Y}$ is normally in a $p_{1/2}$ state and the next higher state available to this proton is a $g_{9/2}$ state, hence a radiative transition between them has a low probability.
35. The neutron cross section decreases with increasing E because the likelihood that a neutron will be captured depends on how much time it spends near a particular nucleus, which is inversely proportional to its speed. The proton cross section is smaller at low energies because of the repulsive force exerted by the positive nuclear charge.
37. (*a*) 71 percent. (*b*) 3.0 mm.
39. 0.087 mm.
41. 0.766 Ci.
43. ${}^2_1\text{H}$; ${}^1_1\text{H}$; ${}^1_0n$; ${}^{79}_{36}\text{Kr}$.
45. 3.33 MeV.
47. 3.1×10^6 m/s; 4 MeV.
49. 4.
51. $E^* = Q + \text{KE}_A(1 - m_A/m_C)$; 4.34 MeV.
53. The neutron/proton ratio required for stability decreases with decreasing A, hence there is an excess of neutrons when fission occurs. Some of the excess neutrons are released directly, and the others change to protons by beta decay in the fission fragments.
55. 253 MeV.
57. The ${}^1_1\text{H}$ nuclei in ordinary water are protons, which readily capture neutrons to form ${}^2_1\text{H}$ (deuterium) nuclei. These neutrons cannot contribute to the chain reaction in a reactor, so a reactor using ordinary water as moderator needs enriched uranium with a greater content of the fissionable ${}^{235}\text{U}$ isotope to function. Deuterium nuclei are less likely to capture neutrons than are protons; hence a reactor moderated with heavy water can operate with ordinary uranium as fuel.
59. (*b*) ~100 percent; 89 percent; 29 percent; 1.7 percent.
61. 2.37 MeV.
63. (*a*) 2.2×10^9 K. (*b*) This temperature corresponds to the average deuteron energy, but many deuterons have considerably higher energies than the average. Also, quantum-mechanical tunneling through the potential barrier can occur,

permitting deuterons to react despite having insufficient energy to come together *classically.*

CHAPTER 13

1. (a) $3.22\ 3\ 10^{222}$ s. (b) The strong electric field of the nucleus separates the electron and positron sufficiently so that they cannot recombine afterward to reconstitute the photon.
3. 74.5 MeV.
7. 60°. (Hint: Use the relativistic expression for KE to find p_p.)
9. (a) B not conserved. (b) Can occur. (c) Charge not conserved. (d) Can occur.
11. Conservation of energy.
13. n_m(mu-neutrino).
15. A negative xi particle, X^2.
17. In order to obey the exclusion principle; no.
19. le.
21. S^1.
23. Only the strong interaction can produce such rapid decays.
25. Since a positron and a neutrino are emitted, the weak interaction is involved. Because this is so much feebler than the strong interaction, the reaction has a low probability even when the protons are energetic enough to overcome the Coulomb barrier.
27. (a) If r is the radius of the balloon, dsydt 5 (1yr)(drydt)s where r and drydt are the same for all points on the balloon at any time. (b) H 5 (1yr)(drydt). If dr/dt is proportional to r, H is constant, otherwise not.

For Further Study

There are many excellent books on every level of difficulty that cover the various elements of modern physics. A number written decades ago still have much of value to say to today's students. What follows is a brief selection of books that either treat the material covered in this book on about the same level but from a different perspective, or are on levels somewhat higher or lower, or give more complete accounts of certain topics. Anyone with access to a college library can easily find still other books that may serve particular needs even better.

Besides books, two periodicals frequently contain both news and review articles on modern physics: the monthly *Scientific American* and the British weekly *New Scientist*. Though specializing in current research, these periodicals regularly include interesting historical and biographical studies as well. The articles in *Scientific American* are usually written by the researchers themselves and are always authoritative; those in *New Scientist* are more often written by science journalists and now and then are rather speculative. These periodicals are nonmathematical and many issues from the past have articles well worth reading by students of modern physics.

General

Other texts at a level comparable with that of this book with similar coverage are:

J. Bernstein, P. M. Fishbane, and S. G. Gasiorowicz. 2000. *Modern Physics*. Upper Saddle River, N.J.: Prentice-Hall.

K. S. Krane. 1996. *Modern Physics*, 2nd ed. New York: Wiley.

R. A. Serway, C. J. Moses, and C. A. Moger. 1997. *Modern Physics*, 2nd ed. Fort Worth: Saunders.

S. T. Thornton and A. Rex. 2000. *Modern Physics for Scientists and Engineers*, 2nd ed. Fort Worth: Saunders.

P. A. Tipler and R. A. Llewellen. 1999. *Modern Physics*, 3rd ed. New York: Freeman.

Three books that give more detail on many of the discussions in this book are:

A. Beiser. 1969. *Perspectives of Modern Physics*, New York: McGraw-Hill.

R. Eisberg and R. Resnick. 1985. *Quantum Physics of Atoms, Molecules, Solids, and Particles*, 2nd ed. New York: Wiley.

F. K. Richtmyer, E. H. Kennard, and J. N. Cooper. 1969. *Introduction to Modern Physics*, 6th ed. New York: McGraw-Hill.

Relativity

A. P. French. 1968. *Special Relativity*. New York: Norton.

R. Resnick. 1968. *Introduction to Special Relativity*. New York: Wiley.

E. F. Taylor and J. A. Wheeler. 1992. *Spacetime Physics*, 2nd ed. New York: Freeman.

Waves and Particles

D. Bohm. 1951. *Quantum Theory*. Englewood Cliffs, N.J.: Prentice-Hall.

R. P. Feynman, R. B. Leighton, and M. Sands. 1965. *The Feynman Lectures on Physics*, Vol. 3. Reading, Mass.: Addison-Wesley.

R. Resnick and D. Halliday. 1992. *Basic Concepts in Relativity and Early Quantum Theory*. New York: Macmillan.
W. H. Wichman. 1971. *Quantum Physics*. New York: McGraw-Hill.

Quantum Mechanics

J. Baggott. 1992. *The Meaning of Quantum Theory*. New York: Oxford University Press.
S. Brandt and H. D. Dahmen. 2001. *Picture Book of Quantum Mechanics*, 3rd ed. New York: Springer-Verlag.
A. P. French and E. F. Taylor. 1979. *An Introduction to Quantum Physics*. New York: Norton.
D. J. Griffiths. 1995. *Introduction to Quantum Mechanics*. Upper Saddle River, N.J.: Prentice-Hall.
M. Morrison. 1990. *Understanding Quantum Physics*. Upper Saddle River, N.J.: Prentice-Hall.
L. Pauling and E.B. Wilson. 1935. *Introduction to Quantum Mechanics*. New York: McGraw-Hill.

Many-Electron Atoms

G. Herzberg. 1944. *Atomic Spectra and Atomic Structure*. New York: Dover.
H. Semat and J. R. Albright. 1972. *Introduction to Atomic and Nuclear Physics*. New York: Holt, Rinehart and Winston.
H. E. White. 1934. *Introduction to Atomic Spectra*. New York: McGraw-Hill.

Molecules

G. M. Barrow. 1962. *Introduction to Molecular Spectra*. New York: McGraw-Hill.
G. Hertzberg. 1950. *Molecular Spectra and Molecular Structure*. New York: Van Nostrand.
L. Pauling. 1967. *The Nature of the Chemical Bond*, 3rd ed. Ithaca: Cornell University Press.

Statistical Mechanics

R. Bowley and M. Sanchez. 1996. *Introductory Statistical Mechanics*. New York: Oxford University Press.
C. Kittel and H. Kroemer. 1995. *Thermal Physics*. New York: Freeman.

The Solid State

C. Kittel. 1996. *Introduction to Solid State Physics*, 7th ed. New York: Wiley.
M. N. Rudden and J. Wilson. 1993. *Elements of Solid State Physics*, 2nd ed. New York: John Wiley & Sons, Inc.
J. Singh. 1999. *Modern Physics for Engineers*. New York: Wiley.
S. M. Sze. 1981. *Physics of Semiconductor Devices*, 2nd ed. New York: Wiley.

Nuclear Physics

M. Harwit. 1998. *Astrophysical Concepts,* 3rd ed. New York: Springer-Verlag.
I. Kaplan. 1962. *Nuclear Physics.* Reading, Mass.: Addison-Wesley.
K. Krane. 1987. *Introductory Nuclear Physics.* New York: Wiley.
M. R. Wehr, J. A. Richards, and T. W. Adair. 1984. *Physics of the Atom,* 4th ed. Reading, Mass.: Addison-Wesley.

Elementary Particles and Cosmology

J. Allday. 1998. *Quarks, Leptons, and the Big Bang.* Philadelphia: Institute of Physics Publishers.
B. Greene. 2000. *The Elegant Universe.* New York: W. W. Norton & Co., Inc.
D. Griffiths. 1991. *Introduction to Elementary Particles.* Upper Saddle River, N.J.: Prentice-Hall.
A. Liddle. 1999. *Introduction to Modern Cosmology.* New York: Wiley.
S. Weinberg. 1992. *Dreams of a Final Theory.* New York: Pantheon.

Credits

Introduction

Page 3: © McGraw-Hill Education/SuperStock; 4: © McGraw-Hill Education/T. Pyle/Caltech/MIT/LIGO Lab; 5: © McGraw-Hill Education; 6: © McGraw-Hill Education; 7: © McGraw-Hill Education / Shutterstock/Andrey Armyagov; 8(top): © McGraw-Hill Education/Rodrigo A Torres/Glow Images; (bottom): © NFRI; 9: ©NASA/-JPL-Caltech/GSFC.

Chapter 1

Page 17: © Gloria NASA; 39: © Gloria NASA; 65: © Shutterstock.

Chapter 2

Page 69: © McGraw-Hill Education/Christine Eckel; 73: © McGraw-Hill Education/Roger Cope/Alamy Stock Photo; 75: © McGraw-Hill Education/Shutterstock/Duncan Andison; 82: © McGraw-Hill Education /Photocrea/Shutterstock; 87: © DUNLEE; 88: © McGraw-Hill Education/Shutterstock/VILevi; 91: Public Domain; 97: © Shutterstock.

Chapter 3

Page 109: © McGraw-Hill Education/Steve Gschmeissner/-Science Source; 119(left): © McGraw-Hill Education /The Center for Electron Microscopy and Analysis at The Ohio State University; (right): © Shutterstock.

Chapter 4

Page 137: © Shutterstock; 146: © McGraw-Hill Education/Ryan McVay/Getty Images; 161: © McGraw-Hill Education/Cultura/Image Source; 167: Courtesy of AEI Fabrication.

Chapter 5

Page 179: © McGraw-Hill Education/Andrew Dunn/Alamy; 205: ©IBM's Research Center; 206: ©IBM's Corporate Archives.

Chapter 6

Page 219: © McGraw-Hill Education/(NASA); 234: © Wooil Media; 235: © Wooil Media.

Chapter 7

Page 247: © Shutterstock; 276: © McGraw-Hill Education.

Chapter 8

Page 285: © Shutterstock/Dominika Zarzycka; 309: © Coherent, Inc.

Chapter 9

Page 315: © McGraw-Hill Education/(NASA); 334: © McGraw-Hill Education/Shutterstock; 347: Public Domain; 350: Public Domain.

Chapter 10

Page 355: Courtesy of Samsung Electronics; 359: © Dr. Stuart McKernan/University of Minnesota; 363: © McGraw-Hill Education/Shutterstock; 367: © McGraw-Hill Education/Shutterstock/Caleb Foster; 381: © McGraw-Hill Education/Shutterstock/Raimundas; 384: © NASA Johnson Space Center; 401: © McGraw-Hill Education/Shutterstock; 404: © IBM Corporate Archives.

Chapter 11

Page 407: © McGraw-Hill Education/AGE Fotostock; 410: © McGraw-Hill Education/Shutterstock.

Chapter 12

Page 439: © Princeton Plasma Physics Laborat; ory; 451: © NASA Johnson Space Center; 459: © McGraw-Hill Education/Image Source; 476: © Nuclear Energy Institute; 486: © NFRI/KSTAR; 488: © Lawrence Livermore National Laboratory.

Chapter 13

Page 495: © Shutterstock/Agnieszka Skalska; 503: © McGraw-Hill Education/xenotar/Getty Images; 504: © Shutterstock; 516: © Shutterstock; 521: Courtesy of NASA.

Index

Physical Constants and Conversion Factors

Atomic mass unit	u	1.66054×10^{-27} kg
		931.49 MeV/c^2
Avogadro's number	N_0	6.022×10^{26} kmol^{-1}
Bohr magneton	μ_B	9.274×10^{-24} J/T
		5.788×10^{-5} eV/T
Bohr radius	a_0	5.292×10^{-11} m
Boltzmann's constant	k	1.381×10^{-23} J/K
		8.617×10^{-5} eV/K
Compton wavelength of electron	λ_C	2.426×10^{-12} m
Electron charge	e	1.602×10^{-19} C
Electron rest mass	m_e	9.1095×10^{-31} kg
		5.486×10^{-4} u
		0.5110 MeV/c^2
Electronvolt	eV	1.602×10^{-19} J
	eV/c	5.344×10^{-28} kg · m/s
	eV/c^2	1.783×10^{-30} kg
Hydrogen atom, ground-state energy	E_1	-2.179×10^{-18} J
		-13.61 eV
rest mass	m_H	1.6736×10^{-27} kg
		1.007825 u
		938.79 MeV/c^2
Joule	J	6.242×10^{18} eV
Kelvin	K	°C + 273.15
Neutron rest mass	m_n	1.6750×10^{-27} kg
		1.008665 u
		939.57 MeV/c^2
Nuclear magneton	μ_N	5.051×10^{-27} J/T
		3.152×10^{-8} eV/T
Permeability of free space	μ_0	$4\pi \times 10^{-7}$ T · m/A
Permittivity of free space	ϵ_0	8.854×10^{-12}C^2/N · m^2
	$1/4\pi_0$	8.988×10^{9} N · m^2/C^2
Planck's constant	h	6.626×10^{-34} J · s
		4.136×10^{-15} eV · s
	$\hbar = h/2\pi$	1.055×10^{-34} J · s
		6.582×10^{-16} eV · s
Proton rest mass	m_p	1.6726×10^{-27} kg
		1.007276 u
		938.28 MeV/c^2
Rydberg constant	R	1.097×10^{7} m^{-1}
Speed of light in free space	c	2.998×10^{8} m/s
Stefan's constant	σ	5.670×10^{-8} W/m^2 · K^4

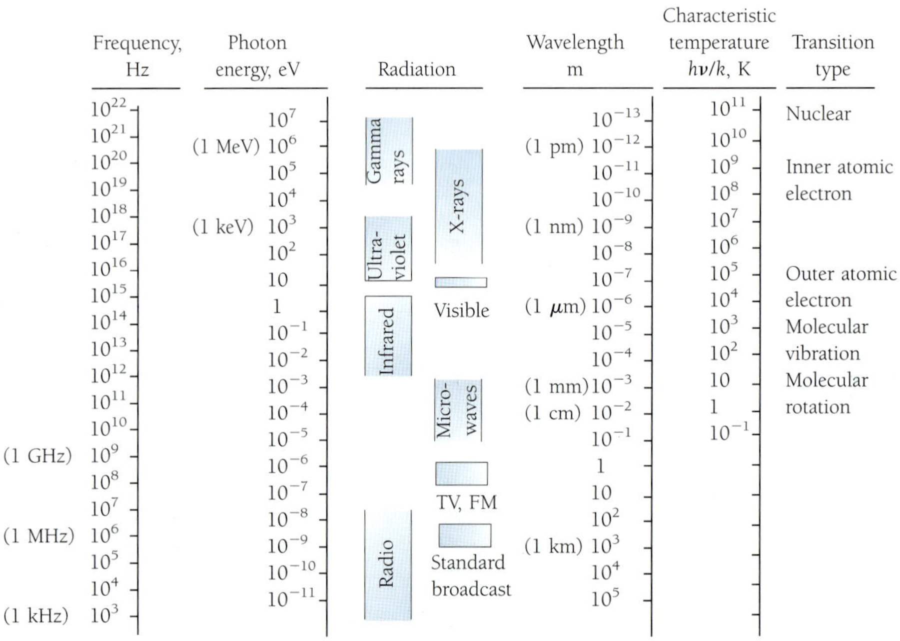
Frequency, Hz
Photon energy, eV
Radiation
Wavelength m
Characteristic temperature $h\nu/k$, K
Transition type
10^{22}
10^{21}
10^{20}
10^{19}
10^{18}
10^{17}
10^{16}
10^{15}
10^{14}
10^{13}
10^{12}
10^{11}
10^{10}
(1 GHz) 10^{9}
10^{8}
10^{7}
(1 MHz) 10^{6}
10^{5}
10^{4}
(1 kHz) 10^{3}
10^{7}
(1 MeV) 10^{6}
10^{5}
10^{4}
(1 keV) 10^{3}
10^{2}
10
1
10^{-1}
10^{-2}
10^{-3}
10^{-4}
10^{-5}
10^{-6}
10^{-7}
10^{-8}
10^{-9}
10^{-10}
10^{-11}
Gamma rays
Ultra-violet
Infrared
Radio
X-rays
Visible
Micro-waves
TV, FM
Standard broadcast
10^{-13}
(1 pm) 10^{-12}
10^{-11}
10^{-10}
(1 nm) 10^{-9}
10^{-8}
10^{-7}
(1 μm) 10^{-6}
10^{-5}
10^{-4}
(1 mm) 10^{-3}
(1 cm) 10^{-2}
10^{-1}
1
10
10^{2}
(1 km) 10^{3}
10^{4}
10^{5}
10^{11}
10^{10}
10^{9}
10^{8}
10^{7}
10^{6}
10^{5}
10^{4}
10^{3}
10^{2}
10
1
10^{-1}
Nuclear
Inner atomic electron
Outer atomic electron
Molecular vibration
Molecular rotation

The Greek Alphabet

Alpha	A	α	Iota	I	ι	Rho	P	ρ
Beta	B	β	Kappa	K	κ	Sigma	Σ	σ
Gamma	Γ	γ	Lambda	Λ	λ	Tau	T	τ
Delta	Δ	δ	Mu	M	μ	Upsilon	Υ	υ
Epsilon	E	ϵ	Nu	N	ν	Phi	Φ	ϕ
Zeta	Z	ζ	Xi	Ξ	ξ	Chi	X	χ
Eta	H	η	Omicron	O	o	Psi	Ψ	ψ
Theta	Θ	θ	Pi	Π	π	Omega	Ω	ω

Multipliers for SI Units

a	atto-	10^{-18}	da	deka-	10^{1}
f	femto-	10^{-15}	h	hecto-	10^{2}
p	pico-	10^{-12}	k	kilo-	10^{3}
n	nano-	10^{-9}	M	mega-	10^{6}
μ	micro-	10^{-6}	G	giga-	10^{9}
m	milli-	10^{-3}	T	tera-	10^{12}
c	centi-	10^{-2}	P	peta-	10^{15}
d	deci-	10^{-1}	E	exa-	10^{18}

Quantum Numbers of an Atomic Electron

Name	Symbol	Possible Values	Quantity Determined
Principal	n	$1, 2, 3, \ldots$	Electron energy
Orbital	l	$0, 1, 2, \ldots, n-1$	Orbital angular momentum magnitude
Magnetic	m_l	$-l, \ldots, 0, \ldots, +l$	Orbital angular momentum direction
Spin magnetic	m_s	$-\frac{1}{2}, +\frac{1}{2}$	Electron spin direction

ATOMIC SHELLS: $n = 1$ 2 3 4 5 . . .
K L M N O . . .

ANGULAR MOMENTUM STATES: $l = 0$ 1 2 3 4 5 . . .
s p d f g h . . .